Thomas Bartholin

Jesper Brandt Andersen

Thomas Bartholin

Physician and anatomist

From Unicorns to Lymph Vessels

AARHUS UNIVERSITY PRESS

Thomas Bartholin

Cover: Carl-H.K. Zakrisson and Tod Alan Spoerl
Cover illustration: Amsterdam Museum
Layout and typesetting: Carl-H.K. Zakrisson and Tod Alan Spoerl
Translated from the Danish by Peter Fisher and Jesper Brandt Andersen
This book is typeset in Newzald and printed on Munken Lynx, 120 g
Repro: Garn Grafisk Produktion
Printed by Specialtrykkeriet Arco, Denmark

Printed in Denmark 2025
1st edition, 1st impression

ISBN 978 87 7219 835 4

Aarhus University Press
Helsingforsgade 25, DK-8200 Aarhus N
unipress@unipress.au.dk
aarhusuniversitypress.dk

Published with the financial support of:
Cand. Pharm. Povl M. Assens Fond
Konsul George Jorck og Hustru Emma Jorck's Fond
Statens Kunstfond

The carbon emission of this book is calculated to be 2,2 kg according to ClimateCalc. CC-000157/DK

PEER REVIEWED
FSC www.fsc.org MIX Paper | Supporting responsible forestry FSC® C010493
Nordic Swan Ecolabel Ecolabelled printed matter 5041 0826

Contents

Preface

I had been intending for some years to write a work on the history of paediatrics. I came across an English book on the subject, including discussion of the Danish physician and anatomist Thomas Bartholin and how he dealt with a diphtheria epidemic among children in southern Italy in 1644. I quickly realized that Bartholin's writings from the time must be the first scientific work written by a Dane on a paediatric subject. Further research on Thomas Bartholin soon opened my eyes to this amazingly interesting and versatile scientist from an exciting period, which was characterized by an astonishing development within the natural sciences, not least in the field of medicine. It was a process where Denmark and the University of Copenhagen – specifically on the strength of Thomas Bartholin's special abilities and enterprise – came to play a leading role.

Thomas Bartholin is a dynamic and colourful representative of an age in which man moved from beliefs, superstition and outdated learning in the direction of observed scientific facts. One moment he writes a book about unicorns, and the next carries out brilliant scientific research that leads to the discovery of the lymph vessels and produces thoughts about medical and scientific questions on a level far ahead of his time. I was quickly carried away by Bartholin's inspiring and alluring world and consequently altered my plans to write this biography of him. It was my original intention to have the Danish edition ready in 2016 – the 400th anniversary of Bartholin's birth – but the longer I engaged with the material, the more it expanded, and therefore the book was not published until 2017. Ever since, it has been my wish to publish the biography in English, since the literature on Thomas Bartholin in English is very scarce, and I have experienced a demand for more from foreign researchers.

In 2021 I published a Danish biography of Bartholin's famous pupil Niels Stensen. The work on that book brought a lot of new material to light about their relationship and Bartholin's enormous importance in Stensen's life and scientific career. I have included this material in the present expanded English edition of my biography of Thomas Bartholin.

With only a few exceptions, Bartholin's colossal literary output is written in Latin, and only a few of his works have been translated into living languages. Reading Bartholin's works demands not only a significant knowledge of Latin, but also a special knowledge of precisely the version of Latin in which Bartholin and other scientists wrote in the 1600s.

During my work on the book, I have therefore been dependent on the exceptional help of senior researcher Niels W. Bruun in translating from Latin into Danish, both where it was a question of understanding difficult passages in Bartholin's productions and in the translation of the titles of Latin works. In the same way Niels has contributed a considerable number of new translations of chosen texts from among Bartholin's works, which have never been translated before into Danish or any other language. For many years, Niels has worked on an English edition of Bartholin's letters and generously given me the opportunity to read a selection of them in translation. Furthermore, Niels has contributed many of the book's illustrations and has continually graced me with his extensive knowledge of Bartholin and his time. A giant thank-you must go to him.

In the same way, I owe Peter Fisher a huge debt of gratitude for his hard work in translating chapters 1-4 and appendix 3 of my Danish biography into English, for proofreading the entire manuscript and for translating difficult Latin texts that have not previously been translated into English or other living languages.

It has also been of crucial importance for the book's realization to be able to draw on the theologian Axel Garboe's outstanding biography of Thomas Bartholin published 1949–50, the philologist H. D. Schepelern's translation of Ole Worm's letters, and the physician Gabriel Tryde's translation of Bartholin's works about his research on the lymphatic system.

Huge thanks must go to associate professor, legal licentiate Jørgen Blomqvist from the Committee for the Protection of Scientific Works (UBVA, ubva.dk), for an assessment and judgement in the conflict between Thomas Bartholin and the Swede, Olof Rudbeck, over the discovery of the lymph vessels (Appendix 3). Corresponding thanks must go to UBVA's president, professor, Ph.D., Morten Rosenmeier, for his interest in the subject and his permission to UBVA's examination and to the member of UBVA, professor and doctor of medicine Mogens Holst Nissen for his introduction of the question to UBVA.

Enormous thanks must go to the staff of the Royal Library in Copenhagen for their kind help and exemplary service in connection with the provision of books and the reproduction of illustrations, and to all the remaining contributors to the book's illustrations.

Special thanks go to the owner of Hagestedgaard, Lord Chamberlain Anders Hvass, for his interest and hospitality.

A big thank-you also to the foundations which have made the production of the book possible through their generous financial support. This includes Cand. Pharm. Povl M. Assens Fond, Landsdommer V. Gieses Legat and Statens Kunstfond.

Finally, thank you to my wife and five daughters for their interest, patience and tolerance during my work on the book.

I hope the book may generate a renewed interest in Thomas Bartholin and the historical perspective to his exceptional contribution to Danish and international natural and medical science.

Nysted, January 2025
Jesper Brandt Andersen

Introduction

"It is not Nature that fails, but humans, and they often fail because they flatter either themselves or their friends."[1] Thomas Bartholin, 1661

The century of the physician and anatomist Thomas Bartholin (1616–1680) – the 17th century – represented within the scientific world a transitional period between the Renaissance, with its cultivation of antiquity, and the era of the Enlightenment, with its focus on reason and science. The 1600s were characterized by a series of scientific breakthroughs, and it was precisely these turning points that formed the basis of the Enlightenment. Medical science was one of the areas that generated the greatest breakthroughs.

In the Renaissance, the development of printing in the first half of the 16th century had paved the way for new publications in Greek and Latin of the medical works of the ancient physicians Hippocrates (c. 460–c. 370 BC) and Galen (c.129–c. 216), and set a new focus on their medical understanding, which had already been mandatory doctrines in universities all over Europe for centuries.

Vesalius, Harvey and Aselli

However, in the mid-1500s something happened that had a decisive influence on 17th-century medical science. With the Flemish anatomist Andreas Vesalius (1514–1564) as a pioneer (Figure 1), people began questioning the doctrines of the ancient Greek authorities. By carrying out dissections himself on human bodies, Vesalius realized that the ancient physicians' anatomical writings did not fit human beings but were based on dissections of dead animals – that is, if they were based on dissections at all.

The ancient physicians' theories about human anatomy were, in other words, just theories and guesswork, but generally not based on practical research and observations. Vesalius' courage in setting himself in opposition to the ancient authorities and his encouragement to others to perform dissections themselves and carry out their own observations generated a new culture of medical science, which did not really gain a foothold before the beginning of the 17th century.

Figure 1. The 28-year-old Flemish anatomist Andreas Vesalius (1514–1564) during the dissection of an arm in 1542. Woodcut by Jan Stephan van Calcar (1499–1546) in Vesalius' epoch-making anatomical folio-work *De Humani corporis fabrica Libri septem*, Basel 1543.

Through his rejection of the doctrines of the ancient Greek physicians Hippocrates and Galen and his call for independent study of the human body, Vesalius revolutionized anatomical research and laid the foundations of modern medicine. He became a great source of inspiration for physicians of the 17th century. The Royal Library in Copenhagen.

Dissections of living creatures, so-called vivisections, making it possible to make direct observations of the functions of the organs, was a part of this development. The tradition of thinking more about structure and anatomy was, in the beginning of the 17th century, gradually replaced by a more dynamic perception. Inspired by the Italian philosopher, physicist and astronomer Galileo Galilei's (1564–1642) new mechanical observations within physics, there was a growing interest in the connection between structure and function, or, in other words, between anatomy and physiology.

Two of the 17th century's greatest discoveries in medical science took place at approximately the same time, at the beginning of the 1620s. One was the finding by the Englishman William Harvey (1578–1657) (Figure 2), that the heart pumps the blood in a circulation round the body, published in the work *Exercitatio Anatomica de Motu Cordis et Sanguinis in Animalibus* (An anatomical exercise on the motion of the heart and blood in animals) in Frankfurt am Main in 1628.[2] This discovery broke with the old opinion that the blood is continuously formed and broken down in large quantities in the liver, which was considered the body's blood-forming organ – an opinion that prevailed until it was found in the second half of the 1800s that the organ that generates the blood is in fact the bone marrow.

Figure 2. The English anatomist and royal physician William Harvey (1578–1657), discoverer of the blood circulation. Posthumous portrait engraving in rococo style by Jacobus Houbraken (1698–1780) from template by Willem van Bemmel (1630–1708). 36.9 × 23.3 cm. Amsterdam 1738. Published in London 1739.

The engraving shows the older Harvey. Symbolizing his profession and specialty, in front of the frame is a tablet with the large blood vessels of the chest and abdomen, the heart and the kidneys. In front of the board is a Rod of Asclepius. Harvey predicted – rightly, as it turned out – great resistance in the medical world against his theory of blood circulation. Thomas Bartholin was one of his earliest supporters. Private ownership.

Figure 3. Gaspare Aselli (1581–1625), discoverer of the lacteal veins. Engraving by Cesare Bassano (1584–1648). 20.5 × 13.6 cm. From Aselli's *De lactibus sive lacteis venis, quarto vasorum mesaraicorum genere*, Milan 1627.

The Latin text in the frame reads "GASPAR ASELLIUS CIVIS CREMONENSIS ANATOMICUS TICINENSIS ANNO ÆTATIS XLII": "Gaspare Aselli, citizen of Cremona, anatomist in Ticinum [Pavia], age 42."

With his discovery of the lacteal veins in the mesentery of a living dog in 1622, Aselli kick-started the 17th-century medical revolution. He was aware that the new vessels transported nutrients from the intestine, but he thought of them as veins and never realized that he had discovered a type of lymph vessel. His publication on the newfound vessels did not appear until two years after his death. Bibliothèque nationale de France.

The second great medical discovery in the 17th century was the finding by the Italian Gaspare Aselli (1581–1625) (Figure 3) of a new kind of vessel, namely the lacteal veins or milk veins. On the dissection of a live dog, after it had eaten, Aselli observed in the mesentery some thin white vessels which, upon intersection, emitted a milk-white fluid. He regarded these vessels as a fourth type of vessel – after arteries, veins and nerves (which at that time were also believed to be vessels – a type of veins) and called them *venæ lacteæ* (lacteal veins). Aselli believed that the lacteal veins ended their course in the liver, and that the milky-white fluid that they contained, rich in nourishment, and which originated from the food in the stomach and intestines, was conveyed directly to the liver to be

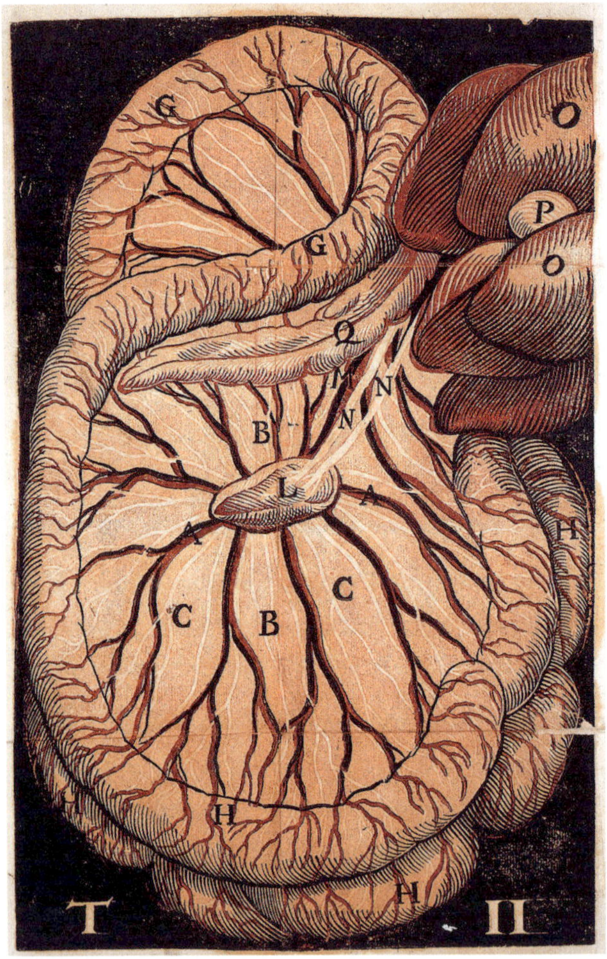

Figure 4. The mesenterial blood vessels (A) and Aselli's lacteal veins (B) on table II in his treatise about his discovery of the latter, *De lactibus sive lacteis venis, quarto vasorum mesaraico-rum genere* (On the milk vessels or milk veins, a fourth kind of mesenterial vessels), published in Milano 1627. It was the first coloured anatomical illustration ever seen.

C: Mesentery.
G: Jejunum (first section of the small intestine).
H: Ileum (last section of the small intestine).
M: Vena cava.
N: Larger lacteal veins with vena cava and the liver.
O: Liver lobe.
P: Gall bladder.
Q: Pancreas.
Wellcome Collection, London.

used in the formation of blood. Aselli's discovery of the lacteal veins was published in Milan in 1627, two years after his death, in a work with the title *De lactibus sive lacteis venis, quarto vasorum mesentericorum genere* (On the lacteal vessels or lacteal veins, a fourth kind of mesenteric vessels) (Figure 4).[3]

These two major discoveries gave rise to vehement discussions and disputes between physicians and other scholars with an interest in natural science until far beyond the middle of the 17th century, but precisely for that reason they also became the area of focus for research in medical science during the same period.

One of those who played a lively part and proved decisive in this debate and research, who continuously promoted the new knowledge, and who generally played a dominant role in 17th century medical development, both in Denmark and abroad, was the main character of this book, Thomas Bartholin. With a maternal grandfather, a father and an uncle who were all highly influential physicians at the University of Copenhagen, he had the best qualifications for pursuing a career in this field. His own abilities, his insatiable curiosity and his almost inhuman capacity for work were additional factors that contributed to paving the way to his status as one of the most famous and influential physicians in Danish history.

The legacy of Tycho Brahe

The scientific environment in Copenhagen in the 17th century, out of which Thomas Bartholin and his mentioned family members grew, was very much established by the Danish astronomer Tyge Ottesen Brahe, better known as Tycho Brahe (1546–1601) (Figure 5). Brahe was a nobleman intended for law studies, but from early childhood he was fascinated by the sky and the stars and ended up being one of the founders of modern astronomy.

In 1570, Brahe had returned to Denmark after many years of astronomical studies in different Lutheran cities in what is now Germany and Switzerland (Leipzig, Wittenberg, Rostock, Basel and Augsburg), during which time he had also learned to construct astronomical instruments. At his uncle Steen Bille's (1527–1586) estate, Herrevad Kloster in Scania (then Denmark, since 1658 Sweden), he had built an astronomical observatory. Here in the evening of 11 November 1572, when Brahe looked up at the clear sky to the north, with his naked eyes he spotted, in the constellation Cassiopeia, a new bright star clearer than Venus, which later turned out to be a supernova (Figure 6).[4] Being familiar with all the visible stars in the sky, Brahe was convinced that he had never seen this star before, and in his book about it, *De nova Stella* (About the new star),

Figure 5. Tycho Brahe (1546–1601) painted in 1596 by an unknown artist. Oil on canvas. 102 × 83 cm.

In the top-left corner is a symbolic depiction of the four elements, earth, water, air and fire. A hand from heaven holds a ladle over a burning obelisk surrounded by a ribbon with the words: STANS TEGOR IN SOLIDO VENTUS STREPAT, IGNIS ET UNDA (When I stand on solid ground, I am safe, even when wind, fire and sea roar). Two putti-like figures spit water, and two hands pour water out of jugs. In the upper right corner is the text: EFFIGIES TYCHONIS BRAHE OTTONI DANI ÆTATIS SUÆ ANNO 50. COMPLETO. QVO POST DIUTINUM IN PATRIA EXILIUM LIBERTATI DESIDERATÆ DIVINO PROVISU RESTITUTUS EST (Portrait of the Dane Tyge Ottesen Brahe painted after his 50th birthday, where after a long exile in his homeland, he regained his longed-for freedom by divine providence).

Brahe was the founder of the scientific environment in Copenhagen at the end of the 16th and the beginning of the 17th century. His empiric method resting upon experience through personal observations and measurements was adopted by later Danish scientists, including Thomas Bartholin and his famous pupil Niels Stensen (1638–1686). Skoklosters Slott. Photo: Jens Mohr. Wikimedia Commons.

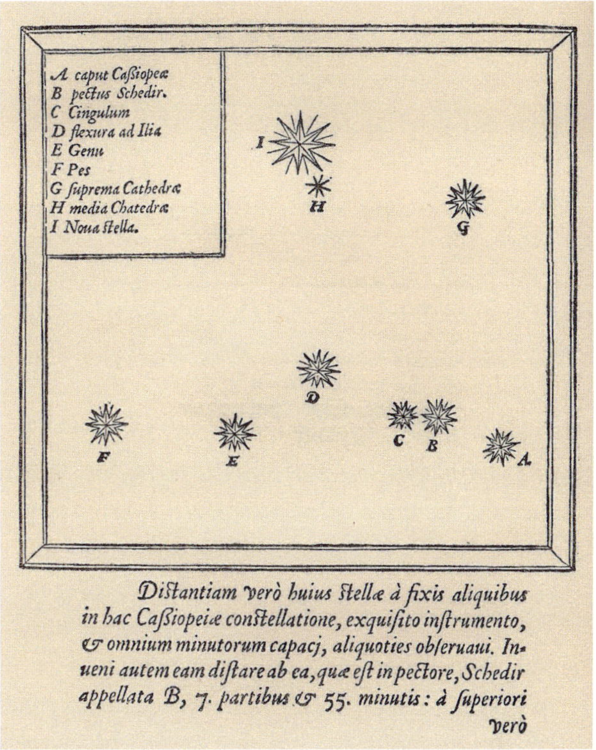

Figure 6. Tycho Brahe's depiction of the new star in Cassiopeia in his treatise about it, *De Nova Stella*, Copenhagen 1673.

The Latin text below the illustration up to the last colon reads: "I measured several times the distance between this star and various fixed stars in Cassiopeia with a specially selected instrument that records every detail. I found that its distance from the star in the chest, called Shedir (B), was 7 parts and 55 minutes." With his discovery of the new star, Brahe buried once and for all the Aristotelian doctrine that the sky beyond the Moon is unchangeable. Wikimedia Commons.

published in Copenhagen the following year, he therefore describes it as "in truth the greatest of all miracles that has occurred in Nature since the creation of the world."[5] In fact, Brahe was so surprised by his discovery that he initially did not believe his own eyes, but summoned witnesses to corroborate it: "I was surprised to such an extent that I could not believe my own eyes, but when I experienced that it could also be seen by others, when the location was pointed out for them, I was no longer in doubt that a new star had really shown itself there."

Until the disappearance of the star in the spring of 1574, Brahe continued to observe it from his observatory at Herrevad Kloster. With his self-produced astronomical instruments, he carried out exact measurements, showing that the star had no parallax,[6] but was a fixed star,[7] which did not change its position in relation to the other fixed stars, the zodiac or the Earth, and which was located much further away from the Earth than the Moon and the other planets then known in our solar system (Mercury, Venus, Mars, Jupiter and Saturn). Brahe concluded: "Therefore, this new star has neither its place in the elementary region below the Moon nor in the spheres of the seven wandering celestial bodies, but in the eighth sphere among the other fixed stars."

The discovery was an astronomical sensation, since it changed forever the ancient Aristotelian idea that the sky beyond the Moon is fixed. Professor in History at the University of York, David Wootton, argues in his 2015 book *The invention of Science: A New History of the Scientific Revolution* that Brahe's *nova stella* was the first crucial discovery in the history of science, and therefore the starting point of modern science, i.e. the Scientific Revolution, since it constituted a proof that there could be change in the heavens, and the discovery was made within astronomy, the only science that at that time worked remotely like a modern science.[8]

Brahe's discovery was noticed by King Frederik II (1534–1588) and the professors at the small and still quite anonymous University of Copenhagen. Despite his lack of academic titles, Brahe was invited to perform a series of lectures at the university. In his opening lecture in September 1574, which was attended by almost all the scholars of Copenhagen, he presented mathematics and astronomy as the basis of all the natural sciences, and he defined astronomy as an empirical science resting upon personal observations and experience.

In 1576 the king granted Brahe the island of Hven, northeast of Copenhagen, where Brahe built up his astronomical observatories Uranienborg and Stjerneborg and constructed the most exact astronomical instruments of his days (Figure 7).[9]

Over the next 20 years, Brahe created on Hven a unique international centre for astronomical science and learning, which was visited by scientists from all over Europe, and he turned astronomy into the first modern science.[10] He established his own printing mill on the island in order to publish and spread his observations and discoveries, which he also did in his numerous letters to leading European scholars and scientists.[11]

Figure 7. Tycho Brahe performing astronomical measurements in the company of assistants with his wall quadrant, Quadrans Muralis. In the background above, measurements are being made with other instruments of his, and in the centre of the image under his outstretched arm is his celestial globe. From the work *Astronomiæ Instauratæ Mechanica* (The renewed mechanical astronomy), Copenhagen 1598.
The Latin text at the top of the image reads "EFFIGIES TYCHONIS BRAHE O.F. ÆDIFICII ET INSTRUMENTORUM ASTRONOMICORUM STRUCTORIS A° DOMINI 1587 ÆTATIS SUÆ 40": "Portrait of Tycho Brahe in the year of our Lord 1587 in his 40th year. He constructed the building's workshop and the astronomical instruments."
Wikimedia Commons.

QVADRANS MVRALIS
SIVE TICHONICVS.

EXPLI-

In 1677 Brahe perforated another Aristotelian idea concerning the construction of the universe. On 13 November, Brahe observed a comet, of which in the following period he made systematic observations. Aristotle (384–322 BC) had considered comets as atmospheric phenomena, i.e. belonging to the atmosphere of the Earth or the sublunar heaven. Moreover, he believed that the planets were surrounded by impenetrable, solid spheres. But Brahe was able to prove from measurements that the comet had no parallax and was therefore located much further away from the Earth than the moon. Furthermore, he managed to show that the comet moved freely through Venus' sphere. He published these findings at Uranienborg in 1588 in the work *De Mundi Ætherei Recentioribus Phænomenis* (On the new phenomena in the heavenly world). Here he states: "It can be established with sufficient certainty that the whole heaven is transparent and clear and not filled with solid and real spheres. For the comets follow as a rule other courses than any heavenly sphere could make possible for them."

In the work, Brahe also explains his own view of the world, the so-called geo-heliocentric or Tychonic system, which was a combination of the Ptolemaic or geocentric system and the Copernican or heliocentric system, with the Earth and the sun respectively as the fixed centre of the universe. In the Tychonic system, the Earth was the fixed centre, around which circled the sun and the moon, while the other planets circled around the sun, which was mobile. Brahe emphasized that the Copernican system could not be proven with the instruments of the time, and by keeping the Earth as the fixed centre of the universe, Brahe's system was not in conflict with the Bible and was therefore accepted by both the Lutheran and the Catholic church.

The activities on Hven took place without interference from the state or the university, but to the benefit of both. Many of Brahe's pupils on Hven later became professors at the university.

Thomas Bartholin's uncle, the physician and antiquarian Ole Worm (1588–1654), writes about Brahe's pupils in a letter of 13 October 1647 to the French scholar Isaac La Peyrère (1596–1676) in Paris: "On my paper are the names of 32, who as students have been his pupils, and who one after another have lived with him on Uranienborg, where they have made astronomical studies. Some of them have later achieved teaching chairs, while others have become bishops and priests."[12]

On Hven, Brahe also had a chemical laboratory, in which he practised alchemy and produced alchemical medicine. His most famous medicine was his plague elixir, *Elixir Pestilentiale Tychonis Brahæi*, used for the treatment of plaque in

Denmark throughout the 17th century and included in the first Danish pharmacopoeia, *Dispensatorium Hafniense* (The Copenhagen pharmacopoeia), published by Thomas Bartholin in 1658:

Plague elixir
made by the Danish nobleman
Tycho Brahe
Take 11 ounces of theriac[13] essence (description follows)
1 ounce of sulphur tincture added to juniper and amber
1 half ounce of angelica extract, absinthe salt and essence of
white crayfish.

> Mix this together and in a retort, you pour as much wine spirits added to theriac, that after extract of theriac essence it covers four finger widths. Place the retort, which is well closed, in a hot water bath, so that the mass can evolve during eight days. Thereafter, it must be filtered through blotting paper and strained. The strained mass should be poured in a new retort and distilled, until there remains in the bottom a boiled down juice, which resembles liquid pitch.
>
> This is what should be used for the purpose. Mark it as above.[14]

Examples of positive mention of the effect of Brahe's plague elixir by Thomas Bartholin and Ole Worm will be given later. Both Bartholin and Worm were inspired and fascinated by Tycho Brahe. Bartholin mentions him in many of his books and letters and even wrote epigrams for him. In his poetry collection, *Carmina varii argumenti* (Poems with various contents) published in Copenhagen in 1669, there is an epigram for Brahe and his famous celestial globe (Figure 8):

For Tycho Brahe's copper globe

A device of cobber indicates the stars in the sky,
and shows, excellent Brahe, your genius.
Just as many years, as the celestial sphere turns,
just as long and for infinite time your fame will last forever.[15]

In his book with a short description of the Anatomy House in Copenhagen, *Domus Anatomica Hafniensis brevissime descripta*, published in Copenhagen in 1662, Bartholin includes another epigram in which he mentions Brahe:

GLOBVS MAGNVS ORICHALCICVS.

EXPLI-

Tycho discovered new stars, a German monk gunpowder.[16]
But a second Bartholin found the clear lymph fluid.
Fire, water and air, these three elements are sufficient
for learned men, let earth cover the rest.[17]

As the epigram shows, Brahe was a man that Thomas Bartholin gladly compared himself with, when he wished to emphasize his own major scientific discovery, the lymph vessels (see later). It could not be any greater. As will later be mentioned, Thomas Bartholin and his youngest brother, the physician and mathematician Rasmus Bartholin (1625–1698), each published a book in 1665 about the comet C/1664W1, which could be seen in the sky over most of the Earth in 1664–5. Rasmus Bartholin was strongly inspired by Brahe and was given the task, by King Frederik III (1609–1670), of publishing Brahe's unpublished handwritten astronomical observations, of which he managed to publish *Specimen Recognitionis nuper editarum Observationum Astronomicarum Tychonis Brahe* (An exploring sample of the nobleman Tycho Brahe's recently published astronomical observations), Copenhagen 1668.[18]

Ole Worm contributed to the French philosopher Pierre Gassendi's (1592–1655) biography about Brahe, *Tychonis Brahei, Equitis Dani, Astronomorum Coryphæi Vita* (The life of Tycho Brahe, Danish knight, astronomical pioneer), Paris 1654, which was the first biography of Brahe and the first of any natural scientist[19]. One of Worm's sources was Thomas Bartholin's other uncle, the professor of mathematics and astronomy Christen Sørensen Longomontanus (1562–1647), who had been among Brahe's pupils on Hven.

The background for Gassendi's biography, which illustrates the significant international interest in Brahe in the following generations, was the greatest quarrel of learned France during the middle of the 17th century, namely the quarrel between Gassendi and his followers about the preferred method of natural science: the Gassendists on the one side, and the great French philosopher and

Figure 8. On his celestial globe, Globus Magnus Orichalcicus (The great copper globe), Tycho Brahe marked the position of the fixed stars in the sky over the years. The globe caused quite a stir at the time. After Brahe's death, it had a wandering existence: it took up residence in a Jesuit monastery in Silesia, at Rosenborg Castle and at the Auditorium Superius of the University of Copenhagen. It ended its days in the observatory in the Round Tower, where it burned during the Copenhagen fire of 1728. From *Brahe's Astronomia Instauratæ Mechanica*, Copenhagen 1598. Deutsche Fotothek.

mathematician René Descartes (1596–1650) and his followers, the Cartesians, on the other. In his famous work *Discours de la Méthode* (Discourse on the Method[20]), Leiden 1637, Descartes had accounted for his method, which was based on rationalism and encouraged discussions and metaphysical theories. This method was criticized by Gassendi, who was a supporter of an empirical approach to science based on practical experiments and observations, or, in other words, experience gained using the senses. Besides a showdown with the new Cartesianism, Gassendi also wanted a showdown with Aristotelianism. Evidently, Brahe was a perfect example of a natural scientist, who fulfilled Gassendi's ideals for the scientific method and at the same time had a certain distance from crucial aspects of Aristotelianism. By writing a biography about Brahe, Gassendi was given the opportunity to use Brahe's example as a promotion for his own viewpoints in the quarrel with the Cartesians. On the other hand, Descartes himself was also interested in Brahe. In his short biography of Descartes, the French physician, chemist and botanist Pierre Borel (1620–1671) writes that Descartes visited Prague at a time with rioting in the city, "but nonetheless he managed to see Tycho Brahe's big instruments and had conversations with his relatives."[21]

In his guide to medical studies, *De Studio Medico*, Copenhagen 1628, Thomas Bartholin's father, Caspar Bartholin the Elder (1585–1629), recommends that medical students read Tycho Brahe's publications on astronomy and alchemy. Brahe is frequently mentioned in Caspar's small 1612 book about stars, *Astrologia seu De Stellarum Naturâ, affectionibus, & effectionibus* (Astrology or about the stars' nature, impacts and their exercise hereof).

In his notebook *Chaos*, written during his four months of literature study in his last year (1659) as a student at the University of Copenhagen, Bartholin's famous pupil Niels Stensen (1638–1686) refers to Tycho Brahe's astronomical letters, *Tychonis Brahe Dani Epistolarum astronomicarum libri* (Books with the Dane Tycho Brahe's astronomical letters), published at Uranienborg in 1696, which shows that Stensen read and studied these letters.[22] There is evidence in *Chaos* to suggest that the young Stensen also adhered to Brahe's model of the universe (The Tychonic system).[23]

Stensen did not carry out actual astronomical research himself, but letters show that he took part in astronomical observations. For example, he writes on 18 May 1668 in Florence about the great comet C/1668 E1, which was first seen in the sky over Lisbon on 5 March 1668 and in the following months over large parts of Europe: "The comet has also been seen by us in Livorno, but only the tail, because the head was either too close to the sun or at least hidden behind some big clouds just over the horizon."[24] On 23 October that same year, in Florence, Stensen took part in telescopic observations of Jupiter.[25]

But what is most interesting about Brahe in the context of Thomas Bartholin and Niels Stensen is his modern approach to science and his introduction to the empirical scientific method. He observed and measured with his own eyes and his own more and more accurate instruments, and he did it more frequently and systematically than any astronomer before him. He wrote down his observations thoroughly in protocols and published his results in books and letters in Latin, so that scientists all over Europe could learn about them, thereby also creating a great scientific network.

Brahe was already dead when Galileo in 1609 pointed his first primitive telescope towards the sky and saw the mountains of the moon and Jupiter and the stars of the Milky Way, published in his *Siderius Nuntius* (Starry Messenger) in Venice 1610. But another two generations had to pass before the accuracy of the telescope could match that of Brahe's non-telescopic instruments.

It has been said about Brahe that he was the first competent mind in modern astronomy to feel ardently the passion for exact empirical facts. His observations were some five times more accurate than the best available observations at the time. Brahe's approach to science, exploring Nature with his own eyes and instruments, is similar to the one practised by Andreas Vesalius and to the one later practised by Thomas Bartholin and Niels Stensen in their research. Bartholin puts it very simply: "We do not trust more, than we can feel with our hands and see with our eyes."[26]

Tycho Brahe's function as a source of inspiration for the physicians illustrates how sciences that we today consider as totally different – in this case, astronomy and medical science – learned from each other in the Renaissance and the 17th century, concerning substance as well as method, and how science, humanistic as well as natural, was considered a confluent unity. This can be seen as a contrast to our days, where we put different sciences into demarcated boxes characterized by a high degree of specialization, resulting in well-known advantages, but also in disadvantages in the form of loss of scientific outlook and broad scientific education.

After an introductory chapter on Thomas Bartholin's most influential source of inspiration, his father, the physician and anatomist Caspar Bartholin the Elder, this book follows Thomas Bartholin's almost ten-year educational journey to the leading European universities in medical science, his period as professor of anatomy and later medicine in Copenhagen, his research into the lymphatic system and discovery and naming of the lymph vessels, his many years as writing professor and administrator with an examination of his most important works, and his impact on his famous pupil Niels Stensen. The book finishes with a chapter on his significance for his own and later eras.

At the end of the book is a timetable of the most essential events in Thomas Bartholin's life (Appendix 1), a bibliography of his works (Appendix 2) and a ruling from The Committee for the Protection of Scientific Works (UBVA) in the dispute between Thomas Bartholin and the Swede Olof Rudbeck concerning the discovery of the lymph vessels (Appendix 3).

Caspar Bartholin the Elder 1

"We do not trust more than we can feel with our hands and see with our eyes."[1] Thomas Bartholin, 1652

Bartholin's textbook of anatomy

In 1611 there appeared in Wittenberg a textbook of human anatomy. Its format was small, and despite its 600 pages it was quite unimpressive. Noteworthy for a book on anatomy it did not contain a single illustration – apart from the head of Janus on the title page, symbolizing that it dealt with the anatomy of both sexes (Figure 9). It paled beside the 16th century's richly illustrated anatomy books, not least the visually impressive deluxe edition by the Flemish anatomist Andreas Vesalius, *De Humani corporis fabrica Libri septem* (On the structure of the human body in seven books), published in Basel 1543 (Figure 10). The paper quality and typography were also on the modest side, and new epoch-making scientific messages were hardly on offer. But it was, at that time, the first teaching book of anatomy, which gave the olfactory nerve (*nervus olfactorius*) its place as the first cranial nerve[2]. But apart from that, the book exclusively contained a presentation of all that was already known. Even the language featured nothing new: it was Latin, the language of learning at the time.

What made the book something special was its high pedagogical value. All unnecessary material was dispensed with, and the presentation was clear and easily understood, and above all the subdivision was logical and straightforward. The book began with the traditional division of the body into four sections with a topographical exposition respectively of the abdominal cavity (and sexual organs), the thoracic cavity (and the neck), the head and the limbs. Then followed, as a new feature, four shorter sections on veins, arteries, nerves, and bones, cartilage and ligaments. Furthermore, the book contained references to all relevant contemporary anatomical works together with mention of their writers and individual excellence. Its title was *Anatomicæ Institutiones Corporis Humani* (The anatomical arrangement of the human body), and it was written by the 26-year-old Caspar Bartholin,[3] Thomas Bartholin's father (Figure 11).

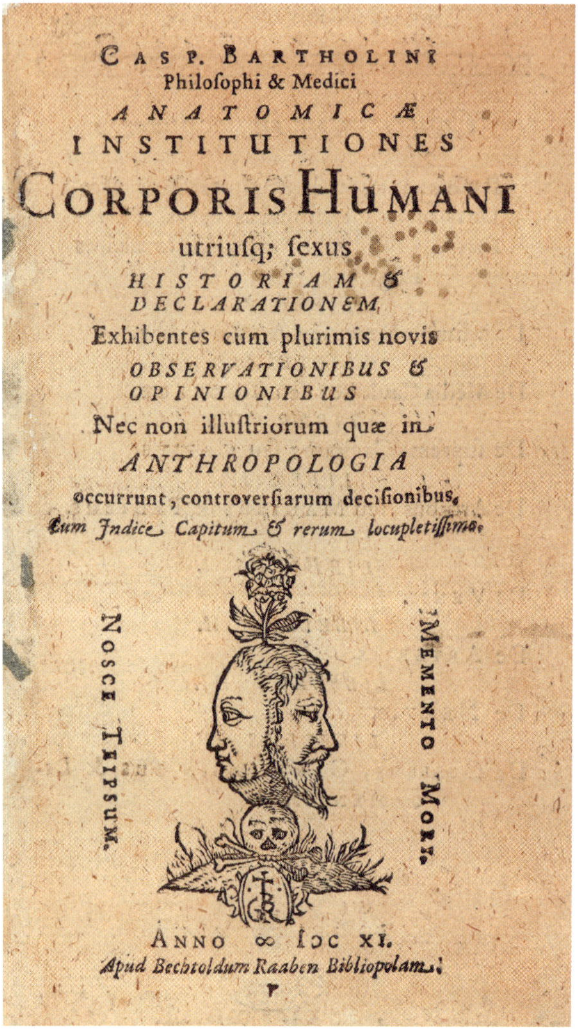
CASP. BARTHOLINI
Philoſophi & Medici
ANATOMICÆ
INSTITUTIONES
CORPORIS HUMANI
utriuſq; ſexus
HISTORIAM &
DECLARATIONEM,
Exhibentes cum plurimis novis
OBSERVATIONIBUS &
OPINIONIBUS
Nec non illuſtriorum quæ in
ANTHROPOLOGIA
occurrunt, controverſiarum deciſionibus,
Cum Indice Capitum & rerum locupletiſſimo.
NOSCE TEIPSUM.
MEMENTO MORI.
ANNO ∞ IϽC XI.
Apud Bechtoldum Raaben Bibliopolam.

Figure 9. Title page of Caspar Bartholin's anatomy book, *Anatomicæ Institutiones Corporis Humani utriusque sexus*, Wittenberg 1611.

The Janus head, which is the book's only illustration, symbolizes that it deals with the anatomy of both sexes. The book became one of Europe's most widely used textbooks of anatomy in the 17th century. The Royal Library in Copenhagen.

Figure 10. The frontispiece of Andreas Vesalius' *De Humani corporis fabrica Libri septem*.

The Latin text in the upper title field reads: "Andreas Vesalius of Brussels, professor at the medical school of Padua, on the structure of the human body in seven books." The Latin text in the lower title field reads: "With the grace and privilege of His Imperial Majesty, the King of France and the Venetian City Council, as stated in their diplomas." At the centre stands Vesalius himself, looking towards the reader, with his right hand in the abdominal cavity of the corpse of a young woman and the index finger of his left hand raised as a symbol of his new teachings, that encouraged making one's own observations through dissections of human corpses. The skeleton with the lance in the right hand was used by Vesalius as a model during his lectures. Above is the coat of arms of the Vesalius family with three weasels, symbolizing the German town Wesel, from which the family originated. The Royal Library in Copenhagen.

ANDREAE VESALII
BRVXELLENSIS, SCHOLAE
medicorum Patauinæ professoris, de
Humani corporis fabrica
Libri septem.
CVM CAESAREAE
Maiest. Galliarum Regis, ac Senatus Veneti gra
tia & priuilegio, ut in diplomatis eorundem continetur.

B A S I L E AE.

DOCT. CASP. BARTHOLIN9 THEOLOG9 MEDICus & PHILOSOPH9 AC SAC. LITER. T. ACAD. REG. HAFN. PROF. PUBL.
A° Chri 1625.
Ætatis 40.
BARTHOLINUS hic est, orbi quem misit Apollo
Philosophum, Medicum, Theiosophumq; pium:
Quod potuit, fecit glyptes; at corporis umbra est,
Et muti tantum: mentis imago deest.
Cernere vis istam? tot libros consule, queis se
Depinxit calamo, gens studiosa tibi!
Serenisß. Reg. Ma. Sculptor Si. de Pas ad vivum delineavit et sculpsit.
Ioan. Stephanus Strobelbergerus Medecinæ Doctor.

The book became one of the greatest successes of all time in Danish scientific literature. It quickly spread across the whole of Europe, where it became one of the most used books of anatomical instruction in the 17th century. It appeared in a further five reprints between 1622 and 1633, but Thomas Bartholin's four revised, illustrated editions are added to this, so that the book appeared in a total of 33 editions or impressions by 1688, including translations into French, German, English, Dutch and Italian.

At the book's first publication in 1611, Denmark was, in a European context, almost a developing country as far as anatomy and medical science were concerned. Medical students at the University of Copenhagen were obliged to make trips to other European universities to study, principally at Leiden, Basel, Paris, Montpellier and Padua.[4]

But the book gave the starting signal for a process that, during the 1600s, gave Denmark and the University of Copenhagen leading international status in anatomy and medical science. Thomas Bartholin more than anyone else personified this process. When his fourth and final revised edition of the book appeared in 1673, the University of Copenhagen had for a long time been an attractive educational institution for students. Its status had changed from a university where Danish medical students needed to go abroad, to one that foreign students sought to attend. With Thomas Bartholin's death in 1680, the days of glory for the Faculty of Medicine gradually faded.

Since Caspar Bartholin had considerable influence over his son Thomas, it is worth dwelling a little first on Caspar.

Figure 11. Caspar Bartholin the Elder (1585–1629), 40 years old. Copper engraving from 1625 by Simon de Pas (1595–1647). 11 × 7 cm. From the frontispiece of Caspar Bartholin's *Systema Physicum*, Copenhagen 1628.

The Latin text in the frame, "DOCT. CASP. BARTHOLIN/THEOLOG/MEDICUS & PHILOSOPH/AC SAC. LITER. I ACAD. REG. HAFN. PROF. PUBL. A°. Chri 1625. Ætatis 40," reads: "Dr. Caspar Bartholin, public professor of theology, medicine, philosophy and holy literature at Copenhagen University. In the year 1625. Age 40." The text under the portrait authored by the physician Johann Stephan Strobelberger (1593–1630) reads: "This is the Bartholin, whom Apollo allowed into the world as a philosopher, physician and devout theologian. The engraver did what he could, but it is nothing more than a shadow of his body and without words; a depiction of his spirit is missing. Do you want to see it? Then read his many books, where it was depicted with his pen for the benefit of you who study!"

Though Thomas Bartholin was only 12 years old when his father died, Caspar continued to be a great source of inspiration to him throughout his life.
The Royal Library in Copenhagen.

Figure 12. Epitaph for Thomas Bartholin's grandfather, the vicar Bertel Jespersen (1555–1613), and grandmother Ane Rasmusdatter Tinckel in Sankt Petri Church in Malmø (at that time Denmark, now Sweden). Oil on wood. Attributed to Peiter Hartmann and dated 1617. Photo: Jesper Brandt Andersen.

Figure 13. Memorial plate in Sankt Petri Church in Malmø for Thomas Bartholin's grandfather Bertel Jespersen, grandmother Ane Rasmusdatter Tinckel and father "Caspar Bartholin, public professor in medicine and philosophy at the Royal Academy in Copenhagen [Copenhagen University]." Photo: Jesper Brandt Andersen.

The influence of Caspar Bartholin the Elder

Caspar Bartholin the Elder, as he is often called to distinguish him from his grandson, the physician and anatomist Caspar Bartholin the Younger (1655–1738), was born on 12 February 1585 in Malmö during the reign of Frederik II. He was the son of the chaplain and parish priest Bertel Jespersen (1555–1613), who may be found on an epitaph for him and his wife Ane Rasmusdatter Tinckel, in Saint Peter's Church in Malmö (Figure 12), beneath which may also be found a memorial tablet, where Caspar Bartholin is mentioned (Figure 13). His paternal grandfather Jesper Pallesen was a farmer in Bigum in Jutland, and from him the family was descended (Figure 14).[5] Caspar Bartholin's boyhood name was Jesper Bertelsen, but as a young man he Latinized it, and from then on Bartholin became the family's new name.

Figure 14. Genealogical table of the Bartholin family.

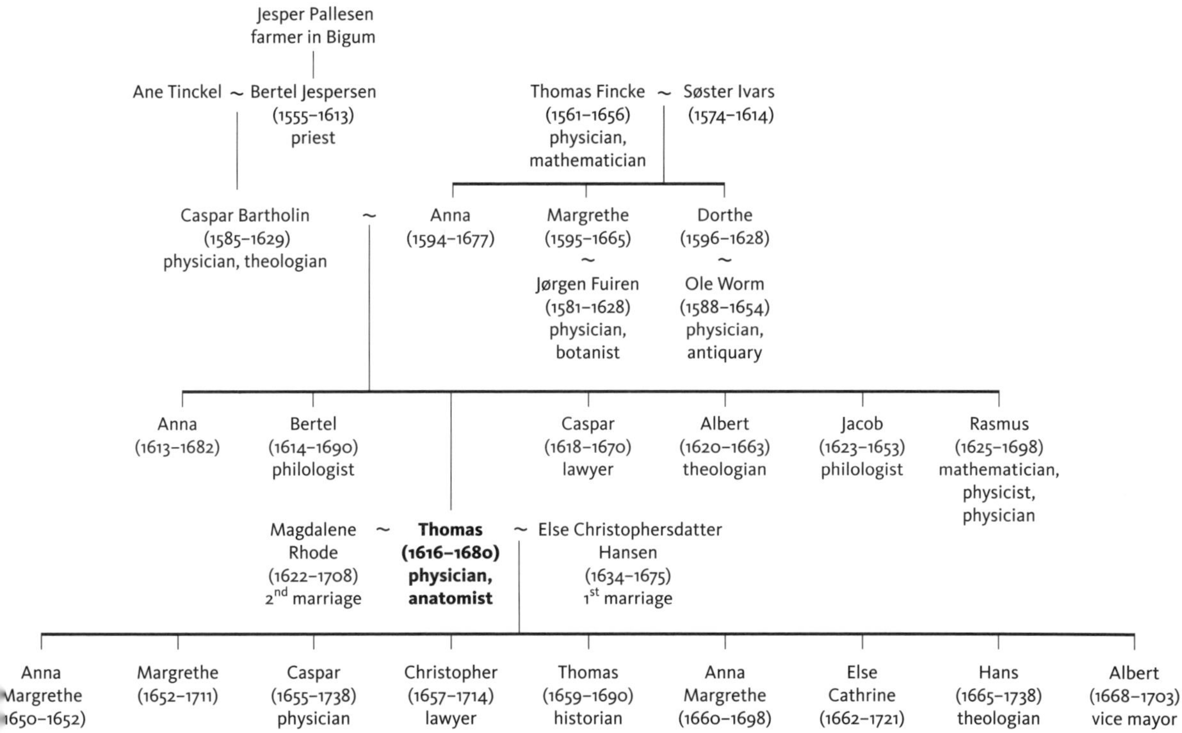

Caspar Bartholin began his studies at the University of Copenhagen at the age of 18 in 1603, but just a year later he absented himself on a seven-year study and educational journey in Europe. Since his economic resources for all essentials were limited to 60 rix-dollars a year for three years from the Chapter of Lund Cathedral, he undertook the journey on foot. Among the countries he visited were Germany, Holland, Belgium, France, England, Switzerland and Italy. He reached as far south as Sicily. His first long stay was in Wittenberg, where he studied philosophy and theology for three years, and in 1605 he took his master's degree.

To pay for his stay and save up for further travel he taught philosophy. After he had visited Belgium, France and England, he reached Leiden in Holland, where a period at the outstanding faculty of medicine inspired him to choose medicine as his main subject. After a short visit home and to Wittenberg, in 1607 he travelled to Basel, whose university with its two famous professors, the physician and anatomist Felix Plater (1536–1614) and his pupil, the physician, anatomist and botanist Caspar Bauhin (1560–1624), was likewise a leading institution in medicine. Caspar Bartholin also taught philosophy here, and in this connection produced a series of philosophical handbooks in logic, rhetoric, ethics, physics and metaphysics, which won great popularity and dissemination and made his name known in many European universities.[6] His first publication was *Enchiridion logicum ex Aristotele* (Handbook in logic according to Aristotle), Strasbourg 1608.

In 1609, Caspar Bartholin visited Rome, where he accompanied the pope's personal physician on visits to the sick, and Naples, where he was offered a professorship in anatomy, which he turned down. During the years 1608–1610 he had three sojourns at the University of Padua, the third among medical strongholds in Europe (Figure 15).

At this university, dissections had been carried out on human bodies since the beginning of the 1300s, chiefly on the corpses of executed criminals, which had been put at the physicians' disposal by the judiciary and those in power. It was here that the Flemish anatomist Andreas Vesalius had worked as a lecturer in anatomy and surgery from 1537 till 1543 and had undertaken the dissections, which constituted the background and the templates for the text and illustrations in his masterpiece already mentioned (Figure 16).

In an epoch-making way, Vesalius settled accounts with the ancient physicians Hippocrates and Galen, whose ideas concerning the anatomy and physiology of the human body and knowledge of diseases were almost statutory, but largely mistaken. Together with the major work of the Polish astronomer and physician Nicolaus Copernicus (1473–1543), *De Revolutionibus Orbium Coelestium Libri*

Figure 15. Palazzo Bo and the University of Padua.
a. Copper engraving in Giacomo Filippo Tomasini: *Gymnasium Patavinum* (The College in Padua), Utini 1654. The Royal Library in Copenhagen.

b. The building remains almost unchanged today and still houses the university and the oldest preserved anatomical theatre in the world. The motto of the University of Padua remains *Universa Universis Patavina Libertas* (The Liberty of Padua is universal and for all). Photo: Jesper Brandt Andersen.

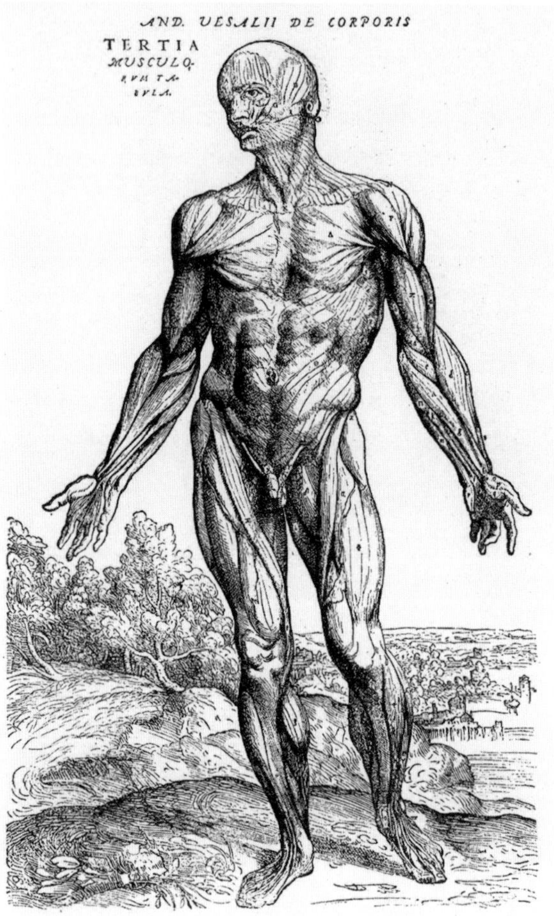

Figure 16. Woodcut of a muscle man on page 218 in Vesalius' *De Humani corporis fabrica*, Basel 1543.

For the preparation of the books' illustrations, Vesalius used some of the leading artists of the time, especially the German-born Italian painter Jan Stephan van Calcar (1499–1546), who was a pupil of Titian (1490–1576). Van Calcar presents the dissected human models in artful allegoric setups of an almost gallows-humoured character with Italian landscapes as background. The Royal Library in Copenhagen.

VI (Six Books concerning the Revolutions of the Heavenly Spheres), which proves that the sun is the centre of the universe, and that the planets (including the Earth) revolve around it, Vesalius' *De Humani corporis fabrica* of the same year is reckoned to mark the beginning of the scientific revolution in the Renaissance. Through his own dissections and demonstrations of dead human beings and living and dead animals, Vesalius was able to show that the ancient physicians' observations were erroneous, since they were solely based on the dissection of animals. When Vesalius began his medical studies, dissections took place with the physician or professor sitting elevated on a chair in the background, reading aloud from one of the ancient physicians' books and pointing out the structures named in the book for a uniformed barber, who freely dissected the specified structures, after which a third person, a demonstrator, pointed them out to the audience (Figure 17). Vesalius found this procedure absurd:

"This detestable procedure, in which someone or other carries out the dissection of a human body, and another produces descriptions of its parts, sitting like jackdaws high up on their tall chairs croaking away with out-and-out arrogance about something they have never investigated themselves, but only learnt by heart from other people's books. The first ones are so uneducated that they do not understand the books of anatomy but tear the corpse to pieces instead of dissecting it skilfully and make a mess of what they should show according to the instructions of the physician, who himself has never set hands on the dissection of a body. The whole lesson is bungled, and days are wasted with ludicrous business in such a confusion, that less is demonstrated there for the audience than a butcher can show in his shop."[7]

If medical science was to advance farther, the physician had to handle the scalpel himself and make his own observations to provide himself with the necessary knowledge through practical experience, rather than obtain it through its being handed down from others. With this new outlook in method, Vesalius founded a new school that paved the way for the 17th century's rapid development in med-

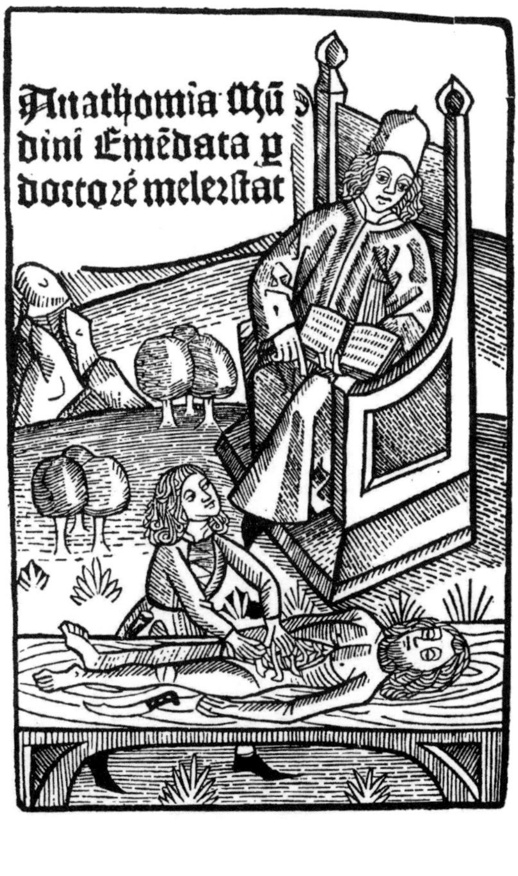

Figure 17. The frontispiece of Mondino's *Anatomy* from 1493.
Mondino de Luzzi (c. 1270–1326) was an Italian physician, anatomist and professor of surgery in Bologna. In the background the professor sits on a chair and reads from an anatomical text in a piece of classical writing while he points out the parts of the body to his assistant, an uneducated barber, who wields the knife during the dissection the body. There would often be a third person, who indicated the structures. Vesalius was highly critical about this form of instruction, which he called "a mess."
The Royal Library in Copenhagen.

ical science. One problem with this method was that, in the Renaissance, it was no easy matter to procure a human body for dissection, not so much on the grounds of resistance from the ecclesiastical and administrative authorities, who simply demanded that dissections should be carried out by experienced physicians in orderly conditions and be followed by an appropriate burial, but more for reasons of resistance among the general population, who had limited understanding of scientific objectives. An extract from a letter of the elderly Vesalius to one of his friends gives an impression of the inconveniences:

"Today I would not at all like to go for many hours and rummage through the bones in the Parisian cemeteries, where I was once at risk among wild dogs. Nor would I like to go alone and expose myself to the place of execution in Louvain to collect bones from the gallows to build up a skeleton. Nor would I trouble the judges by asking them to let people be executed by one or another punishment, or to postpone the execution to a time which was convenient for dissection, and not request the students to keep their eyes on the burials or remind them to take note of any disease in their lecturer's patients, so that they could later get their corpses. Nor will I any more for several weeks store a body in my bedroom, brought to me after a funeral or given to me after a public execution, and not let myself worry about grumpy draughtsmen, who made me more pitiful than those bodies I dissected. All these things, which I bore with an easy mind at my youthful age, with the wish to learn and to promote our joint studies."[8]

Vesalius had carried out his public dissections in provisional wooden anatomy theatres in Padua during winter, when bodies could last longer without decaying. In 1594, the then anatomy professor in Padua, Girolamo Fabrici d'Acquapendente (1533–1619), constructed from his own resources a permanent anatomy theatre, which still exists and is the oldest in the world to have been preserved (Figure 18).

One of the preferred research areas of Fabrici and his pupil Giulio Casserio (1552–1616) was the sense organs. Fabrici had published the work *De Visione, Voce et Auditu* (On the vision, voice and hearing), Venice 1600, while Casserio had published *De Vocis Auditusque Organis Historia Anatomica* (An anatomical description of the vocal and auditory organs), Ferrara 1600. At the time of Caspar Bartholin's arrival in Padua together with his future brother-in-law Ole Worm in the winter of 1608–1609, Casserio was working on a large, illustrated work about the sense organs, *Pentæstheseion*, Venice 1609 and Frankfurt 1610 (Figure 19).[9]

Caspar Bartholin, who must have become quickly trained in the technique of dissection by Plater and Bauhin in Basel and Fabrici and Casserio in Padua,

Figure 18. The anatomical theatre in Padua, which still exists. It was established by professor of anatomy Girolamo Fabrici d'Acquapendente (1533–1619) in 1594 and is the oldest preserved anatomical theatre in the world.

a. Copper engraving in Giacomo Filippo Tomasini: *Gymnasium Patavinum*, Utini 1654, p. 74. The Royal Library in Copenhagen.

b. Section of the theatre seen from the dissection table in the bottom of the room. The dissections were carried out in the light of wax candles. The width of the rows is approximately half a metre, and the height of the railing approximately one metre. Thus, people were standing close to each other, and people who became ill during a dissection could easily fall over the railing. To avoid decay of the bodies, dissections were only carried out during the cold winter months, and the duration of a dissection was limited to one week. In Copenhagen, the colder climate allowed the dissections to last for up to two weeks. Photo: Jesper Brandt Andersen.

helped (according to Thomas Bartholin) Casserio in Padua with the publication by carrying out numerous dissections, which provided a model for the draughtsmen creating the illustrations for the work:

"While he was with Casserio in Padua, he dissected numerous bodies, when on Casserio's behalf he made provision for the illustrations for *Pentæstheseion*. He was also a constant witness to the anatomical demonstrations, which were performed by d'Acquapendente, Bauhin, Plater and the others, who are mentioned in his introduction, and he lent them not merely his learned eyes, but also his scalpel."[10]

IVLII CASSERII PLACENTINI,

PHILOSOPHI, MEDICI ET ANATOMICI, IN VTRAQVE MEDICINA PATAVII VERSANTIS

PENTÆSTHESEION,

Hoc est

DE QVINQVE SENSIBVS LIBER,

ORGANORVM FABRICAM VARIIS ICONIBVS FIDELITER ÆRE INCISIS ILLVSTRATAM,

Nec non

ACTIONEM ET VSVM, discursu Anatomico & Philosophico accuratè explicata continens.

CVM PRIVILEGIO.

VENETIIS APVD NICOLAVM MISSERINVM

As a proof of his collaboration Casserio finishes his introduction to *Pentæstheseion* with a poem by Caspar Bartholin:

> He who does not perceive
> that this book of Casserio
> is written with great power of judgement,
> must himself be without senses.
> He who does not notice Casserio's masterly hand and acumen,
> and can hear, feel, smell, taste and see,
> must be deaf and insensible,
> his nose sluggish, his tongue paralysed and bereft of sight in both eyes.
> *Caspar Bartholin, Dane.*[11]

It appears from the preface to Caspar Bartholin's textbook of anatomy that, despite his young age, he instructed students in anatomy at the University of Padua, and he states the reason for its publication as the prospect of its functioning as a supplement to his lectures. He characterizes the book as a résumé, which does not claim completeness. It builds partly on what he has read and learnt, and partly on his own observations from dissections.

In 1610, Caspar Bartholin received news from the University of Copenhagen that they wanted him back in Denmark. He therefore forsook Padua that same year to journey homeward. On the way he made a stop in Basel, where he received his doctorate in medicine, and in Wittenberg, where he published his textbook of anatomy in 1611. He arrived home at the end of September of the same year and in December took up a post at the University of Copenhagen as *Pædagogus Secundus* (second professor of Pedagogy), the lowest rank of professor.[12] In 1612 he married Anna Thomasdatter Fincke (1594–1677), daughter of Thomas Fincke

Figure 19. The frontispiece of Caspar Bartholin's teacher in Padua, professor of anatomy Giulio Cesare Casseri's (1552–1616) beautifully illustrated work in folio about the sense organs, *Pentæstheseion*, Venice 1609 and Frankfurt 1610.

The Latin text reads: "The *Pentæstheseion* of Giulio Casseri of Piacenza, philosopher, physician and anatomist, who practises both types of medicine at Padua. This is a book about the structure of the organs of the five senses, faithfully illustrated with various images in copper engraving. Also containing their function and use, accurately explained with an enclosing anatomical and philosophical discourse. With privilege. At Venice at the house of Nicolao Misserini." Thomas Bartholin used several of the illustrations of this work in his various revised editions of the Bartholinian textbook of anatomy.
The Royal Library in Copenhagen.

Figure 20. The Worm family 1647.
Oil on canvas. Unknown artist.

Ole Worm is seen to the left wearing a black citizen dress with Christ sitting on his right side and his three wives and two oldest daughters standing on his left side. His children and grandchildren may be seen in the front. The four "apostles" on Worm's right side are his professor colleagues. Number three from left is his father-in-law and Thomas Bartholin's grandfather, Thomas Fincke (1561–1656). The two

men behind the row of women are the husbands of Worm's two oldest daughters. The biggest boy in the middle of the painting is Worm's son Willum Worm (1633–1704). The document to the far left is a poem by Thomas Bartholin saying: "Fruitfully married Worm was blessed with many children. In this painting you see the offspring of wisdom." The Danish National Museum. Photo: Lennart Larsen.

(1561–1656), the patriarch of the so-called Fincke dynasty and the influential leading professor of medicine (*Medicus Primus*) at the University of Copenhagen during the period 1603–1656 and five times its rector (Figure 20).

As a professor of 66 years' experience, Fincke belonged to the university's powerful elite and carefully saw to the progress of his family members' careers. With him as father-in-law it was not difficult for Caspar Bartholin to progress first to *Pædagogus Primus* or *Grammaticus* in August 1612,[13] and afterwards in June 1613 to second professor of medicine (*Medicus Secundus*),[14] succeeding the astronomer Gellius Sascerides (1562–1612), who had been one of Tycho Brahe's assistants at his observatory Uranienborg on the island of Hven.[15] In this job, Bartholin was rector of the university in 1618–1619.[16] When the first professor of theology (*Theologus Primus*), Cort Aslaksen (1564–1624), another of Tycho Brahe's assistants on Hven, died in February 1624, he was replaced by the second professor of theology Jesper Brochmand.[17] Caspar Bartholin then took on – possibly more out of necessity than desire – Brochmand's professorship, while his own was assigned to Ole Worm.[18] Bartholin continued in his theological professorship until his death at the age of 44 on a journey to Sorø on 13 July 1629, the year when he was university rector for the second time.[19]

In just 18 years, Caspar Bartholin managed to hold no fewer than four professorships in three different faculties. Despite his short life he left behind a comprehensive and far-reaching published output covering many subjects and, on account of his almost fifty books, most of which were published in several editions, he is stated to have been the most prolific professor at the University of Copenhagen in the first part of the 17th century and the most read Danish scientific writer in the same period – not least abroad.[20] Besides the aforesaid pedagogical publications and works about medical and theological topics, his works also include books about philosophy, astrology, alchemy and unicorns.[21] Moreover, he was such a popular professor of medicine among the students at the university that it gave grounds for jealousy among his colleagues.[22]

Besides the aforesaid pedagogical publications and works about medical and theological topics his authorship also includes books about e. g. philosophy, astrology, alchemy and unicorns.

As the breadth of his various professorships and writings reflects, Caspar Bartholin was a typical example of a polymath of his time, or in other words: a Renaissance man. He was a scientist, qualified in a broad range of scientific subjects. Perhaps because Bartholin was on the one hand an orthodox Lutheran and theologian, and on the other hand a physician, he saw no conflict between religion and science but conversely subscribed to the philosophy that these two fields hung closely together.[23] The good physician must also be a good natural philosopher, as expressed by Bartholin's own words in Ole Worm's guest book of 1608 in Basel: "The physician who is a philosopher is an equal to the gods;"[24] and a good theologian must have a certain knowledge of God's creation, Nature. It is the physician's duty to turn over every stone and acquire knowledge in all areas of science if he just keeps in mind that, irrespective of what reason and experience show him, the whole thing is the result of God's creation. According to Bartholin, science can be used as a tool to show the world how great God's creation is.

Although Caspar Bartholin as a physician and scientist was based on the doctrines of Aristotle, Hippocrates and Galen, and as a teacher at the University of Copenhagen was obliged to follow these doctrines in accordance with the current charter of 1539,[25] his polyhistoric basic philosophy brought about an openness to theories which it is hard to believe that either a theologian or a physician would be open to. He recognized, for example, astrological theories about the meaning of the heavenly bodies' location with regard to the sickness and health of mankind, the curative strengths of precious stones (which could be further strengthened by the engraving of star signs),[26] the existence of unicorns and their horn's medical effects, together with the disputed theories and ideas about the value of alchemy and iatrochemistry of the Swiss physician and alchemist Paracelsus, alias Theophrastus Bombastus von Hohenheim (1493–1541), and his Danish associate, Peder Sørensen, alias Petrus Severinus (1542–1602), physician-in-ordinary to Frederik II and Christian IV and friend of Tycho Brahe. In some instances, Bartholin associated himself with, or in any case implicitly accepted, certain ideas which, if they had been put forward by ordinary people, might have entailed punishment for heresy. As a recognized scientist, Bartholin avoided this danger, and he was further helped by the fact that Christian IV himself showed interest in alchemy and Paracelsism, and that Bartholin could refer to him as a prominent alchemist.[27] Bartholin sought in vain to deal with his own kidney stone disease with various kinds of nephrite stone, which was brought to him at no small expense from Venice, Nuremberg and Holland, and he explained its lack of effectiveness by saying that the nephrite stone can perhaps drive out a renal calculus and stone, which is loose and soft in consistency, whereas his own kidney stones were unusually large and as hard as flint.[28]

Although Caspar Bartholin as a physician took a starting point that was basically loyal to Galen's teaching and theories, his time in Padua and Basel had drawn him in the direction of the Vesalian teaching of anatomy and opened his eyes to Galen's errors, and at the same time taught him to obtain his own observations and experiences through dissections and experiments. Considering this, it is perhaps not so surprising that Bartholin had a receptiveness to Paracelsus and Peder Sørensen and in certain connections paid direct allegiance to the latter. Paracelsus and Peder Sørensen were both supporters of experience through experiments (Paracelsus' *experimentia*), though with the proviso that the knowledge that one obtains from this is not only revealed through experience and rational thought but also needs insight and penetration. At the same time, they were against dead book learning and out-of-date authorities. In his renowned philosophical masterpiece, *Idea Medicinæ Philosophicæ* (Ideal of Philosophical Medicine) of 1571, Peder Sørensen phrases it in this way:

> "The right to use one's judgement, Nature's illumination, the ability to distinguish and estimate, men should not deprive one another of. The Greeks' absolute sovereignty belongs to the past. It would be shameful to use someone else's diligence in research, effort and results as a cover for one's own lethargy."[29]

A further indication that Caspar Bartholin's change from professor of medicine to that of theology was possibly not his own wish is that, as theological professor, he continued to occupy himself with natural science and medicine. In 1628 he published, for instance, apart from the gargantuan 1300-page book of physics, *Systema Physicum* (The system of physics), a students' guide, *De Studio Medico* (On the study of medicine), written in 1626 for medical students. This was intended and dedicated to his relative Peder Charisius (1608–1685), a grandson of Peder Sørensen, as well as to Bartholin's six sons: Bertel (1614–1690), Thomas (1616–1680), Caspar (1618–1670), Albert (1620–1663), Jacob (1623–1653) and Rasmus (1625–1698).[30] It was therefore obvious that Bartholin wished for one or more of his sons to study medicine, and two of them, Thomas and Rasmus, followed his wishes. That his father's guide influenced Thomas, and that it contributed to his choice of career, is indicated by the fact that he republished it as a part of his own work, *Cista Medica Hafniensis* (The archive chest of the medical faculty in Copenhagen) in 1662. Thomas' title of the guide reads: "Caspar Bartholin's abbreviated and extemporaneous instructions relating to the progress of the medical study for the use of the diligent future physician, who raises himself above the masses." It appears that Caspar Bartholin's personal experience with medical practice also includes "even royal persons, princes, counts and barons, together with an immense number of distinguished citizens and common people." As a good Lutheran he calls first and foremost to piety, a blameless course of life, daily prayers and love of God's word. Everyone who studies medicine must read a chapter or two of the Bible after prayers each morning and preferably go through the whole of it every year. Next, the student must learn Greek and Latin, the languages of medicine. With the sentence: "How will you judge what is good or bad in Hippocrates and Galen, if you do not understand Greek?", Bartholin reveals that he follows the new Vesalian school, which looks upon the ancient Greeks with a critical and nuanced eye, and he invites the guide's readers to do the same. After this, Bartholin goes through a list of disciplines of which the physician must have knowledge, some more important than others, and he indicates relevant authors and their literature, included in which are his own writings (on rhetoric, logic, metaphysics, physics and alchemy).[31] Among the other Danish writers he points to are Christen Sørensen Longomontanus (arithmetic and astronomy), Tycho Brahe (astronomy and alchemy), Christoffer Dybvad (1577–1622, geometry), Peder Sørensen (alchemy and Paracelsian theory) and Christian IV (alchemy). He naturally indicates his own teachers, Plater, Bauhin, Fabrici and Casserio.

Besides the obvious medical subjects like medicine itself, medical questions and disputations,[32] anatomy, surgery, dietetics, diagnostics, "special therapy" and pharmacology, he includes logic (knowledge's ingenious helper), physics (including astronomy, optics and astrology), mathematics (including arithmetic and geometry), mineralogy, metallurgy, botany, zoology and chemistry (including alchemy and Paracelsian theory) among the important subjects, while rhetoric, practical philosophy and metaphysics are named as less important.

For several of the subjects, he demands practical fieldwork. In anatomy, where the first writer he names is Andreas Vesalius, to whom he also gives praise for his illustrations, he requires frequent arrangement of private dissections of animals and the parts of their bodies, and as often as possible the witnessing of public dissections of human bodies. In surgery, where Vesalius also plays a prominent part, you "should not, to improve yourself, refrain from practising 'hands-on' in Italy and France, where you have Europe's best surgeons, and in your daily training pay attention to the essential points, which not every hack or barber is aware of." In pharmacology, one must "often investigate even jars or small spaces at the pharmacies, so that the physician shall not be unaware of something that the apothecary's assistant knows about." In botany you need to "collect plants and establish herbaria with your own hands, rather than reading comprehensive botanical works." And for the person concerned with alchemy Bartholin has the following request: "You must always practise alchemy with its charcoal and glass flasks." Moreover, concerning this disputed science, Bartholin declares:

"I personally am not unaware that some people despise this kind of mystery and do not care very much for it but limit themselves to medical treatment. However, should the latter be carried out, the other may not be neglected. In a desperate situation you can, not infrequently, miss one or another outstanding drug. Indeed, you must never be ashamed to pump away at old crones and barbers about their household remedies, if they are simply effective and have been used successfully."

Caspar Bartholin finishes his instructions with a request for study travels abroad.[33] "On the strength of these few examples everyone – with the exception of fools and imbeciles – will understand the importance that foreign travels hold for the prospective medical doctor."

Hereby, the scene was set for his son Thomas' career in medicine.

Study Travel in Europe

2

"*Semper ardentius*" (Always more ardently)[1] Thomas Bartholin, 1637

Thomas Bartholin was born on 20 October 1616 in Copenhagen, the second eldest son in a family of seven children, among whom, apart from the five brothers already mentioned, was also a sister, Anna (1613–1682). At their father's death, Ole Worm (Figure 21), their uncle, took on the role of parental guardian, and right up to his death he followed, supported and inspired the young Thomas through thick and thin.[2] Several signs indicate that Thomas Bartholin's medical career was planned at a very early stage by either Caspar Bartholin, Ole Worm or both. In any case, in 1631 Thomas was placed for three years in the home of his deceased uncle, the physician and botanist Jørgen Fuiren (1581–1628), and of his maternal aunt Margrethe (born Fincke, 1595–1665), where together with his cousins, Henrik (1614–1659) and Thomas Fuiren (1616–1673), he received personal tuition in medicine from, among others, Frederick III's personal physician, Poul Moth (1601–1670).[3]

On 26 April 1634, Thomas Bartholin matriculated at the University of Copenhagen. At that time, the first three years of medical studies consisted mainly of studies in humanistic subjects (philology, philosophy and theology), which was concluded with a *testimonium publicum*[4] before the student could go forward seriously with the medical subjects, which, as mentioned, required a study trip abroad.

Thanks to his early private tuition Thomas Bartholin probably had a decent advantage in his medical knowledge and insight from the beginning of his studies at the university, and he continued his medical studies in his first years at the university. Among his teachers was Ole Worm. That Thomas Bartholin's medical knowledge must already at this early date have been quite solid is attested by the fact that, on 16 November 1636, he was the respondent in the first of Worm's five disputations during the years 1636 to 1640, published in Copenhagen 1640 with the title *Institutionum Medicarum Epitome* (Epitome of medical instructions). Worm had the intention of forming from these disputations a coherent volume suitable as a textbook for new medical students,[5] as suggested by Bartholin in

Figure 21. a. Ole Worm (1588–1654) c. 1654. Oil on canvas. Attributed to Karel van Mander (1609–1670). See also the text for Figure 96. Aarhus Katedralskole.

b. Portrait engraving by G. Wingendorp after Karel van Mander's painting. From *Museum Wormianum*, Amsterdam 1655. 26.5 × 17 cm.

The Latin text in the frame, "OLAUS WORM D. MEDICINÆ IN ACADEMIA HAFNIENSI PROFESSOR REGIUS A°. M.DC.LIV. Ætat. XVI," reads: "Ole Worm, Royal Professor of Medicine at the University of Copenhagen in 1654, 66 years old."

Thomas Bartholin's epigrams under the portrait, left and right, respectively, can be translated as follows: "Thus, dear reader, he looks, the one you can thank for being enriched with Nature's gifts. He contained so many treasures and was himself a treasure. Therefore, Worm should be included among these great treasures," and "Behold, here is the splendour of the world and the wonder of the fatherland, a proud old man's honourable face. Here hides everything that learning, which is based on a gigantic labour, contains, and all the magnificence that exists in the world."

Worm himself had asked Bartholin for the epigrams and was apparently pleased with them. On 8 July 1654, he wrote to Bartholin in his plague hideout in Roskilde: "Tomorrow I will send to Amsterdam my portrait painted by van Mander. Do you have an epigram for it? What did you give him for the portrait of you?" On 15 July, he writes again to Bartholin from plague-stricken Copenhagen: "Thank you for your beautiful verse for my portrait."

The knowledgeable and versatile Ole Worm was a source of inspiration for many scientists in the following generations. He was married to Thomas Bartholin's maternal aunt. When the 12-year-old Thomas Bartholin lost his father in 1629, Worm took over the role as his guardian and supervised and supported him afterwards through thick and thin. The Royal Library in Copenhagen.

ECCE viri vultum cujus de munere, Lector,
Maxima Naturæ munera munus habes.
Tot cimelia complexus cimelion ipse
WORMIUS hæc inter tanta legendus erat.

CONSTANTER.

ORBIS cerne decus, Patriæ miracula terræ,
Ora triumphantis suspicienda Senis.
Hic latet, immensi quicquid doctrina laboris
Possidet, ingenii quicquid in orbe patet.

TH. BARTHOLINUS D. P. Reg.

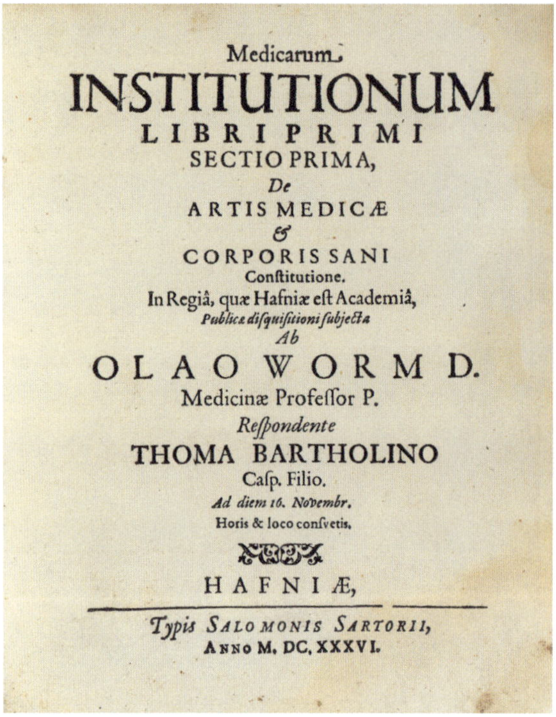

Medicarum
INSTITUTIONUM
LIBRI PRIMI
SECTIO PRIMA,
De
ARTIS MEDICÆ
&
CORPORIS SANI
Conſtitutione.
In Regiâ, quæ Hafniæ eſt Academiâ,
Publicæ diſquiſitioni ſubjecta
Ab
OLAO WORM D.
Medicinæ Profeſſor P.
Reſpondente
THOMA BARTHOLINO
Caſp. Filio.
Ad diem 16. Novembr.
Horis & loco conſvetis.
HAFNIÆ,
Typis SALOMONIS SARTORII,
ANNO M. DC. XXXVI.

Figure 22. Title page of Ole Worm's disputation, *De Artis Medicæ & Corporis Sani Constitutione*, held at Copenhagen University on 16 November 1636 with Thomas Bartholin as opponent. It was published in Copenhagen in 1640 as part of the work *Institutionum Medicarum Epitome*.
The Royal Library in Copenhagen.

a letter to Worm the same year: "I must always remember your services, which the public learns of in your writings, and I personally through your letters. From the former you can expect immortal fame, and from me a grateful mind. Eager for both, I almost persuaded our Marcus[6] (not Tullius[7], however) – if you simply give me the terms for the work – to reprint in a smaller size your basic medical rules, so that the fruits of the tasteful summary may come into use for more people. Once I have myself, together with others, sucked them from your mouth and observed them during public presentations at the university of the fathers."[8]

The 30-page first dissertation *De Artis Medicæ & Corporis Sani Constitutione* (On the art of medicine and the constitution of a healthy body) is about the anatomy and physiology of the healthy body, including the transport of the special nutrient-rich lymph called chyle (Figure 22).

The study tour in Europe 1637–1646 commences

In the summer of 1637, Thomas Bartholin, in company with his brother Bertel, set off on a study tour of Europe, as his grandfather Thomas Fincke, his father Caspar Bartholin and his uncle Ole Worm had done. His destinations were largely the same as theirs, and Bartholin was greatly helped by the fact that he could profit from their good names and connections. The necessary economic support for the journey came – through Ole Worm's mediation – from his grandfather's grant and later from the university.[9] However, with its duration of almost ten years, the study tour lasted substantially longer than anyone in the family expected, and Fincke and Worm put tremendous pressure on Bartholin towards the end to make him come home. Worm followed his nephew's travels closely and continued through frequent letters to give fatherly advice, support and warnings, and his correspondence with Thomas Bartholin is thus one of the most valuable sources for the journey.

The residence in Leiden 1637–1640

Thomas Bartholin's travels began with a voyage to his first goal, the University of Leiden (Figure 23), where he matriculated on 11 July 1637 as a student of philology and medicine.

The University of Leiden offered optimum conditions for medical studies, both concerning learning power and physical surroundings. It had been established in 1575, when the town's population had been given a choice between a university and ten years' freedom from taxes as a reward for having resisted a year's siege by the Spanish.[10]

On the first floor of the little Gothic church, Faliebagijnekerk, at the beginning of the 1590s a university library was set up in one half of the apse (Figure 24), and an amphitheatre for anatomical dissections and demonstrations in the other half (Figure 25).[11]

In January 1594, Pieter Pauw (1564–1617), professor of anatomy and botany, inaugurated the anatomy theatre with a two-week-long dissection of a human body. The previous year, the University had gained permission to receive the bodies of all executed criminals in the Netherlands. At the same time a botanical garden with a broad representation of medical and other plants from all parts of the globe, including the Dutch tropical and subtropical colonies, was established under the direction of the Flemish botanist Charles de l'Écluse (Carolus Clusius, 1526–1609) (Figure 26).

The teachers, some of the most notable of whom will be mentioned below, were reckoned to be among Europe's foremost anatomists. In his first letter home to Ole Worm on 6 March 1638, Thomas Bartholin tells of considerably higher living costs in Leiden than in Copenhagen, and he is worried as to whether the grant provision of 200 rix-dollars a year will stretch to cover his expenses for food, accommodation, washing, clothes, books and private lessons.[12] In addition comes wine, which "serves to keep one's frail health in good shape [...] since

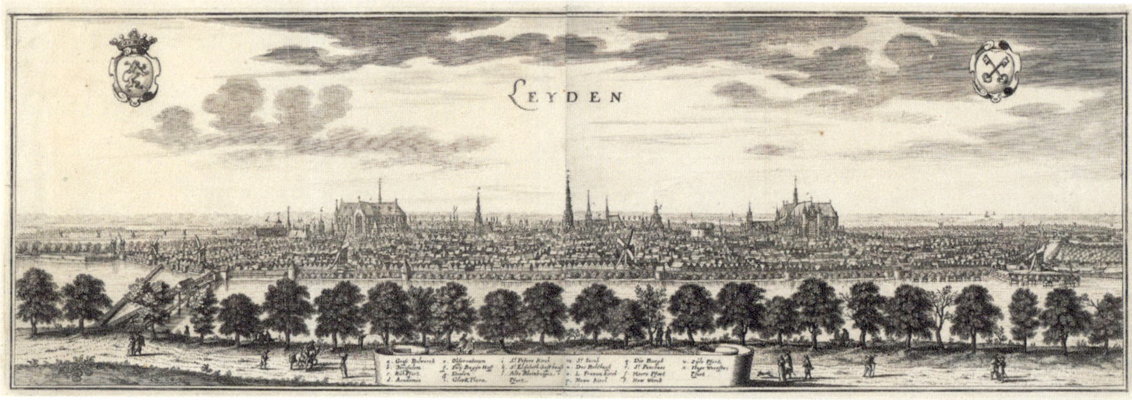

Figure 23. a. Leiden c. 1650. Engraving by Caspar Merian (1627–1686). 21.5 × 63 cm. From Martin Zeiler & Matthäus Merian (eds.): *Topographia Germaniæ inferioris*, Frankfurt 1659.
Rijksmuseum, Amsterdam

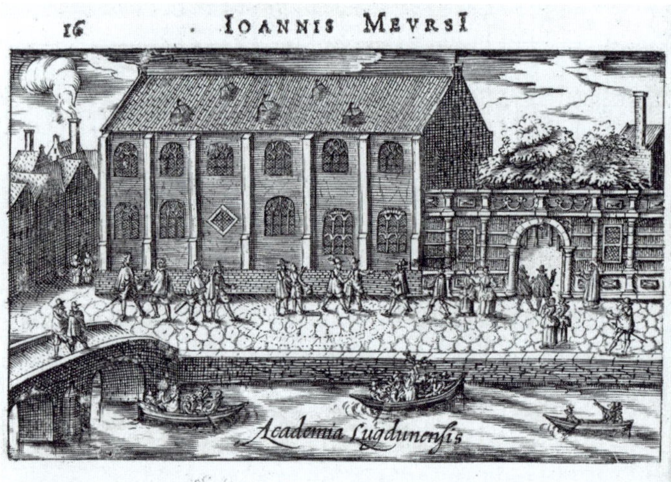

b. Leiden University in 1625. Engraving in Johannes van Meurs (1579–1639): *Athenæ Batavæ. Sive, De Urbe Leidensi, & Academia, Virisque claris; qui utramque ingenio suo, atque scriptis, illustrarunt: Libri duo* (The Dutch Athens. Or, On the City and University of Leiden, and the famous men, who have distinguished it by their intellect and writings: In two books), Leiden 1625, Liber I, p. 16.
Wikimedia Commons.

c. The seal of the University of Leiden on the stones at the entrance to the university. Leiden was one of the first cities in The Netherlands to shake off the Spanish yoke in the 16th century. Consequently, the first university of the Protestant Netherlands was established here in 1575. In 1595 the former Faliede Bagijnhof Church was converted into a university annex, which housed among other departments the anatomy theatre, the library and the botanical garden. This chapel was modified with the establishment of an upper storey as illustrated by the two rows of windows.
Photo: Jesper Brandt Andersen.

Figure 24. Leiden University Library in 1610. Engraving by Willem Isaacszoon van Swanenburg (1580–1612) from a drawing by Jan Corneliszoon van 't Woudt (1570–1615). 33 × 40 cm. Hand-coloured version from Willem Blaeu: *Stedeboeck der Nederlanden*, Amsterdam 1649.

The Latin text, "BIBLIOTHECÆ LUGDUNO-BATAVÆ CUM PULPITIS ET ARCIS VERA IXNOGRAPHIA," reads: "The Leiden University Library, with a true iconography of the pulpit and the citadel."

In the well-stocked library, Bartholin and his fellow students could immerse themselves in medical literature, and on the side wall to the left under the windows, they could view the 11-metre-long panoramic prospect of Constantinople made by the Danish engraver Melchior Lorck (c. 1526–after 1583).
Wikimedia Commons.

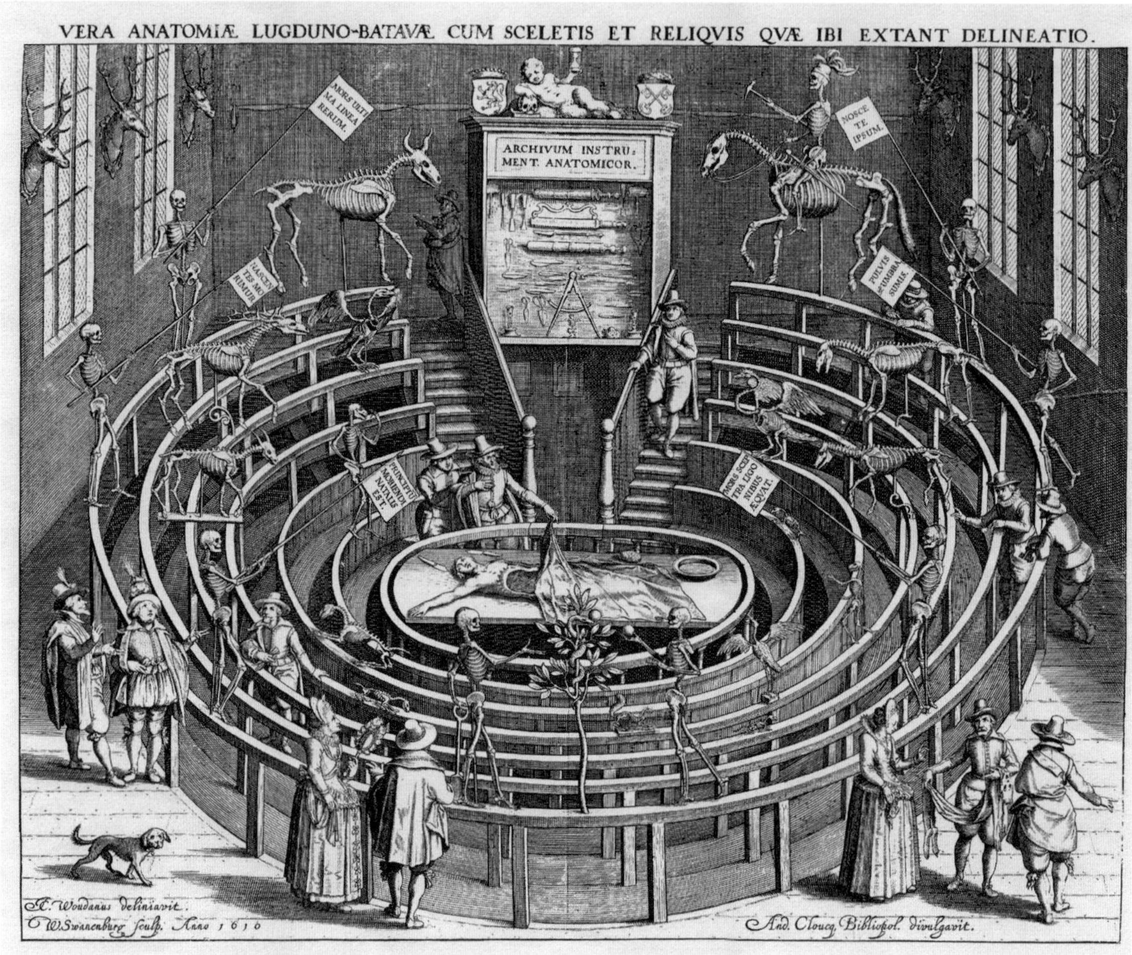

Figure 25. The anatomical theatre in Leiden in 1610. Engraving by Willem Isaacszoon van Swanenburg after drawing by Jan Corneliszoon van 't Woudt. 33 × 40 cm.

The theatre was inaugurated by the anatomist Pieter Pauw (1564–1617) in January 1594. In consideration of the durability of the corpses the dissections took place during the winter. The rest of the year the theatre served as a museum for natural objects with exhibited skeletons of humans and animals as well as stuffed animals. Here you could come to be reminded about your mortality and taught from moral messages on the banners hold by the skeletons. In the front are the skeletons of Adam and Eve (Adam with the spade and Eve with the fruit) separated by the Tree of Knowledge. Over the entrance hangs a cupboard with dissection instruments. On the dissection table an opened corpse has been placed out of season, and in the front of the right side the flayed skin of an executed criminal is shown to shock and warn a noble lady. It seems as if the small dog in the lower left welcomes us to this horror cabinet with a melancholic expression.

The Latin heading, "VERA ANATOMIÆ LUGDUNO BATAVÆ CUM SCELETIS ET RELIQUIS QUÆ IBI EXTANT DELINEATIO," reads: "Accurate drawing of an anatomical demonstration in Leiden with skeletons, and what else is present there." The theatre has been recreated in Rijksmuseum Boerhaave in Leiden.
Rijksmuseum Boerhaave in Leiden.

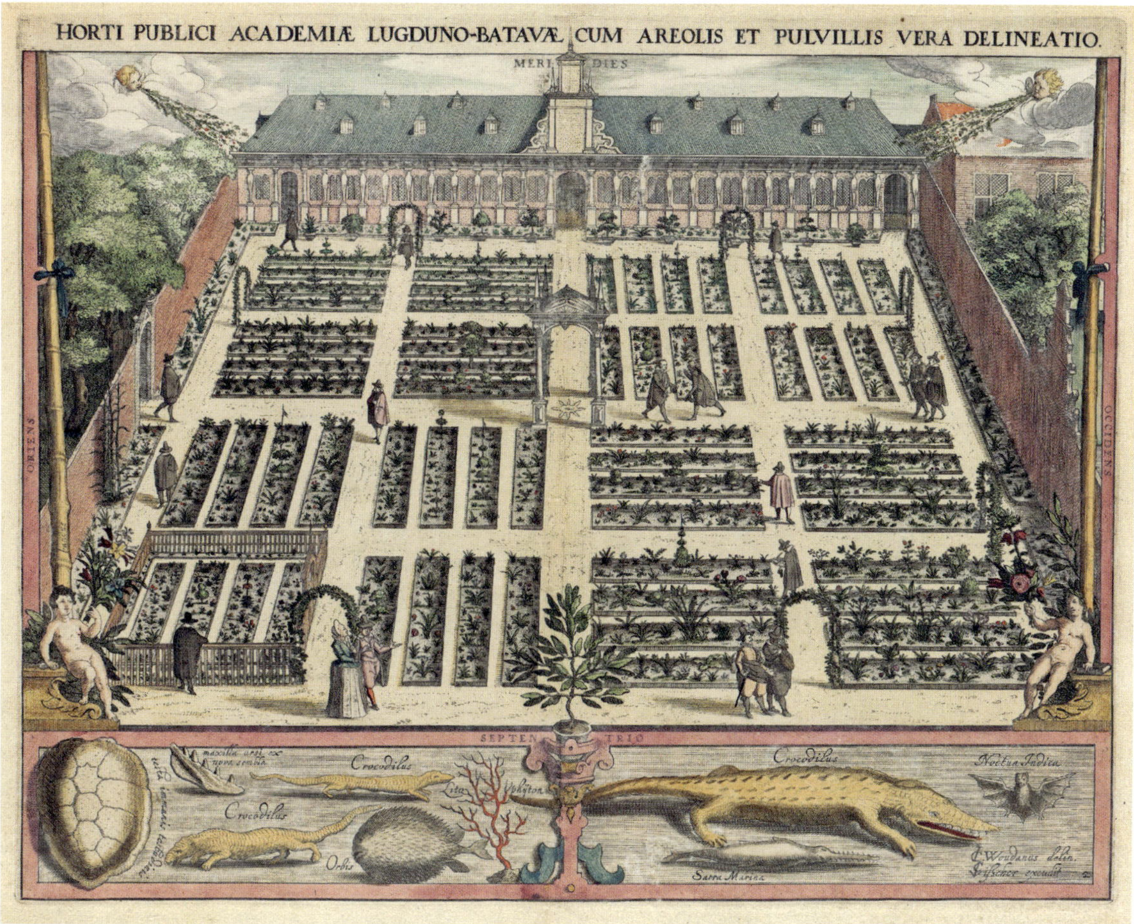

Figure 26. The botanical garden in Leiden in 1610. Hand-coloured engraving by Willem Isaacszoon van Swanenburg from a drawing by Jan Corneliszoon van 't Woudt. 33 × 40 cm.

The Latin heading, "HORTI PUBLICI ACADEMIÆ LUGDUNO BATAVÆ, CUM AREOLIS ET PULVILLIS VERA DELINEAIO," reads: "An accurate drawing of Leiden University's public garden with allotments and flower beds." On the strength of the country's many tropical colonies, the botanical gardens in Holland were well stocked with exotic plants, among them medical plants. To emphasize this, the foreground of the drawing shows various exotic animals, including crocodiles, exotic fish, a tortoise and an Indian bat. Arctic animals are also represented in the form of a piece of the upper jaw of a bear from "nova sembla" (Novaya Zemlya).
Wikimedia Commons.

care of the body is very necessary in these areas, which almost swim in the sea." Bartholin has had fever and is now suffering from constipation. He frequently takes medicines, although the medical school forbids it, but "the familiar physical exercises, for example dancing, boxing, ball games and the like, are unaffordable."[13] About his studies he explains:

"Last winter three dissections of human bodies were arranged. Otto Heurnius[14] dissected only one single individual, one of female sex. But Falcoburg,[15] who with his meticulousness and precision in demonstration, teaching and dissection far surpasses Heurnius, opened and dissected two bodies of male sex. I confess that I attended the first, but the remainder I was unable to because of time; for I do not refrain from other studies to such an extent that I can wholeheartedly occupy myself with medicine."

From the beginning of his journey, Bartholin, as prescribed in his father's study guidance, prioritized acquiring a broad knowledge and a general education, stretching far beyond medicine. After one year in Leiden, he finally chose medicine – to the displeasure of his mother, who had hoped that he would become a theologian,[16] but to the delight of Worm, who on 19 June 1638 writes:

"From your latest letter, which was brought to me first, I could, it seemed, conclude that you had said goodbye to theological studies and turned your mind to medicine, and I want to know if you are serious about it. For if you stick to your decision to devote yourself to it, I shall not stop encouraging such noble endeavours, not merely with advice, but also with action. I shall not use persuasion, but you must follow simply where your heart takes you. No one has ever produced anything of excellence without the consent of Minerva."

When Bartholin has made his final decision, he writes to Worm on 23. August: "But this intention of mine, to which I have been impelled by a particular inner voice, I have very recently submitted to my honourable male relatives." Worm repeats his support on 9 October: "Since it is your decision to devote your energies to medical studies, I shall not dissuade you but pray for complete success and good fortune for you from Almighty God." Bartholin nevertheless continues to devote part of his time to philology and philosophy, "for it will not be amiss to join these to the aforenamed [medical studies] as companions or rather as helpers." Worm also approved of this: "I am very pleased with your plan, according to which you have decided to unite philology and philosophy with the art of medicine; for how profitable these are, not simply with medicine, but also with other branches of science, you have experienced for some time." On 7 January, Worm writes to Bartholin: "However, I invite you, when you have the time, to particularly devote yourself to the study of philology, which you have made a start on, for an excellent style is the greatest recommendation for the subject you have been dealing with, and as the world stands now, it has no taste for anything else than the decorated." Bartholin very much complies with this advice. At the end of his stay in Leiden he studies Arabic with the Dutch orientalist and mathematician Jacob Golius (1596–1667),[17] he discusses Danish antiquities with the German orientalist and philologist Johann Elichmann (1600–1639), and he tries, after Elichmann's death, to secure his collection of Icelandic, Persian and Arabic manuscripts for Denmark.[18]

In a letter to Worm of 23 August 1638, Bartholin relates that he has asked the professor of anatomy, Johannes Walæus (Jan de Wale, 1604–1649) (Figure 27) for advice on which medical writers he ought to read. Walæus' recommendations appear below. Bartholin only states the writers' names, and apart from his father's textbook of anatomy does not mention specific works. For every writer I have inserted the work that is most probably the one in question.

Anatomy:

Caspar Bartholin: *Anatomicæ Institutiones Corporis Humani* (The anatomical arrangement of the human body), Wittenberg 1611.

Adrianus Spigelius:[19] *De Humani Corporis Fabrica Libri Decem* (On the structure of the human body in ten books), Venice 1627, Frankfurt 1632.

Principles of medicine:

Jean Fernel:[20] *Universa medicina* (The world of medicine), Paris 1567, Frankfurt 1577, Geneva 1580.

Daniel Sennert:[21] *Practicæ medicinæ* (Practical medicine) I-VI, Wittenberg 1628–1635.

Physiology:

Cæsar Cremonius:[22] *De calido innato, et semine, pro Aristotele adversus Galenum* (On intrinsic heat and seed, in defence of Aristotle and against Galen), Leiden 1634.

Johann Sperling:[23] *De origine formarum* (On the origin of species), Wittenberg 1634.

Antonio Ponce de Santa Cruz:[24] *De impedimentis magnorum auxiliorum in morborum curatione* (On the hindrances to effective means in the cure of diseases), Madrid 1629.

Caspar Bartholin, *Anatomicæ Institutiones Corporis Humani*, Wittenberg, 1611.[25]

The repeats from Caspar Bartholin's *De Studio Medico* are Daniel Sennert, Jean Fernel, Adrianus Spigelius and Caspar Bartholin himself.

Figure 27. Johannes Walæus (1604–1683). Engraving by Solomon Savery (1594–1683).

The Latin text in the frame, "JOHANNES WALÆUS ANT. FIL. ZELANDUS. MEDICINÆ D. ET PROFESSOR IN ACADEM. LUGD. BAT. ORDINARIUS. NATUS XVII DECEMB. ANNO CIↃ IↃ CIV DENAT. V IVNI A°. CIↃ IↃ CXLIX," reads: "Johannes Walæus, son of Antonius from Zeeland. Doctor of medicine and ordinary professor at the University of Leiden. Born 17 December 1604. Died 5 June 1649."

Through dissections of live dogs, Thomas Bartholin's skilled anatomy teacher in Leiden convinced him of the validity of William Harvey's controversial theory on the blood circulation. Walæus' encouragement of Bartholin to republish Caspar Bartholin's textbook of anatomy in a revised, illustrated edition, which was accompanied by two letters from Walæus in support of Aselli's and Harvey's discoveries, paved the way for Bartholin's fame. Walæus also demonstrated to Bartholin that the heart and vena cava of animals continue to contract, long after the animal has been euthanized. This research inspired and was followed up by Bartholin's pupil Niels Stensen. Leiden University Libraries Digital Collections.

Botanical studies

Early during his stay in Leiden, Thomas Bartholin began to spend time studying in the University's botanical garden. On 15 May 1638, he writes to Worm: "In order to enrich your garden I am sending the last edition of the catalogue of the local botanical garden and ask for instruction on which plants you are short of. This year no plants have arrived from India. I follow zealously the official botanist Vorstius,[26] who teaches about the plants. When I have made progress in this and other parts of the art of medicine, I hope for further instruction from you. [...] For your museum I am sending you two small animals brought from India. In Dutch they are called *Schildtpadden*."[27]

Apart from his garden of rare plants, Worm had begun a collection of scientific, archaeological and ethnographical objects in 1620, the year before he took up the post of professor of physics at the University of Copenhagen. It included mainly rarities from foreign lands. In his answering letter to Thomas Bartholin on 19 June 1638, he himself calls the collection "my collection of Nature's curiosities," but in his own work on the collection, which was published in Leiden and Amsterdam in 1655, the year after his death, he called it *The Museum Wormianum* (Figure 28).[28] Compared with modern-day museums, of which the collection was a forerunner, Worm's collection had a practical purpose. As something new at the University of Copenhagen he introduced it as a tool in the instruction of his pupils by teaching them to identify objects in the collection, which accordingly came to serve as a study-collection – the first of its kind in Denmark. Princely and bourgeois collections of rarities brought home by travellers from foreign parts of the world were not uncommon in the 1600s, but Worm's also belonged internationally among the first scientific collections, and it was one of the first in Europe to become used for practical training sessions in the teaching. Worm continued throughout his life building up his collection, and Thomas Bartholin was a continual contributor of objects to it throughout the whole of his study travel. As will be discussed later, Thomas Bartholin built up his own similar, but significantly smaller collection.

Before the summer vacation, Thomas Bartholin was shown round the botanical garden by the professor of botany, Adolphus Vorstius (1597–1663), and spent part of the summer studying botany. In order that his newly acquired knowledge of plants should not drop out of his memory over the following winter, Bartholin established a growing herbarium. He intended to send plants and seeds home to Worm. At the entrance to the botanical garden a notice board was set up with regulations, which prescribed that one was allowed to look at the plants and smell them, but they must not be damaged. The boughs, flowers and seeds must not be removed, and bulbs not dug up.[29]

Nevertheless, Bartholin succeeded in persuading the gardener several times to let him have some samples that he could either include in his own herbarium or send home to Worm. In a letter to Worm from 1 June 1639, Bartholin regrets that his messenger taking seeds to Worm has been so slow that the springtime has now passed, but he then adds: "While I was thinking over these lines to you, there conveniently arrived, as if on a given cue, a ship laden with costly goods from Phoebus' cradle, in other words the West Indies. For with it the excellent Dr. Piso[30] [Figure 29], personal physician to Moritz, Prince of Orange, governor

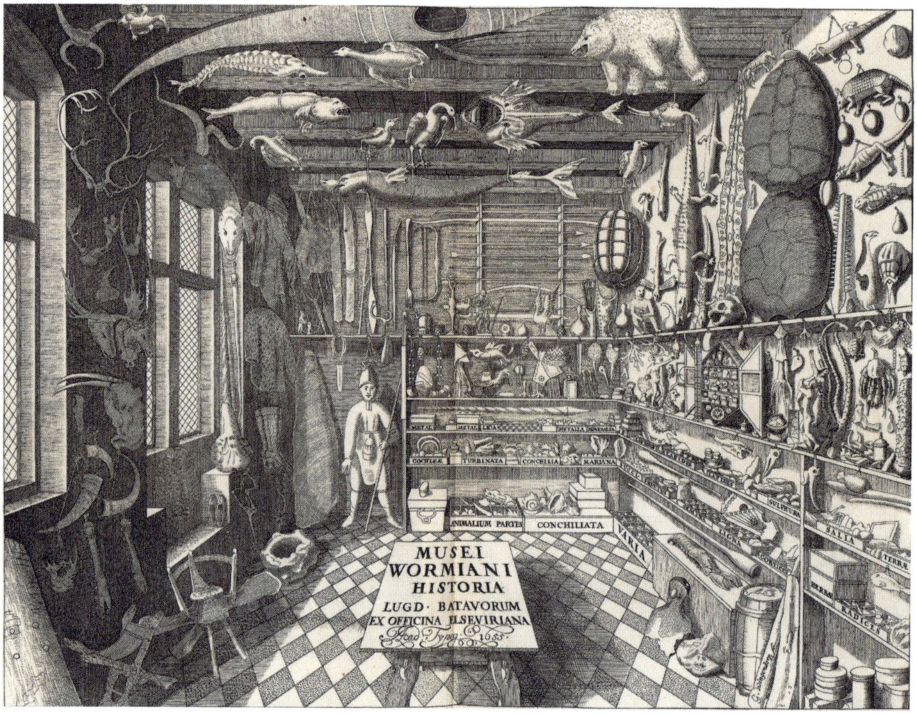

Figure 28. The frontispiece of Ole Worm's catalogue of his cabinet of curiosities, *Museum Wormianum*, Leiden 1655. Engraving by G. Wingendorp.

The illustration shows the museum shortly before Worm's death in 1654. Thomas Bartholin diligently collected objects for the museum during his almost ten-year study trip in Europe. Worm used the objects in the teaching of his students, thereby transforming his museum into a study collection. He was the first teacher at Copenhagen University to do this, and one of the first in Europe. Thomas Bartholin later did the same thing with his own similar collection. Worm's museum was the first private collection of its kind in Denmark and a kind of precursor to present-day museums. Therefore, he is generally considered the founder of the Danish museum service. Upon his death, Worm's collection was included in the King's Art Chamber ("Kongens Kunstkammer"), which later became the core of the National Museum of Denmark.
The Royal Library in Copenhagen.

Figure 29. The coloured frontispiece of Willem Piso's (1611–1678) *Historia Naturalis Brasiliae* (The natural history of Brazil), Leiden and Amsterdam 1648.

Piso was a Dutch physician and natural scientist who took part in a Dutch expedition to Brazil in 1637–1644. During his stay there, he studied the Brazilian fauna and flora and sent home to the botanical garden in Leiden seeds from tropical plants, as mentioned by Bartholin in a letter to Ole Worm. Through his knowledge of tropical diseases, he became one of the founders of tropical medicine. From the text of the frontispiece, it appears that the book does not describe only plants and animals, but also the indigenous diseases, characters and customs. It has more than five hundred illustrations. Wikimedia Commons.

in the West Indies, had sent as an ornament for the University-garden seeds of various rare plants, good-sized fruits of the cassia tree, a slough and a piece of a sugar cane. [...] I am sending a pod of the cassia tree [Figure 30], which I have wrested from the gardener with difficulty. As far as I understand from the renowned Vorstius' letter, it has been made clear through experience that no purgative properties are to be found in this cassia, as opposed to what people generally believe, for they customarily lessen the suggested quantity of this cassia on the grounds that it is supposed to present better qualities than the oriental species. In the same place it is said that amongst the Brazilians there are whole woods planted with these trees, and that their pods, when the wind makes them strike against each other, produce such a noise that the passers-by are terrified. As far as I can perceive with my own eyes, the ones in the west are different from those in the east; the latter are round, the former flatter. For the moment nothing else occurs to me that might satisfy your mind's desire. For I remember that an ostrich egg, one of which is in my possession, is suspended in your very well supplied collection, and the spring is necessary for the care of the rarer seeds, towards which you direct your mind's desire. [...] Our Vorstius is energetic in exhibiting the garden's plants, and in the university lectures about medical issues he deals with simple purgatives."

As the quotation shows, tropical plants in great number were brought home from the colonies in the tropics, planted out in the botanical gardens of the European universities and used as medicines. An example of such an effective medicine was the Peruvian Incas' cinchona bark, which could cure malaria. This will be discussed later.

Figure 30. Four exotic plants in a landscape, including to the left the cassia tree (*Cinnamomum aromaticum*). Engraving from Olfert Dapper: *Naukeurige Beschrijvinge der Afrikaensche Gewesten* (Accurate Descriptions of the African Regions), second edition, Amsterdam 1676, between pages 130 and 131. 21.3 × 33 cm.

The bark from the cassia tree has a scent and taste like real cinnamon and is used as a spice in cookery, particularly in India and China. Cassia cinnamon has been shown to be able to lower blood sugar levels. Wellcome Collection, London.

Instruction in practical medical profession

The University of Leiden had also come to the fore in another area. It was the first university in North Europe to offer clinical tuition to its medical students in a public hospital, the St. Cecilia Hospital. In a letter of 3 October 1638, Thomas Bartholin writes to Ole Worm about this instruction:

"Last Saturday I attended the dissection of a human body in the public hospital, where twice every week Aesculapius' pupils and professors in practical medicine hurry together to carry out investigations in Nature, causes and cures of diseases. We were absorbed in exploring the hidden cause of death, but did not find that any of the essential parts behaved in an unnatural way, and therefore the cause was ascribed to the vitality or the fluids."

As this conclusion shows, people of this epoch had not completely abandoned the ancient physicians' doctrine that illness was due to a weakness in the vitality or libido and an imbalance between the four body humours, blood, phlegm, black and yellow bile, the so-called humoral pathology. The humoral pathology among other things formed the basis of the treatment of a long list of diseases, including the use of bloodletting in the hope of restoring the balance between the body's fluids.

That same winter, Bartholin gives Worm a more ample report about the teaching. The professors instruct the medical students twice a week in practical medicine at the public hospital, where thorough research into diseases and remedies takes place. For this purpose, a catalogue of the hospital's pharmacy and garden has been published. Every week four of the medical students are chosen, who must attend to the patients every day and look after them, so that they receive the medications, which are prescribed after a common consultation between physicians and students. "Indeed, they have even constructed an anatomical theatre in the same building, where the corpses of the dead, infected with diseases, are opened up and inspected by the surgeon under the supervision of the physician," writes Bartholin to Worm on 4 January 1639. Bartholin describes three dissections in which he has participated: an executed criminal with atypical abdominal muscles and an enlarged liver and spleen; a man from Holstein, who after a hard blow on his right side towards his spine had sustained injuries to his ribcage and abscesses in his thoracic cavity; and a man who had died of consumption with excrescences around his spleen, in other words, presumably a cancer. In addition to this practical instruction, Bartholin participates in private theoretical lessons with Johannes Walæus. But Bartholin also benefits from younger teachers: in the botanical garden's colonnade the young Dutch physician Franz de la Boë (Franciscus Sylvius, 1614–1672) (Figure 31) carries out private anatomical demonstrations with the University's permission.

Bartholin writes about this instruction in 1674: "In Leiden it was Otto Heurnius, the hospital physician, and along with him at private dissections Franciscus Sylvius – my contemporary, who at that time had still not gained his professorship, which he so well deserved, but first achieved many years afterwards – who assisted me in the finest manner and opened up many bodies for me and others, who were engaged in this topic, and with thanks and a prayer for both their souls I maintain that this has been a great help to me."[31]

In the letter of 3 October 1638, Bartholin asks for Worm's advice in a case that surprises him. It is a good example of his early capability for critical medical reflection. A Dr. Elichmann in Leiden is apparently able to cure "people who suffer from any kind of illness" with a particular pill. Bartholin would like to know "if such a universal panacea exists, and whether it should consist of vege-

Figure 31. Franciscus Sylvius (Franz de la Boë) (1614–1672). Engraving by Cornelis van Dalen the Younger (1638–1664).

During Thomas Bartholin's stay in Leiden 1637–1640, he and his fellow students were taught about the anatomy of the brain by their peer Sylvius at private anatomical demonstrations in the colonnade of the university's botanical garden. Bartholin struck up a friendship with Sylvius, which later benefited Niels Stensen during his stay in Leiden 1660–1664. Sylvius was one of Europe's leading neuroanatomists. His skilful teaching and dissection technique of the brain formed the basis for Thomas Bartholin's and later Niels Stensen's great insight into the brain's complicated anatomy.
Wellcome Collection, London.

table or mineral material. [...] For the diversity of illnesses, which depend on differences in causes and in patients, occurs to me also to demand a diversity of remedies." Worm answers on 7 January 1639, that he has discussed the content of pills with Elichmann, and that their basic ingredient should be a thick juice of warm ingredients extracted through distillation. They should work against constipation and possibly against "bowel restriction and consumption," and according to Worm doubtless because of the vitality in the minerals. But Worm does not believe they should work against all types of illness, and he knows people who have taken them without success.

In a letter dated 9 October 1638, Worm urges Bartholin to extend his studies to two further areas: "I would be so happy if you would make the acquaintance of one the city's distinguished apothecaries, as well as a surgeon (that is, if it is not possible to reside with one of them!). Through the one, as a sort of variety in your leisure time, you will obtain acquaintance with all the bits and pieces that are in use among physicians. Through the other you will gain the opportunity to observe the subtler operations and cases, whereby you can refresh yourself, when you are tired of the more serious things. France will perhaps bring you into the chemists' sanctuary in due time." Worm knew that Bartholin's next destination would be France.

Revised edition of Caspar Bartholin's's anatomy book, including Aselli's lacteal veins and Harvey's theory of circulation

In the last months of his stay in Leiden in the spring of 1640, Bartholin becomes engaged in a large task, which he explains to Worm 27 March:

> "Among our booksellers there are some good men who are asking for a reissue of my father's textbook of anatomy in a fine print, for no other book has such a great distribution here. According to what the famed Walæus draws my attention to, they have some perfectly accurate, newly made copper-plate illustrations. But with Walæus as intermediary they have asked me if here and there in a few words I might add the new observations and opinions that our age has brought after my father, especially Aselli's and Harvey's, whose illustrations they have also reproduced."

The fact was that Gaspare Aselli's discovery of the lacteal veins and William Harvey's discovery of the circulation of the blood had taken place after Caspar Bartholin had published his textbook of anatomy in 1611.

Both discoveries were controversial, because they finally did away with the Greco-Roman physician Galen's teaching, which largely built on theoretical conclusions in preference to practical experiments, and which had been regarded as a kind of law in medicine for 1500 years. According to Galen, the food underwent a kind of boiling in the stomach, by means of which the nutritional liquid chyle was produced. This passed further to the intestines, became absorbed in the veins of the mesentery and passed further through them to the blood-forming organ, the liver, where it became engaged in blood formation. From here the nutritious blood passed through the vein system, partly to the arms and legs, partly to the right ventricle, where it was divided into two streams: one that passed out into the lungs, and another that passed through invisible pores in the heart's septum to the left ventricle. Here it became mixed with air and was heated up before being sent through the arteries to the rest of the body. According to Galen the pulse is not generated by the heart, but by the arteries themselves.[32] Harvey, on the other hand, emphasized strongly that the blood is not produced continuously in great quantities in the liver, but that it is the heart with its contractions that drives the blood round the whole body through the arteries, and that the blood does not pass from the right to the left half of the heart through invisible pores in the heart's partitions, but through the lungs. Furthermore, Harvey demonstrated the existence of vein valves, which ensure that the blood in the veins can only flow in the direction of the heart. The missing link in Harvey's circulation theory was that he failed to explain how the blood came from the arteries to the veins. This problem was first resolved in 1661, when the Italian physician Marcello Malpighi (1628–1694) with the help of a microscope discovered the capillaries in the lungs of a live frog.[33] Aselli's settlement with the doctrine of Galen was a product of his discovery that the nutritious chyle did not reach the liver through the vein system, but through a hitherto unknown kind of vessel, the lacteal veins, a kind of lymph vessel.

Ole Worm was presumably the first in Bartholin's family to direct his attention to Aselli's and Harvey's discoveries. On 26 May 1631 he writes to medical student Jacob Svabe, who was staying in Leiden: "Since you write that during this winter you have witnessed three dissections of the human body, I would be glad to know whether your anatomists have pointed out Aselli's lacteal veins in the mesentery? Likewise, I would be pleased to know what they think about the circulation of the blood in the body according to the opinion of this Englishman. This winter I have observed it myself in a dog, when I was following the author's instruction. The subject is surely one of great significance and usefulness. Consult your people on this matter."

Bartholin's interest in the lymph vessels was probably already awakened, when during the period 1631–1634 he was instructed by Poul Moth, who demonstrated Aselli's lacteal veins for him,[34] but it was Ole Worm who got him to take up the matter again during his residence in Leiden. In a letter to Bartholin on 7 January 1639 Worm writes: "I am most eager to learn what your anatomists' opinion is about the lacteal veins, and if they demonstrate them." Bartholin answers on 15 February that in the past week he attended Adrian van Valckenburg's dissection of a 20-year-old girl and "you could not see the lacteal veins […] they disappear, before you can observe them thoroughly, and they disappear of course, because the white chyle flows rapidly in great amount to the liver in order to produce the amount of blood that is needed to sustain the spirit of life." Bartholin then goes on to explain what happens to the chyle during dissections of live animals: "When it is consumed owing to the great pain during dissection, the chyle at the animals' death becomes mixed with the amount of blood and easily obliterates these veins. Walæus made this clear to me, but the rest joined the old notion by claiming that the lacteal veins were neither new nor unusual, and this opinion I see also put forward by Plemp,[35] professor in Louvain, in his newly published *Basic principles or instructions in medicine*.[36] He says that it is simply mesenteric veins mixed with white chyle."

Bartholin follows up this topic in a letter to Worm of 26 April 1639: "In the week just gone by, we have twice at our own provision dissected a dog in order to discover the source of the lacteal veins. On the spot was a defender, both energetic and learned, of the elderly, the illustrious Vorstius, by whom the things we sought and demonstrated were regarded as figment of the imagination. However, these whitish veins were clearly seen in the mesentery of the small intestine, separated from the mesenteric veins."

Vorstius conceded that they were not arteries, since the vessels lacked pulsation, but he argued that they were nerves, which Bartholin opposed, since the nerves do not contain a milky fluid and do not disappear from a dead body. Vorstius maintained that he would not subscribe to the idea about lacteal veins before their connection with the liver and their main stem was shown, which Bartholin found difficult, since these vessels disappear rapidly after the commencement of death. However, Bartholin observed during dissection that at the cut through the large gland in the mesentery, where the lacteal veins run together, there runs out a liquid chyle, and he imagines that it will be easier to demonstrate if one dissects without delay immediately after death. He wonders about what he recently heard during a stay in Amsterdam from the Dutch anatomist Nicolaes Tulp (1593–1674), who is known from Rembrandt's famous painting of 1632, *Dr. Nicolaes Tulp's Anatomy Lesson* (Figure 32). Tulp recounted that, during the past winter at a public dissection of a hanged man five days after the man's death, he perceived that some lacteal veins became visible.[37] In a dissected dog the lacteal veins did not become visible, and Bartholin explains this as a lack of nourish-

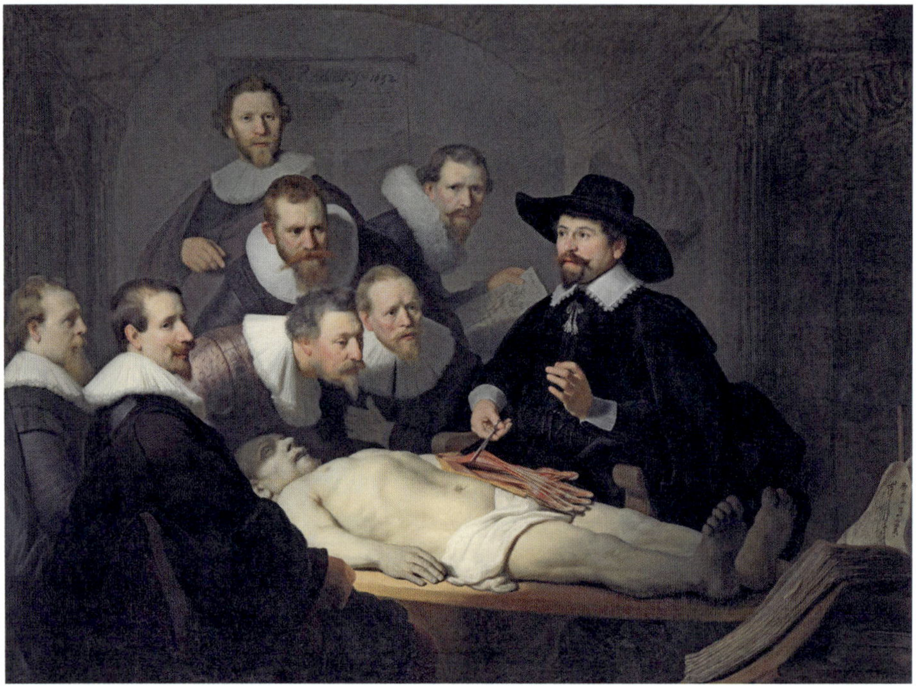

Figure 32. Rembrandt Harmenszoon van Rijn (1606–1669): Dr. Nicolaes Tulp's Anatomy Lesson 1632. Oil on canvas. 169.5 × 216.5 cm. Nicolaes Tulp (1593–1674) was a surgeon and for a period the mayor of Amsterdam. The city's college of surgeons only allowed one dissection per year of the body of an executed criminal. As the city's leading surgeon and anatomist, it was Tulp who, on 31 January 1632, carried out the dissection of the executed thief Aris Kindt. The 25-year-old Rembrandt was present at the dissection and painted the most famous dissection painting ever. In the painting, Tulp instructs his audience in the displayed muscles of the left forearm. Mauritshuis, The Hague.

ment, "unless we take into account the whitish strands which obesity is always associated with." If it is true, as people say, that fat is produced by cold, Bartholin is surprised that in a recent dissection of an apoplectic Tulp found a large accumulation of fat in the left ventricle of the heart, the body's warmest place.

Worm answers on 28 June: "I am surprised that your Vorstius, who is such a learned man, doubts his own senses and eyes to this extent. A topic which is so evident, so splendid, so indispensable to the elucidation of so many entangled anatomic disputes – who would not be tempted by it? I suppose you have adequately opposed his evasions. Or what? If never their source or their introduction could be demonstrated, could you then disown these veins? They certainly lie open for everyone's inspection! Someone or other could perhaps throw doubt on the use which is ascribed to them, until their source and course is established, but as concerns their existence, only the blind can doubt."

He adds, with reference to Aselli's publications, that the latter "has amply satisfied the studious temperament, until time brings better and more certain knowledge," and he urges Bartholin to expand on Aselli's discoveries and with further investigation to "build more accurate data above the well-laid foundation by Aselli." However, for unknown reasons, Bartholin went no further with his own investigations of the lacteal veins until ten years later, as professor of anatomy in Copenhagen, he took them up with his pupils as part of his teaching; but more on this later. As to the lump of fat he found in the left ventricle of the heart, Worm remarks that he himself saw something similar in 1608 at the dissection of a patient with consumption in Basel. This is something not rarely noticed in consumptive and dropsical patients, and he doubts whether, as Tulp observed, it was actual fat, but perhaps a whitish body that resembled fat.

That it was essential for Bartholin to get a discussion of Aselli's and Harvey's discoveries into the revised edition of his father's anatomy book speaks for itself, but it was not without its problems. Owing to their opposition to Galen's doctrine, the two discoveries were certainly frequently discussed among natural and medical scientists in Europe. Harvey's discovery in particular was met with widespread scepticism, and in certain conservative circles even with vehement opposition. Thomas Bartholin himself was still not completely convinced, and it could be risky for a young, ambitious anatomist to take a wrong step. On the other hand, he confides in a letter of 27 March 1640 to Ole Worm that he dares not deny the request to republish his father's textbook of anatomy, "since otherwise they will turn to someone else for that work, which would pose a danger to my father." We shall see later how he solved the problem adroitly.

In the letter, Bartholin names several other authors whose more recent observations he has studied with the purpose of updating the anatomy book. It includes the following writers and presumed works:

Adrianus Spigelius: *De Humani Corporis Fabrica Libri Decem* (Ten books on the structure of the human body), Venice 1627.

Jean Riolan the Younger (1577–1657), French physician, anatomist and professor at l Ècole de Medicine in Paris: *Anthropographia et Osteologia* (Anthropography and Osteology), Paris 1626.

Abraham Zacuto Lusitano or Zacutus Lusitanus (1575–1642), Portuguese court physician and medical historian: *De praxi medica admiranda* (On the admirable practice of medicine), Amsterdam 1634.

Caspar Hofman (1572–1648), German physician and professor of medicine at the University of Nuremberg in Altdorf: *De thorace, eiusque partibus commentarius tripartitus* (Commentary in three parts on the thorax and its parts), Frankfurt 1627.

Peter Lauremberg (1585–1639), German physician, mathematician, poet and professor of these three subjects at the University of Rostock: *Collegium anatomicum* (Anatomical lecture), Rostock 1636.

Bartholin nevertheless lets us know, concerning these writers and their works, that "if they differ now and then from my father, as for example in the case of P. Lauremberg, who as usual is ill-disposed towards my father in his college of anatomy, then I defend my father in a very brief reply." Bartholin has come to an agreement with Johannes Walæus about a perusal of the manuscript before its publication, and he wishes to make the same agreement with Ole Worm.

Concerning the winter's dissections, Bartholin mentions that on a woman "at the lowest point of the intestine, there was a growth, which in size and shape resembled a male penis." The lecturer, Valckenburg, was not able to decide whether the woman was born with this outgrowth, or whether it had developed later. From the description it must have been a Meckel's diverticulum, i.e. a congenital malformation in the form of a remnant of the vitelline duct, a long narrow tube joining the yolk sac to the midgut of the fetus in the early fetal life. It usually disappears before the end of the sixth fetal week, but in about two percent of humans it persists to make a diverticulum protruding near the terminal end of the small intestine. Thus, it is one of the most frequent malformations of the gastrointestinal canal. Meckel's diverticulum often contains gastric mucosa with acid-producing cells and can therefore cause bleeding ulcers in the intestine. In children the diverticulum can cause a potentially lethal condition of invagination, where typically the distal portion of the small intestine slides into the first section of the colon like a sailor's telescope, whereupon the blood supply can become ligated, resulting in gangrene. Meckel's diverticulum was described for the first time in a scientific treatise by the German physician and anatomist Johann Friedrich Meckel the Younger (1781–1833) in 1809.[38] Several sources suggest that a malformation like Meckel's diverticulum had already been mentioned in 1598 by the German surgeon Wilhelm Fabricius von Hilden (Wilhelm Fabry) (1560–1634),[39] while others state that this first occurred in 1699,[40] so Thomas Bartholin's reference here must have been one of the earliest.[41]

Moreover, Bartholin relates that in the current month Leiden has produced two portents: "A woman has given birth to a child with two heads, and a cow has had a three-headed calf. The two-headed child survived for more than a month."

In the next letter to Worm on 10 April 1640, Bartholin has come further along with the revised edition of his father's anatomy book: "Following our renowned Walæus' advice, I have adhered to my father's basic rules and have conveyed them to Hacke,[42] enriched with various discoveries. The renowned Walæus has finally admitted the circulation of the blood, after numerous tests had been carried out on animals, to which he has quite frequently invited us and given us admission, and now it has reached the point where, at my request, he will put down everything he can gather for the illumination and corroboration of Harvey's discovery in an everlasting paper, a burden he wanted to place on my shoulders; but in the end, however, I succeeded in persuading him to summarize the whole matter in a letter, which will be linked with my work."

Six days later came further progress in the story: "They will soon commence the publication of my father's anatomy, if the illustrations for it have been made ready. Our renowned Walæus has the original for evaluation, which I have corrected, and in which here and there I have made some additions. He has called for this work, and we are occupied every day with considering Harvey's opinion concerning the motion of the blood and the movement of the heart. We carry out dissections on various animals while they are alive. We have made many discoveries that support Harvey. I shall send them, when I have the opportunity. They have completely reversed our Walæus, such that recently he has testified as to this opinion in a public debate. The other professors of medicine still hover in uncertainty and are completely ignorant and perplexed as to with whom they must agree." As to this last sentence, Ole Worm himself was no exception. For quite a long time he persisted in his scepticism towards Harvey. On 11 May, he answers Thomas Bartholin: "I am eager to see what observations the renowned Walæus has made about the circulation of the blood, and you will do me a favour in sending them. I have not yet been able to arrive at a belief that things behave in such a way."

At the end of his stay in Leiden, Thomas Bartholin himself still did not feel that he was on safe ground regarding the question of his reliance on Harvey's doctrine of the circulation of the blood. On the one hand he would gladly support the new ideas, and Walæus' experiments had convinced him about the truth of Harvey's discoveries, but on the other hand, like Worm, he could not avoid being influenced by Harvey's influential opponents. As he states in his letter of 10 April 1640, he solved this dilemma skilfully. In the new edition of the anatomy book, he added not one but two letters by Walæus on the circulation of the blood, addressed to Thomas Bartholin himself.[43] By including these letters, Bartholin could be the first to publish an anatomical textbook in support of William Harvey's controversial circulation theory without having directly taken a personal stance on the question.

Figure 33. The frontispiece of Thomas Bartholin's first revised edition, *Institutiones Anatomicæ*, Leiden 1641, of his father's textbook of anatomy from 1611.

Top: Caspar Bartholin flanked by an opened cranium with insight to the brain and a heart.
Left from top to bottom: Hippocrates, Andreas Vesalius, Caspar Bauhin and Pieter Pauw.
Right from top to bottom: Galen, Jean Riolan, Adrian Spigelius and Otto van Heurne.
Bottom: A dissection scene with Thomas Bartholin's teacher in Leiden, Adrian van Valckenburg.
The Royal Library in Copenhagen.

The book appeared in Leiden in 1641 with the title *Institutiones Anatomicæ, Novis Recentiorum opinionibus & observationibus, quarum innumeræ hactenus editæ non sunt, figurisque auctæ ab Auctoris Filio Thoma Bartholino* (Textbook of anatomy, enlarged with new opinions and observations of modern writers, of which a great many have hitherto been unpublished, and with illustrations by the author's son, Thomas Bartholin). It has 496 pages and 73 anatomical plates, of which seven are on folded sheets. After the title page (Figure 33) comes his father's dedication to the learned nobleman, imperial councillor and theologian Holger Rosenkrantz (1574–1642), after which follows Thomas Bartholin's expansive preface and a detailed table of contents. The book's subdivisions follow those of his father, and it finishes with an extensive subject index together with Thomas Bartholin's reminder to the reader. Ole Worm had recommended to Bartholin that he should print his additions in a typeface that was different from the main text, but instead Bartholin puts brackets around his additions.

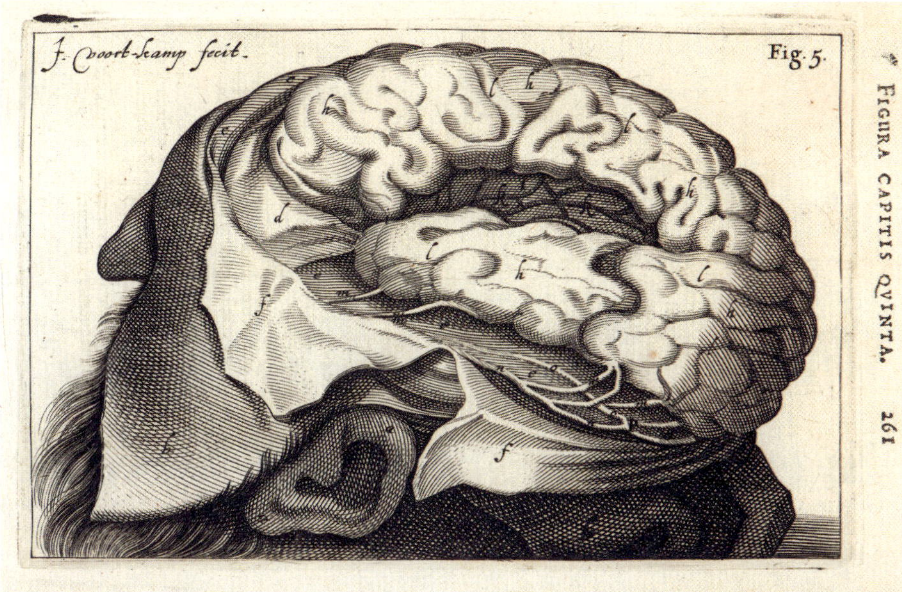

Figure 34. The exposed brain in a man's opened head, where the top half of the skull has been removed and the two outer meninges, the dura mater and the arachnoid, have been cut up and folded to the side. Page 261 in Thomas Bartholin's first revised edition from 1641 of his father's textbook of anatomy.

Bartholin asked the Dutch engraver J. Voort Kamp to produce five new illustrations for the edition, including this one, which is engraved from a drawing by Franciscus Sylvius. The illustration shows the cleft in the brain, marked with a *k*, which separates the temporal lobe from the frontal and parietal lobe, respectively. Bartholin was the first to describe the fissure, which he named *fissura Sylvii* to honour Sylvius, who had shown it to him. The Royal Library in Copenhagen.

The anatomical plates derive, with few exceptions, from other writers' works, mainly Vesalius' *De Humani corporis fabrica*, but Bartholin mentions in the preface to the third revised edition of the book, which appeared ten years later, that the illustrations in the 1641 edition, over and above those from Vesalius, were also taken from works of Bauhin, Casserius, Sylvius and Aselli. The exceptions are figures 4, 5 and 7 in the book's third section (Liber III) concerning the head and the brain, which are all new. They are engraved and signed by the Dutch engraver J. Voort Kamp, while the drawings, which provided the originals for the engravings, are done by Franciscus Sylvius.[44] Figure 5 shows the opened cranium of a man, displaying the cleft in the brain that separates the temporal lobe from the frontal and parietal lobes (Figure 34). The brain's folded surface is drawn back, so that the cleft stands out more distinctly. Bartholin gives Franciscus Sylvius the credit for the discovery of this cleft and names it after him, *fissura Sylvii*. Today the designation *sulcus lateralis* is most frequently employed, but some continue to call the cleft *fissura Sylvii*. Until recently it had been assumed that the illustration in Bartholin's book was the first to show the *fissura Sylvii*, but newer research has revealed that it had already been illustrated in an anatomical plate in 1600 by Caspar Bartholin's teacher in Padua, Girolamo Fabrici d'Acquapendente.[45] Fabrici, however, did not describe the cleft, so Thomas Bartholin probably was indeed the first to give an account of it. For the text to the illustration, he writes:

"If you investigate the furrows in figure 5 carefully, you will observe that they are very deep, and that the brain is divided from one side to the other by the curved fissure, which begins at the front near the nerve roots of the eye and continues back from there along the temporal bone over the base of the spinal cord and divides the upper part of the brain from the lower part."[46]

Bartholin writes in the book's preface about the fissure: "We can all gauge what is remarkable in Sylvius' mind and talent through the wonderful new structure in the brain. [...] On the new illustrations of the brain, the engraver followed the very thorough Sylvius' plan and scalpel." Bartholin writes to Ole Worm on 3 July 1640 from Vliessingen about the new illustrations: "I am sending you the new, very fine depictions of the head, which we have provided, as soon as they are ready; for on them we have reproduced various new convolutions, which were noticed not long ago during a dissection under the supervision of doctor of medicine Franciscus Sylvius; likewise, a new and correct form of the cerebral ventricles, which are not semilunar, as is depicted elsewhere, but round, very much like a horseshoe, as we have seen through personal observation."

Ten years later, Bartholin writes in the third revised edition of the anatomy book: "The convolutions of the brain, which I first learnt about from Franciscus Sylvius, a great anatomist."[47] Worm acknowledges the book on 16 April 1641: "Two copies of your father's textbook of anatomy have come to us, carefully published in a beautiful print. [...] It cannot fail to appeal to everyone's taste, partly by reason of the beautiful print and fineness of the illustrations, but in particular for the splendid discoveries, which through your carefulness have been added succinctly." Concerning the book's containing Johannes Walæus' two letters in support of Harvey's circulation of the blood, Worm adds: "Walæus' letter almost makes me accede to his opinion. There are just some points of dispute remaining, and if they are resolved for me, I shall gladly encounter the breaking truth, as it appears."

Finally, mention must be made of the fact that, in the book's chapter on the heart, Bartholin had also inserted a small two-page supplement about the heart's movements and the pulse, primarily dealing with Harvey's and the French philosopher René Descartes' theories on this subject.[48]

Figure 35. René Descartes (1596–1650) portrayed by Frans Hals (1580–1666) c. 1649. Oil on oak. 19 × 14 cm.

Like many of the other natural scientists of the time, Thomas Bartholin was preoccupied with the great French philosopher and mathematician's thoughts and theories. Thomas Bartholin was first and foremost interested in Descartes' thoughts on the human body, and as the skilful and experienced anatomist Bartholin was, he quickly realized Descartes' mistakes in this area. Unlike his brother Rasmus, he therefore remained critical of Descartes, and this critical view rubbed off on his pupil Niels Stensen. National Gallery of Denmark.

Thomas Bartholin's scepticism regarding Descartes

How burning and controversial the question of the acceptance of Harvey's discovery was, twelve years after he had published it, is illustrated by Worm's letter to Bartholin on 18 April 1640: "Although Harvey's observations are assented to by many people, and though I see that Mr. Descartes recognizes them, they are rejected by others, and in England Primrose has refuted them in a complete dissertation,[49] which I don't know whether you have seen. I have taken care to get it here from England. I point this out not with the purpose of excluding his observations, but in order also to emphasize what Primrose misses in so doing." Thus, it appears that it was not only physicians who took part in the discussion of Harvey's discovery. The great French philosopher of the time and one of the absolute leaders of the 17th century's scientific revolution, René Descartes, was one of Harvey's foremost and most important supporters (Figure 35).

As a scientifically inclined philosopher, Descartes started to become interested in medicine during his stay in Amsterdam from 1629 to 1632. He began comprehensive self-studies, which consisted partly of reading the literature and partly of experiments. From his residence in Kalverstraat in the butchers' area he had easy access for his almost daily visits to the slaughterhouses with the purpose of observing the butchers cutting open various animals and obtaining dead animals for dissection at home.

Ten years later he writes to his friend and colleague Marin Mersenne (1588–1648): "One winter in Amsterdam I went almost daily to a butcher to see him slaughter animals and to have brought to my house the parts of animals that I wanted to dissect in my spare time."[50] In his dissection studies, Descartes relied on the illustrations in Vesalius' *De Humani corporis fabrica*, Caspar Bauhin's *Theatrum Anatomicum*, Frankfurt 1605 and 1621, and Fabrici d'Acquapendente's embryological works *De Formato Foetu* (On the formation of the fetus), Venice 1600 and 1620, and *De Formatione Ovi et Pulli* (On the formation of the egg and of the chick), Padua 1621 – the same writers and works that Harvey had also studied intensively.[51] But as it appears from a letter to Mersenne dated 15 April 1630, Descartes also conducted his own experiments and made observations he could not have read about in books: "I am now studying chemistry and anatomy concurrently. Every day I learn something I cannot find in any book."[52] He elaborates on this nine years later (20 February 1639): "I have not only reflected on what Vesalius and others write about anatomy, but also over many details that are not mentioned by them, but that I have observed myself during my dissections of different animals."[53]

Descartes worked at this point on the book of natural philosophy *Le Monde* (The World), which contains another book, *L'Homme* (The Human Being).

The whole book was published posthumously in French in Paris in 1664, but *L'Homme* had already been published in a Latin translation *De Homine* in Leiden in 1662.[54] Descartes writes to Mersenne about this work in the winter of 1632: "My discussion of the human being in *Le Monde* becomes a little more satisfactory than I had planned, for I have come to grips with explaining all the essential functions in the human body. I have already described the vital functions like the digestion of food, the heartbeat, the dispersion of food etc., and the five senses. Now I am dissecting the heads of different animals, so that I can explain what imagination, memory etc. consist of."[55]

In the work *Discours de la Méthode* (Discourse on the Method), Leiden 1637, which Ole Worm had asked Bartholin on 19 June 1638 to send him, Descartes explains his method in natural philosophy, which aims at seeking the truth in science through a single, clear, easily understood procedure. He outlines the following four points: 1) You must only consider true, that which is evident. 2) You must divide a given problem into as simple parts as possible. 3) You must begin with the simplest and most comprehensible. 4) You must keep account of where in the process you find yourself.

He states, with reference to the unpublished *L'Homme*, his mechanical philosophy, in which human and animal bodies are compared with machines, automata or mechanisms. The physiological processes in both human and animal bodies are explained as matter in movement. Only the human soul differentiates human beings from animals, which have no soul. Descartes rejected the elements and fluids of Aristotle and Galen, but Harvey's discovery of the circulation of the blood fitted excellently into his mechanical philosophy, and in the fifth section of *Discours de la Méthode*, which largely deals with the movement of the heart and circulation of the blood, Descartes gives Harvey credit for his theory:

> "But if anyone asks me how it comes about that the blood in the veins does not drain out, when it runs unceasingly into the heart in this way, or that the arteries do not become overfilled, when all that blood runs out from the heart into them, I can only answer, what has been stated by a physician in England (Harvey: *On the Motion of the Heart*). We must honour him for breaking the ice in this respect. He is the first to have seen that there are many small channels, where the arteries end, and the blood, which comes from the heart, goes through them into the veins' small branches and back into the heart again, so that the blood flows in a continuous circulation."[56]

In the book, Descartes has a didactic discussion about the anatomy of the heart and the passage of the blood through the heart and the lungs, and he explains the heart valves' function in preventing the blood's reflux during the heart's filling (the phase of relaxation, diastole) and emptying (the phase of contraction, systole).

But there is more at stake in Descartes' attitude to the circulation of the blood. His reliance on Harvey concerned the results of this man's experiments, and therefore the indisputable observations of the heart's movements and the circulation of the blood in the heart, lungs and body. But when it came to Harvey's explanatory models, the consensus ceased, and here began the clash between the Aristotelian view of Nature held by the ancient and Renaissance world, which builds on the idea that everything in Nature has a purpose (teleology) and a soul (animism), and the 17th century's new view of nature, in which Descartes led the way. In Padua, Harvey had been taught by Fabrici d'Acquapendente to be dependent on Aristotelian natural philosophy. In *De Motu Cordis*, this is not only seen in Harvey's repeated references to Aristotle and Fabrici but also characterizes his models of explanation to a high degree. In Aristotelian natural philosophy, there was a tradition of using celestial or cosmological metaphors when one wished to give explanations of earthly natural phenomena, including the structure and functions of the human body. The human being was the microcosm in the great universal macrocosm. In *De Motu Cordis*, Harvey already uses a metaphor of this sort in the prefatory dedication to Charles I. He compares the heart with the sun: "The heart of living beings is the basis of their life, the prince of everyone, the sun of the microcosm, on which all living things depend, and from which all power and strength is emitted."[57] As to the teleological aspect, that everything in Nature has a prior purpose, that the competent Nature, including human bodies, is God's creation, and that only God holds the superior explanations for Nature's organizations, there are also many examples in Harvey's writings: "In this way complete and God-given Nature does nothing in vain. She does not endow a heart to an animal, which has no use for it,"[58] and "since consummate Nature creates nothing in vain and in every circumstance fulfils her purpose, the arteries are different all the more in their construction from the veins, the nearer they are to the heart."[59] Harvey explains elegantly that the heart draws itself together like a muscle, and in its contraction presses the blood out into the arteries and, what is more, he compares the atrium's and the ventricle's rhythmic contractions with machines and mechanisms: "And this happens accurately for the same reason as in a machine, where every cogwheel seems to move at the same time, when the one cogwheel moves the other, or as with the mechanical appliance, which has been mounted in a firearm, where you press a trigger, the flint falls down, hits the steel and pushes it forward. A spark now jumps forward, which falls on the gunpowder, this ignites, and the flame creeps forward and explodes, the bullet flies out and hits the target, and all these movements, by reason of their rapidity, give the impression of happening simultaneously, in an instant."[60]

But when he attempts to explain where the vitality of the heart originates, he turns back to Aristotle and the animistic view: "All living creatures are warm, but the dying cold. Inside the body a place is found, which was the origin of the warmth, like a domestic altar and hearth, where Nature's combustible matter and the innate source of fire is contained and preserved, and from which its heat and life fly outward in every feature as if from a fountain, and from which its nourishment comes, and on which its preparation, energy and life strength depend. But that this place is the heart, and that this is the origin of life, and that everything exists in such a way, as I have stated, I hope no one will question. [...] But when the blood, as a result of the cold in the peripheral parts of the outer surroundings, becomes thickened and cooled, and in the same way as the dead are deprived of their animal spirits, it must necessarily turn back to its source and its beginning and there regain its warmth and its animal spirits."[61]

Like Harvey, Descartes also takes as his starting point that it is God who is Nature's creator and makes its laws. He talks about what might pass in a new world "if we consider that God in one or another location within an imaginary space created so much material that from it a world could be made, and that he in this way, without any order of events, set the different fragments of the material in motion, so that he created a chaos, as confused as the poets can imagine, and that God afterwards did nothing else but grant Nature the assistance, with which in the customary way he keeps it and allows it to work according to the laws he has instituted."[62]

But Descartes does not accept Harvey's cosmological comparisons and animistic explanations. He does not compare the heart with the sun, and does not explain the heart's functions as innate animal spirits, which are granted to mankind by God: "And therefore I satisfied myself by supposing that God fashioned a human body entirely like one of us, both outwardly in his limbs and inwardly in his organs, without creating it from any other material than what I had described, and without giving him from the beginning a rational soul or anything else, which could constitute a vegetative or sensitive soul; on the contrary, he lit in his heart a fire without light."[63]

This fire without light, this warmth, Descartes explains as the background of motions of the heart and those of the blood's passage through the heart. When the blood reaches the heart, it is warmed by the heart's heat, with which both the blood and the heart expand, so that the blood is pressed out into the large arteries (the body's main artery, *aorta*, and the lung's main artery, *truncus pulmonalis*), and when the blood comes out here, it is cooled, whereupon both the blood and the heart contract. Descartes concludes that this movement "necessarily follows from the organs' arrangement, which you can see with your own eyes, from the warmth, which you can perceive with your fingers, and from the nature of the blood, just as experience shows, exactly as the movements of clockwork follow from the movements of the weights' and the wheels' strength, position and form."[64]

As much as the physiology in Harvey's explanation of the heart's movement is indisputable, so is the physiology in Descartes' mistaken, for it bears the mark of his not being a physician and not having the practical experience of dissecting live animals. The visionary element in Descartes' explanatory model is that he bases it on mechanics rather than on animism.

Descartes' ideas and theories were highly favoured among contemporary scientists, and Thomas Bartholin was no exception. He followed Descartes' work with interest and discussed it in his correspondence with other scholars. For example, on 11 February 1644, Johannes Walæus informs Bartholin in a letter that Descartes has got himself an adversary in Pierre Gassendi, who has written a large volume, in which he refutes Descartes' philosophy.[65] But for Bartholin, the primary aim was to evaluate the medical content in Descartes' statements, and for good reasons he was not particularly impressed. In his third revised edition of his father's textbook of anatomy, *Anatomia Reformata,* which was published in Leiden in 1651, he comments in four places on Descartes, and in all four places he declares that he is at variance with him. The first place is exactly at the question of Descartes' explanation of the heart's motion. Bartholin writes that the sharp-witted Descartes and his followers explain the heart's movement by saying that one or two drops of blood fall respectively down from the large *vena cava* into the right ventricle and from the lung vein to the left ventricle, where they are heated by the rest of the already warmed blood in the heart and by the heart's warmth. They are expanded through dilution.

In this way, the valves between the ventricles and the atria close, and the heart is enlarged with the diluted blood, which finally cannot be contained in the heart's chambers, but at the same time is driven out from the right ventricle through the *truncus pulmonalis* to all its branches in the lungs and from the left ventricle through the *aorta* to all its branches in the rest of the body. When the arteries are expanded by the blood in this way, the pulse is generated. Thus, the heart's filling phase (diastole) is explained, while the heart's emptying phase (systole) is explained by the fact that the heart is emptied of blood, and the blood is cooled down in the arteries, so that the heart and the arteries relax and decrease in size, and the new drops of blood can be received, and the diastole be repeated. Bartholin does not reject the idea that an easy thinning of the blood can take place because of a mild warming, but he does not accept that a few drops of blood expanding in such a large chamber should be able to cause the heart's movement. He puts forward six objections:

1. The dissections of live animals, vivisections (which he knew that Descartes had not undertaken), show that the blood does not enter the heart drop by drop or in a diluted form, but as a stream of pure blood from the atria to the ventricles.
2. Hearts that are cut into pieces or have a hole pricked in them continue to beat without a thinning of the blood, which is only an illusion.
3. Walæus had observed in strong dogs that, when he cut the tip off the heart, and it was consequently emptied of more than half of its contents of blood, the heart contracted all the same, so that the remaining amount

of blood spurted four feet out. It makes no sense to include the air's coldness as a contributing cause, since the heart does not become cold in an instant, any more than a boiling kettle ceases boiling when you take it away from the fire.

4. The heart's muscular walls, which are hard and strong, are unwilling to rise and fold together at the mere thinning of the blood. A more violent action is required to stir this great mass. The large arteries ought to have a larger entrance, because the thinned blood, which is leaving the heart, will need more room than that which is entering the heart.
5. A confusion between the heart's movement and that of its valves would arise. The diastole of both would happen at the same time, and thus the valves would be useless. Both things are disproved by experience.
6. That the blood should be cooled in the arteries is confirmed neither by common sense nor by examination. The blood in the arteries is warm and differs only little, or not at all, from that which is found in the heart or near to it. In the small arteries there is no pulse, but this must be attributed to the vessels' small size and their distance from the heart, which propels the blood. Nor must the blood thicken and cool, when it runs into the smallest blood vessels to nourish the different parts of the body and turn into the secondary humours. And what is the point of thinning, if it resolves shortly after?[66]

Bartholin maintains that the heart's task is to receive the blood from the large veins, convey it with nutrients and heat, and drive it further through the large arteries round the body. He declares himself at variance with Descartes' followers, in that it is the heart's movements alone that generate the heart's heat.

He believes, on the contrary, that the heart has an innate heat, which is preserved by reason of its movement. He does not think that the heart produces the blood, but that it preserves it in order to convey the nutrients and the heat. Like Harvey, he recognizes that the heart draws itself together like a muscle, but not that it is an actual muscle. He argues that the muscles' contraction can be attributed to control by the will, but that the heart contracts without the use of the will, which is demonstrated by the fact that the embryo's heart beats. He is, however, ready to accept Walæus' theory, that the heat is a special sort of muscle.[67] The eventual adjustment of Harvey's and the ancient physicians' animistic model of explanation and Descartes' unsuccessful attempt at a mechanistic-physiological explanation of the heart's movement first appears with Bartholin's pupil, Niels Stensen, who also stands for the final perception and documentation of the heart as a muscle. As a student, Stensen had already allied himself with both Harvey and Descartes, and he continued to subscribe to Descartes' method of exploring the physical world. Despite this, in keeping with his own experience and knowledge as a scientist, he necessarily had to distance himself from Descartes' manifold errors regarding human anatomy and physiology. Based on his own comparative studies of the skeletal muscles and the heart, which were inspired by Johannes Walæus, he writes the following in a letter from Leiden to Thomas Bartholin in Copenhagen on 30 April 1663: "As far as the material of the heart is concerned, I shall, according to my intention, supply convincing evidence that nothing is found in the heart that is not found in the muscles, nor that the heart lacks anything, which is found in the muscles.[68] In the following year he amplifies this in a treatise on his research concerning the salivary glands, the muscles and the heart: "The heart has been made the home for innate heat, the throne of the soul, yes by some even the soul itself. The heart has been greeted as the actual sun, even the king, but if you examine it properly, you will find that it is nothing but a muscle."[69] The words were addressed to the late William Harvey, who, as mentioned, had made the heart the home of the innate heat and the animal spirits (*spiritus animalis*) and compared it with the sun and a king. Stensen rejected in the most determined way Descartes' explanation of the heart's movement, and this fact was later emphasized by another prominent philosopher and mathematician, Gottfried Wilhelm Leibniz (1646–1716), with the words: "Mr. Stensen has made it clear that Mr. Descartes has made a complete mistake in his understanding of the heart's movement."[70] Nevertheless, Stensen's quoted statement was first and foremost a confrontation with Aristotelian cosmology and animism and consequently an indirect support for Descartes.

Niels Stensen's discourse on the brain's anatomy for a circle of scientifically interested men surrounding the orientalist and scientific patron Melchisédech Thévenot (1620–1692) in Paris in 1665 is legendary.[71] Here he rejected in a convincing and elegant manner Descartes' theory, presented in *l'Homme*, of the anatomy of the pineal gland (*corpus pineale*) and its central role in the functioning of the brain as a meeting point between the soul and the body.[72] But Descartes had already, a year before his death, aired the theory in his 1649 work *Les Passions de l'Ame* (The Passions of the Soul), and Thomas Bartholin had rejected the theory in his *Anatomia Reformata* in 1651 – 14 years before Stensen – with some of the same arguments. In *Anatomia Reformata*, Bartholin writes in the section on the pineal gland that Descartes and his followers are of the opinion that the pineal gland receives motor impulses from the whole body, and that the soul, only in this gland, in this one centre, by virtue of these movements perceives all the impulses and impressions from the five senses, distinguishes the impressions from each other and sends "spirits" (instructions, impulses) out to all parts of the body.[73]

Bartholin puts forward a series of counterarguments, of which the most convincing to modern eyes is that the pineal gland does not receive nerves from either the sense organs or the spinal cord and is too small to accommodate and coordinate so many functions, impulses and impressions without confusing them.[74]

Overall, Bartholin was sceptical of the search for the seat of the soul in the body. He did not believe that such a seat would ever be found. In a letter to Bartholin from Leiden of 3 March 1661, his pupil, the polymath Ole Borch (1626–1690), who was also Niels Stensen's teacher in Copenhagen and his fellow researcher in Leiden, writes that the English courtier, philosopher, astrologist and alchemist Kenelm Digby (1603–1665) places the seat of the soul in a pit in the brain's *septum pellucidum*, the thin pellucid partition, which separates the brain's two frontal horns (*cornu frontale*), which are a part of the frontal lobes (*lobus frontalis*).[75] Digby, who was a friend and warm supporter of Descartes, had already published in Paris in 1644, in his native language, the widely read and controversial work *Two Treatises in the one of which, The Nature of Bodies; in the other, The Nature of Mans Soule; is looked into: In Way of Discovery, of the Immortality of reasonable Soules*, in which he had singled out the *septum pellucidum* as the seat of the soul in the body. Bartholin answers Borch on 9 March 1661 with the wise words: "The famous Digby places the soul in a very fragile house. If it was possible to show that nerves gather around this septum, his sharp opinion would be more likely. I think that the soul's residence will remain hidden to us, as long as the soul itself hides, and that question has not yet been sufficiently seen through by the human intellect."[76]

Like Bartholin, Stensen, in his criticism of Descartes' notions concerning the brain's construction and functions, takes as a starting point Descartes' lack of practical experience in dissections of human brains. In a letter to Thomas Bartholin from Leiden of 26 August 1662 he writes: "In the last few days Descartes' *Tractatus de Homine* with illustrations has been published by Florentius Schuylius.[77] [...] In the book you see figures [Figure 36] that are not inelegant, and that have certainly proceeded from a clever brain; but I very much doubt whether you can conceive of anything similar in the brain."[78]

Nevertheless, Stensen also accedes to the mechanical philosophy of Descartes concerning the brain, when in his brain discourse in Paris barely three years later he calls the brain "this wonderful machine."[79]

The third place in *Anatomia Reformata* where Bartholin declares himself at variance with Descartes is in the latter's assertion that the nerves have valves. Descartes believes that valves in the nerves prevent a reflux of the spirits, since the limbs would otherwise not be able to move. Bartholin stresses that no anatomist has ever seen such valves, and even if they were there, they would not be able to obstruct the impulses.[80]

The fourth difference between the two is in the question of the breathing mechanism. This discrepancy Bartholin deals with in his second revised edition of the anatomy book of 1645, together with which it will be described later.

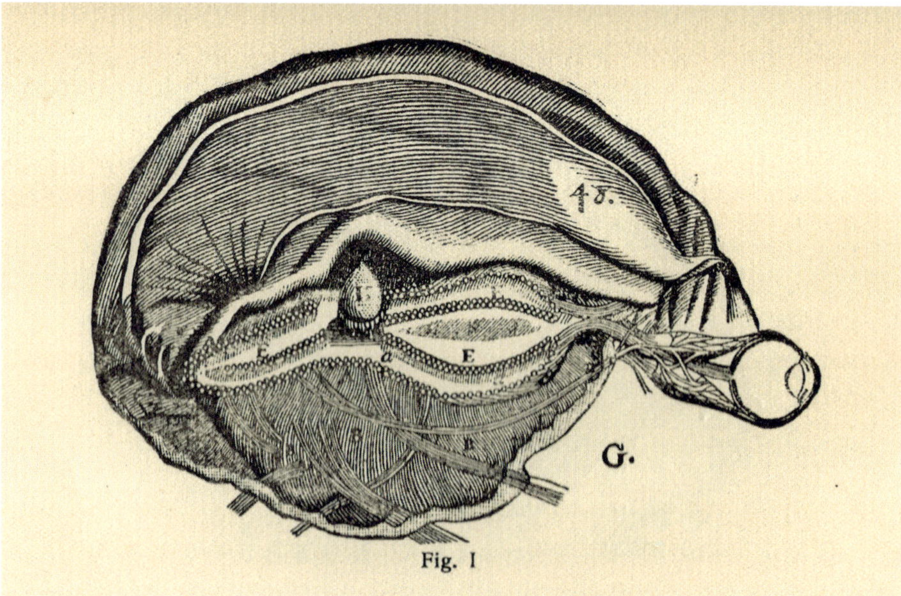

Figure 36. René Descartes' illustration of the brain on pp. 63 and 65 in *L'Homme*, Paris 1664.

The illustration is intended to represent a longitudinal section of the brain in the median plane, where the pineal gland (*corpus pineale*) labelled with the letter I is shown disproportionately large with vertical alignment and as a central pivot point in the brain. The brain's anatomical organization and structures are – as stated by Stensen in a letter to Thomas Bartholin in August 1662 – difficult to recognize. However, it was the illustrations in the Latin translation, *De Homine*, which was published in Leiden in 1662, to which Stensen referred in the letter. They were made by the translator, professor of philosophy in Leiden Florentius Schuyl (1619–1669). As stated in the preface by the publisher of the French edition, Claude Clerselier (1614–1684), he did not believe that the illustrations in the Latin edition matched Descartes' text. He therefore asked the physician and philosopher Louis de La Forge (1632–1666) and the anatomist and mathematician Gerard van Gutschoven (1615–1668) to produce new illustrations for the French edition. La Forge's illustrations are marked with an F and Gutschoven's, as the one shown, with a G. Private ownership.

One can speculate as to whether Thomas Bartholin's scepticism about Descartes was founded in philosophical discrepancies. Bartholin may have had difficulty in accepting Descartes' clash with Aristotelian natural philosophy, which Bartholin's father and many of Bartholin's most important teachers during the rest of his stay abroad had been schooled in. There are, however, no examples in Bartholin's writings that support such speculation. The simple explanation seems to be that Bartholin felt irritated and provoked by the fact that Descartes largely built his anatomical and physiological models of explanation on imaginary theoretical considerations and not on facts obtained through practical research and experience. The fact that Descartes was not a physician, naturally did not improve things. In 1659, Bartholin publishes his clear view on this question, perhaps with a hidden address to Descartes: "Human beings waste their labour in seeking to understand Nature through the chatter of the sophists and the word-battle of the disputations. One must learn to know Nature by personal experience, not by disputing about it. By the sophists' innumerable distinctions, one does not get rid of as much as a flea in high summer. [...] No, it is Nature herself that you must study, otherwise you never learn to know her. We must investigate with our senses, what can be learnt through the senses".[81] Elsewhere he formulates it more directly: "We do not trust more than we can feel with our hands and see with our eyes."[82]

Niels Stensen formulates similar viewpoints in his already quoted dissertation on the brain's anatomy, and without doubt here addresses Descartes: "I, for my part, prefer to confess my ignorance rather than to state with authority opinions, the falsehood of which will some time afterwards be proven by others. We have seen great anatomists who have committed this error, and we have seen others like them, who imagine that one will depend more on these people's insistence than on one's own eyes. I commend this self-righteousness to those who can be satisfied with it, and I try to follow the philosophical rules that teach us to seek the truth by raising doubts about it and not resign ourselves to it before we have gained confirmation of it with a clear proof."[83]

That it was not his medical contributions that Descartes was appreciated for in Thomas Bartholin's family is indicated by the following observation from Ole Worm to Rasmus Bartholin in Leiden on 23 April 1650, scarcely three months after Descartes' death: "We are all deeply pained by Descartes' death, for he was a man of rare ingenuity and born for the advancement of learning, a superb geometrician and mathematician."

In the late 1640s, Thomas Bartholin's brother Rasmus (Figure 37) studied medicine and mathematics at Leiden University. Through his teacher in mathematics Frans van Schooten (1615–1660) he established an acquaintance with

Figure 37. Rasmus Bartholin (1625–1698), professor of medicine and mathematics at the University of Copenhagen, portrayed by an unknown artist. Date unknown. Oil on canvas. 78.5 × 62.5 cm.

Rasmus Bartholin was the youngest and most famous of Thomas Bartholin's five brothers. Through his personal acquaintance with René Descartes in Holland in the late 1640s, he became the leading figure of the Descartes reception in Denmark. He is especially known for his discovery of the birefringence of light in Icelandic calcite, published in *Experimenta crystalli Islandici disdiaclastici*, Copenhagen 1669. In Thomas Bartholin's *De Nivis Usu Medico*, Copenhagen 1661, Rasmus Bartholin explains with accompanying illustrations the structure of snow and ice crystals. As Niels Stensen's teacher in mathematics, he introduced Stensen to Descartes' geometry. In addition, he taught Stensen and his other students to make use of mathematics as a basic science within the natural sciences. Later, he also became a source of inspiration for Stensen in the field of crystallography. In the field of medicine, Rasmus Bartholin was the first to publish a successful treatment of malaria in a patient (himself) with China bark. The Museum of National History at Frederiksborg Castle. Photo: Kit Weiss

René Descartes. Van Schooten introduced him to Descartes' analytical geometry, *La Géométrie*, published as an appendix to Descartes' groundbreaking and controversial work on the scientific method, *Discours de la Méthode*, in Leiden 1637. Rasmus Bartholin had conversations with Descartes, became greatly interested in his philosophical opinions and publications and often wrote about him in letters to Ole Worm.[84] In 1649, Descartes even showed Rasmus Bartholin a great honour, which illustrates the latter's skills as a mathematician. Some years earlier in a Dutch publication, the Dutch mathematician Jan Jansz de Jonge Stampioen (1610–1653) had put forward a difficult mathematical problem, which he was not able to solve himself, and as a result he had lost 600 florins in a bet. When in 1649 Descartes wanted to publish his geometry in Latin, *Geometria*, his French colleagues called on him to include Stampioen's problem and "add comments, so that the road for everyone's very obscure and ingenious solution of the problem in question could stand open."[85] However, Descartes had no suggestion for a solution, but he knew that Rasmus Bartholin had. He therefore asked Bartholin to publish his solution in the book. Bartholin agreed, and his solution can be found on pages 318–23 with the following introductory words by Descartes: "But since the presented solution of this problem includes things that seem to require an explanation, and which could make less experienced people sceptical, I have chosen to put forward here, what the excellent and learned Mr. Rasmus Bartholin, son of Caspar, has set up as an explanation."[86]

Two years later, Rasmus Bartholin had the honour of publishing Franz van Schooten's introduction to Descartes' geometrical method, *Francisci à Schooten Principia Matheseos Universalis seu Introductio ad Geometriæ Methodum Renati des Cartes, edita ab Er. Bartholino, Casp. Fil.* (Frans van Schooten's general mathematical principles or introduction to René Descartes' geometric method, published by Rasmus Bartholin, son of Caspar). After his return to Denmark in 1656, Rasmus Bartholin played a major role in the introduction of Descartes' mathematical ideas at the University of Copenhagen. Whether Thomas Bartholin ever met Descartes is unknown. In his small book *Lidt om Descartes og Danmark* (A little about Descartes and Denmark), the historian of medicine Vilhelm Maar (1871-1940) writes: "When the two editors of Descartes' letters[87] [...] come to think of Thomas Bartholin, where Descartes speaks about 'a Dane', this is in fact extremely understandable. We simply lack a chapter in the 17th century's history of science with the title: René Descartes – Thomas Bartholin. In fact, taking into consideration that Bartholin stayed in Holland from 1637 to 1640 and once again in 1646, and that he had particularly good connections and largely moved in the same scientific circles as Descartes, it is highly unlikely that he and Descartes should not have met one another."[88]

On the other hand, considering Bartholin's great interest in Descartes and his frequent mention of him in his writings, it seems unlikely that a meeting between the two men should have taken place without Bartholin mentioning it even in a single place in his huge literary production, including hundreds of letters – but he never did.

A failed book about pygmies

It was not a discussion about African or Asiatic pygmies in a modern anthropological context, but about various dwarves and gnomes, who at that time were given various supernatural characters. On 10 April 1640, Bartholin writes to Worm: "In order that my leisure hours shall not pass unprofitably, I have read through my father's collected account of pygmies and have expanded it with various additions, which I shall perhaps have published at our printers." Bartholin refers to the fact that the German Jesuit Athanasius Kircher (16021680) claims in his new book on Coptic and Egyptian languages[89] that pygmies are not real human beings, but small trolls, who are usually found around metal mines.[90] Bartholin continues: "Moreover, the Scandinavian people frequently tell stories about these creatures, and I have heard from a Norwegian friend that these small, two-feet-tall manikins move around in the Norwegian metal mines and do no harm to anyone, unless they have been insulted. Such beings are often seen in other places, if one may believe the stories. By the Dutch they are called 'kabuter mancke', by our countrymen 'nisse-god dreng', who now and then pursue their tasks at nighttime. [...] In our fatherland I also remember the old women telling many stories about them and cooking porridge to give them pleasure. Or perhaps they are a kind of demon."

Bartholin asks Worm his opinion about these small "ghosts" and gets the answer in a letter of 11 May 1640, where Worm assures him that pygmies – in contrast to "the phantoms that move around metal mines" – are real people of flesh and blood, who were seen by the Spanish conquistadors on the far side of the Andes and by other explorers on islands in the East. "Who can doubt that the pygmies are real? You see, they are actual human beings equipped with souls and bodies. But one should rather class those 'phantoms' as demons. Just as much as the human being differs from demons, the pygmies differ from these phantoms. Such are spirits, and when they reveal themselves in fantastic shapes to make an impression on mankind, they are dressed in human covering, otherwise they could not just suddenly appear for your eyes and vanish again. But the real pygmies can never discard their bodies at will but keep them coupled with the soul right up to death. They can be touched, they can be killed and treated in other ways like other human beings." However, Bartholin's planned reissue of his father's treatise on pygmies never came to anything, and it thus became one of the rare exceptions to the rule, that almost all the topics that Bartholin occupied himself with throughout his life ended in one or more publications.[91]

Bartholin is struck by tuberculosis

Perhaps the cause of this early exception was that Bartholin's health began to fail towards the end of his stay in Leiden. He himself considered the humid Dutch climate to be a major part of the explanation. At first, he considered whether a stay at a health resort at the warm baths in Aachen might give him back his physical strength, but this was prevented by the Thirty Years' War. On 16 April 1640, he writes again to Worm: "The longer I stay here, the more I feel my health diminishing, so that I have almost decided to visit the spas in Aachen, if the difficulties of travel, which is hindered by the armies of both sides, did not get in the way."

Bartholin's description of his symptoms in the next letter to Worm three weeks later, on 8 May, leaves no doubts as to whether he has been stricken with tuberculosis: "Besides these reasons the consideration for my health dictates me an unavoidable necessity, which follows from a now sufficiently long stay in a dangerous place in this very thick and damp air, and it grieves me that I have incurred a condition involving very serious symptoms, which these years' incessant constipation has inflicted on me, in such a way that I am afraid that both my stomach and my liver have been damaged by consumption arising from injury to the lung [...] and from atrophy. If only the sign of consumption, which I see constantly when I spit blood, might have misled me." The "unavoidable necessity" that Bartholin mentions was his decision to journey southwards to a milder latitude, initially Paris: "Nevertheless, the physicians and Hr. Moth have induced me to prepare myself for the French journey, since the local mildness of the air and the good spas promise – they believe – a change in this harmful susceptibility." Moreover, whereas Leiden had been the right place for philological studies, Paris was, according to what Bartholin had heard from "our illustrious men and others," the best place for philosophical studies, with which he still wished to intersperse his medical studies. Worm is sincerely concerned for Bartholin's health and writes to him on 11 May: "Your frail health causes me deep worry. Be careful not to weaken your frail body's condition by untimely spells of keeping awake researching at night, such that you are destroying your health and depriving your country of the gifts of your splendid soul. Let your father, of blessed remembrance, serve as an example to you, he who, as a young man, wore himself out with fatiguing studies in such a way that in the time when he might have been most profitable to the fatherland, he was overwhelmed to a great extent by the number of advancing sicknesses, so that he could only fulfil the smallest fraction of his intentions. Fortunately, you can make yourself more cautious by regarding others' misfortunes. During your labour, you may safely allow pleasure to amuse your mind. Since the conditions in Germany are so troubled, you will hardly be granted a sufficiently secure passage on a journey to the spas. You must seek other protection for your health through diet and apothecary skills."

Four days later he adds in a new letter that he and Thomas Fincke, with their eyes on Bartholin's health problems, will not oppose his journey southwards to France, but he advises him to continue southwards from Paris to Montpellier, which is rich in able physicians and is a cheap place. Henrik Fuiren had written from Padua not long before that he regretted his long stay in Paris once he had seen Montpellier.

Next stop Paris

At the end of June 1640, Bartholin leaves Leiden at top speed, but in the seaport of Vlissingen in southern Holland he must make an involuntary stop and wait for an available ship and a favourable wind. He employs the waiting time in the collection of exotic rarities, which he tells Worm about on 3 July: One bird of Paradise, Chinese fans of tissue paper, Indian flies, snail shells, Brazilian root (potato), bread baked from cassava root (manioc), reddish-brown beans and finally papaya seeds, which he takes with him to France because of their "excellent cleansing quality." These items he sends to Worm together with Walæus' treatise to the benefit of Harvey's circulation theory. Before his departure from Leiden, Bartholin has arranged with Walæus about doing the final practical work on the publication of the revised anatomy book.

Later in July or at the beginning of August 1640, Bartholin arrives in Paris, from where he informs Worm on the 8 September, that he has been well received by the two Professors of Medicine, Guy Patin (1601–1672) (Figure 38) and René Moreau (1587–1656) (Figure 39): "In particular, I cannot praise enough the kindness of the illustrious, erudite professors of medicine, Moreau and Patin. They have done me all sorts of unreserved services and formed an open friendship, the former solely by reason of my father's merits in medicine, while the latter, partly through Dr. Moth's recommendation, partly guided by common esteem, promises me the greatest progress in my studies. They, as well as others, who are their equals in kindness, soften the tedium of this place." Alongside his med-

Figure 38. The Parisian Doctor of Medicine, physician and Royal Professor Guy Patin (1601–1672). Engraving from 1670 by Antoine Masson (1636–1700).

During his first stay in Paris in 1640–1641, Thomas Bartholin began a lifelong friendship with the competent, sympathetic and faithful Patin, who, as professor of medicine and anatomy at the Medical Faculty of Paris, had his finger on the pulse of everything that was happening in the field of medicine in the city. Patin's extensive Latin correspondence with European scholars, including Thomas Bartholin, as well as other of his writings, have been translated into French and recently made available under the title *Correspondance complète et autres écrits de Guy Patin* on the website https://www.biusante.parisdescartes.fr/patin/. The correspondence is a goldmine of information about what was happening in scientific circles in Paris in the 1600s. It shows that Patin, at Thomas Bartholin's request, acted as a support for Danish students and travellers in Paris, including among others Henrik à Møinichen, Ole Borch, Willum Worm, Peder Schumacher (later Griffenfeld) and Niels Stensen.
Wellcome Collection, London.

Figure 39. Engraving by Michel Lasne (c. 1590–1667) showing "the famous and learned man René Moreau, Doctor of Medicine in Paris, Dean of the Faculty of Medicine and Royal Professor of Surgery" in his library. The engraving seems to be a gift from Moreau's colleague and friend, Doctor of Medicine Guy Patin, who authored the text. The last two lines read: "In the picture, the artist, dear Moreau, shows your appearance, while a library shows your intellect."

Like Patin, Moreau (1587–1656) became a close friend and support of Thomas Bartholin during his stay in Paris in 1640–1641. He contributed to Thomas Bartholin's treatise on a diphtheria epidemic among children in southern Italy, published in Paris in 1646. Wellcome Collection, London.

ical studies, Bartholin wishes to concentrate on philosophical topics. His health has become a little better, and to obtain more improvement he will "try a longer use of asses' milk, whose effectiveness is so highly praised."

In Paris, Bartholin had not exactly reached the stronghold of new thinking within medicine. It was with good reason that his cousin Henrik Fuiren had written to Ole Worm on 28 July 1639: "Frankly, I regret the time I have been wasting in Paris, since it could have been spent to the greater benefit of my studies and less expenditure for me in Montpellier." In the medical faculty, founded in 1369, instruction was given unchangingly – more than in any other places in Europe – according to the pure doctrine of the ancient physicians. To allow the torch of this doctrine to dim or to tamper with its flame would not, according to the French historian of medicine J. Lévy-Valensi, have been a pardonable mistake, but a crime against the divinity. One was expected to conform to the maxim: "Medicine is a religion, the faculty is its temple, and its gods are Hippocrates, Aristotle and Galen" (Figure 40).[92] The dominating figure behind this philosophy, or perhaps rather this regime, was the anatomist Jean Riolan the Younger

Figure 40. French satirical drawing from the 17th century. "The three helpers of life": the physician, the apothecary and the surgeon in their respective armour. From J. Lévy-Valensi: *La médecine et les médecins français au XVIIe siècle*, Paris 1933, p. 235.
The Royal Library in Copenhagen.

Figure 41. Jean Riolan the Younger (1577–1657). Engraving by Gilles Rousselet (1610–1686) after drawing by Pierre Varie (active 1625–1650). From the work *Responsiones Duæ*, Paris 1655, before the title page.

The Latin text in the frame, "DECANUS IO. RIOLANUS PARIS. MED. DOCT. ET REG. PROFFESS.," reads "Dean Jean Riolan, Doctor of Medicine and Royal Professor in Paris."

The two lines under the portrait authored by his colleague, Doctor of Medicine and Royal Professor Guy Patin, read: "This man should not be called Riolan, but Library, for he contains all, that is studied in all medical science."

The Parisian professor of medicine, anatomy and pharmacology was a fierce opponent of almost all innovation. He never recognized the importance of William Harvey's discovery of the circulation of the blood, Gaspare Aselli's of the milk veins, Jean Pecquet's of the thoracic duct or Thomas Bartholin's of the lymph vessels. As the leading anatomist in Paris, he delayed the medical development in the city, and he was Thomas Bartholin's harshest critic. Although he was often unnecessarily polemical, he was right in some of his criticism of Thomas Bartholin, and it helped to make the last two revised editions of the Bartholinian anatomy book better. Wellcome Collection, London.

(Figure 41), who on the one hand enjoyed international recognition as a special authority, and on the other hand had a reputation for being a rough and slanderous polemicist and a hidebound opponent of new thoughts and progress, including Harvey's doctrine of blood-circulation and Aselli's lacteal veins.[93]

For the same reason, Bartholin would have further clashes with him. Riolan had been professor of anatomy since 1608, and in 1620 he opened a new anatomical amphitheatre on the corner of Rue de la Bûcherie and Rue des Rats (present Rue de l'Hôtel Colbert) in the Latin Quarter immediately south of the Seine and Notre Dame (Figure 42). Here dissections were carried out in the old style, where the professor of anatomy – perhaps except for Riolan himself – clad in a long cloak with long arms, a red scapular, cravat and square cap, assisted

Figure 42. a. The old amphitheatre of the medical faculty in Paris on the corner of Rue de la Bûcherie and Rue des Rats (now Rue de l'Hôtel Colbert), just opposite Notre-Dame, but on the other side of the Seine. Drawing by Adolphe Varin (1821–1897) from 1876. From J. Lévy-Valensi: *La médecine et les médecins francais au XVIIe siècle*, Paris 1933, p. 225.
The Royal Library in Copenhagen.

b. The same building in 2024.
The medical faculty in Paris had been located on this site since 1369, but in 1617 construction began of an amphitheatre, which was inaugurated in 1620 by Jean Riolan the Younger. It was rebuilt into the pictured, still existing amphitheatre, which was inaugurated by the Danish anatomist Jacob Winsløw (1669–1760) in 1745. The building still bears Winsløw's name, L'Amphithéâtre Winslow. After the latest major renovation, it looks like this. The Faculty of Medicine is located today in Rue de l'École de Médecine at the southern end of the Latin Quarter.
Photo: Jesper Brandt Andersen.

Figure 43. The botanical garden with medicinal plants in Paris, Jardin Royal des Plantes Médecinales au Fauxbourg St. Victor. Engraving by Adam Pérelle (1638–1695) from around 1660. 18.2 × 27.3 cm. The construction of the garden was begun in 1635 by the physician and botanist Guy de la Brosse (1586–1641) and inaugurated by him in 1640 – the same year that Thomas Bartholin first visited the city. Musée Carnavalet, Histoire de Paris.

by a trained student, directed the course of the procedure without touching the body himself, but giving instructions to a barber with surgical training.

As a supplement to the medical chairs, a chair of surgery and one of botany had been established in 1634, and in 1635 work was started on the establishment of a royal botanical garden with medicinal plants, Le Jardin Royal des Plantes Medecinales au fauxbourg St. Victor, which was inaugurated by the French physician and botanist Guy de la Brosse (1506–1641) in 1640, the year Bartholin arrived in the city (Figure 43).[94] Moreover, Paris teemed with libraries.

It was precisely the city's many possibilities that attracted Bartholin and gave him the opportunity to cultivate studies other than those of a physician. But apart from philosophy, he does not explain which ones. On 27 October 1640 he writes to Worm: "The professors of medicine have still not turned back to their customary lectures, but are still silent, while in the meantime there are other branches of study that absorb me, so that I divide myself alternately. [...] Paris, though dusty, seizes me with its varying possibilities and has beguiled me into becoming its worshipper." It apparently suits Bartholin well that his planned onward journey to Montpellier must be postponed on account of a plague epidemic.

In Paris he also gets the chance to write more in reply on behalf of his father in an old quarrel between Caspar Bartholin and Peter Lauremberg "concerning the blood and water that flowed from Christ's pierced side." During his stay in Leiden, Bartholin had already briefly mentioned this dispute in the preface to the revised textbook of anatomy and at the time asked about Worm's assessment, which he gave on 8 October 1640: "This controversial question, which has been solved by you, you have handled sufficiently well and cleared up, and you have defended your father in an excellent manner. I believe he imagines that Christ suffered on the cross, as the painters are accustomed to depicting him, so that streams of blood and water flow from his side, as were they meant to fill cups. If there had simply been a little drop of blood or serum, that would have been enough."

In Paris, Bartholin finds the time and opportunity to dig deeply into the question in an original article and again asks for Worm's approval; but this time, he gets a somewhat different reply, which shows that the influential family in Copenhagen at that time already had strategic plans to line Thomas Bartholin up for a professorship at the University of Copenhagen and secure him an irreproachable international reputation. It was important to win the favour of the noble lords and at the same time turn down the polemical flame in order not to acquire too many enemies. The first part of the answer comes on 12 December 1640:

"I would like you to send your article on the divergence between your father and Lauremberg to your press or to the Netherlanders and dedicate it to our new chancellor. I dare express my hope that this will prove greatly to your advantage. He entertains great affection for you on account of your father's merits, and in this way you make him particularly obliged to you. So that you will not be without knowledge of his titles, I insert them here: 'Highly exalted and awesome lord, Mr. Christen Thomesen of Stougaard,[95] knight, His Royal Majesty's High Chancellor, the Danish kingdom's chief adviser, the illustrious patron of the Royal University in Copenhagen.' There is no reason for you to be discouraged by the insignificance of the book. In the preface you will insert reference to the work you have done on your father's textbook of anatomy, and of the remaining things with which you can employ his favour. I have discussed the matter with your grandfather and others, and the scheme did not displease him."

The second part comes on 1 February 1641: "My dear Thomas, I have read through your finely written answer and have nothing special to add to or subtract from it. However, I would simply wish to point out to you that, given that your opponent is no longer among the living, it would be suitable if you proceeded against him rather more leniently, so that you do not fall into the same error that you so strongly reproach him of. It is best to delete everything that carries any trace of bitterness, and it will perhaps be enough to mention your adversary by name a single time or not at all. [...] Your grandfather, as well as our bishop, has read it, and with this one exception they took extraordinary pleasure in it. If you alter it, you will gain great praise for its unpretentiousness, which is a young man's grace."

Bartholin responds that he has not considered changing his plans about dedicating the book to Worm himself, but he will gladly dedicate a later book to the chancellor. The book, *De Latere Christi aperto Dissertatio. Accedunt Cl. Salmasii, & aliorum, De Cruce Epistolæ* (A dissertation on the pierced side of Christ. With the addition of letters from Claude Saumaise and others concerning the crucifixion of Christ), was published in Leiden in 1646 on Bartholin's homeward journey to Copenhagen, and as the subtitle indicates, correspondence on the subject between Bartholin and the French classical scholar Claude Saumaise (1588–1653) and others was added.[96] Bartholin also added an epigram to Christ written by himself, and two epigrams to himself written by the Dutch theologian Antonius Thysius the Elder (1565–1640) and the Danish poet and historian Vitus Bering (1617–1675), respectively. It became Bartholin's first book about biblical medicine – a subject that he was the first important physician to address.[97]

He built further on the subject in his work on paralytics in the New Testament, *Paralytici Novi Testamenti*, Copenhagen 1653, second edition Basel 1662, and in his work on different diseases (and their treatment and cure) in the Bible, or more exactly entitled "The Old Testament, Talmud and the New Testament," *De Morbis Biblicis Miscellanea Medica*, Copenhagen 1671, second edition Frankfurt 1672, which, according to the preface, being severely pressed by the publisher (Daniel Paulli), he "wrote more rapidly than one can cook asparagus."

By December 1640, Bartholin had been prepared to journey southwards to Montpellier, but was delayed by a shortage of money, for which he waited until it was sent to him by his mother. The cost of living in Paris was too high, and anatomical instruction too poor. On 10 December he writes to Worm: "We have recently attended a public dissection, but with disgust and the greatest pity." Overall, Bartholin does not receive much medical instruction during his stay in Paris. Instead, he continues to study philosophical topics. 26 January 1641 he writes to Worm again: "I have quite a few topics that I have decided to publish, if time and leisure are granted me: 'Concerning the birthdays of the old', 'superstitious anatomy', 'concerning the idea of nothing', 'the philosophy of the ancients', especially the Platonic (which has enticed me) and so on. [...] I shall soon prepare 'superstitious anatomy', or at any rate a specimen of it, that is, how much superstition and sacredness could be contained in each limb according to the old pagans, and the reason for this superstition."

Moreover, a domestic archaeological topic occupies Bartholin towards the end of his stay in Paris, namely the discovery of the first of the golden horns on 20 July 1639 at Gallehus, between Tønder and Møgeltønder, by the farm girl Kirsten Svensdatter from Østerby. It is Worm who, on 12 December 1640, mentions the finding and his legendary treatise on it for Bartholin: "I am now also fully occupied with the printing of my account of the chosen prince's golden horn,[98] which was discovered here a year ago." Bartholin answers enthusiastically on 26 January 1641: "Concerning your undoubtedly golden description of the golden horn, I have given hope to our philosophical physicians. I shall not hesitate to pay the postman, provided you satisfy our appetite for such a piece of writing. The freshness of the topic and the worthiness of the writer have whetted my appetite to the full." Bartholin must have satisfied his appetite quickly, for before Worm had received his letter, he had sent Bartholin on 1 February his treatise *De Aureo Serenissimi Domini Christiani Quinti Daniæ, Norvegiæ, etc. electi Principis Cornu Dissertatio* (Treatise on the golden horn, which belongs to His Royal Highness Christian V, hereditary Prince of Denmark, Norway, etc.), Copenhagen 1641, "which I wrote recently on the prince's behalf about his golden horn" (Figure 44).

Figure 44. Copper engraving of the long golden horn found on Saturday 20. July 1639 at Gallehus between Tønder and Møgeltønder in southern Jutland by the peasant girl Kirsten Svensdatter from Østerby. The engraving is from Ole Worm's book about the horn from 1641. Thomas Bartholin reused the illustration (the version shown here) in the second edition of his book on the unicorn, *De Unicornu Observationes Novæ*, Amsterdam 1678, p. 149. The Royal Library in Copenhagen.

Arrival in Padua

During the spring of 1641 Thomas Bartholin eventually left Paris together with his brother Caspar to travel south through Orléans and Lyon to Montpellier. From Lyon he writes in a short letter to Worm on 19 May: "Thank you for the treatise you sent about the golden horn. The scholars I meet while travelling find it graceful and first-rate and pester me to be allowed to look at it, but no one has yet achieved this." Despite "the peace for study and the cheaper cost of living" Bartholin does not stay long in Montpellier, and he sends no letters to Worm from there.

By the next letter of 13 November, he has travelled via Florence to reach Padua, where he gains his matriculation on the same day at the university.[99] In a little book about travels for medical study, Bartholin relates many years later that this type of travel, except for soldiers, workers and students with very scanty means, normally took place on horseback, but because they lacked horses, he and his brother Caspar were forced to carry out the journey on foot from Provence to Genoa via Nice, Monaco, Menton, Ventimiglia, Finale and Noli.[100] It is a journey of more than 300 kilometres.

We also get a little information about the stay in Montpellier. Bartholin has not had much response from the town's conservative professors about Aselli's and Harvey's discoveries: "When with some keenness I had begun a public disputation with the professor on behalf of the new opinion about the lacteal veins and the circulation of the blood and was ill regarded by the steady supporters of Galen [...] I made my way to Italy."

The self-illuminating flesh in Montpellier

The stay in Montpellier, however, was not totally unbeneficial. Elsewhere in the letter to Worm, Bartholin writes: "My short treatise on the shining flesh I recently saw in Montpellier shall be dedicated to my grandfather and will perhaps soon be published in this difficult place, if you consider it right."

The story was that Bartholin in Montpellier had become a second-hand witness to a phenomenon from the world of physics – cold light emitted from living organisms (*bioluminescence*, to use a modern word) – which he had never seen before, and which together with light emitted from minerals (phosphorescence) would over the coming decades become an object of intense interest among the world's natural philosophers, including one of Bartholin's later pupils, the earlier mentioned Danish physician, chemist, botanist and philologist Ole Borch.[101] Hitherto it had been the general assumption that light could only be emitted from heated material.

The interest in cold light began in 1602 when the Italian shoemaker and alchemist Vincenzo Cascariolo (1571–1624) found on the mountain Monte Paterno near Bologna, a luminous stone that contained a mineral (barium sulphate) that, like a sponge soaking up water, absorbed light from the sun by day and emitted it at night. The stone, *lapis solaris* (sunstone) or *lapis bononiensis* (Bologna stone) caused a sensation, and alchemical qualities were ascribed to it. Galileo also became interested in the stone and furthermore gave samples of it to the physician and natural philosopher Giulio Cesare la Galla (1576–1624), who was the first to mention the Bologna stone in his treatise *De Phænomenis in Orbe Lunæ* (On phenomena within the moon's circumference), Venice 1612.[102] He declared that the stone was not capable of shining before it had gone through the process of calcification. However, the first great, widely read and discussed treatise on the subject, *Litheosphorus sive de Lapide Bononiensi* (On the Luminous Stone or the Bologna Stone) was published by the Italian physician, philosopher and archaeologist Fortunio Liceti (1577–1657) in Udine in 1640. It gave rise among other concerns to a heated discussion between Liceti, who attributed phosphorescent qualities to the moon and compared its light to that from the Bologna Stone, and Galileo, who held that moonlight was reflected sunlight.[103]

Bartholin's interest in the subject was restricted essentially to light emitted from living organisms. In his book *De Luce Animalium Libri III. Admirandis historiis et rationibusque novis referti* (Three books filled with wonderful accounts and new explanations of light emitted by living creatures), published in Leiden in 1647 and not – as first planned – dedicated to his grandfather, Thomas Fincke, but to Chancellor of the Realm Christoffer Urne (1593–1663), Bartholin picturesquely describes the phenomenon from Montpellier in this way:

> "It was in the month of April. Regenerated after winter the year began to smile on us. The cattle, which came out of their dark cattle-sheds where they had spent the winter, were delighted by the spring. All creatures were anticipating happiness with the ascending sun, apart from the unhappy tribe of donkeys and mules. They awaited, now as before, the lash across their poor backs, and the mules' bells spoke only of the wear and tear for the foreseeable future, such as had been their lot hitherto. But one could not let all the cattle live. Some of the animals had to be slaughtered, and the meat presented for sale in the butcher's shops. Among those who bought some of it, there was a woman who had not time to cook the meat immediately but hung it up in her hut to use it the next day. As she now lies at night on her bed in the little room, which borders on her kitchen, she cannot sleep. Then she suddenly sees the piece of mutton, which has been hung

up in a corner of the hut, shining so brightly, that the gleam from it fills the whole room. The woman becomes terrified and at first believes that the gleam comes from hellfire, or that dead spirits have taken up residence in the slaughtered animal, and in this way are making themselves known – a thousand bewildering thoughts are coursing through the poor woman's head. Soon she believes most of all that she has perhaps been deceived by the one who sold her the meat, and quickly afterwards she hopes that the poor cottage will become famous because of this miracle, or that she now has a constantly burning lamp in her room. As soon as it became day, the woman told her neighbours about the strange thing she had seen. The report of it spread far, and a wondering crowd of a hundred people streamed thither. The light that radiated from the flesh was not like flames, but a white light like the glow from stars, which, however, was not evenly distributed over the whole piece of flesh. Gradually as the flesh rotted, the light was extinguished little by little. Pious people – clergymen – who were eyewitnesses to it, when the light faded away, remarked, that it gradually gathered into the form of the holy cross, until it faded completely."[104]

Bartholin confirms in a later letter (13 August 1643) to Worm that the flesh had putrefied: "In this way many pieces of meat in Montpellier inclined towards putrefaction this spring, and therefore a good deal of it shone in the meat market. Likewise, it is well known that fish and wood shine if they are beginning to rot." Bartholin did not believe that the rottenness produced the light by itself, but that it "releases concealed seeds of light that have been repressed in the mixture of the elements."[105] With modern knowledge it must be assumed that the light was emitted by luminescent bacteria in the rotted mutton flesh. The process is due to the fact that the luminescent organisms contain the pigment luciferin, which with the aid of the enzyme luciferase reacts with oxygen, and thus the light is produced.

De Luce Animalium is, at 396 pages, a bulky work, described by E. Newton Harvey as the first extensive and most important early book about luminescence, and the most distinguished of its kind of the 17th century.[106] Similarly, Newton Harvey finds it probable that the book stimulated contemporary studies of luminescence to a large degree.[107] The book describes all luminous phenomena known at the time, both real and imaginary, organic and inorganic, and it illustrates in the best way the characteristic polyhistoric and natural philosophical access to a scientific subject in the first half of the 17th century. It is divided into three books. The first deals with light phenomena in human beings, including light emitted from the hair of the head, head, face, eyes, teeth, tongue, chest, stomach, abdomen and joints, but also mental and spiritual light. There is also

Figure 45. Carlo I Gonzaga (1580–1637), Duke of Mantua, painted by Daniel Dumonstier (1574–1646) in the 1620s. Oil on canvas. 60 × 50 cm.
It was said that fire appeared when you gently rubbed the duke's skin.
Musée de l'Ardenne.
Photo: Wikimedia Commons.

a short discussion of luminous plants, jewels and minerals, including the stone from Bologna.[108] The second book deals with light emitted from animals, including lions, horses, bulls, wolves, lambs, dogs, foxes, cats, hyenas, mice, snakes, creeping and flying insects, other different insects of uncertain origin, birds and fish. The third book is taken up with the possible reasons for the light emitted from living organisms, including considerations to do with light-emitting matter and the various types and amounts of emitted light. Bartholin refers frequently to more or less fantastic stories from the Bible, from works of ancient authors and from contemporary accounts. Much of the light, he describes, is due to phenomena other than bioluminescence: for example, light reflections from glittering surfaces and sparks from static electricity arising out of friction from clothing and other materials on hair and skin. The light that shone from Moses' face when he descended from Mount Sinai Bartholin explains as a reflection of sun- or moonlight.[109] The flames that poured from the mouth of Tiberius' horse were, according to Bartholin, perhaps simply static electricity, which arose when its mane was combed, and the same explanation is natural for the story, that it was customary for fire to come forth when you gently rubbed the skin of Carlo I Gonzaga (1580–1637), Duke of Mantua (Figure 45).[110]

A more dramatic version of this phenomenon is the story of the woman from Verona, who suddenly found that when she rubbed her body with a piece of cloth, it sent out a glittering, phosphorescent light, so that in a dark room she appeared like a real fireball, an awe-inspiring object, which rendered her superstitious maidservant speechless and terror-stricken and made her run from her mistress in the belief that she had become consumed by flames.[111] Beyond an extensive literary study, it was Bartholin's intention to enrich the book with his own experimental observations. On his later journey in Italy, he caught fireflies in order to investigate them for their light-emitting material, which he assumed must be related to that of the self-illuminating flesh. However, he was not very successful with his investigations. Concerning his attempt to examine whether the firefly's ability to shine vanishes when it dies, he writes for example: "I tried to find out the truth of this conjecture and put a wingless fly aside. But while I awaited the result, it prudently fled and took its light with it."[112] What Bartholin was not capable of demonstrating through his own experiments he could, however, find in a letter from the Sicilian Carolus Vintimilla to the Italian botanist Fabio Colonna (1567–1640).[113] In this letter, Vintimilla gives an account of his experiments with fireflies, which contain three completely correct scientific novelties: 1) Only male flies have wings, 2) the wingless female flies emit a stronger light, so that they may better attract the male flies at night in preparation for mating, and 3) the fireflies' eggs are also luminous. In referring to the letter and allying himself with the results of Vintimilla's investigations in *De Luce Animalium*, Bartholin became the first to publish this new knowledge.[114]

Bartholin himself was of the opinion that light was found in everything and constituted the fifth element after fire, earth, air and water: "I have claimed that innate light is a fifth element, which either shines clearly by itself or with an added motion, like sparks, which arise through scratching on steel or stone, by the punch of a fist in the eye or as sparks from teeth."[115] He elaborates on this opinion in a letter to Ole Worm on 17 June 1643: "That there is implanted in everyone a light, whence its origin is attributed to be, I conclude from the question itself. In other words, I conclude that it is implanted, and that it is transferred to us from our parents, no different from the soul."

As so often, when something is difficult to explain, Bartholin adds that it is God who has created the light, together with the other elements. When the parts of the light are spread as in the heavens, it is weakened or disappears, but when it is gathered and concentrated as in the celestial bodies, it is strengthened. Bartholin acknowledges that "these things seem irrational, but it is in this way we explain almost everything that causes difficulties to others."

Worm looks for Bartholin's answer on 7 July as to "why some living creatures shine, others not [...] even among individuals of the same species. In other words, why did only the one piece of lamb, which you saw in the meat market in Montpellier, shine amongst all of them and not the rest? In that case it seems most important to find the causes, why the innate light shines more in the one than in the other." Bartholin explains on 13 August that the difference can be attributed to either conditions of the shining material in its surroundings, for example the air, or in the light itself: "Ultimately, the light is emitted, if it is plentiful, if for one reason or another it is detached from its restrictions, so that it can discharge itself or is compressed, so that it can be seen. Likewise, not all regions of heaven are illuminated, but only those which are concentrated in stars and comets. Nor does the Bologna stone always absorb the light and retain it, but only when it is prepared by division."

In the year before the publication of *De Luce Animalium*, Worm recounts in a letter to Thomas Bartholin from Leiden on 24 April 1646 a new dramatic tale about the light phenomenon, this time experienced at home in Denmark:

"Thinking about your treatise on light, I cannot omit recounting what happened to me a few days ago: my maidservant recently bought in the fish market among other fish a *Scorpius marinus*, the name used for it by Schonfelde, but by the people round here a sea scorpion. When she had skinned it and thrown aside the skin on the dunghill together with the head, my servant saw several days later, when he walked past the place in the twilight, an unusual light, which shone in the same place in the darkness. This struck him with no small fear, on account of which he sent for the other people in the house and showed them the portent. When the terrified folk had observed this light for some time, one of them, who was a little bolder, went nearer to find out what was at the bottom of it. But he found only the head and skin of the flayed fish, which he carefully kept till the following day. In the morning, he came to me and gave a detailed account of the matter. I told him to bring the shining articles, which I concealed in their half-rotten state in a corner of my museum. But when I several times turned it and handled it in daylight, it was not possible to observe anything of the sort. However, when the darkness was closing in during the evening, I had the experience he described. And when I looked more closely, my eyes could perceive something like four small distinct lights, which gleamed outwards from four places on the head, one over each eye and on each side of the lower jaw. And now, when these places, from which the light shone, were inspected with a light, they did not look black as the rest of the skin, but whitish and slimy. By this sight my eyes were delighted for three to four nights at different times, and I showed it to other people, until the head little by little lost its light, when it dried because of the warmth of the surrounding air. Thus, I think, that this tough and slimy dampness in a white object with a tendency to putrefaction has contributed substantially to the nature of this light."

Contemporary interest in this subject was so great that Bartholin chose to reissue *De Luce Animalium* in Copenhagen in a new edition of over 500 pages with the title *De Luce Hominum et Brutorum libri III* (Three books on light emitted by human beings and animals). It also contained Bartholin's improved edition of the Swiss physician and natural historian Konrad Gesner's (1516–1565) *De Lunariis Herbis, et Rebus Noctu Lucentibus* (On self-illuminating plants, and things which shine at night), Zurich 1555, which was the first book to deal exclusively with luminescence, chiefly in plants.

Despite his lifelong interest in bioluminescence, Bartholin never succeeded – because of the insufficient knowledge of his epoch – in finding a scientific explanation for the phenomenon.[116] To reflect on this, it bears mentioning that the self-illuminating characteristics of phosphorus were first explained in 1974.[117]

Residence in Padua

In Padua, Bartholin had come to the actual end of his journey. It was no coincidence that at this town's university 247 Danish students received their matriculation during the period 1566–1644.[118] One of them, the later physician, professor of medicine and rector of the University of Copenhagen, Christian Ostenfeld (1619–1671), who stayed in Padua from 1651–1655 and studied medicine at its university from 1653,[119] describes the town thus:

"Among Italian towns Padua rises for the fame of its university. While science and teaching were driven out across the whole of Italy on account of the frequent barbaric invasions and wandered about at random, so did the undefeated Venetian republic make redress with the establishment of an asylum for learning. It is fortunate in having built within its kingdom's frontiers this beautiful, famous theatre, whose changing scene is used by actors in all disciplines and is viewed by the world. By reason of the people's friendliness, by reason of the easy way of living, by reason of the refreshing climate, by reason of the freedom to study, students from all corners of the world are drawn here to learn, and they rush here as to a market of scholarship. Nowhere has science stepped more gloriously from darkness to light, nowhere has Astraea[120] triumphed with greater honour, nowhere does anatomy have a more frequent audience, nowhere does one live more pleasantly and have a better reputation. Therefore, those who have once come to Padua are reluctant to leave, others who have spent a long time at foreign universities believe that they have only lived properly when they have lived in Padua."[121]

The University is the third oldest in the world, founded in 1222 by professors from the world's oldest university in Bologna, who had left the latter as a protest over the Papal States' "grave offences to academic freedoms and the failure to observe the privileges solemnly granted to teachers and students."[122] At the time of Bartholin's arrival in Padua – 100 years after Vesalius' days of glory as a teacher in anatomy and surgery – the university was still the leading school in Europe in almost all medical disciplines. It included the earlier-mentioned and still existing anatomical theatre from 1594, where in Bartholin's time dissections took place in the light of torches and wax candles, and where the bones from the dissected bodies were afterwards buried in a particular grave at the front of St. Martin's Church.[123] The university in Padua had found means to solve the fact that it was not always easy to obtain bodies for dissection. Bartholin puts it this way: "A thorough investigation must be made of the bodies of sick persons, and they must be subjected to the anatomical knife in the anatomy theatres. At Padua the anatomists easily see this wish fulfilled, for there they have several anatomical assistants, who seek to provide materials from hospitals as well as from the homes of the dead persons and from recently filled in graves. They enjoy this great privilege thanks to help from the state, and still they must set out with weapons in hand, since they would otherwise fail or not without resistance bring home the prey."[124]

Furthermore, in Padua there was, and still is today, a botanical garden for medical plants, Orto Botanico. Established in 1545, it is the oldest of its kind in the world (Figure 46). This achieved distinction with its great collection of Egyptian plants, which had been collected by the Italian physician and botanist Prospero Alpini (1553–1616), who from 1580 to 1584 had acted as physician for the Venetian consul in Cairo. In 1592, Alpini published in Venice the work *De Plantis Ægypti* (On Egyptian plants), which despite its incompleteness became an important tool for the Swedish naturalist Pehr Forsskål (1732–1763) on the great Danish expedition to "Arabia and other surrounding countries" 1761–1767.[125] The work describes 57 plants and trees and illustrates 49 of them (Figure 47). It gave Alpini's contemporaries, not least medical scientists, important new knowledge about exotic and oriental plants and was accordingly reissued in 1640 by Thomas Bartholin's teacher in Padua, the German physician Johannes Vesling (1598–1649).

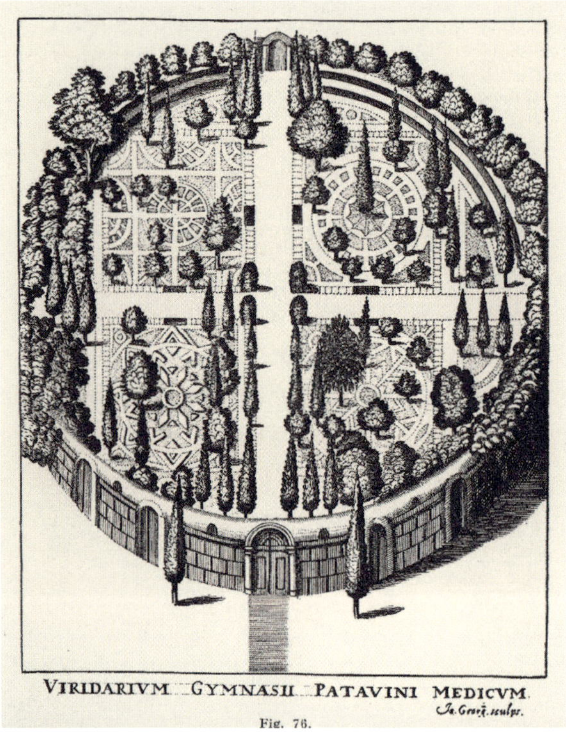

Figure 46. a. The botanical garden of the University of Padova, Orto Botanico, was established in 1545 and is considered the oldest university garden in the world. Engraving in Giacomo Filippo Tomasini: *Gymnasium Patavinum*, Utini 1654, p. 82.
The Royal Library in Copenhagen.

b. Orto Botanico appears today completely intact with its encircling ring wall and its geometrically organized beds of medicinal plants (piante medicinali) as in the 1500s. The garden is tended and cared for daily by numerous gardeners. A visit to the garden is a fascinating journey back in time.

The plant on the left in the picture with the light purple flowers is a *Monarda didyma* (double horse mint).
Photo: Jesper Brandt Andersen.

The medical faculty had at its disposal some of Europe's most skilled teachers and scientists. Apart from instruction in anatomy in the anatomical theatre, the professors of medicine held lectures and lessons in their own homes, and medical education was favoured by the fact that the university was one of the first to emphasize clinical teaching, where the students came along on sick visits in the patients' homes, in monasteries and in hospitals. This had already been introduced by Vesalius' friend, professor in practical medicine Gian Battista da Monte (1498–1551). In the tower of the city's astronomical observatory, Galileo had been the first, in 1609–1610, to discover Jupiter's moons, as well as the craters of the moon, through his primitive, home-made telescope. Bartholin writes in 1665 that he met Galileo once on his travels, when the latter had become an old and almost blind man, but whether or not they talked he does not say.[126]

After the 12th century, Padua had become a free city republic. In the 14th century, it had enjoyed the protection of the powerful Carraresi-family, and at the beginning of the 15th century (1405) it had been placed under the administration of Venice, which was in opposition to the Papal States in Rome. Crucial for Danish and other North European students in Bartholin's time was the fact that the University of Padua practised religious freedom, in contrast to the universities in Bologna and Rome, for example, where you had to be a Catholic to be

Figure 47. Orto Botanico was characterized in the 1600s by its large collection of Egyptian plants collected by the Italian physician doctor and botanist Prospero Alpini (1553–1616). The engraving shows a date palm in Alpini's *De Plantis Aegypti*, Venice 1592, between pp. 11 and 12. The Royal Library in Copenhagen.

Figure 48. Portrait engraving of the physician and polymath Johan Rhode (1587–1659) executed by Albert Haelwegh (1620–1673) c. 1656–1659. 19.5 × 15.5 cm.

The Latin text under the portrait, which was written by his colleague Christian Ostenfeld (1619–1671), reads: "The famous Dane Johan Rhode, physician, polyhistor and philosopher, who earned the favour of his fellow citizens, was honoured by Christian Ostenfeld with this memorial."

From the time he settled in Padua in 1622 and until his death, Rhode's home was a gathering place for Danish and other Nordic students in the city. Thomas Bartholin respected him and benefited from his help, connections and knowledge. Rhode had a conservative scientific nature and did not himself make new discoveries in medical science, but he had an extensive literary knowledge.
The Royal Library in Copenhagen.

able to receive tuition.[127] Furthermore, the general living costs in Padua were lower than in the other Italian university cities and there was an unusual degree of academic freedom.[128] At the beginning, it was only possible to study law in Padua, but in 1399 the university was divided into two departments, *Juristarum* for law studies and *Artistarum* for studies in medicine, philosophy, theology, grammar, dialectics, rhetoric and astronomy. The university was organized as a free union of students, divided into nations from origin in Italy (*citramontani*) or other countries (*oltramontani*). It was the students themselves who approved the university's regulations, elected the rectors, employed the teachers and paid their wages together with the authorities. When Padua came under the administration of Venice, the government in Venice decided that the University of Padua should be the only university in the whole of the Venetian Republic to offer all study disciplines (*Gymnasium Omnium Disciplinarum*) and in addition an unusual degree of religious tolerance and freedom of thought.[129]

Among the Danish students who never left Padua was the physician and polymath Johan Rhode (1587–1659) (Figure 48), who arrived in the city in 1622 and remained there until his death. During his time in Padua, he himself became an institution and his home a meeting-place, especially for Danish and other Nordic students, but also French, English, German and Italian students and scientists.[130] In Rhode's 376-page guest book, which is preserved in The Royal Library in Copenhagen, there are about 800 names, including students, physicians, friends and family of artists and noblemen, who made their way via Rhode's home (Figure 49).

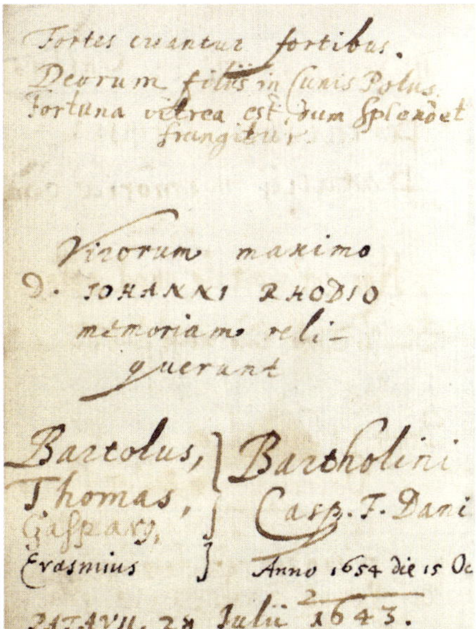

Figure 49. Guests who visited Johan Rhode at his home in Padua signed his guest book. Students, scientists, artists, royalty and many others have, over the decades, written their names in this small, thick book covering the period 1623–1656. On 28 July 1643, Thomas Bartholin and his brothers Bertel and Caspar signed the book, and on 15 October 1654, their youngest brother, Rasmus, added his name on the same page. The verse at the top of the page tells of the fragility of happiness: "Happiness is clear as a glass; when it shines most beautifully, it can be broken apart."
The Royal Library in Copenhagen, Thott 573 octave.

If anyone needed help in his research and travel, he could find it from Rhode. As a newcomer to Padua in the autumn of 1641, Thomas Bartholin needed help and guidance, and since Rhode had in his younger days established a close friendship with Ole Worm and was in continuous correspondence with him, it was quite natural that Bartholin visited Rhode and established acquaintance with him as his first act in the city. Rhode made no new discoveries in medicine and had a conservative scientific nature, but he possessed an extensive literary knowledge.[131] Bartholin came to value him highly, and the sympathy and respect were mutual. As already shown with Caspar Bartholin as an example, Thomas Bartholin was a lifelong industrious publisher not only of his own but also of others' works, including some of Rhodes' works.[132]

Studies of the golden horn

From his letters to Worm, it appears that Bartholin spends a part of his first period in Padua studying archaeological literature and spreading knowledge of the discovery of the golden horn and Worm's treatise on it. From Venice, Bartholin recounts on 21 February 1642 that he has ensured copies for the libraries of Venice, Padua, the Vatican and Milan, "so that the glow of the golden treatise can emanate from every place."

Back in Padua, Bartholin writes to Worm on 20 March 1642 that he has received a letter from Fortunio Liceti with his interpretation of the "hieroglyphs" on the golden horn. Bartholin cites the letter, in which Liceti writes that he interprets the picture descriptions of the Danish golden horn in such a way that "the horn has been made on the order of a very old king of the Jutes, and not for any other use than to prove the state governors' awareness of the management of their provinces and the people's honesty in refraining from robberies. Therefore, the horn was displayed publicly at a crossroad in Ribe. But I do not think the pictures that are found on it concern anything other than an eminent Danish king's character and personality, both in war and peacetime, while he proficiently rules the realm subjected to him."

The widely read Bartholin does not allow himself to be so easily convinced about this interpretation and lets Worm know that he does not intend to support it in his answer to Liceti: "For with what evidence will he convince us that the fidelity of the subjects has been put to the test with such a presentation, since it could be verified with something smaller and less conspicuous? The explanations of the pictures have a greater air of truth; and from his moral presumptions he could easily confine them to the customs of a king." Bartholin thinks that the king to whom Liceti alludes is the saga king "Frode the Great" (Frode Fredegod),[133] but he believes that Liceti's interpretation of the golden horn's hieroglyphs rests on a mistaken basis, namely the ancient Egyptians' and Romans' interpretation and meaning. Bartholin doubts whether "our old Danes have described things with the same images," and he searches for documentation as to who might have brought the Romans' and Egyptians' secret style of writing to the North. If the people of Scandinavia had no knowledge of this style of writing, "it seems likely that they also under the signs, where they have hidden an inner meaning, must have been different, and that thus the golden hieroglyphs can hardly be adapted to the oriental meaning."

Worm urges Bartholin in letters of 16 April and 5 May 1642 to be level-headed and abstain from criticizing "the illustrious Licetus": "He could otherwise take offence, desist from his project and turn his goodwill into revenge." Worm will never regret his work on the golden horn, when a great man like Licetus chooses to offer his labour and time on the business, and it is understandable that there are various interpretations of the hieroglyphs' meaning. Worm agrees with Bartholin, in that they "in many respects do not fit with the Egyptian," but Worm has already given his interpretation in his writing: "Let other people have a different meaning than my own, as long as they are guided by probable considerations."

Since Liceti traces the hieroglyphs to a Danish king, he will not gain Worm as his opponent, unlike a German who has tried to wrench the horn – "the most outstanding among memorials" – from the Danes by claiming that this was the horn that the idol Svantevit on Rügen held in its hands. In order to expose this foolishness to everyone, in a forthcoming treatise Worm will place the golden horn among the other Danish memorials "in increased figure and freed from his [the German writer's] blemishes." For his forthcoming work, Worm has asked Bartholin for a more detailed description of the previously mentioned Florentine horn, including its size, material, form and use together with a description of its hieroglyphs. Is it made of gold? Does it correspond in size to the Danish one? Is it provided with the same hieroglyphs? Can it be blown into? Is it composed of several plates?

Bartholin examines the question with his usual thoroughness, and on 26 June 1642 he informs Worm that there is not just one but two Florentine horns, which he has got an illustrator to draw for Worm's treatise. The horns are made of ivory and probably date from the first king of Portugal, who occupied countries in the East. The incised figures seem to indicate that they were used as hunting horns, but Bartholin believes that they were employed as war trumpets in the Turkish galleys, which had been conquered by the Grand Duke of Tuscany. Crosses and other Christian symbols testify to Christianity, and spurs on the knights' and huntsmen's heels show that the horns can only be a few hundred years old.

Bartholin therefore asks Worm: "But what do these Florentine horns have to do with yours?" He has discussed the golden horn with leading Italian antiquarians and intends to publish their judgements in print. He has also met Liceti who, in his forthcoming book about ancient rings, *De Anulis Antiquis*, describes the golden horn and interprets its hieroglyphs only slightly differently from Worm.

Medical studies in Padua

The letters to Worm from Bartholin's first year's residence in Padua do not leave the impression that it is his medical studies that occupy his thoughts during this period. However, we are given a few pieces of information about them. On 21 February 1642 he writes about a lesson in anatomy:

> "My present friend and greatest patron, the admirable Vesling, has conducted several anatomical dissections with success and great precision. [...] We have performed very painstaking investigations concerning the mouth of the *vena cava*[134] and the *arteria venosa*,[135] but among numerous bodies we have only once found it open, and that was in a particular heart condition, since the normal way through the *vena arteriosa*[136] was blocked by a very large quantity of fat or, more accurately, a slimy fluid, which gushed out from both sides of the ventricles of the heart. We have shown that the observations of this Folius, whom I have mentioned in my introduction, are totally wrong or rather much too particular.[137] I shall discuss it with him personally, while I am called to the Saturnalia here."[138]

On 29 May 1642 he continues his report, informing Worm about the teaching in practical medicine by Professor Giovanni Domenico Sala (1579–1644): "For whole days, I am following the eminent Sala and his practical medical course with various cases of illness and their treatment." Thirty years later Bartholin writes about Sala's teaching: "In the course of three years of visiting the sick in Padua I considered Giovanni Domenico Sala the elder as outstanding and would have enjoyed that advantage for a longer time, had not fate removed from us that great physician, renowned in years, merits, teaching and experience. I shall remember him to the end of my life."[139]

Despite Bartholin's great interest in and contributions to Worm's publications, one notices in the letters from Worm his increased concern regarding Bartholin's focus and time spent on non-medical subjects. Worm attempts to flatter Bartholin through the interest shown by the King's Chancellor Christen Thomesen Sehested (1590–1657) in Bartholin's medical achievements, including the republication of his father's textbook of anatomy, but Bartholin's answer in the letter of 29 May only makes Worm still more worried. Bartholin writes: "If, with his [the chancellor's] approval, you fix my scattered studies to a definite goal, I shall aim at it with greater determination, and I shall gather the spears which, because of the uncertainty of the final purpose, fly dispersed and in different vague directions. When I am more certain of my subject matter, I shall choose a definite direction for my studies. If I am deprived of other sources of aid, I shall offer myself to medicine, which can generally supply one's daily bread."

Worm refers to precisely these "sources of aid" in his answer to Bartholin on 20 June, where he makes it clear to Bartholin how he should prioritize his time: "As regards the purpose of your studies, we have advised you in the direction of medicine, which has hitherto been your purpose, to which end also both stipends are granted to you.[140] These may not be paid out for any other use. If you let yourself be tempted away from it, so that we fail to realize the expectation, we have put in you, you must take care not to be met by compensation claims. Meanwhile, you are behaving properly by working at philology, philosophy and other subjects in your free time."

The last mitigating sentence could perhaps be partially explained by the fact that Worm, despite everything, had a certain realization that it was his own interests that guided Bartholin away from the trail of medicine. After this lashing, Worm tempts Bartholin with the carrot that, after his studies have finished in Padua, a post could be created for Bartholin at home at the University of Copenhagen as his grandfather's deputy, and "if you were brought into the university in this manner, you could not easily be removed from it."

Book about the unicorn

Bartholin, however, did not immediately take the fatherly suggestion from Worm as a piece of advice. Alongside his interest in the golden horn, he had begun to look at a completely different type of horn, namely horns that grow on human beings and animals, and particularly the unicorn's horn. It was again a topic that both his father and Ole Worm had dealt with, and in order to understand this interest it is necessary to look at a little historical outline.

When Caspar Bartholin, as mentioned earlier, published his treatise on the unicorn in 1628, there was still a widespread understanding among scholars and ordinary people that a single-horned animal existed with very particular qualities mainly associated with the horn. Scholars based their opinion on writings both older and newer, but in churches, monasteries, princely treasuries, cabinets of curiosities and pharmacies in Europe were preserved examples of what people took to be genuine unicorn horns.

Figure 50. Stick and drinking cup of narwhal tusk (Greenlandic unicorn horn).

The drinking cup, which was carved by Frederiks III's woodturner Jacob Jensen Normand (1614–1695), is decorated with gold and enamel, and on its lid is a Greenlandic man, symbolizing the Danish king's dominion over the northern seas. On the stick, the rubies of the button conceal holes for compartments with sweet-smelling ointment. The enamel was made by Frederik III's goldsmith Paul Kurtz (d. 1679) c. 1663. The Royal Collection at Rosenborg Castle. Photos: Kit Weiss.

A particularly well-known variety could be found in the Abbey of Saint-Denis, north of Paris. It was so long that a full-grown man could not reach its tip when he stood next to it. It was not particularly thick, slightly decreasing in thickness towards the tip and twisted like a snail's shell.[141] It was the colour of hartshorn, but comparatively heavier. Just about as famous were the two similar unicorn horns that were kept with the other treasures at St. Mark's Basilica in Venice and were shown to the people each year on Ascension Day. Caspar Bartholin had seen these horns with his own eyes during a visit to the city.[142] In many European churches, unicorn horns were used as candelabra and crosiers. The Danish kings also owned staffs, sticks and goblets made from unicorn horns, which were really narwhal tusks (Figure 50).

The anointing chair at Rosenborg Castle, which was commissioned by Frederik III in the 1660s and used for the anointing of kings from Christian V to Christian VIII (1786–1848) during the period 1671-1840, is also made of narwhal tusk, although in Frederik III's time it was said to be made from unicorn horn (Figure 51). Henrik Fuiren's and Thomas Bartholin's later collections (museums) of rare natural historical objects also included so-called unicorn horns.[143]

Literary descriptions of unicorn horns could be found in both the Old Testament and ancient writers.[144] Among those, Caspar Bartholin points out in his book, are Pliny the Elder (23–79) and Claudius Aelianus (175–235), but the famous Italian explorers Marco Polo (1254–1324) and Ludovico di Varthema (1470–1517) also wrote about unicorns.[145] Amongst these writers there was agreement that there were to be found in India, Ethiopia and neighbouring areas one-horned animals, whose horns were the best antidote against poisons and illnesses that were considered poisonous (for example plague and other malignant fevers). Some regarded the effectiveness of the horn as an antidote so potent that when the animal bent over to drink from a river or stuck its horn in the stream, it could purify the water of the whole river. To drink from a cup made from the horn of a unicorn could therefore have both preventative and curative effects, and to be in possession of such drinking cups was considered a protective measure in many European royal and princely houses, not least where one was most afraid of poisoning attempts.

When it came to the animal's exact appearance, agreement vanished, and here one must in particular put one's trust in Ludovico di Varthema's evidence, which was relatively new and partly based on personal observation. Di Varthema had seen with his own eyes two live unicorns at the palace of the Sultan of Mecca, to whom they were sent as a gift by the King of Ethiopia. One was the size of a 30-month-old horse, the other was the size of a year-old foal. They were both brown and had a single, rather long horn in the middle of their forehead. The

Figure 51. The anointing chair, made in the 1660s by the woodturner Bendix Grodtschilling the Elder (1620–1690) of narwhal tusk (Greenlandic unicorn horn), was commissioned by Frederik III with the biblical King Solomon's throne as a model. 252 × 96 cm. The chair was used at the anointing of the Danish kings from Christian V in 1671 to Christian VIII in 1840. The Royal Collection at Rosenborg Castle. Photo: Kit Weiss.

Figure 52. Unicorns often appear on medieval frescoes in Danish churches. Examples are those on the frescoes of the Elmelunde Master, as this one in Elmelunde Church on the island of Møn, dating to approximately 1500. The Elmelunde Master also decorated Keldby and Fanefjord Churches on Møn with frescoes. In all three churches, a unicorn is included in the depiction of the Annunciation as a symbol of purity and virginity.
Photo: Jesper Brandt Andersen.

head was like that of a deer, the neck not particularly long and covered with a short mane, the legs slender and buck-like with hooves. Pliny had a similar description, but with the variation that its feet resembled those of an elephant. Marco Polo, however, described the unicorn as a little smaller than an elephant and with a pig's head, which drooped towards the ground. Gradually a certain consensus formed that the unicorn was an equine animal with a long, spiral, conical, ivory horn in the middle of its forehead, and this consensus was current in Caspar Bartholin's time (Figure 52).

Caspar Bartholin himself was convinced about the existence of unicorns, even though few had seen the animal. "Whoever denies it displays his complete ignorance of both secular and religious writings,"[146] he wrote. He maintained the existence of other one-horned animals, and as long as a single corner of the world remained unscoured, there was hope that a real unicorn could be brought home from a far-off land. When it came to older writers' methods of capturing this timid, wild and powerful animal, which at one time was perceived as a symbol of chastity, purity and strength, Caspar Bartholin was, however sceptical. It was said that the animal would not allow itself to be caught alive but could be attracted by a virgin. It would take its rest and fall asleep in her lap, and in this way the hunters could bind it. When it awoke in its bonds, it would be seized by a lethal wildness. Caspar Bartholin took this tale as a fable, but he was less sceptical about another method of capture, said to be described in a letter to the Pope by the King of Ethiopia. The idea of this method was to lure the unicorn to chase a lion, which, from fear of the pursuing unicorn, fled behind a tree and thus made the unicorn drive its horn into the trunk, so that it stuck, after which the animal could not make its escape. The method is known from the tale *The Brave Little Tailor* by the Brothers Grimm, in which the tailor simply takes over the lion's role.

Narwhal tusks are described in the Icelandic sagas, and it is assumed that they had been brought from the Arctic Ocean to continental Europe in the Viking Age. From the middle of the 15th century, accounts were written suggesting that the unicorn is perhaps not a land animal but a sea creature. The Swedish archbishop and topographical writer Olaus Magnus (1490–1557), in his masterpiece giving an account of Nordic people, *Historia de gentibus septentrionalibus*, Rome 1555, writes that the unicorn is a sea monster with a horn in its forehead, with which it causes ships to sink.[147] Similarly, the Dutch physician Jan van Gorp (1519–1572) writes in 1569: "I have sometimes had a suspicion that the so-called unicorn tusks pertained to a fish. [...] The unicorn tusk found in Antwerp was actually brought back from Iceland."[148] More such reports were written at the beginning of the 1600s, and people began to use the names "sea-unicorn" and "Greenlandic unicorn tusk". In his posthumous 1607 publication *Atlas Minor*, the Flemish cartographer Gerardus Mercator (1515–1594) gives a name to the sea-unicorn: "In the Nordic seas lives the fish narwhal, whose tooth is by some sold as the horn of a unicorn."[149] The Danish admiral Godske Lindenov (d. 1612) brought home narwhal tusks to Denmark from Christian IV's expedition to Greenland in 1606.[150] But a further thirty years had to pass before the final proof came that those horns, which were stored all over Europe in the belief that they originated from the legendary horse-like animal the unicorn, were in fact tusks from a small, harmless whale, which swims about in the northern hemisphere's Arctic seas.

One day in 1636, Ole Worm is sitting conversing with the Royal Chancellor Christian Friis (1581–1639) in the latter's estate, Kragerup. When Worm expresses regret that, in his opinion, the Greenlandic captains do not take enough pains to bring a specimen of the sea-unicorn, so often spoken of, to Copenhagen, the chancellor shows him the skull of a narwhal with a tusk attached. Worm takes the skull home with him, subjects it to detailed investigation, engages an artist to draw it in view of a witness and presents the epoch-making "new" item to the world of learning in a short piece of writing, a corollary to a dissertation with the title,

An os illud quod vulgo pro cornu Monocerotis venditatur verum sit Unicornu? (Is that bone, which is commonly offered as the tusk of a unicorn, really the tusk of a unicorn?), Copenhagen 1638.[151] This piece of writing was published with illustrations of the narwhal's skull and tusk in Thomas Bartholin's book about the unicorn, *De Unicornu Observationes Novæ* (New observations regarding the unicorn), Padua 1645 (Figure 53),[152] in his *Cista Medica Hafniensis*, Copenhagen 1662,[153] without illustrations, and in Worm's own *Museum Wormianum*, which was published in Amsterdam and Leiden in 1655 (Figure 54).[154]

The French envoy Isaac La Peyrère (1596–1676), who was shown the narwhal skull in Worm's treasure chamber, also included the two illustrations in his work on Greenland, *Relation du Groenland*, Paris 1647, and in this way contributed further to spreading news in Europe about the Greenlandic sea-unicorn. Moreover, in *Museum Wormianum* Worm included an illustration of a complete whale with a tusk, which he had had sent by a friend in Iceland, so that the last doubt of this whale's existence was eliminated.

DE VNICORNV. 103

illa quum certam ab experientia & rationibus fidem jure suo meruerint, amplius nec probari debent, nec revocari. Restat ut dérati piscis porrò historię insistamus cujus antea specimē accepimus Sūmi *VVormii* beneficio, cui deinde animalis istius ad balænam propius accedentis figuram, sed rudiorem transmisit Episcopus Islandiæ Borealis D. *Thorlacus Scutonius*, qui eum cum glacie Grœnlandica eo appulsum vidit. Nos verò ejus accuratiorem hic damus ideam non antea visam.

G 4 Qui

Figure 53. Narwhal skull and tooth in Thomas Bartholin's *De Unicornu Observationes Novæ*, Padua 1645, p. 103.

Ole Worm writes on 26 August 1642 to Thomas Bartholin in Padua: "The image of the head I have seen with part of the tooth, I send you reproduced in the form of a smaller picture, a representation of the unicorn's head with its tooth, which I have had in my hand, and I have had accurately depicted in the right proportions by a painter."
The Royal Library in Copenhagen.

Figure 54. The narwhal with its spiralled tooth, which for centuries was sold as a unicorn horn and used as a medicine and antidote. Illustration from Ole Worm's *Museum Wormianum*, Leiden and Amsterdam 1655, p. 282.
Thomas Bartholin writes on 2 May 1646 from Leiden to Ole Worm in Copenhagen: "The very fine whale skull I have previously, in my observations on the unicorn, attributed to you as a source. I wish I had then had the depiction of the whole whale, which you have now shown me the honour to send. In that case, I would indeed have been more deserving of public recognition."
The Royal Library in Copenhagen.

However, the discovery of the narwhal and its tusk still did not convince everyone to abandon the idea of the existence of the "real" horse-like unicorn, or the belief that the narwhal tusk or the Greenlandic unicorn, as it was now called, had potent medical qualities. On 3 October 1636, the same year Worm made his discovery of the narwhal, he conducted experiments with Thomas Fincke, the city physician Joachim Scheel and the apothecary Johan Woldenberg, poisoning two pigeons and two cats with arsenic and mercury sublimate respectively, and using grated Greenlandic unicorn, in other words narwhal tusk, as an antidote.[155] The experiment with pigeons failed, since the pigeon, which was given both arsenic and the antidote, vomited up the antidote and died just as swiftly as the other, which was only given arsenic. The experiment with cats was more successful, since the cat that was given the antidote lived many hours longer than the other. People believed that the experiment confirmed that the Greenlandic unicorn tusk had the power of an antidote. Similar experiments carried out in other places in Europe, even after it was generally known to be a narwhal tooth, confirmed the "unicorn horn's" reputation as an effective antidote. The Copenhagen apothecaries continued, through the whole of the 17th century, to sell Greenlandic unicorn horn as a costly remedy, either in its pure form or as an ingredient in compound remedies, both in liquid form, for example juice and extracts, and in solid form, as powders, salts and the like.[156]

In Thomas Bartholin's letter to Ole Worm on 31 March 1642, he puts questions about horned people, which he has come across in the literature. For example, he has read in book 123 of the French historian Jacques Auguste de Thou's (1553–1617) *Historichen Beschreibung deren Namhafftigsten, Geistlichen und Weltlichen Geschichten* (Historical description of their most famous, spiritual and secular stories), Frankfurt 1622, about the "lineage of the horned Franciscus Troviluvius, who around 1599 wandered about in France." According to the book, Troviluvius had a ram-like horn on his forehead, when as a small lad he was fetched down from the mountains. The horn began to grow in the seven-year-old boy, reaching the length of a finger when he was 17 years old.[157] It grew out from the crown of the head in the median line and bent curved backwards to such a degree that its tip would have grown into the back of the skull, had Troviluvius not regularly filed down the tip. He always wore a hat to conceal it, but at the age of 35 years it was discovered by a villager in his birth town of Firmi in the central part of southern France, whereupon he was banished from the village, because the villagers feared that he was possessed by magic and witchcraft. He then lived from hunting in the mountains. He was "discovered" by a hunting party of French gentlemen, who showed him to Henry IV in Paris and later in 1598 displayed him there in public for citizens and learned visitors. For this "exhibition" a copperplate was made showing the red-bearded, bald mountain man with the remarkable crooked horn on the top of the head and dressed in his usual rustic costume made from fox skin (Figure 55).[158]

In his reply to Bartholin on 5 May, Worm confirms the existence of the horned man in Paris and refers to no fewer than four physicians who touched the horn when they met the man in Paris in 1598. One of these physicians still lives in Copenhagen and has reported the story to Worm several times. Later, however, Worm becomes more sceptical about the phenomenon, when Bartholin on 17 June 1643 sends him a sketch of the horned Frenchman's son, who also has a horn on his forehead: "The French have recently shown in the marketplace this horned man's horned son as a curiosity. I am sending a picture of him." Worm answers on 7 July in the same year: "I express many thanks to you for the picture of the horned Frenchman, but I have a misgiving, that he has been brought into this plight through artifice and a score of tricks. You know that our women, when they make cocks into capons by castrating them, cut off their comb and bring the same birds' spurs in its place. They take root and grow into a form, which looks like a horn. What prevents such a thing happening in this case?" Bartholin gave no answer to this question.

On 20 June 1642 Worm gives Bartholin a supplement with important information about the Greenlandic unicorn: "With regard to the unicorn there was perhaps something that could serve your project, both illustrations and pieces of factual information. [...] For I have held it in my hand, and treated it just as

Figure 55. The red-bearded and balding French mountain man Franciscus Troviluvius (François Trouvillou) from the village of Firmi, with his backwards-bent, ram-like horn on his forehead and wearing his usual fox-skin suit. In his hand he holds the hat he always wore to hide the horn.

When the horn was discovered, he was banished from the village, because the villagers feared that he was obsessed with sorcery and witchcraft. He was later "exhibited" in Paris and presented to Henry IV. Thomas Bartholin mentions Troviluvius in a letter from Padua to Ole Worm in Copenhagen on 31 March 1642. With modern knowledge, Troviluvius' horn undoubtedly represented a case of *cornu cutaneum* (cutaneous horn), a type of tumour that originates from the horny layer of the skin. Woodcut in Ulisse Aldrovandi (1522–1605): *Monstrorum Historia*, Bologna 1642, p. 127. The Royal Library in Copenhagen.

I wished, and ordered this animal's entire skull to be drawn in its exact dimensions, and what people call the horn was still firmly attached to it, even though it is not a horn at all, but a tusk, as I have learnt through personal inspection, and we have proved with strong evidence. Our apothecary has prepared a small piece hereof, which was conveyed to him by His Royal Majesty, with a *calcinatio philosophica*, as they call it (it is prepared with hot steam or fluid), and this has also been put at my disposal. It is dissolved into very thin sheets, which does not happen with horn when it is pulverized in this way. However, I have had the same experience with a walrus tusk, which in the same way protrudes as an outgrowth from the jaw, and the same thing can be seen with ivory or elephant tusk. A certainly quite rough sketch of this creature (for it is a whale) was sent to me by the Bishop of North Island, Thorlak Skulason. He has seen it drift ashore there, with the ice from Greenland. This cranium and this tusk I have examined in its smallest details and described them at length in lectures about the objects, which are to be found in my museum, and at the same time shown illustrations of them to the audience. I also have a fragment of two pounds' weight and similarly small pieces of skin scraped from the skull. If I wanted to describe everything I have perceived and noted down, I would fill all the paper in a medium-sized book. If God grants me life and health, it will be conveyed to you in due course."

Bartholin answers on 7 August in the same year: "I had recently added the appendix concerning the fake unicorn with my own hand, when I saw a similar head in Leiden, and I am now delighted that more proofs of it have been obtained, and that you are in possession of an illustration of it, which I hope to receive easily from you who supports my efforts. I expect to receive a picture of the unicorn horn in Frederiksborg Castle from my brother-in-law, Dr. Jacob Madsen; and from friends in England, France and the Netherlands I expect those, we have either seen ourselves or experienced from reports that they can be found there. I have added, from careful reading, some kinds of unicorn that have been ignored by my father." Afterwards Bartholin asks Worm about the rhinoceros-beetles (chafers), Indian rhinoceros-fish and Ethiopian rhinoceros-birds (Figure 56).

At this time, Bartholin was gathering materials for his book on the unicorn, other fabulous animals, creatures and human beings with horns. The book is dedicated to his grandfather, Thomas Fincke. It covers 304 pages, of which the last ten are an appendix about the golden horn in the form of a small extract from Bartholin's and Worm's correspondence with other scholars on this subject during the period 1642–1644. The book's full title is therefore *De Unicornu Observationes Novæ. Accesserunt de Aureo Cornu Cl. V. Olai Wormii Eruditorum Iudicia* (New observations regarding the unicorn. Added the judgements of scholars on the golden horn of the famous man Ole Worm).[159] Like almost all of Bartholin's books, it has a complete table of contents, an index of names and a comprehensive subject index. 12 illustrations show ancient coins with men having horns on their forehead together with animals and fabulous creatures with horns (Figure 57).

The book builds on Bartholin's literary studies and oral and written reports from others. He did not carry out his own experiments for this book. Its new observations (*observationes novæ*) therefore consist chiefly of the discussion of Ole Worm's new evidence, that what people had understood to be unicorn horns for centuries were in fact narwhal tusks.[160]

With customary thoroughness Bartholin discusses practically all possible angles of the subject. There are chapters about men with one horn, men and animals with two horns, various single-horned animals and fabulous animals, including beetles, reptiles, birds, fish, antelopes, deer, oxen, donkeys, horses, rhinoceroses, the Greenlandic unicorn (the narwhal), and the "traditional" terrestrial horse-like unicorn.

The background for horns and the role of horns in connection with illness are made clear. A series of chapters cover the use of horn, for example as medicine, drinking cups, musical instruments and decorative objects. One chapter deals with the measuring of horns, another with authentic proof of unicorn horns, a third with ivory, and a fourth with fossil unicorns. The theological aspects among others are analysed in a chapter about the unicorn on Christ's cross.

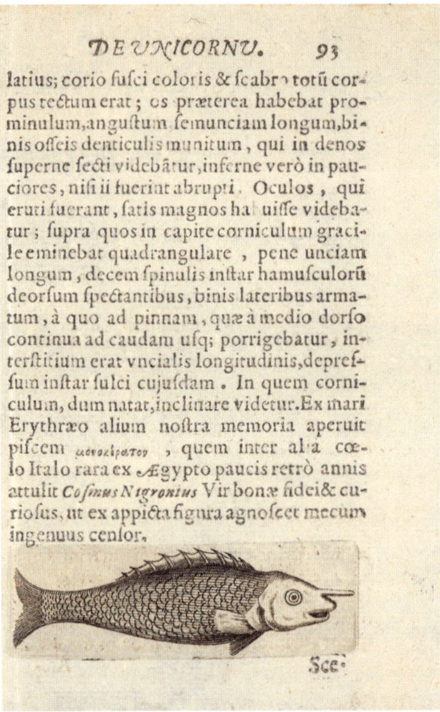

DE UNICORNU. 93

latius; corio fuſci coloris & ſcabro totũ corpus tectum erat; os præterea habebat prominulum, anguſtum, ſemunciam longum, binis oſſeis denticulis munitum, qui in denos ſuperne ſecti videbātur, inferne verò in pauciores, niſi ii fuerint abrupti. Oculos, qui eruti fuerant, ſatis magnos habuiſſe videbatur; ſupra quos in capite corniculum gracile eminebat quadrangulare, pene unciam longum, decem ſpinulis inſtar hamuſculorũ deorſum ſpectantibus, binis lateribus armatum, à quo ad pinnam, quæ à medio dorſo continua ad caudam uſq; porrigebatur, interſtitium erat vncialis longitudinis, depreſſum inſtar ſulci cujuſdam. In quem corniculum, dum natat, inclinare videtur. Ex mari Erythræo alium noſtra memoria aperuit piſcem μονοκέρατον, quem inter alia cœlo Italo rara ex Ægypto paucis retrò annis attulit *Coſmus Nigronius* Vir bonæ fidei & curioſus, ut ex appictà figura agnoſcet mecum ingenuus cenſor.

Sce-

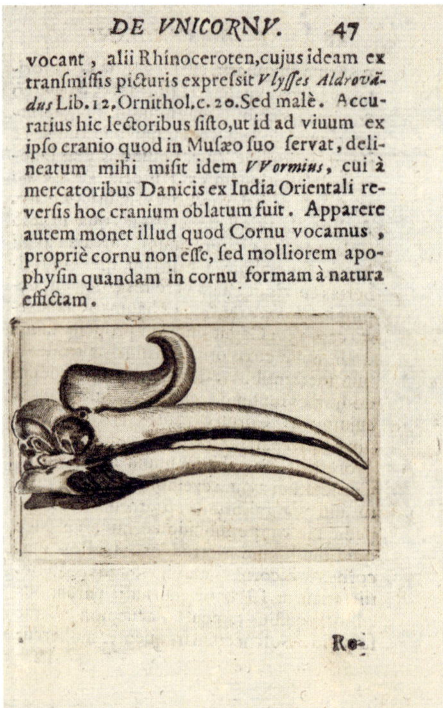

DE UNICORNU. 47

vocant, alii Rhinoceroten, cujus ideam ex tranſmiſſis picturis expreſſit *Vlyſſes Aldrovãdus* Lib. 12, Ornithol. c. 20. Sed malè. Accuratius hic lectoribus ſiſto, ut id ad viuum ex ipſo cranio quod in Muſæo ſuo ſervat, delineatum mihi miſit idem *VVormius*, cui à mercatoribus Danicis ex India Orientali reverſis hoc cranium oblatum fuit. Apparere autem monet illud quod Cornu vocamus, propriè cornu non eſſe, ſed molliorem apophyſin quandam in cornu formam à natura effictam.

Re-

Figure 56. a. The beak of the Indian hornbill Topau. From Thomas Bartholin's *De Unicornu Observationes Novæ*, Padua 1645, p. 47.

Thomas Bartholin writes from Leiden to Ole Worm on 16 May 1646: "But de Laet [Johannes de Laet. (1581–1649)] has promised me numerous observations on one-horned creatures, especially on the one-horned bird brought from India, but which I have not yet seen." Worm replies to Bartholin six days later: "The beak of the bird, which I have called Topau, I have received from a merchant, brought from India under this very name. However, it is not unknown to me that it is called the hornbill by Aldrovandus [Ulisse Aldrovandi] in the *Ornithology*." The Royal Library in Copenhagen.

b. Rhinoceros fish. Ibid., p. 93.
The Royal Library in Copenhagen.

28 T. BARTHOLINI

Sed cur Attilæ cornua? Factum id ejus forsan veneratione. Venerationi erat potentia, quam, ut res secundas, per cornua veteribus adumbratam ex *Plautino* Pseudolo notum Act. 4. sc. 3. vbi querela de Simia:

Nimis ego illum hominē metuo, & formido male,
Ne malus item erga me sit, ut erga illum fuit,
Ne in re secunda nunc mihi obvertat cornua.

Censuit id *Horatius* l. 3. Od. 21. de vini effectibus:

Tu spem reducis mentibus anxiis,
Viresq; & addis cornua pauperi.

Vide Oneirocriticum apud *Achmetem* c. 238.

exstat

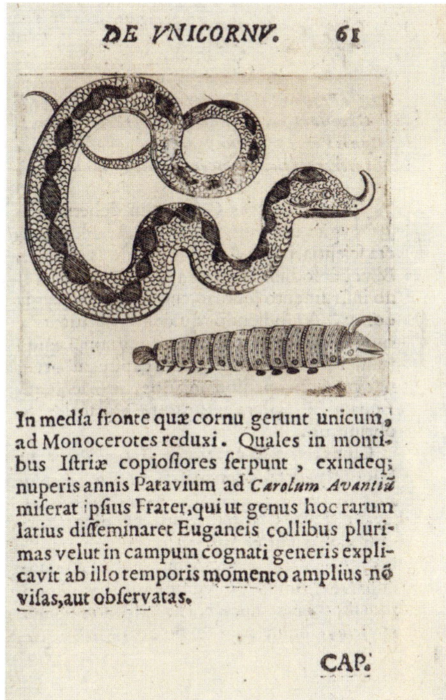

DE VNICORNV. 61

In media fronte quæ cornu gerunt unicum, ad Monocerotes reduxi. Quales in montibus Istriæ copiosiores serpunt, exindeq; nuperis annis Patavium ad *Carolum Avantiū* miserat ipsius Frater, qui ut genus hoc rarum latius disseminaret Euganeis collibus plurimas velut in campum cognati generis explicavit ab illo temporis momento amplius nō visas, aut observatas.

CAP.

Figure 57. a. Antique coins with men with horns on their foreheads. From Thomas Bartholin's *De Unicornu Observationes Novæ*, Padua 1645, p. 28.
The Royal Library in Copenhagen.

b. Fabulous animals with horns. Ibid., p. 61.
The Royal Library in Copenhagen.

In 1678 there came an enlarged second edition of the book, published in Amsterdam by Thomas Bartholin's son, Caspar Bartholin the Younger (Figure 58). It is dedicated to Frederik III and has 381 pages including an introduction, dedication, table of contents and indexes. The chapters are the same 37, covering the same topics as the first edition, but the number of illustrations has doubled to 24. The new illustrations include among others a horned woman (Figure 59), Frederik III's horse with horns protruding from its ears (Figure 60), the rhinoceros with its skull and horn (Figure 61), musical instruments made of horn (Figure 62) and a sketch of the golden horn from Ole Worm's treatise about it (Figure 44).

Bartholin had already begun to collect materials for this edition in 1646, when on 20 April he wrote to Ole Worm from Leiden: "My unicorn has grown not a little since it was published as the result of numerous observations carried out for the second edition, especially concerning a certain man of Tours in

TH. BARTHOLINI
DE
UNICORNU
OBSERVATIONES.
R. de Hooge
fecit

Figure 58. The title sheet of the second edition of Thomas Bartholin's *De Unicornu Observationes Novæ*, published by his son Caspar in Amsterdam in 1678. Engraving by the Dutch painter, sculptor and engraver Romeyn de Hooghe (1645–1708).

Even though Thomas Bartholin lost faith in the existence of the classic horse-like mythical creature with Ole Worm's discovery of the "Greenlandic sea-unicorn" (the narwhal), he did not lose his fascination with the myth about it. In this allegory, both kinds of unicorns are depicted. The top two thirds of the illustration show the classic land unicorn tamed by a maiden in accordance with the myth and surrounded by other unicorns as well as horned people, animals, birds and a snake. In the sea below are two mermen, each carrying a bunch of cut-off "sea unicorn" horns.
The Royal Library in Copenhagen.

Figure 59. "A woman in Amsterdam, who still bears a horn-like projection on her forehead," writes Thomas Bartholin on 20 April 1646 from Leiden to Ole Worm. The horn undoubtedly represented another case of *cornu cutaneum*. Illustration from Thomas Bartholin's *De Unicornu Observationes Novæ*, Amsterdam 1678, p. 5.

Bartholin first published the account and the illustration of this woman in the first volume of his work on rare medical accounts, *Historiarum Anatomicarum Rariorum*, Copenhagen, The Hague and Amsterdam 1654, Cent. 1, account no. 78, pp. 128–29.
The Royal Library in Copenhagen.

Figure 60. Frederik III's horned horse. From Thomas Bartholin's *De Unicornu Observationes Novae*, Amsterdam 1678, p. 154.

Bartholin first published the account of the horse with a depiction of its horns in Volume I of *Historiarum Anatomicarum Rariorum*, Copenhagen, The Hague and Amsterdam 1654, Cent. 2, account no. 10, pp. 185–86. Here he tells us among other things: "In the stables of The Royal Palace in Copenhagen a splendid horse may be seen, on which a horn protrudes from the outer edge of both ears. They look like cockspurs, are curved and glisten on the outside like a pearl. Inside they are hollow and can rightly be counted among the most beautiful horns. Two of these horns have been presented to me by his majesty King Frederik III. Their engravings are shown here."

The horns undoubtedly represented yet another example of *cornu cutaneum*. The Royal Library in Copenhagen.

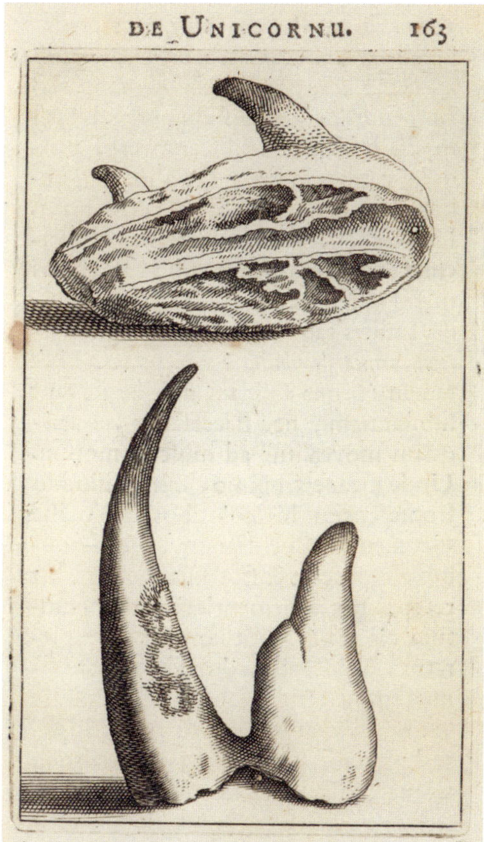

Figure 61. a. The rhinoceros and b. its horns and skull. From Thomas Bartholin's *De Unicornu Observationes Novæ*, Amsterdam 1678, p. 172 and p. 163.

Apart from a few insignificant details, the rhinoceros is identical to the German artist Albrecht Dürer's (1471–1528) rhinoceros from 1515. There is therefore no doubt that Bartholin and his artist have borrowed from Dürer.

As is still the case in many parts of the world, in the 1600s the horn of the rhinoceros was attributed various medicinal properties. In both editions of the book Thomas Bartholin speaks favourably of the rhinoceros' horn. In the third volume of Bartholin's journal *Acta Medica & Philosophica Hafniensia*, Copenhagen 1677, Obs. 89, p. 166, the Swedish physician Herman Nicolai Grim (1641–1711), who served as a medical officer in the Dutch East India Company, reports from Nagapatnam in India to his colleague Ole Borch in Copenhagen that he has cured the Ceylonese Malabar and Coromandel Coast Governor, Richlof van Goes, from a long-standing inflammation of the knee and big toe for which other medicines had failed to provide any relief, with a spirit liquor containing rhinoceros horn.
The Royal Library in Copenhagen.

France, on whom in previous years had grown a goat's horn out from his right cheek, in the same way as a woman in Amsterdam, who still carries a horn-like projection on her forehead, not to mention different accounts, that have gathered among writers."

In contrast to what one would believe with modern eyes concerning a book written about a fabulous animal by a scientist, it was greeted with great interest and recognition among contemporary scholars. Even one of Bartholin's severest critics, Caspar Hofman in Altdorf (see later), was so impressed by the book that he thought the young Bartholin could not have been its real writer. On 31 August 1645 he writes to Johan Rhode: "Are you aware of this work yourself? For you must not believe we are so stupid that we could not notice that it is your spirit that runs through the book. My friend Rupert had scarcely looked at it before he shook his head and said: 'Do you believe that this is the work of a young man of 28 years?' 'Whose do you think it is, then?' I reply. 'Without any doubt the famous Rhode.' A few hours later I agreed with him."[161]

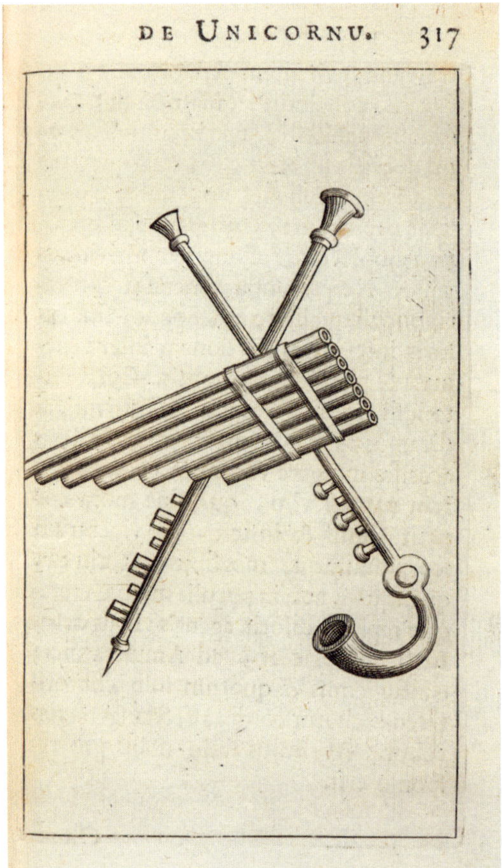

Figure 62. Musical instruments made from horn. From Thomas Bartholin's *De Unicornu Observationes Novæ*, Amsterdam 1678, p. 317.
The Royal Library in Copenhagen.

The Danish zoologist C. C. A. Gosch (1832–1913) gives the following complete assessment of the book: "This book in its various Bartholinian shapes is one of the last beautiful and for its time justly representatives of the simple scholarly natural history, which, if possible, begins its discussions with Aristotle, romps in classical references as its proper element, deals with other people's bare assurances as useful scientific material, and least of all thinks of self-investigation of the natural objects."[162]

If Bartholin had ever had a belief in the myth of the unicorn as a horse-like creature with a horn in its forehead, he lost it with the discovery of the narwhal and its tooth, but he persisted all his life in the belief of the medicinal powers of the "Greenlandic unicorn's horn," and used it in the treatment of his own family and his patients, of which examples will be given later.[163]

Honours in Padua

In Padua on 1 August 1642, Bartholin was awarded the honour of being named the German nations' *consilarius*, a kind of trustee (Figure 63).[164] The more than 1000 foreign students in Padua were at this time divided into 31 nations. Each nation represented an organization with different offers to the students, for example a library, a welfare office and a historical society.[165] Bartholin, however, was apparently more burdened by the duties than he was flattered by the honour. In his letter to Worm of 7 August 1642, he speaks of "the oppressive council duty that has been laid on me, even though I sought to escape it." Because of contemporary disturbances in Italy, on 28 October Worm takes the opportunity to ask Bartholin, through letters to the correct persons of consequence (Chancellor Christen Thomesen Sehested and Royal Physician Jacob Fabricius (1576–1652)) and publication of his works, to prepare for his homecoming to a suitable post in Copenhagen:

> "For the person who wishes to rise high up, it is certainly necessary here to obtain patrons and ingratiate himself with their favour, and this occurs most conveniently through a letter, which cannot be affected by any confusion, and because I know that your elegant style pleases them, see what you can do. You must not be persuaded by anyone to swerve as much as a finger's breadth from your enterprise. Pursue your physician's skill with zeal, as you are accustomed, and now and then when you are tired, make your escape to philology's splendid country house."

Meanwhile, Bartholin was also assigned duties as deputy rector and councillor (*syndicus*), and the combined workload of the three new responsibilities largely robbed him of time for his medical studies. However, he was not slow in solving

Figure 63. a. The memorial plaque for Thomas Bartholin in the loggia in Palazzo Bo's courtyard, Il Cortile Antico, in Padua. At the top it says ALEMANNA (German) which marks Bartholin's position of trust as the German nation's *consiliarius*. At the bottom is the Latin text "THOMAS BARTHOLINUS CASP. F. DANUS" (The Dane Thomas Bartholin, Caspar's son).
Photo: Jesper Brandt Andersen.

b. In 1942, Thomas Bartholin was honoured as one of the 40 most significant students in the history of the University of Padua with a portrait by the Italian painter Gian Giacomo dal Forno (1909–1989). It hangs in La Sala Dei Quaranta (The Hall of the Forty) alongside portraits of Ole Worm and the lawyer and historian Peder Hansen Resen (1625–1688).
Photo: Courtesy of the University of Padua - ph. Massimo Pistore.

the problem, and his way of doing it oozes with self-confidence and demonstrates that, despite his youth, he must have enjoyed considerable status among the important men of the medical faculty in Padua. By the end of the year, he had already had enough and simply renounced his new obligations. On 8 December he writes to Worm:

"The duty as prorector and councillor, with which I have reluctantly been burdened, has enforced silence on me and brought my letters to a standstill. But now, out of humility and to avoid displeasure and greater gathering from applicants, I have renounced all my obligations in the presence of our rector and professors. I have only retained the title and privileges of prorector. Therefore, now a pleasanter freedom once again opens for studies, which my concerns have hindered."

Bartholin was now back on track with his medical pursuits. A publisher in the city was showing interest in yet another new edition of his father's anatomy book, and Bartholin had assured himself of assistance from his teacher, Johannes Vesling. The pages of illustrations had to be augmented with new illustrations from his own dissections at the cost of the earlier ones from older writers' works: "I shall let the old illustrations disappear and seek new ones in their stead, drawn while the dissections are carried out, and afterwards engraved." Bartholin also asks Worm for any proposals he has for changes.

The excretory duct of the pancreas – an epoch-making discovery with fatal consequences?

After this, Bartholin mentions a significant new anatomical discovery, to which he himself bore witness earlier that year:

"I am sending a picture of the new channel that has been discovered in the pancreas. I have seen it myself in the form in which it is here reproduced [Figure 64]. People are uncertain about its function. We believed that it conveyed the gastric juice to the spleen, but since we did not see any entrance, we guessed something else on the basis of the liquid within, namely that the channel was a passage for the bile-containing excrements after treatment in the pancreas. In fact, it stains a slate pencil with darkish bile pigment."

It was a question about the discovery of the pancreas's excretory duct, *ductus pancreaticus* or *ductus Wirsungianus*, which conveys the pancreatic juice containing the digestive enzymes into the duodenum. The story behind the discovery and its consequence is dramatic in more than one sense.

On 1 March 1642, the thirty-year-old murderer Zuane Viaro delle Badia was hanged in the Piazza del Vin in Padua. The following day, the body was delivered to Professor Johannes Vesling or, more correctly, to his prosector, the German anatomist Johann Georg Wirsung (1589–1643), for dissection.

The prosector's role was to carry out the preliminary dissections to prepare the bodies for the professor's lectures in the anatomy theatre. Normally such a dissection in Padua would take place in Fabrici d'Acquapendente's anatomy theatre, but the season for public dissections had just finished at the end of February. Instead, the dissection took place in private, in the hospital of San Francesco, which was part of a still existing monastery. Present at the dissection besides Wirsung were only two foreign medical students, Thomas Bartholin and the German Moritz Hoffmann (1622–1698). According to the Swiss historian of

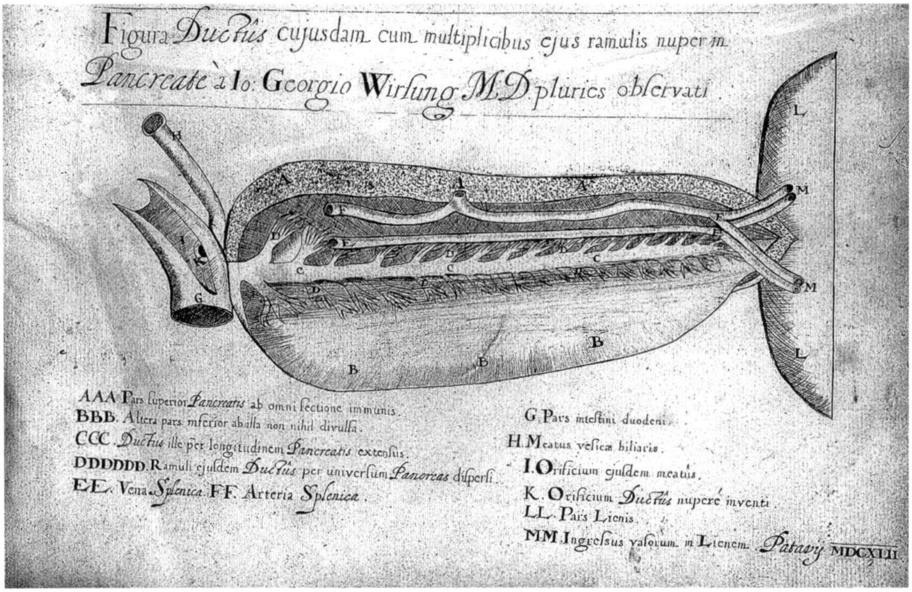

Figure 64. Johann Georg Wirsung's (1589–1643) drawing of the main excretory duct of the pancreas, which he discovered on 2 March 1642 together with his students Thomas Bartholin and Moritz Hoffmann (1622–1698) during the dissection of the 30-year-old hanged murderer

Zuane Viaro della Badia at San Francesco Hospital in Padua. Wirsung had seven engravings of the drawing made which he sent to seven of Europe's leading anatomists to get their opinion on the function of the duct. The first of the engravings he asked Thomas Bartholin to send to Ole Worm. The discovery triggered the exploration of glands. There has been speculation as to whether it was a conflict of priority with Hoffmann regarding the discovery, which led to the murder of Wirsung the following year. Thomas Bartholin did not take an active part in this dispute but honoured Wirsung by naming the duct *ductus Wirsungianus* after him. The duct is marked with a C in the drawing.
Wellcome Collection, London.

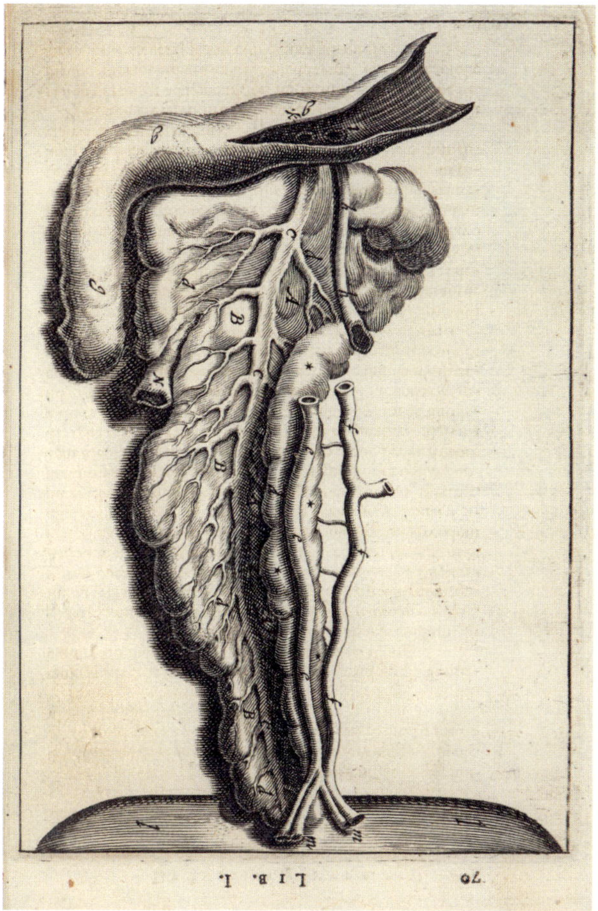

Figure 65. Thomas Bartholin's illustration of the pancreas with the newly discovered main duct C, *ductus pancreaticus* or *ductus Wirsungianus*. To the left is the cut duodenum, into which the pancreas empties its juice containing digestive enzymes. *h* is the bile duct, *l* is the spleen, *e*, *f*, *m* and N are veins and arteries. From Thomas Bartholin's *Anatomia Secundum Locupletata*, Leiden 1645, p. 70. The Royal Library in Copenhagen.

medicine, Alfred Max Schirmer, Thomas Bartholin was the first to carry out and publish thorough studies of the pancreas.[166] According to Schirmer, the latter occurred in 1651 in Bartholin's third revised edition, *Anatomia Reformata*, of his father's textbook of anatomy, but in the previous edition of 1645 the section on the pancreas was already extended considerably and supplied with an illustration of the newly discovered excretory duct (Figure 65). Here Bartholin writes about the finding:

"Besides all this, it also has another membranous passage of a particular nature, which spreads alongside the whole pancreas, sometimes in a straight line, sometimes twisted, and this has so far not been described by any anatomist, but was first observed in Padua when I was there in the year 1642, by Johann Georg Wirsung, an extremely painstaking anatomist; it is remarkable for its hollowness and the strength of its walls."[167]

Wirsung did not publish the discovery himself in text form but drew with his own hand a very detailed illustration of the pancreas with the newly discovered excretory duct, on a copper plate. From this he produced seven engraved copies, which he sent to seven of Europe's foremost anatomists to hear their opinion about the function of the newly discovered excretory duct.[168]

The first of the copies to be dispatched was the one that Bartholin sent with the letter to Ole Worm.[169] Worm's answer to Bartholin comes on 31 December 1642:

"The sketch of the new channel in the pancreas is very welcome to me. I am amazed that no trace has been found of the lacteal veins. What if this channel absorbs the gastric juice from the duodenum to convey it through the lacteal veins, cleanses it of sediment, so that that it can more easily pass through the narrower channels to the liver, and eventually sends the excrements it has received to the intestines in order to increase the expelling strength and colour the excretion with its darkish colour?"

As mentioned earlier, Gaspare Aselli's discovery of the lacteal veins in 1622 had led to the understanding that the food's nutrients pass from the intestines to the lacteal veins in the mesentery, but in the 1640s people still believed that the nutritious chyle fluid in the lacteal veins was conveyed directly over to the liver, which people still thought was the blood-manufacturing organ. Aselli himself thought that the pancreas was a sponge, which absorbed chyle and transported it further to the liver.[170] From this point of view, Worm's observations are immediately intelligible. Many others continued to subscribe to Galen's idea, that the pancreas was simply a supporting cushion on which the stomach could rest.[171] The whole thing was built on speculation. People did not really know the function of the pancreas, and this was the background for Wirsung's asking his seven colleagues for their opinion. The discovery of the excretory duct cleared the way in the first instance for an understanding that the pancreas is a gland, which submits a secretion (pancreatic juice) to the duodenum,[172] and this knowledge appears already in Thomas Bartholin's answer to Ole Worm on 5 February 1643. Here Bartholin, with reference to a simple test with a probe, proves to Worm that the content in the excretory duct cannot, as suggested by Worm, come from the intestine. Because of a valve at the duct's mouth in the duodenum, the content in the duct can only pass the opposite way and therefore must originate from the pancreas:

"Nature shows the channel in the pancreas sometimes single, sometimes double, most frequently single, and the same is then supplied with a small valve, which turns towards the intestine and is open towards it, so that you can scarcely without force stick a pencil into the gland, which serves as no inessential testimony that from the pancreas itself something is conveyed into the intestine, not the reverse."

A letter of 7 July 1643 from Wirsung in Padua to Jean Riolan in Paris, who had not answered Wirsung's first enquiry, gives a few more details:

> "The channel mentioned, whose illustration I had enclosed, has the following qualities: the opening or beginning, if I may call the broadest part the beginning, is to be found in the duodenum near the bile duct. A probe can only be inserted with difficulty from the duodenum into the pancreas, but it passes easily into the intestine from the channel, and the channel runs through the middle of the gland for the whole of its length towards the spleen. It has various lateral branches, which end in smaller branches near the gland's surface, above and beneath the vessels of the spleen, and it sends twisted branches through the pancreas, but does not continue into the spleen. I found that the channel in some people, and also in some animals, is doubled, with a short one in the usual place and a longer one a little lower down. This was found not only in adults, new-born babies and embryos, but also in apes, dogs, cats, pigs, hens, mice, frogs and in all the animals I thoroughly investigated. Should I call it an artery or a vein? I never found blood in it, but a cloudy liquid, which stains a silver probe like bile. What this liquid can be, what function and action it has, is unclear, and I do not know the answer."[173]

Wirsung did not manage to participate in the investigation of the pancreas much longer. Late in the evening on 22 August 1643, while he stood outside his house and spoke with his neighbours, he was shot with a carbine by the Flemish medical student Jacques Cambier. The motive for the murder is unknown, but it became the subject of much speculation and many conspiracy theories. Some sources maintain that the motive was revenge due to a personal feud between Wirsung and his murderer. Others believe that Cambier was engaged by Moritz Hoffmann or Johannes Vesling to carry out the murder. Hoffmann's motive would theoretically be a quarrel about priorities over the discovery of the excretory duct, while Vesling's would be other known cases of unruliness between him and Wirsung.[174] But even with the murder of Wirsung, the drama was not yet finished. Five years later, Hoffmann tried to get the priority of the discovery of the pancreatic duct, for he maintained that he had already discovered the duct in September 1641 in a dissection of a turkey and afterwards shown it to Wirsung, who then began to search for it. Since Hoffmann could not prove his statement, posterity has credited the discovery of the duct to Wirsung, and it is still named the *ductus Wirsungianus*. As the third person present at Wirsung's dissection, Thomas Bartholin did not take part in this dispute, but as stated, he gave Wirsung the credit in *Anatomia Reformata* in 1651. In 1673, in the fourth and last edition of his anatomy book, *Anatome Renovata*, he adds, however, that "although in the previous autumn it had been observed in a turkey by Moritz Hoffmann, who at that time lived in the same house as Wirsung."[175]

Further discussion about Harvey's theory concerning the circulation of the blood

Another topic to which Bartholin returned in his exchange of letters with Worm in the summer of 1643 was the question of Harvey's theory concerning the circulation of the blood. The reason was that Fortunio Liceti had published a work on the subject, and on 17 June Bartholin wanted to inform Worm about it, for he had found a reason to make objections against it "on various grounds. For he [Liceti] imagines a double circulation, one in the veins, the other in the arteries, separately in their own specific vessel, and he believes that the blood is conveyed from the right to the left chamber of the heart, not through the lungs, nor through an opening, and on the whole not through a well-defined passage, but through the *vena coronaria*."[176] While Worm in his answer of 7 July continues to express doubt about the "in my opinion still unsettled question concerning the circulation of the blood," Bartholin speaks his mind in his response of 13 August for the first time on this great question and shows his complete support for Harvey. On the ideas of Liceti he concludes: "How the renowned Liceti can explain his circulation of the blood I do not see, since he refrains from experiments and remains content with simple reflections. But this whole matter depends on dissection of living creatures."

War in Italy and pressure from Worm concerning Bartholin's homecoming

1643 became a disturbed year of study for Thomas Bartholin because of the unrest in northern Italy caused by a war between the allied Italian duchies and the Papal States. The war had an impact on teaching in Padua and reduced the possibilities for travel to other cities. Instruction was apparently confined to "daily visits to the sick together with venerable medical greybeard, Mr. Johannes Sala," and a planned journey to Rome in the spring had to be postponed. Many years of study at night in poor lighting began to give him eye problems. On 17 June he writes to Worm:

"As far as public circumstances are concerned, they remain unknown, for Italy is in confusion. The Venetian army has occupied Ferrara's neighbouring territory on this side of the Po, which is called Poteria, and the Duke of Modena is threatening the city itself.[177] The Pope is daily expected to send a large troop of soldiers, who according to their abilities will hold back the many enemies. In the coming spring I otherwise intended to go to Rome, if it were possible despite these disturbances, for studies here suffer harm through the students' indiscipline, and my eye weakness, which has been brought to its worst, could hardly endure more serious strain."

In his answer of 7 July, Worm exerts serious pressure to get Bartholin home for a position as a substitute for his aging grandfather:

"I have once again had dealings with your grandfather, who wishes dearly for you to come home in order to follow in his footsteps and lecture on medicine, since he is now so terribly weighed down by old age that he cannot discharge his duties, and he wants you, ahead of his other descendants, to be his deputy. If you draw out your delay for much longer, I'm afraid that someone else will snatch this tidbit out of your mouth."

Bartholin's answer, which follows on 20 August, illustrates a couple of his distinctive personal characteristics. He did not allow himself to bow before authority, whether official, familiar or professional, and he did not try to avoid necessary conflicts. In such matters he was self-reliant, obstinate and hugely wilful. He writes to Worm:

"I would not be answering the enclosed note if I were not aware that I do not fulfil your excessive expectations. A teacher's chair demands more mature years and a fuller scholarship, which I lack. It ought to increase with age and greater diligence. However, I entrust my subject to you, who would rather have rest than public unrest, to which my frail body would certainly not be adequate, whereas my mind could perhaps stand the task, since it is closely united with you and the fatherland."

As must be expected, this reply was not well received by Worm, who cuts up rough with a scowl, mingling stick and carrot in his reply of 16 September:

"That a university existence with us does not appeal to you, because it would be filled with outward disturbances, I learn unwillingly, since nothing more peaceful may be found in the whole of our fatherland. We observe nothing to cause disquiet, thank God, and to lecture three to four times a week from the master's chair what can be understood by beginners is definitely only enjoyment and child's play. You will only have a small commitment to disputations, as long as you work in the capacity of your grandfather's substitute, and for your own studies you will surely get no less time here than you have there. Finally, your purpose must be pondered over: Your inheritance is being used up, the possibility of failure threatens, your health is breaking down, and I don't know where in our country you would be able, with greater honour and with less trouble and inconvenience, to manage your life than in our community. I am afraid that you may be among people who give you less good advice. Think through your situation once again for yourself and look to see whether, when you have accomplished your expedition to Rome, you will be able to appear in your homeland next spring."

Happily for Bartholin, this letter, according to him, was "intercepted because of the Ferrarese soldiers' despotism (yet nevertheless preserved for posterity) and when, through his cousin, he gained knowledge of its existence and told Worm of it, Worm partly beat a retreat with an understanding, that the letter did not contain anything in particular, "apart from your grandfather's and your friend's longings, in which they were in need of you and your homecoming; and this is still so, for the propitious opportunity through which we desire that good care will be taken for you, is likely to fail abruptly."

In the same letter, written on the last day of the year 1643, Worm informed Bartholin that the circumstances in Denmark were perhaps not so calm after all:

"The situation is now more troubled than usual, because a few days ago the Swedish general, Torstenson,[178] invaded Holstein unexpectedly with his army, and if he had not encountered obstacles, he would perhaps have occupied the whole of Jutland. Our illustrious king is gathering an army to oppose this attempt, so the bloody fury of the war goddess seems to threaten us in a less lucky way. May God mercifully come to our help and grant peace in the fatherland."

That it was this war against Sweden, the so-called Torstenson War of 1643–1645, that caused Worm to reduce his eagerness a little towards getting Bartholin back to Copenhagen is apparent from Worm's next letter, of 17 March 1644:

"Though our disorganized condition, during which the Swedes, without anyone having expected anything of the sort, have treacherously made inroads both in Jutland and a great deal in Scania,[179] seems to prompt me not to try to write anything to you about hastening your return to us, your grandfather's will and excessive old age demand first priority, for every day threatens us with his demise. But if anything should happen to him, your friends' plans will be overthrown, and no similar opportunity will be presented to see to your needs. So please think over everything more carefully and take the changeability of the human condition more closely into consideration. We are thinking of nothing but your welfare, to which under the present circumstances we are unable to give more consideration."

Further travel to Naples and Rome

Shortly after 4 October 1643, when he wrote his name in Henrik Fuiren's guestbook in Padua,[180] Bartholin must have succeeded in escaping to Naples and Rome. From Rome he writes to Worm on 20 November:

"Returned home from Naples, I am staying in Rome, where everything is filled with the disturbances of war without any sure expectation of peace, which Cardinal Bichi[181] on his queen's behalf[182] presses on for, using all his energies, for he has been sent from France to settle the disturbances in Italy, which threaten ruin. What he can achieve in the face of the princes' obstinacy will be shown by the outcome. If it were omens one should rely on, they would predict ill; for yesterday the Vatican, the Pope's residence, was struck by lightning and two of the Swiss guards were injured. A quite unusual cold weather torments us, and fatigue from travel has not yet abated, so that we are still forced to remain unaware of what we must promise ourselves in the way of studies. In Naples everything indeed seemed very favourable with the leading physicians, who were very obliging towards me, namely Marius Schipanus, my father's friend, and Marcus Aurelius Severinus, the recreator of surgery, who with great friendliness offered to share his home and everything with me, if I returned. However, the great familiarity of the knight Cassiano dal Pozzo,[183] your good friend, keeps us here. The Danish memorials[184] sent to him by Henrik Fuiren have deservedly pleased him greatly."

During his stay in Naples with the surgeon Marco Aurelio Severino (1580–1656) (Figure 66), Bartholin managed to ascend Vesuvius, and through this his fascination with volcanoes was stimulated: "It was in the midst of such a natural bounty, continuing on the other side of the city, that Mount Vesuvius hastened the death of the noble Pliny,[185] and as recently as December 1631 spewed up flames, stones, sulphur, etc., it burned, dried up neighbouring streams, sprinkled ashes as far as Venice, and immersed the nearby towns [Figure 67]. When I passed through the Torre del Greco [...] between Salerno and Naples, it was almost wholly covered with ashes; nor did Vesuvius then cease to smoke, and I saw the sulphur boiling at the bottom of the crater when I climbed the mountain in November 1643."[186]

Bartholin stayed a couple of months in Rome. At the beginning of 1644 he departed on a journey to the southern Italian mainland, Sicily and Malta, accompanied by his cousin Didrik Fuiren (1621–1656) and the young Dutch physician Johannes van Horne (1621–1670). The journey came to stretch across most of that year. To Worm's annoyance he received only two letters from Bartholin

Figure 66. Portrait engraving of Marco Aurelio Severino (1580–1656) by Clemens Ammon 1650. 17.5 × 14.5 cm.

The Latin text in the frame, "MARCUS AUREL: SEVERINUS THARSIENSIS PHIL. MED. IN REGIO GYMNASIO NEAP ANAT: ET CHIR: PROFES: Natus A° 1580," reads: "Marcus Aurelius Severinus of Tarsia, philosopher and physician. Professor of anatomy and surgery at the University of Naples. Born in 1580." The text below the frame reads: "While Nature from its gifts elsewhere provided the anatomy, it will apparently give them all to this man." With his vast clinical experience, Severino was in many ways a significant source of inspiration for Thomas Bartholin. At least three of Bartholin's books were direct products of this inspiration. These are *Anatomica Aneurysmatis dissecti Historia*, Palermo 1644, *De Angina Puerorum Campaniæ Siciliæque Epidemica Exercitationes*, Paris 1646, and *De Nivis Usu Medico Observationes variæ*, Copenhagen 1661. Wellcome Collection, London.

Figure 67. The eruption of Vesuvius on 16 December 1631. Engraving by Matthäus Merian (1593–1650) from a drawing from 1631 by Joachim von Sandart (1606–1688). From Cornelis Danckert's *Historis* 1642.

The text above the illustration in Merian's German edition reads: "The eruption of Mount Vesuvius and the simultaneous fire of the surrounding region drawn after the real event by Joachim Sandart 1631." The letters indicate "A. The crater where the eruption began. B. Stream of boiling water from the mountain. C. Torre del Greco was completely covered in ashes. D. Torre Anunziata was completely covered in ashes. E. Two galleys, on which the rest of the population was rescued. F. The Torre Anunziata bridge. G. Corner of the city of Naples. H. This ravine burned out in Pliny's time. I. Smoke and vapour from the burning ashes mixed with flowing water. K. Snow-capped mountains in the direction of Apulia. L. Beautiful fields full of lemon and pomegranate trees."

One of the geological topics that occupied and fascinated Thomas Bartholin the most were volcanoes. He visited Vesuvius and Mount Etna but also wrote about the Icelandic volcanoes. Peace Palace Library, The Hague. Photo: Wikimedia Commons.

resulting from this journey. The first is from 1 February, when Bartholin and his travelling companions reached Naples:

"Because of the less healthy nature of the climate in Rome, I have, together with my cousin, made off to Naples, to which I have also been lured by friends, whom I have, through our reputation, won in no small number among the scholars. Among these towers Marcus Aurelius Severinus, known to the world for his published work on abscesses and other complaints, which he is hastening to publication shortly, and which includes the learned commentary on the viper. It will soon be published in Padua.[187] I enjoy this domestic intercourse, and I shall not regret that I have done so."

Worm, in his answer on 17 March, tells of his pleasure at Bartholin's friendship with Severino, whom he calls "the foremost among the surgeons of our century," and he therefore hopes that Bartholin's relations with Severino are sustained.

In Naples, Bartholin follows the experienced Severino in his surgical treatment of patients in the city's hospitals. One of the hospitals in particular impresses him: "The Hospital of the Blessed Mary of the Annunciation was magnificent and splendid. All cases of fever were cared for on the lower floor and wounds on the upper, and there was a pleasant open space on the roof where those who were convalescent were admitted so that they might refresh themselves and become accustomed to the air. In the lowest part of the building on the right was a special room to receive the moribund and the noisy, and the reverend fathers lived adjacent to it. On the bed of each fever patient was affixed a stone tablet on which was noted daily when and how frequently he had received his medicine."[188]

Book on the dissection of an aneurism

Bartholin is a little taken aback by the big operations that Severino and his colleagues carry out without using anaesthesia.[189] Among other things, he and Johannes van Horne witness an amputation of a man's arm without anaesthesia. The unfortunate patient had been subjected to bloodletting by a barber, who unluckily had jabbed a hole in an artery instead of a vein, with the result that the artery swelled up into an aneurism (a widening or bulge caused by a weakness in the vessel's wall), which burst, causing heavy bleeding and extreme swelling of the man's forearm, which became blue and cold. Gangrene threatened, and only an amputation could save the man's life. After a prayer, the man's forearm was amputated "in less than a moment."[190] While the surgeons present gave free rein to their feelings with shouts and tears, the patient, according to Bartholin, participated with admirable calm.

After the operation, Severino invited his two pupils, Bartholin and van Horne, to his home to take part in an anatomical examination of the amputated arm, or in Bartholin's words, "to take part in the dead arm's funeral, which had to be carried out with a scalpel, not with a hoe and a spade."[191]

When Bartholin came to Palermo later in the year, he made an arrangement concerning the publication of a little book of 72 pages entitled *Anatomica Aneurysmatis dissecti Historia* (Anatomical account of the dissection of an aneurism),[192] which consists of his 50-page account concerning the pathological anatomical examination of the amputated arm, dedicated to the Italian scholar Angelico Aprosio (1607–1681), an 18-page letter from Johannes van Horne to the Neapolitan physician Josepho Donzello on the same topic, and a list of 20 works published or planned for publication by Thomas Bartholin:

1. His father's anatomy, augmented.
2. The heathen anatomy of the ancients.
3. Three books on the light in animals.
4. Customs in ancient times of various peoples concerning birthdays.
5. New observations concerning the unicorn.
6. Rare anatomical accounts.
7. New controversial questions in anatomy.
8. An explanation of Christ's pierced side.
9. A treatise on fungal diseases.
10. Studies of diphtheria in children.
11. A treatise on petrified tongues in Malta.
12. A treatise on truffles.
13. Medicine from snakes in Bracciano.
14. The formative medical journey in Europe.
15. Ancient customs, which should be known by the physician.
16. Different views on patients with kidney disease.
17. Investigation of the nature of the soal of a deceased person.
18. On bracelets of ancient times. A sketch.
19. A book about the idea of "nothing."
20. A volume of letters and poems, which have been published in various places.

Bartholin's account of the aneurism is particularly interesting, since it is the first scientific treatise known on this subject.[193]

Book on cryoanalgesia by using snow

Severino's smaller surgical interventions such as operations on the arms and legs take place, according to Bartholin, in a heavy-handed way with branding irons, but in this case not without previous analgesia, and here Severino uses a principle several hundred years ahead of its time. He teaches Bartholin to use cryoanalgesia, which is carried out by cooling down the wounded area in question with snow before the operation. In the same way, Severino teaches Bartholin how snow can be used to cool down a catheter, so that its introduction becomes less disagreeable for the patient, and how snow can be used as a means of analgesia before the removal of bladder stones in small boys. Bartholin must have made thorough notes on his newly acquired knowledge about the various possibilities for the use of snow in medicine, but it was not until 1661 that he published a work on the subject in Copenhagen, *De Nivis Usu Medico Observationes variæ. Accessit D. Erasmi Bartholini De Figura Nivis Dissertatio cum Operum Authoris Catalogo* (Various observations regarding the medical use of snow. With a dissertation by Dr. Rasmus Bartholin about snow crystals and a catalogue over the author's works) (Figure 68). Once again, it is the first scientific treatise we know on this

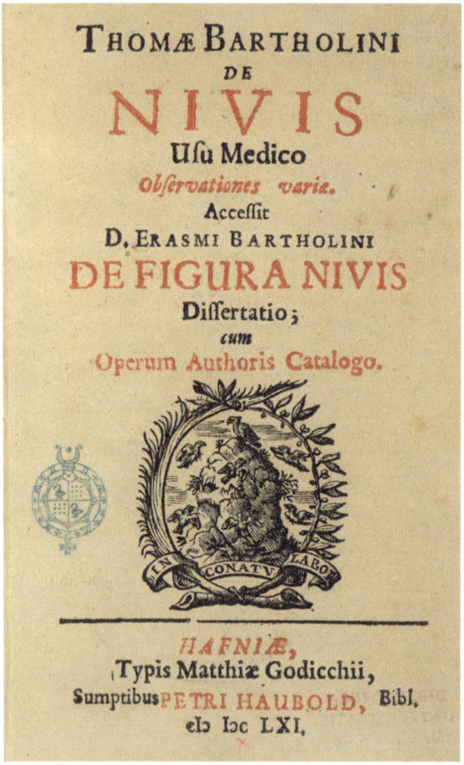

THOMÆ BARTHOLINI
DE
NIVIS
Uſu Medico
Obſervationes variæ.
Acceſſit
D. ERASMI BARTHOLINI
DE FIGURA NIVIS
Diſſertatio;
cum
Operum Authoris Catalogo.

IN CONATU LABOR

HAFNIÆ,
Typis Matthiæ Godicchii,
Sumptibus PETRI HAUBOLD, Bibl.
cIↃ IↃc LXI.

Figure 68. Title page of Thomas Bartholin's *De Nivis Usu Medico Observationes variæ*, Copenhagen 1661. Only in recent times has it been realized that the book was epoch-making, in that it was the first known book to describe the modern treatment cryoanalgesia. Thus, it predates this form of treatment by more than three hundred years.
In the book, Bartholin describes, among other things, how snow can be used as a pain-relieving cooling agent prior to wound treatment with a branding iron. He had learnt about the treatment from the Neapolitan surgeon Marco Aurelio Severino.
The Royal Library in Copenhagen.

subject. The book consists of 232 pages besides the dedication to Reich Count Christian Rantzau (1614–1663), the preface, a poem by the aforementioned Vitus Bering to Thomas Bartholin, a table of contents, a 42-page illustrated dissertation on various kinds of snow crystals written by Thomas Bartholin's brother, Rasmus Bartholin (Figure 69), and a catalogue of Thomas Bartholin's published works as well as those planned for publication.

Offer of professorship in Messina

The report of Bartholin's qualifications has preceded him southwards and ensures that, while he stays with Severino in Naples, he receives envoys from Messina who deliver an offer of a professorship in medicine at the University of Messina.[194] Bartholin has not considered accepting the offer but wants to give the position closer inspection and render thanks for the honour and trust. When, together with his travelling companions, he proceeds with his journey southwards to Sicily in the spring of 1644, he therefore stays in Messina, from where, thankful to have crossed the dangerous Straits of Messina, he sends a longer account in his next letter to Ole Worm on 11 April 1644:

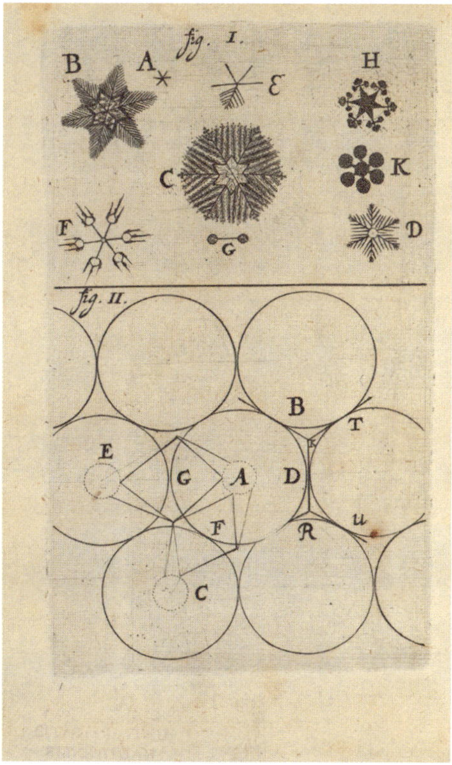

Figure 69. Rasmus Bartholin's drawings of snowflakes and snow crystals were based on his own observations in Holland and later in Denmark during a heavy snowstorm on 3 January 1660. From Rasmus Bartholin's supplement on snow crystals *De Figura Nivis Dissertatio* in Thomas Bartholin's *De Nivis Usu Medico Observationes variæ*, Copenhagen 1661. The Royal Library in Copenhagen.

"I have no doubt that you and our dearly beloved mother are thinking anxiously about the situation of our affairs. However, we are safe, thank God, since for these last few days we have avoided Scylla and Charybdis[195] and have made our entry into the glorious city of Messina, so that I can explore in person the post at the University, which the high council promised me a short time ago with a generous salary, and give my thanks for the offered post of honour. There I have found friendly, gracious noblemen, and I enjoy an intimate relationship with the very pleasant P. Castelli [Figure 70]. This person is fully committed to publishing books on various learned subjects. In his museum he has collected various freaks of nature. He has animal skeletons he has built up himself, of muskrats, swordfish and other creatures. The eyes of the muskrat he has shown with the musk-vessel, as he has described it in an account of that animal. He owns various petrified shells. At his house I have seen an Indian palm fruit, which cannot be crushed even on an anvil. I have given him the catalogue of your museum, which fortunately has fallen in with his taste, apart from the fact that it contained the most common apothecary's seeds. It is, however, certain that they are also essential for Nature's completeness and variety. [...] In the medical area there is only a little that deserves to be written. In the city moats in Messina, Castelli has established a garden with chemical laboratories, a lecture theatre, a chapel, etc. together with fourteen beds for plants, whose contents he has described in a catalogue published in 1640 [Figure 71].[196] In the meantime, it now lies neglected. Only a few others shed a light on medical science. [...] But Castelli raises his head above the rest. It is fifty years since any theriac was prepared here, for Rome, Naples and Alexandria made up for this need. But in 1637, when P. Castelli was prior of the college, Jo. Dom. Cardellus began to sell one in San Giorgio's House. He has given a public description of it, but the same Castelli is its author, and from Sicily and Calabria he has obtained a supply of many single ingredients. It contains an abundant amount of sesame oil, which in Rome is replaced by almond oil. For sesame appears here in such rich quantities that it is sprinkled on coarser bread because of its pleasant taste. Opposite the college may be seen an extensive infirmary consisting of seven small separate blocks for the Capuchins, women and visitors. Stomach complaints are common here, and Castelli fortunately cures them with his dietary electuary, the specifications of which he has given us.[197] Moreover, one ounce is administered an hour before lunch, and it gives splendid aid to the stomach. Choking angina[198] is also frequent in children, and sometimes it attacks adults too, when they have dealings with children. Castelli, who has demonstrated that bleeding is of no use in this disease, treats it solely with a purgative drink of rose syrup and honey. In the

Figure 70. The Sicilian physician, botanist and chemist Pietro Castelli (1574–1662), 55 years old. Engraving by an unknown artist. Castelli's dynamic nature and diverse interests in medicine, surgery, anatomy, botany, chemistry, pharmacology, zoology, geology and curiosities fascinated and inspired Thomas Bartholin. Wellcome Collection, London.

Figure 71. The botanical garden in Messina was created in 1638 by Pietro Castelli and consisted of 14 small gardens or beds, each of which was organized in a specific geometric pattern. 12 of them were named after the apostles of Jesus. Here you can see the garden of St. Peter the Apostle, which is number 1 in Castelli's book on the garden, *Hortus Messanensis*, Messina 1640.
The Royal Library in Copenhagen.

months of May and June, solemn fishing for swordfish takes place. The great white shark fights against it right from the front. With its sharp tail it often cuts the feet of swimmers,[199] and therefore you do not venture far out into the water. I have seen the skeleton of its head in Imperato's Museum in Naples [Figure 72], and I know that its skin is well known and available in your museum. A nobleman in Messina keeps among other rarities two large horns of the rhinoceros, and other horns that are said to come from a unicorn. I cannot give any judgement on this, since people here put gazelle horns on many [animals]."

Bartholin, who was much occupied with Sicily's nature, continues with a mention of different geological phenomena in Messina and its neighbourhood, comprising rock corals, other stone formations and petrified shells, etc. He speculates whether these fossilized sea creatures "have been washed up by the sea floods, through which Sicily was separated and detached from Italy." He finishes off the

Figure 72. Engraving showing the Neapolitan apothecary and collector of curiosities Ferrante Imperato's (c. 1550–c. 1631) natural history collection. From Ferrante's 800-page catalogue of the collection, *Dell Historia Naturale Libri XXVIII*, Naples 1599. Imperato's collection was the direct model for Ole Worm's Museum Wormianum.
Wellcome Collection, London.

letter with legendary histories about Messina's origin, and a disclosure about his next destination: "Messina, en route to Etna, 11 April 1644."

As will later be described, Bartholin's famous pupil, Niels Stensen, took up and proved Bartholin's speculations more than 20 years later: that petrified shells and other petrified parts of sea animals found on land far away from the sea had petrified in the sea and later been washed up by floods.

Book about a diphtheria epidemic among children in southern Italy

Bartholin's discussion of choking abscesses, choking inflammation of the throat or suffocating angina in children is due to the fact that Sicily and the southern Italian mainland since 1618 had been tormented by diphtheria epidemics, which killed, in particular, many children. Bartholin writes many years later that the disease "attacked the children like a plague, infecting some and killing others. The older ones were rarely attacked and easily treated, because they were more willing to take medicines and gargles."[200] In the year before Bartholin's arrival, Severino had published a treatise on this disease, and with customary zeal Bartholin immediately conceived an interest in this subject and ended by making his own observations of affected children, which along with his Italian colleagues' experiences and extensive literary studies of the subject, he included in his own treatise, *De Angina Puerorum Campaniæ Siciliæque Epidemica Exercitationes* (Investigations into a diphtheria epidemic among children in Campania and Sicily).[201] which he published in Paris in 1646 on his journey home to Denmark. This small 140-page book alone deserves a closer discussion firstly because it seems to be the first scientific treatise on a paediatric subject by a Danish hand, but also because it shows that, through his observations and studies of the disease, Bartholin obtained an insight into it that, on a number of points, was far ahead of its time. Paediatrics – the study of children's diseases or children's medicine – was first established as an individual medical specialty in Denmark with the provision of children's hospitals in the middle of the 1800s, and in this period writings and articles on paediatric subjects began to emerge.[202] In other words, Bartholin was a couple of hundred years before his time in this area.

Diphtheria is caused by *Corynebacterium diphtheriae*, which is transmitted through saliva and airborne droplets. In its classic form, the disease produces severe inflammation of the throat with a thick greyish-white coating in the mouth and throat. The throat inflammation can be complicated by a closing up of the throat with a danger of choking, and with shedding of the linings there can occur life-threatening bleeding. The bacteria produce a toxin (diphtheria toxin), which can cause serious disease in the internal organs, especially the heart, and attack the muscles and nervous system resulting in life-threatening convulsions and paralysis. Today diphtheria can be effectively prevented by vaccination, which is also included in modern child vaccination programmes, and the disease can be treated with antibiotics.

De Angina Puerorum also besides a dedication to Bartholin's cousin and travelling companion, Didrik Fuiren, a preface, a testimony from two good friends from his stay in Paris, René Moreau and Guy Patin, a table of contents and at the end in French an authorization of the book's printing and publication from the physicians at the medical faculty in Paris, dated 3 February 1646. After Bartholin's 123-page dissertation follows Moreau's 17-page letter concerning laryngotomy (opening the larynx and trachea by way of a surgical intervention through the neck). Bartholin's treatise consists of seven chapters on the following subjects:

> 1) On the part of the body that is attacked by diphtheria. 2) On the causes of this choking disease. 3) On the nature of this choking disease. 4) On cure through vomiting. 5) On bloodletting from the neck vein, the arteries or with assistance from leeches. 6) On cure by the help of a purple-coloured thread. 7) On laryngotomy.

Bartholin describes in detail the damage done by the disease to the various organs, with particular focus on the neck, the nervous system, the heart, the lungs and the muscles. He describes convulsions and paralysis but also, with reference to Severino, how vomiting can cure the disease. Similarly with reference to Severino he emphasizes viper venom, oil of scorpion poison and bloodletting through opening of the neck vein (*vena jugularis*) or the use of leeches for treatment. According to Severino, bloodletting from the neck vein is a highly effective treatment of both diphtheria and croup in children (and furthermore also effective with prolonged, obstinate headache, prolonged severe hoarseness, difficulty in breathing, asthma, pains in the pulmonary pleura, pains in the spleen, infected wounds, gastric problems and leprosy). Bartholin adds that he and his fellow traveller, Johannes van Horne, in several cases from personal experience have corroborated the curative effects of bloodletting in children's diphtheria. As already mentioned, Pietro Castelli does not believe that bloodletting has any place in the treatment of this disease but recommends violet or rosehip syrup.

According to Bartholin, certain physicians recommend purple-coloured thread for dealing with throat complaints. For example, Paul of Aegina[203] advises throttling a snake with a purple cord and afterwards laying the thread on the neck of a patient who suffers from a throat complaint. One of the theories behind this treatment must be, according to Bartholin, that the colour purple draws blood. Moreover, he mentions that snake or scorpion poison are used by some physicians as medication against infectious diseases.

During his work on the book, Bartholin writes in an undated letter to Moreau: "Together with your friend Severinus I was recently involved in some extraordinarily thorough observations about children's diphtheria, the treatment of which is often a subject of doubt, and we arrived at no satisfactory conclusion. However, it was our common view that witchcraft, which is harmful to this age group, must be fought with a brave and fearless hand. For the famous goat-milking Lilith,[204] and the screech-owl, a hostile company for small children, is not frightened away by shiny spittle or through hanging up of amulets, garlic or mud round the neck or by Alyssum.[205] [...] They need reliable help from the hands of a physician. [...] But when I think of it, my mind gets nervous and disturbed by thoughts of the small child's sufferings. On the one hand the difficulty of the task causes hesitation, and on the other hand a threatening danger does the same thing, and I don't believe that it is suitable for this disease to cut into the neck. Chiron'ssons called it laryngotomy,[206] and it is in this regard that I turn to you, most learned man, our century's great light, so that I may in the end clearly understand, with you as my teacher, what I cannot achieve with my own clouded intellect."[207]

Hereafter, Bartholin mentions that the leading surgeons, Adrianus Spigelius, Marco Aurelio Severio and Nicolaes Tulp, are not opposed to the use of laryngotomy. He is doubtful as to how far his own scepticism over the operation is justified or unjustified and asks for Moreau's decision on this. He explains his scepticism thoroughly:

> "For the disease, which is acute and advances rapidly, scarcely gives time for this slow form of treatment in which one must wait until the skin and subcutis are cut up, the muscles are drawn from each other, and a little tube is put in, adjusted and tied up. The strength of a very young neck, where the ring structures are not clearly detached, and everything is very soft because of the person's age, gives in. Moreover, the difficulties are increased owing to fear of inflammation and concern that the number of infected areas might be increased."

Bartholin is worried in case the operation turns out to be useless, because the cricoid cartilages will fall together, and even harmful because of the inflicted wound and the irritation that will be caused by the metal tube, and since raw, impure and intemperate air is drawn directly down into the lungs without first having been through the nose, mouth and throat. He concludes wisely: "No emergency should force us to an extreme treatment that is riskier than the danger that the emergency itself produces."

Bartholin finishes the letter with a request for Moreau's view concerning the place of purple thread in the treatment of illness – a subject Bartholin has just read about in the ancient French medical writer Marcellus Empiricus' (Marcellus of Bordeaux, b. 395) work *De Medicamentis* (On medicaments).[208] Bartholin finds it particularly hard to understand why purple flax thread should be able to lessen the pains of haemorrhoids, as Marcellus recommended.

Moreau answers Bartholin on 1 March 1646 from Paris in the long 17-page letter already mentioned, with which Bartholin concludes *De Angina Puerorum*.[209] Moreau writes that the disease "is serious and terrible in its appearance, it stops the swallowing process and the breathing, and it must be feared for the fact that, in terminal cases, it spreads to the lungs and often kills the patient within a week. [...] Moreover, it often spreads through epidemics." Moreau refers to such an epidemic, "which made an incursion through Germany in the years 1564 and 1565, and indeed not without reason, for illnesses that hit the whole population have a common cause: polluted and contaminated air; for this air, which because of the necessity of breathing, passes back and f through the neck and trachea so many times that it causes a malignant impression in these areas, and especially in those bodies, which are warmer, more humid and softer than normal, and where the head rests more loosely and delicately, which is the case in children." Moreau refers how Hippocrates wrote how children living in humid areas are affected more frequently by convulsive fits and attacks of asthma. Convulsive fits are due to children's feebler heads, while breathing difficulties are caused by "contraction of the airways, which is due to either inflammation of the trachea and larynx, blocking of the air passage, or overheating of the lungs." Moreau refers to contemporary Spanish and Italian authors' writings about the disease and their great differences on the subject,[210] but praises Bartholin for his insight into the illness: "As far as the nature of the disease is concerned, I see that you are clear in your points of view and do not struggle seriously to understand on what basis and with what preventive initiatives it must be subdued and combatted. The only thing that is a cause for your concern is the question of laryngotomy, where you ask for my assessment, especially regarding whether the operation can be carried out safely and professionally on children with stifling inflammation of the throat."

Moreau endeavours to address Bartholin's concerns by referring to the similar concerns of ancient physicians. He himself has had good experience with bloodletting from the arm veins as a means of treatment, and he has only needed to carry out laryngotomy in the case of two adults, a gardener and a soldier. On both occasions the operation took place without difficulty, and both patients were cured. However, the soldier became persistently hoarse after the operation.

Moreau gives a detailed description of the operation's execution, and he has no doubt that laryngotomy can be carried out with the same success on a child as on an adult. An assistant can easily hold the child gently, keep it calm and divert it by distracting and entertaining means during the relatively short operation, which is neither particularly painful nor bloody, and Moreau does not think that children are afraid either of the surgical instruments or the surgeon's hands. Nor does he believe that there is any substantial risk of any of the complications that Bartholin fears. He is used to performing bloodletting, cupping, setting broken bones in place and removing bladder stones in children, and believes the latter operation is much more dangerous than laryngotomy.

Moreau dismisses Marcellus' claim about the painkilling effect of purple flax thread on haemorrhoids – like many others of Marcellus' claims – as "superstition and gossip from old wives' tales and fabrications, which deserve no credence."

The first English professor of paediatrics, George Frederic Still (1868–1941), most famous for the first description concerning the disorder of attention now called ADHD, writes in his book *The History of Paediatrics* about *De angina Puerorum*:

"In 1646 a tiny volume was published in Paris, in the Rue St. Jacques at the Sign of the Golden Vase, by one whose name lives in the nomenclature of anatomy, Thomas Bartholin. Its value was altogether out of proportion to its size, for it was a piece of careful clinical observation, backed by a knowledge of anatomy which was uncommon in those days."[211]

Still emphasizes Bartholin's insight into the infectiousness of the disease, the character of the poisoning and the effect on the nervous system, and he cites Bartholin for the following correct observations: "It has the insidious character of a toxic disease, in which air contaminated with the virus[212] is inhaled through the nostrils and works its way surreptitiously through the cribriform plate[213] to the brain within."

As one of his proofs of the disease's effect on the nervous system, Bartholin mentions muscle weakness and squint in a surviving child. Cases of sudden death, which others explain by blocking of the lungs, Bartholin explains correctly by paralysis caused by the disease's injuries to the spinal marrow or the nerves that emanate from it. Still continues:

"After these conclusions, so nearly in accordance with much later discoveries, it is rather a shock to find his last chapter devoted to the question of whether a piece of purple thread tied round the neck is of value for affections of the throat. However, this problem gives him the opportunity for a delightful digression on the traditional use of such methods, and especially the effect of colour on health and disease."

Bartholin's interest in childhood diphtheria continued. When Severino published a new book on the subject in 1653, Bartholin, in spite of pressure from his research into the lymphatic system (see later), sent a contribution by way of a commentary from Copenhagen.[214]

Visit to the volcano Mount Etna

Just as the volcano Vesuvius made an impression on Bartholin, so did its three times higher, snow-clad, more active relation Mount Etna in Sicily. Bartholin ascended the volcano, studied its geology and flora thoroughly and mentions it in many of his letters and works. In his guide to medical travel of 1674, the aging Bartholin looks back on his visit to the volcano:

"I devoted letter 52 in century 1[215] of my *Epistolæ Medicinales* to a catalogue of the plants originating around Etna.[216] The apothecaries of Catania compounded the juice of liquorice into pastilles and exported them as a theriac, and in the hot season, when I made the difficult ascent through rocks and ashes to the centre of Etna, I saw *Tanacetum vulgare luteum*[217] flourishing at the summit. There was snow on the summit of the mountain which, carried down to Melita,[218] sold for a higher price than wine; however, it diminished during the season of the greatest heat. It was said that Etna blazes when the sirocco blows, but when returning from Syracuse on a stormy night, I saw its flames. However, Etna always gives off smoke. The last eruption of 1635 displayed a variety of spectacles with torrents of sulphurous liquid sweeping over everything, and overwhelming whatever was in its way, trees, stones and men, and when solidifying it encrusted those things that it had overrun [Figure 73]. I took home a piece of that solidified flow for the Museum Wormianum."[219]

In *Anatomica Aneurysmatis dissecti Historia* Bartholin speaks of "Etna, which quakes when its forge shatters,"[220] in *Angina Puerorum* of "Etna's fire-throat,"[221] and in *De Luce Animalium* he also mentions his purchase of a piece of lava from Etna's solidified streams of lava.[222] In *De Nivis Usu Medico*, he makes Etna a subject for studies of the Aristotelian nature theory of *antiperistasis*. According to this theory concerning the mutual effect of opposite natural forces, physical and biological opposites, for example cold and warmth, enforce each other's self-strengthening ability by being surrounded by their opposite. According to

Figure 73. Etching of the 1669 eruption of Mount Etna by an unknown artist. 18.2 × 20.2 cm. The 1669 eruption, which is the largest recorded historical eruption of the volcano, took place from 11 March to 15 July. It covered several smaller towns and villages on the mountain's slopes in ashes, whereas the strong city walls of Catania largely protected the city and channelled the lava around it, and into the sea. The eruption in 1635, as Bartholin describes, was also powerful. Wellcome Collection, London.

Hippocrates, the body's organs were, as a consequence of this phenomenon, warmest in winter by reason of the surrounding cold at this time of year. Pursuing the same thoughts, Bartholin writes in *De Nivis Usu Medico* about the snow-covered, fire-belching Etna and the underground heat phenomena in Scandinavia:

> "With help from this coldness, the warmth becomes concentrated and preserved by the snow, which, when the warmth is absorbed, even increases it, as Mounts Etna and Hekla, which are greyish from perpetual snow, show, for they are always warm because of their glittering flames, like the cliffs in Norway and Sweden, which are rich in metals. As to why Etna's masses of snow are unaffected by the flames, there are various explanations in literature."[223]

He continues that "the heat in the depths is preserved and increased by the external cold," and the heat "is besieged by the cold outside, just as the roots of the plants, animals and insects are strengthened in winter, for their insides gather strength, which is unable to flow out, and thereby gain greater power."[224]

Paradoxically, however, Bartholin's own pupil, Niels Stensen, came to play a leading role in the dismissal of the theory of antiperistasis by visiting ice caves in mountains in North Italy in the summer of 1671 and explaining correctly how the ice inside is preserved during summer despite the high temperatures outside the caves, namely by virtue of freezing air currents from deep cracks and cavities in the roof, floor and walls of the caves.[225]

Another fallacious theory of the age with which Bartholin becomes engaged during his visits to Sicilian mountain-caves and grottoes is the theory of *succus lapidescens* (stone-producing liquid). The theory was proposed by the German mineralogist and metallurgist Georgius Agricola (a Latinization of Georg Bauer, 1494–1555) in the work *De Ortu et Causis Subterraneorum* (On the origin and causes of underground phenomena), Basel 1546. The idea is that stone is formed when a stone-producing fluid consisting of underground water, rainwater and earth is condensed through heating and evaporation. Inspired by this, Bartholin airs the theory in *De Unicornu Observationes Novæ* that the water that drips down from the roof of the caves can appear to be condensed to stone, and he imagines that the same phenomenon can be the cause of lithiasis in people's bodies, for example kidney stones and bladder stones.[226]

Tongue stones and sealed earth in Malta

At the end of April 1644, Bartholin and his travelling companions journeyed farther south to Syracuse, and shortly afterwards sailed on to Malta. The troubled waters, which were often hit by stormy weather and visited by Saracen pirates, did not make for a safe voyage in Bartholin's time, but this did not hold back the adventurous and curious Dane. As a special attraction the island promised two geological phenomena, which both contained medical characteristics. One was petrified shark's teeth, *glossopetrae* (tongue stones), and the other was Maltese sealed earth, *Terra Sigillata Melitensis*, which was a kind of sacred earth that had been sealed after the apostle Paul's three-month stay on the island, and which possessed certain curative qualities when one took it as a medicine, for example as an antidote against a snake bite. Bartholin and his fellow travellers visited Paul's cave and collected samples from its particularly potent medicinal earth.

Malta's chalk and limestone mountains are rich in fossils, and the most discussed in Bartholin's time were the fossilized shark teeth or snake teeth, which some of the locals considered as originating from giants who had lived in Malta in the distant past, while others considered them to be petrified snake tongues or snake teeth. Among scientists, however, there were some who believed them to be fossilized shark teeth. Irrespective of the understanding of their origin they were an object of considerable superstition and assigned effects against illness, poisoning and magic.

Many years later, Bartholin writes about both phenomena and of his own visit to Paul's cave: "I have visited the cave and viewed the yellowish earth used as an antidote, which still seems unreduced, although every day pieces are broken off for people's use. Moreover, I have myself, with my companions, taken a quantity of the same earth with me out of the cave. This earth, which is so strong an antidote against poison and malignant illnesses, is burnt by the inhabitants and shaped into various forms, whereby it is turned into various glazed vessels, pictures, small tablets and coins, which by no means give way to similar things made from other sealed earth. There are samples of them in my museum. In Syracuse I was told by a person who had been bitten by a snake that he was immediately healed by covering the wound with sand from this earth, which had just been found on a ship in the vicinity. Even the Maltese themselves testify to the forces of this earth and the snake tongues in a public announcement printed in the year 1643, which I have brought with me, that they resist all poisons and not only protect against poisonous bites, but even cure bitten persons when the tongues are hung round the neck and arms, if a little of the said earth is drunk in wine and water, or the bare fluid, when the tongues are dipped in it. If wine or water is scooped up in a vessel made from this earth, and is drunk after it has absorbed its strength, many illnesses are cured. [...] I have seen snake tongues mounted in gold and silver hanging as an amulet in order to ward off witchcraft. In Malta at the home of the Vice-Chancellor of the Jerusalem Order, Johann Fr. Habela,

I have seen an ancient snake tongue dug up from the earth. Attached to it there was a lead ring, a sign that it had hung either in the ears or around the neck. Rondeletius[227] mentions that snake teeth hung around children's necks help when they are suffering from toothache. Moreover, tooth powder is made from them, which retains its hardness in fire and makes the teeth white. A snake tongue held in the hand of a woman giving birth until it becomes warm is said to ease the birth. Snake teeth ground to powder and taken in water are reckoned in Malta to be an antidote against malignant fever and are believed to be effective against boils."[228]

With his customary enthusiasm Bartholin threw himself into the phenomenon. He obtained knowledge about it from experienced people on the island, while he himself hacked out fossilized shark teeth from the limestone and took a collection of them home with him.[229] He began to write a dissertation on the subject, and Henrik Fuiren and Ole Worm put their knowledge at his disposal. On 16 June 1644, Fuiren writes from Padua to Worm in Copenhagen:

"He has plans to write about tongue stones, for the abundance and diversity of those he has dug out from their original location and brought with him gives him the opportunity to do so. At his request I have sent him an excerpt of what may be found written about them in your museum, and if you have later made some observations on them, I have no doubt you will also inform him of these."

Worm answers Fuiren on 25 July in the same year with a longer explanation of tongue stones, ascribed to Thomas Bartholin, and a pronouncement of his great expectation concerning Bartholin's planned publication on the subject:

"I am delighted that Thomas along with your brother, has journeyed back to Naples safely, and even more so as they have come full of very useful and interesting things. Since you have given me occasion, I cannot help but briefly discuss his project, since it is interesting material, he touches on, and it involves considerations in relation to many highly engaging subjects. And I have no doubt that he will dispatch his task with all his characteristic perspicacity."

In his answer Worm maintains, with reference to the Italian botanist and natural philosopher Fabio Colonna (1567–1640) (Figure 74), that the tongue stones are fossilized teeth from the shark *Canis Carcharias*, whose cranium and teeth are to be found in his museum. He also possesses many different fossilized teeth of various forms and sizes (Figure 75), and he believes, like Bartholin, that the fossils came into existence as the result of a petrifying sap, air or evaporation. But regarding the fossils' nature and condition, "the person who has more carefully and more accurately investigated the places where such phenomena occur can

Figure 74. a. The Neapolitan botanist and naturalist Fabio Colonna (1567–1640), 38 years old. Portrait engraving from the book *Lyncei Minus*, Rome 1606.

Colonna concluded in 1616 that tongue stones, *glossopetræ*, were fossilized teeth from ancient sharks. Both Ole Worm and Thomas Bartholin supported this hypothesis, long before Niels Stensen started his research on the subject, but it was Stensen who provided the final proof.
Biodiversity Heritage Library.

b. On page 33 of his treatise on tongue stones, *De Glossopetris Dissertatio*, Rome 1616, Fabio Colonna depicts Maltese tongue stones and teeth of the great white shark side by side to demonstrate the similarity. At the centre of the image on the right he depicts a piece of rock with an embedded shark tooth and at the bottom coral stones shaped like shark teeth.
Biodiversity Heritage Library.

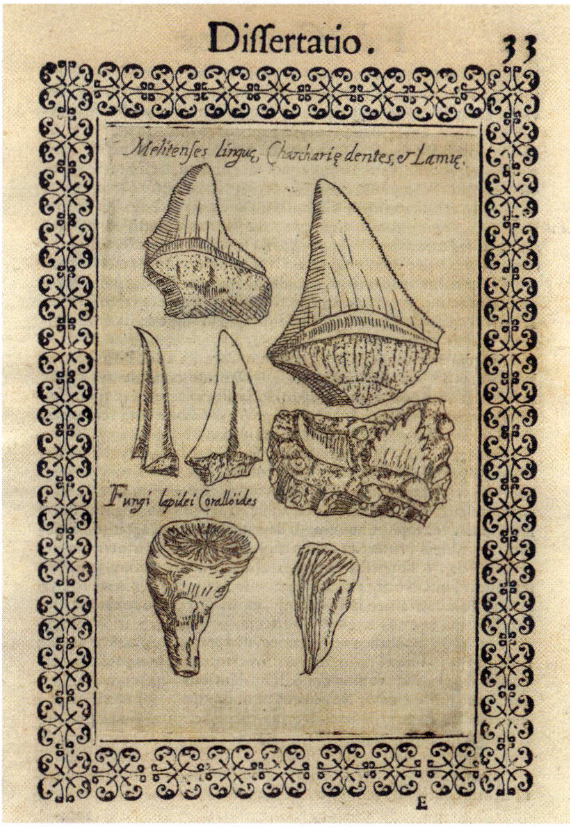
Dissertatio. 33

Melitenses linguę, Charcharię dentes, et Lamię.

Fungi lapidei Coralloides

E

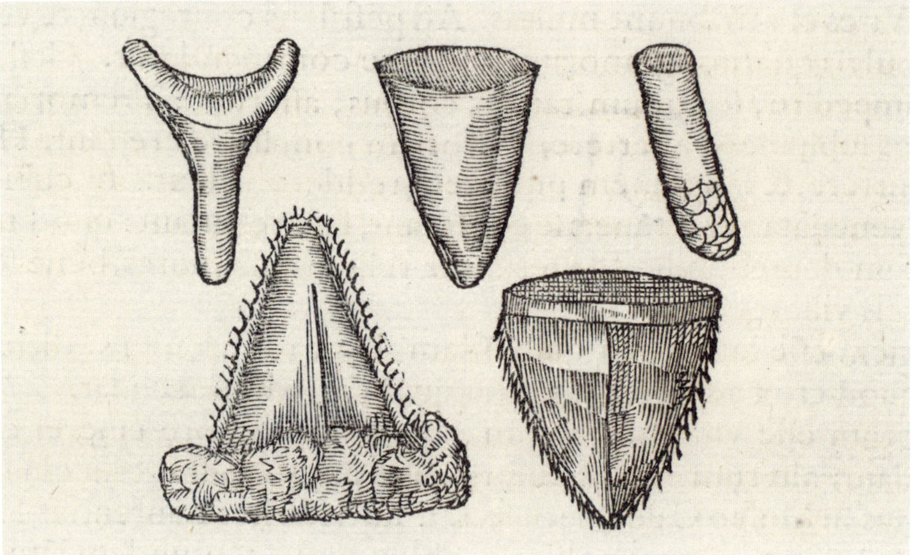

Figure 75. Tongue stones (*glossopetræ*) on page 67 in Ole Worm's *Museum Wormianum*, Leiden and Amsterdam 1655.
The Royal Library in Copenhagen.

be a better judge. I am delighted that my Bartholin is such a person, and I hope that he will produce something that can stand firm concerning these new matters." Worm must have been disappointed, for although Henrik Fuiren dutifully sent his account on to Bartholin, the latter's book on the subject was never published. He only published nine pages about the medical use of tongue stones in the sixth and last volume of *Historiarum Anatomicarum Rariorum* (Rare medical reports) in Copenhagen 1661, from which I have just quoted. In his book from 1670 to his sons about the burning of his library, *De Bibliothecæ Incendio Dissertatio Ad Filios*, Bartholin lists the planned dissertation about Maltese tongue stones, *De Glossopetris Melitensibus Dissertatio*, among the books lost in the fire, and he writes the following about it:

"This dissertation was born to me when I visited the island of Malta, where I collected fossil shark teeth of all shapes and sizes, which the inhabitants of that place brought to me, and I myself searched here and there for every sort for the completion of the study. However, distracted by other studies, I was unable to complete what I had begun, but I indicated the substance of this dissertation in the Historiæ Anatomicæ."[230]

As time would show, Niels Stensen also became the person, who – in the words of Worm – most carefully and accurately investigated the places where such phenomena occur, and who delivered the dissertation with the right answers (Figure 76).[231] Through his dissection of the head of a great white shark (*Canis Carcharias*) caught in the Mediterranean Sea by French fishermen and brought to Florence from Livorno for Stensen's dissection on the order of the Grand Duke of Tuscany, Ferdinando II de' Medici (1610–1670), Stensen in the autumn of 1666 became the first to prove that tongue stones really are fossilized sharks' teeth, and that they have been washed up on land by floods. As a result of periodic changes in the surface of the Earth produced by natural forces like floods, earthquakes, landslides, forceful rainfalls, erosions, volcanic eruptions etc., and by virtue of oozing fluids, the fossils had, according to Stensen, become embedded in layers of earth and rock. Thus, the dissection of the shark's head led Stensen to further paleontological and geological studies of the geologically varied and fossil rich Tuscan landscape and to further fundamental geological discoveries, which eventually made him the founder of modern geology and crystallography.[232] But by then, Worm had long since departed this world.

New stays in Naples and Rome

Bartholin had planned to continue his travels from Malta to Egypt, but this plan was prevented by an outbreak of an infectious disease on the English ship that was to have sailed him there.[233] Instead, in May 1644, he set out once more in a northerly direction towards the Italian mainland and Naples, where he once more stayed in Severino's house.

A couple of weeks later, on 1 July 1644, Worm receives a letter from Severino himself with a fine testimonial concerning Thomas Bartholin: "This man's [Thomas Bartholin's] competence in researching Nature, his enthusiasm, his diligence, which is sufficiently well attested through daily experience, I shall not exert my mind to discuss to any full extent, nor trust my ability to do so."

The fact that Bartholin never managed to publish the treatise he had begun on fossil shark teeth must be ascribed first and foremost to two circumstances. Firstly, he writes in a letter to Johannes Walæus in Leiden on 21 June 1644 that he is exhausted after his fatiguing journey. What is worse is that, during his stay with Severino, he is struck by his first attack of kidney stones – an affliction that came to persecute him for the rest of his life. He has to finish off the letter to Walæus with the sentence: "Kidney stones do not allow me to write more for the moment."[234] The disease must this time have tormented him for over a month, since in a letter to Pietro Castelli on 23 July 1644, he writes that the illness forces

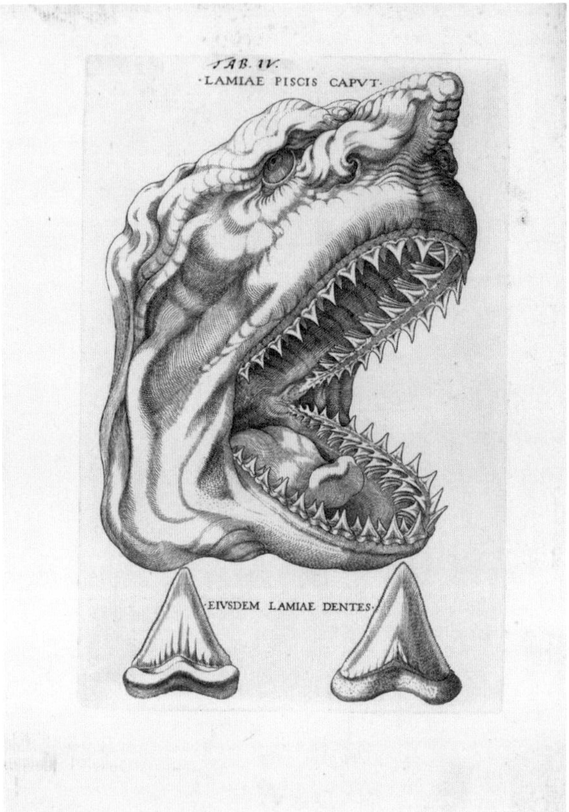

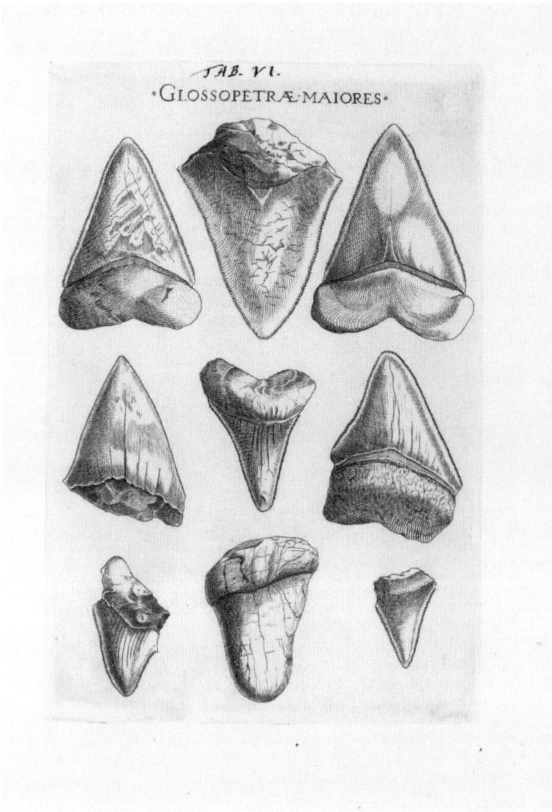

Figure 76. a. The head of a great white shark with its many rows of teeth, and two of the teeth taken out. Engraving from Niels Stensen's dissertation on the dissection of the head of a great white shark, *Canis Carchariæ Dissectum Caput*, Florence 1667, between pp. 70 and 71.
The Royal Library in Copenhagen.

b. Tongue stones (*glossopetræ*) or petrified shark teeth, as Niels Stensen proved them to be. Ibid.
The Royal Library in Copenhagen.

c. First page of Niels Stensen's handwritten manuscript for the dissertation.
The Royal Library in Copenhagen, Vesterlandske håndskrifter, NKS 4019 kvart.
The copperplate (a) does not depict the shark that Stensen dissected. Both plates (a and b) were made for the Italian physician, polyhistor and overseer of the Vatican Botanical Garden, Michele Mercati's (1541–1593) unpublished work *Metallotheca Vaticana*. Thomas Bartholin's friend, the Florentine nobleman, philologist and pupil of Galileo, Carlo Dati (1619–1676), was in possession of the plates and made them available for Niels Stensen's book. It is likely that Stensen, during his student years in Copenhagen 1656–1659, had already seen the great white shark that Thomas Bartholin kept in his collection of natural objects on the first floor of his professor's residence, which he used in the instruction of his students. In *Domus Anatomica Hafniensis*, Copenhagen 1662, p. 57, Bartholin describes the shark as follows: "A Lamia or great white shark. The spine of the same. The jawbones of the same with irregular rows of sharp teeth. The very large teeth of the same, which look like very large tongue stones."

CANIS CARCHARIÆ DISSECTUM CAPUT.

Non dubito quin Lectori fastidiosa fuerit longa, et non interrupta, rerum musculos spectantium expositio; quocirca, cum in varietate oblectationem animus inveniat, materiam priori subjungendam judicavi, quæ variis, nec cohærentibus inter se, observationibus recensendis occasionem præberet. Nec commodiorem aliam desiderare potui, quam quæ mihi sese obtulit, Caput Canis Carchariæ dissecanti. Scilicet Serenissimus Magnus Etruriæ Dux, cum Liburno ipsi significatum esset, ad aliquot milliarum a Portu distantiam captum insolitæ magnitudinis piscem esse, Caput illius Florentiam apportari jussit, mihique tradi dissecandum. Licuit mihi in eo nonnulla videre, animalium partibus cognoscendis lucem haud obscuram afferentia, him to bed.[235] Despite the testing of various remedies, he cannot relieve his pain, and he thinks about the reason for his illness. Is he eating the wrong diet, can strong wine be the cause, or has he inherited the disease from his father, who was also tormented by it? Secondly, Bartholin has still not published several of his other aforementioned dissertations, and shortly after he regains his health, he journeys to Rome following a rare event that occurred on 29 July 1644: the Pope's death.

Pope Urban VIII (Maffeo Barberini, 1568–1644) is perhaps best known as Galileo's patron, and later in the latter's trial in 1633 as the man who forced him to abjure his support of the Copernican theory, that the Earth rotates round the sun, which is the centre of the universe, and not vice versa. But because of his dictatorial form of government and his burdening the people with taxes and dues on, among other things, wine, to finance costly works of art, this pope was so detested at his death that the people attacked his statue in the Capitol. What most attracted Bartholin's attention in connection with the Pope's death was, however, something entirely different, namely the cause of death itself. At the autopsy an unusual discovery was made, which Bartholin describes ten years later in the second of his six-volume work on rare medical reports:

"**A bone in Urban VIII's heart**
The body of Urban VIII, who was a praiseworthy prince, was, in preparation for embalming, opened by the celebrated anatomist Johannes Trullus during August 1644 in Rome. In the left chamber of the heart was found a triangular T-shaped bone. Moreover, there were in the gall bladder five stones the size of a hazelnut, along with many small stones in the kidneys. From the bone in the heart ordinary folk spread various stories about this great prince, and as usual everything was interpreted for the worse. But thanks to Nature's great foresight, it often happens in old people that the otherwise slow-flowing blood is set in motion because there is added, as it were, a mainspring. Plater,[236] in the third book of his *Observations*, pointed out a similar bone consisting of three segments, and we have further confirmed it in the first century, No. 50 of *Hist. Anat.* Colombo,[237] in the seventh book of his anatomical work, states that all this is not inconsistent with Galen."[238]

Bartholin does not express an opinion about the cause of death, which may be suspected to be the "bone" that was found in the heart, which may be assumed to be a calcified blood clot (a thrombus or embolus). How it came about that Bartholin gained access to the information about the pope's post-mortem examination, we do not know.

Return to Padua

At the beginning of autumn, Bartholin was back in Padua, where a worried letter from Worm dated 3 September 1644 awaited him. Worm has learnt from Didrik Fuiren that Bartholin's health is continually poor, and Worm interprets this as due to Bartholin's hard work with all the writings he has prepared.

Bartholin does, however, reassure Worm on 14 October with news from Venice that things are not so bad after all, but that it would help if he would lay off on the unremitting pressure to return home to Denmark:

> "My dissertations flow off by themselves from an easy pen, so that they are put down on paper even before I have thought about it. This must be attributed to fortunate heredity rather than to due determination. Should my country simply grant me the peace I desire, I could well with time dispatch one greater task after another."

The supplication for peace did not work, however, for in his answer on 29 November Worm writes: "Your grandfather seems to take your longer delay abroad to heart, since he had expected your homecoming this winter." Bartholin hereafter tries to placate his grandfather's impatience by dedicating his book about the unicorn to him.

M.D. in Basel

Worm had already been informed about Bartholin's next destination and its aim in the earlier spring. On 23 March he writes to the student Henrik Motzfeldt (1618–1692) in Wittenberg: "Now he and Henrik Fuiren are thinking of a journey to Basel to be honoured with the highest degree in medicine and to return to us in the autumn season."

But Worm is again rejoicing too early. Bartholin lengthens his stay in Padua, and on 1 June 1645, he travels along with the brothers Henrik and Didrik Fuiren to Basel, which they reach in late July with recommendations from Worm and Johan Rhode in their baggage.[239]

Thus, Bartholin had chosen to follow the family tradition and gain his doctorate in Basel, just as his grandfather, father and uncle had done. On 26 September 1645, he made a defence of his doctoral disputation on pleurisy, *De Pleuritide Paradoxa Endoxa*[240] (On acknowledged paradoxes concerning pleurisy) (Figure 77), and at the promotion on 24 October, he gave a talk on deformed creatures – one of the hot topics at the time.[241] Both events must have taken place in the Brabeuterium, the University library dating from 1559 housed

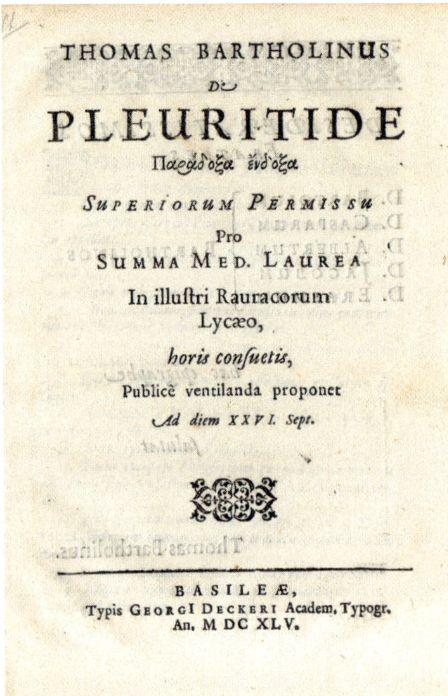

THOMAS BARTHOLINUS

De

PLEURITIDE

Παράδοξα ἔνδοξα

SUPERIORUM PERMISSU

Pro

SUMMA MED. LAUREA

In illuſtri Rauracorum

Lycæo,

horis conſuetis,

Publicè ventilanda proponet

Ad diem XXVI. Sept.

BASILEÆ,

Typis GEORGI DECKERI Academ. Typogr.

An. M DC XLV.

Figure 77. The title page of Thomas Bartholin's doctoral disputation on pleurisy, which he defended at the University of Basel on 26 September 1645 and dedicated to his brothers Bertel, Caspar, Albert, Jacob and Rasmus. It consists of 77 short theses formulated on four and a half pages. The full title of the disputation is "*De Pleuritide paradoxa endoxa. Superiorum permissu. Pro Summa Med. Laurea. In illustri Rauracorum Lycæo, horis consuetis, Publicè ventilanda proponet*": "Thomas Bartholin will with the permission of the authorities and to achieve the highest medical distinction present *Acknowledged paradoxes concerning pleurisy* publicly at the usual time at the famous University of Basel."
Private ownership.

Figure 78. a. The Swiss-born engraver Matthäus Merian's plan of his native city Basel 1615.

The Latin heading, "NOVA ET GENUINA DESCRIPTIO INCLYTÆ URBIS BASILÆ PER MATTHÆUM MERIAN FIDELITER DEPICTA ET SCULPTA ANNO DOMINI MD CXV," reads: "New and authentic description of the famous city of Basel faithfully depicted and engraved by Matthäus Merian in 1615."
The Royal Library in Copenhagen.

b. Section of Merian's plan with the still extant Alte Universität (The Old University) on the banks of the Rhine next to the Mitteler Brücke and the Rheintor. The small chapel-like building closest to the river is the Brabeuterium, where Thomas Bartholin defended his doctoral disputation on 26 September 1645. From 1559 the Brabeuterium housed the University Library, and in 1573 a hall for degree ceremonies, including defences of the medical doctorate, was established, in which public debates and promotions were held, and where Andreas Vesalius's and Felix Plater's manufactured skeletons were displayed. With the establishment of a chair of anatomy and botany in 1589, an anatomy theatre was created on the ground floor of the university's north wing, where public dissections of human corpses could take place. Until 1692 a botanical garden with medicinal plants ("Hortus Medicus") was located on a terrace in the university's southern area.
The Royal Library in Copenhagen.

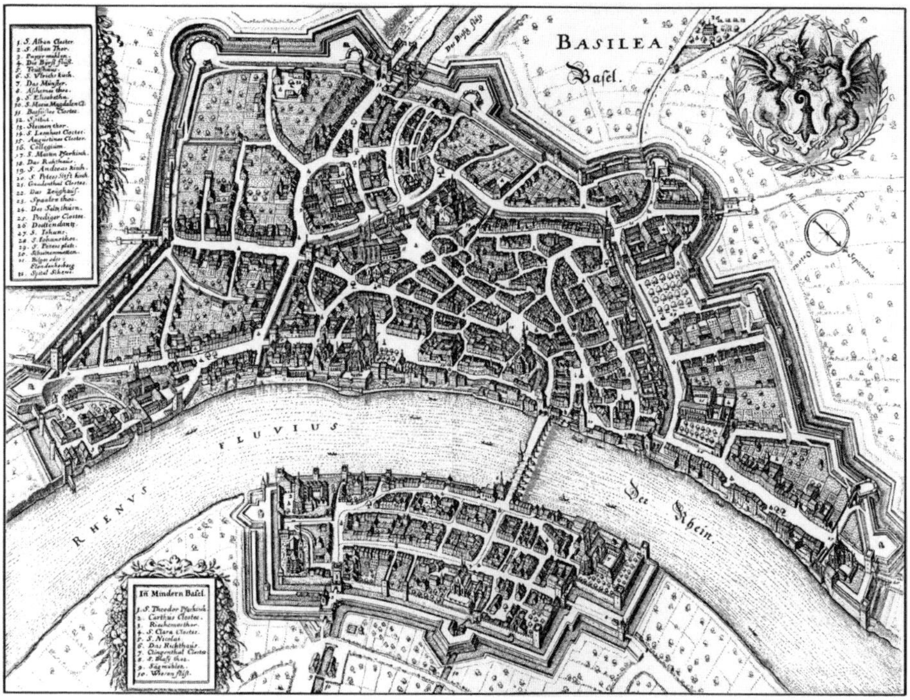
BASILEA
Basel.
RHENUS FLUVIUS
Der Rhein.
In Mindern Basel.

12
16

in a chapel building on the bank of the Rhine (Figure 78), where in 1573 a hall for degree ceremonies of all faculties was established, in which public debates and promotions were held, and where Andreas Vesalius' and Felix Plater's skeletons manufactured from dead humans were displayed.[242]

Bartholin sends no letters to Worm from Basel, but from Orléans he writes on 24 November 1645: "Basel has sent us on our way with the desired titles of honour.[243] We hope they are in accordance with our merits. I have sent my speech on monsters to press at the same place."

Bartholin himself refers to the two dissertations as "trifles" in a letter to Worm of 20 April 1646. Taking Bartholin's own estimation of them and seeing them with modern eyes, one has to admit that he is correct. The doctoral dissertation covers a modest eight pages including the title page, the dedication to his brothers Bertel, Caspar, Albert, Jacob and Rasmus, together with an appendix, while the speech concerning monstrosities runs to eleven pages, inclusive of the title sheet and dedication to the Polish knight Petro Rzeczycky of distinguished status, whom Bartholin had met in August, and whom he showers with compliments. What lies behind all this praise, history does not inform us. In his doctoral thesis, Bartholin exceptionally agrees with Descartes, that pleurisy should be classified as an acute illness.[244]

The second revised edition of the anatomy book

In the letter from Orléans, Bartholin mentions that he is en route to Paris to spend the winter there before his intended homeward journey to Denmark in the spring of 1646. A new edition of his father's textbook of anatomy is on its way to press in Leiden, and in Lyon a physician has already translated it into French.[245] Bartholin explains that "the publishers quarrel among themselves and compete about its publication." On 8 December 1642 he had written to Worm from Padua: "The bookseller Frambot is keen to reprint my father's anatomy book, and everything necessary for it is in hand, so that it can be compared with books from Leiden. I shall omit the old illustrations and have new drawings made on the spot, while the dissections take place, and thereafter engraved accordingly. The anatomist, the renowned Vesling, has also promised his assistance here, so I hope that the work will make good progress. If you notice something that ought to be changed or added, you must reveal it frankly, together with what you believe should be done regarding the dedication. My other works are now on hold because of this."

Worm answers on the last day of the same month:

"I am delighted to hear that your publisher wishes to republish your father's anatomy book. However, I can scarcely believe that it will surpass the Dutch, in neither the beauty of the printing, the quality of the paper nor the precise reproduction of the illustrations. The new pictures will serve as a recommendation for the work to a considerable degree, if only they are accurate and executed by a skilled hand. I have not noticed anything specific that must be changed. As for the six small bones in the *sutura lambdoidea* itself, which were first discovered by me, I have informed you on another occasion. [...] As regards the dedication, I do not know what to recommend, unless you would like to present it to the Lord Chancellor, since Rosenkrantz, to whom your father dedicated it, passed away peacefully here in Copenhagen on 28 October, and the former as a kind of right of inheritance should be his successor."

Bartholin follows Worm's advice and once more chooses Leiden as his place of publication, and he dedicates the book to the king's and the university's chancellor Christen Thomesen Sehested, naturally not without strategic view to his own upcoming homecoming to Denmark. The title of the new edition is identical to that of the 1641 edition, and the same applies by and large to the book's contents. When Bartholin's remaining activities and writings during his student journey are taken into consideration, it is not surprising that he did not find time for a comprehensive revision. His intentions for a full-scale change of illustrations and a replacement with fresh ones taken from new dissections therefore proved not to be realistic either.[246] The edition contains eight pages fewer than the first edition. 14 new pages with anatomical plates have been added,[247] while six of the first edition's pages with anatomical plates have been omitted. More sections have been expanded through Bartholin's newly acquired knowledge, among others from the dissections with Vesling and Wirsung in Padua. This applies for example to an expanded section on the pancreas, which pertains both to its anatomy and its function, including an illustration of it (Figure 65). Three new pages have been added with illustrations of the bones in the cranium; these include the six cranium bones, whose discovery Worm attributes to himself (Figure 79).

Bartholin answers Worm on 26 January 1643 from Padua, saying that he does not recollect receiving anything about the six bones from Worm, but he makes a request about them, upon which Worm explains on 6 April: "In the year 1628, when I showed and explained to a numerous circle of listeners the bones in a human skeleton, I found in the *sutura lambdoidea* itself six small bones, which ran through both of the tabulata of the cranium and, as far as I know, had not been noticed earlier by anyone."[248] After this there follows a longer account of the bones.

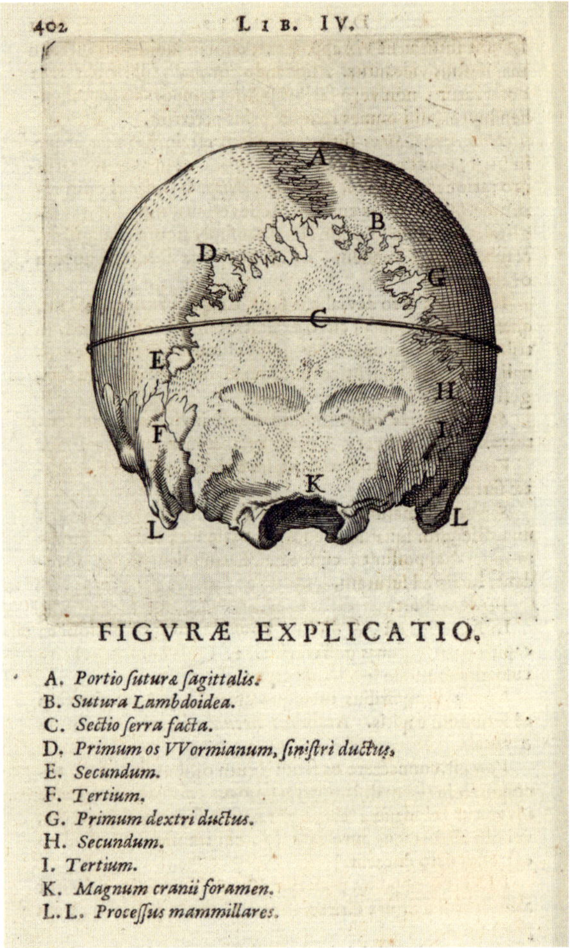
402 LIB. IV.

FIGVRÆ EXPLICATIO.

A. *Portio suturæ sagittalis.*
B. *Sutura Lambdoidea.*
C. *Sectio serra facta.*
D. *Primum os VVormianum, sinistri ductus,*
E. *Secundum.*
F. *Tertium.*
G. *Primum dextri ductus.*
H. *Secundum.*
I. *Tertium.*
K. *Magnum cranii foramen.*
L. L. *Processus mammillares.*

Figure 79. Ole Worm's six skull bones on p. 402 in Thomas Bartholin's second revised edition of his father's anatomy book, *Anatomia Secundum Locupletata*, Leiden 1645. The Royal Library in Copenhagen.

As further emphasis of Bartholin's support for Harvey's circulation theory and Aselli's lacteal veins, Walæus' letters on these topics are extended. There is also an added illustration of the vein valves of the forearm and the lower leg, of which the former is from Harvey's book on the motion of the heart and circulation of the blood (Figure 80), two illustrations of Walæus' dissection of a dog concerning the blood circulation, and an illustration of the mesentery with Aselli's lacteal veins (Figure 81).

A double illustration of the heart with a probe through the fetal window between the atria (*foramen ovale*) serves to dismiss the theory of the Venetian Cecilio Folli of 1639, that even after pre-natal life the blood passes from the right to the left side of the heart by this route, for Bartholin had not been able to detect the aperture in his dissections of numerous bodies with Vesling in Padua (Figure 82). In a letter dated 19 November 1644, this same Folli had sent Bartholin illus-

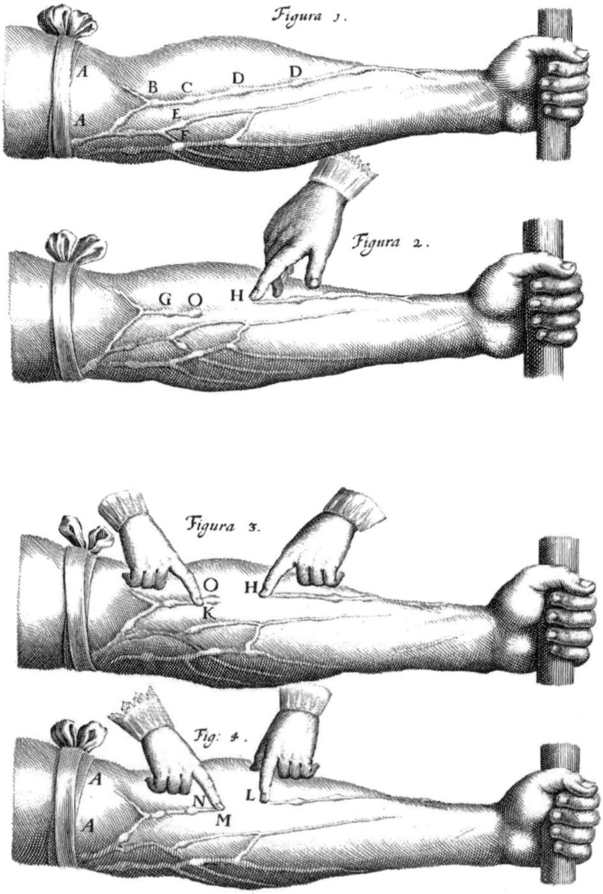

Figure 80. The function of the venous valves of the forearm, demonstrated in illustrations between p. 56 and p. 57 in William Harvey's book on the motion of the heart and the circulation of the blood in animals, *Exercitatio Anatomica de Motu Cordis et Sanguinis in Animalibus*, published in Frankfurt 1628. Thomas Bartholin reproduced the top illustration together with illustrations of the venous valves of the lower leg on p. 310 in *Anatomia Secundum Locupletata*. The Royal Library in Copenhagen.

trations and descriptions of the bones in the middle and inner ear,[249] which Bartholin also reproduced in the book (Figure 83), and which was published in Folli's own small six-page pamphlet on the inner ear, *Nova auris internæ delineatio* (A new sketch of the inner ear) in Venice on 10 May 1640, with the dedication to "that illustrious man, Doctor Thomas Bartholin."

Finally, the book has a portrait engraving of Thomas Bartholin with text by Johan Rhode (Figure 84). The original reason for the engraving was that Bartholin, during his stay in Padua in 1644, was one of the few foreigners who had been received into a distinguished academy for Italy's most significant scholars and artists, Accademia de' Signori Incogniti di Venezia (Venetian academy for unknown gentlemen).[250] Besides French and German translations, the edition was also translated into Italian in verse form.[251]

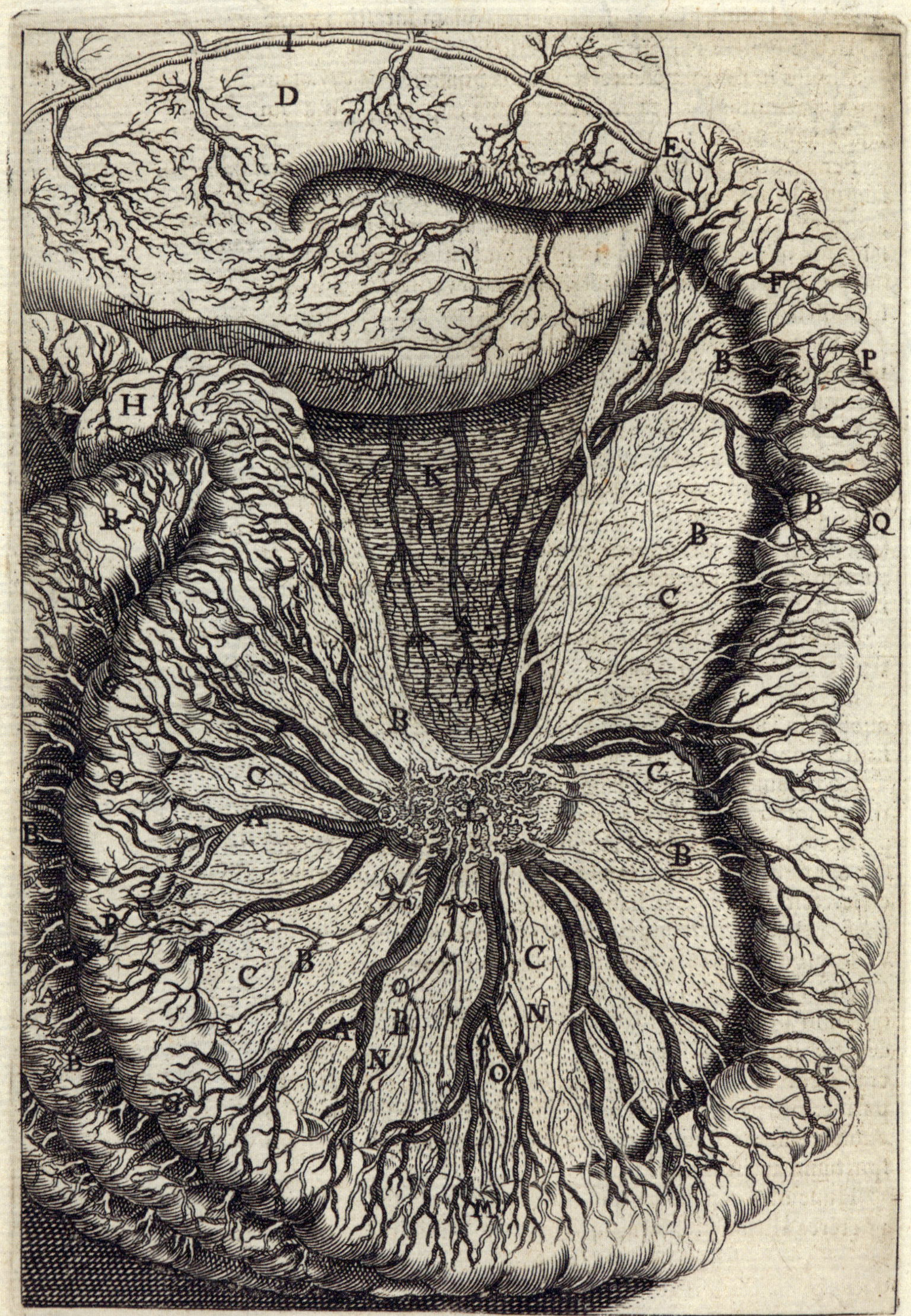
I
D
E
F
A
B
P
H
K
B
Q
B
C
B
C
A
C
B
B
D
C
B
C
O
A
B
N
N
O
B
G
M

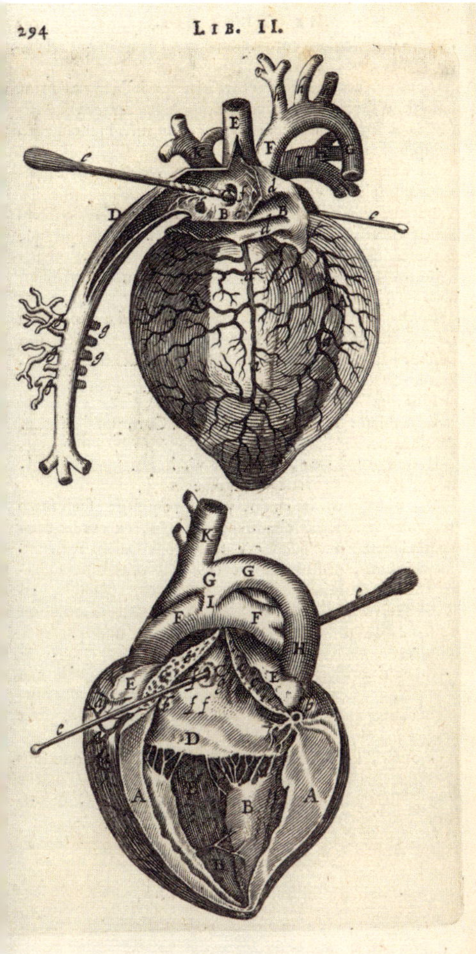

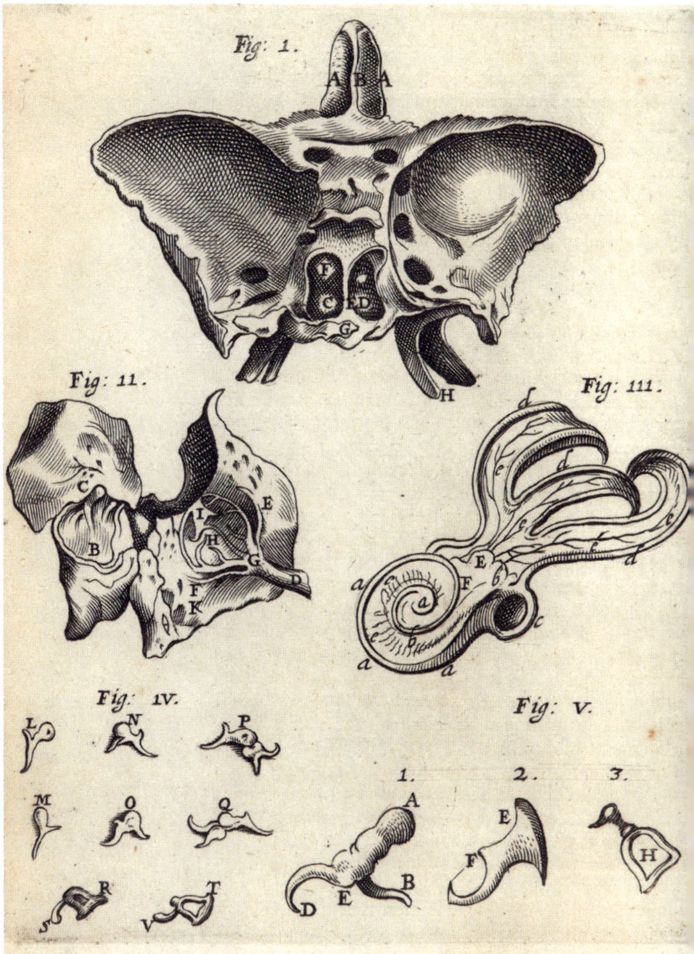

Figure 81. Aselli's milk veins (B) in the mesentery (C), reproduced in Johannes Walæus' letter on p. 474 in Thomas Bartholin's *Anatomia Secundum Locupletata*.
The Royal Library in Copenhagen.

Figure 82. The heart with probes through the fetal oval hole between the atria, the *foramen ovale*. From Thomas Bartholin's *Anatomia Secundum Locupletata*, p. 194 (the page is mispaginated as 294).
The Royal Library in Copenhagen.

Figure 83. The Venetian anatomist Cecilio Follis' (1615–1660) drawings of the bones of the middle ear and the inner ear reproduced in Thomas Bartholin's *Anatomia Secundum Locupletata*, p. 408.

The impressively accurate drawings originate from Follis' small six-page booklet *Nova auris internæ delineatio*, Venice 1645, the first two pages of which are a dedication to Thomas Bartholin.
The Royal Library in Copenhagen.

Aetat. XXIIX
Aº MDCXLIV
D. THOMAS BARTHOLINVS
CASP. F.
I. FINCK. NEP.
Emicat hoc vultu, quae condita pectore Virtus
Et patris et claro nomine crescit aui.
Jo. Rhod.
Jo. Georg
fecit

Figure 84. The first portrait engraving of Thomas Bartholin aged 28 in 1644, while he was in Padua. 14.5 × 9.5 cm. Artist: Giovanni Georgi (active c. 1617–1656). The Royal Library in Copenhagen.

The Latin text on the ribbon and frame, "D. THOMAS BARTHOLINUS CASP. F. T. FINCK NEP. Aetat. XXIIX A° MDCXLIV," reads: "Thomas Bartholin, son of Caspar, grandson of T. Fincke. Age 28, year 1644."

As one of the only foreigners, Thomas Bartholin was admitted to a literary society for unknown (actually known) men, the "Accademia de' Signori Incogniti di Venetia," whose members had their portraits included in a catalogue, in which the engraving is included: Giovanni Francesco Loredan: *Le Glorie de gli Incogniti O vero gli Huomini Illustri dell' Accademia de' Signori Incogniti di Venetia*, Venice 1647, p. 408. Bartholin used the engraving in both *Anatomia Secundum Locupletata*, from which the version shown originates, and in certain reprints of his book on the unicorn. The text under the portrait was written by Johan Rhode and reads: "The face beams with spirit and virtue, thoroughly rooted, the famous name of the fathers increases his reputation even more." The Royal Library in Copenhagen.

Hofman's attack

While in Lyon, Bartholin became aware of an attack on his father by the ageing professor of medicine at the University of Altdorf, Caspar Hofman, who was notorious for his vitriolic pen. In his letter to Ole Worm from Orléans, Bartholin writes:

> "I recently encountered, when I travelled through Lyon, Caspar Hofman's medical doctrines,[252] which were published in the same place, and in which he devoted a whole chapter against Bartholin and his anatomy, simply insulting and not appropriate for an old man's dignity. When I have completed my journey, I shall arrange an answer, perhaps with words of the same weight. For he is presenting various ridiculous statements, which are unworthy of a medical expert."

Hofman puts forward in his book a total of 44 points of criticism of Bartholin's anatomy book, of which two of the most substantial concern the heart and the lungs respectively.[253] The critical point about the heart is totally valid and deals with the fact that Bartholin, despite his agreement with all the other points in Harvey's circulation theory, adheres to his father's mistaken idea that in the partition between the chambers of the heart are found invisible pores through which the blood can pass from the right to the left chamber. The idea originated from the ancient physicians, but Vesalius and many of Bartholin's contemporaries too still agreed with this opinion. "Invisible" here should be understood literally, since curiously no one had ever seen such pores. Bartholin, in both the first revised edition of 1641 and the second revised edition of 1645, makes a curious division of the blood into a thinner, lesser part, which passes through these invisible pores from the right to the left chamber of the heart, where it mixes in with the thicker larger part of the blood, which has been through Harvey's passage through the lungs,[254] which Hofman rightly disputes. Nonetheless, Bartholin sticks to his mistaken opinion in both the third and fourth revised editions of 1651 and 1673,[255] respectively, where he argues that the thin part of the blood nourishes the heart's partition on its way through it and would disappear in the long, dangerous passage through the lungs. When you cannot see the pores in a dissection of the heart, this is, according to Bartholin, because they are twisted and draw together after death, and he brings examples, saying that both the blood and several of the body's other fluids pass across barriers without visible apertures.

In Bartholin's chapter on the lungs, Hofman's point of criticism concerns the physiology of breathing. Here Bartholin has more truth on his side, but through his opinion he does not come into opposition only with Hofman, but also with his earlier teacher in Leiden, Adrian van Valckenburg, and not least with Descartes and his followers. In the second revised edition, Bartholin rejects Descartes' assumption that the air that streams into the lungs when the breast swells does this by reason of the fact, that owing to the breast's displacement of the surrounding air, it can find no other way. Descartes here takes a mistaken starting point, saying that air cannot be compressed, while Bartholin states that "air can easily be compressed, which can be proved a thousand times by experiments", and "at the movement of the thorax or a similar bodily movement we do not see that the lightest thing that exists will be set in motion." Furthermore, he maintains that the lungs "only follow the movement of the thorax to avoid a vacuum, and therefore they only receive the inhaled air, because the thorax in its expansion fills the lungs with air."[256]

The part of Hofman's criticism that most infuriated Thomas Bartholin was the personal attacks on Caspar Bartholin, which mainly dealt with the latter's almost exclusive use of literary sources. Hofman's explanation for this was Caspar Bartholin's lack of experience with dissections. Thomas Bartholin regarded this as hypocrisy, because Hofman himself had, compared to Caspar Bartholin, almost no experience of dissection.

But Bartholin in the first instance had no time to answer Hofman's criticism, since he was on his way to Paris together with Henrik Fuiren. In this city he receives a letter from Worm, dated 24 January 1646, in which Worm congratulates him on his title of doctor and praises the new edition of the anatomy book:

"I acknowledge that the last edition of Bartholin's anatomy is elegant, the one Hacke has sent us, and I don't believe any will appear to surpass the beauty of its printing and illustrations."

As to Hofman's attack he writes: "Hofman seems to take it upon himself to attack everybody. He always appears to assume the tone of command. I have not seen his book of instruction. A person who wants to attack others can expect to be paid back in the same coin."

A new method is used to entice Bartholin home: "If you do not come home in time, you will scarcely find a maiden who is worthy of your embrace."

Worm ends the letter with a request to look up one of the time's legendary Paracelsians and mystics, the Flemish physician, chemist and philosopher, Jan Baptista van Helmont (1580-1644), who without Worm's knowledge had already died on 30 December 1644:

> "There is in Belgium, close to Brussels, if I am not mistaken, a strange paradox-maker, Jan Baptist van Helmont. He considers a physician to be unworthy of that name if over four days he has not cured a patient with fever. He removes with one incision the true reason for all fevers. He boasts that with plaster he has cured hundreds of third-day fevers without any recurrence and says he can perform other marvels in the art of medicine. If you go there, then seek out the man, and try to clarify if there is any lasting truth behind these high-flown words."

Bartholin's answer comes from Amsterdam on 7 March 1646: "The mad Helmont has met his destiny, and I don't think he has managed to fulfil his promises. For he boasted of a unique remedy against the plague, and yet both his sons died of this disease."

Short stay in Paris

In the same letter Bartholin reveals a little about what use he made of his time during his short stay in Paris from December 1645 to February 1646: "My written reply to Hofman I prepared in Paris before my departure, and I would have sent it for printing at the same place, had not the dedication held me back. I'm waiting for this at your advice, so that I may publish it without any delay. I have thought about Gunde Rosenkrantz,[257] to whom it would not seem unworthy, considering that it is an answer to a work of my father. However, in order not to waste my time in Paris without producing something, I have sent my experiences of epidemic angina in children to the press. It is now ready, so I hear, and I shall send it as soon as the printer has sent it to us."

In Paris, Bartholin also finds time to renew the discussion with the Cartesians about the physiology of breathing. He receives a letter in Paris dated 6 December 1645[258] from Lyon, from the French translator of the anatomy book's second edition, Abraham du Prat (1616–1660), who in connection with his translation work has a series of questions about various points in the book, on which he asks Bartholin to shed light. One of the questions concerns Bartholin's rejection of Descartes' explanation of the breathing mechanism. Du Prat does not fully understand this section of the book or Bartholin's arguments, and he asks him for a thorough explanation. Apparently, Bartholin's answer does not satisfy du Prat, who comes back with a new letter on 19 January 1646[259] where, not surprisingly, he makes clear his support for his famous countryman's position, and he asks for evidence from Bartholin for his claim concerning the compression of air. Bartholin answers on 1 February in a four-page letter to du Prat with the title *De motu pulmonis contra Cartesium* (On the motion of the lungs in opposition to Descartes),[260] that he could in the same way demand from the Descartes camp evidence for their point of view, but he does not believe that proofs are necessary since arguments and simple examples are sufficient in this matter. When an almost weightless object like a feather or tuft of wool is set in motion by a breath of air from a person weakened by illness, it must, if it is in the airspace just outside the thorax, also be moved by the air pressure from the thorax's expansion, if it were correct that air cannot be compressed. Descartes and his followers fail to explain how air that is displaced by the extension of the thorax must necessarily sink down into the lungs. Du Prat, who is apparently also keen to be on good terms with Bartholin, attempts a professional mediation between the two parties through an association of their viewpoints, where "the whole breathing mechanism can be explained both by the thorax's and particularly the diaphragm's natural internal movement combined with the effect of the air coming from outside, which, displaced by the movement of the thorax, streams down into its hollow interior."

But Bartholin does not allow himself to be persuaded:

> "Your brilliant attempt to reconcile the disagreeing parties would be very inviting were it not for the fact that the importance of certain factors is more exaggerated than there is any basis for. For why do we seek in the world outside for an explanation of what Nature has given us as an innate principle?"

But let us return to the letter sent to Worm from Amsterdam, in which Bartholin explains that he has hastened his return in accordance with Worm's and the chancellor's wishes, so that he can be home in the spring to take up a vacant professorship in philosophy in Copenhagen.[261] Meanwhile, Bartholin has now been informed by his fellow countryman Peder Lauridsen Scavenius (1623–1685) that the professorship has already been taken by another candidate. Bartholin, who has always perceived philosophy as an auxiliary subject to medical studies, expresses a certain relief about this, but for the sake of his continued travel plans he asks Worm to confirm or deny it.

The journey is concluded in Leiden

The last three months of his almost ten-year study tour brings Bartholin to the city where it began, namely Leiden, where he arrives in the middle of March 1646. Here he receives a letter from Worm dated 16 March with confirmation that the vacant professorship in Copenhagen has been awarded to another. Worm has spoken to the chancellor about another vacant professorship in ethics, and otherwise he reassures Bartholin that he can act as deputy for his grandfather in medical instruction, and he therefore maintains his wish that Bartholin returns home as quickly as possible. But the letter makes Bartholin lose trust in the chancellor and postpone his return home once again. On 11 April, he sends an answering letter, with which he encloses a "copy of our last anatomy book and in a finer binding as intended specifically for our High Chancellor. I would have enclosed a letter to him, if I had thought it worth the trouble of doing so. For with some difficulty, I allow myself to go as far as to fulfil this duty, since I see that the noble lord has entertained other thoughts in his quiet mind than he has allowed himself to express, and that he has abandoned his wish while pretending he had me in mind. Not that I look askance at others' good fortune or pine away when I see them in favourable circumstances, knowing they deserve better than I do; but it is with something quite different from equanimity that I have learnt that it is with empty expectations that I have been duped into hurrying to come home. Therefore, I have now adopted a steadier pace and have begun to curse these ideas about official work. [...] I would like to live an unnoticed life, where I was not in danger of arousing any envy – and it is indeed in my studies that I reap this. Then neither my pen nor my spirit would come to a standstill."

Bartholin builds further on his embitterment in his next letter to Worm from Leiden on 2 May. After yet more thanks to Worm for his love, affection and exceptional solicitude, he continues his crusade against all the others who do not wish him well:

"But as I am so sure regarding your zealous concern, so I commend you as one who is the originator of my welfare, which you could serve best by being eager for my extended stay abroad. I know that yours and my friends' tremendous affection for me does not cease trying to draw me home to the fatherland. But I ask you, my father, why are you thus calling me to my fate? Am I not subjected to sufficient envy here, since it is necessary to endure hostile looks there? I have truly never aspired to envy, but in one way or another I have been given this rough fate. You know that Hofman's and Lauremberg's vengeful spirits will watch over me. Riolan is also angry, because I have not come to an agreement with his father regarding the pudendal nerves. Another one is on the move and argues the mechanism of the breath to Descartes' advantage and against me. Thus, there are stumbling blocks everywhere, and I have noted that there is no shortage of influential persons among you who cannot come to terms with the activities of so many Bartholins at the university, just as if this bond of brotherhood were unheard of there! Consequently, I have had enough of these injustices, though I shall bear with equanimity the deceitful promises that were given to you and me during my absence."

Bartholin asks Worm not to cause the reluctant and deaf "patrons" more trouble for his sake. He has heard rumours that the professorship in ethics has now also been taken by another, which Worm confirms in a letter dated 24 April, and which Bartholin receives after he has sent his own. In his answer of 22 May, Worm promises not to weary Bartholin any more or apply pressure to get him to come home, but he does not refrain from speaking his mind about Bartholin's whining:

"There is no reason for you to complain of envy, since it tends to accompany virtue as the shadow accompanies the body. It was certainly to my mind better what my former teacher Caspar Bauhin, of blessed memory, used to say, that he hoped for envy and scorned pity."

Bartholin's last letter to Worm from his study tour is from Leiden, dated 6 May 1646. Bartholin has entered into a friendship with Worm's friend, the Dutch geographer and director of the Dutch West India Company, Johannes de Laet (1581–1649), who through Bartholin will collect rare exotic objects for Worm's Museum. Bartholin is collecting materials for a second edition of his book about the unicorn, and de Laet has promised him "numerous observations of single-horned creatures, especially the single-horned bird, which is brought here from India." Bartholin's little book about the wound in Christ's side is on its way to press, but he has asked the French philosopher Claude Saumaise (1588–1653) to evaluate it prior to publication. A letter to Saumaise of 28 July 1646 on this topic shows that Bartholin at this time still finds himself in Leiden,[262] while another from Worm to the brother, Rasmus Bartholin, in Leiden, dated 23 October in the same year, shows, that Bartholin is now at last back in Copenhagen after nine years and four months abroad,[263] full to the brim with impressions, knowledge and practical skills and enriched with a broad circle of acquaintances among Europe's leading physicians and scholars. Bartholin never afterwards set foot abroad again – probably never even outside Zealand.

Professor of anatomy in Copenhagen 3

"For one must vary the material, little by little wheedle out of Nature its secrets and not rashly resign oneself to what one has only seen once, when for centuries general experience speaks against it."[1] Thomas Bartholin, 1653

In the first year after Bartholin's homecoming to Copenhagen, no professorship at the University became available. His time was divided between his practical work as a physician and his literary studies. In a letter to his brother-in-law, Bishop Jakob Matthiesen (1602–1660) in Aarhus,[2] on 1 March 1647, he complains that his work of helping other people eats up his time and prevents him from pampering himself with literary activities: "Prescriptions – a terrible word for both the sick and myself – drag me down."[3]

Bartholin's professional skills were also tested on orders from the highest authorities, but unfortunately without success. On 2 February 1647, he writes to Ole Worm, who at this time was living at the court in Nykøbing Falster to treat the sick and drunken Crown Prince Christian (1603–1647), that Royal Physician Jacob Fabricius at the king's command has summoned the physicians in the medical faculty for a consultation about the possibilities for the treatment of hoarseness suffered by the teacher and castle priest, Johan Damianus (1610–1658). Three days later Bartholin informs Worm about the result of the consultation and at the same time tells him of his first visit to the Anatomy Theatre in Copenhagen:

"Your wife said that you will be glad to know the faculty's intention regarding Master Damianus' illness. The origin of his catarrh some people attribute to the head, my cousin to the stomach, Dr. Simon[4] to a laxity of the throat muscles. People agreed about continuing with the treatment begun by Dr. Fabricius, and the younger Dr. Fabricius[5] proposed the construction of a fontanelle.[6] Since the patient is thin, I proposed bloodletting to draw blood away from the lungs. The older Fabricius vetoed this, since he must not be deprived of the heat, and his heart palpitations could be caused by the spleen. We later inspected the Anatomical Theatre together with Dr. Simon, who vainly attempted to wrest the child's skeleton from my cousin."

Despite all his medical duties, Bartholin succeeded in finding time to publish a further two books in Copenhagen. The first, from 14 December 1646, is a 68-page illustrated treatise on midwifery in antiquity, *Antiquitatum veteris puerperii synopsis, opere magno ad eruditos præmissa* (A survey of midwifery in ancient times, a precursor of a large work intended for the world of learning) – a subject that Bartholin, prompted by Walæus, had begun to study during his stay in Leiden. In the book Bartholin gives his opinion on the origin of the Danish word for midwife, "jordemoder", based on a Roman custom (Figure 85).

As the title of the book indicates, during his travels in Italy Bartholin had gathered materials for an almost finished large work on the subject with illustrations of statues, inscriptions, coins, etc., but had postponed publication because he had failed to obtain some promised illustrations from Rome and Gaeta.[7] This large work was never completed because the manuscript went up in flames during the fire at Bartholin's manor, Hagestedgaard, in 1670. Why he had deferred publication so long is unknown, but in his book about the fire he bewails the loss of this "fetus" which he had worked on for more than thirty years. It perished in its own birth. Among other subjects it would have described Jewish, Greek and Roman practices and rituals before, during and after birth.[8] In order to repair the loss, his son Caspar published in Amsterdam in 1675 a 179-page extended edition of the 1646 version with the title: Thomas Bartholin: *Antiquitatum veteris puerperii synopsis. A filio Casparo Bartholino commentario illustrata. Cum Thomæ Bartholini ad Filium Epistola* (A synopsis of birth care in ancient times. Provided with a commentary by his son, Caspar Bartholin. With a letter from Thomas Bartholin to his son). It was dedicated to the lawyer and Royal Councillor Peder Lassen (1606–1681).

Both editions have gained a certain recognition among later writers on the subject.[9]

The second book, published by Bartholin shortly after his homecoming to Denmark, is a 128-page illustrated book about arm-rings in antiquity, *De armillis veterum, præsertim Danorum* (On arm-rings in ancient times, especially in Denmark) with an appendix consisting of Ole Worm's reply to Fortunio Liceti concerning the golden horn (Figure 86).[10] It is dedicated to Christian IV's son-in-law, governor in Norway, Hannibal Sehested (1609–1666). The book's seven chapters deal with the concept of arm-rings and the origin of the word *armilla*, on which parts of the arm the ornaments were worn, the ornaments' materials (gold, silver, copper, etc.), the form and appearance of the rings, both sexes' use of arm-rings and wedding rings and the various use of arm-rings in various peoples.[11]

Judging by its organization and contents, the book seems inspired by Fortunio Liceti's large-scale work *De Anulis Antiquis Librem Singularem* (Single volume

Figure 85. In his work on obstetrics in antiquity, *Antiquitatum veteris pueperii synopsis*, Copenhagen 1646, Thomas Bartholin was the first to give the later recognized explanation of the origin of the Danish word "jordemoder" (midwife). Based on Roman custom, he explains, the midwife is the woman who, as a sign of recognition from the father, takes the newborn child from the ground. He adds the words "unde terræ matres dictæ" (therefore they are called "jordemødre").

On the shown bas-relief from a sarcophagus at Palazzo Sachetti in Rome, reproduced in Caspar Bartholin the Younger's *Expositio veteris in pueperio ritus* (Exposition of ancient customs and practices at births), Rome 1677, p. 65, the new mother is seated on the left, looking at her newborn child, while the midwife sits on her knees, bent over the child, ready to lift it from the bath. A breastfeeding woman stands with a sheet ready to receive the baby, while the two women on the right note the time and date of the birth.
The Royal Library in Copenhagen.

on ancient finger rings), Udine 1645, about classic antique finger rings.[12] In the book's introduction, Bartholin lists his published books, and in addition he mentions a planned anthology of his letters and epigrams already published in various writings, which was never realized.[13] Furthermore, he mentions an apology to the German theologian and Catholic convert Barthold Niehus (1590–1657) concerning differences on points in Bartholin's book about the wound in Christ's side – a debate in which Claude Saumaise also took part, aligning himself with Bartholin.[14] Niehus believed that Bartholin, with his comments on the blood and water that flowed from the wound in Christ's side while he hung on the cross,

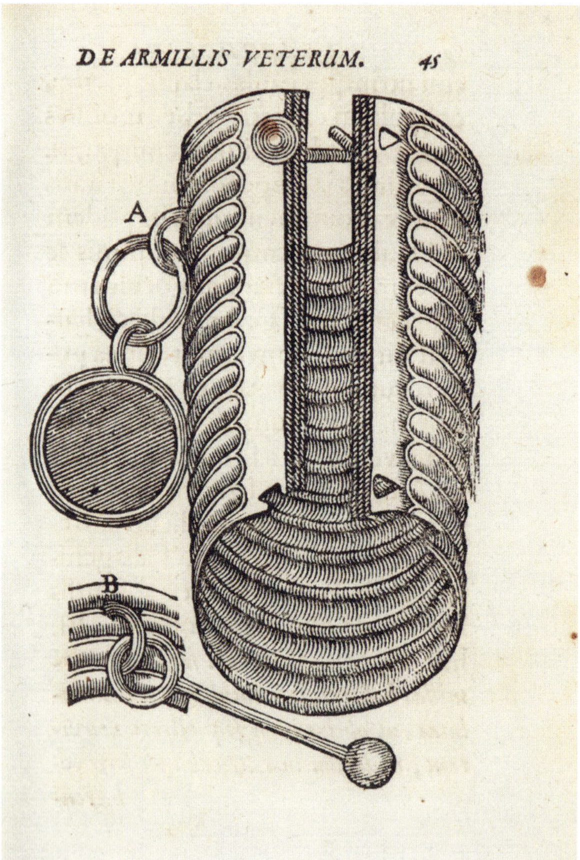

Figure 86. a. Bangles from the end of the Bronze Age, c. 6th–7th century BC.

The bangles originate from the area around Kolding and were originally kept in Ole Worm's Museum.
National Museum of Denmark.
Photo: Lennart Larsen.

b. Engraving of the same bangles on p. 45 in Thomas Bartholin's book on bangles in ancient times, *De Armillis Veterum, præsertim Danorum*, Copenhagen 1647.
The Royal Library in Copenhagen.

shook up the Christian gospels and put a question mark over Christ's martyrdom on the cross, but like the good, orthodox Lutheran that Bartholin was, guided by Worm and Saumaise he allayed the potentially dangerous attack from his opponent with an assurance that everything had been written with the best intention and as an invitation to end the public dispute that subsequently happened. Bartholin had previously, in the aforementioned letter to Jakob Matthiesen, accused Niehus of trying, both publicly in writing and by secret slander, to ascribe impious ideas to him on a false basis.

Differently positive tones about the book concerning the wound in Christ's side sounded from Guy Patin in Paris who, in a letter to Bartholin dated 20 June 1647, calls it an "extremely elegant book" and requests him to write about his travels in Europe and complete his earlier proposed written revenge against the "peevish and barking" Caspar Hofman.[15]

Professor in the Faculty of Philosophy

Soon afterwards, Bartholin had something else to think about. On 17 July 1647 his aged eighty-six-year-old grandfather, Thomas Fincke, demanded a deputy for his professorship in medicine, and ten days later Bartholin was assigned the post with the chancellor's consent and a promise to seek Bartholin's promotion, which in practice would mean the next available professorship.[16] Bartholin did not have to wait long. On 8 October 1647 his uncle (the husband of his father's sister Dorothea), professor of mathematics and astronomy, Tycho Brahe's pupil, Christen Sørensen Longomontanus, died. On 16 October, Bartholin was called "in to the consistory and was ordered to dispute pro professione infima."[17] On 12 December of the same year he disputed in the university's large auditorium about Jesus' healing of the paralysed man at the Poll of Bethesda in Jerusalem (John 5, 1-9),[18] and on 15 December, he took the oath and assumed a professorship in the Faculty of Philosophy.[19] In his opening speech to the disputation, Bartholin recalls Longomontanus, who taught the students about the universe (macrocosm), in contrast to himself, who will now teach them about mankind (microcosm). His aim is to combine medicine with philology. With the story about Jesus' healing of the paralysed man, he wishes to show that God competes with Nature and shows himself as its lord.[20]

The sources give scant information on Bartholin's activities in the new professorship, but he eventually has time to form a response to Caspar Hofman's attack mentioned earlier. In each of the months from March to July 1648 he organizes in the university's large auditorium five disputations in all, answering Hofman with different opponents and publishes all of them during the same year in Copenhagen in a book of 136 pages with the title *Anatomicæ Vindiciæ Cl. V. Casparo Hofmanno* (Anatomic counterclaims directed against the renowned Caspar Hofman). The book is dedicated to Lord Great Chamberlain Corfitz Ulfeldt (1606–1664), the real leader of Denmark during these months between Ulfeldt's father-in-law, Christian IV's death, and Ulfeldt's brother-in-law, Frederik III's assumption of power. The book closes with two letters from Marco Aurelio Severino against Hofman, together with a poem in tribute to Bartholin's book on light sent out by living creatures by his friend, "our time's leading poet" Vitus Bering. In the book's preface, Bartholin praises himself for having been the first to have ventured to attack Hofman,[21] who had died that November and therefore had not had the opportunity to make a reply. For each of the disputations, Bartholin gave a preface, which he published in 1668 in his work including speeches on various subjects.[22] In the large auditorium on 24 January 1649, Bartholin gave a talk to the students about the relationship between theology and medicine and invited them to take both theological and medical studies.[23] Although Bartholin had formally taken over a professorship in mathematics from Longomontanus, there is no indication in the sources that at any point he instructed his students in mathematics or astronomy. For his professorship in medicine, he still had to wait a little.

The Anatomy House in Copenhagen

At the north-west end of what is now Frue Plads, near the corner of Nørregade in the inner quarter of Copenhagen, was the Anatomy House, a two-storey building in Dutch renaissance style (Figure 87). The building was erected at the instigation of Christian III (1503–1559) as a college for students and was finished around 1545. After extensive rebuilding in 1553 it came to house the University Library along with the old auditorium, Collegium Vetus, and additionally, after rebuilding in 1643, the Anatomy House (Figure 88).[24] It was established in 1644

Figure 87. Top: The Anatomy House in Copenhagen. The frontispiece of Thomas Bartholin's *Domus Anatomica Hafniensis*, Copenhagen 1662. Engraving by Albert Haelwegh. 13.5 × 9 cm.

The Anatomy House is seen from the university's courtyard. To the left is a section of the cemetery wall with an exit that led to the former Frue Kirkegaard (the Cemetry of Our Lady), the present Frue Plads (the Square of Our Lady).

To the right is the stair tower, which led up to the first floor and the attic. Above the house hovers its motto: "Here, too, the gods have been."

Bottom: The anatomical theatre on the ground of the house. In the centre of the theatre is the revolving dissection table, and in front of the back wall the skeletons "Adam and Eve" may be seen on either side of the "Tree of Knowledge."
The Royal Library in Copenhagen.

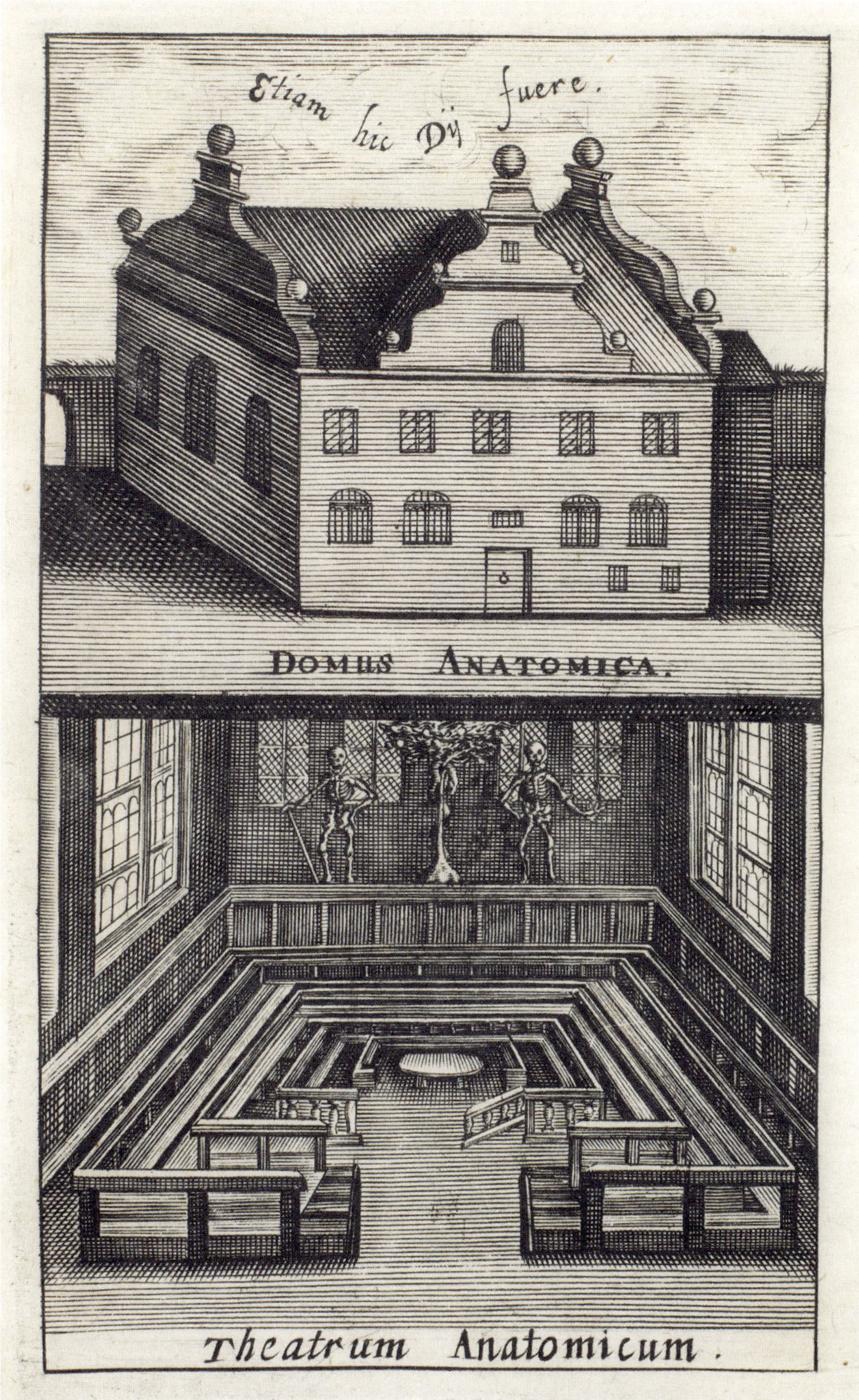
Etiam hic Dÿ fuere.
DOMUS ANATOMICA.
Theatrum Anatomicum.

Mylorg reconst.
J. Hansen del. 1879

by Christian IV, according to his wish that "the young people who flocked to the University of Copenhagen to study medicine should with greater success be able to acquire knowledge of the individual parts of anatomy."[25] To put it somewhat more precisely, it was a building where Danish medical students could learn about human anatomy through dissection of human bodies. Hitherto, Danish medical students and physicians, like Bartholin, had been forced to learn these skills during study trips to the leading European universities, primarily in Holland, France and Italy. In Copenhagen they had no buildings for dissections but taught the students anatomy from illustrations in books. The university did not even have a skeleton for teaching purposes.[26] When the professor Anders Christensen (1551–1606) carried out dissections of human bodies in his own house as part of his teaching, the first in the history of the university to do so, he was met with disgust and contempt and shut out from polite society, so that he soon after had to give up his practice.[27] In his short biography of Anders Christensen in *Cista Medica Hafniensis* Thomas Bartholin writes:

> "He was lucky in the performance of his medical profession and, something that was exceptional at that time: he showed the greatest interest in anatomical practices. When, in his time, he desired to resume these practices at this university, and wished to dissect a male body for the advantage of his young students and for the embellishment of the university, he would, if he had not backed away, almost have been forced to keep himself away from the better classes of society, because people were wholly unaccustomed to dissections. Such great loathing for a highly useful discipline prevailed in his time."[28]

Figure 88. Reconstruction of the university's courtyard as it looked in the early 1600s. Woodcut by J.T. Hansen from a model by R. Majborg in Henning Matzen: *Kjøbenhavns Universitets Retshistorie 1479–1879*, Copenhagen 1879, Vol. 2, Appendix, p. 7.

In the background is Vor Frue Kirke (the Church of Our Lady). To the right is the oldest auditorium building, Collegium Vetus, flanked by the first community building in half-timbering. To the left is the new auditorium building, Collegium Novum. The building in the centre in Dutch Renaissance style – here reproduced from the illustration in Thomas Bartholin's *Domus Anatomica Hafniensis* from 1662 (Figure 87) – was built as a student college during the reign of Christian III (1503–1559) around 1545. After a remodelling in 1553, it housed the university library on the first floor and an auditorium on the ground floor. After a further remodelling in 1644, the building became the Anatomy House in Copenhagen with the university's anatomical theatre on the ground floor and from 1657 the university's museum of natural history on first floor. After Thomas Bartholin's death, the Anatomy House remained largely unused until it, like most of the university's other buildings, perished in flames during the great fire of Copenhagen in 1728.
The Royal Library in Copenhagen.

In his short 1674 piece *De Anatome Practica* (On practical anatomy), or in modern terminology "pathological anatomy," that is to say, the teaching about the causes of illness from examinations of sick or deceased people's organs, Bartholin expresses the scepticism about dissections and post-mortems that he had encountered in the Danish population:

"In Copenhagen I myself was formerly interested in grasping the same favourable opportunity, since I had obtained permission from the authorities, which I later, however, discouraged by gossip and protests, saw myself compelled to abandon. For who will expose himself to the mockery of all the people, when an ice of mistrust and ill-will must first be broken down?"[29]

Later, he makes clear what his dislike often consisted of:

"Nor in our countries is there anything unreasonable in giving latitude to people on this point, when they have the right attitude as to what can be of the greatest advantage, although one quite often meets great difficulties with people who, on the strength of a certain piety or superstition, oppose this dissection in the thought that it goes against the peace of the dead, as if the worms show greater mercy to the corpses. However, it oftens happens that the common good of the survivors is preferred over the private peace of the corpse."[30]

To oppose this scepticism, public dissections and demonstrations were introduced in the Anatomy House from the beginning. The first anatomy professor of the house, the North German Simon Paulli (1603–1680) (Figure 89), published a bombastic invitation programme to his first public dissection in March 1645, that of "the naked corpse of a newly-wed young woman, who was disgraced (for shame!) by an unlawful union."[31] Here he emphasized the dissection's important aim and attempted to settle the prejudices and the objection ("the arrogance of the brutish and spiteful mob"[32]) against them. The young woman was executed by decapitation. For a payment of a single rix-dollar or six Danish marks, "remarkable and noble paragons of all ranks," "lawfully created Doctors of this Royal University of Copenhagen, Masters of Liberal Arts and other men of letters, and also the magnificent Lords Gentlemen, Members of the Senate, distinguished citizens, surgeons and all others devoted to the liberal arts in this glorious municipality of Copenhagen," "the excellent and distinguished gentlemen students of every faculty" and "those, especially, who are dedicated to studies in theology and medicine, together with the older apprentices in surgery of this celebrated municipality" could acquire an entry token, a "tessera anatomica," and take a seat among the audience.[33] The dissection continued for fourteen days.[34] The aim of the fee, which was equivalent to a week's wages for a manual worker,

Figure 89. The North German Simon Paulli (1603–1680) as a 46-year-old professor of anatomy, surgery and botany. Engraving by Albert Haelwegh from a model by Karel van Mander. 17 × 14 cm. From Paulli's *Flora Danica. Det er: Dansk Urtebog*, Copenhagen 1648.

The text on the book with the illustration of a human skeleton, which Paulli holds in his hand, reads: "This is how you will be after death, O human! Remember that you are but a shadow." The Greek text on the shield above Paulli's head is his motto: "He who avoids difficulties, avoids honours." The German text from David's psalm 71:5 under the portrait "Du bist meine Zuversicht, HERR, HERR, meine Hoffnung von meiner Jugend an" reads: "You are my confidence, Lord, Lord, my hope since my youth." Paullis' great interest was botany. In the field of anatomy, he contributed nothing significant other than his role in the creation and inauguration of the Anatomy House. Ole Worm did not like him, and although Thomas Bartholin speaks of him respectfully, their relationship was not without controversy.
The Royal Library in Copenhagen.

was, according to Thomas Bartholin, "to keep away the common crowd."[35] Later he cancelled the fee himself, following the example of Padua.[36]

Since the dissections were generally performed over several days, and in some cases weeks, they had to take place in the winter season, so that the corpses did not rot too quickly. Almost all the bodies dissected during the lifetime of the Anatomy House were those of executed criminals, whose lifeless bodies had been dedicated to that purpose by the king before their condemnation. When Simon Paulli began his public dissections, Christian IV had already sent out a command, that "when any villain in the different fiefs has been condemned after lawful proceedings to be hanged before the end of March, then the provincial governors must send the culprit to Copenhagen to be executed, so that the body may afterwards be delivered to the Anatomy House."[37] To ensure that the dissecting surgeon had as intact a body as possible, the king frequently altered the criminal's punishment from one method to another, for example from decapitation and breaking wheel to the somewhat milder hanging. Whereas otherwise only hanging had been employed as a method of punishment for men who had committed rougher forms of theft, the breaking wheel had become used as a penalty for murder, clandestine childbirth in connection with child murder, crude or repeated theft committed by a woman, robbery, assault on the city's sentries as well as rape committed against girls.[38] By breaking wheel, the actual execution method was beheading with the sword, after which the lopped-off head was placed on a pole, while the body was laid on a wheel. The execution of particularly well-known criminals in public places was often a great draw, where it could be difficult to hold the spectators at a distance. When Frederik III changed a penalty from breaking wheel to hanging, not only did this mean a milder punishment, which could offend people's legal conscience, but the public also missed the theatrical performance of such an execution. In return, you could sign up for the "performance" in the Anatomy Theatre, if otherwise you had the means and the patience. By and large the success of the Anatomy House was due to the king's support, interest and participation.

In 1639 Simon Paulli, on the arrangement of his father-in-law Jacob Fabricius, Christian IV's personal physician, was called from a professorship of medicine at the University of Rostock to an extraordinary professorship of anatomy, surgery and botany in Copenhagen.[39] The chief purpose of this extraordinary (third) professorship was to upgrade the university's efforts in the subject of anatomy, and Paulli was paid directly from the royal treasury with 200 rixdollars per year.[40]

In a letter to Henrik Fuiren in Padua, dated 23 April 1641, Ole Worm describes the circumstances relating to Paulli's appointment: "As to the new anatomist the case is this: As soon as Dr. Jacob Fabricius was called to the court, he contrived with competence (and in this lies his strength), that His Royal Majesty should set aside a new stipend of 200 rix-dollars per annum for a professorship of anatomy. For the discharge of this duty his son-in-law, Dr. Simon Paulli, has been appointed at the same time. This winter he has expounded Galen's book on the bones, but no dissection has taken place, for he puts forward the reason that he lacks a room for such studies."

Although the establishment of the Anatomy House in Copenhagen was largely Simon Paulli's claim to fame, he is not reckoned to have been a particularly outstanding anatomist. His greatest interest was botany, and he is best known for the first edition of the Danish flora, *Flora Danica. Det er: Dansk Urtebog* (Flora Danica. That is: The Danish Herbal), which was published in Copenhagen in 1648 on the initiative of Christian IV, and whose purpose was to describe herbal remedies for the advantage of "the common man," who did not have the means of finding a physician. Furthermore, Paulli must have noticed the resistance shown to this professorship appointment from other academics, who had no wish for him to be there. This applied most of all to Ole Worm, who wished for Bartholin's occupation of the chair as fast as possible, and whose attitude to this affair appears clearly in a letter to Henrik Fuiren, dated 4 February 1643: "We are now fully occupied with the building of the Anatomy Theatre, and when the whole thing is completely finished the director [Simon Paulli] promises something wonderful from it, although in the last two years he has been with us, he has not dissected so much as a dog."

After the death of Christian IV on 28 February 1648, Paulli withdrew from anatomy to concentrate on botany and later gained positions as court and royal physician,[41] and Thomas Bartholin took over Paulli's professorship of anatomy and the responsibility for the Anatomy House and the teaching there.[42] Yet first, on 11 April 1649, Thomas Bartholin obtained the third professorship of medicine (*Medicus Tertius*).[43] Bartholin did not receive a written appointment, but the university's rector reported that it was His Majesty's wish that Thomas Bartholin should be received into the Faculty of Medicine as the third ordinary professor and teach medicine and anatomy, but until further notice he should continue with his salary and his privileges from the Faculty of Philosophy, where Vitus Bering should take over his lectures.[44]

With his new post, Bartholin received a professor's residence, "The Residence," next to the Anatomy House and an adjacent garden, which was located next to the present-day Krystalgade.[45] More precisely, the professor's residence was located between the university's present-day assembly hall and the library building facing Fiolstræde.[46]

Under Bartholin's direction right up to 1661 the Anatomy House experienced its period of glory, and it is no exaggeration when, in his book about this building, *Domus Anatomica Hafniensis brevissime descripta* (The Anatomy House in Copenhagen very briefly described), published in Copenhagen in 1662, he immodestly writes about himself:

"This eminent man [Simon Paulli] was next succeeded by Thomas Bartholin, previously Royal Professor, who also shed radiance to the utmost of his ability on the Anatomy Theatre through a variety of dissections, demonstrations, lectures, etc., from 1648 until 1661, upon which His Most Serene Royal Highness very graciously granted him freedom from his academic labours."[47]

Bartholin's success with the Anatomy House was due not only to his own unquestionable abilities and industry as well as the fact that anatomy enjoyed the status of a fashionable science throughout the whole of Europe in the 1600s, but also largely to Frederik III's support. The shy and introverted king (Figure 90), who was crowned on 23 November 1648, shared Bartholin's interest not only in science and anatomy, but also in books, curiosities and the grey area between science and superstition.

His extensive book collection came to form the basis of the Royal Library, but his collection of various peculiar natural objects, weapons, antiquities, *objets d'art*, paintings, scientific instruments and clocks, items from foreign parts of the world, coins, medals, antique specimens, models of fortresses, ships and new mechanical inventions resulted in his formation of Det kongelige Kunstkammer (the Royal Cabinet of Curiosities) about 1650 (Figure 91), inspired not least by Ole Worm's Museum Wormianum.[48] Moreover, like his father he acquainted himself with alchemy.

Before Bartholin's first public dissection in the Anatomy House in March 1649, he sent out the following prospectus:

"**To the dissection of a male corpse**

May it be auspicious, favourable and beneficial to human beings, that with the blessing of the Supreme Deity, with the bidding of Our Most Merciful King, with the permission of the highly-renowned Chancellor, and lastly with the consent of the Lord Rector Magnificus and the Faculty of Medicine, DR. THOMAS BARTHOLIN, Public Professor, will embark on the dissection of a male corpse tomorrow, on <date>, <year>, at one o'clock in the afternoon.

Figure 90. Frederik III (1609–1670), king of Denmark and Norway 1648–1670, painted by Abraham Wuchters (1608–1682), c. 1645–1650. Oil on wood. 52.5 × 39.5 cm.

The shy and introverted Frederik was Thomas Bartholin's main patron. He shared Bartholin's interest in books, curiosities and natural science and was present at several of his public dissections in the Anatomy House. By supporting the Anatomy House and ensuring access to human corpses for dissection, he created the necessary working conditions for Bartholin's exploration of the lymphatic system. Through his support of Bartholin's early retirement from the practical part of his work as a professor at the university, he provided the ideal conditions for Bartholin's comprehensive literary production. National Gallery of Denmark.

On subsequent days, provided God and his health allow it, he will continue with this so that listeners and spectators who take pleasure in it may come to hear and observe; those who know and like to recognize that they are mortal, who admire the noble study of anatomy and cherish affection for themselves and the dignity of their bodies, he INVITES and SUMMONS to the Anatomy Theatre with the utmost courtesy and an equal measure of affability."[49]

Frederik III had not only put the body of the executed man at their disposal, but "twice appeared in the Anatomy Theatre, observing the proceedings attentively, and thoughtfully exchanged opinions about the principal internal organs of the body with those standing by."[50] About the dissection itself, Bartholin writes the following, showing that even in his time physicians were already aware of the negative consequences of tobacco-smoking on health:

"A formal dissection in 1649

In the presence of Denmark's glorious King Frederik III, our most gracious lord – we sincerely wish, that as God loves him, so may he also love us – and at the same time the realm's most distinguished men and worthy spectators of all ranks – I cut up in March in our Anatomy Theatre at the University of Copenhagen with the aim of a formal dissection the body of a well-built and strong male person. Among various noteworthy observations, I found among other things three small fatty stumps (both glutinous and whitish) in the left ventricle of the heart. They were, I believe, formed as a consequence of the heart's too strong heat and

Figure 91. In the Art-chamber of Frederik III there was an approximately one-metre-high human skeleton made of ivory in an ebony cabinet. Both were made by the first curator of the Art-chamber, Niels Gyntelberg (1626–1661). In the Art-chamber's inventory from 1673, written by Thomas Bartholin and others, the skeleton and the cabinet are mentioned with the following text: "Et Sceleton af Elphenbeen i et Ibenholtzhuus med Windwer for."

Originally, the skeleton held the symbols of death, the scythe and the hourglass, in its hands. It stands on a swivelling plate of ebony, so that you can see it from all sides. On the top of the ebony box is carved in ivory letters a poem in Latin by Thomas Bartholin: "Recognize the exaltation of art, despise Nature's corruptibility: While the latter turns the body to ashes, the former shows how it is joined together. Were it by fate given life and breath, as well as the gift of sense, then nothing more marvellous would be given on the whole earth. Had we still in our grave so great a glory, then every man would wish, that he had no flesh on his bones." National Museum of Denmark. Photo: Kit Weiss.

fat in the blood. Moreover, there was no valve in the right kidney vein, and while all the inner and outer parts were sound, only the lungs were strengthless and crumpled, and they stuck to the ribs. The liver was unusually large, three Danish pounds in weight, and the spleen weighed twelve and a half ounces. It was my opinion that the lungs had suffered damage because a much-too-frequent use of tobacco had conveyed catarrh to them from the head. In the brain there was no damage to be seen or any trace of blackening, though the person concerned was known by many for his heavy smoking of tobacco."[51]

When Bartholin uses the words "the realm's most distinguished men", he means, among others, the King's Chancellor Christen Thomesen Sehested and the Reichshof-Master Joachim Gersdorff (1611–1661). Moreover, Jens Lauridsen Wolf (1583–1660) mentions in his *Encomion Regni Daniæ* of 1654 that the state governor Joachim Gestrup, the treasurers Jørgen Rosenkrantz (1607–1675) and Peder Vibe (1596–1658) and many other nobles and professors were present at this dissection.[52] Frederik III was also present at many of Bartholin's later public dissections.

Bartholin describes the interior of the Anatomy House as follows:

"On the lower floor of this house given over to anatomical purposes is found first of all, in the middle part of the building, an entrance-hall; four doors open from it onto various locations. Through the first door on the right, we are brought into a preparation room, where corpses are made ready on a rectangular revolving table by an anatomist and his assistant. There is a stove here to prevent the dissectors' hands being pinched by the intense winter cold, and besides this a tiny room containing a case and a table, so that the anatomist can store away his instruments, gather himself together in this out-of-the-way corner, and dress properly before a public demonstration.

The second door, adjacent to the first, leads to a kitchen, in which bodies are washed before dissection and, once the demonstrations are ended, the flesh is stripped from their bones, so that these may be united to form a skeleton. For this purpose, they have had to erect a huge stove to provide a fireplace, and in the corner is situated a vast copper cauldron, in which the bones are boiled. There is another door, too, which gives access from this kitchen to the preparation room; through it the corpses are carried, so that there is no need to pass across the entrance-hall with them. The third door on the left opens into the Anatomical Theatre itself, and on it are inscribed the verses of Michael Kirstein:

Here, wayfarer, you see either bones or severed bodies.
Here our skill dissevers and unites the work of Nature.

The Anatomical Theatre or, more properly, amphitheatre is, according to the nature of the place, square and spacious with a very high ceiling, more serviceable than splendid, surrounded on the periphery by four rows of wooden benches gradually rising from bottom to top. The lowest accommodate professors, Doctors of Medicine and the more distinguished visitors. At the top, shelves are arranged for laying the skeletons on. Each row of benches has its own seats, which can be lowered by means of hinges, to enable tired or weak spectators to relax, and raised up for those who prefer to stand. In the middle of the lecture hall for the purpose of exhibiting corpses there is a table, which can be twisted and rotated to any angle, so that every slightest occurrence to do with the body may be shown to all the viewers. The table in the well of the theatre is separated from the lowest benches by a balustrade, to prevent the dissector and his assistant from being hampered during a demonstration by the crowd's repeatedly pressing forward.

Through the fourth door in the entrance hall, you can climb by a spiral staircase to a box above, where a lattice looks over the theatre; from here His Most Serene Highness, KING FREDERIK III, our most sanctified lord, has more than once watched our demonstrations with a gracious eye. There may also be found adjacent in the same place seats for the courtiers and for others who wish to view the proceedings here from a distance."[53]

More recent research has shown that the king's seat in the Anatomy House, rather than being a room on a higher floor, was a two-part locked box set high on the wall, which spanned the auditorium and the entrance-hall, and to which entry was gained via a spiral staircase, which was situated inside the auditorium.[54] The latticework facing towards the theatre ensured that the king and his escort could remain unperturbed in the private box. The king probably had a good view of the theatre from the box's elevated position, but the distance to the dissection table was too great for him to be able to follow the proceedings in detail, so if he wanted to, he would have had to go down among the other spectators, as Bartholin described above.

Simon Paulli, who had devised a method to produce beautiful white skeletons,[55] had fitted out the theatre with skeletons produced from animals and human beings he had dissected there. At one of the end-walls he had placed a man, Adam, and a woman, Eve, and between them an apple tree, which represented the tree of knowledge. Moreover, all around the room were skeletons of a baboon, another monkey, a swan, a goose, a marten, a hedgehog, a cock and a hen (Figure 92). Each of the skeletons was supplied with an epigram, authored by Paulli's assistant, Michael Kirstein (1620–1678).

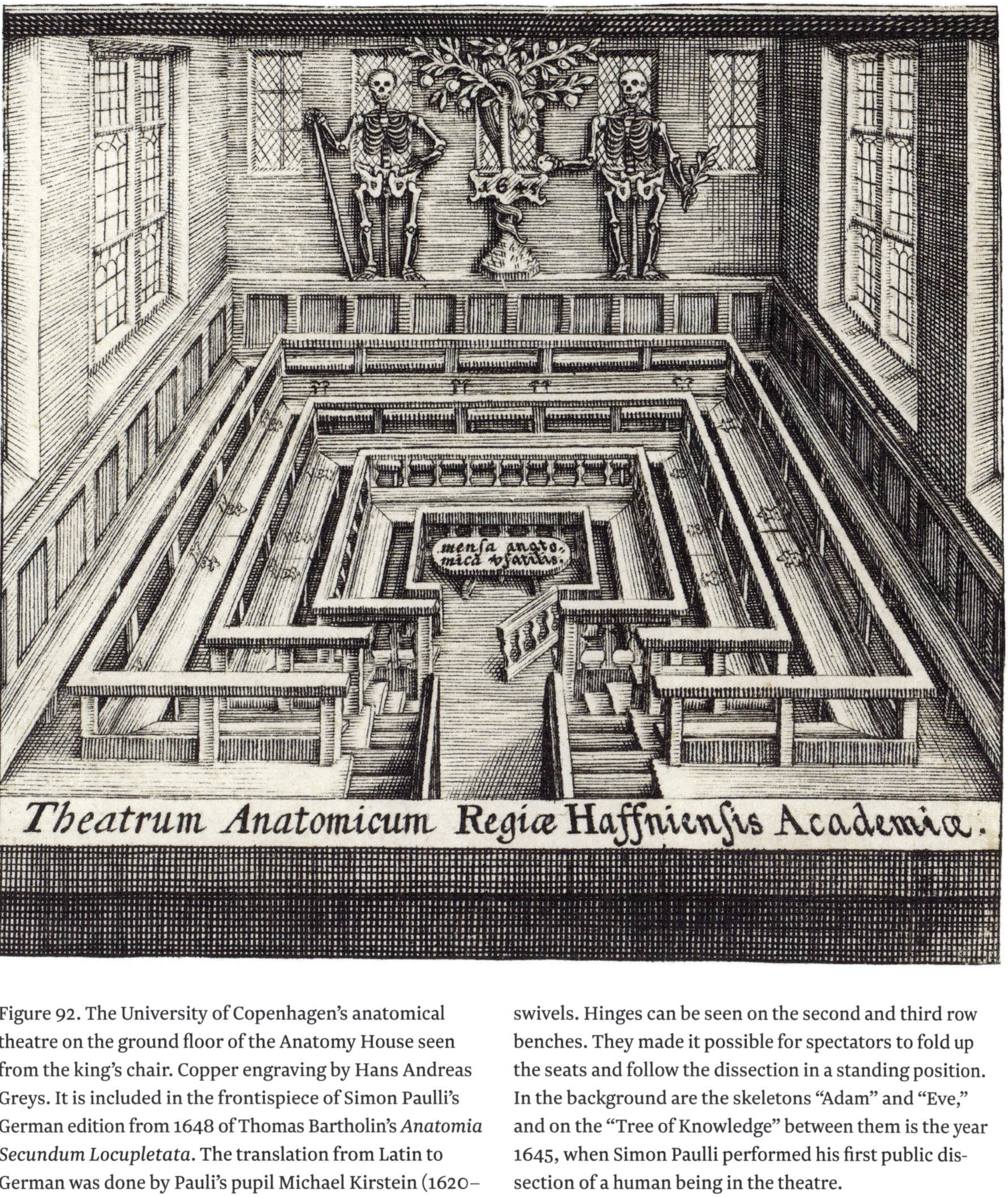

Figure 92. The University of Copenhagen's anatomical theatre on the ground floor of the Anatomy House seen from the king's chair. Copper engraving by Hans Andreas Greys. It is included in the frontispiece of Simon Paulli's German edition from 1648 of Thomas Bartholin's *Anatomia Secundum Locupletata*. The translation from Latin to German was done by Pauli's pupil Michael Kirstein (1620–1678). The Latin text on the dissection table states that it swivels. Hinges can be seen on the second and third row benches. They made it possible for spectators to fold up the seats and follow the dissection in a standing position. In the background are the skeletons "Adam" and "Eve," and on the "Tree of Knowledge" between them is the year 1645, when Simon Paulli performed his first public dissection of a human being in the theatre.
The Royal Library in Copenhagen.

It was not only the bones from the dissected creatures that found further use. The poorly paid anatomical assistants who helped the prosector prepare the bodies before dissection could earn an extra shilling by preparing the skin from the body after the end of the dissection. In 1657 Bartholin describes the procedure for such a preparation of human skin for use as belts for women in childbirth, hysterical individuals, epileptics and suchlike, which he himself had made use of in his practice as a physician.[56]

The wording for the invitation to Bartholin's dissection of a woman needed a particular phrasing:

"**To the dissection of a woman:**
DR. THOMAS BARTHOLIN, Royal Professor of Anatomy, through the amicable permission of the authorities, will tomorrow, ‹ date and year ›, at one o'clock in the afternoon exhibit a naked corpse of the second sex, which he will inspect with chaste eyes and examine with virtuous hands, displaying purely and reverently her breeding ground and the natural vessel of milk for her children; he asks those of a suitable age who have the disposition and knowledge to view with the same attitude and gaze these Floralia, given the kind leave of Cato, and request those who can contemplate modestly the lovely, tender naked body of a Venus, paying respect to the cradle of their own births, with decency and openness, to appear there with all possible delicacy, but also the confidence of worthy feelings, abundant courtesy and propriety in thought and deed."[57]

In March 1650 Bartholin dissected at a public dissection in the Anatomy Theatre a newborn child, who had been murdered by his mother, or as Jens Lauridsen Wolf states it: "A little child, who had been murdered and cast out of the Church of Our Lady."[58] On that occasion the invitation ran as follows:

"**To the dissection of a newborn child**
DR. THOMAS BARTHOLIN, Public Professor, will, with God's blessing, the wish of His Royal Majesty's distinguished Chancellor, and the agreement of the distinguished Rector of the University together with its governing body, tomorrow, ‹ date and year ›, begin the dissection of a newborn child with the necessary dispatch. Every supporter and devotee of dissection for whom it is a pleasure to know in more detail the differences between a newborn child and an adult, he INVITES and SUMMONS to the Anatomy Theatre at one o'clock in the afternoon with all the energy and courtesy he may command."[59]

Four years later Bartholin gives a more detailed description of this dissection in his work on rare medical cases:

"**Dissection of a newborn girl in 1650**
A completely newborn girl had been left in the Church of Our Lady. Her mother, who was just as wicked as Medea, envied the newborn child its life, which she had granted it in the womb, and had strangled it by twisting its neck. In the Anatomy Theatre I dissected the girl publicly in March, while many were watching, so that for my audience I could demonstrate the difference between a newborn and an adult person. In this connection I pointed out a long series of details, which might deserve to be noticed by the readers."[60]

Afterwards Bartholin describes 22 different features in which this newborn child differs from an adult person. Among other details he points out:

"The umbilical cord, which was two feet long, still hung outside the child, together with the umbilical vessels, and dripped the fresh blood that had streamed through it. The embryo's first ureter [urachus] was inaccessible and gave neither above nor below entry to a lancet. [...] The stomach was empty and as large as a walnut. [...] The large intestine was, like the rectum, filled and swelling with black and somewhat coarse excrements, which were yellow along the last piece of the small intestine. These excrements are called meconium. [...] A spleen without strength, of the same size as the stomach. The liver was so large that it filled the whole area under the midriff. The gall bladder was more oblong than capacious and had taken its colouring from the gall. The suprarenal glands were quite large and of the same size as the kidneys, but were triangular, completely hollow and very sizeable. They contained serum and blood and were supplied with less fat than the kidneys. [...] In the breasts there were still no glands, but only fat and an undeveloped nipple. The kidneys were separated by numerous very large glands, but this observation was rather unclear, as it usually is in newborn children. [...] In the heart there was a large, clear connection [*foramen ovale*] between the large *vena cava* and the lung veins.[61] But the small channel [*ductus arteriosus*, see Figure 95] I did not find, even though I sought for it assiduously. Perhaps the reason was that the quite large connection had adequate capability to transport the blood to the right ventricle by itself. The thymus stretched above the heart like a very large gland in three parts, through which the rising *vena cava* had its course. Since in these glands no large vessels have been found during the dissection, it is difficult to believe that the seed for fiery anger is to be found there. [...] As far as the skeleton is concerned, I am not saying anything about the difference between a newborn and an adult human being, because this is demonstrated often enough, and my own anatomy book can be purchased everywhere."

When you consider the dimensions of a newborn baby, you must admire the dexterity, which must have served as the basis, for example, of Bartholin's examination of the pancreas's excretory duct:

"I could only observe the pancreas's duct through some small threads. It clung together with the duodenum at its orifice, which I certainly found when I pushed a lancet in from the duodenum, but because of a small valve I could not force my way in."

Bartholin's aim with the dissection, to show his audience the differences between the child body and the adult body, illustrates his visionary understanding of the importance of these differences in a medical context, and it indicates that at this early stage in the history of medicine he had thoughts about the necessity of an independent specialty in children's diseases, i.e. the subject of paediatrics. There is also evidence of such thoughts in his dissertation on a diphtheria epidemic among children in southern Italy.

The tragic history of the strangled newborn child did not end with Bartholin's dissection of it. In 1653 the child's mother was taken into custody, judged and executed, and Bartholin had her body delivered for dissection in the Anatomy Theatre on 21 December the same year. Later he published his introductory remarks to his anatomical demonstration of her body.[62]

In the following years, Bartholin dissected energetically in the Anatomy House, both at open, well-attended public demonstrations and at closed, private ones, but he also dissected under private conditions in his own professorial residence and at certain occasions in that of his brother Bertel's.[63] Not only human beings but a succession of different animals were dissected, including a deer, a hare, a swordfish, a lumpfish and a porpoise. The latter dissection took place in the Anatomy Theatre in the king's presence. Pigs and cows, which had been delivered to the professor's residence as part of Bartholin's salary from the University, were also submitted to anatomical examinations in connection with their slaughter.[64]

One of Bartholin's most curious dissections took place in the royal chicken run on 10 April 1651 at the direct command of Frederik III. Bartholin gives an account of the incident in the first volume of *Historiarum Anatomicarum Rariorum* in 1654,[65] and mentions it in a letter to professor in logic and physics in Groningen, Martin Schoock (1614–1669), on 15 July 1662.[66] It was the dissection of the nobler parts of a royal cock. It was postulated that the said cock had laid an egg, which was smaller than a hen's egg and larger than a pigeon's egg. Pertinent to the story is the fact that in Bartholin's time it was the general belief, even among scholars, that old cocks could lay an egg, from which could be incubated a so-called basilisk, the king of reptiles (Greek, βασιλεύς – a king), a fierce, poisonous, dragon-like fabulous creature with wings and an appearance like a mixture between a dragon and a snake, or a cock and a snake.

Since both the king and Bartholin were interested in the phenomenon, and the possibility was now offered of getting it examined more closely, the king ordered Bartholin to appear in the royal chicken run and investigate through dissection, whether the cock was at all in possession of the necessary egg-laying organs. As could be expected, Bartholin neither found ovaries, fallopian tubes or a uterus, or anything resembling them, but quite normal male sexual organs. In order not to incur disgrace with the king and from those of his people who adamantly maintained that the cock had laid an egg, Bartholin allowed the question to stay open with a cryptic explanation, that old rotting semen in the cock could have produced the egg, and he even pointed out an enlargement in the cock's intestine, where the egg could perhaps have lain.[67] Although it was not unusual for scientists in Bartholin's time to make a compromise with their belief so as not to fall into disfavour with the authorities,[68] one cannot blame Bartholin's German friend, Ludwig Kepler (1607–1663), son of the famous mathematician and astronomer, Johannes Kepler (1571–1630), for having put to him the question in a letter from Lübeck on 2 October 1657: "I am not completely sure whether you actually mean that a cock can lay an egg, or whether I shall rather believe that you have written on the subject in deference to the court."[69] Noteworthy is the fact that Bartholin chose to publish this letter in Frederik III's lifetime with the possibility that the scientific and literary King might read it, but it is just as understandable that Bartholin chose not to publish his answer, if one was ever written.

Bartholin had, in 1645, in the first edition of *De Unicornu Observationes Novæ* mentioned and depicted the basilisk (Figure 93), and in 1661 he mentions in the third volume of the *Historiarum Anatomicarum Rariorum* Kepler's interest in the animal and expresses his own scepticism about its existence:

> "To the old fables about the basilisk I have nothing to add, since we expect this topic to be discussed in detail and exhaustively treated by Ludwig Kepler, a son of the great mathematician Johannes Kepler. A chicken resembling a basilisk, which he discovered in an egg in Königsberg, has caused him to deal with this matter. On my travels I have here and there seen basilisks, but they were either freaks or created artificially."[70]

Nevertheless, many years later in the apothecary Johann Gottfried Becker's (1639–1711) dispensary in Købmagergade in Copenhagen, Bartholin attempted to incubate a supposed cock's egg to see if a basilisk or a snake emerged from it, but nothing like that occurred.[71]

As a final twist on the story, Simon Paulli, who had witnessed Bartholin's dissection of the fashionable cock in Frederick III's chicken run, ensured its immortalization by preparing its skeleton for Frederik III's cabinet of curiosities. In the chamber's inventory, which was prepared by Thomas Bartholin in collaboration with the two royal councillors Corfitz Trolle (1628–1684) and Enevold Parsberg (1630–1689) in the period April 1673 to August 1674, it is described with the reserved wording: "Skeleton of a cock, which is thought to have laid an egg."[72] Elsewhere in the inventory the following text is found: "Several eggs, which are said to be from a cock."[73]

Figure 93. Basilisks from Thomas Bartholin's *De Unicornu Observationes Novæ*, Padua 1645, p. 51.

In the 17th century it was a common opinion, even among learned people, that old roosters could lay eggs, from which could be hatched a so-called basilisk, the king of reptiles (Greek, *basileos* = king), a fiercely venomous, dragon-like, mythical creature with wings and the appearance of something between a dragon and a snake or a rooster and a snake. The Royal Library in Copenhagen.

The remaining teaching

At the University's annual reformation celebration on 31 October 1649, Bartholin gave a long speech "On the Christian community's illnesses and about the gentle measures that are found to counter them," where he compared human society and the Christian world with a sick body. Obstinacies in religion and religious wars like the recent Thirty Years' War with "towns buried under ashes" and "the beautiful Rhine coloured red with innocent blood," according to Bartholin, have their root in human sin, which must therefore be altered, which only God can do.[74] At the same time Bartholin gives thanks that the Danish church has been spared the infection of heresy. In the same year he published this speech as a 50-page dissertation dedicated to Holger Rosenkrantz's son, Jørgen Rosenkrantz.[75]

On 12 December 1649, Bartholin began a series of disputations, eighteen in total, on miscellaneous anatomical subjects with miscellaneous respondents. The disputations took place in the large auditorium and went on until 30 July 1651. Bartholin published them in a series of eighteen volumes.[76] The subjects deal with almost the entire human anatomy and include the stomach and intestines, the mesentery and pancreas, Aselli's lacteal veins, the liver and bile ducts, the kidneys, the ureters and urinary bladder, the male genitalia, the rib cage, the breasts and chest muscles, the pleura, the endocardium and mediastinum (the walled cavity that separates the left and right lungs, and which contains the heart, the large blood vessels, the oesophagus, the trachea and the thymus), the heart, head and brain, the sense organs, the nerves, the blood vessels, the circulation and the bones.

In the introduction to the first disputation, Bartholin remembers his two anatomy teachers, Johannes Vesling and Johannes Walæus,[77] who had both died earlier that year. He emphasises the great importance of anatomy for the understanding of human health and sickness and God's creative power.

The respondent in the disputation about the ureters and the bladder on 20 February 1650 was Peder Schumacher (later Griffenfeld, 1635–1699), barely 14 years old and later chancellor. In his preface to the disputation, Bartholin put forward a very pertinent prophecy, where he praised the young Schumacher's gifts and maturity far beyond his years, which promised great things for the fatherland.[78] Schumacher gave a disputation once more against Bartholin, on 4 September in the same year, about nerves.

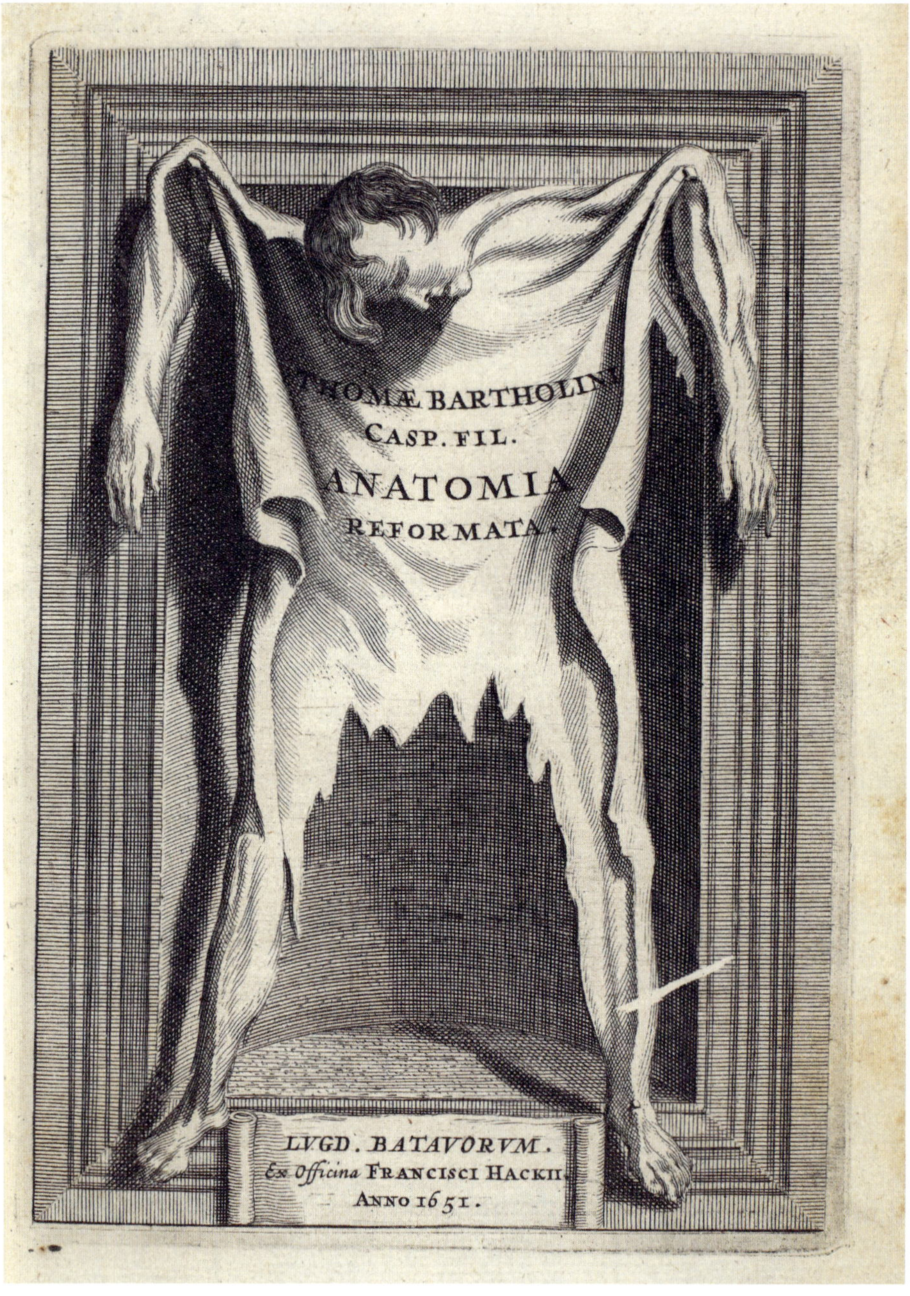

Figure 94. The frontispiece of Thomas Bartholin's third revised edition of the Bartholinian anatomy book, *Anatomia Tertium Reformata*, Leiden 1651.

The text in this edition is reformed to such an extent that, more than the two previous revised editions from 1641 and 1645, it is characterized by being Thomas Bartholin's independent work. The motif of the flayed skin was common in Baroque anatomical title engravings.
The Royal Library in Copenhagen.

The third revised edition of the anatomy book

In Leiden in 1651, Thomas Bartholin published his third revised edition of his father's textbook of anatomy, *Anatomia Reformata* (The improved book of anatomy) (Figure 94).[79]

Although the introduction, the list of chapters and the subjects remain unchanged from the earlier edition, there is, as the title indicates, this time a considerable reorganization of the book regarding both text and illustrations. This edition appears to a much larger extent as Thomas Bartholin's independent work, which among other things builds on his own dissections after his appointment as professor and leader of the Anatomy House in Copenhagen[80]. Many examples of new material from this edition have been given above, and more will be given in the following; for example, Bartholin's discovery of a new muscle, the small lumbar muscle (*musculus psoas minor*), at the dissection of a strong, powerful body in Copenhagen in 1651.[81] The number of pages is increased by 90 to 576, and the anatomical plates (70 in all, of which 6 are on folded sheets) are, with a few exceptions, replaced with new ones, of which a few originals are from Bartholin's own dissections, for example the one shown earlier of the mesentery with the lacteal veins in the lumpfish.[82] But Bartholin makes clear in the preface that there are also illustrations from dissections carried out by his friend, Johannes van Horne, and a physician named Dr. Dionysius à Cruyskircke. Similarly, he writes that the plates for the illustrations in the 1641 and 1645 editions were worn, and that for this reason new ones were prepared for this edition. Among the writers whose illustrations Bartholin uses in this and the following edition of the anatomy book, Casserius and Vesling are the chief ones named (Figure 95).[83]

The book is supplied with a portrait engraving of Thomas Bartholin executed by the Dutch engraver Jonas Suyderhoef (1613–1686) after an oil portrait painted by Karel van Mander (Figure 96),[84] and it is dedicated to "Christen Thomesen Sehested of Stovgaard, the king's, the nation's, and the university's chancellor." The edition was translated into Dutch by Thomas Staffard in 1653[85] and into English by the English physician, botanist and astrologer Nicholas Culpeper (1616–1654), but first published in London 14 years after his death (1668) with the title *Bartholinus Anatomy; From the Precepts of his Father, And from the Observations of all Modern Anatomists, together with his own. With one hundred fifty and three Figures cut in Brass, much larger and better than any have been heretofore. Printed in English. In Four Books and four Manuals, answering to the said Books* (Figure 97).

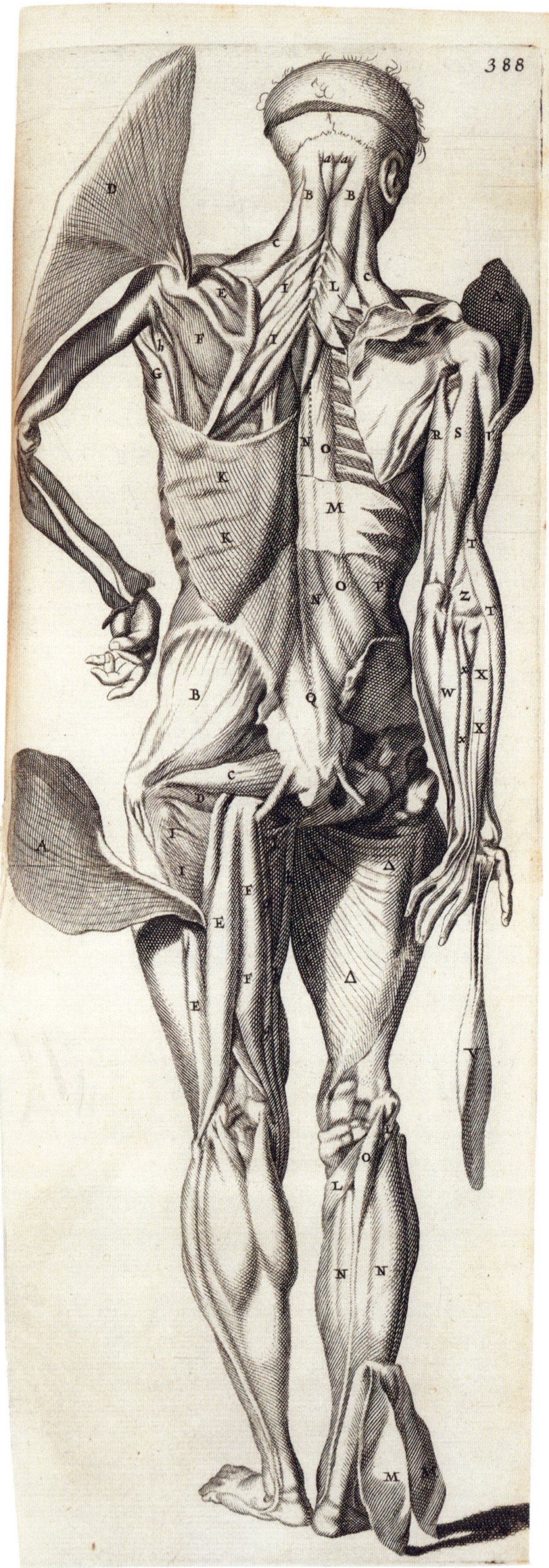

Figure 95. a. Muscle man on p. 388 in Thomas Bartholin's *Anatomia Tertium Reformata*. The Royal Library in Copenhagen.

b. Illustrations of the fetal heart and lungs. Ibid, p. 277.
Figure I: A. The heart. B. Vena cava superior. C. Vena cava inferior. D. Ascending part of the aorta. *e*. Arteria axillaris (shoulder artery). *f*. Descending part of the aorta. *g*. Right atrium. K. Opening in the vena cava.

Figure II: A. The heart. B. Ascending part of the aorta. C. Aortic arch. D. The pulmonary artery. *e*. Ductus arteriosus. *f*. The beginning of the carotid arteries. *g*. The beginning of the right subclavian artery.

Figure III: A, B, C and D. Nerves to the lungs and pericardium. E. The main bronchi. F. The posterior surface of the lungs. G. The pleura. H. Remnant of the pericardium. I. The heart.

Figure IV: A. Trachea. B. Right main bronchus before its divides in two. C. Left main bronchus before it also divides into larger and smaller branches. *d*. Outer bronchial branches.
The Royal Library in Copenhagen.

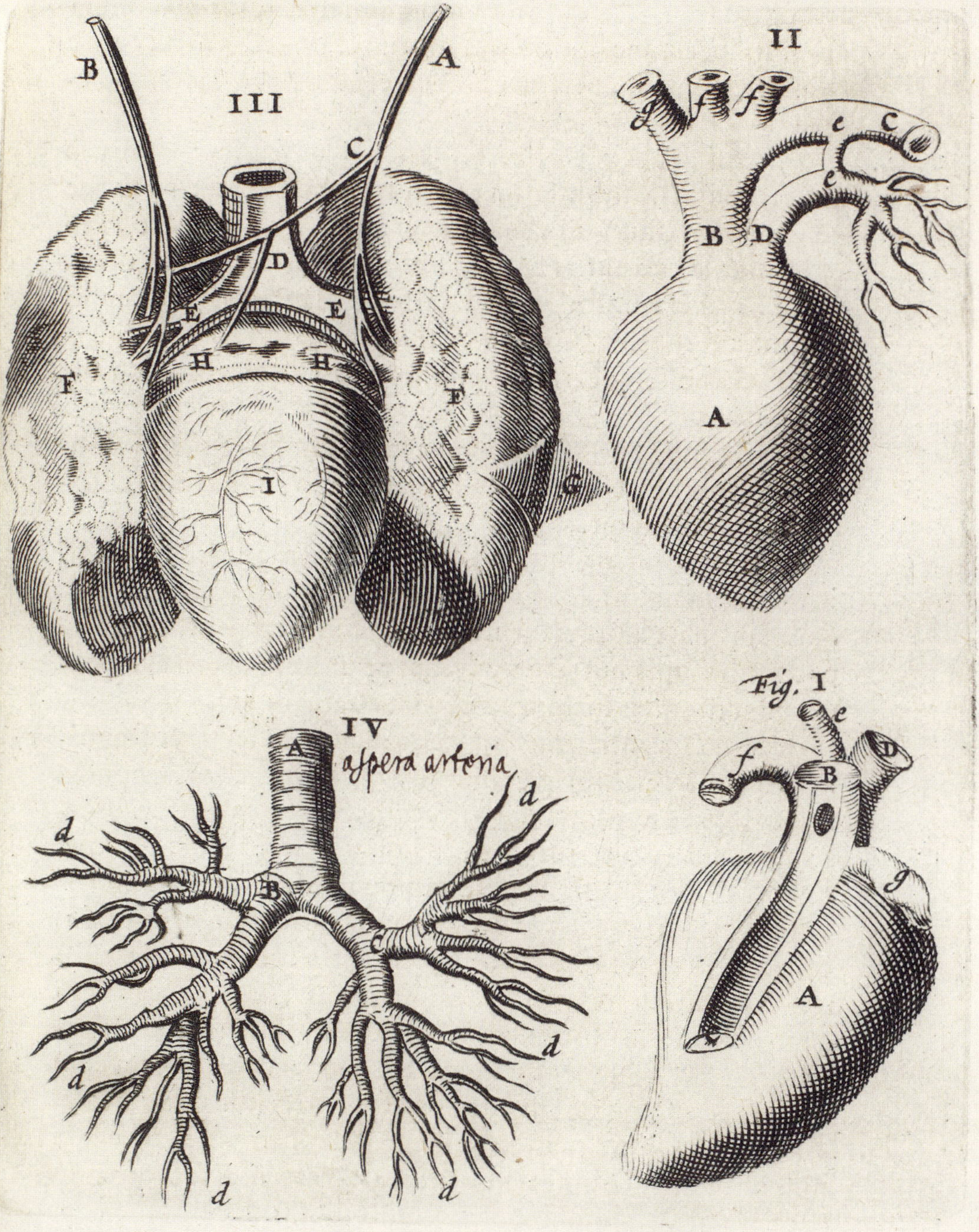
III
B
A
C
D
E
E
H
H
F
F
I
G
II
g
f
f
e
C
e
B
D
A
IV
A
aspera arteria
B
d
d
d
d
d
d
Fig. I
e
f
B
D
g
A

Figure 96. a. Thomas Bartholin portrayed by Karel van Mander around 1651. Oil on canvas 82 × 66.5 cm.

Bartholin's favourite painter and the court's preferred portrait painter Karel van Mander from Delft portrayed the 35-year-old newly appointed professor of anatomy and medicine with a serious, attentive and thoughtful, yet authoritative and calm gaze fixed on the viewer. He has an upturned moustache, goatee, long hair and skullcap and wears a black cloak, white shirt and two small white tassels on the top of the breast. The portrait was made to serve as a template for a portrait engraving (b) for the frontispiece of Thomas Bartholin's *Anatomia Tertium Reformata*, Leiden 1651, and was

therefore sent to the Netherlands with a messenger, presumably as a rolled canvas, as was the case with Karel van Mander's portrait of Ole Worm, which a few years later would serve as the template for a portrait engraving for Worm's *Museum Wormianum*, Amsterdam 1655. The portrait has served as template for several other portrait engravings of Thomas Bartholin and had been considered lost in the literature on Thomas Bartholin (see for example Andrup 1916, p. 57) until now. After stumbling upon the portrait in black and white reproduction in a recent publication (Lea Hagedorn 2022) and subsequently on the website of Amsterdam Museum, I contacted the museum, who informed me that they considered the portrait to be a studio work made as a copy after Jonas Suyderhoef's portrait engraving (b), most likely made to decorate the meeting room of the Collegium Medicum in Amsterdam together with other portraits and probably commissioned somewhere in the second half of the 17th century. After submitting documentation from written sources and arguments in favour of the original portrait, the museum has studied the painting more closely and concluded from its high quality that it may well be the original from van Mander's hand. The painting is not signed. It originally hung in the Collegium Medicum, but in 1865, when the institution that succeeded the Collegium ceased to exist, the painting was given to the city together with other portraits of the Collegium and thus ended up in Amsterdam Museum (personal correspondence with curator at Amsterdam Museum, Tom van der Molen).
Collection Amsterdam Museum.
Photo: Amsterdam Museum.

b. Portrait engraving of the 35-year-old Thomas Bartholin on the frontispiece of his *Anatomia Tertium Reformata*. 12 × 9.5 cm.

As the text says, the engraving is cut ("sculpsit") by the Dutch engraver Jonas Suyderhoef (1613–1686) from a portrait painted ("pinxit") by Karel van Mander (a).

The title below the engraving reads: "Thomas Bartholin, son of Caspar. Physician and anatomist. Royal Professor at the University of Copenhagen. Age 35. Year 1651."

Note that, due to the printing process, the engraving is facing the opposite direction from the painting, while the copper plate, on which the engraving has been printed, is in the same orientation as the painting. This is usually the case for engravings made after paintings, while paintings made after engravings would usually face the same way as the engraving.

A letter from Ole Worm in Copenhagen to his son Willum (1633–1704) on 8 July 1654 reveals that Karel van Mander was, to put it mildly, not satisfied with Suyderhoef's engraving, but actually felt dishonoured by it: "Look my son Willum, here you have my portrait painted quite accurately by Karel van Mander, and Karel himself sends it to Elzevir, so that the engraver can reproduce it all the more accurately; for he does not like Bartholin's, which they left to an unskilled artist, and which he himself thinks dishonours him."
The Royal Library in Copenhagen.

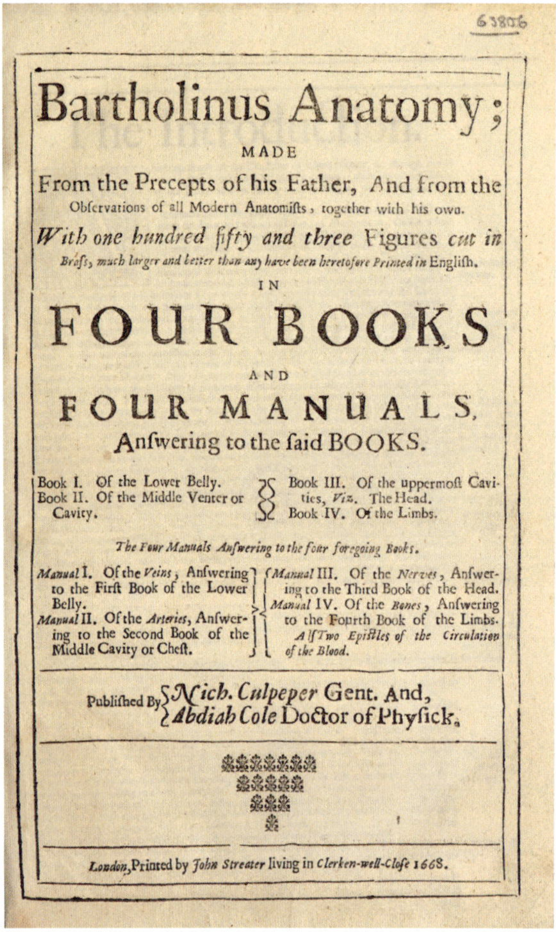

63806

Bartholinus Anatomy;

MADE

From the Precepts of his Father, And from the Obſervations of all Modern Anatomiſts, together with his own.

With one hundred fifty and three Figures *cut in Braſs, much larger and better than any have been heretofore Printed in* Engliſh.

IN

FOUR BOOKS

AND

FOUR MANUALS,

Anſwering to the ſaid BOOKS.

Book I. Of the Lower Belly.
Book II. Of the Middle Venter or Cavity.
Book III. Of the uppermoſt Cavities, *Viz.* The Head.
Book IV. Of the Limbs.

The Four Manuals Anſwering to the four foregoing Books.

Manual I. Of the *Veins*, Anſwering to the Firſt Book of the Lower Belly.
Manual II. Of the *Arteries*, Anſwering to the Second Book of the Middle Cavity or Cheſt.
Manual III. Of the *Nerves*, Anſwering to the Third Book of the Head.
Manual IV. Of the *Bones*, Anſwering to the Fourth Book of the Limbs.
Alſ Two Epiſtles of the Circulation of the Blood.

Publiſhed By *Nich. Culpeper* Gent. And, *Abdiah Cole* Doctor of Phyſick.

London, Printed by *John Streater* living in *Clerken-well-Cloſe* 1668.

Figure 97. a. Nicholas Culpeper's (1616–1654) English translation of Thomas Bartholin's *Anatomia Tertium Reformata*, published in London in 1668, seventeen years after Thomas Bartholin's publication of the book in Leiden, and fourteen years after Culpeper's death, with the title: *Bartholinus Anatomy; MADE from the Precepts of his Father, And from the Observations of all Modern Anatomists, together with his own. With one hundred fifty and three Figures cut in Brass, much larger and better than any have been heretofore Printed in English. IN FOUR BOOKS AND FOUR MANUALS, Answering to the said BOOKS.*

It must be assumed that the co-editor, Doctor of Physick Abdiah Cole (c. 1610–c. 1670), was responsible for the practical publishing. Wellcome Collection, London.

b. The English botanist, herbalist, physician and astrologer Nicholas Culpeper. Engraving by Thomas Cross (active 1644–1682) from Culpeper's *THE English Physitian OR An Astrologo- Physical Discourse of the Vulgar Herbs of this Nation. Being a Compleat Method of Physick, whereby a man may preserve his Body in Health; or cure himself, being sick, for three pence charge, with such things only as grow in England, they being most fit for English Bodies*, London 1652. The republican Culpeper was not only in opposition to the English king but also to the official medical and apothecary establishment, which among other things accused him of witchcraft. Conversely, he criticized their high prices, which the sick poor could not afford, and their use of the Latin language, which ordinary people did not understand. Wanting to make medicine accessible to all people, he translated several major medical works from Latin into English. It is in this light that his translation of the Bartholinian textbook of anatomy should be seen. As a representative of the established medical profession, Thomas Bartholin may have viewed Culpeper's translation with mixed feelings. He includes it in his lists of his works, but he does not comment on it anywhere, and moreover, he does not mention Culpeper's name. Wellcome Collection, London.

In Effigiem Nicholai Culpeper Equitis.

The shaddow of that Body heer you find
Which ſerves but as a case to hold his mind,
His Intellectuall part be pleas'd to looke
In lively lines deſcribed in the Booke.

Croſs ſculpsit

Marriage and children

In the spring of 1649, Bartholin got engaged to the only daughter of the prosperous merchant and Copenhagen mayor Christopher Hansen (1598–1679), Else Christophersdatter Hansen (1634–1675), who was only fourteen years old.

From Leiden, Vorstius congratulates Bartholin on 16 April with his future bride, "a virgin in the bloom of youth, daughter of a great and exceptional mayor with a good dowry."[86] The marriage took place on 8 May 1649 and was celebrated with a great feast, for which the university provided wine, and various poets had composed wedding songs.[87] The couple's first child, Anna Margrethe, was born the following year, but died aged two, during a pernicious epidemic of diphtheria that ravaged Copenhagen in 1652.[88] Later that year, the pair had another daughter, Margrethe (1652–1711).

The discovery of the thoracic duct and the lymph vessels

4

"No discovery, with the exception of the blood circulation, has elicited livelier discussion in the world of medicine than the discovery of the lymph vessels."[1] Thomas Bartholin, 1660

The discovery of the thoracic duct in a human being

19 February 1652, on a slab of slate, which was round at the ends and supplied with an iron ring in each corner, the body of a condemned criminal was carried in and laid on the revolving dissection table in the Anatomy Theatre in Copenhagen (Figure 98). The room was full of spectators. Thomas Bartholin and his assistant, Michael Lyser (1626–1659), took their places around the table. The public dissection could begin. The corpse was the earthly remains of a male child murderer, whose body, according to Bartholin, "was so lean that it could scarcely hang together."[2] He had been condemned to execution on the breaking wheel, but Frederik III had consented to the judgement being converted to hanging, so that the intact body could be given to Bartholin for further examination. Bartholin's mission was to be the first in the world to demonstrate the largest human lymph vessel, the lacteal thoracic duct or simply the thoracic duct (*ductus thoracicus*), which he and others had earlier pointed out in animals. In order to make the task easier, the man was, according to Bartholin's wish, "surfeited with food and wine" five hours before the execution, because the lacteal thoracic duct, as the name implies, gains a milky-white appearance after a meal.[3]

As far as the wine was concerned, Bartholin was forced to go against his principles. Although it was the custom at this time to brace the person condemned to death with wine, so that he could face death with lifted spirits,[4] Bartholin had his reservations. Indeed, he believed, that wine could be useful to get the person condemned to death to tell the truth about his illness as an advantage for the dissection, but nevertheless he could not recommend such drunkenness in the final moments of the approach to death, considering it suitable to keep one's mental state clear.[5] Instead, he referred to the ancient use of vinegar for those being crucified and lemons for robbers condemned to execution by snakebite.

As far as the body's intact condition was concerned, it was barely tolerable. During his detention in the prison cellar under the town hall in Copenhagen's Old Square the man had endured miserable treatment and suffered several illnesses. Besides the physical abuse, which many of the prisoners were subjected to, it was a custom, that the prisoners in the town hall cellar lay on straw, which was only changed once a year. And it could take up to ten years from one thorough cleaning to the next.[6] The quality of the "food and beer,"[7] which was served to the prisoners, one can only guess at, but judged by Bartholin's description of their bodies' miserable condition, there could be no talk of a calorie- or vitamin-rich diet:[8]

"The deceased had apparently been attacked by scurvy,[9] was melancholic and very close to suffering from consumption.[10] Moreover, in prison he had been treated poorly. His toes were in fact badly affected by frostbite, and the big toe on his left foot, which had been attacked by gangrene, had by Nature, which is its own master, begun to separate from the healthy part of the foot. The glands in the groin were hard and swollen and were like buboes;[11] in the intervening

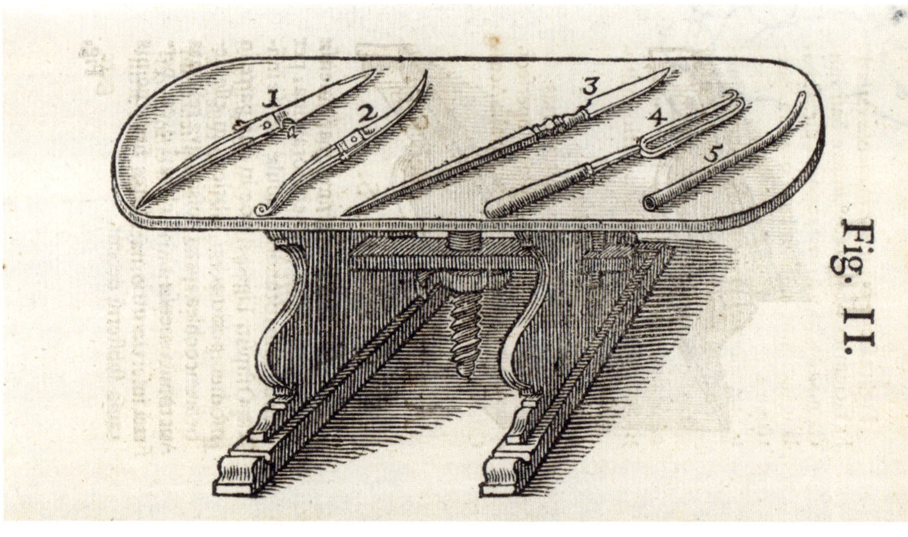

Figure 98. The swivelling dissection table (with table base and wooden spindle) in the Anatomy House's anatomical theatre. Woodcut on p. 5 in the second edition of Michael Lyser's (1626–1659) book on dissection technique, *Culter Anatomicus*, published in Copenhagen 1665 by Thomas Bartholin.

The dissection instruments are: 1. Myrtle-shaped knife with a crescent-shaped incision to support a finger. 2. Carving knife with a curved blade. 3. Knife with an elongated handle. 4. Retractor or holder. 5. Tube for inflating various body parts. The Royal Library in Copenhagen.

spaces of the ribcage's dissected muscles, serum flooded abundantly.[12] The thighs were full of blueish scorbutic stains of a type we also observed in the lungs. Otherwise, the lungs were spongy and slackened and mottled like marble. There was either very little or no fat in the abdominal cavity. The omentum[13] was infiltrated with pus."[14]

Despite the condition mentioned, the lean body of the child murderer revealed itself to be the perfect anatomical specimen. Bartholin and Lyser succeeded in their purpose. The greasy meal before the execution ensured that the lymph vessels (the chyle vessels) in the man's abdominal cavity and thorax stood out clearly because of their content of milk-white lymph, or chyle.[15] The part of the peritoneum carrying vessels (the mesentery) was filled with shining white lacteal veins/chyle vessels, and in the region of the liver there were glittering white strands. Above, in the abdominal cavity on a level with the kidneys, Bartholin and Lyser found three large chyle glands, which were also filled with white chyle, and which were mutually connected by many small chyle vessels. Bartholin and Lyser did not find the chyle container, the receptacle that the Frenchman Jean Pecquet (1622–1674) had described in animals, and which will be mentioned later, but they presumed that the three large chyle glands they found served the same purpose in the human being. From the chyle glands' upper ends, just under the diaphragm, exited single large chyle vessels, which by their convergence at the bottom of the ribcage formed the large thoracic duct. The thoracic duct continued its course up through the ribcage between the spine and the gullet, along the aorta, and the two anatomists were able to demonstrate that the thoracic duct all the way up the man's neck emptied itself through three smaller branches into the left shoulder vein (*vena axillaris*) on the front of the neck just above the left collarbone.

With this, Bartholin and Lyser had not only been the first to find the thoracic duct in human beings, but also shown that the fatty lymph (or chyle), which is formed in the intestine from digested food, passes from the abdominal cavity's lymph vessels (or chyle vessels) and lymph glands (or chyle glands) through the thoracic duct to the blood circulation in the neck, flowing on towards the heart.

Research into the lymph vessel system was at this time a main focus among anatomists throughout Europe, and Bartholin was very conscious of the importance of getting his findings documented in the best and fastest possible way to get ahead of other researchers. Present at the dissection, therefore, was the foremost portrait painter of the court, the nobility and the rich, one of the age's most outstanding artists in Denmark, the Dutchman Karel van Mander (1609–1670) (Figure 99).[16] His task was to carry out sketches from the dissection that would

Figure 99. Karel van Mander (1609–1670). The artist with his family in 1640. Oil on canvas. 136.5 × 111.5 cm.

The Dutchman van Mander was Thomas Bartholin's favourite painter, and the collaboration between him and his fellow countryman, the engraver Albert Haelwegh, led to anatomical illustrations of very high quality in several of Bartholin's works.

A supposedly marvellous work, a life-size anatomical atlas of the human body with copperplate engravings by Haelwegh based on paintings by van Mander from Thomas Bartholin's dissections, was unfortunately never realized, probably due to a combination of delays, the fire at Thomas Bartholin's manor house, and the death of van Mander. Ten of van Mander's paintings for the work were lost in the Copenhagen fire in 1728. National Gallery of Denmark.

afterwards provide a model for copper plates, which should be executed by another of the age's foremost artists, the similarly Dutch-born Royal Engraver Albert Haelwegh (1620–1673). Van Mander presumably had the body of the child murderer delivered to him for further study after the dissection. The result of the two artists' collaboration was three illustrations of very high quality, which show respectively 1) the right side of the child murderer's opened abdominal cavity with the newly-discovered chyle glands and their chyle vessels, 2) the displayed rib cage and abdominal cavity seen in front view with the three large chyle glands, their chyle vessels and the thoracic duct, made clear after the removal of the stomach, the intestines and the abdominal organs, apart from the kidneys, and cutting of the lowest part of the oesophagus and the uppermost part of the aorta, together with 3) the man's throat seen from the left with the entrance of the thoracic duct with three branches into the left shoulder vein (Figure 100).

A good month later, on 24 March, Bartholin gained possession of another body of a hanged criminal for dissection. This time it was a fat thief who also, five hours before his execution, "had been fed in the same way, but had had more to drink." The dissection was carried out under private conditions, but the sources do not make clear if this was also in the Anatomy Theatre or perhaps in Bartholin's private residence. In this man, Bartholin and Lyser also found three large chyle glands in the uppermost part of the abdominal cavity. They were unable, however, to demonstrate the chyle-bearing vessels in either the abdominal cavity or the thorax, but in the thorax they found "the vessels provided for chyle." At the entrance to the thoracic duct's branches into the shoulder veins they made a new discovery. When they blew air or pumped water into the shoulder vein or squeezed it, so that it was distended, there did not occur an answering distension of the thoracic duct's emptying branches. By cutting open the shoulder vein Bartholin and Lyser found the explanation for this, namely that the branches at the entry point were furnished with a valve, so that the blood mixed with chyle in the shoulder vein could not run back into the thoracic duct.

It was now a question of coming out first in the world with these new epoch-making discoveries. Bartholin composed, as quickly as he could, a treatise about his and Lyser's discoveries, *De Lacteis Thoracicis*,[17] which was soon published on 5 May that year, and which was submitted with Michael Lyser as respondent (Figure 101). The publication happened so quickly that, when Bartholin at the end of the book draws attention to some misprints, he adds: "The remainder we entrust to the readers to find, since we have been in so much of a hurry."[18] Bartholin dedicated the treatise to Frederik III with the words: "To the Almighty King of Denmark, Frederik III, the true prince, Thomas Bartholin, his subject, dedicates and inscribes his new discovery regarding chyle, which

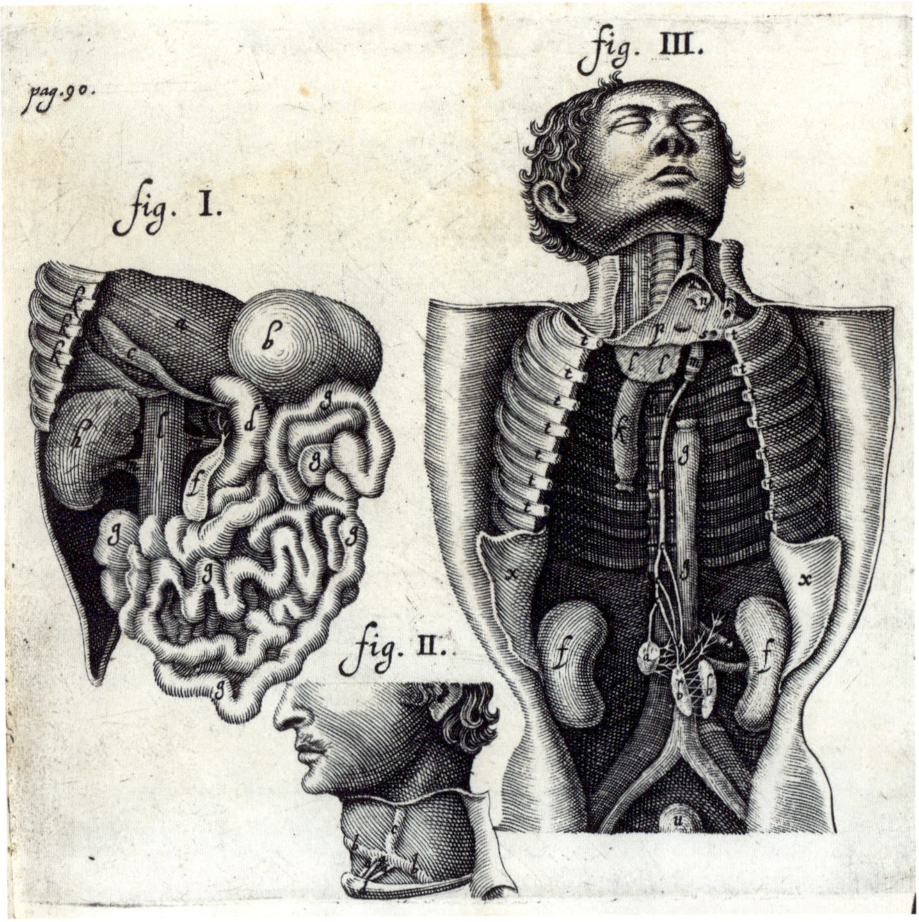

Figure 100. The illustrations from the dissection of the lean child murderer, in whom Thomas Bartholin found the thoracic duct (*ductus thoracicus*) at a public dissection in the anatomical theatre in the Anatomy House in Copenhagen on 19 February 1652. Engraving by Alberth Haelwegh from a template by Karel van Mander. From Thomas Bartholin's *De Lacteis Thoracicis*, Copenhagen 5 May 1652, p. 24. Here reproduced from Bartholin's *Historiarum Anatomicarum Rariorum*, Cent. 1, Copenhagen 1654, p. 90.

The illustrations show: Fig. I: Right side of the man's abdominal cavity with one of the newly discovered chyle glands (*f*) with associated chyle vessels. Fig. II: The man's neck seen from the left with the inlet of the thoracic duct with three branches (*a*) in the left shoulder vein (*vena axillaris*) (*b*), while *c* and *d* mark the external jugular vein (*vena jugularis externa*) and the clavicle (*clavicula*), respectively. Fig. III: The thorax and abdominal cavity seen from the front with the three large chyle glands (*a* and *b*), their chyle vessels and the thoracic duct (*c*, *d* and *i*) visualized after removal of the stomach, intestines and abdominal organs, except the kidneys (*f*), and after cutting off the lower part of the oesophagus (*k*) and the upper part of the aorta (*g*).
The Royal Library in Copenhagen.

never till now since man's creation has been observed in the human being, as he lays himself and his fortune in most humble affection in the hands of His Majesty's completely impartial control."[19] As visual documentation of his new finding, Bartholin supplied his treatise with van Mander's and Haelwegh's three illustrations of the child murderer's dissected body. About these illustrations Bartholin writes:

> "In order that the details should stand out more clearly to the eye, I have enlisted an engraver to help me, following the great man's example. Pecquet has included the picture of a dog in his article, not one that bites, or one that is raving, but one that has been put in chains and eviscerated.[20] [...] But though there were enough materials taken from animals that we could depict, we have preferred instead to use a human being, and with the help of a sketch to convey and put on display for the observant eye the conditions with regard to the receptacle's chyle glands and the thoracic ducts, which, in the two human corpses, we are the first to have searched and found, and we are doing this in the same trustworthy way, as they showed themselves before all the spectators."[21]

By referring to this documentation, Bartholin is pre-emptively guarding himself against subsequent criticism: "If in the meantime they should raise against me a charge of false description, I have for my answer my sketches."[22] Furthermore, he brings as testifiers "the most reputable witnesses of all ranks, who have

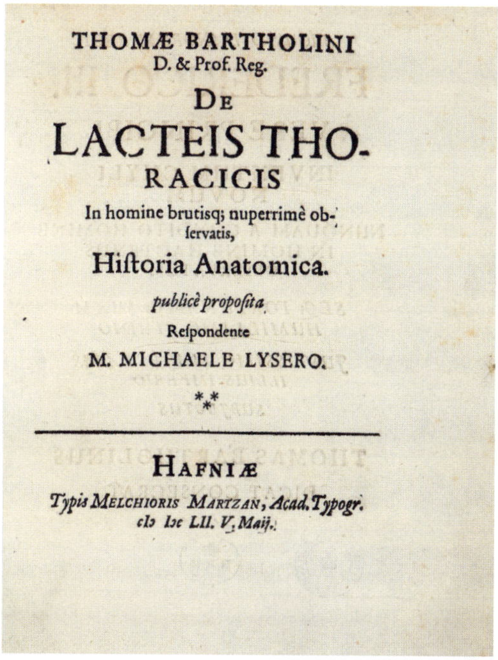
THOMÆ BARTHOLINI
D. & Prof. Reg.
DE
LACTEIS THO-
RACICIS
In homine brutisq; nuperrimè ob-
ſervatis,
Hiſtoria Anatomica.
publicè propoſita
Reſpondente
M. MICHAELE LYSERO.
***
HAFNIÆ
Typis MELCHIORIS MARTZAN, *Acad. Typogr.*
clɔ Iɔc LII. V. Maij.

Figure 101. The title page of Thomas Bartholin's dissertation on the discovery of the thoracic duct in a human being (the lean child murderer), *De Lacteis Thoracicis in homine brutisq; nuperrimé observatis, Historia Anatomica*, presented publicly in Copenhagen on 5 May 1652 with the assistant and pupil Michael Lyser as respondent.
The Royal Library in Copenhagen.

attended our demonstrations and dissections publicly and in private."[23] Here he singles out especially the chancellor, Christen Thomesen Sehested, "the magnanimous patron of the University and of the sciences, and at the same time other noble and distinguished men, whom the new development and their customary friendliness towards me have called to this play of Nature. [...] Furthermore, a great many of the most eminent and experienced physicians in the field of anatomy, Jacob Fabricius, Ole Worm, Simon Paulli, Poul Moth, Henrik Fuiren, Thomas Fincke, other royal professors and a glittering circle of chosen students." As it turned out, Bartholin soon came to need both drawings and witnesses. It was not long before he was met by fierce resistance from a circle of Europe's older anatomists, with "the hyena from Paris," Jean Riolan, at the head, and later he ended up in a bitter contest of priority with the young Swedish student, Olof Rudbeck (1630–1702), who claimed that he had discovered the thoracic duct in a human being before Bartholin and Lyser, but had not got his findings publicized. He accused Bartholin of plagiarism, but more on that later.

Completely in accordance with his basic attitude towards scientific research, Bartholin concludes *De Lacteis Thoracicis* with instructions to other researchers in anatomical procedure, for anyone wishing to investigate the newly discovered lacteal vessels:

"Open promptly the body of a human being or an animal, which has been strangled or hanged four, five, seven hours after a meal, and having removed the breastbone with the attached organs, pull with a forceful hand the lungs and their attached organs on one side to the opposite side in such a way that what is concealed at the back becomes visible. Thereafter dissect with the aid of a scalpel the shining white vessels or the passages glistening with serum, which appear immediately under the aorta and the oesophagus in the middle, or about the middle of the spinal column, carefully free from the adjacent parts and ligate them with a fine thread, either under the *vena axillaris*, or in the middle of the thorax. With another thread one must close off the heart's other vessels above and below, so that the blood does not rush out."[24]

After this follows an instruction for demonstration of the lacteal vessels in the abdominal cavity and the connections from them to the thoracic cage and the blood circulation through the shoulder vein. Bartholin concludes the instruction and the treatise with the words:

"Those people who settle down to rest with something they have not seen, or believe what they have simply heard, will be disappointed. A dissector must seek into the cave of Democritus[25] and the unfrequented places where he will find the

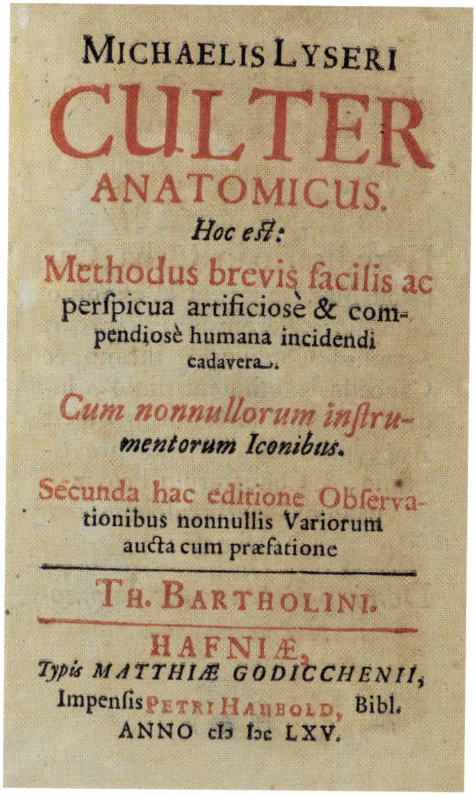
MICHAELIS LYSERI
CULTER
ANATOMICUS.
Hoc eſt:
Methodus brevis facilis ac
perſpicua artificiosè & compendiosè humana incidendi
cadavera.
Cum nonnullorum inſtrumentorum Iconibus.
Secunda hac editione Obſervationibus nonnullis Variorum
aucta cum præfatione
TH. BARTHOLINI.
HAFNIÆ,
Typis MATTHIÆ GODICCHENII,
Impenſis PETRI HAUBOLD, Bibl.
ANNO CIↃ IↃC LXV.

Figure 102. The title page of the second edition of Michael Lyser's (1626–1659) *Culter Anatomicus*, published by Thomas Bartholin in Copenhagen in 1665, six years after Lyser's death and eleven years after the first edition.

The German-born Lyser was Bartholin's most skilled assistant during the dissections that led to the discovery of the lymph vessels in the early 1650s, and it was therefore natural for Bartholin to invite Lyser to prepare an illustrated guide to the art of dissection. As the title page says, Bartholin has enlarged this second edition with his preface and various observations.
The Royal Library in Copenhagen.

Gods. We do not exactly call the places sweet-smelling, nor is it we who disappoint the purchaser, if he does not find anything else there but what is hanging in the window."[26]

On Bartholin's invitation, Michael Lyser published in Copenhagen in 1653 a detailed, 200-page illustrated guide to the art of dissection, *Culter Anatomicus* (The anatomical knife).[27] In *Domus Anatomica Hafniensis* in 1662 Bartholin writes about Lyser and *Culter Anatomicus*: "When he [Bartholin] was occupied with dissections he had extremely painstaking and talented young assistants, whose skills in anatomical dissection he had not so much discovered as created; chief among these was Michael Lyser, who, on the basis of dissections conducted by me, wrote *The anatomical knife* and published it in Copenhagen in 1654 to much acclaim."[28] In 1665, six years after Lyser's untimely death, Bartholin published a second edition in Copenhagen with a preface, an epigram and a letter to Lyser from April 1653 (Figure 102). In 1679, in Copenhagen and Frankfurt, a third edition by Caspar Bartholin the Younger appeared; in 1706 and 1731 in Leiden, a fourth and fifth edition, respectively; and finally in 1740 in London, an English translation was published by G. Thomson entitled: *The Art of Dissecting the Human Body, in a Plain, Easy, and Compendious Method. With the several Instruments necessary, curiously engraved; wherein all the different Parts of the Human Body are described, but particularly the Muscles, in a full and accurate Manner, and their Uses clearly explained.*

Letter to Harvey

On 1 May 1652, four days before the official publication of *De Lacteis Thoracicis*, Bartholin sent a copy of it with a letter to William Harvey,[29] in which he explains his early support for Harvey's discovery of the blood circulation and makes clear that on other occasions during public disputations he has presented the theory for people from foreign countries as well as taught it in his own university. He directs attention to the fact that, in the new edition of his anatomy book, which was published in Holland the year before, according to the best of his capability and on the model of the English surgeon and anatomist Nathaniel Highmore's (1613–1685) anatomical treatise,[30] he has revised the whole anatomy in accordance with the principles of the circulation theory, and he asks for Harvey's evaluation as to whether this work is satisfactorily done. Likewise, he asks for Harvey's views on how the nutritious chyle comes to the embryo within the womb, which despite his numerous investigations in pregnant dogs is a constant puzzle to Bartholin, since he has only found arteries and veins, but no chyle vessels, in the uterus. He wishes to undertake further investigations into this. If the chyle is transported in the arteries, which certainly does not occur in the mesentery, how is it then separated from the blood, and how does it enter the arteries? Bartholin has tried to give an explanation concerning this question in his treatise on the discovery of the thoracic duct in human beings, which he sends with his cousin and pupil, Willum Worm (1633–1704),[31] the son of Ole Worm, and he asks for Harvey's estimation of the treatise.

Fetal science (embryology), including the question of fetal growth and nutrition in the womb, was a topic that attracted great attention among the anatomists of the 1600s. Leonardo da Vinci (1452–1519) had, from his dissections of the pregnant uterus in the late 1400s, correctly concluded that the fetus in the womb does not breathe, since it is deposited in amniotic fluid and would therefore drown if it breathed. Moreover, he had realized that the fetal and maternal blood circulations are not directly connected to each other, but are separated by the placenta, and that the fetus receives its nutrition through the umbilical cord (Figure 103). Harvey's teacher in Padua, Girolamo Fabrici d'Acquapendente, had in 1600 published in Venice the work *De Formato Foetu* (On the formation of the fetus),[32] in which, based on the Aristotelian theory, he describes how nature

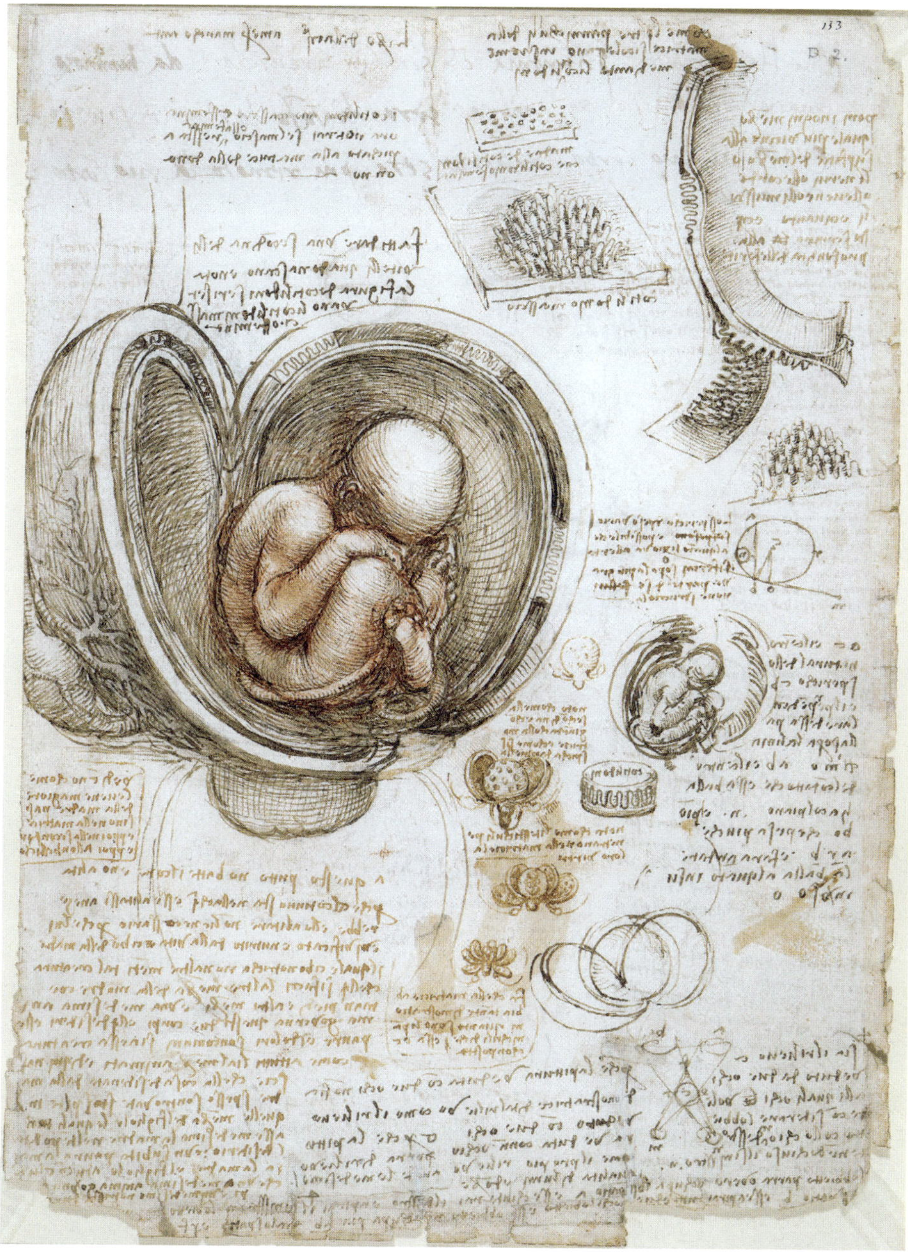

Figure 103. Leonardo da Vinci's (1452–1519) drawings of the fetus in the uterus. Pen and ink over red chalk on paper. 30 × 22 cm. 1510–1512. From the third volume of Leonardo's private notebooks.

Like many of the other artists of the Renaissance, Leonardo made detailed anatomical studies and performed and participated in dissections of both animals and humans to draw inspiration from nature for the representation of the human body. His hundreds of anatomical drawings which were never published and only disseminated widely in the 18th century, testify to the fact that not only was he an exceptionally skilful artist, but he was also in possession of impressive powers of observation. Furthermore, in many of his thoughts about the body's functions, he was ahead of his time: his insight into the fetus' nutrition through the placenta and the umbilical cord, and the fact that fetuses do not breathe in the uterus, are just two examples.

The Royal Collection at Windsor Castle. Wikimedia Commons.

provides for fetal nutrition and growth. The woman contributes matter, and the man contributes form (Figure 104). Inspired by this, Harvey had just published in London and Amsterdam 1651 another controversial work, *Exercitationes de Generatione Animalium. Quibus accedunt quædam De Partu: de Membranis ac Humoribus Uteri: & de Conceptione* (Exercises on the generation of animals, to which are added some concerning parturition, the membranes and the liquids of the uterus and the conception).[33] The last part of the dissertation also includes sections on the functions of the placenta and umbilical cord. This was the background for Bartholin's question to Harvey on this subject.[34]

Finally, Bartholin asks Harvey to explain why he looks with scepticism on Aselli's lacteal veins in his answer to Riolan relating to the circulation of the blood[35] and in his recent treatise on the propagation of living creatures. The paradox was that Harvey, as one of the most prominent sceptics of Aselli's treatise on the lacteal veins and Pecquet's on the thoracic duct and the chyle vessels, was criticized in certain circles for showing the same opposition to the new learning that others, for example Jean Riolan, had shown against his own circulation theory.[36]

As far as one can see, Harvey did not answer Bartholin's letter, but three days before that, on 28. April 1652, Harvey had answered a couple of Bartholin's questions in a letter to the Parisian physician R. Morison.[37] Here Harvey makes clear that he does not doubt the existence of Aselli's lacteal veins, but he rejects the idea that they convey nutritious chyle from the intestine by Pecquet's route through the thoracic duct (not via the liver) to the bloodstream as an important part of the whole of the body's supply of nutrients. He believes that the discov-

Figure 104. The fetus in the uterus. From Girolamo Fabrici d'Acquapendente's (1533–1619) *De Formato Foetu*, Padua and Venice 1600. Plate III, p. 25. Reproduced from Howard B. Adelmann (ed.): *The Embryological Treatises of Hieronymus Fabricius of Aquapendente*, New York 1942.

Fabricius's explanations: "Figure VI shows the position of the fetus in the uterus floating in the amniotic fluid, and the placenta bound together with the chorion. A. The placenta, to which the chorion is bound. B. The umbilical cord. C. The amniotic fluid, in which the fetus floats. D. The cut open uterus. E. The cervix. F. The cut vagina. G. The most prominent parts of the chorionic vessels. Figure VII shows the position of the fetus, which is now about to leave the uterus. A. The head of the fetus. B. The external genitalia. C. The cut abdominal wall."

Fabrici did more thorough studies of fetal nutrition, the placenta and the umbilical vessels than any previous anatomist and performed comparative studies on humans and a wide range of animals. The artist behind the impressive illustrations in his embryological works is unknown. Private ownership.

FIG · VI ·

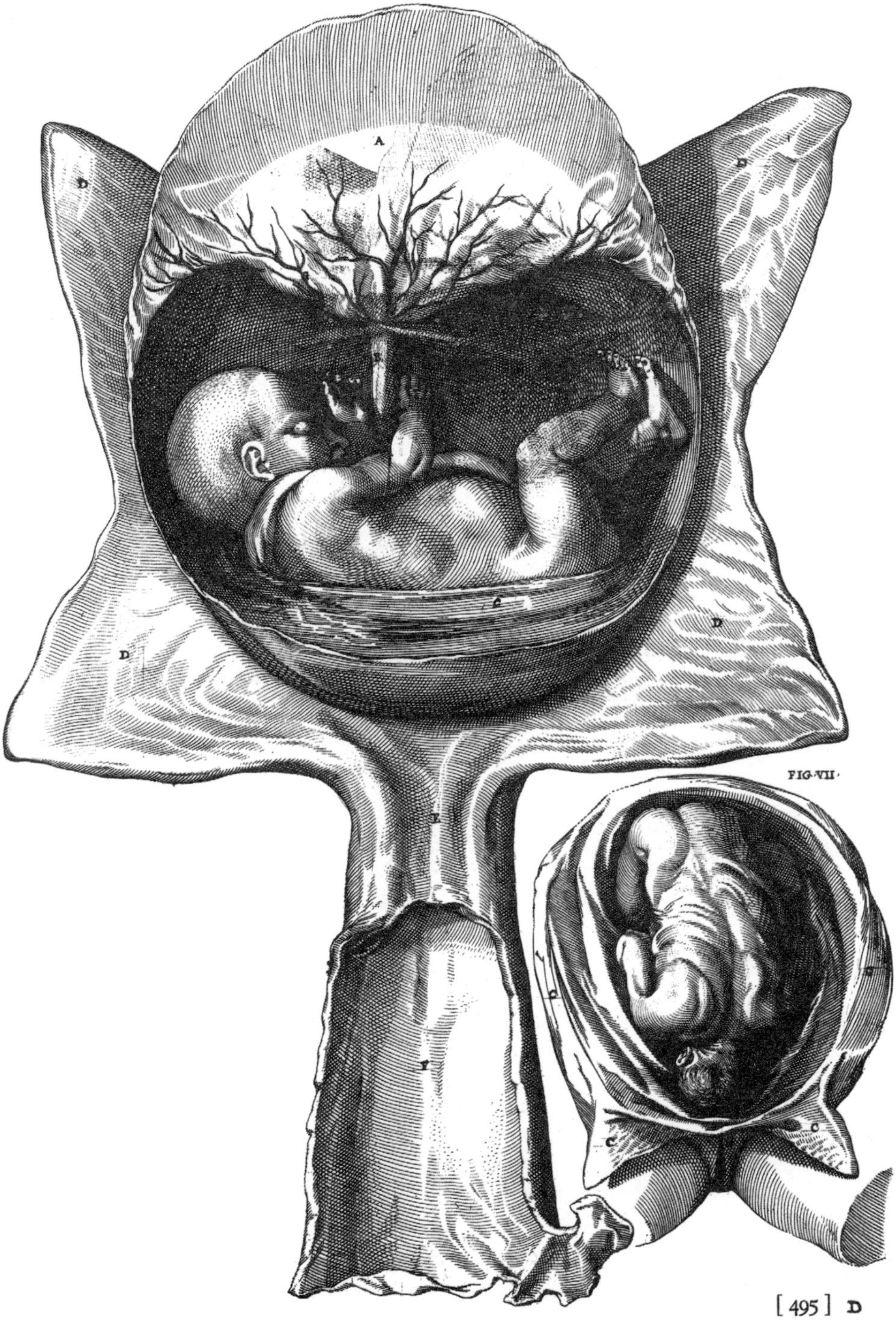

FIG · VII ·

ered thoracic ducts are much too small and narrow to carry all the essential chyle, and that the vessels responsible for the distribution of the body's nutrients should be possible to observe in all creatures at all times, and this has far from been the case with Harvey's experiments with animals. Harvey does not think that Aselli's lacteal veins contain chyle but milk, like the breast's milk ducts, and this milk cannot come from the intestines, which in such a case would themselves have been full of milk, since Harvey does not believe that a liquid can change so radically in its passage through the intestinal wall. Furthermore, if the milk came from the intestine, it should smell as bad as intestinal contents. Harvey believes that most of the nutrients from the intestine pass into the portal vein system and from there directly on to processing in the liver, as evidenced by the intestine's rich supply of blood vessels. As for proteins and sugars, posterity has proven that Harvey was right in this assumption. But as for the fats, he was wrong, as the vast majority of fats pass from the intestine via Pecquet's and Bartholin's path through the thoracic duct to the bloodstream (the jugular veins), bypassing the liver. Harvey was of course also wrong about the postulated milk content of the mesenterial veins and the idea that a liquid cannot change its character when passing through the intestinal wall.

In his answer to Morison about the fetus' conveyance of nutrients from the mother, Harvey is completely right when he declares that the rich supply of blood vessels in the placenta testifies that the fetus does not take its nourishment through the chyle vessels, but through the placenta's rich supply of blood vessels, which join in the vessels of the umbilical cord. Harvey repeats his arguments against Pecquet in two letters to the German physician Johann Daniel Horst (1616–1685), on 1 February and 13 July 1655, respectively.[38]

In the first he expresses regret on account of his poor health, weakened by old age, and explains, "that it requires good eyes, and a mind free from other anxieties, to come to any definite conclusion in regard to these extremely minute vessels; to me, however, as I have just said, neither of these requisites is given."[39]

In the second he mentions Bartholin: "With regard to the letter, which you inform me, you have addressed to Bartholin, I do not doubt of his replying to you, as you desire; nor is there any occasion, wherefore I should trouble you farther on that topic."[40]

On 1 March 1655, Horst had sent a letter to Bartholin from Dresden,[41] in which he expresses doubts as to the lacteal veins leading to the uterus to nourish the embryo with chyle. Moreover, he does not understand Pecquet's conclusion that chyle goes to the heart through the thoracic duct and the jugular veins, when Bartholin has shown that branches of the lacteal veins also lead to the liver. Finally, he does not understand the lymph vessels' transportation of fluid from the periphery to the centre of the body either. He asks for Bartholin's help with the elucidation of these questions.

Bartholin's answer has not been preserved, but in a fresh letter to Bartholin from Darmstadt on 16 July, Horst thanks Bartholin for his reply, which is so "swelling and brim-full with trustworthy solutions to unsolved questions" that he invites Bartholin to publish it together with his answer to Riolan's attack.[42] Moreover, he tells him that he has informed Harvey of his surprise that he does not ascribe any importance to the lacteal veins in the mesentery, even though they can be shown in all healthy animals. He has explained his disagreement with Harvey about the contents of the lacteal veins being the same as milk from the breast, for these two fluids differ in their position, place of production, use and purpose, and there is no doubt that the content in the lacteal veins is chyle from the intestines. Harvey is also wrong in his claim that chyle from the intestines is emptied into the veins of the mesentery, since this can be rejected through personal observation and experiments with ligation. If Horst receives Harvey's answer to Bartholin's letter, he promises to send it, but Horst has no further questions for Bartholin, who explains everything so fundamentally in his writings.

Bartholin's pupil, the German Martin Bogdan (1630–1679), visited William Harvey in London in 1655, and conducted a conversation with him in correct Latin, which according to Bogdan one would not have expected after reading Harvey's writings. During this conversation Harvey maintained his conviction that the thoracic ducts do not contain chyle, but a different serous fluid. On 17 June 1655, Bogdan wrote a letter to Bartholin referring to his meeting with Harvey,[43] and if we are to believe the contents of this letter, Harvey's abstention from answering Bartholin was caused neither by hostility towards him nor by lack of respect for him. Thus, Bogdan writes to Bartholin:

> "He [Harvey] believes that you have benefited the learned world immensely, and he says that because he has one foot in the grave and has reached the age of seventy-eight, he entrusts the completion of all anatomical questions to you."

The background to Bartholin's investigations into the system of lymph vessels

At the time of Bartholin's publication of *De Lacteis Thoracicis* there continued to be great disagreement regarding food's destiny in the human body. All agreed that the digestion of the food began with the chewing and the addition of saliva in the mouth, and that afterwards the food passed through the oesophagus into the stomach. The majority also agreed that the food in the stomach became mixed up with liquid and gastric juice and with the effect of heating underwent a kind of boiling resulting in the formation of a milky substance, chyle.

Furthermore, there was more or less agreement that chyle was included in the formation of blood. The two essential questions were thus: Where did the formation of blood occur, and how did the chyle get there? Here the disagreements began. There were many possibilities, and advocates for all of them. Chyle could be absorbed directly through the stomach or first through the intestines or a combination of the two. It could be absorbed either into the surrounding veins, into Aselli's lacteal veins, or both parts, or through the excretory duct of the gallbladder or the pancreas into the duodenum. And it could be conveyed to the liver, the spleen, the pancreas or a completely different place. Some split the chyle into thin and thick chyle, which they assigned to various purposes. The organ for formation of the blood could be either the liver, the spleen, the heart or blood vessels, or different combinations of these. In brief: there were many questions that still needed answering, and the replies could only be obtained by practical research, which would mean dissections of living and dead animals or of dead human beings.

As mentioned, Thomas Bartholin began his own studies of the lymph vessels at the beginning of 1639 during his residence in Leiden, but he did not resume them until ten years later as professor of anatomy in Copenhagen. In several of his already mentioned disputation exercises for students in the winter of 1649–1650 he talks of Aselli's lacteal veins.[44]

Investigation of Aselli's lacteal veins on the ill-fated servant of the town hall

On 17 February 1650 an extremely unfortunate event occurred in Copenhagen, which gave Bartholin his first opportunity to investigate Aselli's lacteal veins in the mesentery of a human being.

A young man who served in the house of the town hall clerk swiped a piece of ox tongue from a meat platter he was to carry into a dinner. Destiny willed that he should meet the severest punishment for his modest offence: he choked on it. Owing to the circumstances and his unexpected death, the city bailiff, Claus Iversen Ravn (1602–1669), requested that the two surgeons Martin Sixtus and Christian Wineke jointly with "the renowned and excellent Dr. Bartholin, professor of anatomy at the University of Copenhagen, and the illustrious, admirable physician, Henrik Fuiren," should carry out a medico-legal post-mortem of Sven Olsen's dead body. The post-mortem was carried out the day after the death, and the two surgeons write in their report the same day:

"When we had opened the dead body to the extent necessary for our present purpose, we discovered a piece of roasted ox tongue, which had stuck in the topmost part of the trachea and had gone no further. It weighed about 45 grams and had gone past the epiglottis; it had caused a blockage, so that he could neither draw breath nor speak, for breathing and speech are dependent on air being able to pass from the lungs up through the trachea. He had therefore suddenly suffocated despite all his efforts to throw it up, before he was choked."[45]

Later Thomas Bartholin writes thus of the unhappy story:

"Sven Olsen from Varberg was 19 years old and a strong young man, but with a pale face. He was a servant with the Copenhagen city hall clerk, Arnold Bernhardsen, and when he was once, according to the order of the host, serving the guests at a dinner party, he surreptitiously and very quickly crammed a piece of an ox tongue, which was left lying on the dish, into his mouth. As a result, he immediately lost the ability to speak and to breathe. Despite all attempts, and not even under coercion, was it possible to extract the piece of meat that had stuck. Before the surgeon could be called, the man was choked. The next day the body was given a post-mortem examination in my presence and that of Dr. Fuiren and the two surgeons, Martin Sixtus and Christian Wineke together with other medical graduates. It was found that the piece of ox tongue, which weighed an ounce and a half, had been forced into the small opening between the epiglottis and the larynx and had blocked the larynx so badly that the lump of meat could hardly be removed by hand. It is therefore not surprising that his voice deserted him suddenly, and that his breathing stopped. Although the oesophagus had become very narrowed, because it was right next to the piece of meat that pressed against it, it nevertheless gave access to the drink, which came from the stomach and streamed out through the dead man's nose and mouth. The lungs no longer kept their natural condition but were bluish and flabby. Doubtless he would have had only a short time left, even if he had not been as great a glutton. But because on the same day before the fatality in the afternoon he had drunk and eaten everything in abundance, I was confident, that we should manage to see the lacteal veins. I was not disappointed in this hope or my opinion. Spread here and there in the exposed mesentery we saw them with their milky colour and liquid, but there were not all that many of them because of his frugal dinner.[46] Concern for the surrounding audience hindered further research."[47]

In *Anatomia Reformata* Bartholin gives his explanation as to why the lacteal veins in the mesentery of Sven Olsen were visible so long after the onset of death: "Last year in Copenhagen in the body of Sven Olsen from Varberg – he had consumed a piece of ox tongue and promptly choked on it, which was caused by the fact that he had previously eaten and drunk far too much – I have seen the lacteal veins clearly in the mesentery, because the liver needed in no way to draw chyle to itself, when the ox tongue's blockage had prevented breathing, and the heart had been stifled."[48]

At that time Bartholin was, in common with the majority of Europe's other anatomists, of the firm belief that chyle is a nutrient fluid, which is formed from the food in the stomach in a kind of boiling and admixture of gastric juice, in order that it should pass further to the small intestine, from where it is absorbed by the lacteal veins in the mesentery and conveyed by means of these to the liver, which changes it into blood. In his anatomy book he gives a more careful account of his understanding of this process.[49] First and foremost, he states:

"The function of the liver is to make blood. For from the chyle, which is drawn through the mesentery's lacteal veins, the liver manufactures blood, and the blood is made in the branches of the lacteal veins; the hepatic tissue not only sustains the veins, as some believe, but is itself the place where blood is formed."[50]

He explains further that the thick, useless part of chyle, which is first made from the food in the stomach and is completed in the small intestine, then passes further to the large intestine and is excreted with the faeces. The useful thin part of the chyle is, however, drawn into the intestine's lacteal veins and from there, by a thermal force of the liver, into that organ, partly helped by the constriction of the intestines, i.e., primarily by a draw and secondarily by a push.[51] According to Bartholin there exist in the liver invisible connections between the lacteal veins and the branches of the *vena cava*, and between the latter and the branches of the portal vein (*vena portae*). When the chyle content in the lacteal veins' finest branches in the liver has been changed into blood and has changed colour from white to red, it passes via the invisible connections into the branches of the *vena cava* and from there through the *vena cava* itself into the heart. From the heart the nutritious blood is pumped out to the whole body through the arteries, and from the thinnest branches of these the nutrient-poor blood passes back through the vein system to the heart via the *vena cava*. Some of it, however, passes first through the liver via the portal vein. In *Anatomia Reformata* Bartholin still includes the lacteal veins (chyle vessels) in the mesentery as part of the vein system.

He mentions that Aselli does not specify the exact place for the lacteal veins' entry into the liver, but Bartholin has in his dissection studies on the lumpfish discovered that the place is "the third lobe of the liver." In *De Lacteis Thoracicis* he writes:

"Some years ago I myself found repeatedly in the lumpfish, both the male and the female, not merely chyle vessels but the correct, continually disputed, point of insertion in the liver, and I have demonstrated it to the most experienced anatomists, Worm, Sperling,[52] S. Paulli, Fuiren, Pouch[53] and several others, who can reliably bear witness to the truth of what has been discovered. The place was the third, smaller and softer lobe, into which passed a rather large, shining, white branch of a chyle vessel, from the large chyle gland, which does not lie very far away from there, and which was dilated by chyle, and where most of the chyle vessels from the mesentery and the stomach's appendage meet. I have illustrated this with a drawing, which may be seen in my *Anatomia Reformata* [Figure 105]."[54]

In this book Bartholin also mentions these discoveries and explains that the insertion point of the lacteal veins in the third lobe of the liver lies in the gall bladder's immediate vicinity, and that the lacteal veins stretch as a network of branches between the mesentery's two layers from the intestines to the pancreas (named by Bartholin either "the large gland" or "the pancreas") and on to the liver.[55] Furthermore, he explains that the lacteal veins resemble veins in their structure,[56] and that research with ligation has shown that they are supplied with valves on the inside, which ensure that chyle can only run in the direction of the liver, but not back to the intestines.[57] Bartholin ascertains that many scholars are still hesitant in their view of the lacteal veins, because the main stem of these has still not been discovered, but according to his opinion there must be no more doubt about the existence of the lacteal veins than the inhabitants along the Nile must doubt the existence of the river, even though its source is still unknown.[58]

Pecquet's discovery of the thoracic duct in a dog

Until the completion of the work on the new anatomy book in the summer of 1651, none of Bartholin's observations concerning the lymph vessel system had brought any epoch-making news. The great breakthrough came from Paris and reached Thomas Bartholin the same autumn in a letter from his brother Rasmus, who was staying in this city. Earlier that year the young French physician Jean Pecquet had published the work *Experimenta Nova Anatomica* (New Anatomical Experiments),[59] where he gave an account of two groundbreaking discoveries, namely that the chyle from Aselli's chyle vessels (lacteal veins) in the mesentery

TABULA IV. 419

FIGURÆ EXPLICATIO.

Venas lacteas, in Orbe Pisce repræsentat.

AA. *Ventriculus.*
BB. *Appendices ventriculi, in quibus ductus albi conspiciuntur chyliferi.*
CCCC. *Intestina ad latus diducta.*
D. *Intestinum rectum.*
E. *Hepar.*
F. *Tertius lobus hepatis, in quem venæ Lacteæ inseruntur.*
G. *Glandula alba mesenterii chylo turgens, ex qua venæ protenduntur ad tertium lobum.*
hhh. *Venæ Lacteæ.*
iii. *Rami venarum Mesaraicarum.*
k. *Venæ portæ truncus.*
lll. *Mesenterium.*
m. *Vesicula fellis.*

Dd 2

Figure 105. Milky veins (*h*) in the mesentery (*l*) and around the liver (E and F) of a lumpfish that Thomas Bartholin dissected. G marks a mesenterial gland, which is swollen with chyle. A is the stomach, C is the intestine, *m* is the gallbladder, and *k* is the portal vein. From Thomas Bartholin's *Anatomia Tertium Reformata*, p. 419.
The Royal Library in Copenhagen.

is gathered in a single receptacle (*receptaculum chyli*), situated in the abdominal cavity just above the lumbar vertebrae, and that from there other chyle vessels depart, which continue their course up through the ribcage and end up emptying into the blood stream in the left subclavian vein (*vena subclavia*).[60]

Hitherto everyone – like Bartholin and Lyser – thought that Aselli's lacteal veins brought chyle to the liver, where it became converted to blood, and this was in accordance with the ancient perception of the liver as the blood-forming organ. The pivotal aspect in Pecquet's discoveries was exactly the rejection of this perception. Since the bulk of the nutritional chyle converted from the nourishment bypassed the liver through the thoracic duct to the heart, Pecquet's conclusion was that the principal blood-producing organ must be the heart.[61] That this conclusion was also wrong, and that blood production after birth takes place solely in the bone marrow (and in intra-uterine life in the liver and the spleen) would only be discovered in the second half of the 1800s, but more about that later. Pecquet explains in his treatise that Johannes Walæus, in his letters to Thomas Bartholin, had mentioned that chyle in the lacteal veins moves in an upward direction, and that he had observed in large, lean dogs that some of the lacteal veins empty into the mesentery veins (*venae mesenteriae*), some into the portal vein (*vena porta*), some into the bottom of the liver, and a few into the large *vena cava*. According to Pecquet the same thing is concluded by William Harvey, Johannes Vesling, Hermann Conring (1606–1681), Thomas Bartholin and many others – indeed by Jean Riolan himself too – for none of these great men has looked for the hiding-place of the lacteal veins in the thorax.[62] Pecquet, however, believes that it is not due to negligence, but rather to bad luck, that these colleagues have not discovered the receptacle of the lacteal veins. Pecquet's observations were built on three years of studies with dissections of various live animals, among them sheep and dogs, which were dissected about four hours after a meal. His original purpose with the examination of live animals was to study the movements of the heart, but when he took out the entrails and dried the blood, he noticed a milky-white fluid and some shining-white vessels in the thoracic cavity, which he could follow all the way down to the mesentery.[63]

Bartholin immediately perceived the importance of Pecquet's discoveries:

"After I had been notified about his observations in a letter from my brother [...] who had concluded in Blois that Pecquet's experiments were reliable, I became very absorbed with them, and when I had received the whole book, I undertook together with my friend, the physician Michael Lyser, a careful investigation of a live dog, four hours after it had been given a rich meal, to verify the new findings. And I found most of the observations made by him [Jean Pecquet] and shown on the sketch [Figure 106]."[64]

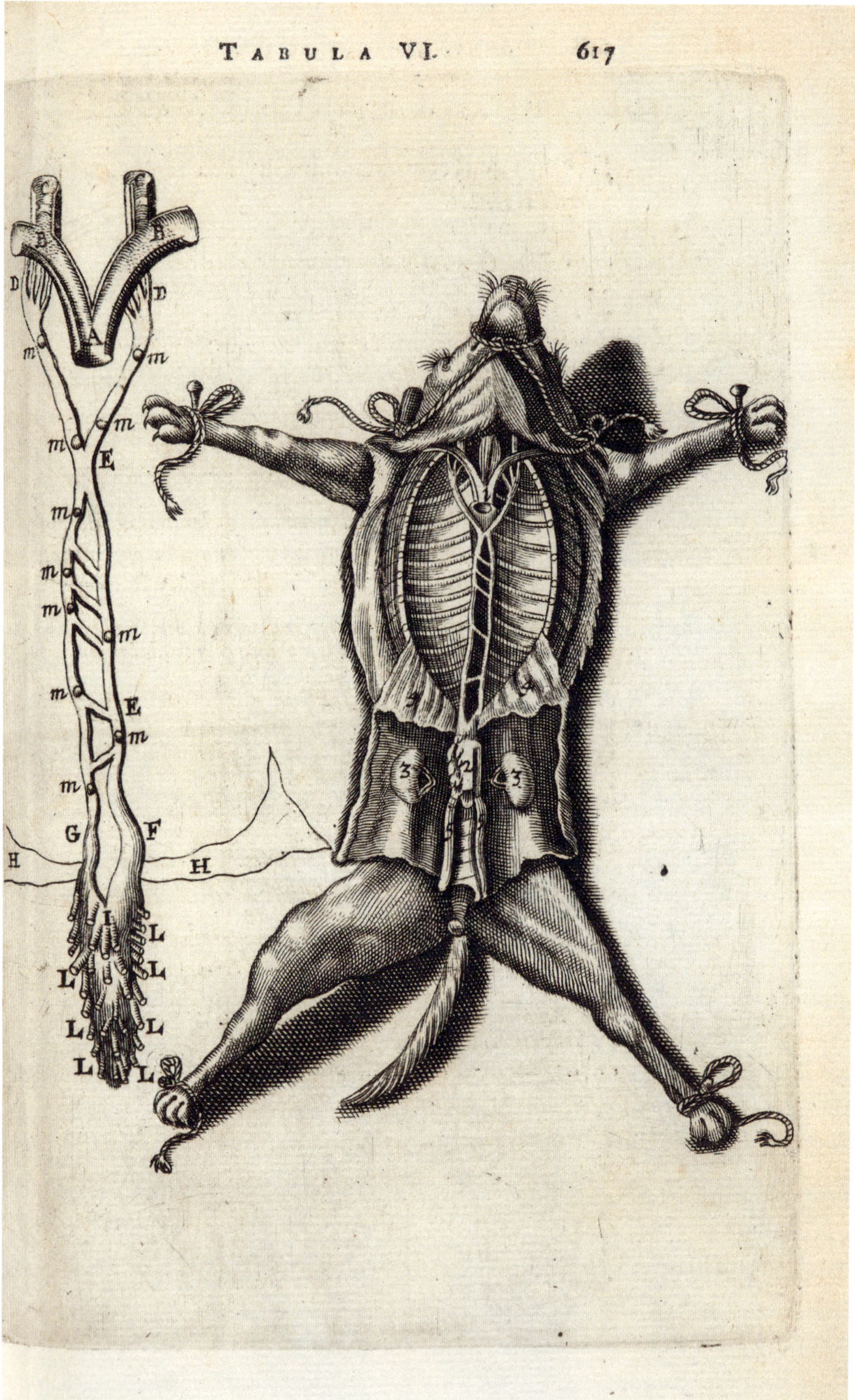
TABULA VI.
617
B
B
D
D
A
m
m
m
m
E
m
m
m
m
m
E
m
m
G
F
H
H
I
L
L
L
L
L
L
L
3
2
3

Bartholin and Lyser undertook this dissection on 15 December 1651 at the Anatomy Theatre in Copenhagen in the presence of, among others, Ole Worm and Henrik Fuiren, along with the French physician Pierre Bourdelot (1610–1685) (Figure 107), "an associate of Pecquet and a very learned man,"[65] who was on a passage to Stockholm visiting the queen, Christina of Sweden (1626–1689) (Figure 108), "the incomparable marvel of her sex and of her century,"[66] who was interested in science, art and philosophy. In the years around 1650 she summoned a swarm of Europe's most learned men with the purpose of transforming the Swedish court into "a centre for knowledge and scholarship of the highest worth," as she herself phrased it in a letter to Pierre Gassendi.[67] Some of the scholars, like Gassendi, were content to correspond with the energetic queen, while others allowed themselves to be enticed to a stay in the cold North, which was not necessarily free from risk. Three months' stay in the penetrating cold of Queen Christina's castle in the winter of 1649–1650 was more than René Descartes could stand. He suffered pneumonia and died in Christina's care in February 1650. The circumstances of the visit and his death are recounted by Rasmus Bartholin in a letter from Leiden to Ole Worm in Copenhagen on 12 April 1650:

"According to reliable information in a letter from the royal French envoy himself, who stays in Sweden, and whose hospitality Descartes enjoyed, together with a letter from the deceased's servant to Hoghelande[68] and the queen, we have come to know that, to comply with the queen's requests, every day at ten o'clock in the morning he [Descartes] instructed her in his philosophy, which she was very keen to dedicate herself to. And it sometimes happened in the bitterest heart of winter that, coming home from the queen's library (where the instruction took place), he was so chilled with cold, that he asked for wine spirits as a remedy. But the bitter cold, it was thought, could not be alleviated in any other

Figure 106. The dog in which Jean Pecquet found the thoracic ducts (E) and the collecting chyle container, *receptaculum chyli* (I). From Thomas Bartholin's *Anatome Quartum Renovata*, Leiden 1673, p. 617.

The remaining letters in the left figure mark: A. Superior vena cava. B. Confluence of the jugular and shoulder veins. C. Valves preventing the chyle from running back from the jugular veins into the thoracic ducts. D. Branches of the thoracic ducts that join the jugular and shoulder veins. F and G: The dilated beginnings of the thoracic ducts that pierce the diaphragm and enter the thorax. H: Diaphragm remnants. L. Mesenterial chyle vessels close to the chyle container that are equipped with valves to prevent backflow. *m*. Various valves in the thoracic ducts preventing the backflow of chyle towards the chyle container. The right figure: 1. The ascending vena cava. 2. Part of the ascending aorta located between the kidneys and adherent to the chyle container. 3. The kidneys. 4. The dissected diaphragm. 5. The psoas muscles.
The Royal Library in Copenhagen.

Figure 107. The Parisian physician Pierre Bourdelot (1610–1685). Engraving by Simon Thomassin (1654–1733) after template by Nicolas de Largillière (1656–1746).

Thomas Bartholin's friend Bourdelot had visited him in Copenhagen in December 1651 on his way to visit Queen Christina in Stockholm. Bartholin had demonstrated on a live dog the newly discovered lymph vessels, and Bourdelot was convinced that they were not milk veins but a new kind of vessel. In Paris, Bourdelot had a private experiment-orientated academy, where medical science was the centre of attention. It was situated in Rue de Tournon in the Latin Quarter south of the Seine and later hosted Bartholin's two pupils Niels Stensen and Ole Borch as frequent guests. Archive.org.

Figure 108. Queen Christina of Sweden (1626–1689) portrayed by Sébastien Bourdon (1616–1671) in 1653. Oil on canvas. 72 × 58 cm.

When the science-loving Swedish queen heard about Thomas Bartholin's research on the lymphatic system from the Frenchman Pierre Bourdelot, she was thrilled, praised him and encouraged her countrymen to be inspired by Bartholin. Bartholin similarly praised Christina for her wisdom and interest in science, but unfortunately they never met. National Gallery of Sweden. Wikimedia Commons.

way than by going to bed afterwards. When the queen had received information about his weakened health, she immediately sent one of her physicians, but when this person wished to bleed him, he would not initially allow it. When he later developed pleuritis and high fever, he agreed to have himself bled three times, but with poor results, so that he departed this life on 1 February to the greatest sorrow of everyone, including the queen herself, of whom we hear that she had a wax effigy made of him. But although the queen ordered that there should be a final honour shown to him with every type of pomp, and that he should be buried in the main church, as a result of some kind of superstition he was interred outside the city in a churchyard where it was customary to bury infants."

That Descartes' fate served as a warning for Christina's future French guest, Claude Saumaise, appears also from Rasmus Bartholin's letter: "Saumaise will start on his way as soon as the weather becomes bearable for his age, and he will not allow himself to be committed as long as there can be cause for him to experience any danger from the cold – he is frightened after the example given by René Descartes." Saumaise, who became one of Bartholin's lifelong friends, stayed with him in the professor's residence in Copenhagen on his way to Stockholm,[69] and perhaps he might as well have remained there, for in a letter from Leiden to Thomas Bartholin in Copenhagen he writes on 11 July 1652: "I do not like the Swedish intellectuals, but prefer those from Denmark, no matter if one seeks erudition, an ingenious scholar or anything that is disposed to create cultivated human beings."[70]

Back in Paris, Saumaise wished nevertheless to pass on the baton for a further visit to Queen Christina, and he had chosen Guy Patin, who, however, was afraid of Descartes' example and declined to take the trip to Sweden, "for he was too fearful of the cold in that country."[71] Instead Saumaise sent Bourdelot in this capacity. Thomas Bartholin, who never met Queen Christina, had this conclusion about Bourdelot's dissection of a live dog:

"But since the chyle vessels very quickly perish during a living animal's strong movements, and towards the end of the painful investigation, and since I had noticed on another occasion with hanged persons that the chyle vessels were preserved, I believed that the investigation would also be successful, if one either strangled the animal speedily, some hours after it had had a meal, or made a dissection on a person who had been condemned to the gallows, or who had suffocated after a copious lunch, because the absorbed liquids become stationary in the same place when the movements of the thorax have come to an end, because respiration has stopped, and they are not set in motion anymore by the pain the dissectors' knife causes, which happens so easily otherwise."[72]

That Bartholin, in contrast to Pecquet, was not completely prepared to write off the liver as a blood-generating organ follows from his letter to another great physician of the age, the German polymath and professor of medicine in Helmstedt, Hermann Conring, on 20 December 1651, five days after the dissection of the dog:

"But since you have taken up with your most learned pen the question of the channels of the chyle and its production in your book *De Sanguis Generatione* [On the generation of the blood],[73] please allow me to ask for your verdict on the new receptacle of the chyle. In a recent dissertation, the Frenchman Jean Pecquet observed, with unprecedented success considering his age, a new receptacle of the chyle beneath the mesentery above the lumbar vertebrae between the psoas muscles, which draws out the chyle from the lacteal veins of the mesentery and pours it back in a continuous movement through the ducts, which have a whitish appearance, to the subclavian branches, into which they are inserted, so that chyle at the level of the external jugular veins is mixed with the blood in the *ven a cava* on its way to the heart. But he saw no lacteal veins leading to the liver, so he only assigns the liver the role of percolation organ. [...] I have made diligent investigations into the truth of the matter, and in the dogs that were dissected and disembowelled approximately four hours after they had eaten, I found everything quite clear, and several things that Pecquet himself had not drawn attention to. For example, a remarkable cavity of the receptacle situated beneath the left kidney, and two tiny, slightly distended tubes hidden beneath the diaphragm, one on either side; sometimes they are joined, and sometimes they are separated by the tiny branches in between, and they go up along the dorsal spine to the left subclavian vein, into which they empty their chyle, as can be seen by pressing with the fingers and with a knife. However, I thought that I observed that the lacteal veins also spread into the liver from the mesentery, and I am therefore unwilling to banish the liver completely. Instead, I believe that it shares the task of blood production with the heart, because the amount of chyle in the mesentery is more abundant than it would be if chyle only streamed through these new, very narrow channels in the thorax. And in this way thinner chyle is perhaps carried directly through these new channels to restore the heart's strength, and what has been brought there causes a sudden renewal of strength, while the thicker chyle, which needs a longer time to digest, enters the liver. In any case, coarser foods remain in the dogs' stomachs, while finer foods are distributed in large quantities to the receptacle and the new ducts. But I await your opinion on these matters, until I can carry out further experiments on a bear, which I am keeping for this purpose, and on a man who is to be hanged after he has eaten. His Royal Highness, our King, has kindly promised us his royal presence at these events. Be so kind as not to delay your reply, so that with your approval I may be sure about all these matters before I publish my planned treatise on this subject. My revised anatomy book has arrived from the press; in it I have more than once mentioned your illustrious name, which stands out here and everywhere like hills on the plains and posies in a garland."[74]

Conring answers Bartholin on 6 February 1652 and states that he did not know about Pecquet's discovery.[75] He is sorry for his late reply and gives the reason, that he did not wish to answer before he himself had carried out similar investigations on live dogs. In this experiment he has made the same discovery as Bartholin, but besides he has observed thin branches going out from the thoracic duct to the large *vena cava*.[76] Conring expects that, in the future, further connections will be found between the lacteal veins and the vein system, since he believes that the quantity of chyle in the mesentery is too large for the thoracic duct's entrance into the left subclavian vein to be the only connection. He maintains, contrary to Pecquet's view, that connections can be found between the lacteal veins and the branches of the portal vein and the *vena cava*, both in and outside the liver, which both he and others (among them Aselli and Bartholin) have seen, but they are also too few and too small to be responsible for the transportation of all the chyle together with the thoracic duct. Conring believes, however, that based on Pecquet's recent discoveries, the liver must be assigned a lesser role in blood formation than hitherto, and he proposes that blood formation takes place both in the veins and the liver, so that eventually it is completed in the heart, whose heat is decisive in the process. He is sceptical about Bartholin's hypothesis that chyle restores the heart's strengths. In the future, he wishes to investigate further whether the small lacteal veins leading to the liver play any role, but he has a firmer belief that Bartholin will first discover the truth about them. It soon turned out that he was right in that assumption.

Bartholin's professional respect for Conring was enormous. His reply comes on 20 May 1652,[77] in which he compliments Conring for his observation, that the thoracic duct supplies small branches to the *vena cava*, for his view on the distribution of chyle and for his arguments in the disagreement about chyle's possible work of restoration on the heart. Bartholin sends along *De Lacteis Thoracicis*, in which he has repeated Conring's answer,[78] and he asks for Conring's assessment of the treatise.

On 15 March and 18 May 1652, Pierre Bourdelot sends from Stockholm a couple of very commendatory letters to Thomas Bartholin. He has told Queen Christina how Bartholin benefits his profession by hard work, experiments and common sense, and the queen has replied that she wished that there were Swedes who were gripped by the same ambition for truth and honour.[79] The queen is delighted to have a man near her who is so proficient in anatomical dissections. Bartholin's excellence, about which she has said delightful things to the professors in Uppsala, has worked as a stimulus on them, and they are now eager to carry out dissections on living creatures themselves. One of them states, that without knowing Pecquet's name, he has made the same observations of the chyle vessels in the thoracic duct, "and he promises a lot of things."[80]

Bartholin answers Bourdelot on 27 May 1652, saying that he feels honoured that his research and anatomical work has won favour with such a great queen. He desires information about what the physicians in Uppsala have observed concerning the lacteal veins in the thoracic duct. He tells Bourdelot that, at the first opportunity, he will send him a treatise on the discoveries he has made on a human being (the lean child murderer) since Bourdelot's departure from Copenhagen. He mentions having discovered three glands in the groin region at the normal place of the receptacle, and that he has only found an entrance of the thoracic duct in the subclavian vein on the left side, not on both sides.[81]

Discovery of a new type of vessel with aqueous content

In a letter to Thomas Bartholin on 23 February 1652, Hermann Conring's pupil Frederik Arnisæus (1621–1654), son of Christian IV's personal physician and coroner Henning Arnisæus (1570–1636), expresses annoyance that Jean Pecquet denies the entry of lacteal veins into the liver, when everyone during every dissection can see it with their own eyes.[82] In Bartholin's answer to Arnisæus on 30 April 1652, we gain insight as to the goal of Bartholin's further research and his great expectations of it. In this letter Bartholin describes in writing for the first time the lymph vessels as a new type of vessel with aqueous content close to the liver and in the limbs:

"However, I still cannot give up the liver, even if I partly support the heart, now that these lactiferous ducts in the thorax have been identified. Perhaps Caecar can share his power with Jupiter, and the two cooks share the production of blood for so large a family, if Aselli's lacteal veins really reach the liver. Pecquet denies it and refuses to assign any importance to them. However, they have been visible to everyone up to now. Or have they seen a shadow instead of Juno?[83]

I have demonstrated them more than once, and I defend them constantly, although my mind has begun to weaken. However, I must not shift the old boundaries so far, so long as there is support for the liver. You consider many new matters, and unless I am making a great mistake, I shall soon reveal a new kind of vessel, which I must still not unveil in public before I have proven my theories through a series of experiments. So, Pecquet can really have a legitimate reason for disallowing lacteal veins close to the liver. For those that are clear do not contain a milky, but an aqueous fluid, and on several occasions I have noticed such things, not only close to the liver but also in other places, including the limbs, which at last spring forth in a new stream.[84] But I am bewildered, and I am not satisfied with myself. In the meantime, I shall now illustrate the known lactiferous ducts in the thorax and work hard on behalf of the liver, so that I may not, if things in the future prove unfavourable for it, and new vessels that are hostile to it come to the fore, be blamed for inconsistency or thoughtlessness. Clouds are forming, which will result in a great torrent of rain and drown the field of anatomy. So eventually we shall await a clear sky from the supporters of truth, whom I shall invite to this play."[85]

The story was that, in their dissections of different dogs during their investigation of Aselli's chyle vessels in the winter of 1651–1652, Bartholin and Lyser had encountered a new kind of vessel in the abdominal cavity, which contained a clear aqueous fluid. Bartholin mentions them a total of three times in *De Lacteis Thoracicis* in connection with three different dissections of dogs, the first of which is in Chapter 6:

"It was remarkable for us to find in a dog, which was fairly large and healthy, and which was opened up seven hours after a meal, vessels that very much resembled chyle vessels, but through which we could sometimes see chyle, sometimes serum, and which meandered from the mesentery or the newly-discovered receptacle, partly up towards the liver together with the portal vein, partly along the kidney veins and partly downwards following the course of the *vena cava* to the pelvis, which they stretched across and were attached to by a thin membrane, until the entrance to the bladder concealed the view of them. When they became closed with an easy ligature, they swelled up below the ligature but emptied themselves up towards the mesentery in such a way that the suspicion arose in our astounded minds that this was a special type of vessel intended for serum, but we suspended our judgement about it until other observations had been carried out."[86]

On the opening of these vessels Bartholin and Lyser ascertained that they contained a volatile fluid that was still not as white as chyle normally is when it is on the point of disappearing. In Chapter 12, Bartholin recounts another dissection of a dog, which was performed on 4 January 1652:

"It is remarkable, what we found in a dog, which was dissected seven hours after a meal, that in the abdomen certain serous vessels followed the veins out to the forepaws on both sides, transparent like those we saw on another occasion in the thorax and the abdomen, whose source as well as mouth we were unable to find, but when they were ligated, they swelled up in the part facing the limbs."[87]

Finally, Bartholin mentions in Chapter 15 the dissection of the dog in the Anatomy House on 15 December 1651, when he gained the opportunity to show this type of vessel to Pierre Bourdelot and discuss them with him. Interestingly, Bourdelot denied both that they were chyle vessels and that it was chyle that was emptied from them, even though at this point Bartholin held to the idea, that when it was not blood, it must be chyle:

"In order to get nearer the truth, I have spent the whole winter seeking facts in the dog's internal organs, and having opened up several, which had been well fed, in our anatomy theatre, I investigated in the presence of many individuals large and small dogs to see if chyle vessels went into the liver; something we had earlier seen repeatedly, but that Pecquet now denies. Then many whitish fibrils became visible, which sought the liver along the *vena portae* recessed into its membranes, in the same way as Aselli and Highmore depict it. Sometimes they were clearly distended, most often by a serous fluid, which you could see through the vascular wall, but if you opened either these or those, they poured out an aqueous fluid, which was still not as white as chyle normally is when it is about to disappear. Dr Pierre Bourdelot [...] therefore firmly denied that they were chyle vessels, and that the discharge was chyle. But if it is not blood, nor the *vena portae*, nor any other vessel known hitherto, I have judged that there is no other course than to take it for a chyle vessel. [...] It is certain that on several occasions we have seen serous chyle of this sort, both in the receptacle and in the chyle vessels, and indeed, even that the serum changed with and followed the white chyle in all the chyle vessels, in the abdomen just as much as in the thorax."[88]

In January 1652, at the deadline for the delivery of the manuscript regarding the treatise on the thoracic ducts, the status concerning the lymph vessels was therefore that Bartholin had a realization, or in any case a clear suspicion, that he and Lyser had discovered a new kind of vessel of another nature with a larger spreading area, another appearance, a more aqueous content and in certain cases a different direction of flow than that of the familiar chyle vessels. But for the aforesaid reasons he could not for the moment regard these new vessels as anything other than chyle vessels. This last conclusion would quickly change.

The lymph vessels – *vasa lymphatica* – the aqueous vessels

When Bartholin and Lyser on 28 February 1652 conducted the customary anatomical exercises, they became very certain that something big was in the offing, and that the vessels with the clear contents that they had perceived earlier were not identical to the chyle vessels already known. The day's programme for dissection was that of a

> "fairly large, healthy dog, on which we were not troubled by an all too thick layer of fat, seven hours after it had had a meal, to notice more thoroughly how the lacteal ducts in the thorax behaved for a longer spell after a meal. The stomach was still full of food. In the mesentery there could be seen as usual the shining white vessels of chyle as in the thorax. A large profusion of vessels went into the liver, along with the *vena portae*, which were heavily filled but certainly not with chyle, but with an aqueous fluid that could be seen through the walls. Several similar water-bearing vessels could be viewed surrounding the *vena portae* in ring formation and further down enveloping the stem of the descending *vena cava* from all sides. [...] When they were ligated in one place or another, they swelled up just below the ligature, but above this in the direction of the mesentery they were emptied, which is why also the branches of chyle vessels around the diaphragm, which were earlier shining white, were now filled with a similar aqueous fluid, as the chyle was driven away, and the strongly filled lacuna lying above the diaphragm and going out from the receptacle appeared very swollen, stretched by this serum. But the thoracic ducts continued to be filled with chyle up to the axillary veins because of the ligation. But what filled us with amazement was that when we ligated the *vena axillaris* at the forelegs, there now appeared the same kind of aqueous vessels as we had seen earlier in the abdomen, swelling and shining like hydatids,[89] stretched across the *vena axillaris*, and when one ligated them, they continued to be filled on the side turning towards the foot end. Much the same thing occurred on the right side, and when one bored a hole in them, a very clear fluid immediately emptied from them, which flowed towards the surface of the breast. The whole thing seemed to us almost like a miracle, because it was so uncustomary, and I did not trust my eyes and a single experience, even though it was clear enough. For one must vary the material, little by little wheedle out of Nature its secrets and not rashly resign oneself to what one

has seen only once, when for centuries general experience speaks against it. Whether this was a special type of vessel specifically for water, we now discussed back and forth, and one conjecture shifted to another, just as we also discussed the question of where they came from and where they were going, if it was after boiling had finished or during it, that the water either disappeared or was absorbed by them; but I allowed that it should remain undecided. I therefore deemed it necessary to defer judgement on this new circumstance until we had had further experiences."[90]

When a little later a small dog became available for dissection, Bartholin and Lyser did not waste the opportunity to confirm their discovery, which they succeeded in doing. As Bartholin phrased it:

"In order not to render so many martyrs useless, we strangled it quickly with a cord, seven hours after it had had a plentiful meal, and eviscerated it. Without wasting time and with disregard for everything else, we immediately went into the extremities and the liver, and saw, as we had earlier, these aqueous vessels, here connected with the *vena portae* with numerous rings and, as it were, sunken in its wall and attached to *ramus axillaris*[91] like ivy."[92]

The two anatomists proceeded energetically with dissections on a long line of other dogs, both fasting and full dogs, pregnant and milk-producing dogs. They discovered the same thing in all of them. In the same way as they had detected a valve at the entry of the thoracic duct in the shoulder vein, they noticed valves at the entrance of the new vessels in it.

The next thing to be resolved was the old question as to whether the chyle vessels went into the liver, or whether the vessels that Bartholin, Lyser and many others had so often seen around the liver were some of the new vessels. If this were the case, Pecquet was right in refuting the idea that chyle vessels went into the liver. A definite distinction between the new vessels and the chyle vessels would, according to Bartholin's investigation, require that a living stream of fluid flowed through them, and in order to settle the question he and Lyser therefore dissected a live dog, which had been fed four hours beforehand. They saw no shining white chyle vessels in the area around the liver and gall bladder, but only the newly discovered transparent vessels containing a clear fluid. Only in the mesentery did they see shining white chyle vessels. After having ligated the portal vein and the surrounding new vessels with a ligature close to the liver, the two physicians saw that the new vessels swelled up on the liver's side of the ligature, while the portal vein, as expected, swelled up on the opposite side of the ligature.

With this, Bartholin and Lyser had found the final proof that the vessels they had often seen in the vicinity of the liver were not chyle vessels, which brought chyle from the intestines to the liver, but the newly discovered vessels, which brought fluid *away* from the liver. After this, Bartholin had to support Pecquet in abandoning the explanation of the liver as a blood-producing organ.

Since Bartholin had great expectations of quickly being able to demonstrate the new vessels also in human beings and getting this into his publication, he chose in the first instance to spend time on these matters and postpone the publication of his and Lyser's discoveries in animals. He states soberly:

> "However, I neither could nor should settle with what I have discovered before I have involved human bodies too in my deliberations. For the honour of making a discovery is less if it is not extended to include all types of living creatures. It increases tremendously when it also includes human beings, which would contribute a great deal to its future importance."[93]

He emphasizes that the greatest discoveries of the age, namely Aselli's of the lacteal veins, Pecquet's of the receptacle and thoracic duct, and Harvey's of the circulation of the blood, were all carried out during experiments on animals, and that none of these researchers found correspondences in human beings, though none of them doubted that the same features may be found in human beings.[94] They were therefore dependent on others to complete the work. The problem with the new vessels was that, even more quickly than is the case with chyle vessels, they disappear after death. Bartholin writes that their walls are made of fine membranes, which resemble cobwebs, and which disappear completely when the transparent fluid has been drained out. The dissection of live human beings was very much more than the God-fearing Lutheran Thomas Bartholin wanted to throw himself into: "The organs of living human beings are forbidden to us by our humanity and our God-fearing."[95]

Besides, in Bartholin's time there was a real fear that the bodies of hanged criminals would wake up again if the dissection were begun while the body was still warm. Bartholin writes on this subject:

> "I have no doubt that they [the lymph vessels] are more easily found in a newly deceased human body that is still warm shortly after hanging. In a woman, whom I recently opened, the heat had still not completely disappeared. No-one surely wishes to dissect anyone fresher than that. The experiences are alarming, and we fear, that the dead will wake up again. Riolan reports quite a few examples of people who have come to life again after being brought to the anatomy theatre. In the last years it was rumoured among one or two people in Oxford that some-

thing similar occurred, and this winter the Hamburg physician Schlegel broke down completely, because when he had stuck his knife into a criminal who lay on the anatomy table, he was terrified to observe that this person moved."[96]

Bartholin also has an explanation of the phenomenon:

"The breathing is actually brought back easily, provided the closure – either through the fault of the executioner or because of ossification of the larynx cartilage – has not been complete, if the body is either washed in warm water, which is the normal custom, or provoked by a small irritation from the knife."[97]

According to the story, the Hamburg physician Paul Marquart Schlegel (1605–1653) (Figure 109) did not simply fall to the floor at the sight of the awoken body but died two to three weeks later from shock at the experience. The story, however, appeared unable to withstand further investigation.

Michael Lyser left at the beginning of May 1653 on a longer scientific journey to Germany and Italy. With him he had copies of Bartholin's recently published treatise (see below) concerning their discovery of the lymph vessels in dogs for distribution among leading researchers in Bartholin's circle of acquaintance. It was in this treatise that Bartholin brought in the account of Schlegel. One of the first to receive a copy was Schlegel's close friend, Hamburg's medical officer Johannes Placcius (1605–1656), who most certainly denied the account. Lyser writes to Bartholin from Wittenberg on 15 June 1653:

"In Hamburg I gave the same treatise to Dr. Placcius for his perusal. He was very satisfied with everything and read it through twice. Only the remarks about the blessed Schlegel he liked less. He gave his solemn word that the rumour that has been circulating is untrue. [...] It was on the third day after the hanging that Schlegel set his dissecting knife into the criminal's body; therefore, I ask how any residue of life could have remained hidden so long? Poor Schlegel had earlier complained about his weak health, and the day after he started the dissection, he began to bend under a malignant fever with a rash, which took his life on the seventeenth day. Placcius therefore gave the friendly advice that on occasion you should adjust the passage, so that no-one could accuse Your Highness of falsehood."[98]

Bartholin's pupil Henrik Møinichen (1631–1709), who in the following year went on a similar journey and visited this same Placcius in Hamburg, confirmed to Bartholin in a letter from Augsburg what Lyser had written:

"The physicians in Hamburg, particularly the medical officer Dr. Johannes Placcius, told me that Your Eminence made a mistake in drawing the conclusion

Figure 109. The Hamburg physician Paul Marquart Schlegel (1605–1653), who according to rumour fell to the floor and died of fright when the body of a hanged criminal he was dissecting suddenly moved. Oil on canvas. 1653. Artist unknown. 90 × 65 cm. The Latin text "PAULUS MARQUARDUS SCHLEGIL. Philosoph. et Medic. Doctor. Physicus Civitatis Hamburgensis Ordinarius. Christ. Anno. 1653. Die February. Ætatis 48" reads: "Paul Marquart Schlegel, Doctor of Philosophy and Medicine. Ordinary city physician of Hamburg. February 1653. Age 48." Staats- und Universitätsbibliothek Hamburg: Gemäldesammlung: 6.

that the illustrious Schlegel died of fright. [...] Placcius solemnly maintained that Schlegel died of a malignant fever. He said that the rumour about dismay having caused his death had been invented by ill-natured people, and that no-one had observed any sign of the least movement in the body during the dissection. This man, Placcius, said that he had passed on these pieces of information to the excellent Michael Lyser, and that he had asked him to make Your Eminence aware of them in a letter. Whatever the truth may be, the rumour circulates constantly among the inhabitants of Hamburg that he died of fright, but all the physicians know the story and agree that the cause of death was a malignant fever."[99]

Whether due to Bartholin's indisputable fondness for good stories, or the fact that he felt himself too important to allow himself to be influenced by this type of suggestion from his assistants and a medical officer in Hamburg, or some other explanation, he waited a very long time before he followed up the request. In the collected work of his own and other selected writers' discoveries concerning the system of lymph vessels, *Opuscula Nova Anatomica De Lacteis Thoracicis et Lymphaticis Vasis* (New anatomical writings about the thoracic ducts and the lymph vessels) from 1670,[100] Bartholin produces a new version of the story. Schlegel's cause of death has now been corrected to the true version, but as an extra seasoning to the proper story the body's movement is now changed, so that the body simply hit the horror-struck Schlegel to the floor with a smack in the face:[101]

"In the later years, rumours have circulated in Oxford about something similar, and this very winter the Hamburg anatomist Schlegel is said to have fallen ill, because when he laid his knife on a criminal, who lay on his anatomical table, to his alarm he saw that the body moved. But in reality, this man, who was created to cast lustre over anatomy, died of a pernicious fever, causing great harm to the republic of medicine. This happened several weeks after the arm of the corpse, which had been bent backwards and now straightened itself, had given him a punch in the face, which Johann Daniel Major bears witness to in his *Anat Chil. § 25. ch. 2.*"[102]

In *De Peregrinatione Medica* (On medical travels) from 1674, Bartholin contents himself with noting: "[...] the hand of a dissected cadaver cast terror into Schlegel."[103]

When throughout 1652 Bartholin and Lyser, with "deep attention" to the lymph vessels, had carried out dissections on a total of six strangled or hanged, fasting or fed criminals without finding with certainty these vessels in any of them, they had to give up their project for a while. The otherwise optimistic Bartholin concludes with barely concealed vexation: "But I am as sure as anyone that the mentioned lymph vessels constantly conceal themselves in the human being, and that the honour of finding them completely will fall to others. Someone will come, who in a human being will put the final touch to our investigation, and to him we shall come to owe the same large debt as Pecquet does to us.[104]

Instead, Bartholin now prepared at top speed a treatise about his and Lyser's discovery of the lymph vessels in animals. He called the new vessels "lymph vessels" (Latin, *lympha* = clear water) or water-bearing vessels, and the treatise was given the name *Vasa Lymphatica, Nuper Hafniæ in Animantibus inventa, Et Hepatis exseqviæ* (The lymph vessels, which were recently discovered in animals in Copenhagen, and the abandonment of the liver) (Figure 110).[105]

It was published on 1 May 1653, and Bartholin dedicated it, not without strategic purpose, to his worst critic, Jean Riolan in Paris. Karel van Mander and Albert Haelwegh were again appointed to illustrate, this time the dismembered body from the vivisection of the large, healthy dog – Bartholin's own dog named Molossus – which had had a meal seven hours previously (Figure 111). Bartholin later honoured Molossus with a poem (see later).

In this treatise he contented himself by honouring the abandoned liver with the following epitaph after a long introduction in his usual ironically humorous manner:

STOP WANDERER!
IN THIS TOMB IS HIDDEN THE ONE
WHO HAS BURIED SO MANY,
YOUR BODY'S FIRST COOK AND LORD,
THE LIVER, KNOWN BY CENTURIES,
BUT UNKNOWN BY NATURE,
WHO BY RUMOUR ALONE FORTIFIED THE MAJESTY
OF ITS NAME AND DIGNITY,
AND PRESERVED IT BY THE POWER OF POPULAR OPINION.
SO LONG DID IT BOIL,
UNTIL TOGETHER WITH ITS BLOODY KINGDOM
IT COOKED ITSELF TO DEATH.
GO AWAY WITOUT A LIVER, WANDERER,
AND LET THE LIVER RETAIN ITS BILE,
SO THAT WITHOUT BILE YOU SHALL BOIL WELL
FOR YOUR OWN BENEFIT. PRAY FOR IT![106]

Figure 110. The title page of Thomas Bartholin's dissertation on the discovery of the lymph vessels in a dog, *Vasa Lymphatica Nuper Hafniæ in Animantibus inventa, et Hepatis exseqviæ*, Copenhagen 1 May 1653. The Royal Library in Copenhagen.

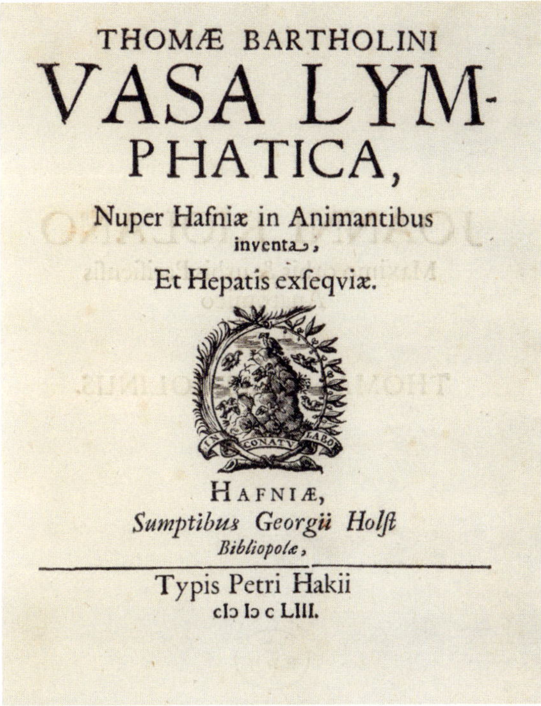
THOMÆ BARTHOLINI

VASA LYM-
PHATICA,

Nuper Hafniæ in Animantibus
inventa,
Et Hepatis exſeqviæ.

HAFNIÆ,
Sumptibus Georgii Holſt
Bibliopolæ,

Typis Petri Hakii
cIↄ Iↄ c LIII.

The dissection of Anne Michelsdatter

After this, Bartholin moved on to another question prompted by the discovery of the thoracic ducts, which he had planned to investigate further, namely the question of whether the thoracic ducts are connected to the nearby mammary glands in the breasts, and whether chyle from the thoracic ducts in this way is included in the formation of the breast milk. Moreover, as he mentioned in the letter to Harvey, he wished to investigate whether chyle vessels go directly from the mother's mesentery to the fetus in the womb in preparation for nourishing the fetus with chyle.

In *Anatomia Reformata* Bartholin had argued that the breast milk is produced from the blood and not from chyle,[107] but the discovery of the thoracic duct had once again made him doubt:

> "It seems to be not entirely impossible that these thoracic ducts, which lie so near to the mammary glands, contribute something to the milk being deposited in these. [...] There have certainly been some people who have maintained that they have seen that in human thoracic cages some branches went out to the mammary glands; but I am afraid that it is the nerves to the said glands, which have deceived them. But there are very many indications which, in agreement with Hippocrates, suggest that chyle or milk is conducted to the mammary glands."[108]

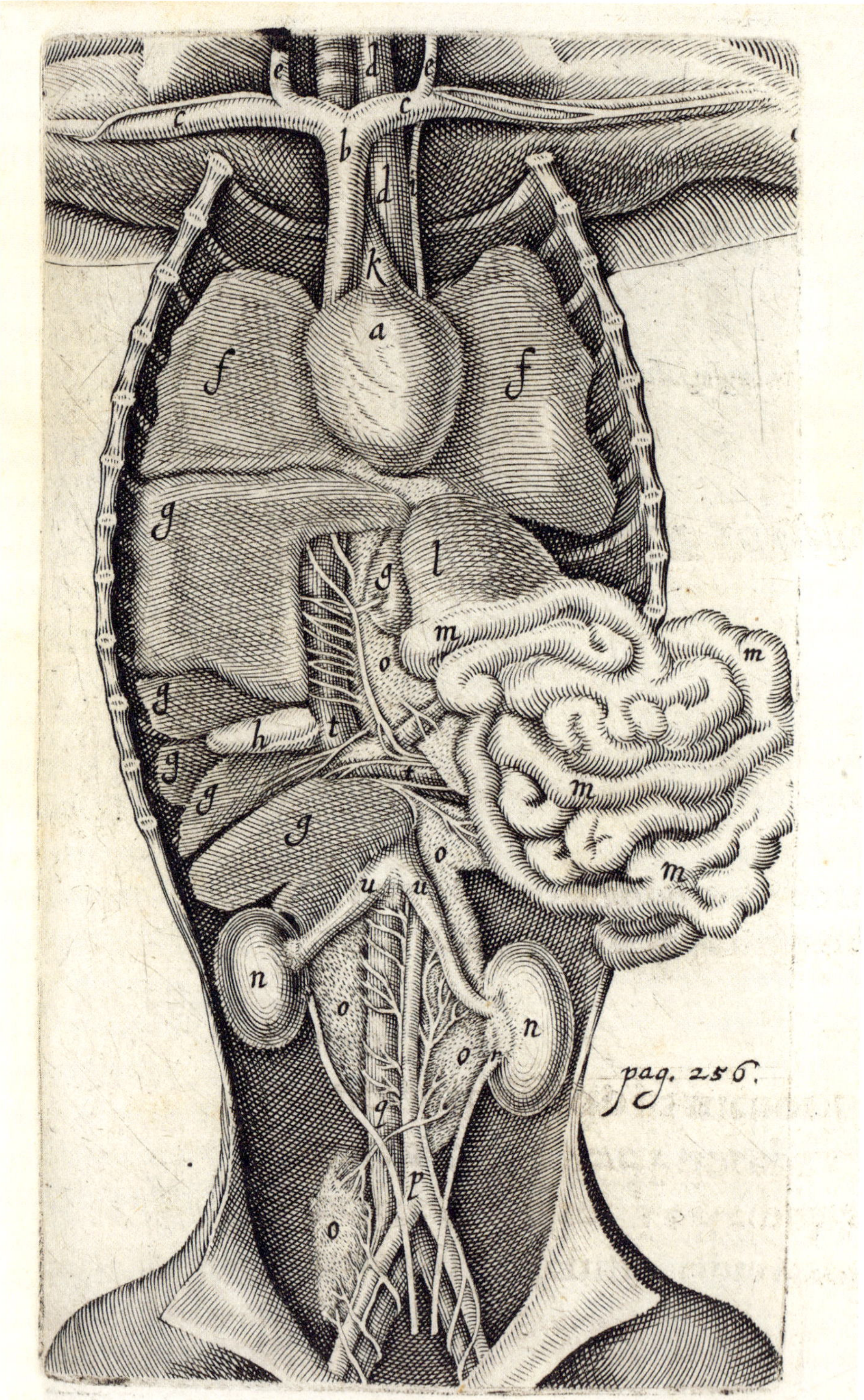
e
d
e
c
c
b
d
i
k
a
f
f
g
l
g
m
o
m
g
h
t
g
g
t
g
m
o
m
u
u
n
o
n
o
g
p
o
pag. 256.

Pecquet had denied that chyle vessels lead to the mammary glands, but the Englishman, Nathaniel Highmore, for whom Bartholin entertained the greatest respect, had, during the dissection of a live pregnant dog, made observations that supported such a connection. Bartholin was reserved towards these observations and concluded as usual that there was need for further examination: "However it may be, I have collected for this purpose some bitches, both pregnant and lactating, to which, after the completion of this writing [*De Lacteis Thoracicis*], I shall devote my whole attention."[109]

However, in February 1653 another unhappy incident occurred in Copenhagen, which gave Bartholin and Lyser the possibility of carrying out the desired examinations on a woman's body. A very young serving-maid, who seemingly could not cope with the fact that she was about to become a mother out of wedlock, had in her desperation killed her child and buried it under her employer's bed; she ended up being revealed because of the stench. At the district court she was condemned to beheading by the sword. By royal permission, Bartholin was given the body for dissection, after which it was to be handed over to Karel van Mander, so that he could illustrate it. Bartholin recounts it thus: "A rather young girl, healthy and not with a bad-looking body, but who had shown her cruel side in killing her child, had recently, six weeks after the birth, been arrested, punished by beheading and handed over to me for solemn dissection."[110]

Copenhagen's city bailiff, Claus Iversen Ravn,[111] whose duty it was to pay the executioner for his trouble, writes under items of expenditure, on 12 February 1653 in his annual account for 1 May (Philippi Jacobi day) 1652 to 1 May 1653: "Given the executioner, who according to judgement of the district court beheaded with sword Anne Michelsdatter, who served Arent Boemand in Vimmelskaftet, and there in their absence gave birth to her child and murdered it and in their own bed, in which they lay, buried it, and because of the stench that came from it, it was revealed. [...] Then after the execution, in accordance with the King's command, Karel van Mander was given her body to illustrate it" (Figure 112).[112]

Figure 111. The opened thorax of the dog Molossus on which Thomas Bartholin found the lymph vessels. From *Vasa Lymphatica*, between the preface and page 1, here reproduced after Thomas Bartholin's *Historiarum Anatomicarum Rariorum*, Cent. 2, Copenhagen 1654, Hist. 48, p. 256. Engraving by Albert Haelwegh from a template by Karel van Mander.

The newly found lymph vessels are the fine white strands, that stretch between the liver *g* and the intestine *m* and across the vena cava *t*. Lower down in the abdominal cavity, they appear as a network of strands stretching between the large lymph nodes *o* and across the large vessels, vena cava *q* and the aorta *p*, from and to the hind legs.
The Royal Library in Copenhagen.

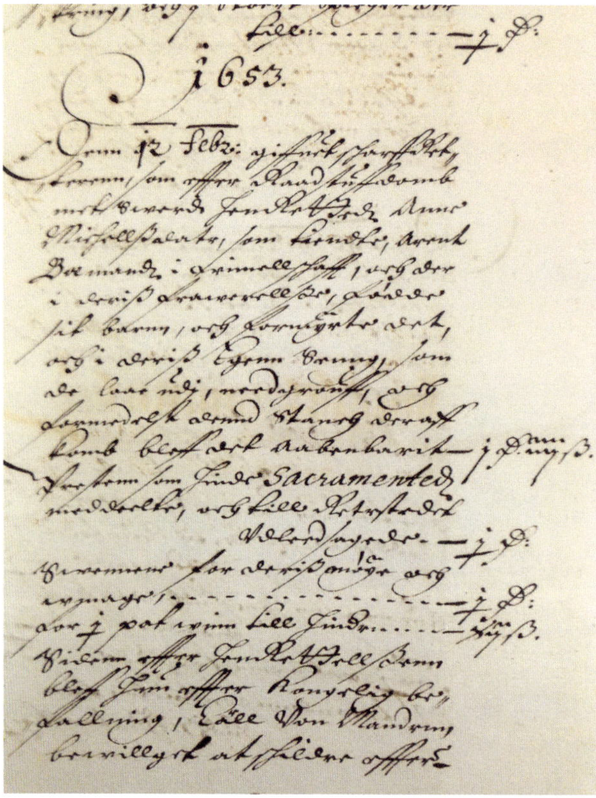

Figure 112. City bailiff Claus Iversen Ravn's (1602–1669) note in the city bailiff's accounts about the execution of Anne Michelsdatter on 12 February 1653. The Danish National Archives. Photo: Jesper Brandt Andersen.

Anne Michelsdatter's beheaded body was handed over to Bartholin for dissection on the day of her execution. He writes: "On the 12 February 1653 in the biting cold I began the dissection. When I had separated the skin and fat, I cut open the muscles of the abdomen across to the peritoneum, since originally, I had only had in mind to demonstrate the female genital organs."[113] Bartholin found the chyle vessels and chyle glands in the mesentery, but neither chyle vessels to the womb nor any direct or indirect connection between the thoracic duct and the mammary glands and their milk ducts. The young girl, who was small in stature, was well nourished. The large amount of fat made it difficult to distinguish the thoracic duct, and there were no larger vein branches to be seen around the mammary glands and their milk ducts. However, Bartholin succeeded in getting the milk ducts stylishly dissected away from their surrounding adipose tissue, so that Karel van Mander could illustrate their anatomy in a new quickly written treatise, *Dubia Anatomica de Lacteis Thoracicis* (Anatomical doubts concerning the thoracic ducts),[114] which Bartholin presented publicly on 27 June 1653 in Copenhagen with his student Henrik Møinichen as respondent, and which he dedicated to Guy Patin with the words: "To the erudite, the earlier dean, my old friend." On the dissection of Anne Michelsdatter's breast Bartholin writes:

"Since we were also concerned with the question of the milk in the mammary glands, we tackled these with the same sharp knife. After we had cut through the layer of fat, in which the glands lay hidden, there came to light a whole collection of smaller glands, in many places sprinkled with silvery strings. Ten or more very strongly filled milk ducts surrounded the papilla as though in a ring. They were filled with milk, and every single one of them was divided into several branches and continued outwards and ended in the outer edge of the glands, in the way that the reader can see it in the sketch that I append"[115] (Figure 113).

Supplying dissections of pregnant and lactating bitches did not bring Bartholin any nearer to a resolution of the two unanswered questions about the thoracic duct's possible connection with the mammary glands and the fetus's possible supply of chyle vessels and of chyle from the mother.

Demonstration of the lymph vessels in an illegal dissection of a Norwegian tailor

Bartholin's and his pupils' observations of the new vessels during the vivisections of dogs, and Bartholin's designation of the vessels as lymph vessels, won international acknowledgement during 1653, and Bartholin therefore became even more determined in his wish to get his work completed, and to become the first in the world to demonstrate the lymph vessels in a human being. He largely got others to undertake his obligatory teaching duties and entrusted the rest of his obligations at the university to others, so that he could concentrate on further research:

"We must struggle with the task that we are undertaking. [...] We shall truly exhaust our energies in vain with the meticulous dissection of living animals if we do not show on ourselves and other people that what we have found in them is true. [...] Truly every work of dissection would be needless if we did not in the end set foot on human beings. To sharpen our knife with this in mind for the benefit of the present and future will deserve the highest praise."[116]

Despite sharpened knives and great industry, after the dissection of seven further human bodies throughout 1653, Bartholin and his pupils were not successful in reaching the target of their project. According to Bartholin there was an unusual shortage of thieves and consequently of hanged criminals, leading to a delay in successful discovery.

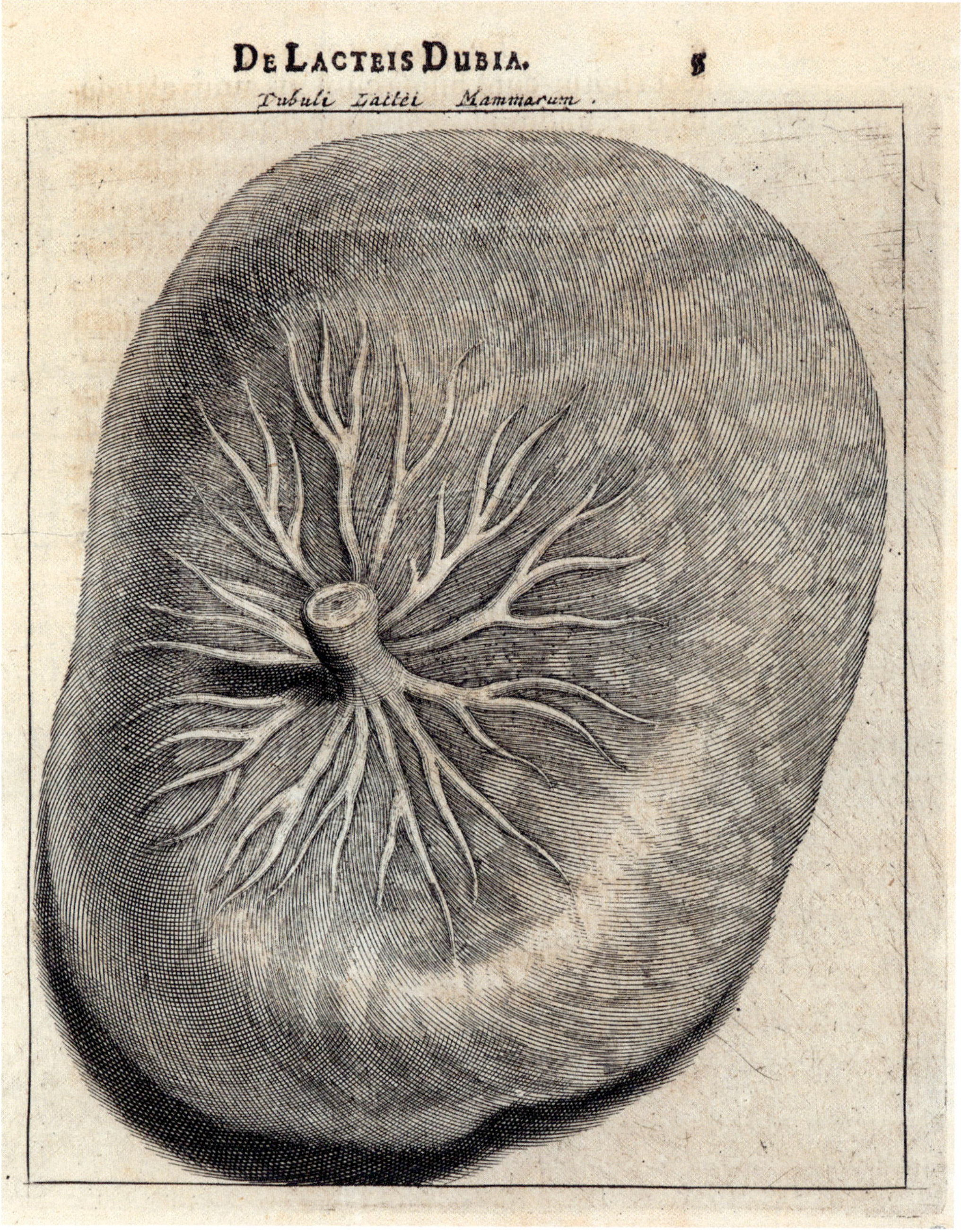

Figure 113. Anne Michelsdatter's breast with the dissected filled milk ducts radiating out from the the nipple. Engraving by Albert Haelwegh from a template by Karel van Mander. From Thomas Bartholin's *Dubia Anatomica de Lacteis Thoracicis*, Copenhagen, 29. June 1653, p. 5. The Royal Library in Copenhagen.

The thought that others should beat him to it had gradually become so unbearable that the otherwise quite loyal and law-abiding Bartholin for once found himself compelled to resort to shady methods. He had long had his own alternative plan in reserve, and at the end of March 1654 he found the opportunity to realize his plan.[117] Ole Worm had made him aware of a down-at-heel, drunken tailor living alone, who had been struck with a wasting illness, which was making him slowly wither away.[118] A lean body without superfluous fat that could obstruct the discovery of the lymph vessels was exactly what Bartholin needed. He made use of sources who were keeping the languishing Norwegian under observation, so that Bartholin and his dissection team – like a flock of vultures patiently awaiting a weak animal's demise – could intervene when Nature had completed its work. According to Bartholin it was, however, neither he nor Worm who had pointed out this man as a future model for dissection, but Nature herself, and this was also Bartholin's explanation: early in the course of his illness, the man had lost his voice, so that he "could say nothing but one small word, which must have meant yes", and so he was not in a condition to disclose for others the treasure his body enclosed, to which Bartholin would soon be given access:

"But our consumptive patient appeared to have lost his voice, not because of some obdurate submission to his pastoral care, but because of a physical defect. Something lay hidden in his body, which, for the time being, one had to be silent about, and which must not be disclosed by anyone before the hour of his death. He was silent, while he lived, so that we might find in his dead body, what we would have been able to speak about loudly. [...] He was silent, until we made our discovery in the animals, but at his death he declared, as it were, his praise of us for this discovery, when others insidiously tried to rob it, and in every feature he awarded our diligence with the palm of glory. Thus, by keeping silent he reserved for us the greatest treasure."[119]

Whether the Norwegian would also have spoken that little word meaning "yes" to this interpretation is indeed doubtful. But it is true that, after death, his body came to serve a higher scientific purpose. In addition, Bartholin writes about the man:

"A certain Jens Lauritsen, a Norwegian, who during his remarkable existence pursued his livelihood through tailoring, thirty years of age, somewhat stout, athletic in build and of fairly irascible temperament, was two years ago attacked by consumption, and in the course of some years he became emaciated across his whole body to such a degree that he could scarcely hang together, and is now as transparent as a Punic lantern.[120] [...] When he had been attacked by the wasting illness, he changed his residence on various occasions. He never settled down in peace but had more lodgings than friends. Finally, during this reticent and pitiful inactivity in life he moved to a miserable residence, which lay right up against my father-in-law's garden here in the city,[121] where in cramped physical conditions and wretched housing conditions he dragged out his life alone in a miserable manner with the assistance of drink, until, completely deprived of his remaining powers of speech, he passed over to where they say no one returns. Four hours after his death I hurried secretly to his residence, together with the scholarly medical candidate Henrik Martin à Møinichen, who was accustomed to helping me with my dissections, and my colleague at the Copenhagen school, Peter Rasmussen. Prior to this, I had been informed by a woman who had lived in the house that the place was not tainted with, nor had the sick man died of, anything pernicious. For more than a few people had succumbed around the city, because of the year's terrible weather, to a contagious, pestilential fever, and a dissector is seldom out of danger, when bodies that have been infected in this way are opened too early."[122]

Shortly beforehand, Copenhagen had been hit by a plague epidemic, which only ended after the loss of 9,000 Copenhageners' lives, i.e. approximately a third of the city's 30,000 inhabitants. Bartholin himself gives a picturesque description of the sickness, the epidemic and its consequences, and reveals at the same time his own great insight into the concept of infection more than two hundred years before anyone knew about bacteria and seriously understood the meaning of hygiene, infection control and epidemiology:

"The typhus, which ravaged in winter, was gradually succeeded at the beginning of spring and throughout the summer by a violent plague epidemic, which in Copenhagen sent more than nine thousand people to their graves. The infection was spread to us from Dutch ships, which in Danzig had loaded grain, hemp, flax, etc. and through fear of the English fleet had put into our port. Several of the sailors in private lodgings met their deaths. Their clothes were hung out in the sun and infected the children who touched them. In this way the disease spread and took possession of the whole city and the neighbouring villages together with other localities, to which the deceased's clothing, papers, or other possessions were brought. The venom even adhered to bread, spirits and similar items that came from a plague-ridden house. Many breastfeeding women gave their infants the infection in their milk, while they themselves survived. Many who wore the same clothes that infected persons had taken off, escaped safely, while others had to pay with their lives for their foolhardiness. It was also found that there died more young women, pregnant, unmarried and newly married girls as well as children of both sexes, than younger and older men, and more people of poor status, who did not seek to avoid socializing with those who had caught the plague, than those of higher social rank. The plague manifested itself for the most part by a massive shivering, and next there came backache and headache, followed by high fever, then pain in the left leg, and sometimes there were blisters on the soles of the feet. When they disappeared, and the venom entered the heart, deadly blisters appeared. With their outbreak all pain stopped up to the last breath, and thus there were many who immediately went away and laid themselves down in their coffins. They normally died on the third day. In my opinion the venom contained a narcotic ingredient, which numbed the sense of pain. But hope of a cure was sometimes aroused when buboes appeared in the groin, but this was crushed if the buboes one moment broke out and the next moment disappeared, when Nature's ability to develop the buboes had weakened. In the neighbouring villages, those stricken by the plague were seized by so violent a madness that they pitched themselves into rivers and drowned and even ended their lives using nooses or knives. Few of the infected became healthy again, and only if they immediately took antidotes, which provoked violent sweating, and provided they promptly used it a second time. Many patients profited from Tycho Brahe's plague elixir, which is often sold in our chemists besides the others, which the Copenhagen doctors have provided, when they gave advice to the public. Those who buried the dead smeared the rope with pitch because of the smell and thereby kept themselves free of the disease."[123]

Bartholin and his family survived by shortly afterwards staying with the provincial governor and member of the national council Niels Trolle (1599–1667) at Roskildegård in Roskilde[124] until the epidemic had ceased, and the danger of infection was dispelled in February 1655. The university closed, and only three professors, including Ole Worm, remained in the city. On 29 April 1654 he writes to his son Willum Worm in Leiden: "The plague has now begun to cause devastation here; may God look on us with merciful sight!" Worm mentions in a new letter on 22 July in the same year an antidote, which works by dispelling the disease's poisonous materials. Nonetheless he himself and several members of his family died during the epidemic.

One understands readily that Bartholin wished to secure himself and his collaborators against the risk of plague infection from the deceased Norwegian. Partly reassured by the interviewed resident that there was no danger of infection, and with female guards concentrated outside the house, Bartholin and his assistants then began their secret, illegal dissection of the dead, emaciated Norwegian, who lay wrapped in grave clothes in a narrow corner of the dark room. In the abdominal cavity they found, separated from a partly corroded omentum and a here and there slightly dark-coloured small intestine, sound organs, which did not show anything abnormal. The mesentery had no fat or chyle vessels. But suddenly the long-awaited discovery emerged: two lymph vessels around the mesentery's large vein and the spleen vein, whose convergence creates the portal vein, and furthermore three smaller lymph vessels around the portal vein:

"But see! Then appear around the mesenteric and splenic branch two lymph vessels, so much the more shining in that they were very welcome to our eyes. The radiance and novelty of the discovery now dispelled every fear, and since we regarded the supposed danger to be compensated for by the great delight, we rejoiced, that what mankind had failed to perceive hitherto had now been seen by us in man. The vessels mentioned, which were slightly wrinkled, meandered across the mesenteric branch. They were elongated, like a pigeon's feather in thickness, shining because of the clear water they contained, and perfectly resembling those we had found earlier in animals. Near the liver appeared smaller ones, three in number, which followed down the length of the *vena portae*, and which were filled with and shining throughout from the same gleaming liquid. We could not discover any more lymph vessel branches, because neither the inconvenient space, which turned away from the light, nor the anxious impatience of the women waiting outside, for whom information is quite often to be withheld in such secret sections, gave us in our hurry the opportunity for a closer investigation. Yet, in passing, we quenched our scientific thirst a little as to the movement of the liquid contained in the lymph vessels, like the dog when it came to the Nile."[125]

By compressing with his fingers the lymph vessels he had found, Bartholin and his assistants ascertained that the lymph in the two large lymph vessels, which followed the mesentery's large vein, flowed in an upward direction towards the liver, while the lymph in the three smaller lymph vessels, which came from the liver, flowed away from this, as with the animals. Moreover, they were able to conclude that the lymph vessels in human beings, as in animals, are supplied with internal valves, which prevent the lymph from flowing backwards. In the lungs they found the cause of the man's illness and death: a lung infection which in its description is consistent with tuberculosis. "What more is there to do?", asks Bartholin, and with this he brings an end to his investigation of the lymph system:

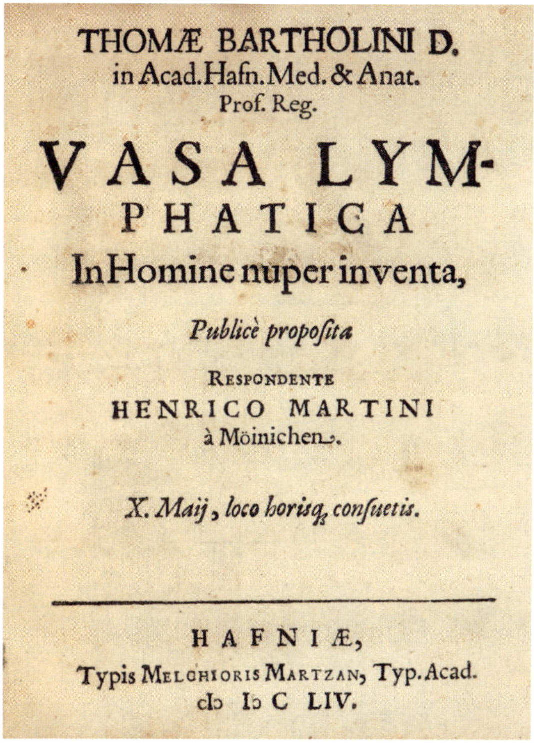

THOMÆ BARTHOLINI D.
in Acad.Hafn.Med.& Anat.
Prof. Reg.

VASA LYM-
PHATICA
In Homine nuper inventa,

Publicè proposita

RESPONDENTE
HENRICO MARTINI
à Möinichen.

X. Maij, loco horisq; consuetis.

HAFNIÆ,
Typis MELCHIORIS MARTZAN, Typ. Acad.
cIↄ Iↄ C LIV.

Figure 114. The title page of Thomas Bartholin's dissertation on the discovery of the lymph vessels on the Norwegian tailor Jens Lauritsen, *Vasa Lymphatica in Homine nuper inventa,* presented publicly in Copenhagen 10 May 1654 at the usual place and time with Henrik Martin à Møinichen as respondent. The Royal Library in Copenhagen.

"To put a rest to our enquiring mind it seems for the moment sufficient to me to inform the world of learning that the same lymph vessels both are contained in and can be shown in the human circulation."[126]

And he did so in a hurry with the publication of the treatise *Vasa Lymphatica in Homine nuper inventa* (The lymph vessels, which have recently been discovered in man),[127] which he presented publicly on 10 May 1654 in Copenhagen with Henrik Møinichen as respondent (Figure 114). About Møinichen, Bartholin writes in *Domus Anatomica Hafniensis* eight years later: "Henrik Møinichen, whose acknowledged expertise in anatomy later became known to the Italians and delighted his fellow-countrymen."[128] Bartholin dedicated the treatise and the "water's sources and their hitherto unknown channels" discovered by him in human beings to "Frederik III, the muses' glorious protector, the great patron of the natural sciences and the sciences' generous friend."

Illegal private dissections – in miserable conditions with danger of infection – of poor people deceased through illness was not an unusual means for Bartholin in the quest to gain new knowledge. It could occur "with a stolen knife without the knowledge of the relatives"[129] or with a suitable payment to the relatives. Whether Bartholin had paid any of the Norwegian's relatives for his project he does not mention, but in other instances he did not refuse to pay compensation. In his writing on pathological anatomy in 1674, he says:

> "However, I have not cast off the burden nor fled with the load, but in order that the world should see that I was ready to serve the public interest, even at the cost of my fortune and my health, I have arranged private autopsies for people in poor circumstances, who preferred profit to piety, as I quite frequently offered and paid an amount of money, either when the deceased had wasted away through a prolonged illness or had been carried away by sudden death. In the case of acute and pernicious illnesses this sought-after and purchased contribution of knowledge is not without danger of infection. In the event of plague, it is surely certain death for the anatomist, unless he is protected by *alexipharmaca*."[130]

Bartholin goes into detail as to what he means by *alexipharmaca*: "In relation to the infected corpse whose dissection is to be undertaken, anatomists secure themselves with the aid of God, who favours an honest prayer, through a consumption of theriac, by smearing *oleum rutaceum*[131] in the nostrils and protecting the mouth with *acetum calendulæ*[132] and at the same time putting on such clothing as we have depicted in our "*Historiæ Rariores,*" as used by physicians and surgeons during the plague in Rome [Figure 115].[133] [...] In poor homes and hovels the anatomists protect themselves against the stench of the diseased corpses with sweet-smelling substances such as *balsamum apoplecticum* or *rutaceum* or the one made from amber or through other substances."[134]

After this, Bartholin includes notes for a series of means to disguise the stench from corpses, but ends with recommending "smoke from juniper trees, pastilles of angelica root, almond confection or preserved citrus fruit, scorzonera or marigold [...] in order that the mouldy hovels of the poor or the stench of corpses during autopsies should not trouble sensitive noses."[135] About his own procedure he informs us: "Personally I have never myself liked ointments in the execution of these actions, nor have I ever felt at all unwell in looking at rotting entrails, but merely been able to enjoy snuff, with which I protected my nostrils."[136]

Bartholin's thoughts on the functions of the lymph system

It is not difficult to argue that anatomy and physiology are two inseparable topics in medical science. An anatomical observation first gains its full value when it is followed up soon afterwards by considerations of its use and function. In his treatise on the thoracic duct Bartholin formulates this succinctly as "the honour

Figure 115. Plague doctor in protective suit. From Thomas Bartholin's *Historiarum Anatomicarum Rariorum*, Cent. 5, Copenhagen 1661, Hist. 71, p 143. The Royal Library in Copenhagen.

is worth nothing, if what we perform is of no use."[137] In accordance with this he has many thoughts, as a researcher should, about the functions of the thoracic duct and lymph vessels and their significance in illnesses.

As mentioned, Bartholin considered chyle to be a nutritious fluid that was included in blood formation, which he at first believed took place in the liver, afterwards in both the liver and the heart, and finally only in the heart.

Summing up, he thinks that the function of the thoracic duct is to shorten the route to the heart for the ingested, heart-strengthening foods, fluids, medicaments, etc. in the stomach by bypassing the liver: "For through the liver's parenchyma[138] and the network of veins it slows down, and the force of the heart-strengthening substances is either diminished or distributed to more places."[139] He supports this point of view with more examples. A hanged criminal, whom he had recently dissected, had some hours before being hanged had a good meal and been plied with "an excessive quantity of wine", but on his dissection Bartholin found no places with chyle vessels, and the thoracic duct contained no chyle. From this Bartholin concludes that the criminal's fearless mind, contemptuous of death, can only be explained by the fact that the wine had very rapidly gone the short distance through the thoracic duct to the heart.[140] Among other examples he mentions the palpitation of the heart and the increasing pulse rate after the consumption of a large meal or a large amount of liquid and "the unexpected death, which comes after the enjoyment of an abundant amount of food, which has certainly not been able to traverse the detour through the liver with the requisite speed."[141]

As far as the aqueous lymph is concerned, Bartholin states in his treatise on the lymph vessels in animals that he finds it probable that it is a product of a particular metabolism in the body's different tissues and organs. Thus, he believes that the lymph is a part of the blood's fluid phase, and is what remains behind when the arterial blood has delivered its nutrients to the tissues and metabolism has taken place here:

> "The water is conveyed mixed with the arterial blood, and thus not freed from the untreated blood's contamination, through the arterial tubes out to the nourishment of the tissues, and when afterwards the profitable nutritious materials have been extracted and absorbed by the tissues, the water alone remains to be emptied into one or another flood by particular routes."[142]

Thus, one must understand Bartholin to mean that fluid, together with the nutritious materials, seeps out from the bloodstream in the tissues, and that the superfluous fluid, which remains after metabolization in the tissues, is afterwards absorbed by the lymph vessels, resulting in the formation of the lymph. As in the case of the chyle, he states that the lymph ends up in the thoracic duct and by that route comes via the shoulder vein to the heart. Bartholin's understanding of the lymph vessels' function can be summarized in the following five points:

1. To take part in the absorption and transport of nutrients (fatty substances) from the intestine to the bloodstream and onward from there to the body's organs and tissues.
2. A draining function, which consists of "relieving the tissues, which must be nourished, from the load of water they have no use for,"[143] and to help the body to get rid of surplus fluid, for example in the case of dropsy (the earlier word for abnormal accumulation of fluid in the body) to bring

> fluid to the kidneys. Bartholin explains this by describing what happens when the lymph vessels cannot perform this function, because they are clogged, compressed or burst: "I understand it thus, that when, as results from obstruction of the lymphatic vessels, the thin liquid does not find an outflow from the tissues, anasarca or oedema arises,[144] or when the passage through them is blocked, and the fluid is forced to go into the veins, it causes ascites,[145] or finally, when the vessels, because of the fineness of the walls, are burst and destroyed, as a result of the quantity of fluid, or its sharp nature, or the greater pressure of the surroundings, the fluid empties itself out into the nearest cavities or the abdomen or the ribcage, and thereby gives cause for serious illness."[146] He gives almost the same explanation and formulation in the treatise concerning the lymph vessels in human beings: "For when those newly discovered vessels have been attacked in different ways, it follows that, little by little, water comes out into the skin, or hydrops appears in the firm parts.[147] Thus, what happens is that: either, if the lymph vessels have become blocked, the inner organs, the joints and cavities are distended by a large quantity of fluid and swell up, because the water cannot be separated from the tissues, which have been nourished, and from the blood; or, if the lymph vessels for one or another reason burst, that the water, which should run away through these channels, is poured out into the cavities and the abdominal spaces, and so the terrible hydrops grows, which does not put any restriction on itself."[148]

Bartholin adds that if the accumulation of liquid takes place in closed cavities, tissues or organs with surrounding, close-fitting membranes, fluid-filled blisters, so-called hydatids of different sizes are formed.

3. To supply the heart with water in order a) to thin and cool the blood in the heart, which would otherwise become too thick and too warm, b) to increase the heart's completion ("boiling") of the blood and c) to thin chyle in the mesentery and the thoracic duct, so that it becomes more free flowing. In other words, Bartholin attributes to the lymph the function of diluent. Moreover, he considers whether the lymph's constant circulation, which is comparable to repeated distillations, extracts something that could be of further use, and he believes that the lymph, so as not to accumulate, is removed with the serum (the blood's watery phase), either by being excreted as urine through the kidneys or by being included in the body's composition.

4. To contribute to the sweating: "I would venture to say that almost all sweat seeps out from the lymph vessels, which follow the blood-carrying veins, with the exception of the one, which, before the lymph vessels intercept it, evaporates through the skin's pores and the interstices of the muscles from the tissues themselves and their nourishment."[149] Bartholin argues that if the sweat were secreted from the veins, blood would come out or, leastways, the sweat would be coloured by blood.
5. To moisten the tissue and extend life in the case of illness, where the body is worn and dried up. Here Bartholin uses the example of the tuberculous Norwegian: "We saw this in our Jens, whose worn-out body could thank the lymph vessels that his life had lasted so long."[150]

In addition, Bartholin considers whether the lubricating synovial fluid comes from the lymph vessels, and he calls for investigations of this.

What do we know about the functions of the lymph system today?

The lymph system's functions can in all essentials be summed up in the following three statements:

1. A draining function, which consists of bringing the filtered tissue fluid (filtered out through the blood stream's capillary walls) and proteins back to the blood from the tissue space outside the bloodstream.
2. An immune defence function, which consists of supplying the blood with immune defence cells (lymphocytes) and antibodies (immunoglobulins) from the lymph glands.
3. A fat transport function, which consists of transporting fatty materials absorbed from the food in the intestine to the bloodstream.

Re. 1: The exchange of oxygen, carbon dioxide, nutrients and waste products between the bloodstream and the tissues occurs in the human organism through the capillaries, which are the connecting link between the arterial system and the venous system. As mentioned, the capillaries had not been found at the time of Bartholin's exploration of the lymph vessel system. About 90% of the fluid (plasma) which is filtered out from the bloodstream into the tissues through the arterial end of the capillaries is reabsorbed at the venous end. The remaining approximately 10%, which in an adult person amounts to 2–3 litres a day, is absorbed together with waste products, proteins, foreign bodies, dead cells and microorganisms (bacteria, viruses and parasites) in the smallest lymph vessels, the lymph capillaries, which have a greater permeability than the blood capillaries, and seeps as lymph back into the bloodstream, approximately 75% (the lymph from the lower part of the body and the left side) via the thoracic duct, and as far as the rest is concerned (the upper right quarter) via the *ductus lymphaticus dexter*, which is a large lymph vessel in the upper right side of the thorax. The thoracic duct, *ductus thoracicus*, is a tube 40–45 centimetres long with a diameter of 2-3 millimetres. It begins its course in the abdominal cavity with a broad extension or container of 4–5 centimetres in length and 4–5 millimetres in width, the *cisterna chyli*, which corresponds with the already mentioned *receptaculum chyli*, which the Frenchman Jean Pecquet noticed for the first time in a dog. From the abdominal cavity the thoracic duct (or as Bartholin prefers to write, thoracic ducts, for often there are two to three parallel tubes) passes up in the thorax through the same aperture in the diaphragm as the *aorta* and follows first this and afterwards the oesophagus to the neck. Here it flows (with certain individual variations, as Bartholin has described) into the bloodstream on the left side in the angle between the two neck veins, the *vena jugularis interna* (the inner neck vein) and the *vena subclavia* (the collar bone vein). The right lymphatic duct enters the neck veins on the right-hand side. The diameter of the lymph vessels increases gradually from the lymph capillaries, which do not contain valves, over the lymph vessels to the thoracic duct. Because of the regular occurrence of valves, which prevent the lymph flowing backwards, the filled lymph vessels' appearance becomes like that of a pearl necklace. The lymph is driven forwards through the compression of the lymph vessels by their surroundings and the contraction of the smooth musculature in the lymph vessels' walls. Around two thirds of the body's lymph derive from the intestines and the liver.

Re. 2: Before the lymph is emptied into the bloodstream, it passes through the lymph glands, which function as a filter, where the foreign bodies, dead cells and microorganisms are sorted out from the lymph, and lymphocytes (a kind of white blood cell, which function as immune defence cells) and antibodies are added. The lymph glands or lymph nodes, of which a human being has 500–1000, which vary in size from a few millimetres to over one centimetre, are thus a strong part of the immune defence, which is the reason why they are enlarged during illnesses that activate the immune defence, such as infections, rheumatic or autoimmune diseases and cancers.

Re. 3: About 95% of fat from food is taken up in the small intestine. In contrast to the carbohydrates and proteins, which after absorption in the small intestine are dispatched directly with the portal vein's blood to the liver, a large proportion of fats (the long-chained fatty acids, triglycerides, cholesterols and phospholipids) is absorbed in the small intestine's lymph vessels and are dispatched with these in Pecquet's and Bartholin's way via the thoracic duct to the bloodstream in order by this way to reach their further metabolism in the liver. Normally the lymph is clear as water, but after a meal the lymph from the small intestine (chyle) is coloured milk-white and the blood pink by the long- chained fatty acids. This effect over a few hours gradually decreases with the metabolism of the fats in the liver and the adipose tissue.

The significance of Bartholin's discoveries

It must be concluded that the results of Bartholin's research into the lymphatic system are largely in harmony with modern knowledge. His anatomical descriptions correspond very closely with what we know today. The same goes for our understanding of the lymphatic system's function regarding the absorption and transport of nutrients (fatty substances) from the intestine to the bloodstream and his description of the lymphatic system's important function as a drainage system for the tissue fluid. His detailed understanding of the various forms of liquid accumulation that happen when the lymph vessels' drainage function is compromised in connection with different processes of illness appears impressive when seen by modern eyes.

These ideas have, without doubt, had significance for the understanding of the trajectory of illness in connection with cancers, rheumatic complaints, inflammatory conditions, problems of the heart and blood circulation, parasitic infections and other illnesses with local or disseminated accumulation of fluid. The discovery of a completely new fluid circulation with separate functions from the blood circulation was epoch-making and resulted in an adjustment of the earlier understanding that these functions were attended to by the blood circulation. Bartholin's discovery of the thoracic duct and the lymph vessels in human beings is therefore reckoned among the 17th century's most substantial discoveries in human anatomy and physiology.[151]

For good reasons Bartholin could not argue in favour of the substantial importance of the lymphatic system for immune protection. Almost 120 years had to pass before anyone discovered, with the help of the microscope in 1771, that white blood cells are produced in the lymph glands and the thymus, and that they arrive in the bloodstream through the thoracic duct,[152] and almost another 120 years before it was discovered in approximately 1890 that these white blood cells (lymphocytes) played an important role in immune protection.[153] That blood production in human beings takes place in the bone marrow (in prenatal life, however, also in the liver and the spleen) was first discovered in 1868.[154]

Bartholin's assertions of the lymphatic system's importance for the secretion of sweat and the moistening of the tissues were disproved about ten years later by his pupil Niels Stensen, who found that these functions are due to sweat glands in the skin and other exocrine glands attached to the internal surfaces of the body. In 1664, Stensen became the first to describe how sweat is produced in sweat glands in the skin.[155] His observations built on his studies during his stay from 1660–1664 at the University of Leiden, where Franciscus Sylvius was his teacher, but more about this later.

A couple of strong pats on the shoulder, which sets Bartholin's discoveries in relief in his own time, came from Jean Pecquet in Paris. The first came in a letter to Bartholin on 13 December 1654:

> "Take care of yourself; I beg you to remain in good health, something so important for the world, for you have improved anatomy with such resplendent and miraculous proficiency that I no longer believe that anything can be found, however well-hidden it may be, that cannot be brought into the light of day if you draw your knife from its sheath or cast your expert eyes on the matter."[156]

The second came in a new letter dated 17 April 1656:

> "Moreover, there is a great deal that is continually waiting to be discovered, and that, I tell you, is waiting for your attention. Therefore, seeing that you did not hesitate to follow my guidance in the discovery of the lacteal veins, do accept my offer when I keep on pushing you. I too feel myself continuously inspired to further research, because of the distinction you have won through your discovery of the lymph vessels, which no one had previously heard of."[157]

The attacks from Riolan

There were others who did not show the same chivalrous view of Bartholin's discoveries as Jean Pecquet. From two sides in particular Bartholin was subjected to fierce attacks; one was perhaps more to be expected than the other.

Like a watchful bird of prey, Jean Riolan had long sat on his perch and waited for a suitable opportunity to dig his sharp claws again into the young Dane, who pushed forward with points of view that challenged the medical worldview of the old Greek authorities, otherwise so unassailable (Figure 116).

In 1649, while Bartholin was setting about his investigations into the system of lymph vessels in Copenhagen, Riolan had published in London the book *Opuscula Anatomica Nova* (New anatomical opuscules).[158] As a bitter opponent of all new thinking, he directed his polemics against Harvey's new teaching about the blood circulation and against the two letters that Johannes Walæus had written in support of this teaching, and that Thomas Bartholin had included in his revised editions from 1641 and 1645 of his father's textbook of anatomy.[159]

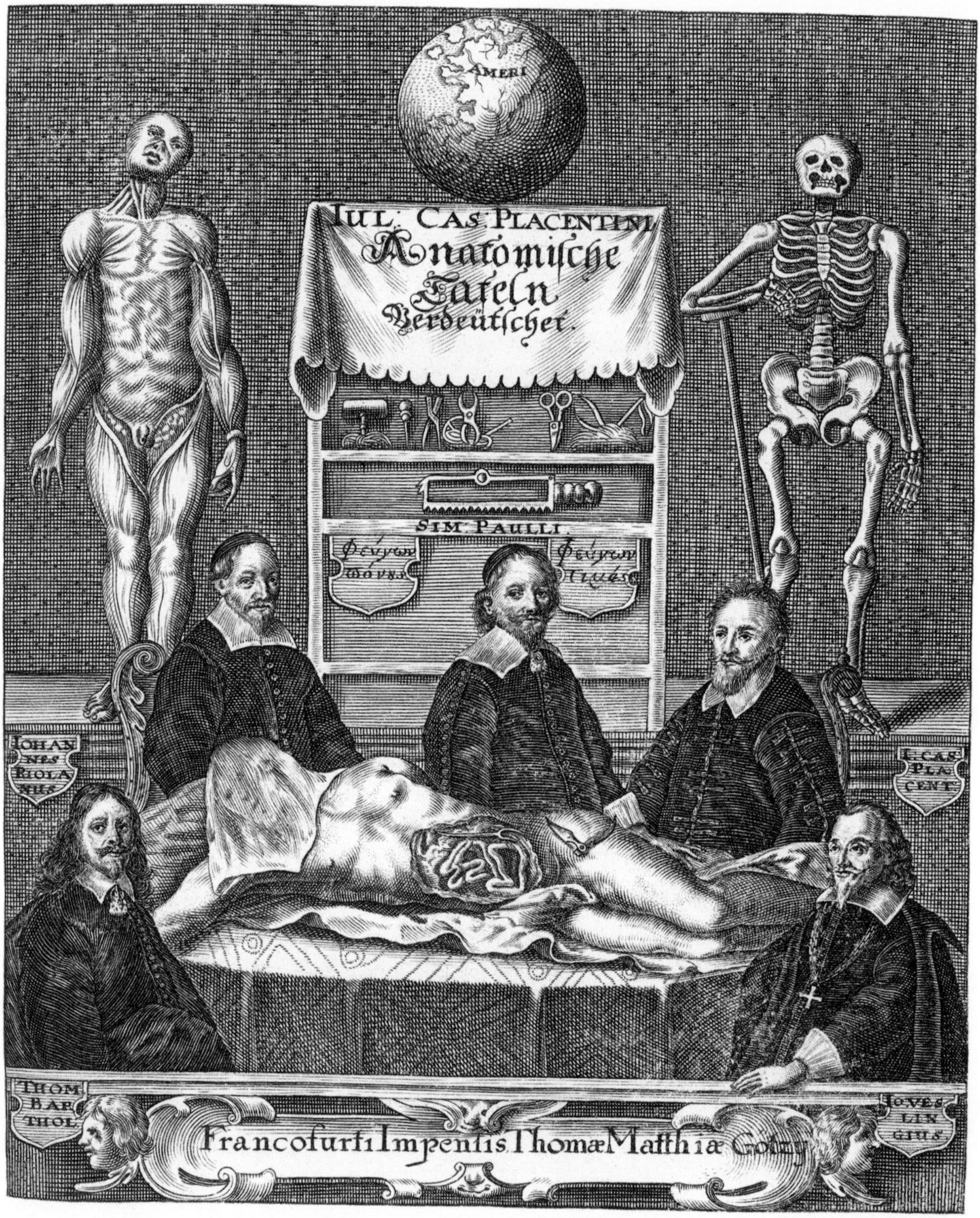

Figure 116. Here, Simon Paulli has depicted himself at the dissection table surrounded by four of the leading anatomists of the 17th century. From top left are Jean Riolan, Simon Paulli and Julius Casserius. In front from left are Thomas Bartholin and Johannes Vesling. The frontispiece of Simon Paulli's German edition, *Anatomische Tafeln,* Frankfurt 1656, of Julius Casserius' impressing anatomical atlas in folio format, *Tabulæ Anatomicæ*, Venice 1627. The Royal Library in Copenhagen.

Against these two publications, moreover, Riolan put forward a series of critical points directed at both father and son.[160] Among other things Thomas Bartholin was criticized for embellishing himself with borrowed feathers from his father and from Riolan himself, while his father was attacked for lacking experience of practical dissection. Moreover, Riolan pointed out a number of factual mistakes, of which some were justified. Thomas Bartholin essentially met Riolan's criticisms with silence and confined himself to correcting selected critical points and mistakes in the next edition of the anatomy book in 1651. If Bartholin had expected that Riolan would afterwards fall silent, he was mistaken. Riolan was not satisfied with Bartholin's corrections, and in his *Animadversiones Secundæ ad Anatomiam Reformatam Thomæ Bartholini* (Further criticisms of Thomas Bartholin's improved anatomy book), which was published together with his *Opuscula Nova Anatomica* in Paris in 1653, a series of new critical points followed, besides a reiteration of those that Bartholin had not corrected or, according to Riolan, had wilfully misunderstood.[161] Riolan writes that he would gladly have refrained from publishing this new criticism if Bartholin had taken all his earlier criticism as good advice, but according to Riolan, Bartholin has shown himself ungrateful and attacked him more bitterly than before. Attentive as he was to everything that came from Bartholin's hand, Riolan also refers here to some attacks that Bartholin had raised against him in his published lecture and disputation series, *Collegii Anatomici*, for the medical students in Copenhagen in the years 1649–1651. Riolan, like a schoolmaster, takes Bartholin to task and calls him more pretentious than his age and presentations can warrant, and he advises him (with a degree of justification) not to spread himself over so many different subjects (medicine, the humanities and theology) if he wishes to be a proficient anatomist.

Although many of Riolan's points of criticism were needlessly polemical, pedantic or simply mistaken, there were also many of them that were justified, and that helped to make the third and fourth editions of Bartholin's anatomy book better than the two previous ones. An example of one of Riolan's most valid criticisms, to which Bartholin paid no attention but repeated all the way through to the fourth edition of 1673, is a tale handed down of an embryo produced by fertilization in a woman's stomach, developed there and born through her mouth.[162] It will be discussed in more detail later.

Conversely, there were other points of criticism from Riolan, where posterity would prove that Bartholin was right. An interesting example is the question of the function of the adrenal glands. Here Bartholin has a fine description of the organ's anatomy, position, and blood and nerve supply, which he supplements with one of the book's new, original illustrations, representing the organ in human beings and certain animals (Figure 117). He states that the Italian physician Bartolomeo Eustachi (c. 1500–1574) was the first to describe these small organs and designate them as glands, and later they were described by Bauhin and Casserius, of whom the latter called them adrenal glands, "as they seem like small kidneys that lie on the large ones." Bartholin describes them as capsule-like with a central cavity, which contains a thick, black gall-like fluid. He, therefore, like his father, calls them *capsulæ atrabilariæ* (capsules with black gall). He states that their function is unknown but guesses that they cleanse the blood of waste products, which are afterwards excreted with the urine. The adrenal glands' different functions as a hormone-producing organ first began to be understood at the end of the 1800s,[163] so Bartholin could not have known about them. Bartholin stresses that the organ is larger in newborn children than in adult human beings, which causes Riolan to conclude that it has no function after birth but only has the purpose of supplying the embryo's kidneys with fat. But Bartholin argues convincingly against this point of view, for:

> "1. Their hollowness, veins, arteries, fluids, etc. allow us not to say, that they wither away in adulthood. 2. The use of the umbilical veins stops, because the child no longer needs to be united with its mother or receive nutrients from her. And that these capsules have a function in adults is proved earlier, for otherwise their veins, arteries, etc. would have no purpose. 3. That the kidneys in embryos must always be laid bare of fat, I have found, is false. 4. The kidney fat is never made from the whey-like black juice, and no one has ever seen an oily juice in these capsules."[164]

It is thought-provoking that C. C. A. Gosch, 220 years later, brings to light this critical point from Riolan against Bartholin as one of the justified ones with the remark: "which [capsulæ atrabiliares] he [Riolan] rightly designates as mere fetal organs in human beings."[165]

Riolan himself had often seen Aselli's lacteal veins during his dissections of hanged criminals, but he ascribed no particular significance to them.[166] This was why Bartholin's writings on this subject inevitably sooner or later gave rise to new attacks from the old man in Paris, although Bartholin sought to prevent

Figure 117. Differently shaped adrenal glands in humans: round (top), triangular (middle), square and oval (bottom). From Thomas Bartholin's *Anatomia Tertium Reformata*, Leiden 1651, p. 123.
The letters mark: A. The intact adrenal glands. B. The same dissected to make the various shapes of their internal cavities visible. C. Arteries and veins emanating from aorta, vena cava or the kidneys' vessels. D. Vena cava. E. Aorta. F. Kidney vein. G. Kidney. The Royal Library in Copenhagen.

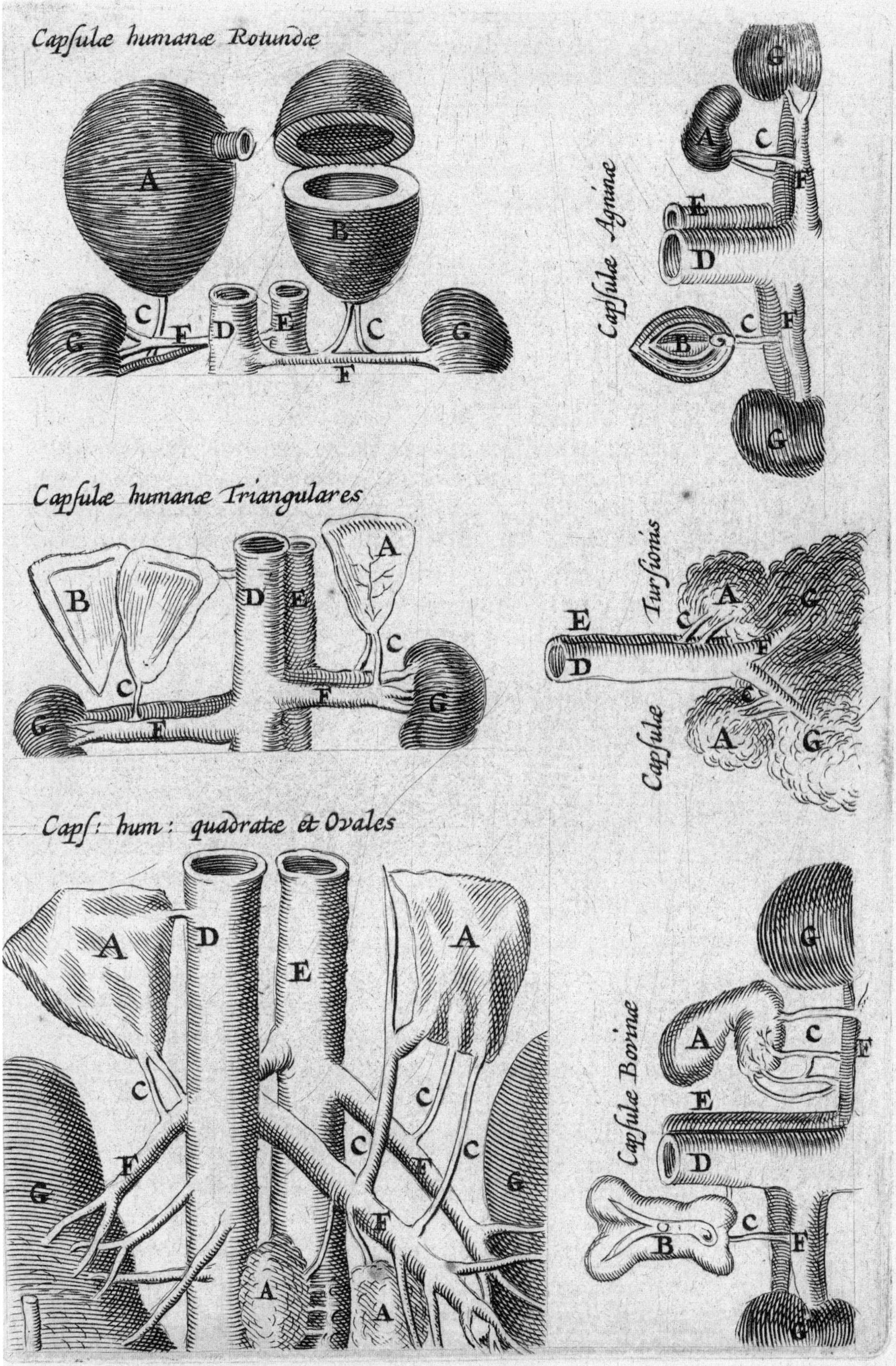
Capsulæ humanæ Rotundæ
A
B
C
C
G
G
F
D
E
F
Capsulæ Agninæ
G
A
C
F
E
D
B
C
F
G
Capsulæ humanæ Triangulares
B
D
E
A
C
C
F
G
F
G
Capsulæ Tursionis
E
D
A
C
C
F
C
A
G
Capſ: hum: quadratæ et Ovales
D
E
A
A
C
C
C
C
F
F
F
G
G
A
A
Capsulæ Bovinæ
G
A
C
F
E
D
B
C
F
G

them with a deft strategy. Bartholin's treatise on the discovery of the thoracic duct in human beings contained only a few, less serious criticisms of Riolan and, as mentioned, he attempted to flatter him into silence by dedicating his treatise on the discovery of lymph vessels in animals to Riolan, "the World's and the city of Paris' greatest anatomist". This treatise, moreover, contained no criticisms of Riolan at all. But "the Parisian hyena," as the Swiss anatomist and botanist Caspar Bauhin the Younger (1605–1685) called him in a letter to Bartholin,[167] did not allow himself to be flattered into silence indefinitely. So when, in the treatise about anatomical matters of dispute, Bartholin again took the old man in Paris by the collar, a new thorough attack from Riolan was inevitable. Bartholin criticizes Riolan for not wishing to recognize that chyle flows through the thoracic duct to the heart in order to be converted into blood there: "Although the truth about the thoracic ducts is sufficiently clear and becomes absolutely trustworthy by what you have seen with your own eyes, it has not been possible to convince the highly praised Riolan that chyle is conducted through them to the heart."[168] Instead Riolan thinks that the admixture of chyle is directed at making the blood thicker, and together with the admixture of fibrinous (solidifying) material this causes the blood to slow down and remain longer in the heart, where heat and acuity are supplied in a kind of fermentation process.[169] Bartholin comments on this theory with scarcely veiled sarcasm in this way:

"Naturally this most eminent man, worthy of every crown, must deny that chyle is converted in the heart, since he has indeed wished to preserve the liver's old task unharmed and untouchable. Therefore, one must commend the rapidity of this great spirit with which he has discovered and shed complete light on this new purpose. If without regret it must be admitted to have a different view than that of the old anatomist, I shall assume that the blood does not require any holding up, but that it must rather have a great impetus, both in order not to suffer any harm, and so that the tissues are not cheated of the suitable nourishment that the unbroken movements of the heart and the arteries convey."[170]

After this Bartholin delivers an extremely far-sighted addition, which reflects the fact that he was not merely a scientist, but also a practising doctor with his own patients, in a position to transfer theories into a practical, clinical context: "But if the circulation of the blood is hindered owing to one or another defect, either in the blood's degree of concentration, in the heart's capacity for expulsion, or finally in the tissues that must receive it, the physician attempts in every way to compensate for this deficiency."[171] As modern examples of such treatment one can mention blood-thinning, blood-clot-dissolving, heart-fortifying and diuretic medicine, together with operations for varicose veins and balloon dilatations and bypass operations for constricted arteries.

Bartholin answers Riolan's dismissal of Pecquet's discovery – as a kind of revenge – with the same accusation as Riolan himself had employed earlier against Bartholin's father: "But if he, as we did, had more often consulted people's and animals' carcasses with reference to this discovery, he would, such being his way of thinking, have declared a friendly opinion in favour of Pecquet, which we trust will occur sometime in the future, if God will, as we pray, extend his illustrious age with further years."[172]

After this, Bartholin answers point by point fourteen critical arguments posed by Riolan against Pecquet's discovery,[173] but apart from Bartholin's points concerning the anatomical facts, most of his answers are no less characterized by the erroneous opinions of the time than Riolan's arguments. Bartholin finishes the article with a section in which he states that, although Pecquet's and his own discoveries have abandoned the liver as the body's blood-producing organ, that does not mean that the methods for dealing with diseases of the liver and heart must be changed. In this almost poetical section Riolan gets (with a small mitigating passage at the end) a concluding broadside for his stubborn insistence on the liver as being the blood-producing organ and his everlasting dogmatic emphasis on the old medical authorities' version of the truth as an argument against everything new:

"Riolan, the keen champion of the old traditions, but who has never been persevering, when it was a question of examining and acquiring new discoveries, where he saw Nature's purpose, has stood as a maintaining God on behalf of the liver against Pecquet. [...] But in the meantime, let us see, without any enmity, but only out of love for the truth, what the old master into anatomy has sent in the way of help to the liver. There is complaint that the ancient medicine is overthrown, and that we are attacking our standard-bearers of science, Hippocrates, Aristotle, Aretæus,[174] Galen and others, who have vouched for the fact, that the production of blood is a function of the liver. [...] But we shall not detain ourselves with the old authorities, who in the infancy of anatomy we know have built up the art of medicine, starting from the organs they knew about and their supposed functions. It is beneath an anatomist's dignity either to stand or fall with the old authorities. The knowledge of anatomy has grown tremendously, and even if the old practitioners were called back from the underworld and brought into our anatomy theatres, they would imagine themselves removed to another world, and would, after having seen Harvey's circulation, and Aselli's and Pecquet's chyle vessels, our lymph vessels, the pancreatic duct and many of Riolan's observations, with which, to the undying reputation of his name, he has advanced the cause of anatomy, reject most of their conclusions and bequeath us a volume of abandoned observations, before they went home to the Stygian waves."[175]

Bartholin sent the treatise on the anatomical questions of doubt to Guy Patin, o whom, as already mentioned, he had dedicated it. Patin thanks him on 12 September 1653 with a notice to Bartholin, that Riolan has a riposte in the press in three parts, where in part 1 there is an answer to Bartholin's treatise on the thoracic duct, in part 2 there is an answer to the treatise on the lymph vessels in animals, and in part 3 there is a commentary on the revised third edition of the anatomy book.[176] As far as the treatise on the anatomical questions of doubt is concerned, Patin does not know how Riolan will react. For the sake of the reader, Patin has requested Riolan to include in his publication those of Bartholin's treatises that he has commented on. On 26 September 1653, Patin writes again to Bartholin.[177] He has shown the treatise about the anatomical questions of doubt to Riolan, who immediately threw himself into investigating it, and Riolan will now extend his answer to include four parts, which, after persuasion from Patin, will have the addition of Bartholin's three treatises concerning the system of lymph vessels.[178] Patin regrets this professional quarrel between his two friends with the words: "I am very unhappy about it, but what am I to do? I am not strong enough to bring such powerful billows to rest. Let the truth win, and the future will pronounce a clear judgement, provided there is the necessary and sufficient knowledge to settle such great problems; scholarship strengthened by positive, clear experience."

Riolan acknowledges in his riposte to a large extent the anatomical data from Aselli's, Pecquet's and Bartholin's research, but on the other hand not their physiological consequences. He does not recognise the lacteal and lymph vessels as an independent vascular system but believes them to be ramifications of the veins. Thus, he believes that the thoracic duct is a connecting vessel between the portal vein (*vena portae*) and the large *vena cava*. Moreover, he maintains that the liver is the blood-producing organ.

Before Bartholin grabbed his pen to formulate a reply to Riolan, on 16 June 1654, from his plague-free refuge in Roskilde, he sent Riolan's texts to Ole Worm in plague-ridden Copenhagen asking for his judgement. Worm's estimation and advice was not to be mistaken:

> "I have glanced quickly through Riolan's jottings, which I return with thanks, so that I shall not deprive you of them for too long. But I have always regarded as true, what appears in the schools: the obedient sons of learning gently proceed forwards! They have laid all coarseness to one side. Meanwhile, I am experiencing the opposite with our Riolan, who is indeed well-informed and renowned for his anatomical writings, but whose manners neither the muses nor the court nor the years have been able to soften. I believe he has been brought up with old women, among whom there is entertainment derived from squabbling, so that

they compete in taunts and sarcasms and tearing apart the reputations of decent people. And this old-fashioned schoolmaster is moved neither by friendly appeals nor proof of errors, unless he overturns a whole dustbin of insults and curses over well-deserving men, who try hard with their efforts to benefit the world of learning. He wishes only to be worshipped and be revered as Our Lord of Anatomy. What kind of insolence is this? It is up to you, my Bartholin, to stand with your chin up and in silence to disregard the grumbling old fellow's insults. Scholars and orderly people who read the writings of both of you and weigh up the matter with impartial assessment scorn his worthless chatter. Indeed, he is only seldom concerned with the substance of the matter, but most often focuses on what is inessential. Believe me, he will get more annoyed, because his nonsense is scorned than if he were attacked with sharper words. What is the use of threshing long straw on this kind of wrangling? If you want to mess with a drunken man, you are just driving a madman madder. May God preserve you unharmed for greater and more profitable purposes."

Worm died two and half months later and thus did not know that his nephew did not follow his last piece of advice. Bartholin knew that he had the scientific data on his side, and the temptation to use the opportunity to put the Parisian fossil in his place once and for all was too strong. Thus, in 1655 Bartholin published the refutation *Defensio Vasorum Lacteorum et Lymphaticorum adversus Joannem Riolanum* (Defence of the lacteal and lymph vessels against Jean Riolan),[179] where with reference to his own and others' dissection discoveries and simple tests with ligation of vessels he could refute Riolan completely.[180]

The battle with Riolan, however, was still not fully over. In 1654 Jean Pecquet had published a new edition of his *Experimenta Nova Anatomica*, where with flattering expressions he had dedicated a treatise on the lacteal veins to Thomas Bartholin. Consequently, Riolan made a new attack on the new doctrine with the same arguments as previously in his *Responsiones Duæ: Prima, Ad Experimenta nova Joannis Pecqueti, Doctoris Medici Montpeliensis: Altera, Ad Pecquetianos duos Doctores Parisiensis, adversus sanguificationem in Corde, etc.* (Two answers. The first concerns the new experiments of Jean Pecquet, Doctor of Medicine in Montpellier. The second is against blood formation in the heart, addressed to two Parisian doctors and followers of Pecquet, etc.), Paris 1655.[181] Shortly afterwards Bartholin was ready with a last response against Riolan in the form of *Spicilegium ex Vasis Lymphaticis* (Aftermath of the Lymph Vessels),[182] which, besides a new refutation of Riolan and a great acknowledgement of Pecquet also contains a criticism of the Englishman Francis Glisson (1597–1677), who in the work *Anatomia Hepatis* (Anatomy of the Liver) of 1654 had stated the erroneous view that the lymph vessels were formed of an emanation of the arteries with an addition of material from the nerves.[183]

Bartholin sent the last-named treatise of his to Pecquet, who in his answer of 17 April 1656 writes about Riolan: "I feel sorry for Riolan, who, after giving nourishment to so many serious insults, will find that his life fades away before his derision does. I have decided to tone down his inborn inclination to slander by remaining silent; especially because the discoveries that you and I have made can stand by themselves, and countless people are proud to follow in our footsteps and compete with one another to spread the knowledge they have learnt from us."[184]

After this, the criticism from the old grump in Paris fell silent, and when not long afterwards (on 17 February 1657) Riolan drew his last breath, Bartholin was finally rid of his worst critic.

The striving for priority with Rudbeck

A much more serious attack on Bartholin came from a completely different and unexpected quarter. This time it was not a question of an old, established anatomist, but a very young, unknown researcher in Denmark's sister country to the east. The dispute was not simply about professional differences, but the very right to call oneself "discoverer of the lymph vessels." As was mentioned earlier, Pierre Bourdelot informed Thomas Bartholin in a letter dated 18 May 1652 that one of the professors in Uppsala maintained, without knowing Pecquet's name, that he had made the same observations of the chyle vessels in the thorax, and he promised a lot of things. This did not, however, concern a professor, but a highly talented, 21-year-old medical student called Olof Rudbeck, son of the bishop of Västerås in Sweden (Figure 118).

In the summer of 1653, a couple of months after Thomas Bartholin brought out his first treatise on the lymph vessels, Rudbeck published in his hometown of Västerås a little dissertation of 24 pages on the same subject: *Nova Excercitatio Anatomica, Exhibens Ductus Hepaticos Aquosos, & Vasa Glandularum Serosa, nunc primum inventa,* æneisque *figuris delineata* (A new anatomical examination of the liver's limpid excretory ducts and the gland's serous vessels, now discovered for the first time and illustrated with sketches).[185] In this treatise Rudbeck gives an account of the process of his discoveries of respectively the thoracic duct and the chyle receptacle in animals and the lymph vessels in animals. He leaves no doubt that he understands his own discovery of the lymph vessels to be the first, for on the title page he adds "nunc primum inventa" (now discovered for the first time).

Figure 118. Olof Rudbeck (1630–1702) portrayed in 1687. Oil on canvas. 84 × 67 cm. Attributed to Caspar Kenckel (1650–1724).

The priority dispute between the established and famous Copenhagen anatomy professor Thomas Bartholin and the unknown Swedish student Olof Rudbeck over the discovery of the lymph vessels is legendary in the history of 17th-century natural science.

The dispute became long and bitter, and Bartholin let his student Martin Bogdan take the pen for him, but the relationship between Rudbeck and Bartholin ended many years later with amicability and mutual respect.

The Swedish National Portrait Gallery at Gripsholm Castle. Wikimedia Commons.

Concerning the thoracic duct and the chyle receptacle he states that on one occasion, no more clearly defined than in 1650, he happened to attend the slaughter of a calf. The blood vessels of the neck, trachea and oesophagus were already removed, but when the women who were responsible for the slaughter wished to remove the entrails and addressed the intestines, Rudbeck noticed that there ran from the calf's neck a milky liquid. On closer inspection he found that the milky fluid came from a tube in the thorax behind the aorta, and in following it to just under the diaphragm he found a receptacle containing the same milky fluid. Since the neck vessels, the intestines and the remaining bowels were removed, he had no possibility of investigating the matter further, and because of other duties he could not undertake dissections for the rest of the year. On 8 February 1651, he demonstrated the thoracic duct on a calf and a sheep. In the same year on an unidentified date, he followed up the discovery in the calf with the dissection of a cat five hours after its last meal. With the help of systematic ligation tests, he found that the chyle in the intestine of the cat also emptied itself into a vesicle lying upwards and posteriorly in the abdominal cavity. From here he could, as with the calf, follow the flow of chyle up through the thorax in a tube, which he now observed emptied into the bloodstream at the shoulder vein's mouth into the neck vein. From here he could follow the white chyle down into the right-hand chamber of the heart. Why Rudbeck does not already at this point publicize these epoch-making observations, or in other ways share them with others, he makes no mention. This appears first in connection with a dissection he carried out for Queen Christina in April (or May) the following year.[186] Concerning the dissection Rudbeck states:

"In 1652, I demonstrated the aforementioned chyle container during an anatomical dissection, which was undertaken not only at the command of, but even in the presence of Her Majesty, before whom the most outstanding of physicians, Professor of Medicine Stenius,[187] gave an account of the subject in a lecture. On this occasion the royal physicians informed me that, shortly before, a message had come to Stockholm, that Pecquet had published a book on the same subject, and this I learnt about in June. A short time later I obtained, in Janson's bookshop, van Horne's treatise published in May on the newly discovered chyle duct.[188] And finally, I obtained in the September-market in Stockholm Thomas Bartholin's newly published treatise on the discovery of this container in the human being. There is therefore no reason for me to deal with this question at greater length, since it has already been dealt with by so many writers; for from their experience and considerations it appears quite clearly that the liver is not the most important organ in the production of blood."[189]

One may wonder a little about Rudbeck's piece of information, that he did not gain knowledge before April (or May) 1652 of Pecquet's book, which was published almost a year earlier. As already mentioned, in Copenhagen on 15 December 1651, Bourdelot had witnessed Bartholin's vivisection of a dog completely according to Pecquet's description. Therefore, Bourdelot doubtless had knowledge of Pecquet's discovery and publication when he came to Queen Christina and the physicians in Uppsala at the beginning of 1652, and he explains in a letter to Thomas Bartholin on 15 March 1652 that he had already at this point informed the queen about Bartholin's work and experiments, and the queen had responded by calling for the same pursuit of truth and fame among the Swedish physicians.[190]

In Bourdelot's letter to Bartholin dated 18 May 1652 we come to know directly that Queen Christina has told the physicians in Uppsala, including apparently Rudbeck, about Bartholin's excellences, and that Bartholin has inspired them to perform vivisections: "Your Excellency, about whom the queen has given a fine report to the professors in Uppsala, has functioned as a stimulus to them, and they are now plunging enthusiastically into the dissection of living creatures; one of them [Olof Rudbeck] made the same observation of the chyle-carrying vessels in the thorax without knowing Pecquet's name, for this is what he claims, and he promises a lot of things."[191] One perceives Bourdelot's scepticism in reference to Rudbeck's alleged ignorance of Pecquet and his discoveries.

So much for the chyle vessels, the receptacle and the thoracic duct. As far as the lymph vessels are concerned (the crystal-clear or serum-filled vessels), Rudbeck gives an account with precise dating for the chronology of his discoveries,[192] which cannot be understood in any other way than that he wishes to claim his priority.

On 27 January 1651, he noticed behind the large intestine and rectum of a dog some serum-filled vessels, which emptied their contents into the chyle container (Pecquet's *receptaculum chyli*). On 6 March 1652, he found in the abdominal cavity of a sheep numerous very transparent and knotty serum-filled vessels, which also emptied their contents in the chyle container, and on 19 October of the same year in the dissection of a cat he found the same type of vessel, which followed the loin veins beneath the kidneys and sent branches along the length of the muscles of the abdominal wall. On the 30 October he found in the dissection of a dog's thorax a pipe connection between two glands in the mediastinum and the thoracic duct. When on 27 April 1653 Rudbeck demonstrated the thoracic duct's entry into the shoulder vein for some friends, he also found some small glands behind the breastbone at the collarbones, and these glands sent two small serum-filled branches to the thoracic duct.

Rudbeck's treatise is supplied with two fine illustrations, drawn by himself, from the dissection of a dog. One of them shows the mesentery's chyle vessels, the chyle container (*receptaculum*) and the newly discovered serous vessels (lymph vessels), between the chyle container and the liver (Figure 119). The other shows the chyle vessels, the chyle container and the newly discovered serous vessels in the abdominal cavity together with the thoracic duct in the thorax (Figure 120).

Rudbeck divides the crystal-clear vessels into two types. From investigations with ligation of a long series of different creatures he demonstrates that the one type emanates from the liver, follows the portal vein and the bile duct in the direction of the pancreas and empties its crystal-clear contents into the chyle container between the kidneys. He calls these vessels the liver's crystal-clear excretory ducts and explains their knotty appearance by their regularly occurring valves, which open in the direction away from the liver and ensure a flow direction away from this organ. The body's remaining crystal-clear vessels, which are inserted between their matching glands, he calls the glands' serous vessels. Rudbeck writes that he demonstrated the liver's crystal-clear excretory ducts at the dissection for Queen Christina in April (or May) 1652. He had established that the vessels left the liver and before his discovery of their emptying into the chyle container he was under the impression that they contained a fluid

Figure 119. Olof Rudbeck's drawing of the chyle vessels or lacteal veins (*c* and *d*) and the receptacle (R) in a dog. The newly discovered serous vessels (lymph vessels) (*e* and *f*) are seen between the receptacle and the liver lobes (I). The letters mark: A: Stomach. B. Pylorus. C. Duodenum. D. Jejunum. E. Ileum. F. Coecum. G. Colon. H. Rectum. I: Five liver lobes. K. Gall bladder. L. Kidneys. M. Kidney veins. N. Vena cava. O. Vena portae. S. Mesentery. T. The weak part of the mesentery, where ligatures can be conveniently attached to the hepatic ducts. *a*. Glandular part of pancreas in the middle of the mesentery. *b*. Fleshy part of pancreas attached to the duodenum and beneath the stomach. *c*. Lacteal veins between the intestines and the pancreas. *d*. Lacteal veins departing from the glandular part of pancreas and entering the receptacle. *a*. and *e*. Liver ducts, which originate from outside the liver. *f*. The same running to the gland *m* and continuing from there to the receptacle. *g*. Two branches of bile ducts. *k*. The bile duct (choledochal duct) running into the duodenum. *i*. Mesenteric veins. *m*. Gland below the vena portae receiving liver ducts. *n*. One of those ducts most often seen crawling around the gall bladder. *o*. Branches of vena portae to the liver. *r*. Liver ducts, which wrap around the vena cava and enter the receptacle. *t*. Mesenterial veins. P. Valves in a serous (lymph) vessel.

From Rudbeck's *Nova Exercitatio Anatomica Exhibens Ductus Hepaticos Aquosos, & Vasa Glandularum Serosa, nunc primum inventa, æneisque figuris delineate*, Västerås in the summer of 1653. The dissertation was published a few months after Thomas Bartholin's *Vasa Lymphatica*.
Reproduced after Thomas Bartholin's *Anatome Quartum Renovata*, Leiden 1673, p. 629. The Royal Library in Copenhagen.

p
p
p
p
p

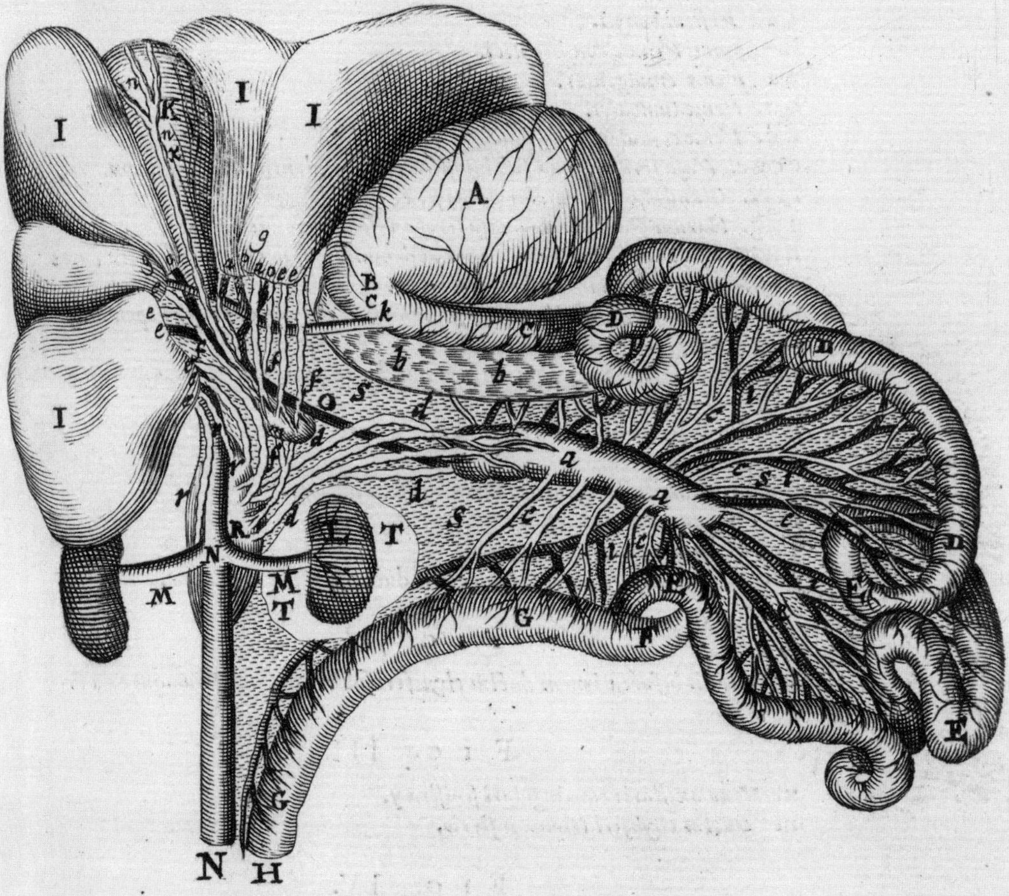
I
K
I
I
A
I
B
C
k
D
D
b
b
O
S
d
a
a
L
T
N
M
M
T
G
F
F
E
E
N
H
G

that was of no use for the nourishment of the body and that, after being carried to the pancreas, was excreted through this organ's excretory duct to the small intestine as a waste product, like bile from the liver. His further investigations had not brought him to any certain clarification of these vessels or the function of their inside fluid, but he mentions as possibilities that the fluid can perhaps contribute to fermentation and/or diluting of chyle and towards the chyle's movement. With reference to the vessels' significance in illness, he mentions that their eventual blockage can cause accumulation of fluid and the development of hydatids in the liver and abdominal dropsy with exudation of fluid from there. He refers here to Chapter 14 in Thomas Bartholin's *Anatomia Reformata*: "In the liver's surrounding membranes occur water blisters, from which, according to Plater, dropsy occurs."[193] To the glands' serous vessels Rudbeck ascribes the function of conducting fluid from parts of the body and from the glands to the chyle container and the thoracic duct. He ascribes to the glands the function of absorbing serum, which for the purpose of warming is sweated out from the inner organs, and conveying it to the chyle container, so that later it can be separated from the blood in the kidneys. Alternatively, he suggests, the glands have the purpose of drawing chyle out from the blood and preparing it by dissolving

Figure 120. Olof Rudbeck's drawings of the thoracic duct, the chyle receptacle, the chyle vessels and the lymph vessels in the thorax and abdominal cavity of a dog. From Rudbeck's *Nova Exercitatio Anatomica*, Västerås 1653. Reproduced after Thomas Bartholin's *Anatome Quartum Renovata*, Leiden 1673, p. 631.

The letters mark:

Fig. I: A. Left side of the heart. B. The lungs turned to the right. C. Vena cava. D. Vena axillaris dextra (right shoulder vein). E. Vena axillaris sinistra (left shoulder vein). F. The same opened to make visible the inlet of the thoracic duct. G. Sternum drawn in points only. H. Vena jugularis sinistra (left neck vein). I. Aorta. K. Chyle receptacle. L. Branches of vena cava to the liver. *a*. Kidney veins. *b*. Lumbar veins. *d*. Crural veins. *e*. Serous vessels, which extend below the rectum to the receptacle. *f*. Glands situated adjacent to the crural veins. *g*. Serous vessels extending from the glands to the receptacle. *h*. Similar vessels progressing between the abdominal muscles to enter the chyle receptacle. *i*. Lacteal veins (chyle vessels) crawling between the pancreas and the chyle receptacle. *k*. Pancreas. *l*. Lacteal veins between the pancreas and the intestines. *m*. The thoracic duct. *n*. The inlet of the thoracic duct into the shoulder vein. *o*. Glands of sternum. *p*. Confluence of a chyle duct branch and a serous duct. *q*. Branch of a serous duct for the ribs. *r*. Cardiac glands. *s*. Serous vessel inserting into the thoracic duct beneath the heart.

Fig. II: *t*. Shows the bifurcation of the thoracic duct below the heart.

Fig. III: *x*. The shoulder vein with the left neck vein *y*. *u*. Triple insertion of the thoracic duct into the shoulder vein.

Fig. IV: *z*. Various ramifications of the thoracic duct found in animals.

Fig. V: α. Oesophagus. β. Glands attached to the oesophagus. γ. Serous vessel extending from its origin in the gland to its inlet into the thoracic duct. δ. Thoracic duct.
The Royal Library in Copenhagen.

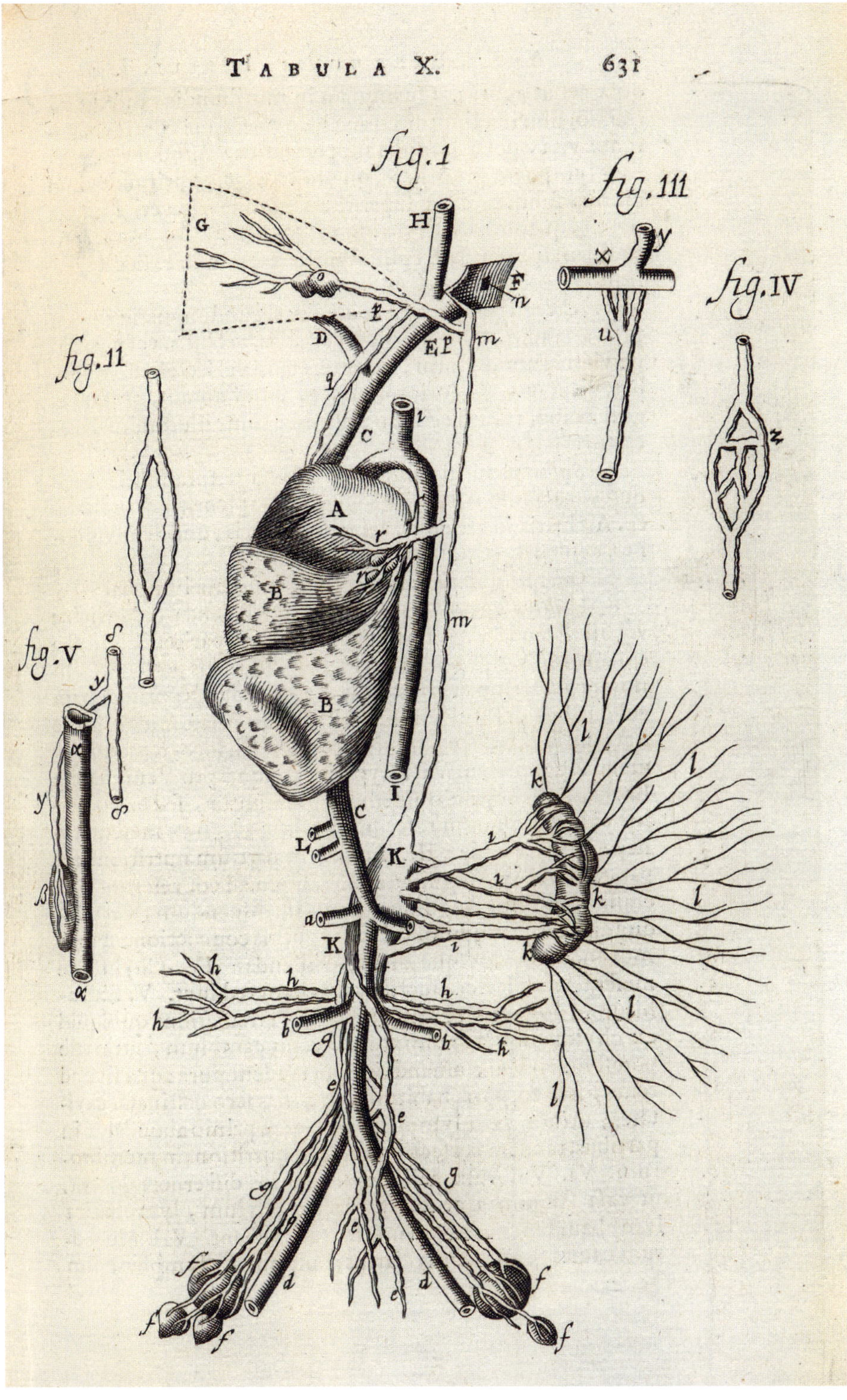
TABULA X.
631
fig. 1
fig. 11
fig. 111
fig. IV
fig. V

and sifting it, after which it is conveyed back to the chyle container and from there in a kind of cycle further back into the blood, so that it is transformed into nourishment for the body. In a similar way he mentions that the glands' task is to take up fat, which through the warmth of the body, for example during a fever, has become fluid, and to convey it farther to the chyle container, so that it can function as nourishment for the body. As for the liver's serous excretory ducts he also describes how obstruction of the glands' serous vessels can cause infiltration of fluid (hydrops) and leaking of fluid into the abdominal cavity (abdominal dropsy), the thorax and other cavities of the body.

Rudbeck's discoveries only build on the dissection of animals, and he gives instruction about how you take out both the liver's crystal-clear excretory ducts and the glands' serous vessels in animals. On the other hand, he again gives no explanation as to why he did not at a much earlier point choose to publish his results or just parts of them.

He finishes his treatise with the following request: "But nothing will be more welcome than if there should be someone, helped by more abundant experience, who will more specifically elucidate the meaning of these developments and give their results to the anatomical world."[194] It would soon appear that someone or other with some justification could question the sincerity of this statement. When Rudbeck first heard of Bartholin's *Vasa Lymphatica*, published in Copenhagen on 1 May 1653, is unknown. Shortly after the publication of his own writing in Västerås, he went on a study trip to Leiden with economic support from Queen Christina, and for the time being did not say any more in the tale of the lymph vessels. In Leiden, however, Rudbeck struck up a friendship with Thomas Bartholin's cousin, Willum Worm, who in a letter to Bartholin, dated 3 November 1653, informs him of Rudbeck's treatise, gives a resumé of it and encloses Rudbeck's two illustrations.[195] However, he does not give Rudbeck's name or the times of his discoveries but describes him simply as the author and his close friend, who has demonstrated all these matters with great thoroughness for himself and Bartholin's earlier travelling companion, professor Johannes van Horne. Thomas Bartholin explains later in a letter dated 31 July 1654 to one of his pupils, Martin Bogdan, that he received Rudbeck's treatise from his bookseller around Christmas,[196] and that it did not happen earlier is supported by a letter from Ole to Willum Worm, dated 17 December 1653, in which he writes: "The name of the Swede and his treatise we have been expecting with our merchant Erik Munk. He mentions that he talked with you in Amsterdam but received nothing from you."

At the beginning of 1654 something happened that raised the young Rudbeck from his chair and caused a storm between him and Bartholin to break loose. It was common at the time to reprint others' works with the purpose of publishing a composite volume. This is what the Dutch medical student in Leiden, Sibold Hemsterhuis (1629–1695), did, including the last three years' important discoveries of the new vessels. This composite volume, which he called *Messis aurea triennalis* (A three-year golden crop), consisted of Pecquet's treatise on the thoracic duct in animals (1651), Bartholin's on the same in human beings (1652), Bartholin's treatises on anatomical questions of doubt (1653) and the lymph vessels in animals (1653) and Rudbeck's on the same subject (1653).[197] Hemsterhuis dedicated the book to Thomas Bartholin, "anatomy's eye," and in the preface assigned priority to Bartholin in discovering the lymph vessels. Bartholin doubtless knew of the book and sanctioned it before its publication, for he availed himself of the opportunity to clarify some of the dates of his discoveries of the lymph vessels in the new imprints.[198] In the first edition of his treatise on the thoracic duct, he had not supplied the date of his demonstration for Ole Worm, Henrik Fuiren, Pierre Bourdelot etc. of several small vessels around the portal vein and the liver with serous content, which Bourdelot most definitely denied was chyle; but in Hemsterhuis' publication Bartholin made clear that the date was 15 December 1651. For another dissection with the discovery of the lymph vessels around the liver in a dog Bartholin supplied a date of 9 January 1652, and in Chapter 2 of the treatise on the lymph vessels in animals, where he had in the first edition simply given 28 February as the date when he and Lyser finally observed the lymph vessels, he now also made clear that the year was 1652.

Rudbeck, who found himself in Leiden and was a contributor to Hemsterhuis' collection, at the eleventh hour received for inspection the correction with Bartholin's precise dating and now flew into a rage. He wrote in great haste a six-page letter to Hemsterhuis dated 23 December 1653, which he got Hemsterhuis to put into *Messis aurea triennalis* just before his own treatise. In this letter, which is characterized by an uncompromising tone, he accuses Bartholin directly of fiddling the dates with the purpose of usurping the priority of the discovery of the lymph vessels after having read Rudbeck's treatise on the same topic. He particularly contests Bartholin's addition of the year 1652 to the date 28 February, where Rudbeck believes that the correct year is 1653, when the treatise in question (Bartholin's *Vasa Lymphatica*) was published.

Rudbeck's accusation was, however, groundless: partly because Bartholin had still not read Rudbeck's treatise at the time he supplied the precise dates of his undertakings, and partly because Bartholin had multiple witnesses that the dissection in question actually took place on 28 February 1652.

More or less at the same time as Hemsterhuis' publication, which Bartholin received from Willum Worm in February 1654,[199] Bartholin published the first two centuries (volume containing 100 reports) of the work concerning rare medical accounts, *Historiarum Anatomicarum Rariorum.*[200] Bartholin specifically calls account 48 in Volume 2 "The first discovery of the lymph vessels" (*Lymphaticorum Vasorum Inventio Prima*). Here he reproduces the drawing of the dissected dog from *Vasa Lymphatica* and repeats information about the precise dates of his and Michael Lyser's discoveries of the lymph vessels in animals.[201] Last, but not least, he finishes off with a thorough broadside against Rudbeck:

"In this way it occurred that the new type of lymph vessels, which had been unknown and hidden for so many centuries, and which were very useful when one had to explain certain illnesses, revealed themselves to me distinctly for the first time in Copenhagen, because Nature was kind and obliging. For the honour due to the God of creation, for the extension of Nature, which has been well-disposed towards me, for the health of mankind and the reputation of our fatherland and without any wish at all for personal fame or desire to advance myself I allowed this discovery in short to be made public for the first time in Copenhagen on 1 May 1653, during this year, that has passed so rapidly. The description was composed with a very swift pen, partly because it ought not to appear that I despised kind and compliant Nature, partly because I wanted to avoid my discovery being treacherously snatched by some unknown, uninstructed thieves. Its reputation had indeed already spread all around, and I had suggested it in a few words in Chapters 6, 12 and 15 of *De Lacteis Thoracicis*, which was published on 5 May 1652. It is in fact proof of a morally tarnished period when you attribute praise to yourself for the work of others and advance yourself with the help of a discovery you could not have achieved yourself. Whatever should happen in future, I shall wrap myself in my honesty and integrity and depend on what my illustrious audience will think. They were present in the Anatomical Theatre with us and observed on many occasions the preliminary practices that led to our findings, and they sincerely wished us good fortune with our successful discoveries. Among them I mention for honour's sake and to fulfil my promise of reliability the renowned men and medical doctors Ole Worm and Henrik Fuiren, Master Michael Lyser and Willum Worm, Jacob Holst, Peder Schumacher, Georg Seger and Henrik Martin à Møinichen, who are medical candidates, and others, who could be assembled in a tolerably long line, should the need arise."[202]

As already stated, one must give Bartholin credit, for in the chapters mentioned in the treatise on the thoracic ducts he does indeed refer to these vessels, which in the following year he names "lymph vessels," and he is even modest when he writes that he only indicated them in a few words.

Regarding the account in *Historiarum Anatomicarum Rariorum*, Bartholin writes in a letter to Martin Bogdan on 31 July 1654:

> "I have written an account concerning the discovery of the lymph vessels without any deceit or pretence in *Historiarum Anatomicarum Rariorum*. If it contains anything that is invented or untrue, I will invoke the wrath of the gods upon myself."[203]

This report became Bartholin's only official contribution to his priority dispute with Rudbeck concerning the lymph vessels. The remainder of the quarrel came to a head between the two young fighting cocks Olof Rudbeck and Martin Bogdan, while Thomas Bartholin adroitly kept in the background.

The official explanation of this, which Bartholin made a remarkable effort to convince the world of, was that Bogdan became so piqued at Rudbeck's false accusations against his mentor, that on his own initiative and without Bartholin's knowledge he launched a counterattack against Rudbeck. Although the smoking gun is absent, one must be more than commonly naïve if one does not agree with Rudbeck, that Bartholin from his place of concealment staged and orchestrated everything. As the cases involving Hofman and Riolan show, Bartholin was an able and shrewd polemicist who did not put up with anything, least of all that someone should try to deprive him of the honour of his greatest discoveries hitherto. Such a serious attempt as Rudbeck's could not go unheeded. The question was simply how the matter should be tackled. To take part in public battles against other established and acknowledged scholars such as Hofman and Riolan was not a problem, but to enter into public retaliation against a Swedish student was below the dignity of the internationally well-reputed Danish professor of anatomy, Thomas Bartholin.

We get the first indication that something is awry with Ole Worm's letter to his son in Leiden on 21 January 1654:

> "I have given Dr. Bartholin the Swede's ivy wreath. There is here a German medical student [Martin Bogdan], who is thinking about some remarks against his treatise; perhaps they will be dealt with quickly. But I am keen that you should keep quiet about this until they have been printed."

It is very difficult to imagine that Worm, Bartholin's uncle and closest confidant, could have possessed this knowledge about Bogdan's imminent counterattack against Rudbeck, without Bartholin knowing it himself. Worm's wish to keep the counterattack secret speaks for itself.

Bogdan's first 44-page defence of Bartholin, or rather, as the title indicates, counterattack on Rudbeck, *Insidiæ Structæ Thomæ Bartholini Vasis Lymphaticis ab Olao Rudbeckio Sveco in suis Ductibus Hepaticis et Vasis Glandularum.*

Arosiæ editis (The treacherous ambush on Thomas Bartholin's *Vasa Lymphatica* instigated by the Swede Olof Rudbeck, in his writing about the liver's excretory ducts and the glands' vessels published in Västerås) was published in Copenhagen and Frankfurt in February 1654. In it, Bogdan accuses Rudbeck of having stolen the most substantial part of his treatise from Bartholin.[204] Shortly after its publication, Bartholin writes in a letter to Willum Worm on 10 March 1654:

"I also send Bogdan's apology, which he wrote and published without my knowledge. [...] I would never have written a single word in reply. Now that I have sufficient fame, I avoid disputes, and I am satisfied if my conscience is clear. Moreover, I do not feel that it is worth beginning much work in finding out who was the first to discover this or that. It is enough that it has been discovered and can contribute to the good of the community."[205]

In the same letter Bartholin sends his thanks for receiving one of Rudbeck's sketches of the chyle container in a human being. Apart from a single lymph branch to the underpart of the body, which is mistakenly marked as being a lacteal vessel, Bartholin praises the sketch, which he calls an elegant supplement to his own, and he compliments "the excellent" Rudbeck for his thoroughness. However, he adds: "If only certain people preferred to become famous through suchlike or other additions instead of attacking others' opinions and discoveries."

You could easily be lured into taking the letter to Willum Worm as proof of Bartholin's innocence, magnanimity and good intentions. Certainly, Willum Worm had established friendship with Rudbeck, but why should Bartholin write falsehoods to his confidant and cousin?

One should, however, remember at this point, that Bartholin's letters were published and edited by Thomas Bartholin himself, and were his most important tool for giving a picture of himself. In 1663, when this letter was published, it was always important for Bartholin on the one hand to maintain his priority over Rudbeck in the discovery of the lymph vessels, but on the other hand to paint a picture of himself as the great anatomist, who dispenses magnanimity, is willing to share his fame with others and wishes to promote the good of the community. How creative Bartholin was in the self-editing of his letters no one knows, for only a few of the original letters survive, but there are no extant complaints from Bartholin's contemporaries about the editing, as far as is known. Self-editing of published letters was at this time by no means unknown, and Bartholin defends the process by saying that letters that are to be made public must be composed in a way that can be followed by everyone.

Rudbeck's answer to Bogdan's apology did not take long to arrive. A couple of months later he published in Leiden a 164-page piece of writing, *Insidiæ*

Structæ, Olai Rudbeckii Sueci Ductibus Hepaticis Aquosis, & Vasis Glandularum Serosis. Arosiæ editis, a Thoma Bartholino Médico Regio *Professore in Academia Hafniensi* (The treacherous ambush on the Swede Olof Rudbeck's *The liver's crystal-clear excretory ducts and the glands' serous vessels*, published in Västerås, instigated by Thomas Bartholin, Royal Physician and Professor of Anatomy at the University of Copenhagen), in which he denies the accusation of plagiarism with the same arguments as earlier and brings critical remarks on Bartholin's treatise concerning the lymph vessels. It appears that he regards Bartholin as the true author of Bogdan's apology.[206] One cannot accuse Bogdan – nor perhaps, more justly, Bartholin – of lacking industry or perseverance. A few months later he was ready in Copenhagen with a new apology in Bartholin's defence, this time covering no fewer than 284 pages: *Apologia pro Vasis Lymphaticis D. Thomæ Bartholini Medici Regij Professoris & Anatomici, conscripta a Martino Bogdano Drifna Marchico contra Insidias Secundo Structas ab Olao Rudbeck Sueco* (Apology in defence of Royal Professor of Medicine and Anatomy Dr. Thomas Bartholin's *Vasa Lymphatica*. The defence was composed by Martin Bogan from Driesen in Neumark and directed against the hidden ambush, which for the second time was instigated by the Swede Olof Rudbeck). In this piece of writing, which he dedicated to Henrik Fuiren, Bogdan does not include any new information in the striving for priority but repeats with more bitterness his earlier accusations against Rudbeck and takes the opportunity to reply to the latter's personal attack on Bogdan himself.[207]

The most interesting fact about this long apology is that it is enclosed with a 15-page letter from Thomas Bartholin in Roskilde to Martin Bogdan in Elsinore, dated 31 July 1654.[208] Naturally, it is unthinkable that Bogdan would have published this letter without Bartholin's knowledge and acceptance. The letter is a long self-production on Bartholin's side. He attempts to draw a picture of Bogdan as a capable and talented pupil, who on his own initiative has undertaken the chivalrous attempt to defend his beloved teacher against spiteful allegations from "those who like to neigh and kick, because they have still not been tamed." In this errand Bogdan himself has become the victim of a malevolent attack, but Bartholin gives consolation:

"Thus it is, my dear Bogdan: Envy also grows side by side with the discoveries of Nature. It is the good fortune of our age to abound in plenty of examples of diligence and to bloom with the untiring efforts of unknown observations. But no age has ever been more fruitful of vices or unjust accusations than this one, where morals are overturned, wickedness is in control, and human affairs are going downhill."

The fact, that Bogdan puts his thoughts in a totally different way than Bartholin demonstrates, according to Bartholin, that Bogdan is the true writer of the apology. Bartholin regrets all the attacks he himself has been the victim of:

> "No line has flowed from my pen without being tainted by the spiteful tongue of the envious. I have written nothing that they, who observed what I wrote with sick eyes, did not dare to condemn. If they could interrupt even the lightest strokes of the pen, if they could tear up the paper, there would be no lack of people who would promise themselves undying fame as a result of my ruin. [...] The old men insult me out of envy, the young ones tear at the back of my garments out of malice, while the ones in between, motivated by secret rivalry, hurl weapons while maintaining a friendly expression, and if I calculate correctly, there is plotting against me from all sides."

Although God has given him strength to grow higher like a palm the more attempts have been made to crush him, his wellbeing no longer tolerates this type of dispute, and as for the rest he has already obtained abundant fame and recognition:

> "Those who value things honestly have caused more praise to be accorded to me, than the malicious critics have stolen from me. I, who considered many people's adverse judgements of me to be of little significance, have always regarded the approval of good men, the consciousness of right and truth, and public usefulness, to which I have devoted myself, as the best means of increasing one's reputation. The wicked and corrupt behaviour of certain people has never led me astray."

Intertwined periodically in between all this whining, Bartholin directs one well-placed liver-blow after another at Rudbeck's body:

> "And so, I think that one should forgive the disputes of rising minds, unless they seek to gain their reputation by means of abuse or by threats of violence, or by using cunning, which results in the disgrace of illustrious men. I have noticed not without admiration that this is done with more boldness than modesty by certain rivals, whom I have understood are unjustly snatching my observations from me, your book from you, and our reputation from us both. [...] My age and experience did not seem unsuitable for new observations. [...] I leave to their untimely arrogance the question of whether they are gaining the same thing, who without a guide, without any practice, are keen to attain the learning and reputation of an anatomist. For we are amazed by the dreadful characters of those who teach before they have learned, and who make every effort to invent new anatomical things for themselves, although they do not have the knowledge of the ancients and do not even have a superficial acquaintance with the ancient writers."

Rudbeck did not reply to Bogdan's new apology, and with that their quarrel was settled for the time being.

Three years later, a 72-page open letter emerged in Uppsala from Rudbeck to Bartholin.[209] Here Rudbeck regrets the tone in his aforesaid letter to Hemsterhuis and explains that he did not wish to answer Bogdan's shameless apologies since he did not regard him as his proper, worthy opponent, but simply as a spokesman for Bartholin. But since in his letter to Bogdan, enclosed in the latter's second apology, Bartholin swears he is not the author of the apologies, Rudbeck finds cause to write this letter to him. In the letter, Rudbeck assigns priority to himself for the discovery of the lymph vessels (which he continues to call the serous vessels) in and around the liver, the heart, the pericardium, the lungs, the mediastinum, the sternum, the oesophagus, the stomach, the spleen, the testicles, the ovaries, the uterus, the rectum and the region of the loins. He includes new data for many of these discoveries, some of which were first made in the period 1653–1655. Only for the discovery of the lymph vessels of the shoulder does he give Bartholin priority.

Likewise, he points out that he gained insight into the function of the lymph vessels and took exception to the idea of the liver as a blood-producing organ before Bartholin. As evidence, he points to his earlier-mentioned public dissection for Queen Christina and various physicians in April (or May) 1652, where he demonstrated the thoracic duct and the lymph vessels around the liver on a live dog, as well as his drawings in his *Nova Excercitatio Anatomica* in the summer of 1653.

Moreover, he now indicates that he had already, in October and November 1650 and in January, February, March, April, September, October and November 1651 (i.e. before Bartholin's demonstration for Pierre Bourdelot and the others previously mentioned in the Anatomy House in Copenhagen on 15 December 1651) demonstrated the new serous vessels around the liver several times for various people at private dissections. As examples he mentions five persons "of the highest distinction and celebrity", namely his own brother, professor of theology Johannes Rudbeck (1623–1667), professor of mathematics, astronomy and medicine Olaus Stenius (1597-1660), the rector of the university Johannes Tenstad, the treasurer Claudius Laurentius Sæbenius and the medically interested Eric Bernhardsson Bougt. It is worth remarking that the only one of these witnesses who could be said to have the slightest medical insight was Stenius, but even that seems to have been limited. Rudbeck's biographer Gunnar Eriksson writes thus about the one-eyed Stenius: "The other professor, Stenius, clearly encouraged him [Rudbeck] in practical work, but it seems less probable that this astronomer, without any actual medical studies behind him, can have imparted any deeper insights into the practicalities of anatomy."[210] At the same time Rudbeck tries, with a series of cryptic arguments and insinuations, to prove that Bartholin is speaking untruthfully in Hemsterhuis' *Missa aurea triennalis* about the data for his discoveries of the lymph vessels mentioned in the treatise of May 1652 on the thoracic duct. For example, he believes that when Bartholin, in Chapter 6 of this treatise, during his first mention of the new serous vessels, says "It was remarkable that in a dog that was large and healthy, and that had been opened up seven hours after its meal, we found vessels that were very similar to chyle vessels," this is proof that this was the first time Bartholin saw these vessels, for otherwise he would not have used the word "remarkable". Based on his curious and illogical arguments, Rudbeck concludes: "From these proofs, which are clearer than the midday sun, I do not think that only men equipped with all their senses, but even the blind themselves can distinguish which of us was the first to discover these vessels, unknown to all earlier ages." Rudbeck's letter seems strained, almost desperate, and Bartholin did not answer it. Neither did he react to the fact that, in 1661, Rudbeck reiterated the statements about his priorities in the treatise *De sero ejusque vasis* (On serum and its vessels).[211] On the contrary, in 1662 Bartholin put in a letter to Rudbeck's professorial colleague in Uppsala, Johannes Schefferus (1621–1679), the germ of a reconciliation and fraternization with Rudbeck. Bartholin proclaims that he is eager to hear what Rudbeck is doing, addresses him respectfully, sends him greetings and proposes that Rudbeck should be the illustrator for the next edition of the Bartholinian anatomy book.[212] In Scheffer's reply to Bartholin he sends Rudbeck's most respectful greetings and passes on his pleasure at having been mentioned in so friendly a fashion in Bartholin's letter.[213]

In the period 1679–1702, Rudbeck published the strongly nationalistic four-volume work *Atlantica*. Here he attempts to prove that Sweden is identical with the lost Atlantis, that Scandinavia is the cradle of civilization, that Swedish is the world's original language, that the ancient gods and heroes were Swedish, and much more. Not surprisingly the work later became the object of Ludvig Holberg's (1684–1754) piercing satire.[214] In the work's second volume of 1689, nine years after Thomas Bartholin's death, Rudbeck mentions a letter from Thomas Bartholin dated 20 January 1666[215] in which Bartholin, in very deferential and respectful phrases, thanks Rudbeck for another letter and for his book on horticulture, which Bartholin praises for its great tastefulness.[216] He encloses his own treatise on Danish medicine and Danish medicines, *De Medicina Danorum Domestica* (see later), Copenhagen 1666, in which he also praises Rudbeck's gardening book. He appreciates Rudbeck and their friendship deeply. In an epilogue he mentions his planned fourth edition of the anatomy book, and he invites Rudbeck to enrich the book with his comments, just like Niels Stensen had already contributed with many new additions.

The compendium mentioned earlier about the system of lymph vessels, *Opuscula Nova Anatomica* of 1670 (Figure 121), brought nothing new in the striving for priority with Rudbeck, but in the said fourth edition of the Bartholinian anatomy book of 1673 Bartholin includes his last chirp in the tale:

"In various writings I have published a complete account of the new vessels, and in written rejoinders against hostile attacks I have defended my rights and in special editions strengthened it further. Likewise, in two written rejoinders Bogdan has awarded me these vessels.[217] Almost at the same time in Sweden, Olof Rudbeck discovered similar liquid-filled channels and later described them. In England, Joyliffe's friends boast that he has also made the discovery.[218] That these vessels have been seen by others apart from me I shall not dispute, but no one blazed the trail before me, the first person to step out into the arena. I have, however, never ascribed such great importance to this discovery as to stop me acknowledging that Nature stands available to everyone."[219]

I agree fully with Bartholin's conclusion. On the one hand, despite the defective documentation for Rudbeck's early observations, there is no basis for denying that he discovered the lymph vessels at about the same time as Bartholin. Perhaps the best proof of Rudbeck's early insight into the subject are his detailed, masterly drawings, which according to his own statement he presented at Queen Christina's court at the end of May 1652,[220] and which constituted the model for the two copper plates in his treatise concerning the lymph vessels from the summer of 1653. The drawings impressed Thomas Bartholin, who incorporated them on pp. 629 and 631 in the fourth edition of his anatomy book.

On the other hand, there is no doubt that priority for the discovery of the lymph vessels according to the rules of scientific priority of the present day – and in all probability of that time too – must go to Thomas Bartholin. According to these rules priority is awarded to the researcher who first makes his results accessible to independent expert witnesses, either in a public printed publication or an equivalent verbal presentation. Bartholin preceded Rudbeck on both counts. He demonstrated lymph vessels in a dog for Pierre Bourdelot and several others in Copenhagen in December 1651, four or five months before Rudbeck's verbal demonstration for Queen Christina, Pierre Bourdelot and several others in Stockholm in April (or May) 1652. Thomas Bartholin refers to lymph vessels in a letter to Frederik Arnisæus on 30 April 1652, and in three places in his treatise on the thoracic duct's first demonstration in the human being in May 1652. He made public his treatise on the discovery of the lymph vessels in animals in May 1653, about two months before Rudbeck. Moreover, in this treatise he gives the lymph vessels the name, they have today. He is the first in the world to point

VASA LACTEA
et
LYMPHATICA.
TH BARTHOLINI
Hafniæ Sumtibus DANIELIS PAULLI.

out the lymph vessels in human beings and publicizes the information on 1 May 1654 in a treatise on the subject.

The idea that Bartholin must have plagiarized Rudbeck lacks any form of documentation. Since Rudbeck first publicized his discoveries after Bartholin, there is no documentary evidence that Bartholin knew their content. Pierre Bourdelot's mention of an unnamed Swede's research in a letter to Bartholin on 18 May 1652 is the only proof, that Bartholin had any idea that someone in Sweden was researching in this area. Conversely, it appears from Rudbeck's treatise about the lymph vessels from the summer of 1653 that he received Bartholin's treatise on the thoracic duct in September 1652, and that he had detailed knowledge of Bartholin's section on the lacteal veins in the third edition of the anatomy book of 1651, from which he quotes. On the strength of my reading through their respective treatises I consider that Bartholin also had a substantially more nuanced and detailed understanding of the lymph vessels' important draining function and its significance for the processes of illness than Rudbeck, which is to be expected when you consider Bartholin's greater medicinal knowledge and his far greater clinical experience. In his own mention of dropsy and hydatids Rudbeck thus refers to Bartholin's anatomy book.[221]

Many writers, most of them Swedish, have over time awarded Olof Rudbeck priority to the discovery of lymph vessels.[222] The most modest of these is Gunnar Eriksson, who is content to give priority to Rudbeck for the discovery of the lymph vessels of the liver in animals,[223] but this too must be disputed since, as already mentioned, Bartholin demonstrated exactly these lymph vessels in a dog for the same Pierre Bourdelot and several others in the anatomy theatre in Copenhagen

Figure 121. The frontispiece of Thomas Bartholin's *Opuscula Nova Anatomica De Lacteis Thoracicis Et Lymphaticis Vasis*, Copenhagen and Frankfurt 1670. Engraving by Johann Philipp Thelott (1639–1671). 14.6 × 8.7 cm.

Thomas Bartholin published this collection of his own and other selected authors' works on the lymphatic system. The frontispiece depicts a paradisiacal universe, where water (lympha) and milk (lactea) are the life-giving elements of nature and the human body. Two plump, naked naiads with flowers in their hair sit on opposite mountain peaks, holding a laurel wreath in one hand over the head of Bartholin, who is growing out of the ground in the middle of the picture. In the other hand, one of them holds a palm branch and the other a bowl to collect the milk that springs from their breasts and continues in winding streams made like lymph vessels down between animals and plants to a sea with two mermaids along with water that emerges from springs in the mountains and the ground. The animals are an owl, symbolizing wisdom and learning, and three mammals, of which the cow and the sheep are milk-producing, and the dog was the test animal, on which Bartholin found the lymph vessels. Bartholin wears a skullcap and neck frill, and the portrait is inspired by Figure 96 a + b. Bartholin dedicated the book to Christian V, and in the dedication he writes that the two faithful and humble naiads pour their lymph and milk in abundance over his majesty's feet. See also Lea Hagedorn and Reinhard Hildebrand 2022.
The Royal Library in Copenhagen.

in December 1651, four months earlier than Rudbeck's showing of them to the same Pierre Bourdelot, Queen Christina and others, likewise in a dog.

Eriksson writes: "Rudbeck's early anatomical achievements are known almost exclusively through his own accounts of them. For the most part these reports seem truthful, and we must stick to them for want of any better ones, as long as they are not contradicted by other undeniable facts."[224]

I believe with this survey of the documents of the case that I have provided such facts, and think it appropriate that modern historians of medicine cut through and award Thomas Bartholin what is rightfully his.[225] Among these is the Swedish-British historian of medicine Jan Bondeson, who writes about Bartholin: "He had a real scientific genius and was one of the founders of modern anatomy through his discovery of the lymphatic system."[226]

One can always, as Bartholin himself says, discuss the importance of establishing priority in scientific observations. Does this giving of credit have any purpose other than to serve the honour of the individual researcher and their closest associates, and their country's national pride? The striving for priority in medical science was common in the baroque era and are not unusual today. In both Bartholin's time and today many researchers work simultaneously with the same topics and make the same observations. For humankind it is important that observations are carried out, but not by whom this is done.

Lack of respect for the credit of the individual researcher and for their rights will nevertheless be strongly demotivating and therefore harmful to research, and it is in this light that one should see Bartholin's and Rudbeck's battle for priority in the discovery of the lymph vessels. The system of lymph vessels held a prominent place in 17th century scientific research, and the honour for discoveries in this area was therefore correspondingly pursued, something beautifully illustrated by the following statement from another researcher into the lymph vessel system, Johannes van Horne, in a letter to Thomas Bartholin dated 17 August 1654:

> "As far as the history of the lymph vessels is concerned, it is not unknown to me to what extent other individuals exert themselves to plead priority in the discovery of these vessels. But this is a vice, in which we all share. In the meantime, it is our opinion that you are the primary authority, although it must be allowed that it was Master Rudbeck who was the first to demonstrate them for me."[227]

In 1660 Bartholin stated: "No discovery, with the exception of the blood circulation, has elicited livelier discussion in the world of medicine than the discovery of the lymph vessels."[228]

See appendix 4 for UBVA's (Committee for the Protection of Scientific Work) decision concerning the priority dispute between Bartholin and Rudbeck.

Administrator and author 5

"Only accordance in our knowledge will bring us closer to true wisdom."[1]
Thomas Bartholin, 1668

Bartholin becomes second professor of medicine and rector of the university

With the death of Ole Worm on 31 August 1654, at the pinnacle of the ravages of the plague in Copenhagen, the university lost its rector and second professor of medicine. Thomas Bartholin took over the professorship,[2] while the professor of theology Thomas Bang (1600–1661) temporarily took over the rectorate. Due to bad health, Bang wished to give over the rectorate to someone else as soon as possible, and he therefore gathered his fellow professors to a meeting in the Convent House in Roskilde on 4 November 1654 to discuss the question.[3] Because of his scientific work, Thomas Bartholin did not want to take over the rectorate and its administrative duties, but equally could not decline it, and was duly elected as rector for the rest of the current period 1654-1655.[4] A little over a month later, the plague had calmed down so much that Bartholin called the teachers at the university to a meeting in the consistory on 13 December, and two days later he held a memorial speech for Ole Worm in the large theatre.[5] Bartholin and his family moved back to Copenhagen, and on 1 February 1655 he sent out a new Latin university program. In the beginning of April, teaching and the other activities of the university were resumed.[6]

Probably Bartholin's greatest achievement during his rectorate, which continued until 6 July 1655, was his collection of funds for a museum for medicine and natural science – a cabinet of curiosities with medical and natural objects and rarities, which could be used in teaching. The largest such collection in Denmark was Ole Worm's, to which Bartholin had contributed so energetically during his nine-year study tour. Besides natural objects it also contained ancient relics, antiquities, objects of art and ethnographic objects. Worm's great scientific work and catalogue of the collection, *Museum Wormianum* – a principal work in older European museum literature – was in press when he died, and was published in Leiden and Amsterdam in 1655, the year after his death. After Worm's death, the collection was bought by Frederik III and, during July 1655, transferred to Københavns Slot (Copenhagen Castle), where great parts of it were included in Kunstkammeret (the Art Chamber), by means of which it finally ended up in Nationalmuseet (the National Museum).[7]

As a collector, Worm was strongly inspired by the Neapolitan apothecary and collector of rarities, the earlier-mentioned Ferrante Imperato, whom he had visited in 1609.[8] Imperato was interested in zoology, botany and geology, and his collection included stuffed animals as well as minerals, corals and petrified animals, sea animals and plants. His almost 800-page illustrated catalogue of the collection, *Dell Historia Naturale Libri XXVIII* from 1599, was the direct model for Worm's *Museum Wormianum.*[9] As the title indicates, Imperato's catalogue consists of 28 small books about different topics within natural history, which besides the already-mentioned topics also deal with alchemy. In the book, Imperato makes himself a spokesman for the opinion that petrified shellfish and fish teeth originate from real animals.[10]

As mentioned, Thomas Bartholin had his own collection of natural objects in his professor's residence in Copenhagen, and so did Henrik Fuiren. But the university also owned the beginning of such a collection, which in *Acta Consistorii* 27 January 1654 is called "Raritetskammeret" (Chamber of Rarities),[11] and which was located in the old library, *Bibliotheca Vetus*, in an extension to the south side of Vor Frue Kirke (the Church of Our Lady).[12] It appears from the documents that the collection, according to the chancellor Christen Thomesen Sehested's will, should be moved and accommodated according to the best model, and the cost of this was to be covered as needed. A royal ordinance of 18 March 1654 instructs the community's economist Thomas Fincke to set aside 1000 rix-dollars from the community treasury for the purpose of preparing the first floor of the Anatomy House to be able to house the university's collection of rarities, which has grown in size, after "several professors and others have presented the university with some rarities, and no place for it exists, except above the Anatomy House."[13]

The project did not get any further at this time. The plague temporarily halted the plans, and when they were resumed in January 1655, it turned out that the community could not allocate the necessary funds. Rector Thomas Bartholin then took the matter into his own hands. As *Medicus Secundus* he held the patronage of Grevinge Church in Odsherred in North Zealand, which in 1571 had been transferred from Roskilde Cathedral to the University of Copenhagen's medical faculty. As the church's patron, Bartholin disposed of its financial means, and as its fixtures and building were in fine condition, he decided to use its equity capital of 468 rix-dollars for the arrangement of "Academiets Antiquitet Kammer" (The academy's antiquity chamber).[14] In this way, the road to the establishment of the museum was secured. At the death of Thomas Fincke in 1656, Bartholin persuaded the other heirs to donate a further total of 1000

Figure 122. Memorial plaque for the establishment of the Natural History Museum of the University of Copenhagen on the first floor of the Anatomy House in 1657.

Thomas Bartholin erected the plaque to honour his grandfather Thomas Fincke, who had died the year before, aged 95. Fincke's name is seen at the top and Bartholin's at the bottom. The plaque was destroyed during the Copenhagen fire in 1728, but fragments of it were found during excavations at the university's main building on Frue Plads in 1905. It was reconstructed and is now mounted on the building that houses the Zoological Museum in Copenhagen. Photo: Jesper Brandt Andersen.

rix-dollars to the cause,[15] thanks to which the necessary capital was provided. The museum, which was completed in 1657, was in 1659 enriched with Henrik Fuiren's collection, and in 1674 with that of his nephew Theodor Fuiren.[16]

Whether Thomas Bartholin's collection became part of the museum, or what happened to it otherwise, is unknown. The museum with all its contents burned down during the great fire in Copenhagen in 1728. However, during excavations for the university library in 1905, some fragments were found belonging to the memorial tablet with Latin inscription put up by Thomas Bartholin some years after the museum's inauguration to honour his grandfather.[17] The tablet was reconstructed and put up in the university courtyard. Today it is located on the outer wall of Biological Institute in Copenhagen (Figure 122). In his memorial speech for Henrik Fuiren on 7 June 1659, Bartholin airs his view on the important role of museums of natural objects as places where you can study nature's diversity with your own eyes. He expresses that, by studying natural objects in such museums, you can "learn more than from a thousand books or from two thousand disputations."[18]

Bartholin becomes first professor of medicine and stops dissecting publicly

Upon the death of Thomas Fincke on 24 April 1656, Thomas Bartholin took over his chair as first professor of medicine (*Medicus Primus*).[19] The newly vacant chair as second professor of medicine (*Medicus Secundus*) was taken over by Christian Ostenfeld,[20] who had just returned from study tours abroad, and who married into the Fincke family the following year.[21] When the chair as third professor of medicine (*Medicus Tertius*) was taken the following year by Rasmus Bartholin,[22] the Bartholin-Fincke family occupied all three chairs of medicine. As the conductor of this orchestra, Thomas Bartholin had obtained significant influence within the medical field of the university. In the chair as second professor of medicine he had not been obliged to perform public anatomical demonstrations or to give lessons in anatomy, but according to his own wish he had continued to do so until a couple of months before Thomas Fincke's death. According to his own statement, his failing health could no longer withstand the long winter days at the dissection table, and the desire for more time for planned literary works became more and more pressing. On 9 February 1656, Thomas Bartholin therefore carried out his last public demonstration in the Anatomical Theatre of Copenhagen in the presence of Frederik III: the dissection of a hanged male thief.[23] In his opening speech to the dissection, Bartholin said, among other things:

"If my body's strength were as great as my mind is willing, I would much more often offer you to witness a complete dissection of a human being to get an insight into this God's workshop. But my failing health and other duties prevent it, now that I am no longer young. Therefore, I now step forward to speak my last words to you from this place."[24]

After a mention of his many years of research through dissections, he adds – in the same way as Vesalius – what impact this work has had on his health:

"Nights and days we have – unnoticed by others – been bent over the preparations. We have exhausted our hands, worn out our eyes, spoken ourselves hoarse, when the discoveries were to be presented and discussed without preparation, so that we had hardly any time to sleep, eat or rest."[25]

After thanking Frederik III for his support, Bartholin then declared that after this dissection he would leave this work to others.

Bartholin could now concentrate on his literary endeavours but, as he stated in *Domus Anatomica*, he continued as the daily leader of the Anatomy House until 1661, and as we shall later see, he continued to teach his pupils in anatomy and to carry out dissections on both humans and animals in private settings for many years to come. He continued his function as preceptor for medical students for the rest of his life.

Work on rare medical cases

Bartholin had already begun to publish his first larger multivolume work a couple of years earlier. The ink from his pen could barely have dried after the publication of his discovery of the lymph vessels in a human being before he published in the same year (1654) the first volume of a new three-volume work on rare medical cases, *Historiarum Anatomicarum Rariorum*, in Copenhagen, the Hague and Amsterdam.[26] The second and third volumes were published in 1657 (Copenhagen and the Hague) and 1661 (Copenhagen) respectively. Each volume consists of two centuries. A total of 33 of the cases are illustrated across the three volumes: the majority with engravings, the rest with woodcuts.

The first volume has 360 pages besides the title page with Thomas Bartholin's portrait (Figure 123), the dedication to Frederik III, the preface, a list of Bartholin's 22 works so far, the almost mandatory poem by Vitus Bering to Thomas Bartholin and the index. 17 of the volume's 200 cases are illustrated. The second volume, which ends with a description of a private dissection in the Anatomical Theatre in Copenhagen on 19 December 1656 of an African lion from the royal menagerie, has 430 pages besides the title page, the dedication to councillor of the realm Tage Thott (1580–1658), the preface and the index. Seven of the cases are illustrated. To this volume is added the Dutch botanist and anatomist, Leiden's first professor of anatomy, Pieter Pauw's (1564–1617) *Observationes Anatomicæ Seclectiores* (Selected anatomical observations). The third volume has 386 pages besides the title page, the dedication to Chancellor of the Realm Peder Reetz (1614–1674), the preface, a poem by Vitus Bering to Thomas Bartholin and the index. Nine of the cases are illustrated. To this volume is added Johan Rhode's *Mantissa Anatomica* (Anatomical addition).

The work deals with a wide range of medical topics. There are reviews of Bartholin's dissections leading to the discovery of the thoracic duct in man and of the lymph vessels, objective medical cases, and descriptions of epidemics of plague and dysentery in Copenhagen. It also contains cases of malformations, petrified fetuses, horned people, a horned horse, fantasy creatures and mythical beasts like mermaids, basilisks and unicorns; not to mention cock-and-bull stories, for example a pregnant woman who is cut in two after being shot by a cannonball, while her baby remains unharmed; a soldier who is shot through the mouth and excretes the bullet with the stool five days later; a cock and a

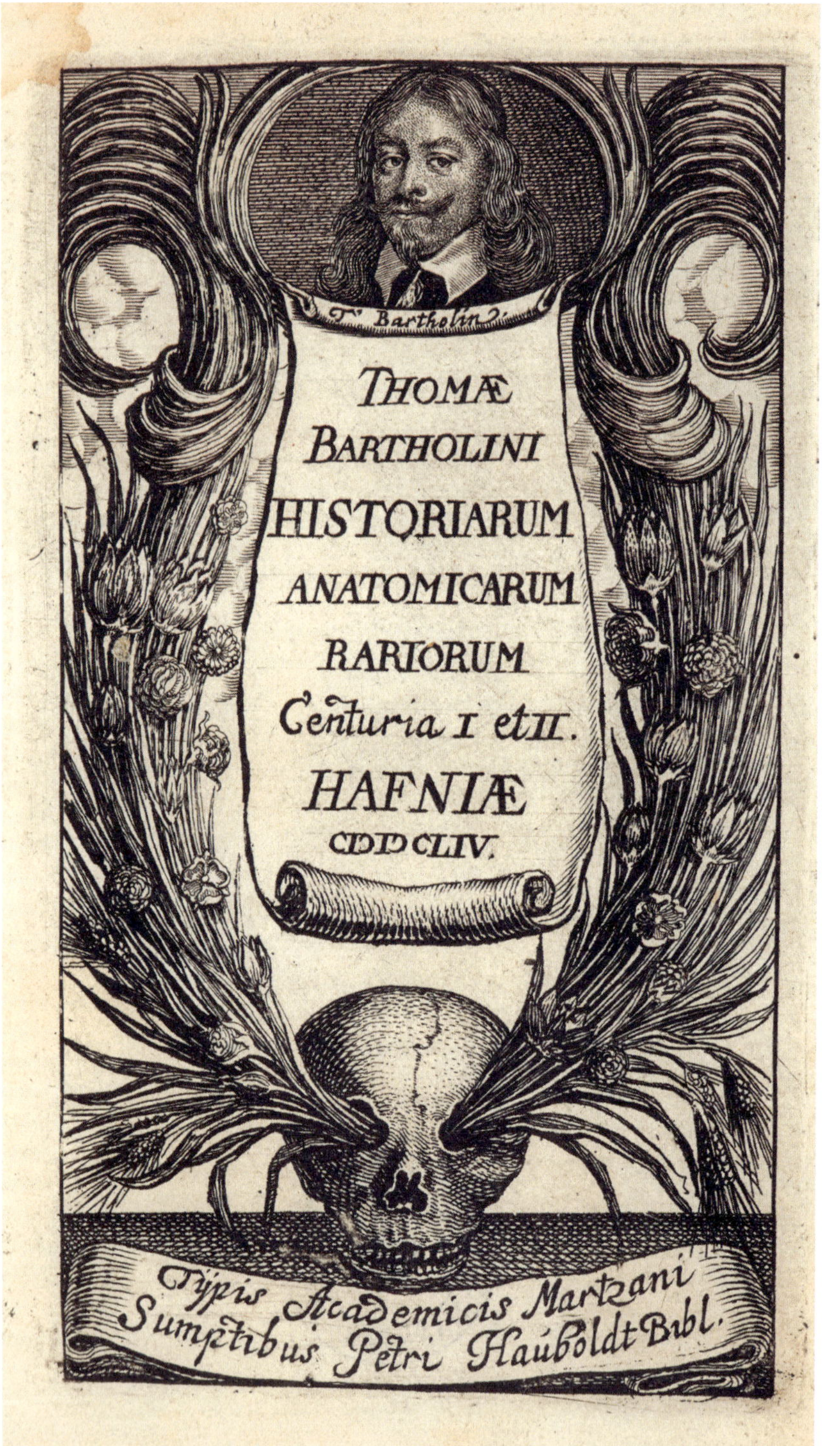

Figure 123. The frontispiece of the first volume of Thomas Bartholin's *Historiarum Anatomicarum Rariorum*, Copenhagen 1654. Engraving by Albert Haelwegh. 13 × 7 cm.

The volume contains two centuries with 100 accounts each. The second and third volumes, also with two centuries each, were published in 1657 and 1661 respectively. The portrait of Thomas Bartholin is inspired by Figure 96 a + b. The Royal Library in Copenhagen.

woman laying eggs (Figure 124); a woman giving birth through the mouth; and a schoolboy who swallows a six-inch-long flute and excretes it three days later (Figure 125). In other words, the work is a strange mixture of sober science and regular superstition. Although this was by no means unusual in Bartholin's time, one cannot help wondering why he chose to include many of the most improbable accounts and only exceptionally expressed his doubt about their truthfulness, but on the contrary in many cases tried to give them rational explanations. In the foregoing, examples of the work's accounts and illustrations have been given in relevant places. Further examples will be given here to illustrate the wide-ranging content.

Bartholin-Patau syndrome

In 1960, the German physician Klaus Patau (1908–1975) and his associates discovered that a known malformation syndrome characterized by lacking or underdeveloped eyes, a wide and flat nose without nostrils, a wide malformed mouth with double cleft lip and palate, a small head circumference, malformed and low-set ears, a small chin, fused and excess fingers and toes, and a four-finger furrow (transverse groove across the entire palm) is caused by an extra chromosome number 13.[27] Moreover, the children are severely brain damaged, and severe malformations in the brain, the spinal cord and the heart cause many of the children to die within the first days or weeks, and only 5–10 percent survive for more than a year. After Patau's discovery, the syndrome was called trisomy 13 or Patau syndrome. Likewise, in 1960 the Danish ophthalmologist Mette Warburg (1926–2015) became aware of the fact that Thomas Bartholin was the first to describe a child with this syndrome in the literature, and the name of the syndrome was thereafter changed to the present Bartholin-Patau syndrome.[28] Bartholin's description of the child can be found in the work's second volume:

"**A monster without eyes**
In the town of Væ, in Danish Scania, respectable people have had a malformed child. Everything went as it should during the delivery, but the newborn had no eyes. The open forehead was reddish in depth. The nose, which lacked nostrils, was broad and oblong, and on it grew a tumour, which was covered on the one side by bone and on the other side by skin. The mouth was broad and malformed. The upper jaw consisted apparently of one single bone. Both hands had six fingers, and in the same way there were six toes on the left foot. With its pitiful and continuous whining, the child evoked the pity of everyone, until it expired."[29]

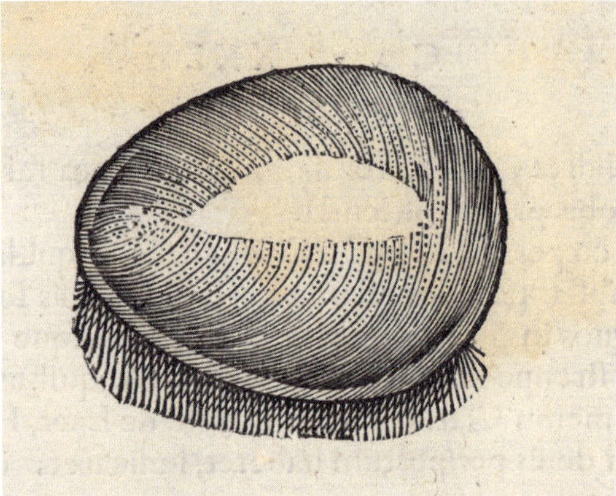

Figure 124. The magic egg, Ovum Magicum, the second of two eggs that according to several witnesses was "born" by the Norwegian woman Anna Omundsdatter in Sundby in Stavanger in the spring of 1639. Illustration from Ole Worm's folio work, *Museum Wormianum*, on the objects in his "collection of the curiosities of Nature," as he used to call it, Leiden 1655, p. 312. The Royal Library in Copenhagen.

100 THOMÆ BARTHOLINI

1646. tibiam inflat ſeu buxum, ſeptem octove digitos transverſos longum. Accedens commilito, tibiæ poſſeſſor, inſtrumentum repetit. Ille verò non commotus monitis, in incepto perſiſtit. Hinc commilito aſtu fiſtulam apprehenſurus, forte manu vehementius adacta in gulam per fauces tibiam intruſit, ut extrahenti digito amplius apprehendi non poſſet. Sangvinis ex faucibus ſtillicidium ſeqvutum. Poſtea qvum vario conatu nec extrahi poſſet, nec deglutiri, tandem tamen in ventriculum delapſa, dextrum latus videbatur obſediſſe. Unde doloribus feſſus lectum petiit, & ſine alimentis tres dies noctesq; inſomnes tranſegit, latere dextro dolente, cui tibia inhærebat. Tertio die circa qvartam horam pomeridianam ſine medicamentorum ſtimulo, per infe-

Figure 125. The flute of boxwood that according to Ole Worm was six inches (15.7 cm) long and according to Thomas Bartholin eight cross-fingers long. In August 1647 it was accidentally swallowed by a 17-year-old schoolboy in Køge south of Copenhagen, when he was playing the flute in front of his master's door and got into a scuffle with a friend, who knocked it down in his throat. The flute came out with the stool three days later with no damage to it or the boy. Bartholin had the account from Worm, who was in possession of the flute. Illustration from Thomas Bartholin's *Historiarum Anatomicarum Rariorum*, Cent. 1, Copenhagen 1654, Hist. 59, pp. 99–101. The Royal Library in Copenhagen.

The fused brothers

An account in the first volume of *Historiarum Anatomicarum Rariorum* deals with one of history's most famous and unusual cases of conjoined twins. In the 1630s and 1640s the Genovese Lazarus Colloredo (1617–1646) travelled around Europe and earned his living by exhibiting his conjoined twin brother Giovanni Baptista, who was fused to his chest. In 1637 they visited among other places Copenhagen and Charles I's court in England. In 1638 they were in Paris, in 1639 again in England, and in 1640 in Poland and Turkey. In 1642 they were in Scotland, in 1645 in Strasbourg and Basel, and in 1646 they were on a tour in Germany and Italy. Thomas Bartholin first met them in Copenhagen in 1637 and later in Basel in 1645, and both times they made an indelible impression on him. For his description of them, he asked Karel van Mander to make a drawing, from which an engraving was made by Alberth Haelwegh:

"**A brother, who was grown together with his brother's breast**
I first saw the Genovese Lazarus Colloredo in Copenhagen when he was 28 years old, and again later in Basel, but in both places with horror. This Lazarus had a little brother born with him on his breast, if I guess right, fused with the sternum.[30] Only the left leg hung down on him, two arms, only three fingers on each hand. Traces of the birth parts appeared. He moved his hands, ears, lips; the pulse was in the chest. The little brother's excrement came out of the mouth, nose and ears, and he is nourished by what his bigger brother eats. He has the vital parts for himself and sleeps, sweats and moves, when his brother is awake or is quiet or fasting. Both have been given a Christian name: the bigger is called Lazarus, the smaller Giovanni Baptista. The entrails, such as the liver, the spleen etc. they share. The eyes of Giovanni Baptista are almost closed, the breath weaker; only by holding your hand before his mouth do you feel a small warm breath. His mouth is gaping open and is almost always oozing with saliva. Only the head seems to have consumed all the food to its host; for it is much larger than the brother's but badly formed; and the hair hangs down with the head. Both have a beard, but Baptista's is not trimmed. Lazarus was of normal height, of proper physique and manners and dressed like a fashionable courtier. He wore a cloak, with which he concealed the brother's body and kept it warm, and from the first conversation one would not believe that a monster was hidden inside. He seems to be quite content, except from the fact that he is sometimes worried about the final outcome and fears for the brother's death, since he predicts that he himself will die from stench and putrefaction, for which reason he takes better care of his brother than of himself. I here present this very rare monster's exact appearance for the reader in the form of an engraving [Figure 126]."[31]

Figure 126. The conjoined brothers from Genoa, Lazarus and Johannes Baptista Colloredo (1617–after 1646), whom Thomas Bartholin saw "with horror" in Copenhagen in 1637 and in Basel in 1645. Engraving by Albert Haelwegh from a template by Karel van Mander. Illustration from Thomas Bartholin's *Historiarum Anatomicarum Rariorum*, Cent. 1, Copenhagen 1654, Hist. 66, pp. 116–17. The Royal Library in Copenhagen.

One gets a further impression of the contemporary amazement at the sight of the two brothers from the English poet Martin Parker (1600–1656), who saw them in London in 1637 and begins his ballad from that year "The Two Inseparable Brothers" in this way:

> I many Prodigies have seene,
> Creatures that have preposterous beene,
> to Nature in their birth,
> But such a thing as this my theame,
> Makes all the rest seeme but a dreame,
> the like was nere on earth.[32]

Bartholin's description of the brothers is confirmed by numerous independent contemporary reports, flyers and illustrations.[33] One of these reports is found in one of the most famous works of the time about malformed humans, animals and plants, the earlier-mentioned Italian physician Fortunio Liceti's richly illustrated *De Monstris* (On monsters), published posthumously in Amsterdam in 1665 by the Dutch anatomist Gerard Blasius (1627–1682) with an appendix written by Blasius with his "careful account of new and rare monsters" from among other sources Thomas Bartholin's *Historiarum Anatomicarum Rariorum*.[34] It includes a letter about the brothers, when they were only a week old. It is dated Genova 19 March 1617 and is from the Genovese physician Augustino Pinceto, who writes that on the 12 March 1617 in the parish of Bartholomæus de Costa in Genova two conjoined brothers were born. Their parents, Pellegrina and Baptista Colloredo, are healthy and wealthy and have several well-formed sons.[35] The brothers are fused in the abdomen, while the other body parts are separated. One brother is twice the size of the other and well-formed, while the other is missing a leg and has closed eyes. They will be baptized as two individuals. Only the big brother sucks, but when you drip milk on the little brother's lips, he licks it. The brothers have common excretory pathways for faeces and urine. The description is supplied with an illustration of the brothers later in the childhood.[36] Apart from the information about the excrements and the localization of the fusion in the abdomen, this description corresponds quite closely to Bartholin's much later description of the brothers in their adulthood. Another report from the brothers' early life can be found in the Italian physician Paolo Zacchias (1584–1659) work on medico-legal questions in three volumes, *Quæstionum Medico-Legalium Tomi Tres*, Frankfurt 1688,[37] while further descriptions of the adult brothers agreeing with Bartholin's description are given by the French lawyer and historian Henri Sauval (1623–1676), who met the brothers in Paris in 1638,[38] and the Scottish historian John Spalding (1609–1670), who met them in Aberdeen in 1642.[39]

Treatment of third day fever with Greenlandic unicorn horn

In the summer of 1652, Copenhagen was hit by third day fever or malaria, which was by then a quite common disease in Denmark. It was called second day fever or quotidian feber (*febris quotidiana*), third day fever or tertian fever (*febris tertiana*) and fourth day fever or quartan feber (*febris quartana*), depending on whether the fever, which changes with the cycle of the various malaria parasites in the red blood cells, occurred every day, every second day or every third day. Later in the summer an epidemic of dysentery and acute diarrhoea, which killed thousands of the city's inhabitants, began.[40] Bartholin treated himself and his family with unicorn horn:

"**An epidemic of third day fever**
In the summer of 1652, which, in contrast to our usual climate, was very hot and dry, a third day fever attacked many families. [...] Also, I, my wife, daughter and the whole family were attacked. During the seizures diaphoretic medicaments helped the most when they were administered with theriac after previous medical treatment. In this regard our Greenlandic unicorn horn showed its power. Just half a drachma of it taken with an adequate amount of water before the seizure not only resulted in a large amount of sweat, but also mostly abolished the fever. I have tried this new medicament with great success in many cases and with other physicians in our city, the most experienced men in the art of medicine."[41]

Against the subsequent dysentery, the unicorn horn apparently had no effect, since Bartholin's 2-year-old daughter Anna Margrethe died during this epidemic.

Treatment of fourth day fever with China bark

In the fifth century of his 1661 work, Bartholin gives the first known account of the treatment of malaria with quinine in the form of China bark imported from Peru, where it was used by the Incas as a medicament against fever. Bartholin describes his treatment of a young woman with fourth day fever, but unfortunately, he did not have enough of the rare and expensive drug to cure the patient:

"**Peruvian antipyretic powder**
I was also strongly interested in this drug, and when some of my friends had returned from Rome and had given me three doses of the bark, I tried it in a case of *Febris Quartana*. A nice young girl had for eight months been tormented by *Febris Quartana*, which had first been simplex and then duplex, and had been

Figure 127. China bark tree (*Cinchona*) from Thomas Bartholin's account of his attempt to treat a young woman for malaria with the bark of this tree. The patient got better, but Bartholin realized that it would take several doses to cure her, and he did not have enough of the bark to succeed. *Historiarum Anatomicarum Rariorum*, Cent. 5, Copenhagen 1661, Hist. 50, pp. 107–16. The Royal Library in Copenhagen.

changing. No drugs could help her. At last, she agreed to take Peruvian powder, although she did not like it. She took it macerated with wine but soon threw it up again. The double fever, however, became simple and milder, but soon after again double. I still had two doses. In order that they should not provoke nausea, I gave her them in another way: I macerated them for forty hours with white wine, strained it, and gave her the strained wine twice, without causing nausea, because the drink tasted pleasantly bitter like lemon seeds. The fever again turned simple, but it did not completely cease, and it was not possible to make more attempts because of lack of bark. I will now try to have more of it sent up here, for we have long needed such a powerful drug against *Febris Quartana*. [...] Through my colleague, Dr. Møinichen, the famous Hieronymus Bardus in Rome[42] has kindly sent an image of the three doses, which I reproduce here for those whom it may interest [Figure 127]."[43]

As we shall later see, his brother Rasmus Bartholin had better luck almost twenty years later.

Operation of a noblewoman for breast cancer

By virtue of his high status as anatomist, physician and professor and his connections to the royal court and the nobility Bartholin often served as physician for the nobility and its children, of which he gives several examples throughout his writings. In 1656 he became involved in the medical treatment, and in 1658 in surgical treatment, of breast cancer in the noblewoman Elisabeth Lunge (1610–1659) of Krenkerup Estate on Lolland (Figure 128).

He describes the treatments in the fifth century (third volume) of his 1661 work:

"**Removal of an ulcerated breast cancer**

The honourable Elisabeth Lunge was about 46 years old and had a melancholic temperament, which had worsened after the death of her famous husband Palle Rosenkrantz (1587–1642). In the year 1656 she discovered a nodule in her right breast, close to the nipple. At first it was quite small but then started getting larger day by day with all the signs of a cancer, because her menstruation had ceased. The patient consulted me and Dr. Paul Moth as well as Dr. Henrik Fuiren, and because she resisted a surgical intervention, we tried with all means to stop the growth of the tumour. We tried to divert the melancholic fluid by the aid of medical wine and pills, which should be taken one right after the other, but first and foremost we restored the absent menstruation, which immediately appeared, because we used pills made from common orchid with decoction of mushroom, castor, long snakeroot, savin tree, red myrrh and distilled oil of amber and cinnamon, whose curative effect we have often experienced. Each time she had menstruation as wished, she noticed that the tumour became smaller. Therefore, she began to hope that it would completely disappear with time. In that conviction she travelled home to her lovely estate on Lolland, where she monitored her health according to advice given.

In the following year her sadness increased, because of her daughter's death and the calamity that hit the state and the country.[44] Her health and the tumour therefore began to deteriorate critically, and although she consulted the local physician, he could not prevent the formation of a festering sore and the onset of severe pain. She therefore felt sorry for her miserable life and at the same time feared for greater disasters. Hence, in the summer of 1658 she travelled back to Copenhagen to consult us physicians about her health once again. Only a surgical intervention now left any hope, for the cancerous tumour had spread to the whole breast, and approaching death threatened every day with the ravages of

Figure 128. Thomas Bartholin's breast cancer patient, the noblewoman Elisabeth Lunge (1609–1659) of Krenkerup Manor on Lolland, flanked by her husband Palle Rosenkrantz (1587–1642) and her son Jørgen Rosenkrantz (1637–1660). Oil on canvas by Heinrich Dittmers (1625–1677). Part of the family epitaph in Radsted Church near Sakskøbing. Photo: Nils Roland.

cancer. We palpated the cancer carefully and were able to see that a surgical intervention could be carried out without any great risk, for the tumour was removeable and did not extend so deeply that it was attached to the ribs. Only in a single place near the sternum was it attached, but together with the breast the tumour could be removed completely. This expectation gave her new courage, and she therefore preferred to die at the hand of the surgeon than continue living a tortured life.

Fearless and prepared for death, she now wanted to submit herself to our surgeon and demanded an operation. While Dr. Moth, Philip Hacquart[45] – royal surgeon and a specialist in his field – and I made all the necessary preparations, the patient was reconciled to God by the priest. On the 16 June we gathered. Different contracting medicaments were prepared, and in a side room the fire was also kept ready. The patient, who lay in a four-poster, was wrapped in linen and had a quilt over her. We asked her to be confident and said a prayer to God together with the priest, before we started the work. With the left hand the surgeon gripped the whole cancerous tumour, and with the right hand he very quickly directed his broad and sharp knife around it. He began the procedure at the sternum, where part of the tumour had attached itself, and after having cut deeper at the side he rapidly made three incisions and removed the whole breast together with the cancer. To make the tumour appear more clearly, the bystanders lifted the patient's right arm in the air. In this way the tumour could be separated from the large breast muscle, and the arm's ability to move was not harmed. [...] The patient endured the operation with a calm mind, but full of sighs. The blood, which flowed in copious quantities from the vessels of the chest, we stopped after some time, so that the dirty and melancholic blood from the ill place could be completely removed. [...] The massive wound was then cured by the application of haemostatic drugs, while the blood was stopped with tight ligatures, so that it was not necessary to use the irons. To prevent new material from flowing to the operated site, the upper arm was fitted with a bandage. Hereafter, we used healing medicaments, which during six weeks gradually almost closed the wound and caused the skin to be restored. Traces remained of a gland on the sternum, where the cancerous tumour had taken root and threatened to spread. With powerful irons we removed the excess tissue, and the wound was cleaned. But this small wound could not be closed completely, because there was a fear that pain would arise as a result of the natural influx. For this reason, the patient preferred to use this place as a draining hole in the future. In this way the honourable lady was by the blessing of God happily freed from a cancerous tumour, and before my departure she arranged for a thanksgiving service be held in the church. I have since learned that, in the following severe winter during the Swedish siege of Copenhagen, she was suffocated by a catarrh, which arose when she heated her house up with coal, the use of which she was not accustomed to."[46]

As mentioned, the operation of Elisabeth Lunge was carried out 16 June 1658, and she died on 7 January 1659, almost 49 years old. Although it is not difficult to imagine that a weakened woman could die from a catarrh and its possible complications during a very cold winter, as Bartholin writes, it seems on the other hand more probable that the underlying breast cancer and its presumed spread to other parts of the body, including the lungs, was the real cause of death. The thanksgiving service arranged by the patient, however, suggests, that the operation liberated her from the worst torments for a while, and the operation may well have prolonged her life.

The case report is one of many examples of Thomas Bartholin's teamwork with the most skilful barber-educated surgeons in the country. They were often educated abroad and immigrated to Denmark from countries with a higher expertise in surgery, typically Germany or France. Philip Hacquart, to whom Bartholin usually refers with superlatives, was of French origin and came to Denmark as a military surgeon during the last years of Christian IV's reign.[47] In accordance with tradition and the Danish Medical Ordinance of 1619 (and later that of 1672), surgeons were obliged to consult the university educated physicians concerning the treatment of complicated, dangerous and severe diseases. The indication for such operations was decided by the physician, who also oversaw the operation and guided the surgeon in anatomical questions. On the other hand, the physicians did not carry the knife themselves.

As will later be shown, many examples of this kind of teamwork are given in Bartholin's medical and natural philosophical journal, *Acta Medica & Philosophica Hafniensia*. As will also be shown, Bartholin made an important contribution to the education of the surgeons in Denmark.

The execution of the wholesale murderer Jens Fuglefænger

The brutality of the methods of punishment and execution in the 17th century are well known. In the third volume of *Historiarum Anatomicarum Rariorum* Bartholin reports death of the wholesale murderer Jens Fuglefænger (Jens Bird Catcher) and the punishment that preceded it:

"**Movement without the heart**

Jens Fuglefænger, that hateful man, who in recent years had murdered many people in isolated houses by Kongevejen outside Østerport, at night, on his own with an axe, had now finally been arrested and sentenced to a dreadful and painful death. He was driven on a cart through the city and taken out to the place of execution; on the way there, he was torn apart in the most terrible way with red-hot pliers all over his body. First his legs and arms were crushed, then his tongue was torn out of his mouth, and strips of skin were torn from his back. Thereafter the executioner cut open his chest with a quick hand, ripped out the heart and threw it at his head. He bore all this calmly, with a sensible and fearless disposition. When the pulsating heart had been thrown at him and lay by his side between different instruments of torture, he quickly moved his head, regarded the bystanders with a sharp look, and at the same time observed his heart closely, before his head was chopped off."[48]

In support of his report, Bartholin refers to article 15.32 in the English philosopher, lawyer, politician and author Francis Bacon's (1561–1626) writing *Historia Vitæ et Mortis* (The History of Life and Death), where Bacon writes:

"I remember that I saw a Traytor embowled, whose heart beeing cast into the fire, leaped five foote high, and afterward lower for the space of seaven or eight minutes. Also the old tradition of a sacrified Oxe, that in embowelling lowed, deserves to be beleeved, thogh it be more certayne that a man executed and embowelled, after his heart was pluckt out, and in the hangmans hand, was heard to utter three or foure words of his prayers."[49]

Notice that, unlike Bartholin, Bacon refers to this incredible report as his own personal observation. As I shall later discuss, this represents an important difference between the two of them.

Dissection of a mermaid

In the spring of 1646, on his way home from his study trip in Europe, Bartholin made the acquaintance of Ole Worm's friend, the Dutch geographer and director of the Dutch West India Company, Johannes de Laet (1581–1649). They came to an agreement that de Laet would collect rare exotic objects for Worm's Museum. On 1 July 1646, de Laet writes from Leiden to Worm in Copenhagen about a hand and a rib from a mermaid:

"Famous man! I have been longing to resume my writing pursuits, which have been interrupted for some time, and now the illustrious Thomas Bartholin gives me the most pleasant opportunity, since he has returned to us from his journey to Italy. From him I have also received the catalogue of your museum published last year, and at the same time I have understood that you plan to publish the same for public use supplied with information about the objects in the museum. After having seen the catalogue, I have started to look around for which of the things that I have collected during a long time and at different occasions I could add to your collection, although they are lying unprocessed and scattered in different places here. [...] However, I have put in a box certain things that I have not found mentioned in your catalogue, and I will pass it on to Mr. Bartholin, so that it can reach you with greater certainty. Among these things I hope that you will not neglect the skeleton of a hand and a rib from a sea monster, which is common in the sea off the coast of Africa close to Angola. The Portuguese call it in their language *Perxe de Moliher*, which is the same as a siren. Balls made from the ribs are praised as a unique remedy against haemorrhoids, which I have heard from a learned Portuguese who had frequently experienced their effect. I hope to get the picture of the living creature soon."[50]

It ended up being Bartholin, who with the help of Royal Engraver Albert Haelwegh eight years later, in the first volume of *Historiarum Anatomicarum Rariorum*, delivered the picture of the "live" mermaid, both out of the water and swimming in it, together with illustrations of de Laet's mermaid's hand and ribs (Figure 129).[51]

In the corresponding report Bartholin discusses the topic carefully with references to relevant literature and reports from learned people. He introduces it by stating that some of the older reports about mermaids are false and others true, but he does not doubt that a human-like creature exists in the sea. He is also sure that new species of sea animals have arisen through the movement of terrestrial animals into the sea: "Marine foxes, wolves, calves, dogs, horses etc. exist. Why then, do we reject the picture of the marine human sea monster? It is certain that on land you find apes, which, apart from their way of thinking, appear as humans in external appearance and attitude. All such sea monsters are related with the seal."[52] One could argue that with these thoughts Bartholin anticipates the evolutionary thoughts of Charles Darwin (1809–1882) in *On the Origin of Species* by two hundred years.[53]

Then Bartholin reports about a mermaid, which at the end of the 16th century was caught close to the coast of Brazil by merchants from the Dutch West India Company and dissected in Leiden by the Dutch anatomist Pieter Pauw under the supervision of Johannes de Laet. The head, the chest and the area down to the navel looked like the human species, while the area from the navel to the feet consisted of a deformed body with no sign of a tail. In order not to deceive his reader, Bartholin keeps the mermaid's hand and rib in his museum thanks to the courtesy of the laudable de Laet. Moreover, he attaches an illustration of the mermaid in both standing and swimming position, so that all doubters can be satisfied. Then follows a detailed anatomical description of the dissected mermaid's hand and rib in accordance with the illustrations. Bartholin ends his report with the following conclusion:

> "There is such a great difference between the old and new authors concerning the appearance of the mermaids, that it comes as no surprise that some consider them fabricated. As far as we are concerned, we have hands as proof, and we can refer to mermaids, which have in fact been seen. Neither hands nor ribs can deceive, and we have included completely genuine illustrations of them."[54]

The hand and rib of de Laet's mermaid are included in Bartholin's 1662 catalogue of his collection on the first floor of his professor's residence in Copenhagen.[55]

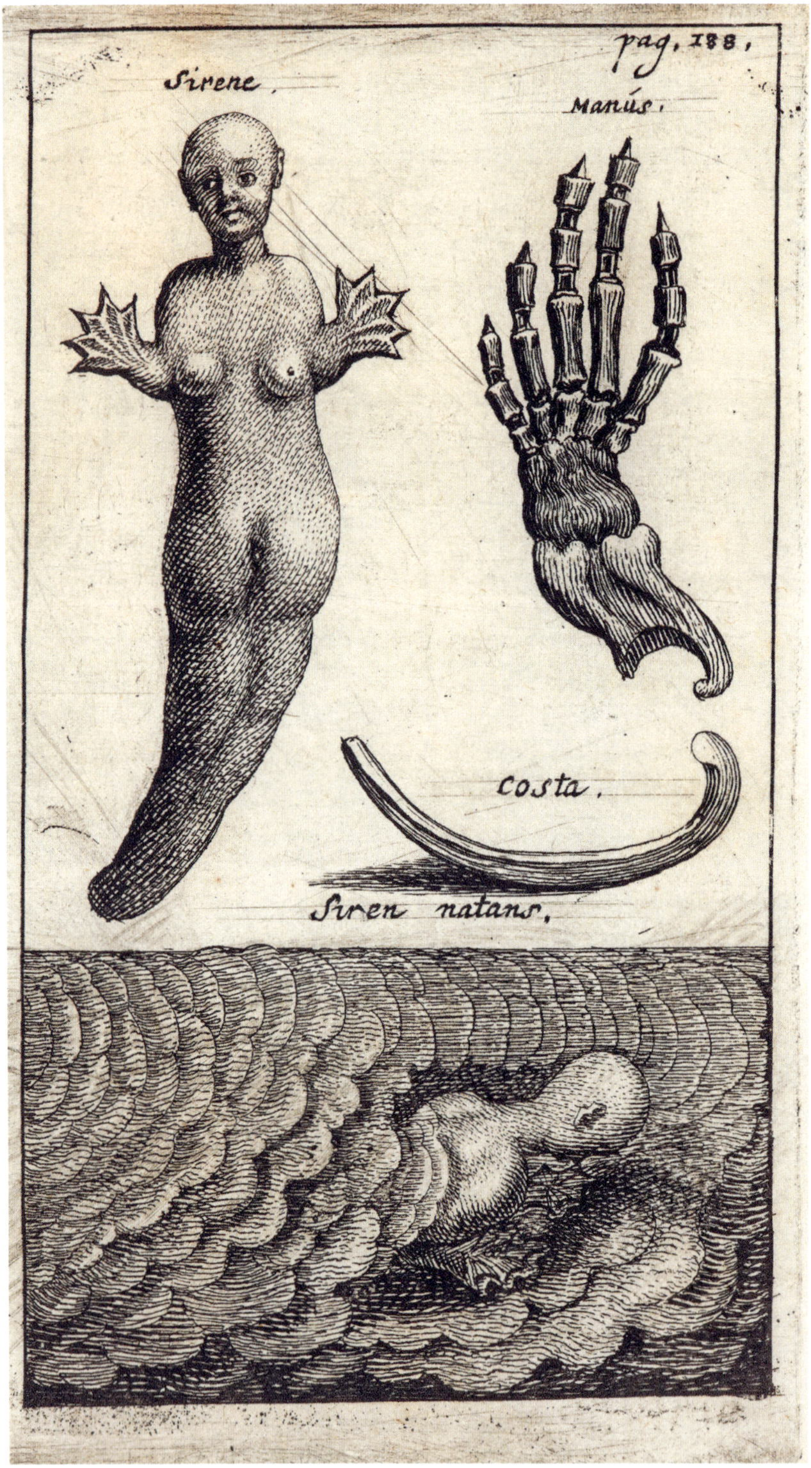

Figure 129. Albert Haelwegh's engraving of the mermaid who in the late 1500s was captured by merchants from the Dutch West India Company in the sea near the coast of Brazil and later dissected in Leiden by the Dutch anatomist Pieter Pauw (1564–1617) in the presence of his compatriot, the geographer Johannes de Laet (1581–1649), who sent the mermaid's depicted hand and ribs to Ole Worm in Copenhagen. Thomas Bartholin was given the opportunity to study both in detail. He published the result of these studies and his account of the mermaid in the first volume of *Historiarum Anatomicarum Rariorum*, Copenhagen 1654, Cent. 2, Hist. 11, pp. 186–91, where the engraving can be found on page 188. Bartholin later included the hand and rib in his collection of curious natural objects. The Royal Library in Copenhagen.

Birth through the mouth

In the third revised edition of 1651 of his father's anatomy book, in the chapter about the stomach Bartholin added a reference – indeed as an example that lies outside Nature's intentions – to a narrative handed down from the personal physician to the princes of Anhalt, Philip Salmuth (1571–1626), about a fetus which is born through the mouth through vomiting after fertilization and growth in the stomach.[56] Salmuth was notorious for his credulity.[57] The narrative with the title "Infans per os exulsus" (child secreted through the mouth) is included in his posthumous work *Observationum Medicarum Centuriæ Tres* (Medical observations in three centuries), Braunschweig 1648.[58]

According to a "reliable eyewitness", a young woman had given birth through the mouth to a fetus about the length of a finger. The neighbouring wives had buried it, but when the authorities learned about the matter, the fetus was excavated and examined. It was concluded that it was a well-formed fetus. The simple-minded woman claimed that her husband had had oral sex with her, and that she did not know of any other way to have sex until she was instructed by the neighbouring wives. Bartholin's reproduction of Salmuth's copious narrative was noticed by Jean Riolan in Paris, who added it to his numerous other points of criticism of the aforementioned edition of Bartholin's textbook of anatomy in his *Animadversiones Secundæ ad Anatomiam Reformatam Thomæ Bartholini* (Further remarks to Thomas Bartholin's improved textbook of anatomy), Paris 1653. Here Riolan calls Salmuth's narrative absurd and Bartholin's reproduction of it ridiculous and incomprehensible, and he adds: "But Bartholin does not care whether what he takes from others is true or fictitious, as long as he can fill up his anatomy book with this kind of nonsense and unbelievable tales."[59]

Bartholin, however, did not take the criticism into account in the least. On the contrary, in the third volume of *Historiarum Anatomicarum Rariorum* in 1661 he resumes the narrative from Salmuth and declares:

> "I dare not doubt the truth of this disgusting story, since Salmuth certifies that he wrote it after a letter to Johannes Komeler from his relative and compatriot Gottfried Hoffmann, who once lived in the house of the consul in the town, where it happened, so he was a reliable eyewitness. Every day many things happen that we must condemn but still believe. Why then, could this story not be reliable?"[60]

Bartholin argues that, although the uterus is the ideal organ for fertilization, the growth of the fetus and birth, "there is nothing to prevent a fetus from forming in any cavity in the body," including the stomach, as long as the cavity can receive the semen and offer the fetus the necessary heat, shelter and vascular supply:

"Who would then be able to deny that a lost fetus in this strange place could be clothed with the membrane, inside which it could eventually thrive in its new home. For it is certain that it is not from the uterus, but from the semen, that the fetus is created."

Bartholin supplements Salmuth's narrative with a similar one from the animal world, which was told to him by his brother Albert Bartholin:

"My brother has a similar report about a cat. It was pregnant and threw up its kittens in a property in Frederiksborg to the astonishment of those present and gave birth in this way. That they were swallowed in advance, or that they came out of the abdomen, was not observed by any among the curious audience. They are willing to testify under oath that the event really took place as described. Whoever wants to can believe it. If in the future others should be less inclined to believe it, it will not matter. In that respect everyone has their free will, and we exercise no tyranny over the observations. On the other hand, people who report what they have seen or heard should not be silenced any more than those who, based on reports given by reliable people, investigate the reason for rare things independently and carefully."[61]

In his book about unusual means of birth in humans, *De Insolitis Partus Humani Viis Dissertatio Nova* (New dissertation about unusual birth paths in humans), Copenhagen 1664, Bartholin includes a whole chapter on birth through the mouth,[62] where he repeats the narratives of Philip Salmuth and Albert Bartholin, and in 1673 in the fourth revised edition of his textbook of anatomy, Salmuth's narrative is still present in unchanged form.[63]

Witchcraft, magic and demonic possession

In the time of Thomas Bartholin, witch burnings still took place in Denmark. His opinion about his contemporaries' belief in witchcraft, magic and demonic possession is not unambiguous, but the tendency is that as a younger man he generally gives the impression of believing in that sort of thing, while in his older days he takes more distance from it.

"**Pellet excreted through vomiting**
A noblewoman in Scania had in November 1653 suffered from pain of unknown cause in the whole body for a long time. The pain was accompanied by frequent nausea and the feeling of vomiting. Finally, the nausea increased so much, that she threw up a wax-ball with the shape and size of a nutmeg. As soon as the pain had ceased, a rumble was heard in the house, and four cats were seen fighting in the bedroom. All this clearly showed, that witches had a hand in the game. The pellet was sent to us in Copenhagen, and I have seen it myself and held it in my hands. It is smooth on the outside, and when it had been cut in half, we found linen, down and a pin mixed in the wax. Either this pellet had been stuffed into the mouth of the lady with the devil's cunning, while she was asleep, or kneaded into the stomach by sorcery. The cunning of the evil spirit and the deeds of the witches surely appear here. Her virtuous spouse, who wants to avenge the harm inflicted on his wife, has summoned the instigators of the witchcraft, who have teased the god-fearing and honourable matron with their enchantments."[64]

In 1655 Bartholin writes: "That different diseases can be inflicted upon human bodies by the devil is so well known by us that it does not need my confirmation."[65] In contrast, he writes in 1673: "Usually, I attribute little or nothing to sorcery, since I attribute to Nature or illness the rare or unusual cases whose causes we do not understand. Especially, hypochondriacs are exposed to several things that the common man attributes to a supernatural cause. The devil and the witches are blamed for bringing about the miracle, and the credulity of the human being attributes so much power to him and them that one should believe that they control our destiny, which is after all guided by God and Nature."[66]

Historiarum Anatomicarum Rariorum was received with great interest abroad, and the second volume was translated into German by Bartholin's former pupil Georg Seger (1629–1678)[67] as well as into Dutch.[68] 107 of the 600 case reports were completely or partly translated into Danish by the Danish literate Odin Wolff (1760–1830) and published in his *Journal for Politik, Natur- og Menneskekundskab* (Journal of politics and the knowledge of Nature and the human being) in 1828–1829. In 1827 Wolff writes among other things:

"It is curious that none of our scientists before me have noticed the treasury of knowledge and reports about Nature, the human being and the epoch in Denmark contained in the mentioned work of our compatriot Thomas Bartholin, one of Europe's sharpest and most famous learned men of his time. Perhaps this forgetfulness can be partially explained from the fact that the work was written in Latin."

Wolff's subsequent suggestion for publishing a beautiful work with Danish translations of "the publicly useful and popular" content of Bartholin's Latin works, and his invitation to interested subscribers to approach him in order to prepare such a work, was unfortunately never realized.

Contemporary and later criticism for verbosity

As presented above, scattered throughout Bartholin's voluminous authorship are curious case reports for which Bartholin, as might be expected, was criticized and accused of verbosity and lack of scientific judgement by contemporary scholars and later medical historians.[69] In a 2018 analysis of 15 such case reports I concluded that much of the criticism, especially during the era of positivism at the end of the 19th century and the beginning of the 20th century, is unjustified and rests upon misunderstandings and ignorance.[70] As such, some of the criticized reports have content that seems surprising but can be confirmed by present-day knowledge. Examples of such reports are the ones concerning horned people (cutaneous horns or *cornu cutaneum*), conjoined twins and petrified fetuses (*lithopædion*). Other reports have content that, with present-day knowledge, we must reject, but which are understandable from our knowledge of that time. Examples of this are egg-laying women and the existence of unicorns and mermaids. This leaves a third category of reports, the veracity of which we must reject with present-day knowledge, and which even with insight into the knowledge of Bartholin's time is difficult to understand. It was mainly reports in this category for which Bartholin was criticized by his contemporaries. Examples are the French surgeon Ambroise Paré's (1510–1590) report about the embalmed body from which hair and nails continued to grow after 24 years, which was criticized by Caspar Hofman; Salmuth's report about the woman giving birth through the mouth, which was criticized by Jean Riolan; the court's report about King Frederik III's egg-laying cock, which became the subject of Ludwig Kepler's ridicule; and the report about the Hamburg physician Schlegel, who died from fright when the body he was dissecting moved, which was criticized by Schlegel's colleague Johannes Placcius and by Bartholin's pupils Michael Lyser and Henrik Møinichen. Of course, the reports about the execution of Jens Fuglefænger, the schoolboy swallowing a six-inch-long flute and the cat giving birth through the mouth also belong to this category.

These reports have in common that they were all handed down from others. In contrast, I have not succeeded in finding one single example indicating that Bartholin, when practising science himself, did not reproduce his observations correctly and truthfully. In that context, he was not satisfied by making conclusions based on the knowledge or observations of others. As mentioned, he puts this very clearly in *De Lacteis Thoracicis*: "We do not trust more than we can feel with our hands and see with our eyes," and "Those who settle for something they have not seen, or believe what they have only heard, will be disappointed."

That leaves us with the question of why Bartholin, a highly respected and famous anatomist, chose to take the risk of losing his good reputation as a skilful, well-informed and serious scientist by including in his books incredible reports from others, in which he could not possibly have believed himself, and some of which he was even told were fictitious. Indeed, one of Bartholin's most faithful friends, the Parisian anatomist Guy Patin, asked him this very question. Thus, Patin writes in a letter of 19 May 1662 to Bartholin that he has read all six centuries of *Historiarum Anatomicarum Rariorum*, which he praises for many fine topics, but that he would have preferred that Bartholin had not included two of the reports in the sixth century.[71]

In the first of these, Bartholin refers to a verdict passed by the court in Grenoble in 1637 concerning a case of adultery and mysterious pregnancy.[72] The husband of a certain Madame d'Auvermont from Grenoble had been posted permanently as a captain in Germany for four years, without visits to his home, when his wife gave birth to a son. His wife claimed that she had had no sexual relations to other men, but after she had dreamed that she had had intercourse with her husband, she became pregnant. Several witnesses swore an oath to her honour, after which the court acquitted her. Bartholin doubts the truthfulness of the account, but includes it nevertheless, and Patin informs him in the letter that the judgement is a forgery fabricated by the child's true father to ridicule the judges in Grenoble and to show his contempt for the physicians at the University of Montpellier.

Patin's other criticism is interesting because it is an example of unjustified criticism from not just someone who was a friend, but one of the period's most respected French anatomists. The report deals with a case of a stone child situated outside the uterus. A sixty-year-old childless woman from Pont-à-Mousson in Northeast France committed suicide on 6 August 1659 by throwing herself out of a window.[73] Through her 30 years as a widow, she had often complained about the feeling of an unpleasant heavy mass in her abdomen. Therefore, it was decided to perform an autopsy on her, upon which an estimated six-month-old incapsulated stone child was found in her abdomen. It was attached to the peritoneum and the intestines, while the uterus was empty and looked normal. Patin concludes that pregnancy cannot take place outside the uterus and calls the report a fable based on deception and mendacity. In contrast, Bartholin's description of the report testifies to an impressive insight into ectopic pregnancy – a topic about which he published a whole book, *De Insolitis Partus Humani Viis Dissertatio Nova*, in Copenhagen a couple of years later:

"If it really is as described, I can only imagine that this fetus must have been conceived in the fallopian tube. The semen can easily get in there, and undoubtedly it hides there, until it unites with the female semen. But as soon as the access back to the uterus is blocked, the conception takes place in the tubes, which are expandable. When the fetus grows in the narrow tube, it suffocates and drags the mother with it to death, if the wall of the tube is strong. But if the wall is thinner, the fetus breaks through it by force and crashes into the abdominal cavity. The latter scenario, I think, must have taken place in the case of the child from Pont-à-Mousson. I would have thought that the fetus had made its way through the uterus if the autopsy had not convinced me that the uterus was completely without traces of lesions and scars, which the coroner would possibly have found in the tube if he had examined this small part. Surely the fetus breaks more easily through a cracked tube than through the uterus."[74]

Bartholin answers Patin in a letter of 29 August the same year:

"I have myself always doubted about the truth of the judgement from Grenoble. Anyhow, I added it to the truthful reports to reward the readers with a pleasant variation. Just like poets, historians also wish to be of use and to please."[75]

Concerning the stone child from Pont-à-Mousson, Bartholin states that there have been different opinions. He has had concerns about the authenticity of the report but nevertheless found it probable that it might be true. Whoever its originator is, he deserves trust. After giving further examples of Nature playing games with human reproduction, Bartholin adds:

"In my defence I will say that I have added these and other reports to my medical rarities in the same form as I have received them, for I do not want to break my promise in this matter for the sake of truthfulness. The cases that I have seen myself, I can report as true – as far as the other cases are concerned, the reader must rely on the authority of others and not mine."[76]

Riolan was thus largely right in his criticism when he wrote that "Bartholin does not care whether what he takes from others is right or wrong." When Bartholin published case reports originating from other people, he did not feel responsible for their truthfulness, and frequently he did not believe in them himself.

In Enlightenment France a hundred years later, the physicians and medical historians had an eye first and foremost for Bartholin's qualities and therefore looked with forgiving eyes on his penchant for repeating stories with questionable content. In *Collection Académique*, which published French translations of a great number of the volumes of Bartholin's journal of medicine and natural philosophy, *Acta Medica et Philosophica Hafniensia*, it was put this way in 1766:

"He was a highly industrious scholar, who combined the qualities of a beautiful soul with spiritual endowment: even his faults were due to virtues. Incapable of deceiving, he believed that others were just as trustworthy as him, and often he was deceived by charlatanism and superstition. Some of his own observations bear the stamp of this excessive confidence. Moreover, it has happened to him more than once that, after making great efforts to explain certain facts, he has subsequently been forced to withdraw them and admit that he had been fooled. [...] If he lacked anything, it was perhaps the philosophical spirit, which is the torch of all arts, and without which one is not sufficiently wary of prejudice, superstition and the love of the marvellous"[77]

Bartholin's reasons for including the case reports in question were to present new and interesting material to his readers, to entertain them, to arouse their curiosity and to challenge their own judgement. These purposes were obviously of such a great importance to him that he continued to include the same incredible case reports over and over again despite criticism from his colleagues and advice from his friends and students to remove them from his works.

Part of the explanation for this conduct could be Bartholin's own curiosity and the curiosity he expected from his readers to find explanations for rare phenomena in nature. Many things in the 17th century were not as clear as they are today, and Bartholin was by nature open and curious, but by no means cocksure, and there are many examples in his authorship of his leaving the question open in case of doubt. Some medical historians have attacked Bartholin as a scientist for being only interested in curiosities, rarities and exceptions, but not in common things that are more easily explainable.[78] That is by no means a fair accusation. A characteristic feature of natural science in Bartholin's era was precisely its interest in exploring Nature's diversity, or in other words, the richness and diversity of God's creation. Within medical science this manifested itself, among other things, in a great interest in and fascination with deformities, at that time often called monsters. This interest is illustrated by the large number of comprehensive works on this topic, including Ambroise Paré's *Des Monstres & Prodiges*, Paris 1573, Ulisse Aldrovandi's posthumous *Monstrorum Historia*, Bologna 1642 and Fortunio Liceti's and Gerard Blasius' *De Monstris*, Amsterdam 1665. The interest in and the publications about deviations and rare cases – casuistic reports or case reports – have always been and continue to be an important tool for medical science to acquire new knowledge and a prerequisite for setting the boundary between the normal and the abnormal.

Publication in Danish under a pseudonym

In 1658 Bartholin published in Copenhagen one of his few books in his mother tongue. It was a translation from Italian of the book *Discorsi della vita sobria*, published by the Venetian nobleman and patron of the arts Luigi Cornaro (1464–1566) in Padua a hundred years earlier. It was about how, through moderate and sober living, you live to be 100 years old. Cornaro had, at the age of 40, caused his health to break down through immoderate consumption of food and drink, promiscuous sexual behaviour, etc., but on the advice of his doctors he changed to a healthy lifestyle, including a low-calorie diet, a restricted consumption of wine, balance between work and rest, and country life with fresh air. Cornaro's book became overwhelmingly popular both in his own time and later and has been published in numerous editions in numerous languages all the way up to present time. Bartholin published his 64-page translation on the 100th anniversary of the original under the pseudonym Tienst Berede and gave it the title *Et Edrue Lefnets Gafn oc Nytte* (The benefits and usefulness of a sober life). As I will later elaborate, healthy living was a subject that interested Bartholin greatly.

Denmark's first pharmacopoeia

In 1619, Christian IV introduced Denmark's first medical ordinance with an associated medicine tariff. According to the text of the ordinance, the reason for it was the pharmacies' poor supply of medicaments, exceeding of the shelf life of the medicaments, too-high drug prices and a desire to stop the disputes between university-educated authorized physicians, barber surgeons, apothecaries and unauthorized practitioners by defining their individual tasks.[79] The ordinance stipulated that physicians from the medical faculty and the city of Copenhagen should twice a year carry out inspections of the pharmacies, which meant that the latter now became subject to the administration of the former. It also meant that the apothecaries' preparation and storage of medicaments had to be supervised by physicians, and that all prescriptions had to be written or approved by physicians. Only authorized physicians were allowed to prescribe medication for internal use, and in cases of treatment of serious external injuries a physician should also be consulted. In return, the physicians were obliged to carry out sick visits, when required, except during epidemics of plague or other dangerous infectious diseases. Thus, the 1619 ordinance was the first serious attempt at quality assurance in the Danish healthcare system.

In 1645, a new tariff was introduced, which raised the prices slightly, and which now applied to all pharmacies throughout the country. Since this was not complied with, and there were many other complaints about disorder in the pharmacy system, Frederik III in 1651 wrote to Ole Worm and Simon Paulli to draw up a new tariff and new rules that could bring order.[80] However, nothing happened, and in 1654, the year of Ole Worm's death, the task was given to the Royal Physician Peter Bülche (1605–1671), who drew up a new tariff, but due to complaints from the apothecaries it was not enacted.[81]

Instead, the work was placed in the hands of the newly appointed dean of the medical faculty, Thomas Bartholin, who soon thereafter prepared Denmark's first pharmacopoeia, *Dispensatorium Hafniense* (Copenhagen Pharmacopoeia), and published it at bookseller Joachim Moltke's Publishing House in Copenhagen 1658.[82] As the title indicates, it is a collection of recipes with the purpose of guiding the physicians and apothecaries in Copenhagen in their preparation of medicaments. Several cities abroad had long had pharmacopoeias, for instance Augsburg, where in 1564 they got *Pharmacopoeia Augustana*, from which Bartholin drew inspiration. The main purpose of *Dispensatorium Hafniense* was to ensure better control, quality and harmonization in the pharmaceutical field. Bartholin's way of achieving this was to reduce the number of medicaments from around 2000 in the 1645 tariff to around 800. In order to lower the prices and facilitate access, he prioritized Danish medicaments over foreign ones, and he removed obsolete ones. The book has 283 pages in addition to the frontispiece (Figure 130), title page, a dedication to Frederik III, preface and an alphabetical index of the medicaments. According to the preface, a further 10 physicians contributed to the work, including Peter Bülche, Simon Paulli, Henrik Fuiren, Poul Moth and the medical professors Christian Ostenfeld and Rasmus Bartholin. The ingredients in the medicaments are mainly plants, including herbs, leaves, roots, seeds, berries and fruits. Chemical and animal ingredients were used to a lesser extent. Among the more famous medicaments are Tycho Brahe's earlier mentioned plague elixir, *Elixir Pestilentiale Tychonis Brahæi*.[83] Of course, the book also has a comprehensive recipe for the famous all-purpose remedy theriac,[84] which was credited with being effective against snakebites, poisoning, plague and other serious diseases. Among the ingredients were viper meat, castor oil,

HAFNIA:
Cochlearia Danica
Chamæmorus Norweg.
DIS-
PENSA-
TORIUM
HAFNI-
ENSE
1658
Gentiana Norweg.
Angelica Islandica
Serenißː Reg: Daniæ, Sculptor Alb: Haelwegh Sculpsit.
FLORA

rhubarb, aloe, myrrh, saffron and opium mixed with honey and wine. It could be obtained in Danish pharmacies right up until the 20th century.

But Bartholin also had another intention with the book. He used his dedication to Frederik III to appeal to the king to secure better conditions and esteem for the physicians. Among other things he wrote: "We, who day and night carry out our work for the common good, are mocked by the most illiterate people," and "Few appreciate wisely ordered conditions, and hardly a single quack can be cleansed of his madness by a thousand pounds of hellebore."

Bartholin also had a hard time with the intrusion of the apothecaries into the physicians' territory. He wanted the introduction of sharper professional boundaries. On 17 July 1647, for example, he wrote to Guy Patin in Paris about the apothecaries:

> "I do not know why their skin is so itchy almost everywhere that they cannot stay on their own turf. Often, in empty boasting, they practise the art of medicine in secret, and thus they use their seal in the physicians' harvest, and while they forget the submissive servitude to which they are bound, they spin malice against the reputation and happiness of their masters, the physicians."[85]

Later, Bartholin took a milder and more cooperative view of the apothecaries. As will be mentioned later, in *Acta Medica* there are many examples of Bartholin's experiment-based research collaboration with the apothecaries Johann Gottfried Becker from Copenhagen's *Elefantapoteket* (Elephant Pharmacy) and Christopher Heerfordt (1609–1679) from *Svaneapoteket* (Swan Pharmacy).

For the time being, *Dispensatorium Hafniense* only applied to Copenhagen's four pharmacies, but as will be elaborated on later, its content was imposed on every pharmacy in the country 14 years later.

Figure 130. The frontispiece of Denmark's first pharmacopoeia, Thomas Bartholin's *Dispensatorium Hafniense*, Copenhagen 1658. Bartholin reused the engraving of Simon Paulli's *Flora Danica*, Copenhagen 1648. Engraving by Alberth Haelwegh. With Copenhagen as background, the Roman goddess of flowers and spring, Flora, sits with a wreath of flowers in her hair and wears a flower-decorated antique costume. In front of her is a vase of flowers, with more loose flowers on the floor. In the four corners, four medicinal plants from the then Danish kingdom are shown on oval plates: top-left *Cochlearia Danica* (Danish scurvygrass), bottom-left *Gentiana Norwag* (Norwegian Gentiana), top-right *Chamæmorus Norwag* (Norwegian cloudberries) and bottom-right *Angelica Islandica* (Icelandic Angelica). The Royal Library in Copenhagen.

The war against Sweden and the introduction of absolute monarchy

From the beginning of the Swedish siege of Copenhagen in August 1658 until the early summer of 1659, Thomas Bartholin and his family again took up residence in safety in Roskilde. He describes his impressions from other people's reports about the fallen Swedish soldiers in the snow in front of Copenhagen's ramparts in the night of the Swedish assault on the city on 10–11 February 1659 in his work on the medical use of snow (Figure 131):

> "Some were frozen with angry expressions on their faces, others were lying with open eyes; still others bared their teeth as if in a rage and had their mouth wide open. Some had their arms outstretched and threatened with their swords, others lay in the most various positions. When the ice thawed in the spring, a Swedish rider appeared, sitting on his horse. I do not remember what he held in his hands."[86]

That autumn, Bartholin had two soldiers from the Dutch auxiliary troops quartered in his home.[87]

1660 saw the introduction of an absolute monarchy in Denmark and the subsequent celebration in front of Copenhagen Castle of Frederik III's ascension to the throne as the first absolute king of Denmark on 18 October (Figure 132). In connection with this, the consistory of the university decided on a meeting of 13 October that the university would take part in the celebration in the form of a party, where Thomas Bartholin, accompanied by music, was to make a tribute to the king, who would be present. On 23 October, Bartholin then delivered his tribute before King Frederik, Crown Prince Christian, the Dutch ambassador, the councillors of the realm and deputies from all estates.[88] In the speech, Bartholin said among other things: "We have a king, who is greater than Caesar, happier than Augustus, better than Trajan. We build our happiness on such a king's grandeur, bliss and mildness."

Figure 131. The morning after the Swedish assault on Copenhagen in the night between 10 and 11 February 1659 painted by Christian Mølsted (1862–1930) in 1919. Oil on canvas. 147 × 236 cm.

In the foreground is the Danish King Frederik III on horseback in front of the small frigate Højenhald that played an important role in the defence of the city. With a force of 180 men, the Swedes tried in vain to capture or destroy the ship. The frigate's cannons killed and wounded a great number of Swedish forces along the coast south of the city. The Swedes' attack came after they had besieged the city for six months. The Copenhageners had been warned by spies and were therefore well prepared. The Swedes suffered heavy losses and had to retreat after a few hours of fighting. The Copenhageners' victory was crucial for Denmark's continued existence as an independent nation.
Photo: Bruun-Rasmussens Kunstauktioner. Wikimedia Commons.

Figure 132. The inheritance tribute in the square in front of Copenhagen Castle on 18 October 1660.

Painting from 1879–1880 by Heinrich Hansen (1821–1890). Oil on canvas. The large-scale ceremony, where Frederik III was hailed as Denmark's first absolute king, marked the transition from elective kingship to hereditary kingship and absolute monarchy in the Kingdom of Denmark-Norway. This state upheaval ended centuries of power struggle between nobility and royalty in the latter's favour. The nobility was deprived of significant privileges, while the bourgeoisie was rewarded for their efforts during Copenhagen's defence against the Swedes. Representing the citizens' intellectual elite and the high-ranking state officials, Heinrich Hansen depicted Professor of Medicine and Dean of the Medical Faculty at Copenhagen University, the famous Thomas Bartholin, in the painting. He can be seen immediately to the left of the canopy in the centre of the image wearing black civic clothing with white shirt and shirt frill. The Museum of National History at Frederiksborg Castle. Photo: Ole Haupt.

Niels Stensen – the ingenious pupil 6

"I, for my part, prefer to confess my ignorance rather than to state with authority opinions, the falsehood of which will some time afterwards be proven by others."[1] Niels Stensen, 1665

In November 1656, a very special 18-year-old man began his studies at the University of Copenhagen.[2] He was the son of Royal Goldsmith Sten Petersen (d. 1645) and grew up in the family's gold-smithy on the corner of Købmagergade and Klareboderne in central Copenhagen. It was a few hundred metres from the university, on what is now Frue Plads, and two hundred metres from the astronomical observatory Rundetaarn with the extended Trinity Church, including the university library on the first floor – a building complex whose construction had begun shortly before the young man's birth (Figure 133). His name was Niels Stensen (1638–1686) (Figure 134). As a boy, Stensen learned craftsmanship and instrument handling first from his father, whom he lost at the age of seven, and then from his stepfather, who was also a goldsmith. At the age of 10, Stensen began eight years of schooling in Copenhagen's oldest school, Vor Frue Skole (the School of Our Lady), right next to the university. His two most important teachers were the school's rector, the mathematician and philologist Jørgen Eilersen (1616–1686), who taught Stensen Euclid's geometry, and the physician, philologist, chemist and botanist Ole Borch (1626–1690) (Figure 135), who mainly taught him botany and chemistry (Figure 136). Eilersen was an acquaintance of Thomas Bartholin, and Ole Borch had been Thomas Bartholin's pupil in anatomy. They both later became professors at the university.[3]

Most of Stensen's years at the School of Our Lady coincided with Bartholin's research into the lymphatic vascular system in the Anatomical Theatre just a stone's throw away, and with Borch and Eilersen as his teachers, there can be no doubt that, during his school years, Stensen followed the famous anatomy professor's epoch-making research and discoveries with interest. In addition, during the same years, Bartholin had worked as a tireless and high-profile lecturer, written communicator and administrator and as such, together with his reputation as a successful researcher of international stature, he had raised the Faculty of Medicine and the entire University of Copenhagen to unprecedented international heights. From being a place that Danish students sought to move

Figure 133. a. The Round Tower in Copenhagen drawn by Holger Jacobsen (1650–1701). Holger Jacobsen's travel diary 1671–1679. The tower was built in the period 1637–1642 as the first part of a building complex consisting of an astronomical observatory (The Round Tower), a church (Trinity Church) and a university library. The Royal Library in Copenhagen, NKS 375 kvart.

b. Peder Resen's map of Copenhagen 1642 – the year when the four-year-old Niels Stensen was able to observe the completed Round Tower a few hundred meters from his birthplace. Resen has marked the university's auditorium building as number 12 opposite the Church of Our Lady, just to the left of the centre of the image. From the book *Kobberstikkene fra Peder Hansen Resen Atlas Danicus 1677*, Copenhagen 1974. Private ownership.

Figure 134. Thomas Bartholin's most important and famous pupil Niels Stensen (1638–1686) portrayed during one of his stays in Florence. Since he is depicted in secular clothes, the portrait is presumedly painted before his ordination as a Catholic priest in 1675, i.e. from the period 1666–1672. It is the only surviving portrait of Stensen that was with certainty painted with himself as a live model. It is unsigned but attributed to the Medici family's Flemish court painter, Justus Sustermans (1597–1681). Oil on canvas.

Sustermans has depicted Stensen with a gentle, calm, intelligent and thoughtful expression, that awakes confidence and meets the viewer with friendliness. This fits well with the impression you get of Stensen from his letters and works, as well as other people's descriptions of him.
Galleria degli Uffizi, Florence. Su concessione del Ministero della Cultura.

Figure 135. Ole Borch (1626–1690), another of Thomas Bartholin's important pupils, painted by Johan Jepsen (1655–1700) in 1691, probably from a sketch from shortly before Borch's death. Oil on canvas. 76 × 60 cm.

The highly versatile and gifted Borch became Niels Stensen's first important teacher and source of inspiration in Copenhagen and later his research partner in the Netherlands and Paris. His letters from Leiden home to Thomas Bartholin in Copenhagen are an important source for Niels Stensen's anatomical research in that city in the early 1660s, and his travel diary is an important source for his and Stensen's activities in Paris in 1664–1665 and for the environment and activities in the private scientific academies in Paris.
The University of Copenhagen.

Figure 136. a. Ole Borch on a botanical excursion with his students at the School of Our Lady or the University. Hand-coloured drawing by Holger Jacobsen. From Holger Jacobsen's travel diary 1671–1679. The Royal Library in Copenhagen.

b. Ole Borch's chemistry laboratory. Hand-coloured drawing by Holger Jacobsen. From Holger Jacobsen's notebook. The small house, in which Borch later set up a similar laboratory, survived the Copenhagen fires and therefore still exists. It is situated in the garden of Borchs Kollegium in Store Kannikestræde between The Round Tower and Frue Plads. The Royal Library in Copenhagen, NKS 309 aa quarter.

away from, he had made it a desired educational institution, which foreign students sought to attend. It is therefore not surprising that Stensen, like so many other students at the School of Our Lady, chose Thomas Bartholin as his preceptor at the university. That Bartholin was a desired teacher and mentor is evident from the following statement from Stensen's schoolmate Johan Monrad:

"[...] and I then went out to study at the Academy under the professors, and as the choice was free to all, I chose as my schoolmaster at the academy Dr. Thomas Bartholinus, for he was famous among all the others because of his piety and virtue."[4]

Although the students were formally free to choose their preceptor, there must have been limits to how many students Bartholin could take under his wing. He was undoubtedly involved in selecting his students, and the highly gifted, talented and ambitious Stensen had just as certainly received the best recommendations from Eilersen and Borch. According to the university's register, 23 pupils from the School of Our Lady, including Stensen and Monrad, were enrolled at the university on 27 November 1656. Seven among them gained Thomas Bartholin as preceptor, while this was only the case for one pupil among the 83 enrolled from various other schools. In other words, there are indications that Eilersen and Borch encouraged many of the school's pupils to request Thomas Bartholin as preceptor, and that Bartholin largely met the wishes of the pupils of that school, knowing that those from Eilersen and Borch had received a solid basic scientific education.

As the rules at the university stood at the time, new students had to come forward before the dean (in this case Thomas Bartholin) and present a certificate from the Latin School, which contained both a *testimonium eruditionis* (testimony of knowledge, learning and education) and a *testimonium vitae* (testimony of living, faith and morality). As will be shown in the following, Niels Stensen, interested in natural science and medicine, was fully aware of his future teacher's style, and Bartholin, on the other hand, was soon to realize that he had brought a brilliant student into his stable. A unique symbiosis in Danish scientific history was about to take shape.

Most of the first two of Stensen's three years at the University of Copenhagen, up to the siege of the city by the Swedes in August 1658, proceeded calmly with regular teaching. In addition, Stensen participated in private chemical and physical exercises and experiments with Ole Borch, who, as a tutor to the sons of Reichshof-Master Joachim Gersdorff, had access to Gersdorff's private chemical laboratory on Kongens Nytorv. The last study period during the Swedish siege was, however, marked by unrest and disturbances. Stensen, along with the other students, participated in the defending of the city from the ramparts, but it is not known whether he participated in actual fighting.[5] For long periods, the teaching at the university was more or less suspended, and only a few professors remained in the city. Thomas Bartholin was not one of them. As mentioned, he stayed with his family in safety in Roskilde from the beginning of the siege in August 1658 until the early summer of 1659,[6] and during this period his brother Rasmus Bartholin was the only medical professor who taught students.

Because of this and the fact that Thomas Bartholin withdrew from performing public dissections in the Anatomy House in February 1656, it has been questioned how much instruction in anatomy Niels Stensen actually received from Thomas Bartholin during his student years in Copenhagen.

However, it is beyond any doubt that the teaching of Stensen in anatomy and medicine was mainly conducted and executed by Thomas Bartholin. Several facts bear witness to this: Stensen's sublime dissection technique when he arrived in Holland in the spring of 1660; their close teacher-student relationship; and Stensen's detailed knowledge of Bartholin's achievements and approach as an anatomist, as evidenced by their mutual correspondence during Stensen's stay in the Netherlands. Moreover, we know from Bartholin's writings that, during Stensen's student years in Copenhagen, Bartholin performed private dissections on animals and humans, in which it is likely that Stensen participated. For example, on 19 December 1656 in the Anatomical Theatre Bartholin carried out a private dissection of an African lion from Frederik III's royal menagerie.[7] The following year he carried out a private dissection of a muscular man with only 11 ribs on one side,[8] and in the severe winter of 1658 he was called to Simon Paulli's home to attend the dissection of a female owl, which had died of cold (Figure 137).[9]

Rasmus Bartholin also became an important teacher for Stensen. As previously mentioned, Rasmus Bartholin played a leading role in the introduction of Descartes' mathematics and geometry in Denmark. In March 1656 he had been appointed professor of mathematics, and in June 1657 he also became professor of medicine.[10] His teaching of mathematics and geometry was undoubtedly of great importance for Niels Stensen's later use of these disciplines in his research in anatomy, geology and crystallography.[11]

In 1657, Rasmus Bartholin gave a speech for Niels Stensen and his fellow students in favour of the use of the Danish language as a language of science in order to achieve greater scientific dissemination. In this speech, *De Studio Lingvæ Danicæ* (About the study of the Danish language), Rasmus Bartholin emphasized the mathematical disciplines as the basis of all art and science, without the help of which no one has ever conceived or accomplished anything excellent[12] –

106 THOMÆ BARTHOLINI

mentum utrinq; membranoſum teres ad renum adipoſam tunicam adſcendebat. Os vulvæ prima fronte quadrangulare apparebat, latius verò diductum oblongius viſebatur. Sub hoc vaſculi Civetici oſculum eminet, in quo vaſculo ſeu clauſo ſeu diſſecto nihil ſingulare notavimus, nec ulla parte à maſculo differens, unde illi diutius non immorabimur, contenti in tabella ante oculos ponere ſitum partium & formam.

FIGURÆ PRIMÆ

Explicatio.

a. a. Renes, dexter & ſiniſter.
b. b. Renes ſuccenturiati, dexter major, anguſtior ſiniſter.
c. c. Venæ & arteriæ deſcendentis truncus.
d. d. Ligamentum teres membranoſum.
e. e. Teſticuli glandulis infinitis diſtincti.
f. f. f. f. Vaſa ſpermatica præparantia, partim ex truncis, partim ex emulgentibus orta.
g. g. Cornua uteri utrinq; protenſa.
h. Uterus ipſe.
i. Os vulvæ quadratum.
k. Veſica ad latus reflexa, ut melius cætera veniant in conſpectum.

l. l. l. l. Ur-

Figure 137. The female owl dissected by Thomas Bartholin in Simon Paulli's home in the winter of 1658, probably with Niels Stensen's participation. From Thomas Bartholin's *Historiarum Anatomicarum Rariorum*, Cent. 5, Copenhagen 1661, Hist. 49, pp. 105-107.
The letters mark:
Fig. I: *a*. Kidneys. *b*. Adrenal glands. c. Vena cava and aorta. *d*. Round ligaments. *e*. Testicular glands [ovaries]. *f*. Spermatic vessels [Blood vessels for the ovaries]. *g*. Horns of the uterus. *b*. Uterus. *i*. Square bone of vulva. *k*. Urinary bladder. *l*. Ureters. *m*. Rectum. *n*. Anus. *o*. Small vessels. *p*. Mouth of a small vessel.
Fig. II: *a*. Vulva. *b*. Openings of small vessels. c. Fissures of small hair vessels.
The Royal Library in Copenhagen.

almost an echo of what Tycho Brahe had said in his speech at the university in September 1574. Like Brahe, Rasmus Bartholin also advocated the empirical research method in his speech.

There are indications that Rasmus Bartholin also inspired Niels Stensen in the crystallographic field, or that at least a mutual inspiration existed.[13] In *Chaos* (see below), Stensen studies, describes and draws various forms of snow crystals, and he suggests further investigations.[14] Thomas Bartholin's *De Nivis Usu Medico* from 1661 added an illustrated dissertation by Rasmus Bartholin about various kinds of snow crystals. In 1669, the same year as Stensen published his basic geological and crystallographic work, *De Solido Intra Solidum Naturaliter Contento Dissertationis Prodromus* (Forerunner to a dissertation on solids naturally embedded in other solids) in Florence, Rasmus Bartholin published in Copenhagen his most famous work, the physical and crystallographic treatise on the birefringence of light in Icelandic limestone, *Experimenta Crystalli Islandici* (Experiments with Islandic crystal), which was published in English translation in *Philosophical Transactions* in London on 16 January 1670.[15] In the treatise, Rasmus Bartholin presents the results of his thorough physical and chemical studies of the crystals of pure, transparent Icelandic limestone, which was brought to Denmark from the quarry in Helgustadir by an expedition in 1668. Throughout the treatise he uses geometry and geometric illustrations in his presentation and explanations. His most important observation was the limestone's birefringence of light. He observed that only one of the two light rays obeyed Descartes' law of refraction in *la Dioptrique*, which was explained by his friend and colleague, the Dutch mathematician, physicist and astronomer Christiaan Huygens (1629–1695), in his *Traité de la Lumière* in 1690.[16]

In his treatise, Rasmus Bartholin also notes the perfect embodiment of the intersection surfaces of the limestone, and that crystals are only easily broken along the directions of the cutting planes. Furthermore, he notes that the breaking of a larger crystal produces smaller rhombohedrons, all of which have the same shape and – as Stensen establishes as a general law in *De Solido intra Solidum* – constant angles between the surfaces (measured by Rasmus Bartholin to 101° and 79°).[17] In other words, teacher and pupil beautifully complemented each other the same year in two epoch-making treatises.[18]

The young Stensen's enthusiasm for Descartes and his method of exploring Nature, formulated in *Discours de la Méthode*, was undoubtedly also due to the influence of Rasmus Bartholin.

From March to July 1659, when teaching at the university was suspended, Stensen made a comprehensive literature study of 77 authors' more than 100 works concerning mainly natural philosophy and medicine. He recorded his

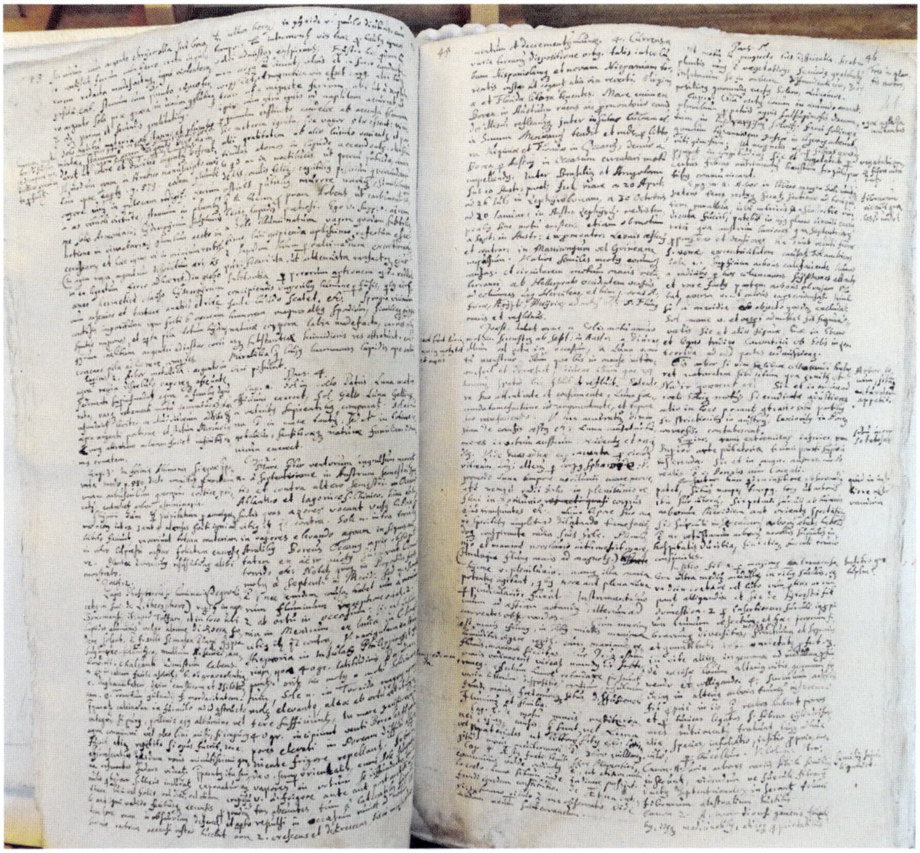

Figure 138. Double page from Niels Stensen's notebook *Chaos* from his four months of literary studies in Copenhagen from March to July 1659. His neat writing testifies to his excellent eyesight and fine motor skills, two important qualities for a skilled anatomist.
Gal. 291, Biblioteca Nazionale Centrale di Firenze Manoscritti e Rari. Photo: Jesper Brandt Andersen.

reading fruits in the form of excerpts, notes, comments, and mentions of self-performed exercises and experiments in a 92-page handwritten notebook called *Chaos* (Figure 138), which was rediscovered in 1946 by the Austrian Stensen-researcher Gustav Scherz (1895–1971) in the Galilei-archive at the department for manuscripts in the National Library in Florence.[19] Stensen's main supervisor in this study of literature was Ole Borch, who had access to Joachim Gersdorff's well-stocked private library, made chemical experiments with Stensen in Gersdorff's chemical laboratory and had frequent conversations with him during this period.[20] However, towards the end of the period, when Bartholin had returned to Copenhagen, there are several indications that he had a hand in Stensen's choice of works to review. From this point on, more anatomical and medical works begin to appear in *Chaos*, not least several of Bartholin's own works and Jean Pecquet's *Experimenta Nova Anatomica*, which Stensen quotes from and writes notes about for five days between 19–23 June 1659. Stensen read the second edition from 1654, which begins with a six-page letter of dedication, dated 1 January, from Pecquet in Montpellier to Thomas Bartholin in Copenhagen. In this letter Pecquet praises Bartholin's *Vasa Lymphatica*, which has had great importance for him and reassured him of the correctness of his research. Furthermore, he denounces Riolan's furious manner and compares him to an *andabata*, a Roman gladiator whose helmet was not equipped with holes for his eyes, so he fought blindly without hitting his opponent.[21]

Since Bartholin owned a copy of this edition, which he had received from Pecquet himself,[22] and since it was not among the books in Gersdorff's library,[23] it is most likely that Stensen had borrowed it from Bartholin, and that his reading it was due to Bartholin's instruction.

After three years of study at the University of Copenhagen, thoroughly trained in both theoretical and practical skills, in humanities, natural sciences and scientific method, by teachers of international standing, Stensen was ready for further study abroad.

Stensen in Amsterdam – the discovery of a major salivary duct

In early 1660, several months before the final peace treaty with Sweden was signed, Stensen managed to escape the siege of Copenhagen to begin his compulsory medical study trip. He was provided with financial means from an inheritance and private and royal support. Like other Danish medical students, his first destination was Holland, though not Leiden as was usual, but Amsterdam, where Thomas Bartholin's acquaintance Gerard Blasius lived (Figure 139). Educated in Leiden, Blasius had recently begun teaching medical students in anatomy at Amsterdam's *Athenaeum Illustre*, and later the same year he became the city's first professor of medicine.[24] He is particularly known as a neuroanatomist, and his most important contributions to anatomical research are his works on the central nervous system (brain and spinal cord), especially the work *Anatome Medullæ Spinalis et Nervorum inde provenientium* (Anatomy of the spinal cord and the nerves emanating from it), published in Amsterdam 1666, which is based on dissections of 119 animals and explains anatomy using 78 illustrations.

Gerard Blasius was the son of Christian IV's master builder Leonard Blasius (d. 1644), who had had several major construction projects in Copenhagen from

Figure 139. Gerard Blasius (1627–1682), Niels Stensen's anatomy teacher and host in Amsterdam in 1660. Engraving by an unknown artist, c. 1660.

Blasius shows a human skeleton in a textbook of anatomy. Although skilled, he was not a great anatomist himself but strived for recognition and fame. When Niels Stensen, during a dissection of a sheep's head in Blasius' private anatomical theatre discovered the main duct of the parotid gland, Blasius could not resist the temptation to claim the discovery himself. He did not expect his young Danish pupil to bite back so insistently and skilfully that the "theft" would end up bringing dishonour to himself.
Stadsarchief Amsterdam.

1640–1644, while the family lived in the Danish capital.[25] However, Gerard had apparently made the acquaintance of Thomas Bartholin through letters of recommendation from others.[26]

When Stensen arrived in Amsterdam at Easter in 1660, bearing letters of recommendation from Thomas Bartholin,[27] Blasius was in the middle of a teaching program in the form of three weeks of private anatomical demonstrations in his home. As a new guest and lodger in the home, Stensen followed the last six days of dissections of the brain, eyes and ears, before Blasius concluded the final day with a cursory review of known excretory ducts of the salivary glands of the mouth.

Spurred on by this and with one eye on his opportunities as a guest in the house, Stensen asked Blasius for permission to purchase animals that he could dissect on his own in the small dissection hall of the house. He got permission, but when he soon after bought a sheep's head with the intention of dissecting the brain, he had no chance to anticipate where his very first dissection abroad would lead him. The dissection took a completely unexpected turn, which sowed the seeds for Stensen's fame as an anatomist and immortalized his name within this profession, but which at the same time brought him involuntarily into his first drama as a scientist, a bitter priority dispute with his host – one of the most famous of its kind in the 17th century. A year later, on 22 April, Stensen explains what happened in a letter to Thomas Bartholin from Leiden:[28]

"A year ago, as I had been received by Blasius as my host, seeing a good opportunity of having anatomical subjects in that series of lectures, which he finished in the third week of my arrival, I asked the very famous gentleman to allow me to dissect with my own hand what I would provide for myself. Having obtained permission, I had a stroke of luck when, at the first opportunity on 7 April, when I was dissecting alone in the small study, I found in a sheep's head a duct that nobody had described, as far as I know. Having removed all the common wrappings, I considered dissecting the brain, when casually I decided to examine first the vessels passing through the mouth. Thus, exploring to this end with a probe pushed in the paths of the arteries and veins, I observed that its tip was no longer retained inside the straits of the tunics but roamed in a large cavity. As I immediately afterwards pushed the metal probe further in, I heard it clinking against the teeth."

What Stensen had found and inserted his probe into from the outside of the sheep's upper jaw inward towards the mouth was not, as he thought, a severed blood vessel, but the excretory duct of the parotid gland into the oral cavity. Since, as he further writes to Bartholin, he had not previously seen or heard of this duct, he consulted Blasius:

"Amazed by the novelty of the thing, I called my host to hear his opinion. He at first accused me of having forced my way, then he resorted to the frequent whims of Nature, and finally he invoked Wharton's book.[29] But since nothing turned up there, and the rather carelessly treated vessels did not permit further investigation, I decided to examine them another time with more attention. Some days later I succeeded in doing so in a dog's head, although less clearly."

Despite the fact that the affirmative finding on the dog's head must have made Stensen reasonably sure that he had enriched anatomy with a significant new discovery, he did not allow himself to be carried away by the enthusiasm that would inevitably strike any young scientist in such a situation. His further actions in the case show that even at the beginning of his fledgling career he understood the importance of scientific humility and caution with hasty conclusions. He informs Bartholin about a letter to his friend Jacob Henrik Paulli (1637–1704) in Copenhagen to hear his opinion about the discovery and about his consultation with the able and skilled Franciscus Sylvius after having left Amsterdam that summer to study in Leiden:

"Since thereby a great relationship with the lower salivary duct suggested the function of this vessel, in the same month I told my very close friend Mr. Jacob Henrik Paulli that I had discovered a small salivary duct and added a description of it. But since something similar had already been discovered, and I could not guess whether this had been observed by other people, I kept silent, until I had an opportunity to consult the very famous Sylvius on the matter. After hearing me, he thought that this duct had to be looked for in man, and having found it, he showed it several times to onlookers. This is, very illustrious Sir, the story of the discovery."

Stensen's findings show that he must already have been an exceptionally skilled dissector at this early stage, as these salivary ducts are very small, delicate structures. Since it was his first known dissection abroad, he must have acquired these skills during his study years in Copenhagen, and most likely from private dissections with Thomas Bartholin, who generally had enormous experience with the dissection of both animals and humans and particularly with the almost microscopic, thread-like and fragile lymph vessels.

The letter from Niels Stensen to Thomas Bartholin is the earliest preserved letter from Stensen and the first of a total of 14 letters, seven from each correspondent, from Stensen's stay in Holland during the years 1660–1664, where he made a string of epoch-making anatomical discoveries, especially concerning glands, muscles and the heart. Among the more than 500 surviving letters from and to Niels Stensen,[30] only the first hundred or so are from his time as a natural scientist up to 1675. The correspondence from Holland with Thomas Bartholin gives the best picture of Stensen's unique abilities as an anatomist, and Stensen's seven letters from this correspondence are therefore traditionally included in his 34 works of natural science. The correspondence is also an important testimony to Thomas Bartholin's crucial role as Stensen's teacher and mentor and as a catalyst for his scientific career. Stensen's awareness of and gratitude for this fact shines through in his introduction to the first letter:

"When I consider how well you, very famous Sir, are disposed towards me, and since I believe I do not deserve such consideration, I wish I could honour your extreme kindness with an equal respect. The proofs of your love indeed appear so numerous that it makes clear to me that the ancients rightfully wanted teachers to take the place of a venerable father. I think that the day when I was adopted as one of your disciples must deservedly be regarded as equal to my birthday."

An interesting sidelight on Stensen's research activities in Leiden can be obtained from Ole Borch's voluminous letters to Thomas Bartholin from the same city in the same period. Borch left Denmark in November 1660 to begin a six-year study trip in Europe.[31] He honoured Stensen with his company in Leiden from January 1661 to May 1663, and during that period he served as Stensen's research partner at the dissection table. Bartholin published both correspondences in the third and fourth centuries of his medical letters, *Epistolarum Medicinalium*, in Copenhagen in 1667.

The first letter from Borch in Leiden to Bartholin in Copenhagen is from 3 March 1661. Among other interesting things, it shows how important letters of recommendation from a famous and influential anatomist like Bartholin were for students like Borch and Stensen. Borch writes to Bartholin:

"You can judge what profit I gained from your letters of recommendation to Vorstius, Sylvius, van Horne and Blasius from the fact that I was received by all of them with the greatest kindness, and was most considerably let into the secrets of their art. Thus, at the public disputations I was given a seat on the professor's benches, above all those who did not have the title of professor, and exactly the same happened in the anatomical theatre of the hospital and in the other public theatre."[32]

Of the activities at the dissection table, Borch reports that van Horne has dissected a girl and a strong thief, during which he made frequent respectful references to Bartholin. Sylvius has dissected four corpses in the hospital, each time thoroughly examining the brain's cavities and showing the pit in the *septum pellucidum*, the thin transparent partition that separates the brain's two frontal horns (*cornu frontale*), which are part of the frontal lobes (*lobus frontalis*). Borch adds that this is where the Englishman Kenelm Digby places the seat of the soul.

The era's interest in glands

As explained, almost 20 years earlier Bartholin had been a co-discoverer of the main excretory duct of the body's largest salivary gland, the pancreas, and he had participated in the further exploration of this gland's smaller salivary ducts, among other things with the help of introducing small probes into the ducts. On the whole, the discovery of the pancreatic duct boosted anatomists' interest in and exploration of the body's many different glands and their functions.

The first major work on the subject was that of the English physician and anatomist Thomas Wharton (1614–1673) (Figure 140), *Adenographia: sive glandularum totius corporis descriptio* (Adenography or description of the glands of the whole body), which appeared in London in 1656, and which was the one

Figure 140. The English anatomist Thomas Wharton (1614–1673). Oil painting by unknown artist c. 1650.
With his work on all the human body's known glands, *Adenographia: sive glandularum totius corporis descriptio*, published in London in 1656, Wharton provided the starting point for exploration of glands. He was therefore also an important inspiration for Niels Stensen.
Royal College of Physicians, London.

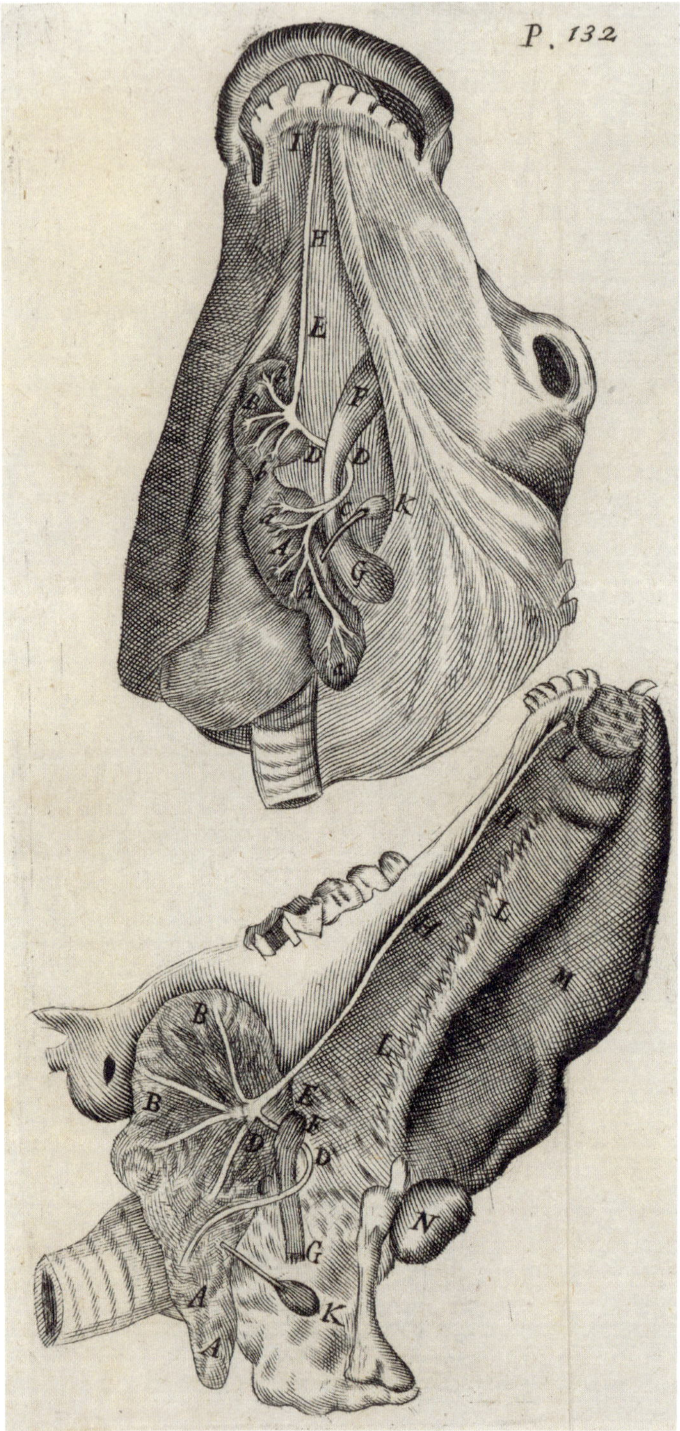

Figure 141. The maxillary salivary gland and its main excretory duct, *Wharton's duct*, of a calf in Thomas Wharton's *Adenographia: Sive Glandularum Totius Corporis Descriptio*, London 1656, p. 132.

The letters mark: A. Posterior part of the gland. a. Posterior branches of the salivary duct. B. Anterior part of the gland. b. Anterior branches of the salivary duct. C. Posterior trunk of the salivary duct ascending the tendon of the biventricular muscle. D. Return of the same duct and union with the anterior duct. E. Common trunk of the salivary duct. F and G. Biventricular muscle. H. Progression of the mentioned duct towards the anterior and inferior teeth of the maxilla. I. Opening of the salivary duct under the tip of the tongue close to the incisors of the maxilla. K. The round maxillary gland. L. Area of roughness hidden under the tongue. M. The tongue drawn back from its place to better visualize the departure of the vessels. N. Tonsil. O. Opening of the ear.
The gland does not exist in humans.
The Royal Library in Copenhagen.

Blasius referred Stensen to. In 40 chapters, Wharton describes all the glands of the body known at the time, from the genitals up through the abdominal cavity and chest cavity to the neck and head including the brain. He divides the glands according to type, appearance and function, and describes known diseases related to them. He compares glands with organs, for example the pancreas with the spleen, explaining the differences. In addition, he describes and illustrates his discovery of the excretory duct of the upper jaw's salivary gland in a calf (Figure 141).[33] This gland does not exist in man, but it is similar to the salivary gland of the lower jaw in man, and Wharton was therefore given priority to the latter (*ductus submandibularis, ductus Whartoni* or *Wharton's duct*). Wharton also describes the parotid gland,[34] but as Stensen writes to Bartholin, there is no mention in Wharton's book of the excretory duct of the parotid gland, which Wharton had not found, and which had not previously been described by others either.

Wharton insightfully explains in the book the importance of saliva for chewing and the digestive process, but he has not realized that the parotid gland is a salivary gland, and he attributes the following three functions to it: 1) to take up excess fluid from the nerves and carry it on to the veins, 2) to warm the ear, and 3) to fill out and level the pit around the ear.[35] Wharton's thoughts about the functions of the glands in general are characterized by the limited knowledge of the era. For example, he believes that one of the functions of the thyroid gland is to make the neck rounder and more beautiful, especially in women.[36] The book makes frequent reference to Thomas Bartholin's *Anatomia Reformata*, not least concerning the milk veins, lymph vessels and pancreas, but also to Bartholin's point-by-point rejection of Descartes' theory that the pineal gland (*corpus pineale*) of the brain is the meeting point between body and soul.[37] In his treatise *Spicilegium Secundum ex Vasis Lymphaticis* (Second Aftermath of the Lymph Vessels) from 1660 (see later), Bartholin praises Wharton's thorough work, which, according to Bartholin, has raised Wharton's fame as an anatomist to the peaks.[38] Bartholin certainly had a point there, for the work achieved great recognition in its time and was reprinted six times in Europe.[39]

Stensen's priority dispute with Blasius

About Blasius' theft of the discovery, which consisted in Blasius publishing it as his own in his *Medicina Generalis* (General Medicine) in Amsterdam in 1661, long after Stensen had moved on to Leiden, Stensen wrote to Bartholin in a letter of 22 April 1661:

"I will tell you about the whole matter like a disciple to his teacher, and I will leave to your judgement what must be decided from that. This duct had already quite often been shown to people by the most illustrious Sylvius and by the most famous Mr. van Horne when, some weeks later, the very illustrious Gerard Blasius, who had become hostile to me for a reason that I do not know, yielded so to his feelings that, through his brother, he notified the very famous gentlemen Sylvius and van Horne that this discovery was due to him. He did so even though he had never applied his hand to the investigation of this duct, even though he had not succeeded in indicating what it was when I showed it to him cut in half, even though in a letter to Eyssonius his own brother attributed the discovery of it to me some days before these reproaches were notified to me,[40] and finally, even though he himself, in his last writing, the *Medicina Generalis*, assigned neither the true origin nor the true orifice to the thing he says he has discovered."

Blasius' blatant theft of Stensen's discovery put Thomas Bartholin in an uncomfortable position between his friend and his student, and it would prove difficult for him to act as a mediator in the resulting protracted and bitter priority dispute. Although his professional insight and knowledge of the nature of the two combatants did not allow him anything other than to place his sympathy on Stensen's side, on the other hand he did not want to end up in enmity with Blasius. Bartholin must, along with Stensen himself, have been annoyed that Stensen had not followed the advice he had given him precisely to avoid such theft. Thus, it appears from Stensen's mentioned letter to Bartholin of 22 April 1661 that Bartholin, in an unpreserved letter, had urged him to publish a paper with an illustration of the discovered salivary duct as soon as possible:

"Since, however, by the same letter you advise me to publish a picture of the external salivary duct, I cannot but expose to you briefly also the envy which this small discovery caused against me and, along with it, what profit I drew from this envy, not to seek fame in trifles but to reject the hateful crime of plagiarism. I indeed regret that I am forced either to speak much of a matter of so little importance or to be the victim of an infamous stamp of ignominy."

As this book shows, Bartholin was a particularly diligent publisher of his works, and he tried hard to transfer this quality to Stensen, who did not by nature possess it. This is just the first of many examples of Bartholin trying to press Stensen to publish his findings, but the releases often took longer than Bartholin liked. It is not surprising, therefore, that 16 of Stensen's 29 anatomical works ended up being published by Bartholin.[41] Stensen's steadfast reluctance to publish his discovery of the salivary duct became too tempting for Blasius, who himself hankered after an anatomical discovery that could make him famous. He had hardly imagined that his young Danish guest would fight so zealously and skilfully for his discovery.

But Stensen probably had many good reasons for postponing the publication, which is presumably why Bartholin's call was only for a small preliminary publication with a drawing of the salivary duct. Among these reasons were undoubtedly Stensen's carefulness and desire to achieve complete certainty of the find, also in man, which was the same reason that Bartholin had for postponing his publication of *Vasa Lymphatica*. Added to this was his change of place of study from Amsterdam to the University of Leiden, his preoccupation with further exploration of the salivary glands, the lacrimal glands and the tear duct (see later), and his obvious desire to include these results in the publication. Finally, a few months after his arrival in Amsterdam, he had published a small disputation on hot springs, *Disputatio Physica de Thermis* (A physical disputation on hot springs) – his first scientific publication – which he defended on 8 July 1660 before professor of philosophy at Amsterdam's Athenæum[42] Arnold Senguerd (1610–1667) and Gerard Blasius.[43]

From Stensen himself not a peep is heard of the discovery until the mentioned letter to Bartholin, more than a year after it took place. The first source to mention the discovery is a letter to Thomas Bartholin from Ole Borch, who after a month's stay with Blasius in Amsterdam had arrived in Leiden in the beginning of January 1661 to join Niels Stensen.[44] On 3 March 1661, eight months after Stensen left Blasius in Amsterdam, and 11 months after Stensen's finding of the salivary duct, Borch writes in his first letter to Thomas Bartholin:

> "There have also been two public demonstrations of Steno's duct ["ductus Stenoniani"] (for so the distinguished van Horne always calls it in honour of its discoverer); that is to say, the two upper salivary vessels, which were discovered here by our countryman Niels Stensen. These originate from the parotid glands, pass by various branches below the cheekbone straight to the part of the cheek, which is opposite the third upper molar, and there pierce the buccinator muscle and discharge saliva into the mouth. Here a small protuberance sticks out in the mouth on each side. If a narrow stylus is introduced through it, this will go straight down to the beginning of the parotids and then gradually follow the vessel, once discovered, to the cheek. Subsequently, the aperture of the mouth is cut wider open and a stylus introduced there."[45]

What we get here is an accurate anatomical description from a man who clearly knows what he is talking about, and who has dissected with the discoverer of the excretory duct, his own former student, now a budding master – as well as with the experienced anatomist Johannes van Horne, who had been Thomas Bartholin's travelling companion in the southern Italian mainland and on Sicily and Malta in 1644. Van Horne became one of Niels Stensen's most important teachers in Leiden. He had become an extraordinary professor of anatomy at Leiden University in 1651, and in 1653 became ordinary professor in both anatomy and surgery.[46] Like Bartholin, he was interested in the lymphatic vascular system, and at the same time as Bartholin he had published his discovery of the thoracic duct in a human, *Novus Ductus Chyliferus. Nunc primum Delineatus, Descriptus & Eruditorum Examini expositus* (The new thoracic duct, now for the first time illustrated, described and presented to the examination of the learned), Leiden 1652 (Figure 142). The publication's dedication is dated 6 May 1652,[47] while Bartholin's *De Lacteis Thoracicis* appeared on 5 May 1652.

Moreover, van Horne had a special interest in salivary glands. In 1656, the same year as Wharton's mentioned work on all the body's known glands was published, van Horne had published in Leiden a small dissertation titled *Disquisitio Anatomica de Ductibus Salivalibus* (Anatomical examination of the salivary ducts), and he was among the first of the anatomists of the 1600s to thoroughly grasp this topic.

As Borch argued, it was also van Horne who ensured Stensen's immortality in anatomical nomenclature with the eponym *ductus Stenoniani*. It happened in 1662 in the second edition of van Horne's anatomy book *Microcosm seu Brevis Manuductio ad Historiam Corporis Humani* (Microcosm or brief guide to the knowledge of the human body), where van Horne on page 23, after reference to the excretory duct of the submandibular salivary gland, adds: "To this must be added Stensen's new salivary duct [*novus ductus salivalis Stenonianus*], which spreads out its branches in the parotid gland, runs obliquely downwards under the skin and ends in a large hole in the cheek region." The names *ductus parotideus, ductus Stenonianus, ductus Stenoniani* and *Stensen's duct* are still used indiscriminately for the excretory duct of the parotid gland.

Bartholin's response to Borch's letter of 3 March 1661 is dated just six days later, that is 9 March. Of the discovery of the duct, he writes: "You and our friend Stensen deserve praise for having noticed the new salivary ducts. This is how Nature divides up her ages for us. She does not disclose all her sacred mysteries at once; instead, our own lifetime sees one thing, that which follows us will see another."[48] On the same day, Bartholin wrote to his colleague and former pupil of many years and Stensen's fellow student in Copenhagen, Georg Seger in Basel, and here Bartholin's pride in the anatomical accomplishments of his two students shines through more clearly:

"I believe you've heard about the enlarged salivary ducts, which the honourable van Horne calls Stensen's ducts, and which he publicly demonstrated twice recently in Leiden. Certain ducts were discovered in a private capacity by that same Stensen of ours, once a pupil of mine (on whom I congratulate myself), along with Ole Borch. [...] Moreover, there were other discoveries, which I have not time to repeat. Thus, Nature discloses herself more and more each day to her investigators."[49]

With both Sylvius' and van Horne's recognition and confirmation of Stensen's discovery and van Horne's honourable award of its priority to Stensen by public demonstrations, one would think that this matter was definitively settled. But when Borch again wrote to Bartholin on 20 March, he reported that rumours were circulating that the zealous and ambitious Blasius was plotting to assign the priority for the discovery to himself in an imminent publication. Borch writes:

"I hear that Blasius is claiming for himself the discovery of the upper salivary duct, but if we are to believe our friend Stensen, who is thoroughly modest and as yet quite ignorant of ambition, the Muses will have none of it. Stensen is quite firm in his assertion that he himself discovered the vessel at Blasius' house. He called Blasius into the theatre, showed him his discovery and, when Blasius was unsure what kind of vessel it was, pointed out to him the stylus going into the mouth. Stensen makes it clear that he will defend his position against Blasius in public, if need be – not in the hope of winning any plaudits by this, but to avoid giving the impression of having made Sylvius and van Horne tell an untruth when they publicly referred to "Stensen's duct." I shall not interfere in this quarrel."[50]

Borch's ink had barely dried before Blasius came out in Amsterdam with his *Medicina Generalis* (Figure 143). The book's dedication is dated 22 March 1661. It is a concise textbook of medicine with the main emphasis on anatomy and written based on other people's works. In the book's ultra-short section on the parotid gland, no mention is made of the newly discovered duct, and Blasius does not attribute other features to the gland than the three mentioned by Wharton.[51] The equally short section about the saliva on pages 63–64 mentions that saliva is a liquid predominantly consisting of serum, which by the virtue of heat is secreted from the blood in glands in the upper jaw. From here it is let through a salivary duct, visible in both the upper and the lower jaw, to the glands at the front of the mouth, and after being squeezed out from there by movements of

Figure 142. a. Frontispiece of the second edition of Johannes van Horne's (1621–1670) introduction to surgical methods, *Methodica ad Chirurgiam Introductio*, Leiden 1668.

A mixed crowd of what appear to be well-dressed citizens, children and patients have lined up as spectators in the surgical theatre, where a surgical assistant is amputating the left lower leg of a horrified patient with a saw, while being guided by a standing surgeon to the right of the image. On the table to the right and on the wall at the back of the theatre are surgical instruments (drills, saws, knives, a syringe), a bone and a jug (with wine for anaesthesia?). On a pedestal stands an antique figure with a Rod of Asclepius in the right hand and a rooster in the left. The stage is set, and the performance can begin.
The Royal Library in Copenhagen.

b. Johannes van Horne's illustration of the thoracic duct in a human in his treatise *Novus Ductus Chyliferus* etc., Leiden 1652, the page between the dedication and the preface. The illustration is reproduced in Thomas Bartholin's *Anatome Quartum Renovata*, Leiden 1673, p. 619.

The letters mark: *a*. Mesenterial veins and arteries. *b*. Lacteal veins are disseminated throughout the mesentery. C. Pancreas. *d*. Two extraordinary branches connected with the middle branch from the pancreas proceeding to the kidney. F. Gland laying over the left kidney. *g*. Curvature around the aortic trunk. H. The new thoracic duct ascending longitudinally through the thorax. *i*. The bifurcation of the same between the fourth and the sixth vertebra of the thorax. K. Part of the thoracic duct that leans against the oesophagus with the heart above it. L. The insertion of the thoracic duct in the left jugular vein. The inlet of the same in the beginning of the shoulder vein or the subclavian vein. N. Part of the vena cava above the heart. O. The left bifurcation of the same. P. Oesophagus. Q. Heart. R. Lungs. S. Trachea. T. Descending aorta. V. [Lower L] Dissected parts of the diaphragm. X. Stomach positioned on its side. Y. Part of the rectum removed from its natural place. Z. Left kidney with its vein and artery. α. Venule crossing the renal gland and terminating in the lumbar muscles. β. Intercostal branches emanating from the aorta. δ. Part of the jejunum. ε. Liver lobe resting on the stomach.
The Royal Library in Copenhagen.

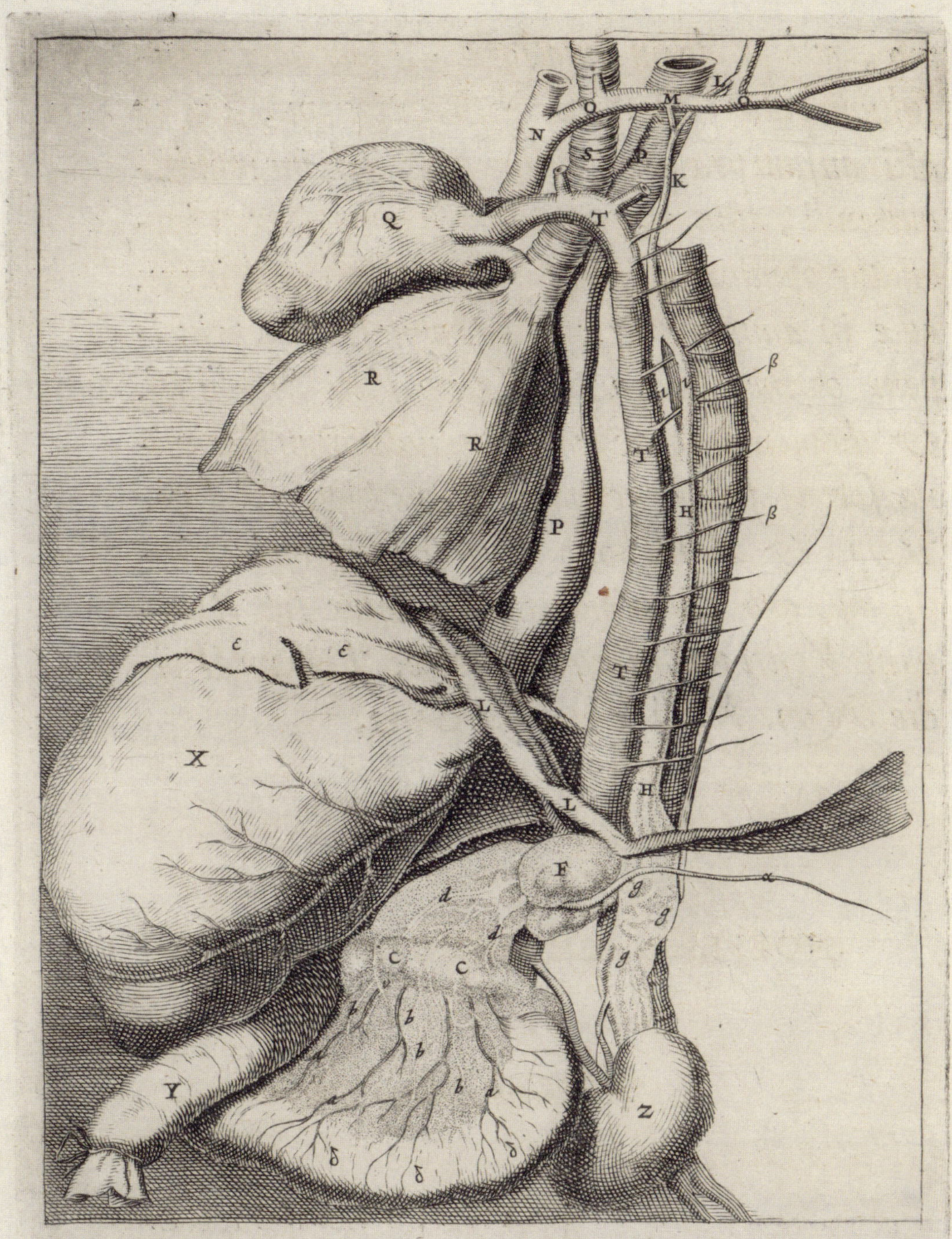

N
O
M
L
Q
S
P
K
T
R
P
ε
X
L
H
β
F
d
C
b
δ
Y
Z
α

the tongue it serves the taste, etc. However, the precise source and mouth of the salivary duct and the parotid gland are not mentioned here. But in the book's preface, Blasius writes:

"In the matter itself, there is hardly anything deserving attention. What has been added, you will easily spot yourself, when you compare what we have written with the writings that others have published so far. However, I need to draw attention to one thing, and that is that the salivary ducts, which appear in the upper jaw, and which I have mentioned on p. 64, must be classified among new

Figure 143. The frontispiece of Gerard Blasius' *Medicina Generalis*, Amsterdam 1661. In this work, in which Blasius credits himself with his student Niels Stensen's discovery of the main salivary duct of the parotid gland, he has – hardly without a strategic motive – presented himself as the experienced teacher, who teaches ignorant and inexperienced children. Wellcome Collection, London.

discoveries, for when, a year ago or more, I was preoccupied with private anatomical exercises, I first found them in a calf's head. In addition, the skilled young man Niels Stensen, my diligent student, later, when a favourable occasion appeared, demonstrated them for the spectators on a human head in the hospital in Leiden a few months ago."

Stensen had written in *Chaos*: "I should never be disturbed by any evil nor become angry because of insults against me or my friends or become distressed because of misfortune of my relatives. God sees and foresees everything, and all that happens is from him and for the glory of his name."[52] But Stensen could not possibly let Blasius get away with the theft of his first major anatomical discovery. As Borch predicted, Stensen took up the fight.

In a letter to Bartholin on 22 April 1661, his indignation shines through very clearly:

"Blasius in private letters did not defend himself with arguments but with insults and called me a liar, being unfair, malevolent, swollen by the ferment of envy and I do not know what else, so that he has deserved that I not only point out everything in my favour, but also that I should turn upon him everything he charged me with; though I know that he will never prove what he accused me of, that what must be rebutted is so well known by most people that they do not need me as a public crier. But nevertheless, if my name had not been associated to this duct by the very famous Mr. van Horne in such a noble place in such an assembly of all kinds of learned people, again and again, I should have willingly withdrawn my claims. Yet so that modesty should not be considered as a sign of guilt, I am forced to complain, although reluctantly, of the insult thrown upon me. Finally, to get on to other matters, let me add one more piece of evidence that I consider the strongest. Blasius himself in the *Medicina Generalis* reveals clearly that he never investigated this duct. Indeed, he assigns to it neither its true origin nor its true orifice. Even to the gland itself, from which the duct has its origin, he ascribes such an obscure function that, if I had not been sure of having shown it to him myself, I should assert that he had never seen it. This will appear more clearly in that disputation, which I have at hand."

The scrap into which Bartholin had thus been brought between his Dutch colleague and his pupil was not lessened by the fact that he had not seen through the ruse that Blasius had set in motion to win Bartholin over to his side. Unaware of Blasius' shady plans, he had at his invitation written a poem of tribute to him, which Blasius had placed immediately after the preface of *Medicina Generalis* with the purpose of using Bartholin as a guarantor of the book. Furthermore,

Blasius had just republished both parts of Bartholin's *Spicilegium Ex Vasis Lymphaticis* in Amsterdam in a single volume with the title *Spicilegia Bina Ex Vasis Lymphaticis* (Two aftermaths of the lymph vessels),[53] with a new portrait of Bartholin on the title page (Figure 144) and a flattering tribute poem from Blasius and his brother Johannes to "Thomas Bartholin, the leader of anatomists, Nature's unique examiner, the discoverer and describer of the lymph vessels, my incomparable best friend and benefactor."

Bartholin had thanked in a letter to Joan Blasius on 10 April 1661.[54] He tried to quell the dispute by taking on the role of mediator. To Stensen, however, he clearly indicated where his sympathy in this case lay, and that he knew very well to whom the priority of the salivary duct belonged. On 10 May 1661 he answers Stensen:

"Learned Denmark and I cannot praise enough your eagerness to explore the secrets of the human body and your luck in examining them; your fatherland is proud of such a great citizen, and I am proud of such a great pupil, by whose skills anatomical knowledge increases daily, and the knowledge about our lymph vessels increases enormously.[55] You will indeed oblige the whole world to give thanks if you, as I recently advised you to, would give a figurative presentation as well as a detailed description of the external salivary duct. You can share the credit with Wharton, when to his inner duct you add an outer one, and in addition you discover the sources of the saliva, about which many have so far advanced numerous conjectures, but none before you has been able to provide the tangible proof. Go on, my Stensen, striving after the immortal reputation that the perfect anatomy promises you. It pains me that on the threshold of your fame, you procure yourself adversaries, even among the best of friends. It is bitter to see the famous Blasius quarrel with you, since, as a friend to both of you, I rather saw concord than discord between you. May you both consider the interests of the medical profession and ensure that the competitive spirit does not turn into jealousy. One must keep the bile away from the writings of the scholars, so that, when the pen exceeds the limits of moderation, it does not pour itself out more to the suffering than to the benefit of the reader. By the way, it is hardly necessary to admonish you; I know your modesty and your calm mind. Nor do I doubt the wisdom of the famous Blasius. I do not want to add more, because I care about both of you. I only ask in the interest of the public that the truth will suffer no harm from the dispute. Search your shelves to get the promised discoveries forward to the public as quickly as possible – for the honour of your fatherland, for the advancement of natural science and for the praise of you, whom the learned will cheer."[56]

Figure 144. The frontispiece of Gerard Blasius' reissue of both Thomas Bartholin's aftermaths on the lymph vessels, *Thomæ Bartholini Spicilegia Bina Ex Vasis Lymphaticis*, Amsterdam 1661. 11.5 × 6.5 cm.
Blasius has portrayed Bartholin as the experienced, thoughtful, well-read and literary prolific anatomist in the process of authoring a new work in his library. As a template for the illustration of Bartholin, Blasius has used Figure 96 a + b. In his priority dispute with Niels Stensen over the discovery of the parotid duct, Blasius tried in vain to win Bartholin to his side by using flattery.
The Royal Library in Copenhagen.

One could easily get the impression, that Bartholin had completely forgotten his own disputes with Hofman, Riolan and Rudbeck, but besides the fact that it was in his own obvious interest to avoid discord between Stensen and Blasius, it was understandable, based on his own experience, that he wanted his talented pupil to be spared such battles and instead devote his time and energy to research. His new invitation to Stensen to publish his findings testifies to the fact that he did of course not think that Stensen should let Blasius get away with the theft. The letter's conclusion more than suggests that Bartholin had already at this early stage made plans to make Stensen his successor in Copenhagen:

"But you, my Stensen, continue to immerse yourself in anatomical study, so that you can happily return as the jewel of the fatherland. Rest assured that by your skill, you enjoy the recognition of your country. Live well. I wish you happiness and also in the future ask for your love. Wishing you well from the bottom of my heart, I am unreservedly your Thomas Bartholin."

In the correspondence between Borch and Bartholin you sense the embarrassment of the case. Borch soberly states to Bartholin on 21 April 1661: "I will not touch the dispute that has arisen between Blasius and our Stensen over the salivary ducts."[57] Bartholin answers almost as briefly on 19 August of the same year: "I wish Blasius and Stensen would resume their former friendship. The contentious issue hardly deserves such a disagreement."[58]

Not quite as cautious in expressing his opinion on the matter was the clear-sighted Georg Seger. On 4 June 1661, he thanks Bartholin for the information (in a letter of 9 March the same year) about Stensen's new discovery, which has inspired him to resume the art of dissection, in which he was trained by Bartholin. The previous week he had begun dissecting a sheep's head to find "Stensen's ducts" when he was disturbed, but he does not doubt their existence. He continues:

"Added to that, not so long ago I happened to come upon a reference to these ducts in the preface to the *Medicina Generalis* of the honourable Blasius. Nevertheless, I suspect that there it had no other purpose than to display his belief that he was responsible for their initial discovery. I certainly wondered, while I was reading it, about the warped habits of our present age, but I will not add anything concerning them now, because they are well known to you through your lymph vessels.[59] We have nearly arrived at the point where inquisitive persons perceive the discoveries more readily than their discoverers."[60]

Stensen's treatise on the salivary glands of the mouth and their excretory ducts is published

In Leiden, July 1661, Stensen's treatise on the salivary glands of the mouth and their excretory ducts finally arrived, with an account of the new finding of the parotid gland's excretory duct and other new observations made by Stensen in Amsterdam and Leiden. Its title is *Disputatio Anatomica de Glandulis Oris, & nuper observatis inde prodeuntibus Vasis* (Anatomical dissertation on the salivary glands of the mouth and their newly observed excretory ducts) (Figure 145).[61] As the title page says, it is a dissertation, which "the Dane from Copenhagen Niels Stensen" presented publicly on 6 and 9 July 1661 at "Leiden's famous

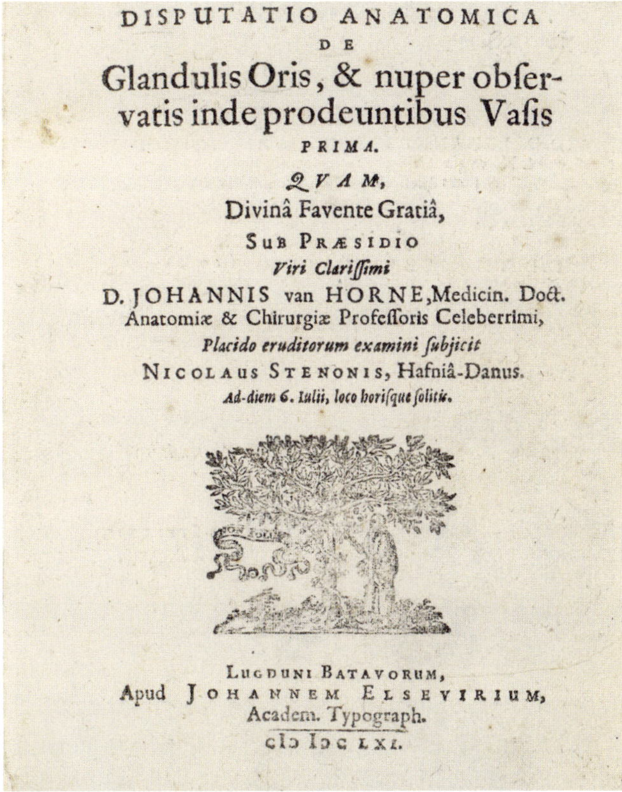
DISPUTATIO ANATOMICA
DE
Glandulis Oris, & nuper obſer-
vatis inde prodeuntibus Vaſis
PRIMA.
QUAM,
Divinâ Favente Gratiâ,
SUB PRÆSIDIO
Viri Clariſſimi
D. JOHANNIS van HORNE,Medicin. Doct.
Anatomiæ & Chirurgiæ Profeſſoris Celeberrimi,
Placido eruditorum examini ſubjicit
NICOLAUS STENONIS, Hafniâ-Danus.
Ad-diem 6. Iulii, loco horiſque ſolitis.

LUGDUNI BATAVORUM,
Apud JOHANNEM ELSEVIRIUM,
Academ. Typograph.
CIↃ IↃC LXI.

Figure 145. The title page of Niels Stensen's first anatomical treatise on the salivary glands of the mouth and their excretory ducts, *Disputatio Anatomica de Glandulis Oris, & nuper observatis inde prodeuntibus Vasis*, Leiden 1661.
With this treatise Stensen introduced himself to the international anatomical scene.
The Royal Library in Copenhagen.

University" for a learned audience and under "the famous man and Doctor of Medical Science, the renowned professor of anatomy and surgery and venerable preceptor Johannes van Horne's defence."

After the title page follows Stensen's dedication to councillor of the realm Otte Krag (1611–1666), archbishop Hans Svane (1606–1668) and "the very famous and experienced gentleman, Dr. Thomas Bartholin, very illustrious Honorary Professor of the illustrious Majesty King of Denmark and Norway at the University of Copenhagen, and Dean of the Faculty of Medicine, and to the Maecenas, Patrons and Teachers [I am] strongly attached. With respect for ever, Humbly and obligingly Niels Stensen D.D.D."[62] The slightly expanded second edition, which was published in Leiden in 1662 as the first of four treatises in the anthology *Nicolai Stenonis Observationes Anatomicæ* (Niels Stensen's anatomical observations) has a seven-page preface with Stensen's natural philosophical reflections. Here Stensen shows for the first time his abilities and his insight as a natural philosopher to the learned world. He makes a tribute to man's imagination and capacity for reasoning, but his message is that these characteristics are not sufficient for an accurate and comprehensive exploration of Nature's diversity and details. Such an exploration also requires academic knowledge and practical experience gained by using the senses.

He highlights his teacher Thomas Bartholin as an example to be followed:

"Who, even the most perceptive, has for so many centuries thought even in dreams about the lymphatic vessels, before you, most famous Bartholin, presented them to the eyes of everybody, after you had observed them them, not with the eyes of your mind, but with those of your body?"[63]

Stensen then transfers this to his own research:

"The origin of the saliva confirms the same need of observations: reasoning deprived of the work of the senses did not find the paths carrying the saliva into the mouth. Neither will the paths that carry its substance to the glands ever be certain for us, as far as I can see, if they are to be expected from thinking alone. Observation must lead to reasoning, and the object itself must be examined, as far as possible from all its sides, so that its figure be imprinted in the mind corresponding truly to the object itself."[64]

Based on numerous dissections of various animals (Figure 146), Stensen explains in detail the anatomy of all the salivary glands of the mouth, including their salivary ducts and their supply of blood vessels and nerves, but also the anatomy of the lymph nodes and lymph vessels of the glands and mouth, and he makes suggestions for the functions of saliva. He even explains the anatomical basis of the taste buds in the form of accumulations of nerves and their stimulation of saliva formation by the influence of flavours.

He convincingly rejects Blasius' claim to the discovery of the excretory duct of the parotid gland with the same arguments as in the letter to Bartholin, the main argument being Blasius' lack of insight into the anatomy of the excretory duct in his description of it in *Medicina Generalis*:

"Yet the famous gentleman would have written very differently, and he would not have imagined glands from which saliva is expressed by movement of the tongue, if he had ever introduced a probe in the duct through its conspicuous exit either in animals or in man.[65] [...] Among those who saw that the salivary duct continued up to this gland, and who observed the considerable roots of the same duct distributed in great number very delicately in this gland, who of them will believe that Mr. Blasius would ascribe such low functions to the parotid glands, if he had ever put his hand to investigating the salivary duct?[66] [...] The true function of the parotid conglomerate glands is to prepare that saliva, which is excreted through the external salivary duct into the external cavity of the mouth."[67]

Stensen explains elsewhere in the treatise that the rather large orifice of the salivary duct in the oral cavity is found next to the second molar in the upper part of the mouth. Stensen found and examined the salivary ducts by cutting the salivary glands with different incisions and inserting small probes into the ducts and, as far as the smallest of them are concerned, by pressing the glands together and observing the appearance of saliva. Thus, he also challenged the common assumption of the time, that the mucus in the nose, mouth and palate originated from the brain, from where it ran down through small holes in the skull bones. In his arguing that the components of saliva originate from the arteries, which means that not all the fluid in the arteries passes on to the veins, Stensen again refers to Bartholin: "The arteries have something they do not transmit to veins; the veins have something they did not receive from arteries. This has been proved by the famous Bartholin in *Spicilegium Ex Vasis Lymphat*icis *I. c. 7*, where he shows that there is normally more copious serum in arterial blood."[68]

The reaction from Blasius

Scarcely had Blasius read Stensen's treatise before he sat down at his desk and wrote a 26-page letter to Thomas Bartholin dated 16 July 1661.[69] From the beginning to the end, it is a letter of defence against the charges in Stensen's treatise. However, the tone is so harsh, implacable and in many places sarcastic that the letter takes on the character of a personal attack on Stensen rather than a defence. He calls Stensen impudent, shameless and vicious and claims that, long before Stensen's arrival in Amsterdam, Blasius had demonstrated the excretory duct of the parotid gland to his students, while the demonstrations of the salivary glands, which Stensen himself witnessed, were only superficial. Similarly, his descriptions of the salivary ducts in *Medicina Generalis* are only superficial, because this book only reviews the anatomy in general terms, which fact Stensen has misunderstood. Blasius appeals throughout the letter to Bartholin's sympathy, in some places more directly than others: "My dear Bartholin, please consider the arguments Stensen uses against me in the disputation, consider my reply, and you will see how badly he treats me." The letter has the character of a convulsive attempt by Blasius to save himself from a most embarrassing affair in the hope of at least maintaining his friendship with Bartholin. As also appears from the letter, Blasius' big mistake was that he had never expected that Stensen would defend his rights, but that the student would bow to his teacher in reverence and gratitude for the favours that Blasius had undeniably granted him. As might be expected, there was no sympathy to gain from Bartholin, who, on the other hand, did not want the case to cost him his friendship with Blasius either.

I
II
III
IIII
V

Figure 146. a. Niels Stensen's illustrations of the salivary glands on a calf, which conclude the first part of his dissertation on the salivary glands of the mouth.
Stensen's figure explanation:

Figure I: *a*. The parotid gland. *b*. Lymph node belonging to the parotid gland. *c*. Lymph vessels running downwards from *b*. *d*. Branches of the main duct of the parotid gland. *e*. The main duct (*ductus parotideus* or *ductus Stenoniani*), which is pierced at its mouth by a probe. *f*. Outer branches of the jugular vein. *g*. Nerves, that are both inside the gland and the head, and which also outside these places, for example in *h*, form an interlacing network. *i*. Nerve fibres, that follow the salivary duct.

Figure II shows the salivary ducts of the mandibular salivary glands: *a*. The mouths of the ducts leading from the lower salivary gland to the cheeks, three of which have a hair inserted. *b*. The opening of the external salivary duct. The other points indicate the other holes, from which sticky fluid comes out, when they are pressed together.

Figure III: *a*. The sublingual gland. *b*. Its ducts. *c*. The mouths of the ducts. *d*. A depression observed on the side of the tongue. Figure IV: *a*. Holes in the palate, through which a viscous fluid can be expelled. *b*. The tonsils. Figure V shows one of the small vessels coming from the lower part of the salivary gland in the cheeks.
The Royal Library in Copenhagen.

Speicheldrüsen. 497

542. Rechte Ohrspeicheldrüse, von rechts.

(Das platysma und die Nerven sind vollständig entfernt, die Gefäße teilweise.)

Die **glandula submaxillaris** *(Unterkieferdrüse)* (s. auch Figg. 540 u. 541) ist abgeplattet-eiförmig. Sie liegt in der Grube, die oben lateral von der fovea submaxillaris der mandibula, sonst von den mm. digastricus und stylohyoideus begrenzt wird. Oben medial berührt sie die untere Fläche des m. mylohyoideus und entsendet um den hinteren Rand desselben einen verschieden starken Fortsatz auf seine obere Fläche, der die glandula sublingualis erreicht und medianwärts an den m. hyoglossus stößt; aus diesem Fortsatz geht der Ausführungsgang hervor. Hinten reicht die Unterkieferdrüse oft bis an die glandula parotis und berührt den m. pterygoideus internus. Ihre freie Oberfläche wird von der v. facialis anterior gekreuzt und ist vom platysma überzogen. Der Ausführungsgang, *ductus submaxillaris [Whartoni]* (s. Figg. 517, 539 u. 540), läuft an der medialen Fläche der glandula sublingualis nach vorn und aufwärts und mündet in der caruncula sublingualis.

Die **glandula parotis** *(Ohrspeicheldrüse)* (s. auch Fig. 541) ist an ihrer lateralen, leicht konvexen Oberfläche ungefähr dreieckig. Sie füllt mit ihrem medialen, verschmälerten Abschnitt, *processus retromandibularis* (s. Fig. 540), die Grube aus, welche vorn vom hinteren Rande des ramus mandibulae und des m. pterygoideus internus, hinten von der pars tympanica oss. temporalis, von der cartilago meatus acustici, sowie vom vorderen Rande des proc. mastoideus und des m. sternocleidomastoideus begrenzt wird; in der Tiefe stößt sie dabei an den proc. styloideus, die mm. stylohyoideus und styloglossus, die a. carotis interna und die v. jugularis interna. Lateral schiebt sie sich auf dem ramus mandibulae und m. masseter nach vorn und reicht bis nahe an den arcus zygomaticus. Unten endet sie zugespitzt nahe unter dem angulus mandibulae und berührt oft die glandula submaxillaris. In der Drüse verlaufen streckenweise die a. carotis externa und mehrere ihrer Äste, die v. facialis posterior, sowie der n. facialis und seine peripheren Äste. Der Ausführungsgang, *ductus parotideus [Stenonis]*, geht vom oberen Abschnitte des vorderen Randes ab, verläuft auf der lateralen Fläche des m. masseter annähernd horizontal nach vorn, biegt dann medianwärts, durchbohrt schräg den m. buccinator und mündet im vestibulum oris dem zweiten oberen Mahlzahn gegenüber (s. Fig. 544); auf dem m. masseter liegt dem Gang oft eine verschieden große *glandula parotis accessoria* an.

b. The parotid gland and its excretory duct in Werner Spalteholz: *Handatlas der Anatomie des Menschen*, Vol. 3, Leipzig 1909, p. 497. Note the use of the term *ductus Stenonis* for the excretory duct. Private ownership.

As this book testifies, Bartholin was endowed with a high social intelligence, and he mastered the art of both polemic and diplomacy. His four-page reply to Blasius on 1 September 1661 shows that in matters such as this he was capable of balancing on a knife edge. Officially, he assumes the role of neutral mediator between his two friends, but reading between the lines, through his praise of Stensen he clearly signals that his sympathy is with him and herein lies his rebuke of Blasius. Bartholin begins the letter by expressing his annoyance at a dispute between two of his best friends. He does not want to argue with his friends, cares for both, is unwilling to take sides, and begs Blasius not to let anyone persuade him to believe that Bartholin will take sides. But then his smackdowns to Blasius follow:

> "Our friend Stensen was kind enough to inform me by letter of the new salivary ducts he had discovered. I was glad that my cherished pupil had cast new light on the art of anatomy, and in my reply encouraged him to write a precise description and have an exact illustration made of the vessel he had found, and to give the world the benefit of this important addition to knowledge. I could not possibly know that you too had seen or recognized this duct. Neither you nor anyone else informed me of that until the quarrel between you burst into the open with unrestrained attacks in print. However, Stensen had complained about this in a letter, saying that someone else was robbing him of the praise due to his diligence and successful research. I urged him in a friendly way towards reconciliation. Now that he is beginning to acquire a reputation – which I think will be a great one – he should not, I said, allow himself to be worn out by excessive wrangling with opponents; he is destined for greater things, that will win him greater glory. In case it is necessary to prove the innocence of what I wrote, I then begged him with all my heart to treat his teacher with his customary restraint, which I know so well; in this way he would avoid damaging the truth or giving the impression that his success had cajoled him into losing some of his usual modesty. Up to this point I can think of nothing I did wrong: I praised the hard work of an excellent pupil and encouraged him to write about his latest discovery to the great benefit of anatomy."[70]

Then follows Bartholin's barely concealed invitation to Blasius to throw in the towel and drop the weapons:

> "You are so well endowed with good sense that, having won fame in the world of letters with other learned works, you can regard the disputed discovery of this little duct as a mere trifle. Even without it your reputation is solid and everlasting. I shall not attempt to judge the matter over which you have entered dispute.

I do not want either of you to be in the wrong, that is, whoever I might find to be wronging his friend and mine. May each man's conscience assure him of his innocence."

Bartholin ends the letter by concluding that the discovery is more important than the identity of the discoverer, and he urges Blasius to restore his friendship with Stensen.

Here, Bartholin's role as mediator in the dispute ended. The dispute, on the other hand, did not. When Stensen had republished his treatise as part of his *Observationes Anatomicæ* and thus reasserted his priority to the discovery of the salivary duct, the dispute flared up again. Blasius made three of his former students testify that he had shown them the said salivary duct before Stensen's arrival in Amsterdam. Blasius sent these testimonies to his friend, the newly appointed professor of medicine and philosophy in Utrecht, Nicolas Hoboken (1632–1678), together with a cover letter dated 29 August 1662 and Blasius' long letter to Thomas Bartholin of 16 July 1661. The purpose was to make Hoboken publish this material, which he did with the addition of a letter to Blasius, in which he assigns to him the priority of the salivary duct, which according to Hoboken should be called *ductus Blasianus* and not *ductus Stenonianus*. The letter is dated 16 August 1662, so it appears as if Blasius' sending of the mentioned documents is a response to Hoboken's request.

Hoboken's publication appeared in Utrecht in 1662 with the title *Novus Ductus Salivalis Blasianus, in luceum protractus* (Blasius' new salivary duct brought to light). It contains a mocking anagram of Stensen's name, which is changed from STENONIS to IS NON EST (It is not him), SI NON EST (If it is not him) and SINON EST (Is he a Sinon?),[71] after which the book ironizes over these meanings at Stensen's expense.

In January 1663, Stensen was ready with his answer, *Apologiæ Prodromus quo demonstratur, judicem Blasianum et rei anatomicæ imperitum esse, & affectuum suorum servum* (Introduction to a defence which shows that Blasius' judge is ignorant of anatomical matters and a slave to his passions),[72] in which, as the title indicates, he regards Hoboken as Blasius' ignorant pen-master and Hoboken's publication as a trial, in which Stensen is the accused, Blasius the prosecutor and Hoboken the judge. Beyond that, Stensen's response brings nothing new to the matter. He maintains that the salivary duct described by Blasius does not exist, and he therefore calls it "Blasius' imaginary duct." From Blasius came nothing more. Hoboken subsequently corresponded privately with Bartholin and backtracked and hoped for reconciliation between Stensen and Blasius.[73]

In the work on his previous research into the heart, muscles and glands, which was published in Copenhagen in 1664 (see later), Stensen aims his last broadside at Blasius. According to Stensen, Blasius tried to render the discovery of the excretory duct of the parotid gland disgraceful to its discoverer, and Stensen says that the duct described by Blasius is scarcely to be found in any other creature than himself, "except perhaps in the inhabitants of the moon, or in those creatures who are creeping around in the spaces between the worlds of Epicurus."[74] With the publication of the last 200 of his *Epistolarum Medicinalium* in 1667, Bartholin cracked his whip on Blasius' back by giving the learned world insight into the proper context of the priority dispute, but as will be shown later, he had already given his unequivocal verdict in two other writings in 1666.

Stensen's further research into the lacrimal glands and ducts, the sweat glands, the thoracic duct, the lymph and chyle vessels, the muscles and the heart

Having finished his research into the glands of the mouth, Stensen continued his research by exploring the glands and vessels of the nose as well as the lacrimal glands and the drainage of the tears from the lacrimal points in the inner corner of the eye and down to the nose through the nasolacrimal duct (*ductus nasolacrimalis*), which he discovered.

Bartholin followed the results attentively and gave advice and guidance through his correspondence with Stensen himself and with Borch. Again, the first mention of Stensen's research in the lacrimal gland is in Borch's letter of 3 March 1661 to Thomas Bartholin:

"The same Stensen, while working on dissections with me in private, found certain ducts, which I call hygrophthalmics,[75] because they seem continuously to install a clear fluid into the eyes, being designed by Nature to allow freer movement and to prevent the eyes from getting dry. They are situated at the outer corner of the eyes within the orbits, and discharge into the eyelids above. They can be found if the whole eye with its appendages is carefully removed from the orbit and closely inspected up to a lamp, while the eyelid with the attached vessels is stretched by the hand. Their origin is in certain minute glands, for the ducts themselves are also quite tiny. In some cases, we saw two such ducts in each eye, in others only one. I would imagine that what Nature does with these ducts is the same as engineers, carriage drivers and others do, when they smear their rotating axles with either water or thick oil to prevent them from being damaged by continual motion and either getting too hot or drying out and no longer moving properly."

In a later letter to Bartholin on 23 July 1661, Borch continues with a description of the nasolacrimal duct, also found by Stensen during their joint dissections:

"In addition, there will be some observations on the vessels that keep the eyes moist; and especially on a duct that starts in the *puncta lachrimalia*, goes off to the nostrils, and then runs down the length of the nose and discharges on the inside right at the end, where the breath emerges. This explains why, when people cry, a quantity of fluid streams out, not only through the eyes, but also through the nostrils. This is a channel perhaps never before observed by anatomists."[76]

Bartholin replies on 19 August 1661 to Borch: "We are eager to get your observations of the brand-new vessels in the eye, and we shall give the discoverers the cordial thanks they deserve for their hard work and success."[77] Stensen himself first mentions these findings to Bartholin in a letter on 12 September 1661:

"But when I often during my work seek proofs through dissections, some things not yet seen present themselves to me. Besides various salivary ducts and those lacrimal vessels observed previously in the external gland of the eye, I indeed observed also two vessels of the same kind in the lacrimal caruncle, one on each side of the cartilage. They admit a probe without difficulty. On the very obvious continuation of the lacrimal point, I will add nothing, since I believe it has been described for you by the famous Borch, and you are to see more of this shortly in the treatise *De Glandulis Oculorum* [Figure 147].[78] However, on the excretory vessel, which I observed not so long ago in the tunic of the nose, some doubts arise, on which I should like to have your opinion. To examine more accurately the continuation of the lacrimal point in a sheep, I had broken the bones of the nose, and after having removed these, I saw in the tunic of the nose a small vessel, glittering like lymph vessels, running downwards and forwards, and therefore I immediately took it for a recurrent lymph vessel. But after having cut open the nostrils above the open orifice of the lacrimal points' continuing duct, I noticed a narrow and dark pit or furrow, through which a hair found an open path into the vessel, and when I examined more accurately the place beyond which the hair was not able to penetrate, it appeared from the obvious roots there to be an excretory lymph vessel. It is amazing that such a delicate duct proceeds in the thin tunic from these minute portions of glands. Yet I wonder even more to what end the humour secreted there is carried to the extremity of the nose, as its orifice is at a distance of hardly a finger's breath from there. I could say that it is a waste substance, but in other parts of the body we see that humour excreted through vessels of the same kind is destined to a well-

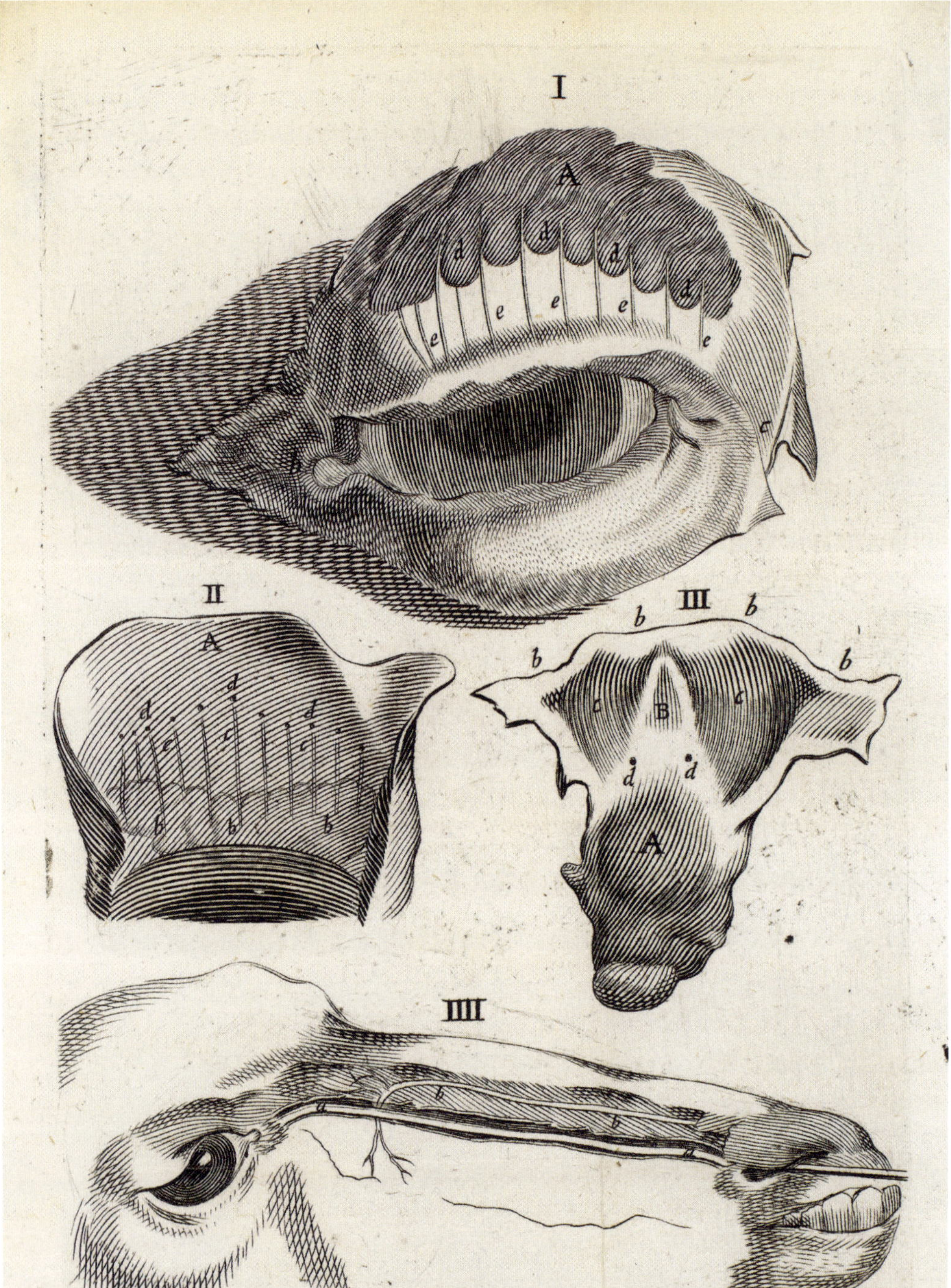
I
A
d
e
b
c
II
A
d
c
b
III
b
B
c
d
A
IIII
b

Figure 147. a. Niels Stensen's illustrations of the lacrimal glands and tear ducts on a calf.

Figure I shows the left eye of a calf: A. The lacrimal gland. *b*. The large corner of the eye. *c*. The small corner of the eye. *d*. The lobes, into which the anterior margin of the lacrimal gland is divided and in between which the ducts *e* exit.

Figure II shows A. The inner surface of the upper eyelid. *b*. The lacrimal gland, which, with its small ducts *c*, shines through the thin mucous membrane. *d*. The mouths of the ducts.

Figure III shows A. The lacrimal gland in the inner corner of the eye. B. The cartilage emerging from the gland itself. *b*. The cartilaginous rim. *c*. The membrane. *d*. The two orifices, one on each side of the cartilage.

Figure IV shows *a*. The continuation of the tear ducts to the outer part of the nose, i.e. the nasolacrimal duct, which is pierced by a probe at the nostril. *b*. The specialized nasal excretory duct.
The Royal Library in Copenhagen.

b. The title page of the dissertation *Nicolai Stenonis De Glandulis Oculorum, Novisque earundem vasis Observationes Anatomicæ, Quibus Veri lacrymarum fontes deteguntur*, which was part of the collection *Nicolai Stenonis Observationes Anatomicæ*, Leiden, 1662, and from which the illustrations in a can be found after page 100.

In the dissertation, Stensen proved that tears originate from the lacrimal glands, and he definitively quashed all previous assumptions that they come from the brain.
The Royal Library in Copenhagen.

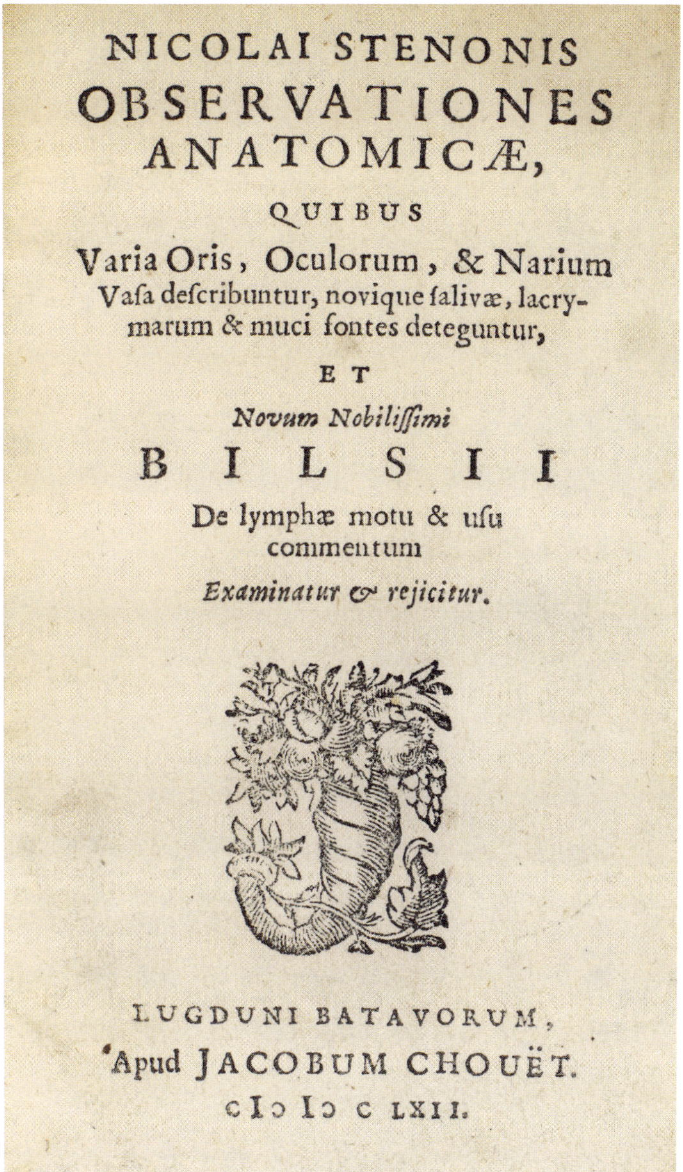

NICOLAI STENONIS
OBSERVATIONES
ANATOMICÆ,
QUIBUS
Varia Oris, Oculorum, & Narium
Vaſa deſcribuntur, noviqueſalivæ, lacrymarum & muci fontes deteguntur,
ET
Novum Nobiliſſimi
B I L S I I
De lymphæ motu & uſu
commentum
Examinatur & rejicitur.

LUGDUNI BATAVORUM,
Apud JACOBUM CHOUËT.
cIɔ Iɔ c LXII.

determined function and is not simply a waste substance. However, here, where the humour is hardly outside its canal when it is also outside the body, I do not see what function it has, and therefore I do not find a name for it. If thus you would be kind enough to let me know your opinion on this matter, you would greatly increase my debt of gratitude to you."[79]

In the same letter, Stensen recounts his discomfort with performing dissections on living animals (vivisections) and his scepticism about the Cartesians' comparison of animals to soulless machines with no sense of pain. In other words, it is an example of Stensen's scepticism towards Descartes' thoughts, which grew with his practical experiences as an anatomist and consequent attention to Descartes' many mistakes regarding animal and human anatomy. As mentioned earlier, this was a scepticism that Stensen shared with Thomas Bartholin. He writes to Bartholin:

"As I was in Amsterdam, I have repeated the experiment of Bils on the movement of the chyle.[80] But I did not find this difference in the blood, although I kept alive to the third hour the dog that had spent the whole day in this torment. Since, however, one experiment is not sufficient to conclude something reliable, at the first opportunity I will repeat this painstaking work, although I must say that I do not torture living animals in such extended torment without horror. Cartesians boast about the certitude of their philosophy. I wish they would convince me as certainly as they are convinced themselves that in animals there is no soul, and that it makes no difference whether you touch, cut or burn the nerves of a live animal or the cords of a running machine. Then I should indeed search more often and more willingly, for several hours, the viscera and vessels of living animals, as I see that much needs to be investigated that cannot be expected to be found in another way."

Finally, Stensen mentions that, during the dissection of the same dog, he found that the thoracic duct, together with a lymph vessel from the glands of the neck and several other lymph vessels, unites in a small canal, which discharges itself in the blood in the vicinity of the axillary and jugular veins. Bartholin answers Stensen on 10 November of the same year:

"So, every day brings you renewed reason to increase your reputation, because your pen and your sharp eyes do not know about holidays. [...] The excretory vessel in the mucosa of the nose leading from the back of the nose to its anterior parts is new and will make your, the discoverer's, reputation immortal. What to think of it, I have not dared to say openly, until I have seen it with my own eyes.

[...] Of the animals on which you have observed this nasal vessel, from which the roomier bony part of the nose is wetted, I can speak more freely.

Indeed, it is easy to find here the reason why the nose of animals, such as oxen, sheep, dogs, etc., is constantly wet. In addition, it must be considered whether this has anything to do with the animals' excellent sense of smell. The animals are certainly distinguished by their sense of smell, which they need for the choice of healthy nutrition. [...] Nevertheless, many people can do without the ability to smell without harming their health; some do not notice smelly things because they choose their food purely by intellect. The animals, on the other hand, cannot do without the sense of smell, and their noses are therefore always overflowing with moisture. In this connection, I have observed that those people who have no sense of smell have dry noses. What is understood about this lymphatic moisture in no way applies to another strong catarrh, which impedes the ability to smell and weakens the noble senses."

In humility towards his skilful pupil, Bartholin concludes his reflections with the words: "You will make a more correct judgement of these things based on what further observation will teach you. After all, I judge here like a blind man about your colours."[81]

Bartholin, who, as mentioned, during his exploration of the thoracic duct and the lymph vessels had found different variants of their entry into the veins of the neck, concludes his letter to Stensen by answering this question:

"I would like to see more certain experiments of the mouth of the thoracic duct. I must confess not to have noticed the little channel that you have seen. I will not, however, dispute the matter, if by constant observations you can find a regular occurrence, for Nature often plays its game with individuals, and not infrequently deceives us. Continue to explore the hidden in Nature to find the truth, and be eager to increase the knowledge of our lymphatic vessels, so that finally, after so many glorious anatomical discoveries – if the world is not, as usual, ungrateful – it may be said of you in the words of the comedian in the Bacchides: "This man must be made up for with gold, of him a statue of gold must be erected."[82] In any case, if men cannot or will not sacrifice the gold, the laurels of the learned will not be denied you. Live well, my Stensen, and carry on with the work that is happily begun. Copenhagen, 10 November 1661. Utterly yours, Thomas Bartholin."

The excretory duct, which Stensen had found in the nose of a sheep, and in dogs, and which he at first assumed to be a lymph vessel, was the excretory duct of a small gland newly discovered by him in the same animals and later named after him *glandula nasalis Stenonis*.[83] It is not found in humans.

It seems that Stensen agreed with Bartholin's ideas on the importance of this gland for the sense of smell in the animals concerned, for in his little appendix on the vessels of the nose, *Appendix De Narium Vasis*,[84] which appeared shortly afterwards as part of *Observationes Anatomicæ*, Stensen ascertains that the nasal mucosa is constantly moist in both animals and humans, and he rejects Hippocrates' statement, that the sense of smell only works in a dry nose with three arguments: 1) If the sense of smell required a dry nose, there would be no reason for the nasal mucosa to be filled with fluid-filled vessels, 2) dogs, praised by everyone for their strong sense of smell, have a wet nose, and 3) fluid is needed to facilitate the transfer of the steamy scent-particles into the pores of the nerve endings.

With his next letter to Bartholin on 9 January 1662, Stensen sends the recently published *Observationes Anatomicæ*, which, in addition to the slightly expanded version of the treatise on the salivary glands, also contains the treatise on the tear apparatus, the appendix on the vessels of the nose, and a response to a dispute about the role of the liver as a blood-forming organ, which will be discussed later. The publication had been delayed, partly because of a trip that Stensen had made around Holland with Borch and some other friends in the summer of 1661. Stensen writes to Bartholin:

"What I had destined to you for a long time, famous Sir, finally, after more than six months, presents itself to be subjected to your judgement. Since the disputation, I had decided immediately to prepare a new print to transmit the integrity of what I had proposed in short at that time. [...] And if narrow circumstances at home had not retained me, I should have proceeded further in the examination of glands, wishing to survey the vessels of them all and to investigate the paths of the lymph visible in various cavities of our body. Indeed, I have no doubt that the task of all the glands, which the famous Sylvius calls conglomerate,[85] consists of moistening adjacent regions through peculiar vessels. But I am also close to believing that all aqueous humours occurring in various cavities must be derived from similar glands. Does not sweat also seem to have the same origin? Several times indeed, in the hospital, in patients emaciated and completely exhausted by the long tyranny of an illness, I saw just underneath the skin, either in the abdomen or in other places, countless small glands connected to the skin through very thin fibrils, which seemed to have been considered as vessels. A few days ago, the famous Silvius, who during this trimester teaches the practice of medicine, gave me the opportunity of examining anything of my choice. In a subject in whom he had demonstrated the parts that served his purpose, after the skin had been opened in the completely swollen thighs, I saw glands, which otherwise are usually hardly visible, so well visible, that their origin, by which they must be reckoned, could not be concealed."[86]

Hereafter, Stensen returns to his latest observations concerning the inlet of the thoracic duct in the great veins of the neck, where he has now discovered a valve that allows the passage of chyle into the blood stream, but no passage of blood or chyle the opposite way. Again, you get an impression of Stensen's elegant working method and craftsmanship:

"If I am not mistaken, I wrote last time something on the inlet of the thoracic duct, which I have been busy examining more than once. But recently I observed that its opening into the *vena cava* is provided with a fairly large and delicate valve opposing regurgitation of the blood. This same valve also closes the orifice of the lymphatic duct descending from the neck. At first, when I had introduced a hair through the said lymph vessel all the way to the *vena cava* itself, and, from there through the thoracic duct, trying to find a path into the same axillary vein, I hardly penetrated it. Thus, as I noticed the hair in the *vena cava* itself, I opened this vein and made apparent two small orifices very close to each other and covered by a common valve. The junction of lymph vessels near this division of the vein is amazing. But the thoracic duct itself deserves even greater admiration: a probe introduced into it leads into the *vena cava* and, if from there you deviate, it enables you to ascend the probe in the descending lymph vessel all the way to the neck, as, however, a different way opens from the opening of this, through the same lymph vessel, also into the sanguineous canal."

The letter again illustrates the mutual sympathy that existed between Stensen and Bartholin, and how important Bartholin's support was to Stensen:

"Certainly, this delay in addressing myself to you would have embarrassed me very much, if this kindness, so rare in others, most familiar to you, of which your latest letter testified, had not given me more courage. I rejoice and consider myself happy and someone to whom providence has given such a great gentleman as teacher. And if the things that have been started by my insignificance, please you, I will judge that I have obtained the largest fruit of my labour. [...] I thank you very much for taking pains to commend me to the promoters of science, and, among them, to the Magnificent Mister Chancellor.[87] Since I have no means of deserving your indulgence, I will recommend you to the most holy Divine Power, and I will harass Him with the most ardent prayers, that He give you with the new year new favours of complete happiness and make it so, that you will return to heaven late and remain prosperous with us for a long time. I end with this wish and ask strenuously, that you continue to find me more worthy of your favour. Do well and excuse my hasty writing. Leiden, 9 January 1662. Yours devotedly, Niels Stensen."

The contemporary correspondence between Bartholin and Borch shows how impressed they both were with Stensen's abilities as an anatomical researcher, and that Thomas Bartholin already had big plans for Stensen at this early stage. Since Bartholin himself had left the professorship of anatomy in Copenhagen upon the death of Ole Worm on 31 August 1654 to become second professor of medicine after him, the professorship of anatomy had been vacant. Bartholin's former student and skilled prosector Henrik Møinichen had returned to Copenhagen in 1660 from a 6-year study trip, during which he had spent five years at the University of Padua and obtained his medical doctorate in Angers on the way home. He had been offered the professorship but had declined, in order to later become court physician to Frederik III. Stensen was the obvious candidate for the professorship of anatomy, and Bartholin was already preparing the way for him. Shortly before Stensen's publication of *Observationes Anatomicæ* in January 1662, Ole Borch wrote from Leiden to Thomas Bartholin in Copenhagen on 24 November 1661: "Stensen is now publishing a second edition of his discoveries and adding a defence against Deusing."[88]

Bartholin himself had been appointed honorary professor by Frederik III in March 1661,[89] so that he could retire from his daily duties at the university to devote himself to the task of authoring scientific works, and he therefore no longer contributed to the university's daily teaching of anatomy. He replied to Borch 30 November 1661, airing his future plans for Stensen:

> "I cannot find sufficient words to praise our friend Stensen's hard work and success. Although I have still not yet seen any of his published works, I recently praised him to some dignitaries as an excellent man, who could one day be put in charge of our anatomical theatre – which at present is still covered in dust. Møinichen has turned down the post, so the art of medicine is on vacation at the university."[90]

On 9 January 1662, the modest and loyal Borch agrees with Bartholin's praise and plans for Stensen:

> "Stensen's short treatise has now been finished at last. I hope, it may meet with your approval and that of his other patrons and mark a happy beginning of its author's reputation. Certainly, his abilities deserve to reach maturity, to fulfil the hopes of our country; and if, after devoting some years to these mysteries, he was given public responsibility for anatomy in Copenhagen, then I confidently expect that this beneficial study would make still further advances."[91]

In the same letter, and thus the same day, that Stensen wrote to Bartholin, Borch supplements Stensen's account of the new observations concerning sweat glands. Borch writes about van Horne's dissections in his own home and in the anatomical theatre and about Sylvius' most recent dissections at the hospital in Leiden. About Sylvius' dissection in the hospital of a 50-year-old man, who had died during a febrile illness with jaundice, Borch recounts, among other things:

"As the lips were found to be swollen, they were cut open and revealed a lot of glands immediately under the skin. The next day the legs were opened in a private dissection and again showed large numbers of glands concealed beneath the skin. So perhaps it is not implausible that all over the body there are minute glands attached to the inner surface of the skin, from which sweat is expelled, so that the body's pores would simply be the tiny ducts of these glands; and that these glands are so small in a healthy body as to be almost invisible, whereas they are more conspicuous in the diseased."

Bartholin's efforts to promote Stensen in Copenhagen were not limited to high-ranking dignitaries. When in his next letter of 14 February 1662 he thanked Stensen for receiving *Observationes Anatomicæ*, he was also able to inform him that he had passed the book on to the king himself, together with his personal recommendation of Stensen:

"I have received your quite new and excellent observations, which in two ways are very dear to me: partly because they have strongly promoted anatomy to the glorification of your name, partly because you have also dedicated them to me. For the first good deed, you will harvest the recognition and applause of the learned world; the second I will try through every possible amicability to deserve. Hardly had I read them, hardly had I shown them to my colleagues, before a favourable opportunity presented itself, as you had written, to present them respectfully to our most gracious King. His Majesty mentioned with benevolence your zeal, and after perusing the little book, he interrogated me about its content and your person. I have, as befits friends, respectfully commended you and your efforts to the ruler's favour; I suggested that they might one day be of use to the Anatomical Theatre. I likewise mentioned the various functions of your new ducts: that of the salivary ducts to transport the saliva, the second duct, which stretches through the nose, to correct the frailties of the eyes by nasal cleansing, and whatever else occurred to me. [...] In any case, of the king's grace and the applause of the scholars you can be sure. Continue with the eagerness with which you have started to win everyone's recognition. Promise me that I may expect another copy at the first possible opportunity, since I do not want to be without this small but powerful book."[92]

On the same day Bartholin writes to Ole Borch:

"I have also received at last *Observationes Anatomicæ* of our friend Stensen, a work of great learning, original and erudite, and shown them to our most serene monarch. His Majesty was delighted by the careful research of his subject. I explained to him the practical applications of the new discoveries and heaped praise on the author as a man who will one day do much for the renown of our anatomical theatre."[93]

How important Bartholin's support was to Stensen's professional confidence, drive and appetite to continue his research is illustrated by his response to Bartholin on 21 May 1662:

"The more your last letter delighted me with pleasure, the more the hindrances that prevented me from answering until now made me embarrassed. I hope nevertheless that my very close friend Matthias will perform the task of intermediary, if this intercession is needed by the one whose kindness competes with his erudition, which is highly praised the world over. By commending the writings of a disciple to our colleagues and even to our most clement King, you have shown in deed as well as in word that these writings have pleased you. It is not so much your eloquence in praising me – which makes me blush, conscious of my own insignificance – that made me happy, as your well-disposed feeling, from which this most friendly judgement resulted. It is true praise to be praised by somebody who is praiseworthy, and who has deserved more titles for praise in medicine than you? Although I know that all that is feigned is very alien to our frankness, I know, however, that in this matter you accommodate yourself to the tradition of the teachers. They praise the works of their disciples, not so much because these are praiseworthy, but to encourage and incite the disciples to do something even better. Thus, also, my efforts and labours are spurred on to greater effort by your praise, which nourishes the once-awakened hope, although you, in your dexterity at dissecting, which is an object of admiration to all, easily see that I have been devoted to an insignificant matter, which was previously neglected by almost everyone.[94] But your works, too, which deserve preservation, promise a better fate to my small vessel than that of Wharton. It seemed to you insufficient to have praised me to our men, so you made my name known to the universal literary community.[95] So that your extreme benevolence would not leave anything untouched, you even acted so that I would be assured the clemency of the King. Well, every time I turn all that over in my mind, I cannot but have mixed feelings: should my soul perceive greater delight from this most happy message or greater pain by considering its own weakness? One solace only remains to me, that of hoping that heaven will return the favours which it is not in my power to return, heaven will recompense so many of your merits by utmost glory and all kind of happiness."[96]

Bartholin answers Stensen on 5 August 1662:

"For your excellent observations, which among others were also dedicated to me, I have not yet given the thanks, which the present cause, your diligence and your honest friendship deserve. May there finally be an opportunity for me to show publicly and privately how deeply I feel obliged to you for advancing anatomical science with your propitious knife and judicious pen, and for bestowing upon me your affection."[97]

In his letter to Bartholin of 21 May 1662, Stensen had, among other things, discussed the mechanism behind the increased production of tears when holding strong substances such as snuff tobacco and ginger in front of the nose, and the shadow images that bright light leaves on the retina. Bartholin, who, as mentioned, himself used snuff to soothe the stench of corpses at his dissections, comments on the first as follows:

"The observations you make about the nasal ducts do not seem to be far from the truth and show us the way to solving many problems, one of which, I reckon, is that the snuff tobacco, when inserted into the nose, helps against bad eyes. We see that when this powder is sniffed, plenty of liquid drips out, the eyes fill with tears and after a lapse of time become clearer, as I myself have experienced. The reason for this is clear from your discoveries, and as I see, you explain it in your learned letter that you sent me."

Stensen's experiments with sunlight in his eyes give Bartholin paternal concerns:

"As for the luminous images that remain in your eye after staring into the sun, I ask you to be careful, that the innate light does not fade, while your reputation shines ever brighter. You see the disadvantages of great fame, which harm our health as well as our peace of mind. So, keep your way and spare your eyes, so that they can withstand many years of study. I speak from experience, as I have fallen into the same trap. In no small measure are the eyes, brain and lungs molested by the smell from the stinking corpses which we breathe in fully by frequent animal dissections."

Bartholin ends the letter by confiding to Stensen the dramatic story of the death of his sister-in-law and her newborn child. The tragedy was a direct contributor to Bartholin's later efforts for better education and examination of midwives in Denmark:

"Now I must deal with a funeral, because I have to mourn the unexpected death of my dear sister-in-law, Karen Bartholin. After a case of colic, she had become pregnant and recently happily gave birth to a male fetus, which was entangled in the amniotic sac, while the placenta surrounded the body in the middle, in which place numerous veins appeared in the fetus. However, the midwife was so inexperienced that she did not know what it was about, until I advised her to cut open the covering membrane, through which she saw the healthy fetus fall out. He barely breathed, because he had been enveloped in the amniotic membranes for far too long. But he started breathing, screaming and sucking again, so that he survived for up to four hours. The next day the mother fell into a cold sweat, her limbs grew cold, and at noon she exhaled her blessed soul, having complained of nothing but that on her right side she felt pain, which seemed to come from the uterus. Others had the groundless suspicion that the uterus had fallen out along with the fetus. They allowed themselves to be deceived by the fact that the placenta immediately came with the fetus, which not infrequently happens without harm to the woman giving birth. This is why our women also speak of 'the child with the house', that is, a fetus in its holster. Others believed that the fatal symptoms came from an overly abundant enjoyment of strawberries, since strawberries as a food at the time for some had unfortunate consequences. It would be no wonder, as strawberries are diuretic and drive out the fetus. Moreover, since due to the unusually hot summer they contained a sharper juice, in one case they could provoke colic, in the other diarrhoea. I did not doubt, however, that it was a malignant disease, as all over the city we heard of serious feverish cases. Be that as it may, we have lost an excellent woman, who leaves a sad loss in her surviving husband, my brother Albert Bartholin, and in the rest of us. Forgive my complaint, forgive me for sending you this page full of sadness. I know that you feel our sorrow, which is why I have not been able to conceal it from you, my lovingly participating friend. Live well, my Stensen, and keep on pursuing the study of anatomy, which you have embarked upon, and to which you seem to have been born, with customary zeal, for the benefit of sick human beings. However, do it with moderation, which will not be harmful to the long life I wish you, or to your eyes."

After the publication of *Observationes anatomicæ*, Stensen had hoped for a break from strenuous and bloody dissections in favour of the more peaceful and pleasant pursuits of geometry. However, various circumstances, including his limited financial means and the attacks of Blasius and two other Dutch anatomists (see later), forced him back to the dissection table or "this bloody task", as he puts it in his next letter to Thomas Bartholin on 26 August 1662.[98]

He threw himself into comparative anatomical studies of various animals. In the letter he shares with Bartholin the details of this research. First, he investigated the variations of the mouth of the bile ducts and pancreatic ducts into the small intestine of a water raven, a duck and some dogs. Next, he began one of his upcoming major research topics, the heart. In the company of Sylvius, the previously mentioned Dutch tropical physician Willem Piso and the young German physician Christoph Gottwald (1636–1700), he examined on the raven how the heart continued to beat with the same force and speed for up to two hours after the bird's chest was opened. As the heart began to beat more weakly and less frequently, Sylvius drew the attention of the others to the fact that the atria and the parts of the *vena cava* close to the heart continued to contract vigorously, and that was still the case when, three hours after the opening, the heart stopped. When the atria also stopped, the *vena cava* continued its contractions up to five hours after the opening of the chest, and "after the right ventricle of the heart had been resected, and all the blood, which was in the *vena cava* between the liver and the heart, had poured out, a swelling rising upwards of the transverse fibres, one following the other, was nevertheless seen in the tunics of the *vena cava*, which were about to collapse."

Subsequently, Stensen dissected four rabbits and made various ligature experiments on the large veins. He found, that here, too, "[the vena] cava's movement was the last of all." and continued its contractions for several hours, for as long as it was supplied with blood, but "as soon as the blood was taken away, they all [the heart, atria and *vena cava*] ceased beating." Stensen asks for Bartholin's "valuable" assessment of the "causes of these movements and their nature," which Bartholin gives in his next letter to Stensen on 7 September 1662.[99] He refers to the first of Johannes Walæus' two letters on the circulation of the blood in support of Harvey, printed in Bartholin's first revised edition of his father's anatomy book from 1641.[100] Bartholin himself had made similar studies and observations in vivisections of dogs in Leiden in the late 1630s with Walæus, and it was probably these studies that Sylvius knew about and had been inspired by. Bartholin writes:

"From your knife, which you often sharpen to the downfall of animals, but to the growth of science, a great hope of certain experience shines towards us. Your raven has brought a lot to light that was worth knowing. [...] Walæus had already observed in *Epistolæ 1* that the origin of the movement comes from the *vena cava*, even when the heart is cut off, and therefore around the heart there are muscle-like fibres which are not otherwise seen in the *vena cava*. Nor is it surprising that it is here that one should seek the beginning and the end of the

Figure 148. Marcello Malpighi (1628–1694) shows illustrations of the bronchial tree of the lungs. Attributed to Giovanni Bernardo Carbone (1614–1683). Oil on canvas, 85 × 72 cm. Date unknown.

The Bolognese Malpighi revolutionized anatomical science through the systematic use of microscopes he made himself. With this method, he found the missing link of William Harvey's circulatory theory, the microscopic capillaries, and he was the first to describe the microscopic air sacs of the lungs, the alveoli, where the exchange of gases (oxygen and carbon dioxide) between the air and the blood takes place. Thomas Bartholin quickly realized the epoch-making perspectives of Malpighi's method and discoveries. He helped raise awareness of them by republishing Malpighi's relatively unnoticed letters to Alfonso Borelli in his own treatise on the lungs, and he encouraged Niels Stensen and Ole Borch to acquire a microscope and verify Malpighi's findings. This was the starting point for Niels Stensen's use of the microscope in his anatomical and geological research and thus a significant part of the foundation for his success as a natural scientist.

movement, since all the blood in the body enters the heart this way, as long as the living creature is healthy, or tries to enter, when it dies. And so that the liver should not be inflated, Walæus has indeed observed the same movement of the *vena cava* from the clavicle on live dogs, as you write that you have observed on your raven."

It is characteristic that the well-read and eager correspondent Bartholin's letters to Stensen are filled with references to the research and literature of ancient and contemporary physicians. This is also the case for this letter, which Bartholin concludes by referring to the Italian physician, anatomist and biologist Marcello Malpighi's (Figure 148) previously mentioned epoch-making discovery of the pulmonary capillaries of a living frog using a microscope:

"I had almost forgotten Malpighi's observations, about which I have written to the erudite van Horne and to Borch." And then, as usual, follows Bartholin's invitation to Stensen to take a closer look at Malpighi's results and begin researching on the subject: "Research eagerly what Nature in this area hides from your eyes, and determine what your opinion is on it."

Malpighi's discovery, Harvey's missing link – the pulmonary capillaries

With his discovery of the microscopic pulmonary capillaries, Malpighi had found nothing less than the missing link in Harvey's theory of the blood circulation, the link between the smallest arteries, the arterioles, and the smallest veins, the venules. Through the walls of the capillaries, the exchange of fluid, salts, gases, etc. takes place between the blood and the surrounding tissue. As the absolute pioneer of the use of the microscope in medical research, Malpighi produced in his laboratory in Bologna some of the best microscopes of the time for his own use, and he was a master in the manufacture of the best preparations for microscopy. In this case, he took advantage of the great translucency of the frog's lungs, which enabled him to see the flow of the microscopic red blood cells through the pulmonary capillaries. In Bologna in 1661 he published his discovery of the lung capillaries and the adjacent microscopic air sacs, the alveoli, in a small pamphlet in the form of two letters to his friend and mentor in Pisa, Galileo's pupil, the mathematician, physicist and physiologist Giovanni Alfonso Borelli (1608–1679) entitled *De Pulmonibus Observationes Anatomicæ* (Anatomical Observations of the Lungs). It was Malpighi's first ever publication.

The discovery remained relatively unnoticed for the first few years, but after Thomas Bartholin, aware of its importance, republished Malpighi's two letters in 1663 in Copenhagen as an appendix to his own treatise on the nature and movement of the lungs, *De Pulmonum Substantia & Motu Diatribe, Accedunt Cl. V. Marcelli Malpighii de Pulmonibus Observationes Anatomicæ* (Treatise on the substance and movement of the lungs. In addition, the famous man Marcello Malpighi's anatomical observations of the lungs) (Figure 149) it became more widely known.[101]

In the preface, Bartholin writes that he received Malpighi's writing with the two letters sent from Florence by his good friend, the Florentine nobleman, philologist and pupil of Galileo, Carlo Dati (1619–1676), who would later play a role in Stensen's life in Italy. Bartholin decided to republish the paper for use by medical students and add new remarks to show his gratitude for the work and lucky discoveries of others – just like he himself had been met with gratitude when he discovered the lymph vessels. He emphasizes that the world owes a great debt to Italy's gifted scientists and their humanism, and that he will never allow the fame of this important people to be silenced in Denmark. It was the University of Bologna that made these discoveries in the lungs through Malpighi, and it was the splendid Pisa that recognized them through Borelli, while elegant Florence, through Dati, ensured they came to the attention of many people including Bartholin.

Malpighi's two letters of 16 and 8 pages, respectively, supported by four pages of figures and figure texts, complete the book. It is in the second of these two letters that Malpighi, using microscopy of the lungs of a live frog and a live turtle, describes a new annular network of microscopic vessels, "which is so abundantly

Figure 149. The illustrations from Marcello Malpighi's *De Pulmonibus Observationes Anatomicæ*, Bologna 1661, collected on a single page in Thomas Bartholin's *Anatome Quartum Renovata*, Leiden 1673, p. 417.

Top: Figure I shows the network pattern of blood vessels and air sacs in a piece of dried lung. Figure II shows the outer surface of small air sacs and a section of their interior which cannot be seen with the naked eye. Figure III shows differently arranged lobes over the trachea and the pulmonary vessels.

Bottom: Figure I shows the lungs of a frog, where A is the laryngeal cartilage, B is the vocal cords, C is the location of the heart, D is the outer surface of the lungs, E is the network structure of the lungs, and H is the small pulmonary veins. Figure II shows in the frog a cross-section of a lung segment with small air sacs and blood vessels. The Royal Library in Copenhagen.

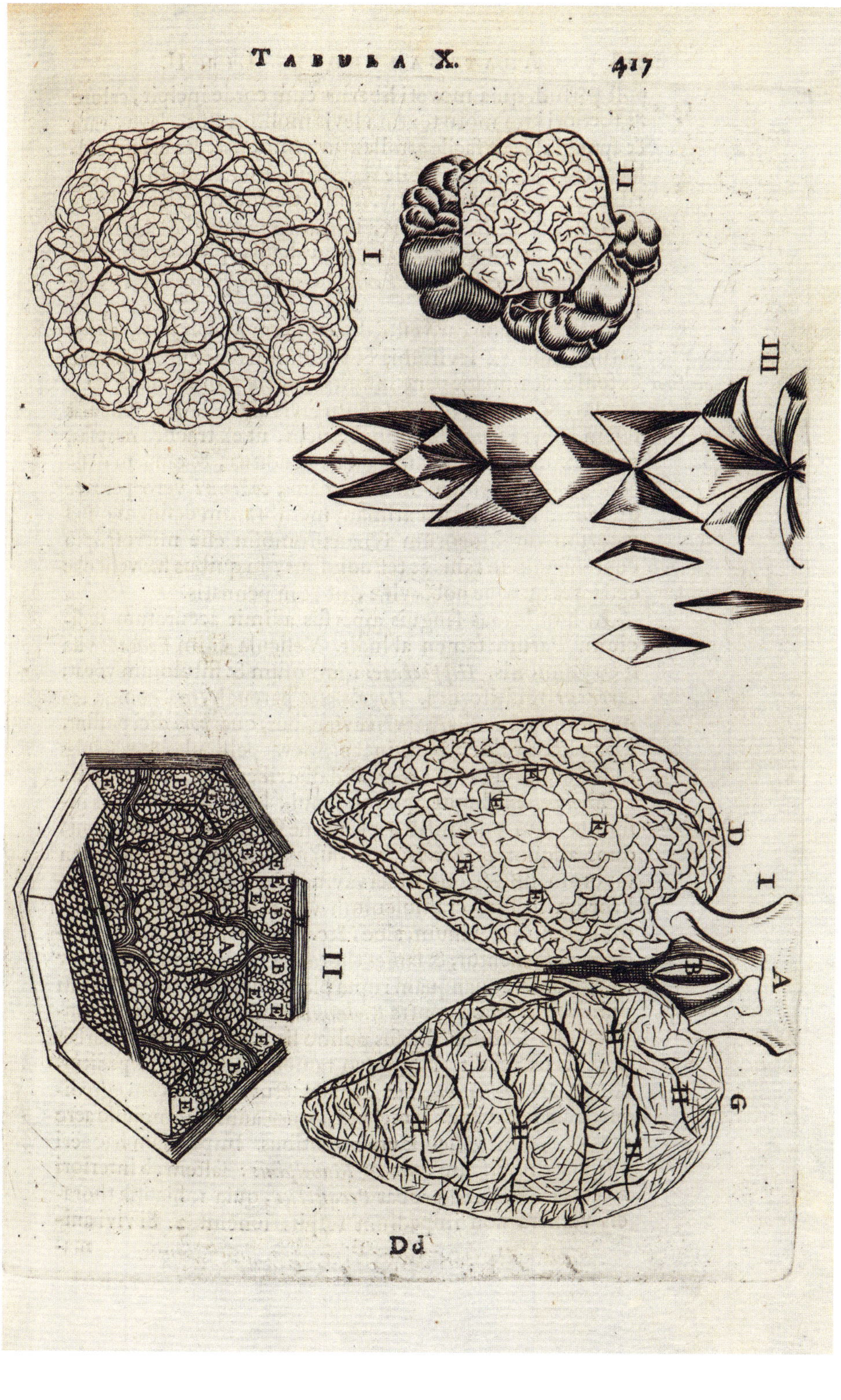
TABULA X.
417
Dd

branched, where they arise from the veins and arteries, that they no longer have the character of a certain kind of vessel, but seem to constitute a network formed by the continuations of both kinds of vessels, which not only occupies the entire outer wall of the lungs, but even extends to the internal partitions; [...] a vascular system interwoven in the vesicles and cavities of the lung, designed to direct blood from the arteries to the veins. And although it sometimes appears in the lungs of the higher animals, as if the vessels of the centre of the network's rings end wide open, it is nevertheless likely that, as is the case in the lungs of frogs and turtles, they continue in these smallest vessels, despite the fact that these, due to their incredible smallness, cannot be directly observed. The question of mutual union and confluence of arteries and veins can, due to these circumstances, with the highest probability be considered resolved."[102]

Bartholin's introductory dissertation of 103 pages is divided into five parts: 1) The substance of the lungs, according to Hippocrates, is not fleshy,[103] 2) the substance of the lungs consists of membranous blisters, 3) the movement of the lungs contrary to the Flemish nobleman, physician and mythical philosopher Jan Baptista van Helmont's view, 4) the movement of the lungs is not due to the movement of the heart, but the thorax's or diaphragm's movement, 5) contrary to Descartes' view, the air does not enter the lungs through the skin. As previously mentioned, Bartholin had discussed the latter point with the Cartesian Abraham du Prat during his stay in Paris in the winter of 1645/1646.

As mentioned in the letter to Stensen, Bartholin had written to van Horne on 8 September 1662 about Malpighi's discovery. The letter is a foretaste of the contents of Bartholin's forthcoming book on the lungs and an illustrative example of Bartholin's and his contemporaries' way of conducting anatomical research. Based on a synthesis of his own observations from dissections etc. (empiricism), other people's observations and thoughts in the literature from antiquity to the present (literature review) and comparison of human anatomy with animal anatomy (comparative anatomy) Bartholin draws his own conclusions by reasoning. Further, the letter shows that Bartholin, one year after his honorary pension was awarded, had in no way ceased dissecting and active research:

"But by way of bringing you and your judgement back from this part of the body to the 'instruments of respiration,' I will explain to you, as you are most keen on these matters, what new information Italy has given us recently on the subject of the lungs. At Bologna, Marcello Malpighi, the professor of medicine there, has discovered, by careful dissection of animals, that the stuff of which the lungs are composed is by no means fleshy, but some sort of aggregate formed of membranes which, being spread out as they are and twisted, form endless circular and curving vesicles. These are positioned and connected as follows: From the trachea to the vesicles themselves there is soon open access from one to another, and they finally come to an end with the membrane of which they are composed, which has the form of a net, loosely woven of various fibres, which he discovered to be the mouth-openings of the pulmonary vessels. This is a summary of a new observation contained in two letters to Jo. Alph. Borellius,[104] a mathematician from Pisa. You should look further into the truth of the matter!

I consulted the entrails of animals – as if for purposes of prophecy[105] – particularly the entrails of domestic animals, which were readily to hand; not because I had any doubts about the reliability of the learned man, but so that my eyes too could believe it with greater certainty. In the case of the lungs of cattle and sheep we saw the interweaving of vesicles on the external surface, especially in the part where the lobe is less thick, so that it is translucent when set before a light. Furthermore, when a pair of bellows is applied and inserted into the trachea, the vesicles are quite clearly seen to rise in the outermost membrane. In fact, in the case of frogs, which Malpighi too has heard croaking, I clearly saw everything in a delightful spectacle. When dissected, their lungs appear inflated on both sides, shining like crystal and satiating the eyes of beholders with their elegant interwoven texture. For quite a long time they stay inflated with their own air, but gradually, as they contract, they become darker, unless, on being thrown back again into water or subjected to aeration, they display the diverse shapes of their vesicles. All lungs present a pouch with a capacity for receiving moisture and air and retaining them for a long time. Harvey, too, in *Exerc. 3 de Gen. Anim.*,[106] has noticed these vesicles in frogs and toads, and I have observed them clearly in birds, especially in geese and ducks, whose lungs are, even according to Aristotle 1.3 *Part. Animal.* c. 6, rather spongy and like membrane. In their case, however, there is a single membranous pouch or sac through which they breathe, extending from the bronchial tubes to the abdomen, although the lung adheres to the ribs. Harvey (1.6) observed this in the ostrich, the turkey-cock and the common cock. Gassendus in the *Vit. Peireskii*[107] observed in a large tortoise a rugged windpipe producing some branches to the kidneys and bladder, which, if a particularly dexterous hand were to demonstrate their equivalents in man, would open up for us a convenient way of purging the pus of consumptives through the kidneys.

Certainly, Corn. ab Hogelande[108] demonstrates along with Malpighi more or less the same construction in human lungs, for he affirms that the parenchyma consists of very tiny vesicles, and that there are present, as it were, small cavities, which might be called 'pores', distinct from or bordered by the aforesaid little vesicles. Hippocrates (1 *De Carnibus*) gives fame to a probable observation: He seems to have divined through a haze something of the sort."[109]

From the trachea via the bronchi and bronchioles all the way to the alveoli the bronchial tree divides into finer and finer branches, in humans up to 28 times. Similarly, the lungs are divided into lobes, segments and lobules. The branches of the pulmonary artery, which carry the dark red, unoxygenated venous blood from the right half of the heart, follows the bronchial tree all the way to the alveoli, where they continue in a dense, round meshed capillary network around the alveoli. This proximity between alveoli and capillaries enables exchange of oxygen and carbon dioxide by diffusion between alveoli and capillaries. The capillaries continue in venules and then larger veins, which follow the connective tissue between segments and lobes and carries the pink oxygenated blood through the two large pulmonary veins (*venae pulmonales*) back to the left atrium of the heart and from there via the left ventricle to the aorta, from where it is distributed throughout the body. Lymph vessels and nerves follow the branches of the bronchial tree like the arteries. The diameter of the human trachea is 1.5–2 cm. The alveoli, of which there are approximately 300 million in a person's lungs, have an average diameter of 0.2 mm, which is exactly the smallest size that the human eye can distinguish. Capillaries have an average interior diameter of just 8 micrometres, i.e. 0.008 mm, and are therefore not visible to the human eye.

Bartholin apparently did not use a microscope in his examinations of the lungs of cattle, sheep, frogs and the birds mentioned, so when he talks about the interweaving of vesicles on the surface of the lungs and their swelling, when bellows are used, he cannot possibly mean the individual microscopic alveoli, which Malpighi describes by microscopy of the lungs of the living frog, but the end surfaces of the lobes, which give the surface of the lungs a humped appearance. In humans, the lobules have a volume of 1–2 cubic centimetres and represent millions of alveoli. Curiously, Malpighi himself regarded the structure of the lungs with microscopic air sacs surrounded by a network of microscopic blood vessels as a well-constructed mechanism for ensuring the mixing of chyle with blood and not to ensure the mixing of air with blood.[110] Bartholin, who had countless times seen with his own eyes that the mixing of blood and chyle takes place in the veins of the neck and ultimately in the right half of the heart, expresses in the letter to van Horne his amazement at Malpighi's point of view:

"But the function of the lungs designated by the same Malpighi seems to me doubtful. He thinks they contribute to the mixing of the substance that constitutes blood. Will not blood, chyle and lymph more rightly be mixed in the heart, where there is a larger cavity, in which chyle and lymph meet in confluence with blood?"

But with characteristic Bartholinian openness, he leaves a door ajar for Malpighi's thoughts:

"Perhaps, however, this mixing will become more complete because of the innumerable tiny vesicles, so that the lungs are the heart's sieving device, or a place where humours are mixed, like the liver in the abdomen. I will not deny that increased liquidity is brought to the blood from the "white humour", as this same man writes, to the extent that it has lymph mixed in."

Malpighi mainly built his assumption based on the light colour of the air-filled lungs and the lighter colour of the blood that had been through the lungs. This indicated, according to Malpighi, that in the lungs there was a mixing of the blood with the milky-white chyle. He in no way denied the fact that the thoracic duct empties its chyle into the large veins of the neck, but he believed that the final mixing of chyle and blood took place in the lungs, and that the blood was therefore finished there. On the last page of his second letter to Borelli, he explains it thus, that back to the lungs "turns the troubled blood, which during its course through the body has been deprived of its nourishing constituents, and which from the subclavian vein has received a liquid that through the continued activity of Nature must be perfected. Here, therefore, the blood must be transformed and prepared, so that it preserves the nature of the small parts of the flesh, bones and nerves. While it flows through the countless pulmonary vessels, it is distributed, so to speak, into different extremely thin jets, and in this way the small parts of the blood acquire a new character, a new state and a new movement. Thanks to this, meat, bones and life spirits are created."[111] Recognizing the enormous importance of Malpighi's discovery for medical science, Bartholin concludes the letter to van Horne with the following short conclusion: "Whatever the case may be, many phenomena may be deduced from the observation of Malpighi."

Stensen and Borch did not wait for Bartholin's book with Malpighi's letters before following Bartholin's call to try to reproduce Malpighi's discoveries. However, a prerequisite for this was access to a microscope of a satisfactory quality, and such were not yet so easy to get hold of. Malpighi's was the first major discovery in medicine that had been obtained using a microscope, and it heralded a new era in medical research. With the microscope used by Malpighi, he would not have been able to see alveoli or capillaries in mammalian or human lungs, but the great translucency of the lung tissue on the examined live frog made it possible to see the capillaries at a relatively low magnification. Malpighi's success in the new discipline of microscopic anatomy (histology), of which he is considered to be the founder, was not only due to the magnifying power and accuracy of his microscopes and his technical ingenuity in operating them. His various methods of preparing the best possible preparations also played an important role. He used drying, boiling, inflating (of the bronchial tree and the blood vessels), flow (vascular perfusion), air discharge, etching, etc.

Neither Stensen nor Borch had personal experience with microscopes in anatomical research. Had it been just ten years later, they would have been in just the right place in the world, because then another of the great pioneers of the microscope, the Dutch craftsman, microscope maker and microbiologist, Anton van Leeuwenhoek (1632–1723) (Figure 150), was manufacturing the best microscopes of the time (Figure 151) in Delft, just 20 kilometres from Leiden.[112] But now they had to wait about a month for the loan of a microscope from one of Stensen's unnamed friends, who may well have been his new anatomical dissection and research partner, the Dutchman Jan Swammerdam (1637–1680), to whom Stensen later in the letter refers as "the in anatomical exercises very

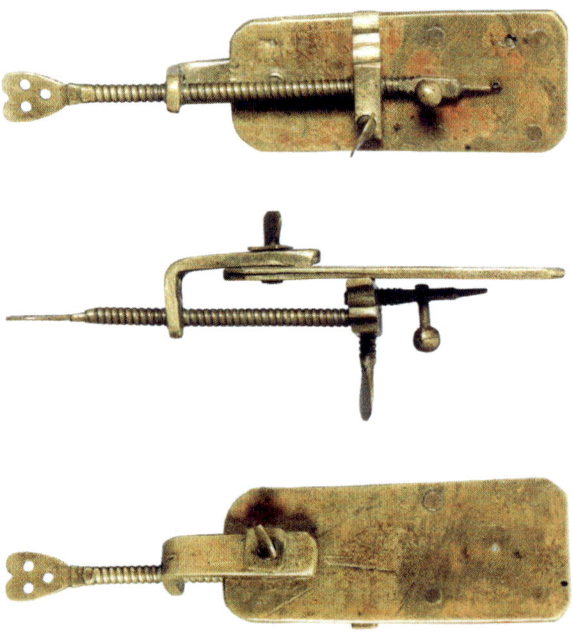

Figure 151. Microscope manufactured by Anton van Leeuwenhoek in brass with a hand-cut lens (the tiny hole at the end of the screw tip to the left).

The microscope shown has a linear magnification power of 266 and a resolving power of 1.34 micrometres. A red blood cell has an average diameter of 7 micrometres and could easily be seen with van Leeuwenhoek's microscopes. The resolving power of the human eye is around 0.2 millimetres or 200 micrometres. Rijksmuseum, Amsterdam.

Figure 150. Anton van Leeuwenhoek (1632–1723) from Delft painted by Jan Verkolje (1650–1693) probably in 1680. Oil on canvas, 56 × 47.5 cm.

With the background of a craftsman's education, van Leeuwenhoek produced some of the best microscopes of his time. He used them himself in his research and became one of the pioneers of microscopy and microbiology. He is portrayed here at his desk with a compass in his hand. He diligently shared his results with the Royal Society in London, which in 1680 honoured him with a membership. The portrait was probably painted on this occasion, because on the large, folded document at the front left of the picture you can make out the words "Caroli Secundi," i.e. the English King Charles II (1630–1685). Rijksmuseum, Amsterdam.

capable and diligent young Mr. Swammerdam." Swammerdam later became a famous anatomist and zoologist and is also counted among the pioneers of the microscope. He was already at the forefront of this technique, and in 1658 he was the first to see red blood cells under a microscope. However, there is also the possibility that Stensen's friend with the awaited microscope may have been a completely different person much more well known in the world of philosophy than in the world of natural science. A few kilometres west of Leiden towards the coast, in the village of Rijnsburg, lived the later-famous rational philosopher Baruch Spinoza (1632–1677) (Figure 152). He was born in Amsterdam in a Jewish family. In 1656, at the age of 23, he became expelled and banned as a heretic by the Sephardic congregation in Amsterdam because of critical statements about Moses and the Mosaic law and accusations on the denial of God and the immortality of the soul.[113] Thereafter, he Latinized his name to Benedict[114] de Spinoza and trained as an optician and glass and lens grinder in Amsterdam. By 1660 he had settled in Rijnsburg to earn a living from this profession, and here he remained until 1663. He quickly became known for producing excellent lenses and microscopes, and Ole Borch mentions him in his diary on 10 September 1661: "There is close here at Rijnsburg a certain Baruch Spinoza, converted from Jew to Christian and soon almost atheist. [...] This man lives a pure and immaculate life and engages himself in manufacturing binoculars and microscopes."[115]

Whether the microscope was provided by Swammerdam or Spinoza, it reached Stensen and Borch in February. On 1 March 1663, Borch writes to Bartholin:

"Malpighi's book is also in print here. I must admit that although we have continued to examine the small vesicles in the lungs with more than ordinary care, using inflation and our eyes, aided by the most accurate microscope, we have so far been unable to present them clearly. Since Stensen writes more in detail about this, I will not elaborate further on the topic. Neither shall I review what will soon be published about the muscles and cardiac action. Stensen is again working on an excellent dissertation."[116]

Four days later, on 5 March 1663, as promised by Borch, Stensen follows up with a letter to Bartholin:

"A long-desired microscope offered by a friend finally provided me with a further opportunity to reflect upon the observations of Malpighi concerning the

Figure 152. Baruch Spinoza (1632–1677) painted by Barend Graat (1628–1709) in 1666 in an antique setting. Oil on canvas.

Spinoza is best known as a courageous and far-sighted philosopher with a tolerant view of faith and religion, which brought him into conflict with the Sephardic community in Amsterdam and the strong believer Niels Stensen. But he was also an excellent optician and microscope maker. Ole Borch mentions this in his diary in the autumn of 1661, and Niels Stensen's friend, the Dutch physician and anatomist Theodor Kerckring (1638–1693), writes in 1670: "My excellent microscope, made by the noble philosopher and mathematician Benedict Spinoza, with which one can see the lymph vessels spreading in different threads, before they enter their conglobate glands and continue, until they again depart from there and unite into a single vessel." It is likely that Stensen's first microscope, which he mentions to Thomas Bartholin on 5 March 1663, came from Spinoza's workshop in Rijnsburg, a few kilometres from Leiden.
Photo: Wikimedia Commons.

parenchyma of the lungs. Having read your letters addressed to Borch and me, I did not doubt that various vesicles could be shown in the lungs, and not only in one way. It is beyond any doubt that several cavities are to be seen in dried-up lungs, if all blood is first washed off, as Malpighi does, before the lobes are dissected. Many small branches of the vein, as well as the two arteries, are indeed dissected at the same time, and it is clear to everyone that they are all embedded in obvious cavities. If, before they are dried, the bronchi and the lungs are distended with air, how many small cavities is it possible to see in the dissected specimens, and of what size? The subdivisions of the trachea, which are scattered in countless numbers throughout the substance of the lungs, will certainly, if you dissect them during a violent distension with air, display an appearance of countless vesicles. But it does not seem necessary for that reason to state that the parenchyma of the lungs is made of pure vesicles. From this it does not even result that the extremities of the trachea solely end in vesicles, not to mention the extremities of the vein, artery and lymph vessels. How much did the microscope reveal to me in the dry lung of a dog? Besides the round orifices of divided vessels and myriads of very minute filaments running without any law, absolutely nothing was seen, no round, small cavities covered on all sides by a membrane, nothing at all that could be called vesicles. Neither did the intact surface of a dried lung display any sign of vesicles under the microscope. There were protuberances, but these were so large and unequal, not bulging at their middle, nor circular in their perimeter, that they do not at all deserve to be called vesicles. But I am far from withdrawing my trust in the experiment of the most intelligent gentleman because I did not succeed in one or two attempts, perhaps because I did not use the same procedure."[117]

In the latter assumption Stensen was quite right, for even if one assumes that his microscope was as good as Malpighi's or even better, it would not have been of a sufficient quality to see the alveoli and capillaries in a dried lung from a dead dog. An almost transparent lung on a live frog or turtle was a completely different matter, and this essential point of Malpighi's approach Bartholin had not emphasized in his letters to Borch and Stensen, but only in the letter to van Horne. All the more unfortunate for Stensen and Borch, for Malpighi in his second letter to Borelli had in fact, explicitly to be helpful to other researchers, given a recipe for his approach:

"In order to be more easily convinced of the correctness of what I have mentioned, one constricts on a frog the protruding and swollen lung after the opening of the body cavity. This is done with a thread in the place where the lung unites with the heart, and while the blood is still flowing through the lung from all sides. For then in its dried state the blood-filled vessels are still preserved, which one can very easily observe using a microscope with a single lens and horizontally incoming light, or as follows: one puts the lung on a glass disc, illuminates it from below with a lamp through a tube and observes it through a microscope with two lenses; then the annular vessels appear. With the same instrument and illumination one can also observe the movement of blood in these vessels, and at different brightnesses detect several other things that cannot be described with the pen."[118]

Stensen's lack of knowledge of this guide is supported by the following sentence in the letter to Bartholin, who must have regretted his inadequate information: "But this will suffice on Malpighi, until it is possible for me to examine his letters myself."

It was soon to be. In his next and last surviving letter to Thomas Bartholin dated Leiden 30 April 1663,[119] Stensen thanks him for the receipt of Bartholin's book including Malpighi's two letters, but in his letter Stensen does not write anything about his further investigations of the lungs. Ole Borch, however, does so in his last letter to Thomas Bartholin from Leiden on 12 May 1663:

"Since Malpighi's book appeared, we have been hard at work here examining the inner organs of frogs. We have often found in the place of the lungs two vesicles marked with a pattern of lines or gratings on the inner surface, but not the huge number of tiny vesicles he delineates. It seemed to me that these two pulmonary vesicles had some similarities with bees' honeycombs, as their projecting interstices gaped with open hollows towards the inside."[120]

About the frogs' lungs Borch does not report any more, and nothing at all about the vessels of the lungs, but he adds the following about the vitality of this animal: "I do not remember ever having seen a creature that was livelier during its torments than this one. When the heart was cut out together with the lungs and placed separately on the table, it continued with its systole and diastole quite visible for more than an hour. At the same time the rest of the body, put in a different place, also moved at intervals, especially round the base of the back, where it looked as if a new heart was producing a new systole and diastole."

Stensen continues his letter of 5 March 1663 to Bartholin with reporting on other observations he has made concerning the lungs. He has cut the fetuses out of the womb of a pregnant rabbit and a pregnant dog by premature caesarean section and observed that the premature baby rabbits breathed and then showed other signs of life. On one of the baby rabbits, who had been breathing, he subsequently found that the lungs were whitish and spongy, while on a stillborn baby rabbit they were quite different – dense and bloody – and he wondered that such a big change could happen in such a short time.

In this book, Bartholin also writes himself into the history of forensic medicine. He writes that anatomists agree that the lungs of fetuses in the uterus are red and blood-filled and sink to the bottom in water, while the lungs of living newborns and adults are pale and air-filled and float on top of water. He points out that William Harvey has suggested using this difference to prove whether a woman has given birth to a living or dead child.[121] In 1681, Johannes Schreyer (1631-1694), the city physician in Zeitz, used this very proof to acquit 16-year-old Anna Voigt of killing her newborn child. When Schreyer placed the dead baby's lungs in water, they sank to the bottom, proving that it was a stillbirth and not infanticide. In 1690, Schreyer published an account of the case and method, referring to Thomas Bartholin's *De Pulmonum Substantia & Motu Diatribe*, among others.[122]

Research on muscles and heart

One of Niels Stensen's most significant efforts in anatomical research concerns muscles, including the heart. As mentioned by Ole Borch in the letter to Thomas Bartholin on 1 March 1663, Stensen was by then working on a dissertation on this subject, and he himself unveils these plans to Bartholin in his letter four days later:

> "Various experiments and other reasons, which at intervals presented themselves to me while I was doing or looking for something else, provided me with the opportunity to examine more thoroughly the movement of the heart, as I then mentioned.[123] To that end I investigated more carefully into the fibres of the heart. On these, with the help of divine grace, I will soon publish a small treatise."

The human body has three types of muscles, namely skeletal, cardiac and smooth muscle. Based on the appearance of the musculature in the microscope, these types are again divided into striated and non-striated. Skeletal and cardiac muscle belong to the former, and smooth muscle to the latter. Skeletal muscles constitute, as the name suggests, all the musculature attached to our bones, which is responsible for those movements of the body that are subject to the dominion of the will. Heart musculature occupies a special position, in that it is in constant activity throughout life, regardless of the influence of the will. Smooth musculature is found everywhere in the body in those structures and organs that can move beyond the dominion of the will, for example the vessel walls, the bronchi, the gastrointestinal tract, the urinary bladder, the uterus and the sperm duct. The average thickness of a heart muscle fibre is 15 micrometres and thus far less than the resolution capacity of the human eye of around 200 micrometres, while the thickness of skeletal muscle fibres varies between 10 and 100 micrometres.[124] A few very thick skeletal muscle fibres can be seen with the naked eye, but when Stensen, Bartholin and Borch in the following mention the course of muscle fibres that they have seen with the naked eye, they mean bundles of fibres, which are also called fascicles. To ensure the complex movement pattern of the heart with alternating contraction (emptying, systole) and relaxation (filling, diastole), as well as coordination of the movements of the ventricles and atria, the course of the fibres of the heart is complex. In addition to running in several layers, some fibres run straight, others oblique or circular, and still others S-shaped.

In his letter to Thomas Bartholin on 1 March 1663, Ole Borch reports in detail about the findings from numerous dissections on humans and animals which he and Stensen in the past winter have either witnessed or participated in. Among other things, he mentions the findings from Sylvius' dissections of three women on Leiden's hospital. Borch's rather detailed descriptions of the course of muscle fibres in various bodies testify that there was a focus on this topic in education and research. For example, Borch describes how he made up his mind to examine on his own the uterus and bladder more closely on one of the three dissected women, who was old. About the bladder neck and the bladder sphincter muscle, which is a smooth muscle, he writes:

> "The bladder neck was as long as the width of two fingers. The sphincter was small, but it could not be distinguished by circular fibres, as is often said, but by fibres, which are extended along its length – except that they ran a bit at oblique angles, though it could hardly be discerned. The sphincter was on the outside surrounded by a thick sinewy layer, just like on the penis. Here the fibres were circular, except that they ran a bit like in a spiral. There is almost nowhere in our body where fibres describe a perfect circle, which has many notable reasons. This was seen in a woman. Further studies of the course of fibres in the sphincter of men are required."

Borch then describes the course of the fibres in the bladder neck of a man with bladder stones, who died the previous summer, and then on a pig. While Borch, according to his letter to Bartholin on 12 May 1663, continued his research in the smooth muscle of the bladder wall, Stensen threw himself into comparative studies of the striated musculature of the heart and skeletal muscles. In his letter to Bartholin on 30 April 1663, he gives the first detailed insight into his research in this area, this time accompanied by explanatory drawings:

"Moreover, I am busy with thorough examination of the heart and muscles, and I hope, if fate will leave me the necessary time, soon to complete a description of the structure of both, illustrated with figures. Whether that which I have observed in the muscles was already noticed by others, your boundless reading will teach me. I will not mention the vessels, which did not yet display much of importance. However, as far as their fibres are concerned, I cannot admire their delicate structure enough. Anyone who once studies a single fibre will see that in the middle part it is fleshy, and at both extremities it is tendinous, a fact that is well known. But the whole composition, how they are connected to each other, I have more rarely seen. Thus, the fleshy portion does not extend in a straight line from one extremity of the muscle to the other but traverses the muscle between the broad sinewy strings in such a way that the single fibres of the fleshy parts run parallel with each other. When making a section from one extremity of the muscle to the other along the direction of the fibres, this structure is clearly seen [Figure 153]:

A B. The tendon extending over one surface of the muscle.
C D. The other tendon extending over the opposite surface.
E E. The fleshy part of the fibres between the two tendons.

Displayed in a simple way, the whole thing has the following structure [Figure 153]:

G H. One tendon extension.
I K. The opposite tendon extension.
L L. The fleshy belly.

There are also compound muscles, and not only of one kind. The ones that seem most elegant to me are those in which the fibres are disposed in the following manner [Figure 153]:

P. One of the tendons, which is split into two expansions, of which the upper one extends towards Q, and the other, the lower one, extends towards R.
M. The other tendon, which distributes its fibres through the middle of the belly.

From N to O and farther between the middle and the two opposite external expansions, parallel to the fleshy parts of the fibres. The membrane peculiar to every muscle must not be overlooked. The transverse orientation of its fibres and their distribution between the fleshy fibres seem to contribute not a little in the expla-

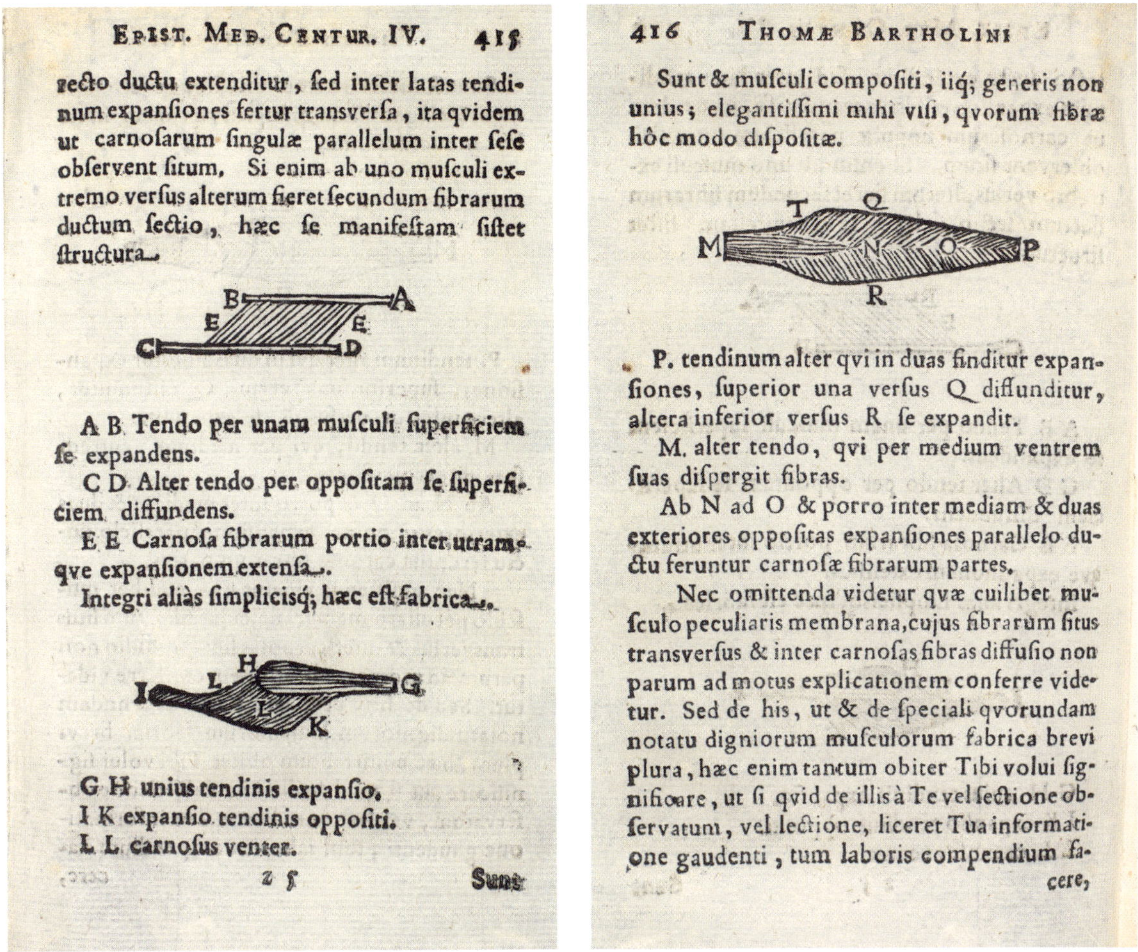

EPIST. MED. CENTUR. IV. 415

recto ductu extenditur, ſed inter latas tendinum expanſiones fertur transverſa, ita qvidem ut carnoſarum ſingulæ parallelum inter ſeſe obſervent ſitum. Si enim ab uno muſculi extremo verſus alterum fieret ſecundum fibrarum ductum ſectio, hæc ſe manifeſtam ſiſtet ſtructura.

A B. Tendo per unam muſculi ſuperficiem ſe expandens.

C D. Alter tendo per oppoſitam ſe ſuperficiem diffundens.

E E. Carnoſa fibrarum portio inter utramqve expanſionem extenſa.

Integri aliàs ſimplicisq; hæc eſt fabrica.

G H. unius tendinis expanſio.

I K. expanſio tendinis oppoſiti.

L L. carnoſus venter.

Z ſ Sunt

416 THOMÆ BARTHOLINI

Sunt & muſculi compoſiti, iiq; generis non unius; elegantiſſimi mihi viſi, qvorum fibræ hôc modo diſpoſitæ.

P. tendinum alter qvi in duas finditur expanſiones, ſuperior una verſus Q diffunditur, altera inferior verſus R ſe expandit.

M. alter tendo, qvi per medium ventrem ſuas diſpergit fibras.

Ab N ad O & porro inter mediam & duas exteriores oppoſitas expanſiones parallelo ductu feruntur carnoſæ fibrarum partes.

Nec omittenda videtur qvæ cuilibet muſculo peculiaris membrana, cujus fibrarum ſitus transverſus & inter carnoſas fibras diffuſio non parum ad motus explicationem conferre videtur. Sed de his, ut & de ſpeciali qvorundam notatu digniorum muſculorum fabrica brevi plura, hæc enim tantum obiter Tibi volui ſignificare, ut ſi qvid de illis à Te vel ſectione obſervatum, vel lectione, liceret Tua informatione gaudenti, tum laboris compendium facere,

cere,

Figure 153. Niels Stensen's three geometric muscle drawings to Thomas Bartholin in a letter dated 30 April 1663. From Thomas Bartholin's *Epistolarum Medicinalium*, Cent. 4, Copenhagen 1667, pp. 415–416. Stensen's figure explanation appears in the body text. The Royal Library in Copenhagen.

nation of the movement. But about this and about the particular structure of some muscles more noteworthy, more will be said later on. But I wished to point out these matters to you incidentally, so that if you observe something about them either by dissecting or by reading, it would be possible to the one delighted by your information on one hand to save work and costs and on the other hand to orientate the explanation of the description according to your precepts."

Stensen does not write it directly, but it seems clear that he used a microscope in his studies of skeletal and cardiac muscle fibres. In view of the fact that, immediately prior to his muscle studies, he had acquired and applied a microscope to his and Borch's studies of lung tissue, it would be highly illogical if he had not made use of this obvious advantage in his studies of the finer and microscopic details of the tiny muscles fibres (cells). Remarkable in this context is also Borch's letter to Bartholin of 1 March 1663, where immediately after the mention of his and Stensen's use of "The most accurate microscope" for examining the dried lung of a dog, he mentions Stensen's upcoming publication on muscles and the heart. Furthermore, a much later letter from Stensen to G. W. Leibniz states that it was precisely a desire for investigation of the substance of the lungs that led to Stensen's exploration of the heart fibres and comparison of these with skeletal muscle fibres.[125]

Stensen continues the letter to Thomas Bartholin with one of the most famous phrases in the history of medicine:

"As far as the substance of the heart is concerned, I will, so I believe, provide the striking proof that nothing is found in the heart that is not found in a muscle, and nothing is absent in the heart that is found in a muscle, if you consider what makes the essence of a muscle. This will appear obvious for the atria as well."

In other words, Stensen states here that the heart is a muscle. Stensen continues with his observations of the course of the heart muscle fibres and its significance for the contraction of the heart and with his observations of the septum of the heart. Both are inspired by Bartholin's previous studies and chapter on the heart in the latest edition of the Bartholinian anatomy book:

"Hence it is obvious that there are no straight or circular or, as described by Vesalius, oblique fibres in the heart. All are of the same kind. They are simple like those that are found in any other muscle. In the heart, however, they are different from some of the others only in that they do not run straight, but where they go down obliquely from the base towards the apex, they return upwards. But I will demonstrate, I hope, from my own observations, both extremities of all the fibres, together with the function of the small eminences or muscles, the origin of the valves, their number and movement, and much of what concerns the vessels. Your opponents, who were of the opinion that the figures in your *Anatomia Reformata* [published in Leiden 1651], p. 245, showing us the systole and diastole of the heart, are directly against common sense will see from this that they have claimed a triumph before victory, since it is expressed very clearly that the function depends on the structure of the fibres. But the less accurate opinion on the parenchyma of the heart, according to which people believed that the heart behaves like a small bladder, thus that it dilates when it shortens and swells, and becomes narrower when it lengthens and collapses, misled them. For what you have demonstrated with ligature, dissection and palpation makes the structure of the fibres obvious and proves almost everything you present on p. 249 about the shape of the heart during its movements. However, it will also be obvious that the conus is not the densest part, and that the apex, where the fibres return upwards, in the heart of an ox is thinner than the tip of a needle. It was not yet possible to see whether the septum affords a passage. Following the small cavities, I found them closed. The fact that the external walls also have small cavities not smaller than the others, as well as the countless passage of fibres through the septum, seems to argue against it."

As Stensen states, Bartholin had on page 245 of *Anatomia Reformata* added new illustrations of the heart's systole and diastole in a living individual. This was partly to show that the contraction takes place by means of the tip of the heart, the apex or conus, as Stensen calls it, drawing up against the broader base, partly to illustrate the direction of the circulation of the blood by depicting the lungs during binding of the large vessels (Figure 154). He had also studied the course of the heart fibres (but without characterizing them as muscle fibres) and thought about the significance of these courses for contraction. Thus, Stensen was once again inspired by and, in a sense, continuing Bartholin's previous research in one of the hot anatomical subjects of the day – obviously by taking advantage of the possibilities offered by the microscope, the new technical aid of the epoch. In the text on page 249, to which Stensen refers, Bartholin writes:

"As regards the shape of the heart in systole and diastole, it is characterized by three things, 1) the anatomy of living beings, 2) an adequate movement and rest, 3) the course of the fibres and location of the parts. During the systole, the tip of the heart pulls up against its base or wide end and becomes wider, because it releases the blood during a simultaneous change of its elongated shape to a wide shape, since the base or the wide end is motionless, unlike the tip, which is not attached to any vessels. [...] Vivisections show that the heart is shortened during the systole."

In a section with the subheading "Is the heart a muscle?" Caspar Bartholin writes in the original text of the work that Galen rightly rejected the claim that the heart is a muscle, because it has different types of fibres, and because a muscle moves by the force of the will, but he then adds that "it may be true when Hippocrates says that the heart is a muscle, for he defines a muscle to be flesh that is built up in a circular form."[126] Immediately hereafter, a square bracket follows with Thomas Bartholin's added text:

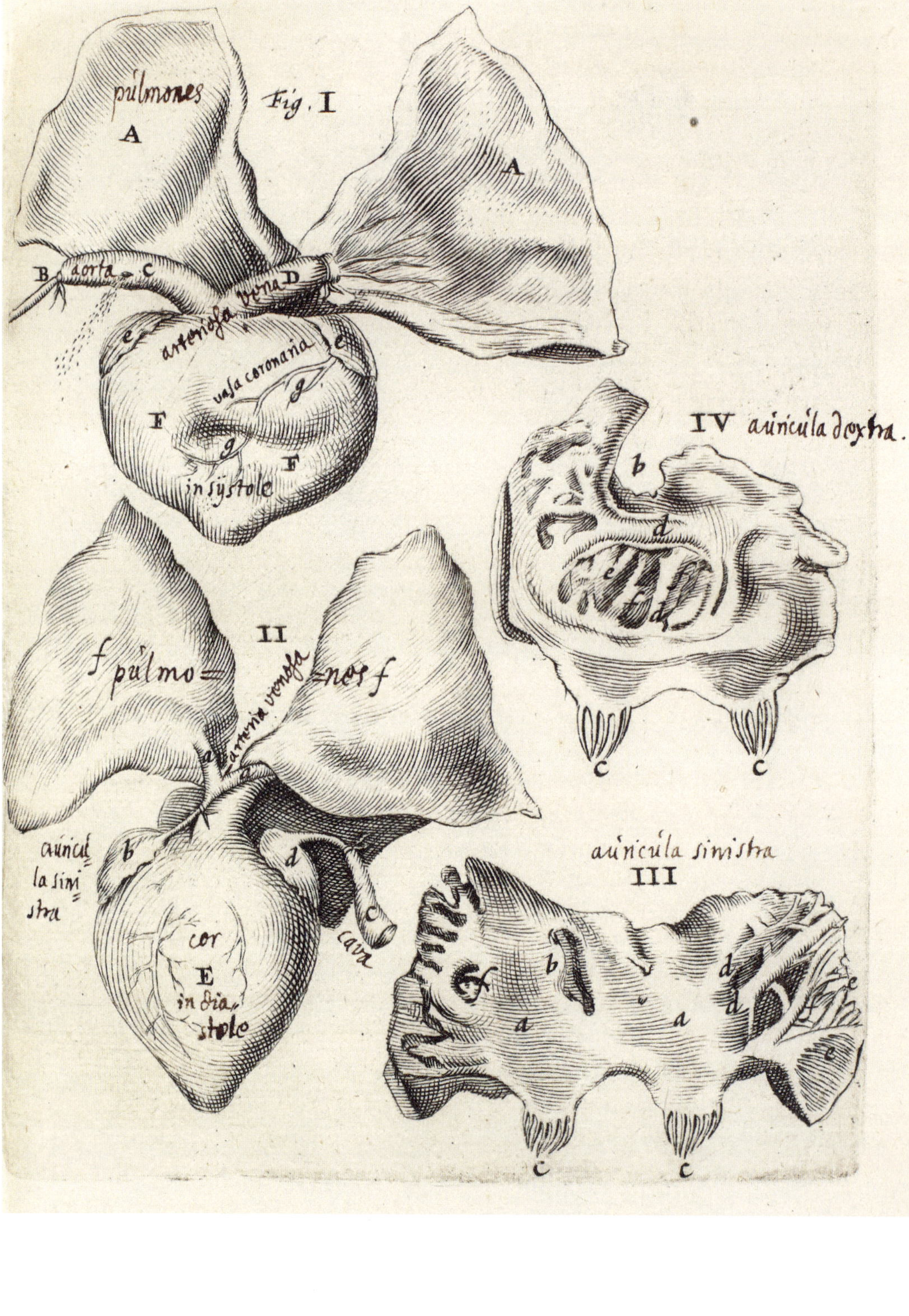
pulmones
Fig. I
A
A
B aorta C
D vena
arteriosa
e
e
vasa coronaria
g
g
F
F
in systole
II
f pulmo=
=nes f
arteria venosa
a
a
auricula sinistra
b
d
c
cava
cor
E
in diastole
IV auricula dextra
b
d
e
f
f
d
c
c
auricula sinistra
III
b
f
a
a
d
d
e
e
c
c

"Others argue that when the heart boils for a long time, it resembles a muscle, or rather several muscles, which at different movements are contracted in between each other. Still others believe that it is an independent muscle that, like the diaphragm, moves constantly. Walæus, who is more right than everyone else, does not call it a muscle, but says that it contracts with a movement like a muscle by means of fibres interwoven in the flesh, especially in the ventricles, like the temporal muscles when chewing."[127]

In other words, Johannes Walæus had already carried out comparative studies of the heart with skeletal muscles, but not – like Stensen – dared to take the controversial step of calling the heart a muscle.

Stensen's examination of the heart's septum for leaks requires a small explanation. Before Harvey's detection of the blood circulation through the lungs, it was the common assumption from Galen's teachings that the blood in the right ventricle was divided into two streams: one that passed into the lungs (but not from here on to the left half of the heart), and another that passed through invisible pores in the cardiac septum and into the left ventricle. Here it was mixed with air and warmed up before it was sent through the aorta to the rest of the body via the arterial system. According to Galen, the pulse was not produced by the heart, but by the arteries themselves.[128] Many decades after Harvey's discovery of the pulmonary circulation, which made the existence of such pores superfluous and in fact provided evidence directly against them, many anatomists were still convinced of their existence. This was so, even if no one ever had been able to detect them by either dissections or experiments with attempts of penetration of liquids through the septum. Thomas Bartholin had, as previously mentioned and shown, in 1645's *Anatomia Secundum Locupletata* published a double illustration of the heart with a probe through the oval foramen between

Figure 154. Thomas Bartholin's illustrations of the heart's systole (Figure I) and diastole (Figure II) on page 245 in *Anatomia Tertium Reformata*, which Niels Stensen refers to in his letter to Bartholin on 30 April 1663.

The lungs are in Figure I marked with A and in Figure II with *f*. The atria of the heart chambers are labelled *e* in Figure I and *b* and *d* in Figure II. Figures III and IV show the inside of the left and right atrium, respectively, with their valves *c* and their structure of a network of muscle fibre columns *d*, *e* and *f*. To indicate the direction of the blood flow in the large vessels in Figure I, ligatures have been placed on the aorta B and the pulmonary artery D, and in Figure II on the pulmonary veins *aa* and the vena cava C. In Figure I, a small incision C has been made in the swollen part of the aorta to give an impression of the increased pressure produced by the ligature. The Royal Library in Copenhagen.

the fetal atria and an accompanying text to refute the Venetian physician and anatomist Cecilio Folli's alternative theory from 1639, that the blood of healthy people, even after the fetal life, passes this way from the right to the left ventricle. But despite his early adherence to Harvey's theory of the blood circulation, and despite the fact that he himself never succeeded in detecting such pores in the cardiac septum, Thomas Bartholin never rejected their existence. On the contrary, as already mentioned, he maintained their existence through all of his revised editions of the Bartholinian anatomy book right up to the fourth and last edition, *Anatome Renovata* from 1673, even though he continued to have doubts about their existence. Thus, he writes in his letter, mentioned below, to Niels Stensen on 25 July 1663: "As for the septum, I have more than once been in doubt as to whether, against Nature, it is permeable." Harvey (and several others) had rejected pores in the cardiac septum. Stensen was downright sceptical and never endorsed the assumption of their existence. As will be mentioned later, Stensen's pupil, Thomas Bartholin's son, Caspar Bartholin the Younger, finally rejected the existence of such pores in 1678.

Bartholin's reply to Stensen on 25 July 1663 is his last surviving letter to Stensen, and the last surviving letter to or from Stensen from his four-year long stay in Holland. It further shows Bartholin's considerable insight into Stensen's new area of research, the anatomy and physiology of the heart and muscles. Bartholin begins by thanking Stensen for the forwarding of his apology against Hoboken's writing and the first part of Sylvius' *Disputationum Medicarum Decas* (Ten Medical Disputations), Amsterdam 1663, in which Sylvius himself for the first time describes the previously mentioned fissure in the brain, which Bartholin had depicted and named after him 22 years earlier in the Bartholinian anatomy book. Bartholin continues:

"Your situation pains me, because so many and such significant opponents surround you. But it is beneficial for excellent talents to have this honourable rivalry. I would have liked for peace between you and the famous Blasius to be restored, but I have renounced my promise, for I do not know how to avoid appearing biased, and that people shall think that I am more favourable to you. Be that as it may, rest assured that I am your friend, and that I praise your indefatigable anatomical studies, by which you excel and make your name highly famous. Truly, your observations of the muscles and the heart are excellent and worthy of dissemination. The spirit of Hippocrates will praise you, because through your excellent observation you revive his observations of the heart, which had been abandoned, and provide conclusive proof that the heart really is a muscle. Galen and other successors will thank you because you have ascertained that the fibres of the heart are of one and the same kind. A short time ago, no small number of anatomists have, in defence of Hippocrates, assumed a similarity of the heart with the muscles. Through your work, the heart is deftly opened again, and the individual fibres of the perpetually moving heart properly matched together. And because the contraction of the fibres, according to your Linden's incomparable opinion in Chapter 6 of *Medicina Physiologica*,[129] takes place in a screw motion, it is carried out with lower power consumption. The physicist learns that the special thing about the screw is that this movement uses comparatively less force than any other machine. All other muscle peculiarities you have been able to easily apply to the heart, but the movement causes difficulties since the movement of the heart is different from that of the other muscles. The latter is a movement subjected to the will, while the former is a natural movement independent of our will, which, under the constant influence of blood, persists even if we do not want it or think about it. Therefore, many believe it to be a special muscle group, akin to that of the diaphragm, in which the movement is now voluntary, now natural, which I cannot endorse, because after all, the movement of the heart excludes any voluntariness and is independent of the will, even though the flow of blood is often either delayed or accelerated, which we see when physical exercises are either rapid or moderate or cease altogether."[130]

We see from this that Bartholin throws his own previous doubts overboard and unconditionally acknowledges Stensen's evidence that the heart is a muscle, which is why he also added a section on this subject in *Anatome Renovata*, which begins with the observation: "Our Stensen has recently proven by the whole new structure and the anatomy of the fibres that the heart is indeed a muscle" (Figure 155).[131] The following passage suggests that Bartholin has misunderstood Stensen. In his letter Stensen did not write that the heart muscle fibres do not have an oblique course; on the contrary, he wrote precisely that they *do* have an oblique course. On the other hand, he wrote that the anatomy of the individual heart muscle fibres is similar, regardless of whether their course is oblique or straight. Stensen is also right that the fibres run around the apex of the heart. Bartholin also seems to recognize this, but why he perceives these fibres to be double is unclear. He writes:

"The opinion that only simple fibres appear in the heart muscle contradicts your own experience. Surely, according to your own confession, they will double through each other's adversarial trajectories. Indeed, you say that when they have run obliquely down from the base towards the tip, their course again straightens upwards. Since, however, they run obliquely downwards, it is not

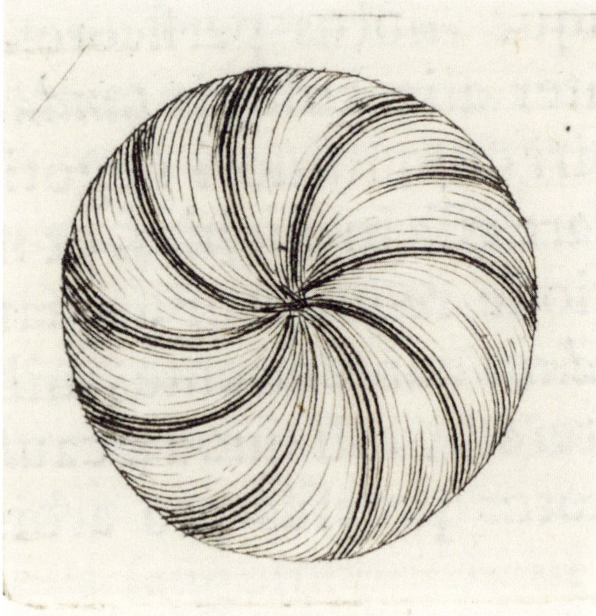

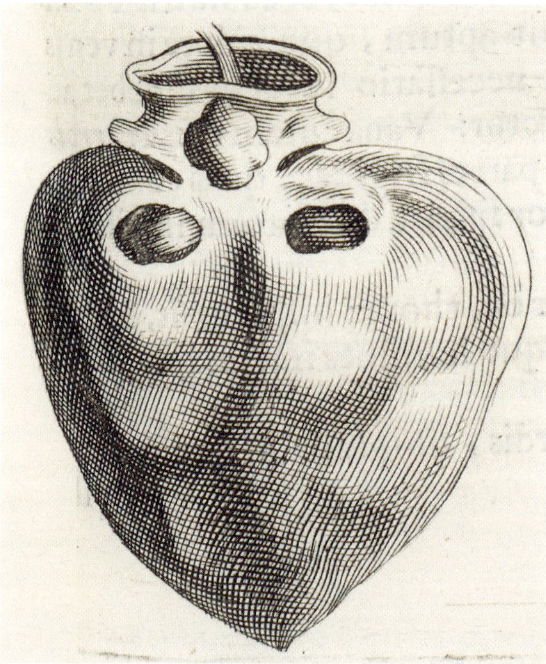

quite right that you, unlike the great Vesalius, deny the existence of oblique fibres in the heart. Their existence is a necessity when we consider the activity of the heart. Its movement, consisting of systole and diastole or of contraction and dilatation of the ventricles, requires different fibres, so that different and opposite movements can be performed. However, you will be able to easily eliminate these and similar questions of doubt by providing the prepared observations on the muscles and the heart with a fuller commentary. But I wonder why the perception of the heart as a muscle is attributed to Hippocrates. In the book about the heart, he calls it a strong muscle, not because of the nerves, from which the name muscle mainly originates, but due to the compression of the flesh. The highly learned Linden calls in chap. 6 of the *Medicina Physiologica* §71, in accordance with Hippocrates' opinion, the heart for a muscle – because of its activity, not because of the tendon.[132] Just as muscle activity consists in contraction and relaxation and is carried out through this, thus also is the activity of the heart."

Bartholin then questions Hippocrates' authorship of the said book on the heart, since, according to Galen and the Greco-Roman philosopher and writer Plutarch

Figure 155. Illustrations of the course of the heart muscle fibres in accordance with a demonstration of a boiled heart, that Niels Stensen performed for Thomas Bartholin, among others. From Thomas Bartholin's *Anatome Quartum Renovata*, p. 361.

All three illustrations are from the frontispiece of the Copenhagen edition of Niels Stensen's book on muscles and glands from 1664 (see Figure 166). Bartholin explains the illustrations as follows: "That the heart is indeed a muscle, our Stensen has lately proved by the whole new structure and organization of the fibres, because: 1. Like other muscles, it has arteries, veins, nerves, fibres, membranes, and no other parenchyma. 2. All cardiac muscle fibres are fleshy in the middle and sinewy at the ends, which all muscle fibres have in common. 3. In the heart, as in other muscles, the fibres run uniformly. They usually form an arc about the centre of their fleshy part, sometimes almost like angles and ascend with the tendon on both sides. Some of the fibres in the left half of the heart have a twisted course, so that the tip of the heart appears star-shaped like this: [A]. The fibres first converge in the opening on the left side, then on the right side and then run from the base slightly downwards. Immediately afterwards, they are twisted under the remaining fibres in a leftward direction. It follows that the preceding fibre always runs a little further down, before it curves, and thus curved, it bends in the same way under the following fibre, that is, under the first twisted fibre below [B and C]. Consequently, the fibres resemble a snail shell that winds from right to left down towards the apex of the heart and then bends back upwards towards the right, as Stensen has demonstrated to me on a cooked heart. Around the base of the heart, the fibres begin and end their course." Bartholin then mentions three more of Stensen's arguments in favour of the fact that the heart is a muscle. The Royal Library in Copenhagen.

(c. 45–120 AD) the author of the book is a Platonic philosopher. Bartholin concludes the letter with the following observations: "But the people who, against our anatomy, dare to compare the heart with a bladder are refuted by your highly accurate discoveries. The heart's substance is condensed flesh, says the author of the Book of Hearts in the above place, where the heart is not equated with a bladder, but is said to be *in* a bladder by means of a smooth membrane or pericardium. [...] What news the greatest troop of learned people has found of these and other things, we would love to learn. Your works appeal to me more than the works of anyone else, because they deal with Nature, because they are learned, and because they come from our fellow citizen. Congratulations on your zeal, my Stensen, keep conquering reputation by such writings and be assured that I am your friend. [...] Live well, be happy and keep liking me. Your Th. Bartholin."

There were no more letters or publications from Stensen's hand during the remaining six months of his stay in Leiden. During this period, however, he was examined by the professors of the Faculty of Medicine of Leiden University for the medical doctorate, which he was awarded in absentia about a year later.[133] Upon his stepfather's death in October 1663, Stensen was recalled to Copenhagen and must have left Leiden in late 1663 or early 1664.

Honorary professor and landowner 7

"I think that the soul's residence will remain hidden to us as long as the soul itself hides, and that question has not yet been sufficiently seen through by the human intellect."[1] Thomas Bartholin, 1661

Bartholin becomes honorary professor

The retreat from the public dissections in the Anatomy House did not necessarily reduce Bartholin's workload. As *Medicus Primus* and Dean of the medical faculty he was instead given several administrative tasks. On top of that came teaching obligations and a never-ending literary production, which was often carried out at such a breathless speed that the printers were unable to keep up. Many nights must have been spent on this. Since his youth, he had suffered from recurrent kidney stone attacks, which further taxed his health, and about which he later wrote to Guy Patin in Paris: "May God save you and my other friends from falling into the hands of this and similar executioners."[2] After almost five years as *Medicus Primus* he must have felt tired and worried as to whether his health could hold out for all the years that were required for him to finish all his planned literary works.

In the beginning of 1661, he therefore applied to Frederik III to become an honorary professor for the rest of his life, so that he could complete his literary works, including useful medical and antiquarian works. Bartholin had listed the works in question in a catalogue added his book *De Nivis Usu Medico*, which was published on New Year's Day that year. The king granted the application on 13 February 1661,[3] and on 6 March Bartholin took leave of his colleagues as ordinary professor,[4] although he formally retained the position of first professor of medicine until his death.[5] Bartholin could now, at the age of 44, retire and write, but at the same time retain his privileges as a professor at the university and his professorial salary. However, contemporary documents indicate that he effectively went down to half his salary, from 400 to 200 rix-dollars per year.[6] On the other hand, there is no indication that Bartholin, despite a large family, had significant financial problems during his time as honorary professor. Bartholin also retained the deanery in the medical faculty, and the duties associated with it, for the rest of his life, as well as the professor's residence in Copenhagen. He was elected rector three more times, in 1665/1666, 1671/1672 and 1680.[7] Many students continued to choose him as their preceptor.[8]

The paper war with Jonkheer Bils

In 1660, Thomas Bartholin got involved in another paper war.[9] This time his opponent was the Dutch miracle-monger Jonkheer Lodewijk de Bils (1624–1669), or Louis de Bils, Heer van Koppensdamme, Boonem, & c., as he used to call himself in his writings. He came from a Flemish noble merchant family from Rouen and Amsterdam.[10] He never studied medicine at a university, but he had an interest in anatomy, and during the 1650s he made himself famous among anatomists in both Holland and abroad. There were two reasons for this: his particular and apparently effective art of embalming, where the bodies and their finest anatomical structures could allegedly stay preserved forever, and his ability to carry out almost bloodless dissections of live animals. The anatomists' interest in exactly these two abilities was not surprising, since long-lasting bodies were of great importance to the teaching of anatomy, and a method for vivisections that could prevent profuse blood flow from obscuring the view would be highly valuable for anatomical research.

Bils was, in the family tradition, educated as a merchant. He had no academic education, and he did not master Latin. However, as a boy he took an interest in anatomy and dissections. He started dissecting dogs and collecting bones, and in Rouen he was arrested during an attempt to steal body parts from a hanged thief who was still hanging on the gallows. The established anatomists started to take an interest in Bils when in the spring of 1651 he granted Leiden University's anatomical collection a gift consisting of several expensive animal skeletons, four human skeletons and a dried-up human skin, which, according to his own statement, he had ripped off the dead man without making incisions in his head and arms. Johannes van Horne, who was the custodian of the university's collection, was impressed by the quality of Bils' preparations. On behalf of the university, he praised Bils for his anatomical skills in a letter of thanks, and he arranged for his preparations to be exhibited in the collection with information about the donor's name. In the following years Bils continued his work developing a method of embalming bodies and organs. He presented examples to van Horne, who was impressed and encouraged Bils to continue the work. Bils also contacted Franciscus Sylvius and the Amsterdam surgeon Paul Barbette (1620–1665), who were initially sceptical, because Bils neither knew the anatomical nomenclature nor had the faintest idea about the function of the organs. Barbette therefore gave Bils a copy of his anatomy book to make him study. By virtue of his gradually established name, the unauthorized Bils quite extraordinarily succeeded in persuading the government in his home city of Sluis to deliver the bodies of executed criminals to him for embalming. Since he was not satisfied with the damaged carcasses from hanging and beheading, he even encouraged the city government to execute the criminals by suffocation from inhaled smoke from burned sulphur. In his home he kept an exhibition of embalmed bodies cut up in different places at different angles and prepared in such a lifelike way that it looked as if they had just died. For a small fee, people could come and see the exhibition.

When Bils performed publicly with his bloodless dissections, for a high entrance fee, he tied the dog to a dissection table and started with a cut in its throat, so that it could not bark. After this, he cut up the dog and showed its organs to the audience. According to witnesses, he could make the blood in the animal flow in the direction he wished. Thus, he could make the arteries on one side of the body fill up until they bulged, while at the same time those on the other side of the body became flat and empty. He could keep the dogs alive for five to six hours while he demonstrated their anatomy to the audience.

Over time, however, Bils was met with increasing scepticism from the established anatomists, because he would not reveal the secrets of his methods. Instead, he tried to sell them for outrageous sums to state institutions, academies and foreign patrons of science. His price for an embalmed body was 8,000 golden florins and for his recipe for embalming 150,000 golden florins,[11] while an ordinary annual salary was a few hundred golden florins. His professional opponents accused him of fraud and quackery, and in 1658 he tried to defend himself in an 11-page pamphlet in Dutch, written as a letter to the physician Laurens Jordaan, with the title *Waarachtig Gebruik der tot noch toe gemeende Gijlbuis, beneffens de verrijzenis der Lever, voorheen zoo lichtvaardig in't graf gestooten* (The real function of the thoracic duct, in contrast to its hitherto ascribed function, and the resurrection of the liver, which has previously been so superficially referred to the grave), published in Rotterdam. As the title indicates, the pamphlet was a direct attack on Bartholin's *Vasa Lymphatica* and *De Lacteis Thoracicis* as well as on Pecquet's and van Horne's treatises on the thoracic duct. In the pamphlet Bils rejects the new doctrine that the chyle passes from the intestine to the thoracic duct bypassing the liver. Instead, he postulates that the chyle passes directly from the intestine to the liver together with the blood in the mesenteric veins, and he thus maintains the Galenic doctrine about the liver being the blood-forming organ. He acknowledges the existence of the thoracic duct, but believes that it only transports lymph, which is then carried around the body through small channels to act as saliva, mucus, tears, sweat etc. In contrast to Bartholin, who in experiments with ligatures had shown that the lymph vessels carry the lymph in the direction from the periphery towards the centre, Bils postulates that it flows in the opposite direction. He explains the mistakes of the established anatomists by asserting that they do not, like him, master the art of performing bloodless vivisections.

The following year the pamphlet was published in Latin,[12] and in the same year he followed it up with another little six-page pamphlet in Dutch with the title *Aan alle ware Liefhebbers der Anatomie* (To all true lovers of anatomy). Here he states that Pecquet, van Horne, Bartholin, Rudbeck, Sylvius and all other scientists have been wrong, and that no one except him has understood the origin or flow of the lymph. He claims that medical students can learn more from him in half an hour than they can learn from van Horne in a couple of years. Van Horne answered this insult with sarcasm in the writing *Waerschowinge aen alle Liefhebbers der Anatomie teegens de gepretendeerde Weetenschap derselver van Joncheer Louys de Bils* (Warning to all lovers of anatomy against Junker Louis de Bils' claimed knowledge), Leiden 1660, where he ironizes Bils' self-overestimation, boasting, tricks and empty promises. He notes that Bils claims to be the only person with true anatomical knowledge, but can it really be true that, in this century, where anatomical knowledge has reached unprecedented heights, the last seal in Nature's book has now been opened, and heaven has saved this moment until the noble Bils' arrival, so that physicians can see the structures of the body "not through a cloud and a dark mirror, but through a clear glass in brilliant sunshine?"

Paul Barbette followed up in the writing *Aenmerckingen van Paulus Barbette, der Medicynen Doctor, Op d'Anatomische Schriften van Jonckheer Lodowyk de Bils* (Remarks to Junker Louis de Bils' anatomical writings) with, among other things, this sarcastic remark: "Now an uneducated jerk, who recently admitted to me that two years ago he still did not know the names of the body parts, wants to play the great master."

At home in Denmark, Thomas Bartholin also did not neglect the opportunity to answer the young, self-imposing, arrogant Dutch nobleman's provocations. One thing was that it was both easy and, in the light of Bils' growing fame, necessary for Bartholin to reject the Dutchman's mistakes and expose Bils' ignorance on Bartholin's home ground. Another thing was that Bartholin himself was one the most diligent, fast and generous scientific mediators of his time, and he truly disliked people who were secretive and closed-fisted with their knowledge. Bartholin published his answer to Bils in *Spicilegium Secundum Ex Vasis Lymphaticis* (Second Aftermath of the Lymph Vessels), Copenhagen 1660. In the seventh and last chapter with the title *Ludovico de Bils nova experimenta ventilantur* (Louis de Bils' new experiments considered) he takes Bils to task. He calls him a poor dilettante, who wants to play scientist and blames him for his arrogance and secrecy concerning his artifices, which according to Bartholin's opinion is just as unworthy for a true scientist as his efforts to earn money on his discoveries. Hereafter, Bartholin rejects Bils' vivisection of a dog as the basis of his mistaken conclusions about the transport of the chyle from the intestine. Bils had opened the belly of a live dog, shortly after it had consumed a big meal. After having ligated the intestinal arteries, he had sewn the dog back together. When he opened it four or five hours later, he insisted that the blood in the mesenterial veins was more pink than usual, and from this he concluded that it was mixed with chyle. Bartholin rejects Bils' method, conclusions and theoretical considerations as support of his views on the lymphatic system and its functions. Aware of his lacking abilities concerning the Latin language and of the fact that he could not match Bartholin in writing, Bils asked his friend, the experienced surgeon from Rotterdam, Nicolaas Zas (1610–1663), to write for him in the paper war against Bartholin. Zas did not wait long. He published in both Dutch and Latin a pamphlet starting with an eight-page letter to Thomas Bartholin dated 30 January 1661, with the breathtaking title *Epistola Apologetica Ad Magnum Th. Bartholinum Regium in Regia Hafniensi Professorem Anatomicorum Coryphæum De Calumniis Nobiliss. Ludovico Bilsio à Coppensdamme, Bonem etc. Arcanorum Naturalium Scrutatori Subtiliss. nec non vivorum ac mortuorum Anatomico Incomparabili, perperam impactis* (Apology letter to the great Thomas Bartholin, Royal Professor and leader of the anatomists at the University of Copenhagen, concerning the false accusations that are improperly directed against the noble Louis de Bils to Koppensdamme, Bonem etc., who is a very sharp investigator of Nature's secrets and an incomparable anatomist, who dissects living and dead).

The letter was published together with Bils' mentioned publications in Dutch and Latin, van Hornes mentioned letter of thanks and the coat of arms of the Bils family. It is written in a sharp tone and contains nothing new to refute Bartholin's attack or to explain Bils' theories, but only personal attacks on Bartholin. Zas writes, among other things, that Bils in his and others' presence has said that if Bartholin will accept the challenge to meet Bils in a public dissection in Holland, Bils will give him a lecture in the art of dissection, so that Bartholin will understand how much he still needs to learn. And if the Danish king will invite Bils to Copenhagen, he will be happy to travel up there and show – not by word, but by deed – that he is better at dissecting than Bartholin, although the latter is probably better at speaking and writing. Ole Borch writes in his letter to Thomas Bartholin 3 March 1661, that the Danish resident in the Netherlands Peder Charisius has sent a copy of the Dutch edition of Zas' "biting epistle" to Copenhagen, and Borch now sends Bartholin a copy of the Latin edition, which he has received from Bils himself.

Borch, Stensen and Møinichen visit Bils in Rotterdam

As soon as Borch heard that "this spider's web of a work" had been published, he travelled together with Niels Stensen and two other Danish students to Rotterdam on behalf of Bartholin to confront Bils personally with the content. They wanted to know if Bils could really accept everything that Zas in his publication tried to make the world believe, and if he would really accept an invitation to Copenhagen. Bils pointed out that he had not written the letter himself, and that he could not accept everything in it, but he was surprised by Bartholin's fierce attack on him, if Zas had translated Bartholin's writing correctly for him. Borch explained Bartholin's viewpoints and assured him that Bartholin had expressed himself in a kindlier way than Zas had reproduced. Borch assured him that Bartholin would be ready to fearlessly meet him "face to face and scalpel to scalpel," if he was ready to travel to Copenhagen and demonstrate his skills, but this was rejected by Bils as Zas' idea. However, he would consider it, if the Danish king would offer him a generous reward and protect him against his enemies, and even then, he would only reveal his secret for a payment of 150,000 golden florins. He refused Borch's suggestion about meeting Bartholin halfway in Hamburg, but if Bartholin would take the trouble to travel to Holland, Bils would be more than happy to demonstrate publicly how much more skilled he was than Bartholin and the anatomists in Leiden. Finally, he accepted a possible challenge from the anatomists in Leiden, and he invoked the professor of medicine at the University in Groningen, Anton Deusing (1612–1666), as a witness of his skills.

Borch notes here in parenthesis to Bartholin that Sylvius has told him that Deusing himself has never opened a human body and therefore praises Bils for something in which he himself has no experience. Moreover, a medical candidate in Groningen has told Borch that Deusing's colleague, Professor of Medicine Henricus Eyssonius (1620–1690), in support of Bartholin's views, has recently demonstrated a mathematical calculation showing that all the chyle and sufficient nutrition for the whole body can easily pass through the thoracic duct, and while he demonstrated this, the listening Deusing grimaced. Bils then showed the four Danes his collection of embalmed corpses. He argued that his own anatomical drawings were correct and those of the anatomists wrong, and he concluded that a reform of practically all of anatomy depended on his system. He dissected a live dog, and Borch admits that Bils with a single cut with the scalpel from the groin to the clavicle on the left side almost cut the whole dog open. In a moment he demonstrated the thoracic duct and the receptaculum, cut open the thoracic duct and collected some of its clear, thick liquid on a spoon to demonstrate to his onlookers. He also demonstrated that a lacteal vein enters the liver, but Borch rather got the impression, that it was a lymph vessel. The dissection was, according to Borch, far from bloodless. At least a pound of blood oozed out, and with his quick cut Bils had damaged part of the liver. After a three-hour visit, Borch and his companions were politely accompanied to the ferry by Bils, and Borch's general impression is that Bils is friendlier than his reputation, and his episodes of excitement are probably caused by animation from others. Bils will before the summer retreat to Sluis and wait there until someone will pay him the sums he demands for the revelation of his secrets.

In his answer to Borch on 9 March 1661, Bartholin thanks him for his and his companions' visit to Bils and directs his anger towards Zas for failing in his duty as a good and honest translator.[13] Bartholin does not understand from Borch's letter that Bils has brought new evidence that chyle is transported directly to the liver or lymph through the thoracic duct, but his curiosity concerning Bils' claimed special abilities is intact: "Please inform me in a few words, if you can nose out which methods Bils uses to preserve his corpses against decay, whether he stops blood without ligatures, and whatever else you managed to observe when you were with him." In his answer to Bartholin on 20 March, Borch writes that he has no news from Bils' camp, but then he adds:[14]

"Our friend Stensen and I have recently observed in a healthy dog, which had died from poisoning, three or four places where the thoracic duct splits, not far from where it empties into the shoulder veins, at the location where Bils places his bulge or labyrinth. But we chose not to examine it closer due to suspicion of poison present. One day will teach the next, and if I am not mistaken, this Troja will at last be revealed, even without Bils. It is hardly probable, that he will ever be able to demonstrate that the lymph runs from the body to the extremities. All experience clearly rejects this, and not long ago we clearly saw in a beheaded killer, who was still not cold, that clear liquid flew quickly in the direction from the hands and the feet towards the body. Bils will never succeed in misleading with his assumptions based on the use of ligatures, and when I was with him, he could not change my opinion with his example with a treadmill. [...] Stensen himself was present, together with Walgesten[15] and Quernes,[16] the carrier of this letter, when I discussed with Bils what Bartholin had written. Therefore, I invoke them as my witnesses, for what I said and did, and I wish you could hear from their mouths and not mine how honestly and as well as possible I defended Bartholin's reputation."

Bartholin answers on 29 March:

"The brilliant fame attached to Bils and his companions is diminished and weakened each day, since we know that this professor is not such a great a man as his reputation seems to promise. He thinks that the rest of us are blockheads, as if no one before him has ever consulted the entrails. By the weakness and immaturity of his judgement he is so much lured into crediting himself too much. The entire long string of works I have published to date by virtue of a good portion of work and God's generous mercy – I, who have been used to performing publicly with the privilege of age – all this he would like to blow away with a single breath, if he were as good at acting as he is at boasting. With all his skills, even the most learned of them, he will never make me believe against the testimony of my eyes that the lymph moves from the centre of the body to the periphery. If he succeeds in bewitching my eyes with his sorcery, he will to me be the great Apollon. But kidding aside, I have more serious matters to attend to. How will he ever change the direction of the valves in the lymph vessels and in that way make disorder in the whole system? With which trick will he loosen the ligatures that I, as the Gordian knot, have rudely placed in his way?"[17]

On 31 March Borch answers Bartholin's letter of 9 March. He repeats his earlier reproduction of Bils' dissection of the live dog and adds among other things:

"Ligatures, he said, had no influence on the matter. They did not show the right flow of the liquids but rather disturbed it. He tried to demonstrate this with a water pump as example. When we came forward with various protests, he just opposed them with his authority. He repeatedly insisted that it was necessary to modernize the anatomy from its present fundaments to his own and said, that he could demonstrate the flow of the chyle, which was the opposite of what everybody believed, and the blood flow and other things that Nature had hitherto kept secret."[18]

Borch adds that the physician Anthony Everaerts (1627–1679) from Middelburg has just published a work about the origin of the human being and the animals, where he agrees with Bils' view that semen, milk, fetuses and liquids originate from chyle and are distributed in the body with the lymph through the lymph vessels. According to Borch, William Harvey does not disagree with this view in exercise 56 in his earlier mentioned *Excercitationes de Generatione Anmalium*, where he writes that a milk-like liquid is present all over a healthy and viable human fetus, especially in the thymus, which is located almost where Bils places his bulging or labyrinth.

Borch has apparently succeeded in seeing through Bils' methods to achieve bloodless vivisections and long-lasting embalmings:

"By applying a pressure, he presses three times the amount of blood normally contained in the vessel into any desired part of the body, and to avoid reflux he uses ligatures. With the vessels greatly distended in this way, it looks as if, by using both his hands and knives of bone, he slowly and carefully divides the muscles in such a way that no blood vessel is damaged. To embalm a body, if I have observed him correctly, he carefully dissects the body, after which he boils it in a large brass pot, by which, through foaming and evaporation, he removes everything watery or oily. The rest he boils in pitch and other substances, which prevents decay. This is what I have understood, partly from what he told me, partly from a thorough inspection of the corpses."

Borch adds that the corpses have an unpleasant stench of conditioner,[19] and he believes that Bils' method of embalming is just as poor as the German-Dutch alchemist Johann Rudolf Glauber's (1604–1670) method with a particular salt (Glauber-salt = sodium sulphate), which it does not resemble although the two men have in common a penchant for bragging. According to Borch's opinion, wine alcohol is an equally good or maybe even better preservative. Finally, Borch acknowledges Bartholin's reply to Zas' letter, which, in addition to its publication in Copenhagen, has just been published in Amsterdam in both Latin and Dutch:[20] "Your very appropriate answer to Zas' word puzzles has also been printed in Amsterdam. We all read it here and admire your pen's great speed and force."

Møinichen and Stensen visit Bils

As soon as Bartholin had received Zas' letter from Gerard Blasius,[21] he grabbed his pen and wrote his reply with the title *Responsio de Experimentis Anatomicis Bilsianis et Difficili Hepatis resurrectione, ad Clarissimum Virum Nicolaum Zas* (Answer to the famous man Nicolaas Zas concerning Bils' anatomical experiments and the difficult resurrection of the liver), Copenhagen 18 February 1661. In the 24-page letter, Bartholin repeats his earlier viewpoints and adds that he has heard about Bils' embalmed corpses from the Danish diplomat and councillor of the realm, Otte Krag,[22] who had seen them and referred to them positively. Therefore, Bartholin had asked his former pupil and prosector Henrik à Møinichen to look in on Bils in Rotterdam on his way home to Denmark to take a closer look at the embalmings.

Møinichen first visited Bils as a regular paying guest. As an experienced dissector, Møinichen had described the embalmed corpses as curiosities, unusable for dissections since the finer anatomical structures were not preserved. He had encouraged Bils to demonstrate his new doctrine of the chyle by a dissection, and Bils had then asked him to return a week later. So he did, bringing a letter of recommendation from an acquaintance in common in Amsterdam. While Møinichen tried to persuade Bils to carry out the promised dissection, Niels Stensen and another medical student showed up in the expectation that it would happen. But when Møinichen repeated his request to Bils, the latter once again rejected it and suggested a demonstration with figures. Møinichen pressed further for a dissection by stating that Bils' assertion about the direction of the lymph flow was not possible because of the valves in the lymph vessels. Bils answered that he would publicly withdraw everything, if Møinichen could show him these valves, but when Møinichen then offered to do so, Bils would not be part of it anyway, and Møinichen and Stensen had to leave Rotterdam with their business unfinished. By maintaining his secrecy and refusing to submit his discovery to a scientific test, Bils had, according to Bartholin, shown that he was not a real scientist, but a charlatan. Hereby, Bartholin meant to have pacified Bils. To Borch he writes on 19 April 1661:

"I do not care about Bils' opinions. I believe that I have eliminated his efforts or his translator's efforts in my answer to Zas. If he wants to continue puffing himself up, be tyrannical and full of arrogant spirit, then let him be aware of Megara's thunderbolt in *Hercules furens*.[23] Arrogant people are pursued by the god of vengeance. The man has a lack of taste, but someone will rub him in lots of salt if he is not, as I believe, in possession of Glauber's miraculous salt. Neither Bils, Everaerts nor anyone else will be able to turn the chyle and lymph flow, even if as many lined up as there are gates in Thebes or mouths in the Nile. [...] I cannot be convinced that Nature contradicts itself, and that we are participating in a battle with shadows. It is not Nature that fails, but humans, and they often fail, because they flatter either themselves or their friends."[24]

It would turn out that Bartholin was right about the last sentence, but he had not pacified Bils. When Bils had had Bartholin's answer translated, he was inflamed with new anger. According to Borch's letter to Bartholin of 21 April, Bils gathered a group of his friends, and by dissecting a dead dog he once again convinced them of his skilfulness as dissector and his rightness concerning the lymphatic system. He persuaded the Dutch physician Laurentius Jordaan to write a new libel against Bartholin in Bils' name.[25] Bartholin answered under the pseudonym Nicolaus Stephanus "to the famous bully Louis de Bils."[26] Thus Bartholin ended his paper war with Bils, and to Borch he wrote the following closing remarks on 6 July 1661:

"I have sent Bils the old divorce form[27] and have no more concerns regarding him. I will not take any further part in this bickering. As emeritus I cannot afford it.

[...] Let everyone who wishes to admire this man, follow him, praise him, kiss his shadow. I will not consider that to be something of my concern. Let it be the mob's business: fools, who often give honour to the undeserving; senseless worshippers of reputation, who are impressed by titles and images."

Niels Stensen takes over Bartholin's dispute with Bils and begins a new one with the anatomists Deusing and Everaerts

Bartholin had barely stepped out of the dispute with Bils before his loyal pupil and friend Niels Stensen stepped into it. His two visits to Bils with Møinichen and Borch, respectively, had not been useless. As mentioned, Borch had realized that Bils' embalmings were not so epoch-making as first assumed, and in his letters to Bartholin he had stated that van Horne had developed a method that seemed to give results at least as good.[28] He had also explained Bils' bloodless vivisections with his use of ligatures. Borch, Møinichen and the discoverer of the lymph vessels, Bartholin, had further established that Bils' asserted observations of the lymph flow from the centre of the body to the periphery, against the direction of the valves in the lymph vessels, could not be possible. Therefore, they concluded that Bils' observations rested either on mistakes or simply on manipulations. But Niels Stensen's curiosity had been awoken, and it turned out to be him who, by virtue of his excellent powers of observation and analytical reasoning, brought the solution to the riddle and thereby the deathblow to Bils' new theories on the chyle transport and the lymph flow.

The story was, in short, that in animal dissections Stensen had observed a new vessel, which differed from the thoracic duct in both location and content. He wanted to examine this more carefully and had therefore dissected a dog, shortly after it had consumed a meal. When he opened its abdominal cavity, he found, as expected, the lacteal veins of the intestines filled with chyle, while this was not yet the case for the thoracic duct or the chyle container. To also make them appear clearer he had ligated them and put the dog aside until the next day. When he opened it again, both the thoracic duct and the chyle container was full of chyle, but moreover, he observed surprisingly that a long lymph vessel, which stretched from the chyle container in the top of the abdomen to one of the lymph glands in the groin, had also been filled with milky white chyle. Hence, this chyle must have flowed from the centre to the periphery and therefore against the direction of the lymph valves, just as Bils had observed. Stensen's astonishment and further observations brought him forward to the explanation. In a letter to Thomas Bartholin 22 April 1661, he explains the whole thing:

"Amazed as I was in the beginning, I asked myself why the chyle, which had been carried on to the gland, had not been detained by the obstruction made by the valves. But soon I saw without any difficulty that the valves, which often collapse when death occurs, could make no resistance, especially when, after the vessel has been pressed together, they lie close against its wall, as I have heard from the famous Sylvius. Nor do I doubt that it is this appearance that deceives Bils, since in order to press out the liquid, he squeezes the vessels together, so that the valves lie up against the vessel walls and in this way allow the liquid to pass both forward and backward."[29]

Stensen continues the letter with the following demystifying remarks about Bils' embalmings and bloodless dissections:

"As concerns his other secrets, they gradually seem to lose value, since the famous Mr. van Horne's embalmed hand surpasses his mummies, and dissection without bloodshed is easily performed. Since he has told us that he uses ligatures, it is easy to see why the blood can be directed from one part of the body to another, while the animal is still alive. If you ligate the descending trunk of the aorta near the heart, as well as the ascending *vena cava*, I do not doubt, that the heart by dragging all the blood from the lower part of the body will fill the upper part, and in this way it will be easy to dissect parts when they are empty of blood, while on the other side you will clearly be able to see vessels that otherwise escape the eye, emerge, bulging with fluid."

In his mentioned first anatomical treatise on the salivary glands of the mouth, Stensen in thesis 42–44 attacks Bils' doctrines about the lymphatic system. He rejects that the lymph functions as nutrition for the body's spermatic parts[30] and as the source of all the body's watery fluids, including the saliva. Stensen adds that this erroneous opinion about the saliva is supported by professor of medicine at the University of Groningen, Anton Deusing (Figure 156), on page 191 in

Figure 156. Anton Deusing (1612–1666). Engraving by Crispijn van de Passe the Younger (c. 1594–1670) in 1652.
The text in the frame states that Deusing is Professor of Medicine and PhD at the University of Groningen and the personal physician to the governor of the province. The engraving is from 1652, when he was 40 years old. The epigram under the illustration reads: "This is Deusing's hospitable face, and this hospitality was given to his beautiful spirit, the goddess of wisdom and the muses. A portrait of his spirit he himself gives in learned works and through his impeccable behaviour. Show esteem for this distinguished man's glance, whose wit is admired by celebrities and the supreme gods of the Muses." Wellcome Collection, London.

ANTONIVS DEVSINGIVS. M.&PH.D. IN ACAD. GRON. & OML. MED. PROF. PRIM. ILL.ss GVBERNATORIS & PROVINC. ARCHIATER. A° 1652. Æt. 40 An.
Hæc illa hospitii facies, pulcherrima quod mens
DEVSINGI nacta est, et Sophia, et Charites.
Effigiem mentis per docta volumina, mores
Perq. inculpatos, exhibet ipse suæ.
Herois vultum venerare, ad acumina cujus
Lumina et Aonidum numina summa stupent.
IOHANNES WITTEN Nob. Gel.

his *Exercitationes Physico-Anatomicæ De Nutrimenti in Corpore Elaboratione* (Physical-anatomical exercises concerning the processing of food in the body), Groningen 1660.[31] By referring to his own, Bartholin's, Pecquet's and others' experiments Stensen then rejects Bils' ideas about a direct transport of the chyle from the intestines to the liver, regardless of whether this should be through lacteal veins, lymph vessels or veins,[32] and about the direction of the lymph from the centre to the periphery of the body against the direction of the lymph valves. Stensen mentions that Bartholin, in his *Spicilegium Secundum Ex Vasis Lymphaticis*[33] and in *Responsio de Experimentis Anatomicis Bilsianis*, has encouraged Bils to demonstrate his ideas about the direction of the chyle and the lymph, as Bils had described in *Epistolica Dissertatio*, and as he had promised to do by dissecting a live dog in the presence of Stensen, Borch and Møinichen. Stensen is therefore surprised that Deusing in that context accuses Bartholin of being unfairly demanding, since Deusing at the same time explains the meaning of the chyle from Bils' way of thinking. Stensen documents the existence and effect of lymph valves by referring to his own experiments with ligatures and probes during dissections, including a dissection of a live dog in the presence of Ole Borch.

In the treatise Stensen also rejects the earlier mentioned Bils-supporter Antony Everaerts' assertion that the chyle supplies the breasts with material for breast milk and is otherwise carried around to the whole body through a richly branched network of chyle vessels (lacteal veins) – a view he had gained through dissection of milk-fed baby rabbits and published in *Novus et Genuinus Hominis Brutique Animalis Exortus* (A new and genuine origin of humans and animals), Middelburg 1661.[34] Stensen refutes it by referring to his own dissection of the teats of a dog that had just given birth. Stensen found no chyle vessels in the dog's teats but, on the other hand, "large and countless blood vessels." As mentioned, Thomas Bartholin had already obtained the same result in his dissection of Anne Michelsdatter in 1653.

Ole Borch writes on 23 July 1661 to Thomas Bartholin about the effect of Niels Stensen's refutation of Bils' new doctrines: "Our friend Stensen has with his newly published treatise, dedicated to your distinguished name, inflicted a serious injury to the Bilsian heresy."[35]

A week later, on 30 July, Paul Barbette writes to Bartholin, that Bils is on his way back to Flanders and will bring his secrets with him, which is not regretted, "for they are indeed of little importance, though the ignorant man values them so highly. [...] His embalmed bodies are sometimes very wet and other times drier than they ought to be, so that they turn into dust; how long they ultimately last must be assessed by those who understand salt and alum. His bloodless dissections and their progress from place to place (which no one among even his best friends has yet seen) confuse the structure of the human body more than they enlighten it, for all structures thinner than blood vessels are destroyed."[36]

Bils never responded to Stensen's criticism, which apparently surprised Stensen. On 12 September 1661 he writes to Bartholin: "Bils is still silent, and I do not know if it is the dog days, that also let his swear words take a vacation, or if he wants to take a vacation from this kind of business forever."[37] As regards Stensen's criticism of Deusing, Borch writes to Bartholin on 23 July 1661:

> "If he has attacked Deusing, the liver's new sustainer, he did it out of love for his teacher Bartholin, and in my opinion perhaps not without reason. Deusing's hand is always cramping to write something, but he has not devoted much energy to the study of anatomy or Bils' mysteries. Accordingly, Stensen found it intolerable that he should triumph at the expense of other people's hard work without having to sweat himself."[38]

Borch believes that Deusing's publications testify more to hard work and petty criticism than to thorough and intelligent scholarship, and among others he mentions "Deusing's treatise on the function of the liver (in which he criticizes the excellent Bartholin more covertly)."

When it comes to the years 1661–1662, Borch definitely has a point in his remark about Deusing's joy in writing. Unlike Bils, Deusing became infuriated by Stensen's criticism, and his response became the beginning of a bitter dispute with Stensen, which again involved Bartholin and Borch.

To begin with the latter part of the dispute, Deusing had earlier in 1661 published a new treatise in support of Bils' ideas about the lymphatic system. As mentioned by Borch, the book had an appendix in the form of a critical letter dated 15 April 1661 to Thomas Bartholin about the function of the liver, *Exercitationes Physico-Anatomicæ De Nutrimento Animalium Ultimo. Ubi De Sanguinis Usu, Ac Commento Nutritii Succi per Nervos influentis. Accessit Dissertatio Epistolica De Hepatis Officio, ad Celeberrimum Virum, D. Thomam Bartholinum, Medicum et Anatomicum Excellentissimum, Professorem Regium Hafniæ* (Physical-anatomical exercises concerning animals' last food, including the function of the blood and considerations about the nutritional juice flowing from the nerves. Added a dissertation in the form of a letter about the function of the liver to the famous man, excellent physician and anatomist, professor at the University of Copenhagen, Doctor Thomas Bartholin).

Bartholin wrote a quick reply in the form of an 84-page dissertation as a letter to Deusing dated 1 August 1661, *Dissertatio Anatomica de Hepate Defuncto, Novis Bilsianorum observationibus opposita* (Anatomical dissertation on the abandoning of the liver against Bils' new observations).[39] In the letter, Bartholin rejects Deusing's and Bils' claim about the thoracic duct's lack of capacity as an argument that part of the chyle from the intestines passes directly to the liver. Interestingly, Bartholin refers to Descartes' geometry about the movements of liquids and bodies through circles and tubes in *Principia Philosophiæ*, Amsterdam 1644.[40] Without saying anything new, Deusing responded the following year in *Disquisitio Ulterior De Chyli Motu Atque Officio Hepatis Ad Excellentiss. & Inclytum Virum D. Thomam Bartholinum, etc.* (Further investigation of the movement of the chyle and the function of the liver to the excellent and famous man Thomas Bartholin, etc.), which is an appendix to *Resurrectio Hepatis Asserta; Contra Socium Larvatum Vincentium Slegium, sub Personati Blottesandæi cohorte furiosa Signiferum* (Defence of the resurrection of the liver against Vincent Schlegel, the bewitched confederate of the hidden Blottesandæus, the leader of the mad mob).

In a collection of letters, *Deusingius Heautontimorumenos sive epistolæ selectæ eruditorum, quæ immaturis Antonii Deusingii medici Groningensis, scriptis larvam strictim sed, sincere detrahunt etc.* (The self-tormentor Deusing or letters by selected scholars, which degrade the immature physician in Groningen Anton Deusing's bewitched and superficial, but genuine writings etc.), which was published in Hamburg in 1661 under the pseudonym Benedictus Blottesandæus (the exposer of the truth), Ole Borch had displayed Deusing's incompetence as a natural scientist with the aim of defending Bartholin and his viewpoints. Deusing guessed that Borch was the author, but Borch denied it, both in letters to Bartholin[41] and in a letter to Deusing on 6 May 1662, published in the above mentioned *Resurrectio Hepatis Asserta*, and thus it was never proven.

In a letter to his former pupil Georg Seger in Lüneburg on 11 October 1662, Bartholin writes about this Blottesandæus: "Whoever this author is, he seems to be learned. If only he had proceeded more kindly towards Deusing, who defended himself with a sharp rebuttal."[42]

But back to Deusing's reply to Stensen's criticism. It quickly appeared in a published 40-page letter dated 9 July 1661, i.e. on the second day of Stensen's defence of the dissertation. Remarkably, it was not addressed to Stensen, but to Johannes van Horne, whom Deusing considered to be the dissertation's true author. The title, *Appendix Ad Dissertationem De Hepatis Officio Seu Vindiciæ Hepatis Redivivi, Leni correctione tangentes sequiorem Interpretationem Clarissimi Viri D. J. van Horne, Med. Doct. Anatomiæ & Chirurgiæ in Acad. Lugd. Bat. Professoris* (Appendix to the dissertation about the function of the liver or defence of the liver's resurrection, which considerately corrects the famous doctor and professor of anatomy and surgery at the University of Leiden Johannes van Horne's less good interpretation) shows that Deusing considered the letter as an appendix to his earlier letter to Bartholin about the function of the liver.

Stensen became aware of the answer on a summer trip with Borch and a small company of other friends to the northern Dutch provinces in July-August 1661. Back in Leiden, he writes to Bartholin in the earlier-mentioned letter of 12 September:

"On the journey we found in Groningen instead of Deusing himself, whom I sought in vain, his *Vindiciæ*. Although in this he fights my words, he directs the criticism against van Horne and pours his bile on him. But he must be very hot-tempered, since he cannot bear so light a touch. I have said that he is friendly towards Bils, I have said that on the question of the origin of the saliva secreted from the glands, he supports Bils' view, I have said that I am surprised that in his explanation of the word chyle from Bils' point of view, he attributes an unfair demand to Bartholin. But otherwise, I have said nothing about him. These words of mine, however, outraged him, as if I had said the most unpleasant things. Among other things, he wriggles strangely in the dispute that he has inflicted on you, and taking things you have said in another connection, out of their context, he claims that it is a matter about a controversy between you and Bils about the movements of the chyle through the blood vessels of the mesentery. It seems as if either he wants to conceal or he does not want to know the reason for *Exsequiæ Hepatis*,[43] and that he has only perused your *Spicilegium* very superficially. The real movement of the lymph, as first observed by you, has however dethroned the liver, and you yourself say in *Spicilegium Secundum*, where you speak about the lymph movement advocated by Bils, that with this cornerstone all your observations stand and fall, and you ask him to keep his word as soon as possible. Without taking this into account, Deusing sticks to Bils' chyle explanation and calls your demand that Bils should demonstrate the movement of the chyle unfair, as if Bils had not promised this, although he says in *Epistola Dissertatio* that the liquid contained in the chyle vessels is transferred to the liver and partly to the thoracic duct, and promises that he will prove it in an obvious way. But this and all the rest you will find more fully dealt with in *Responsio ad Vindicias*, which you, God willing, will soon receive, together with the dissertations, and a short treatise on the glands of the eyes, which you might have had long ago, if various other obstacles than those already mentioned had not got in the way."

Although Stensen is not mentioned by name in Deusing's letter addendum, there was no doubt that it was aimed at him, and as proclaimed in the letter to Bartholin, he soon replied with the retort *Responsio Ad Vindicias Hepatis Redivivi* (Response to the defence for the resurrection of the liver) in the form of a 22-page letter to "the famous and experienced Dr. Anton Deusing, celebrated physician and professor of medicine," dated 18 November 1661.[44] It was published in Leiden 1662 as part of the earlier-mentioned anthology, *Nicolai Stenonis Observationes Anatomicæ*. The retort is more a defence of van Horne's and Bartholin's views in the dispute with Bils about the circulation of the chyle and lymph than a defence of Stensen's own consistent views, and rather than including anything new, it is a rhetorical display in which Stensen, with his characteristic eloquence, tears Deusing's accusations and arguments apart one by one.[45]

As in Bartholin's disputes with Hofman and Riolan and in Stensen's own with Blasius, Stensen's main complaint against Deusing is that the latter draws his (wrong) conclusions from rational considerations, instead of from secure evidence from experience gained through practical experiments. In this context, he refers to Hippocrates: "What is accomplished by reason alone, one can draw no advantage from, but one can from what can be proven by deed." Besides the fact that Deusing does not perform experiments himself, he rests his views on the dubious observations of self-taught researchers like Bils, while he ignores the results of learned and professional people like Pecquet and Bartholin, whose writings on the lymphatic system Stensen frequently refers to and quotes from in his argumentation against Deusing and Bils. Stensen could hardly have received the next letter from Bartholin, dated 10 November 1661, before he wrote the retort to Deusing, dated eight days later.

In the letter, Bartholin appeals to Stensen for a gentle treatment of Deusing:

> "So, every day brings you renewed occasion to increase your reputation, because your pen and your sharp eyes do not know of vacation. But I could, among so much well-deserved praise, also wish you applause from Deusing, who – I do not know if it is your fault – seems to have an unfavourable attitude towards you. Towards me, at any rate, everywhere he shows an unusual kindness, which is why I would like to see you treat him gently in your *Vindiciæ*. What you in such a kind intention adduce in my defence, I regard as a special proof of friendship towards me, which I will try to earn by every possible service. That I have not attributed to Bils an opinion that he does not have, I am as sure of as I am of the fact that he cannot prove the immediate transition of the chyle from the chyle vessels to the liver."[46]

Bartholin did not comment on Stensen's response to Deusing, but in his letter to Stensen on 14 February 1662, he refers to his own mentioned response to Deusing and Bils, the 84-page letter to Deusing of 1 August 1661: "You can read the dissertation *De Hepate Defuncto* at our good friend Borch and rejoice at the man's sympathy."

Stensen's retort elicited a further quick response from Deusing in the form of a 13-page appendix, *Vindiciarum Hepatis Redivivi Supplementum* (Appendix to vindication for the resurrection of the liver), to the apology, *Apologeticæ Defensionis Pro Oeconomia Corporis Animalis Prodromus, Quo Personata Cuidam Benedicto Blottesandeo Larva detrahitur* (Forerunner of a protective defence of a system of life expressions in the body of the living being, and where the personal mask is torn from a certain Benedictus Blottesandeus), which, as the title indicates, was a response to Blottesandæus' (Ole Borch's) attack. It was published in Groningen at the beginning of 1662.

Undoubtedly also as a consequence of his friendly correspondence with Bartholin, Deusing is here on the retreat. He even highlights Stensen's youth and acknowledges his work, which warrants great expectations. He says nothing new, and Stensen found no reason to reply. He contented himself with a small comment in a letter to Bartholin on 21 May 1662:

> "Nor do I care about Deusing's abuse, which aims to teach me modesty. I am obviously supposed to learn it from his *Prodromus*, in which he not only reveals himself as a follower of Bils, but also as a completely obvious imitator of his errors and slanders."[47]

In a new letter to Bartholin on 26 August 1662, Stensen sends the last volley at Deusing:

> "With what zeal, however, our Deusing has attacked Mr. van Horne, and with what titles he has adorned me, should not be unknown to you. If he were as equal with the mathematicians in the use of truth-proofs as he is eloquent when firing at others, he would be a veritable terror to everyone who does not belong to his camp. [...] Indeed, if every effort were not made to prevent him from interpreting the silence of others as his own victory, I would let him make accusations with impunity."[48]

Bartholin's own last refutation of Deusing (and Bils) was published in Copenhagen 1666 in the treatise *De Hepatis Exautorati Desperata Causa* (On the abandoned liver's hopeless cause). Deusing, who died in January the same year, did not live to read it. In the treatise Bartholin diligently refers to Stensen's dissections and observations in his argumentation, and he also refers to Simon Paulli's son, Jacob Henrik Paulli's ditto in the treatise *Anatome Anatomiæ, Bilsianæ* (An anatomical examination of Bils' anatomy), Copenhagen 1663. The dispute between Bartholin and Deusing took place in a friendly and respectful manner. The same applied to their concurrent private correspondence of a total of 11 letters between 1 May 1661 and 20 November 1663.[49] Unlike Blottesandæus (Ole Borch), Niels Stensen is not mentioned in these letters.

Antony Everaerts' response to Stensen's criticism was at the more restrained end of the scale. Contrary to Deusing, he recognized Stensen as the true author of the dissertation. In his letter to Bartholin on 26 August 1662, Stensen writes:

> "To fully complete the number of intrigues, Anthony Everaerts, a distinguished Middelburg physician, has now also joined in. He is more modest than the others, avoids swear words, but also abstains from proofs, though not from experiments. A few days ago, he published a dissertation with the title *Lux e Tenebris Affulsa ex Viscerum Monstrosi Partus Enucleatione* [The light shines in the darkness, according to the explanation of the entrails of a deformed fetus], in which, after describing the deformed fetus, which lacked head and breast, he first complains that the liver is thought by some without reason to be a useless mass, while it is as necessary as the heart (he had indeed seen in this malformation two small heart-shaped organs, which were attached to the liver on both sides as appendages). Then he lets me appear and believes to have demonstrated before the eyes of his friends that chyle vessels are supplying the parts usually called the spermatic parts, which I had refuted both by reason and by experiments. But either my eyes deceive me, or has Nature given the Zeelandic rabbits what it has denied the Dutch. I have opened the stomachs of older rabbits, and I have examined the entrails of younger animals, I have examined the udders and the uterus of pregnant animals, but I have still not been allowed to be sharp-sighted enough to observe other chyle vessels than the mesenteric ones, the thoracic one, and those vessels that originate in the mammary glands and carry the milk to the nipple; in these, however, no matter if they are in the liver, the neck, within the fat capsule or in the groin, I have never in vain searched for lymph, which can be witnessed by the famous men Sylvius and Borch, since they almost always enriched my dissections with their presence. Not long ago I ligated all the vessels that run down under the chyle container with a single band, since I was of the opinion that if a chyle vessel should be found at this place, it would reveal itself through a reaction; but after the ligation, all lymph between the band and the fat capsule disappeared immediately, while a great number of vessels above the band gradually swelled up to a greater volume and contained absolutely nothing but a totally clear lymph. But I should let myself be called a fool, if I would add even more to a matter that, since the first discovery of the lymph vessels, is fully familiar to you."

Bartholin answers Stensen on 7 September 1662:

"I would like to know if the light that shines from the darkness in Middelburg can withstand a gentle breath. As far as I could see from the report, it can be extinguished by a single breath or gust of wind. What does the author want to deduce from the monstrous birth? Surely nothing else than monstrous conclusions. That both heart chambers were sunk in the liver of the malformed fetus is sufficient proof that it is a matter of the activity of the heart and not of the liver, which usually has no cavities of this kind."

After a mention of different authors' descriptions of possible small lacteal veins (chyle vessels) to the breasts and the uterus, Bartholin continues:

"To tell the truth, I believe I have seen more than once small white, glistening fibres here and there, without daring to call them lacteal veins. However, I will not condemn the care of others, when they have seen with the sharp eye of the lynx what has escaped our eyes. I wonder why you have not observed in the rabbit what Everaerts with such great certainty establishes to his readers. However, it is difficult for this reason to throw a man from the saddle, because an insignificance can often change the whole situation, and it often happens that we come across something by chance that afterwards cannot be retrieved, despite much zeal. [...] But from your knife, which you often, to the destruction of the animals, but to the growth of science, sharpen, a great hope of certain experience shines towards us."[50]

Time would prove that Bartholin was right. Stensen's sharp knife and brain were to bring many more epoch-making anatomical discoveries in the years to come.

The medical faculty's archive chest and a book about the Anatomy House

As Dean of University of Copenhagen's medical faculty, Thomas Bartholin was responsible for its archive chest. He wished to secure the contents of the archive for posterity and therefore worked in the period 1656–1661 to collect and organize the archive's documents. As a result of this work, he published in Copenhagen 1662 the book *Cista Medica Hafniensis* (The archive chest of the medical faculty in Copenhagen), which is bound together with the earlier mentioned book about the Anatomy House, *Domus Anatomica Hafniensis brevissime descripta* (The Anatomy House in Copenhagen briefly described).[51]

Cista Medica has 645 pages and is dedicated to the Grand Duke of Tuscany Ferdinand II de' Medici. Besides the frontispiece, title page with information about the co-bound *Domus Anatomica Hafniensis*, dedication, preface and index the book contains a poem by Vitus Bering to Thomas Bartholin. The frontispiece was made by Albert Haelwegh. It has a portrait of Bartholin in an oval frame consisting of laurel branches and placed between six similar frames with portraits of men, who played a significant role in Thomas Bartholin's life (Figure 157). At the bottom is the archive chest of the medical faculty.

Cista Medica is an important source document for the Danish medical history of the 16th and 17th centuries. Among other things, it includes descriptions of the education of physicians, the pharmacy service, drugs, physician's instructions for medical students, biographies about important physicians, the advice of the university, the professors and leading physicians to the population about prevention and treatment of plague, forensic reports and royal ordinances for physicians, surgeons (barbers) and apothecaries. Thomas Bartholin states in the preface that he has "curiously collected the physicians' scattered papers as if they were scraps from a shipwreck, and arranged them chronologically, without giving priority to anyone." He has done it "to avoid our physicians, few in number but considerable in merit, being deprived of an enlightened posteriority as well as their well-deserved immortality; [...] in some places I have also included historical documents, in the sure hope that what, in my opinion, could secure the legacy of the future, shed light on the University of Copenhagen and bring the art of medicine forward, would be to the taste of righteous and scientifically interested people."

In the book you find Caspar Bartholin's earlier mentioned study guide to his sons and Peder Charisius, as well as Johan Rhode's three directions for education in the art of medicine (*Johannis Rhodii de artis Medicæ exercitatione, Consilia tria*). The forensic reports, which cover the period 1621–1659, are written by the university's, the king's and the city of Copenhagen's leading physicians. In the beginning the declarations were solely based on written case material and oral statements and interviews, but later also on autopsy findings.

An example is the aforementioned autopsy of Sven Olsen, who was suffocated by a piece of beef tongue. Another case, in which Thomas Bartholin took part himself, shows that some of the forensic reports were more in the vein of what we would today consider as patient complaints. It took place in August 1659 and concerned the physician Jacob Svabe's and the surgeon Godfred's treatment of the patient Torben Petræus for an illness in the nose. Petræus had complained about the treatment, which had no effect, and as Dean of the medical faculty, Bartholin was requested by the rector and the other venerable professors to answer the complaint together with his brother Rasmus and professor Christian Ostenfeld. On 6 August, the three professors had a meeting in Bartholin's residence, where they questioned the patient and the two treating physicians and

Figure 157. The frontispiece of Thomas Bartholin's *Cista Medica Hafniensis*, Copenhagen 1662. Engraving by Albert Haelwegh using Figure 96 a + b as templates. 13.5 × 7.5 cm.

Above the letter chest are Bartholin and six important men from his life framed by laurel wreaths. The six men are, from left: His cousin Henrik Fuiren, his uncle Ole Worm, his grandfather Thomas Fincke, his father Caspar Bartholin the Elder, and his two colleagues Simon Paulli and Johan Rhode. To his friend in Venice, Aprosio Vintimiglia, Bartholin wrote that he was not content with the portrait, which did not quite look like him. The Royal Library in Copenhagen.

examined the medicine used by Jacob Svabe. Bartholin and his colleagues concluded in their report of the same day:

"We have come to the conclusion that Torben Petræus' nose illness, an ozæna,[52] is so old and malignant that it can only with difficulty be cured with the drugs, which doctor Svabe showed us and assured us that he had used. However, we do in fact believe that these drugs are excellent and useful against this illness, and we have no doubt that the treatment would have had a better outcome if only the doctor had been summoned immediately, while the illness was still in an early stage. Therefore, as far as we can judge, one should not blame the doctor and the patient as much as the obstinate and malignant illness itself, for over time so many small bones have come out of the nose and although it – flattened as it is

– looks healthy enough, so much pus keeps coming out of it that he soils several handkerchiefs daily, as we have been shown. This is our opinion in brief, according to rector magnificus' and the gentlemen professors' kind request."[53]

Other cases dealt with more fundamental questions, such as whether someone can die from drinking brandy, even if there is no flame coming out of his throat, or if milk in women's breasts is a sufficient proof of pregnancy. A recurring figure in the reports up to 1636 is Christian IV's personal physician, the clear-sighted and righteous Henning Arnisæus (1570–1636), the founder of forensic medicine in Denmark. You get the impression of a man who, despite his time, had an admirable understanding of the basic legal principle of giving the accused the benefit of the doubt, and who was very aware that his statements could mean the difference between acquittal and death on the scaffold for the accused. All medical arguments that could be in favour of the accused had to be put forward. This did not least apply to cases where young women were accused of murdering their own fetus or newborn child, but where there was no evidence of anything other than spontaneous abortion or natural stillbirth.

An example of this is the case of Margrethe Johannesen's abortion in 1633.[54] Margrethe Johannesen from Elsinore was accused by the city council and the district council of having killed her fetus. On 4 April, some women had given testimony in the city council that there was milk in Margrethe's breasts, and one of them had noticed that there were scars from skin ruptures on her belly, as if she had given birth. Margrethe had on 29 April admitted that she had been with a man three weeks before Easter, and that she had become pregnant as a result. Around Midsummer's Day she had lost the child, which was not larger than a goose egg, and she had thrown it on the dunghill, because she did not know what it was. The 14 men, who were appointed as judges in the case, believed that the milk in the woman's breasts and the skin ruptures on her abdomen, which were found upon the examination of her a year after the intercourse, were more weighty evidences that she had given birth to a living and full-term child and then killed it than the woman's own explanation and denial, and they therefore wanted to convict her of murder. The king's chancellor therefore asked the physicians in Copenhagen about their professional opinion in the case and, according to the chancellor's wish, their representative Ole Worm asked Henning Arnisæus to give a statement.

Arnisæus states on 24 September that the question of the woman's guilt on the present basis "must be answered clearly in the negative. For although it has been stated that the milk in the breasts was white and flowed more abundantly at the first than at the second examination, it only means that she has been pregnant. And the ruptures on the abdomen do not mean anything else either. Whether she has given birth to a full-term and living child after a normal pregnancy or has aborted a very small undeveloped fetus is completely impossible to determine and establish from these signs, which appear and show after both kinds of births. Indeed, physicians have their diagnostic signs, which makes it possible to distinguish an abortion from a natural birth. [...] But they provide no absolute certainty – and there must be no doubt about that – but a probable, doubtful and uncertain certainty, and they cannot distinguish an abortion from a natural birth, except in the case of a woman who is still warm after a recent birth. [...] Thus, based on these signs, not the slightest objection can be raised against the explanation of the accused."

This is Arnisæus' official statement, which he asks Worm to pass on to the chancellor as his judgement in the case. However, in some unofficial remarks to Worm, Arnisæus casts doubt on the accused's explanation and regrets the judges′ poor handling of the case. He asks Worm to advise the judges not to give the woman the very harsh sentence for infanticide, if they should anyhow decide to judge her for this crime based on her own explanation of birth in the fourth month of pregnancy. The physicians of Copenhagen state on 30 September, that it is sufficiently clear and obvious, that the woman has been pregnant. "But the documents that we have been presented with, do not allow us to further conclude that it was her own fault that the fetus died, or that it was expelled because she had taken some medication or something of that nature. Finally, we also cannot determine whether the fetus has been more developed than the woman indicates. To confirm this, we have attached the seal of the Faculty of Medicine."[55]

The royal ordinances from Frederik II's and Christian IV's time essentially deal with the supervision of the healthcare activities carried out by physicians, apothecaries, barber surgeons and other practitioners. Some ordinances obliged the physicians to carry out supervision with the country's pharmacies twice a year to control the composition, effect, durability and the prices of the medicaments and mixtures as well as the breadth of the pharmacies' product range. Other ordinances concerned regulations for the training and business of surgeons and apothecaries. Many of the ordinances have visionary content, which today's health authorities could profitably learn from. This applies not least to the control of and sanctions against the paid treatment of sick people by unauthorized persons ("empiricists"). An example of this is "Christian IV's ordinance concerning physicians, apothecaries and surgeons and especially about physicians' fees and fines for empiricists." It is from 10 January 1619 and says, among other things:

"And as we have most graciously learned that various empiricists dare to give the sick internal medicine, we wish for the future to have this most graciously prohibited. However, they may prescribe wound potions for a fall, sting or wound, but in this as in other dangerous external illnesses an authorized physician should be consulted, unless the surgeon himself wants to bear the moral and legal responsibility for any wrongful treatment. Therefore, only authorized physicians and medical doctors will be allowed to prescribe medicine for internal use. All others – barber surgeons, apothecaries, chemists, oculists, hernia cutters and quacks – must refrain from it and stick to their own business and educate themselves therein."

The penalty for violation of the rules was significant: "If anyone should be caught in violation of this, he shall be fined 100 rix-dollars each time."

Domus Anatomica Hafniensis has its own pagination with 64 pages and is provided with its own frontispiece (Figure 87), which is normally placed opposite *Cista Medica*'s frontispiece. The title page has an illustration of a pelican, which pecks itself in the chest with its beak, so that its offspring may drink of the flowing blood (Figure 158).

On page 27 and 35 there are illustrations of both the front and back of the admission card (*tessera*) (Figure 159) which were used by both Simon Paulli and Thomas Bartholin in their era at the Anatomy House. After the final page, there is a page with the seal of the Faculty of Medicine, *Sigillum Facultatis Medicæ Regiæ Acad. Hafniensis*, which belongs to *Cista Medica*. Besides what has been described, *Domus Anatomica Hafniensis* also includes mention of a new hall on top of the Anatomical Theatre (the earlier-mentioned museum of natural history), established in 1657. The hall was to house Henrik Fuiren's collection of rare natural objects (The Fuiren Museum), which he bequeathed to the university. A six-page inventory of a selection of the items in this collection follows and then follows a 12-page "Inventory of the anatomical and zoological natural objects in Thomas Bartholin's Museum," which is located on the first floor in his professor's residence. The objects "attract the curiosity of foreign visitors; [...] some were collected by him on his travels, some purchased, some sent by friends, and some left behind after public anatomical dissections."[56] Finally, the book has a section with that kind of Bartholinian self-praise that has provoked so many Danish historians of medicine and contributed to their unjustified and unvarnished labelling of Thomas Bartholin as a self-asserting braggart with limited value as a researcher and natural scientist.[57] The section reads:

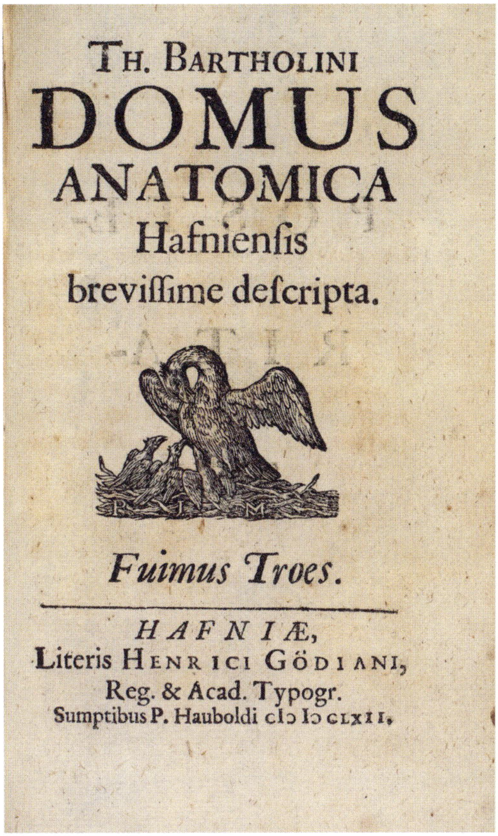

TH. BARTHOLINI
DOMUS
ANATOMICA
Hafnienſis
breviſſime deſcripta.

Fuimus Troes.

HAFNIÆ,
Literis HENRICI GÖDIANI,
Reg. & Acad. Typogr.
Sumptibus P. Hauboldi cIↃ IↃ CLXII.

Figure 158. Title page of Thomas Bartholin's short description of the Anatomy House in Copenhagen, *Domus Anatomica Hafniensis brevissime descripta*, published in Copenhagen 1662.

The illustration shows a pelican pecking its breast with its beak, so that its dead fledglings can be brought back to life by drinking the outflowing blood. The motif originates from an ancient fable.

The printer Henrik Gøde also used the illustration on the title page of Volume III of Bartholin's *Historiarum Anatomicarum Rariorum*, Copenhagen 1661.
The Royal Library in Copenhagen.

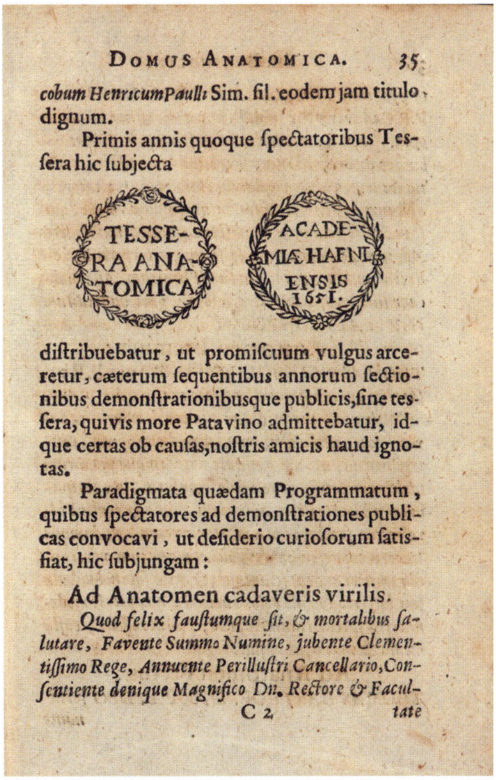

DOMUS ANATOMICA. 35

cobum HenricumPaulli Sim. fil. eodem jam titulo dignum.

Primis annis quoque ſpectatoribus Teſſera hic ſubjecta

TESSERA ANATOMICA

ACADEMIÆ HAFNIENSIS 1651.

diſtribuebatur, ut promiſcuum vulgus arceretur, cæterum ſequentibus annorum ſectionibus demonſtrationibusque publicis,ſine teſſera, quivis more Patavino admittebatur, idque certas ob cauſas,noſtris amicis haud ignotas.

Paradigmata quædam Programmatum, quibus ſpectatores ad demonſtrationes publicas convocavi, ut deſiderio curioſorum ſatisfiat, hic ſubjungam:

Ad Anatomen cadaveris virilis.

Quod felix fauſtumque ſit, & mortalibus ſalutare, Favente Summo Numine, jubente Clementiſſimo Rege, Annuente Perilluſtri Cancellario, Conſentiente denique Magnifico Dn. Rectore & Facul-

C 2 *tate*

Figure 159. Access badge (*tessera*) for the audience to the Anatomy Theatre in Copenhagen in the early years of Thomas Bartholin's days as head of the theatre, here from 1651. *Domus Anatomica Hafniensis*, Copenhagen 1662, p. 35.

According to Bartholin, the purpose of the badge was "to keep the mob away." However, after a few years, the sign was removed, following Padova's example, and the doors were opened to everyone at the public dissections and demonstrations.
The Royal Library in Copenhagen.

"In the meantime, while he scrupulously examined the internal organs in different human and animal bodies, he was the first to discover, demonstrate and explain a man's thoracic lacteal vessels in a published work, which was offered and dedicated to his great monarch, FREDERIK THE THIRD, under whose kindly protection these rich new lacteals flowed forth. In this connection he amused himself as follows:

The lacteal vessels creeping through the quivering breast
Have claimed the heart as a lord for their own guidance.
So may your mountains flow with milk, when you regain
The golden sceptres of the North, O sovereign Frederik.

T.B.

More good fortune shone forth there, when he first revealed in this theatre the lymphatic vessels, never seen or known before, initially in dumb animals, then in human beings, too; with these discoveries he roused the learned world partly to approbation, partly to disputes and varied exertions, and through this new finding shed important light on the practice of medicine. In memory of that accomplishment his colleague and relative, Hr. Dr. Ostenfeld, with excellent ability and excellent learning has raised this tablet:

FOR HER FAMOUS LEADER AND RESTORER,
GLORY OF THE DANISH NATION,
SPLENDOUR OF THE WORLD OF LETTERS
AND GREAT CONTRIBUTOR TO MEDICINE,

THOMAS BARTHOLIN,

WHO, WHEN HE HAD REVEALED, PURGED AND DESTROYED
THE UNHAPPY ERRORS OF SO MANY GENERATIONS AND OF SO
MANY LEARNED MINDS, AND HAD SUBDUED THE SPITE OF
AN ENVIOUS AGE
DURING THE REIGN OF THE BEST OF KINGS,

FREDERIK THE THIRD,

IN THE FIFTY-SECOND YEAR OF THE SEVENTEENTH CENTURY,
FIRST FOUND AND DISPLAYED THE LYMPHATIC VESSELS,

THOSE UNFAILING RUNNELS OF LIFE-GIVING LYMPH,
UNKNOWN AND UNSEEN IN EVERY CENTURY,
STREAMS AND CHANNELS OF TRANSPARENT DEW,
WHERE THROUGH ALL PARTS OF THE HUMAN BODY,
ITS HOLLOWS AND DEEPEST RECESSES, THERE FLOWS
THE TRUE, HABITUAL NECTAR OF HEAVENLY KIND,
THE INDIGENOUS NOURISHMENT OF THE BODY,
THE MOST CELESTIAL PORTIONS OF ITS FOOD,
LIKE A FIFTH ELEMENT,
IN THE HONOUR OF THIS BRILLIANT DISCOVERY

ANATOMY HERSELF

DECREES AND OFFERS
LAURELS,
RENOWN,
MONUMENTS
AND IMMORTALITY
WITH GLAD RECOMPENSE,
THROUGH

CHRISTIAN OSTENFELD,

DOCTOR OF MEDICINE AND PUBLIC PROFESSOR
IN THE ROYAL UNIVERSITY OF COPENHAGEN.

[Here follows the previously quoted poem about Tycho Brahe and Thomas Bartholin]. Tycho Brahe chose exile to observe the wandering planets, and the one who [...] was the German inventor of gunpowder, drew the gunfire and curses of posteriority; however, let these not be a true presage that Bartholin with his lymphatic vessels should either drink only water, or through the depravity of a wicked age himself become demented. May God avert all misfortune, for till now, like *Nepos'* Epaminondas, he has received nothing from his nation but renown. And although mankind itself might be totally unwilling to owe any obligation to him or to his diligence and good fortune, he himself has been grateful towards his dog, in whom he first discovered those vessels, and for that reason supplied its skeleton in that same location with this inscription:

FOR MOLOSSUS
BENEFICENT IN HIS VIOLENT DEATH,
SHINING WHITE NOT WITH HIS COAT,
BUT WITH HIS BONES,
WHO WITHOUT GRUDGE FIRST
SHED HIS UNSEEN
LYMPHATIC FLUIDS
FROM HIS EEXPOSED AND SUFFERING ENTRAILS,
SO THAT BLEACHED, IRKSOME TO NONE
AND THE ENVY OF HIS FELLOWS,
HE MIGHT SURVIVE AFTER DEATH
AS A MEMORIAL FOR MORTALS,
DR. THOMAS BARTHOLIN
HAS PLACED THIS TABLET IN GRATITUDE."[58]

The Bartholinian letters are published as a three-volume work

Despite Bartholin's great desire to write and his writing ability, he never published an autobiography, nor does he mention anywhere in his writings plans for such. On the other hand, he enriched posteriority with an extensive three-volume work, *Epistolarum Medicinalium* (Medical letters), which contains his entire correspondence with the scholars of Europe over 25 years in roughly chronological order from January 1639 to July 1664. In total there are 400 letters in the form of four centuries with 100 letters each. Nine of the letters have illustrations. Bartholin made drafts of all his own letters and saved all incoming letters from the scholars of Europe.

The first volume with the first two centuries was published by Peder Haubold's Publishing House in Copenhagen in 1663 with the title *Epistolarum Medicinalium à Doctis vel ad Doctos scriptarum Centuria I & II. Cum Indicibus necessariis* (Medical letters written by scholars to scholars. With the necessary indexes). It covers the period from January 1639 to February 1660 and has 739 pages, besides a dedication to Johan Georg III of Saxony (1647–1691), who three years later married the eldest daughter of Frederik III, Anna Sophie of Denmark (1647–1717), a preface, a list of the learned senders and recipients of the letters and a keyword index. The volume contains the earlier-mentioned illustrations of the bones of the middle and inner ear in a letter from the Venetian Cecilio Folli of 19 November 1644 and illustrations of a ship from antiquity, the horns from a horned cock, the heart valves from Johannes Trullus' autopsy of cardinal

Camillo Melzi (1590–1659) on 21 January 1659 (sent to Thomas Bartholin in a letter from Henrik Møinichen in Rome on 12 November 1659) and abnormally long fingernails on a human (Figure 160).

A second volume with the third century was published in Copenhagen in 1667. It covers the period from September 1659 to June 1662 and has 442 pages besides a dedication to the learned French diplomate and advisor of Louis XIV, Antoine de Courtin (1622–1685), a preface, a list of the learned senders and receivers of the letters, a poem to Thomas Bartholin by bishop in Trondheim Erik Pontoppidan the Elder (1616–1678) and a keyword index. Besides an illustration of a large bladder stone from a man, the last page of the volume has an illustration of a stone from the stomach of a dog, and then follow two unnumbered pages with two rather primitive illustrations of a monster from Ubberød (*Monstra Fionensia*) with partially missing brain (anencephaly), cleft lip and palate, low-set ears, umbilical hernia containing abdominal organs (omphalocele) and club feet (belonging to letter no. 75, June 1662, from professor in theology at the high school in Odense, Jacob Bircherod (1624–1688).

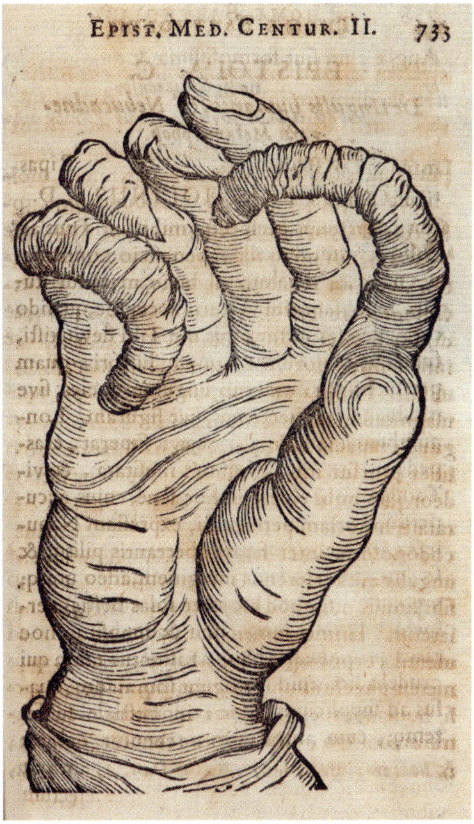

Figure 160. Grotesquely long fingernails. Illustration from Thomas Bartholin's epistolary work *Epistolarum Medicinalium*, Cent. 2, Copenhagen 1663, Epist. 100, pp. 732–39.
The Royal Library in Copenhagen.

A third volume with the last century was also published in Copenhagen in 1667. It covers the period from May 1662 to July 1664 and has 568 pages. It is dedicated to Peder Schumacher (1635–1699), who had just returned from a study trip abroad and had become the director of Frederik III's library and archive. Then follows the preface, a list of the learned senders and receivers of the letters, a poem by Professor Ole Borch to Thomas Bartholin and, after the letters, a keyword index. Only letter no. 70, from Niels Stensen in Leiden on 30 April 1663 about the structure of the muscles and the heart, has illustrations.

Epistolarum Medicinalium is an extremely valuable source of insight into and understanding of the history of science and natural philosophical thoughts of the 17th century. In this book many examples have been given of the letters' broad content of natural science. The reason why the last published letters are from 1664 is of course not that Bartholin's letter correspondences ceased at this time. He had prepared further three centuries with letters ready for printing, and Haubold, the publisher, had made the necessary preparations for the edition, when the entire manuscript perished in the flames of the fire of Bartholin's manor Hagestedgaard in 1670. In his book about the fire, Bartholin writes the same year:

"I am greatly distressed that the letters of good friends concerning a variety of medical and natural matters, as well as my replies offering many observations were destroyed. May readers believe me without oath that replies sought by erudite friends did not perish through my fault or forgetfulness. Although I am by no means unaware that occasionally there was an exchange of letters with those whom public opinion either had condemned or has held suspect (it does not concern me whether rightly or wrongly), yet may everyone know that it was always my intention that Chapter II of Seneca's *De tranquillitate animi*[59] should be observed: never be ashamed of a good thing in a bad author. Also, I was prepared for the decrease of my reputation, especially through the increasing authority of greater ones, to test the pens of men famous for whatever erudition or experience and to search in forgotten places, if thereby secrets could be elicited for human perception and health. If in this way I have sinned, may it be imputed to a zeal for aiding mankind, to which I have hitherto consecrated myself."[60]

It is almost chilling to think what scientific and cultural treasures were lost to posteriority with the loss of this manuscript. For example, a further exchange of letters with Niels Stensen has certainly been part of the loss. Only a few Danish letters from Bartholin's hand have been preserved. As the owner of a manor like Hagestedgaard he was obliged to deliver a few young farmhands for military service. A letter to the mayor of Holbæk on 25 April 1664 is about this (Figure 161).[61]

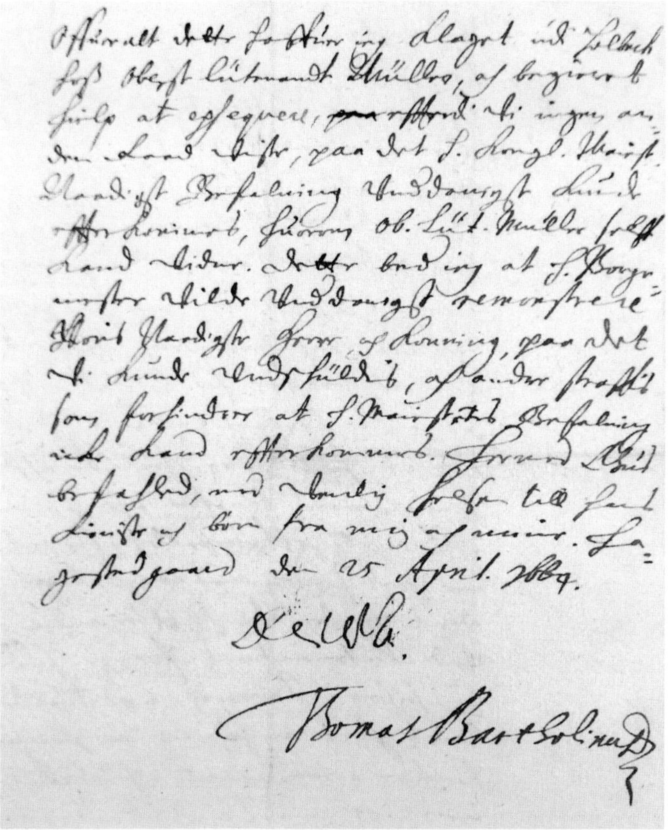

Figure 161. Last page of a letter in Danish from Thomas Bartholin to the mayor of Holbæk on 25 April 1664 regarding military service for Hagestedgaard's male staff. The Danish National Archives (Rigsarkivet) in Copenhagen. Photo: Jesper Brandt Andersen.

Bartholin buys the manor Hagestedgaard close to the town of Holbæk

The Danish nobility had largely failed to defend the country during the so-called Carl Gustav wars of 1657–1660 between Frederik III's Denmark-Norway and Carl Gustav X's Sweden. The first of the two wars was declared by Frederik III in June 1657 in the hope that Denmark could regain the territories lost in the Torstensson War of 1643–1645. But the ice winter of 1658 put the superior Danish fleet out of action, and the Swedish troops were able to cross the frozen belts and force the Danes to a humiliating peace treaty in February 1658, in which Denmark had to cede large areas of land east of the Sound and in Norway. But Carl Gustav was not content. He wanted to conquer all of Denmark-Norway. In the beginning of August 1658, less than six months after the peace agreement in Roskilde, he broke it and landed an army in Korsør on the west coast of Zealand hoping that he could conquer Copenhagen and thus the rest of the Danish Kingdom with a surprise attack.

However, this time it turned out that Carl Gustav had miscalculated the situation and underestimated Frederik III's and the Copenhageners' willingness to defend themselves. When Frederik was notified of the Swedish landing in Korsør, he refused to follow advice to flee to safety in Norway or Holland.

Unlike the nobility and the government, who largely fled, he stayed in the city prepared to defend it to his death, uttering the famous words from The Book of Job: "I shall die in my nest." By promising them privileges he incited the citizens to take up the fight against the Swedes, and within a few days a strong defence was built on the ramparts of Copenhagen. Approaching the city, Carl Gustav realized that this time he would meet tough resistance. He therefore decided to besiege the city – a decision that would prove to be fateful. During the autumn, a Dutch rescue fleet under the command of admiral Jacob van Wassenaar Obdam (1610–1665) managed to slip past the Swedish fleet and reach Copenhagen after a short battle with the Swedes in the Sound (Figure 162). With the help of the Dutch troops and supplies, the Copenhageners managed to beat back the main Swedish storm on the city in February 1659, with heavy Swedish losses as a result. The war ended with a peace treaty in Copenhagen in May 1660, in which Denmark only got back the island of Bornholm and a part of the lost land in Norway, Trondhjems Len.

Domestically, however, the war took on far greater significance. The fact that Frederik III and the citizens of Copenhagen had saved Denmark from extinction as a sovereign and independent country, while the nobility had failed to take part in the defence, quickly had significant political and economic consequences. Based on the new alliance between the king and the citizens, the king's absolute power and sovereignty were already adopted that autumn through the introduction of the Act of Succession and absolute monarchy in Denmark. Through new provisions, the political influence of the nobility was reduced, and that of the citizens increased. The nobility's privilege of tax exemption was abolished, and it became possible for citizens to acquire manors and land owned by the nobility.[62] Courtiers, civil servants and wealthy citizens began to buy the nobility's manors and pleasure estates, and the areas around Copenhagen were particularly attractive.

A typical example is Thomas Bartholin's purchase in 1663 of the manor Hagestedgaard near Holbæk in North Zealand from the nobleman Henrik Thott (1606–1676) (Figure 163).[63] Like many of his peers, Thott was a former wealthy Danish nobleman and landowner with large landholdings, who within a few years fell into financial trouble as a result of the destruction and plundering of the Carl Gustav wars and the new absolute monarchy's removal of significant economic privileges from the nobility.[64] That Thomas Bartholin should have benefited from the new rights of the citizens through his purchase of this manor

Figure 162. Thomas Bartholin's manor house, Hagestedgaard near Holbæk, which he bought in 1663. For a number of years Bartholin was also the owner of the neighbouring smaller manor Ellingegaard which he bought from his brother Caspar's father-in-law, Minister for Finance Henrik Müller. Through his marriage to Magdalene Rhode in 1678, he became the owner of a third manor, Fjellebro on Funen. Photo: Jesper Brandt Andersen.

can rightly be challenged as rather ironic, since, as mentioned, he did not take part in the defence of Copenhagen during the siege, but, like the nobility, and like he did during the plague in 1654, sought safety with his family in Roskilde. When it came to bravely putting his life on the line, Bartholin was no hero. The only wars he fought were fought with a pen. Nevertheless, Bartholin somehow managed to stay on good terms with the Danish kings, and the list of gifts, honours, properties, financial benefits and rights granted to him by Frederik III in the 1660s and Christian V in the 1670s is long.[65]

When he had been appointed honorary professor, Bartholin began to look around for a place where he could retire to quiet, rural surroundings and clean air, in order to write his books far away from the noise, stench, dirt and infectious diseases of the narrow streets and alleys of Copenhagen. On 7 April 1663 he wrote to Niels Stensen in Leiden: "But you should know that I am preoccupied with thoughts of leaving the noise of the city and trying to become a better person in the tranquillity of the countryside." Only half a year later, he had found his rural refuge, where he could retire to peace and quiet with his books and manuscripts in scenic surroundings. Some years later, he wrote in his book about the burning of his library, that the charm of the place was hardly surpassed by any other place in the world:

"My peaceful Tusculum, which had given me pleasure scarcely to be equalled elsewhere in Borea, was thrown into confusion by the fire. Here and there washed by the neighbouring sea, refreshed by a salubrious climate and shaded by three woods, with fertile fields and flowering meadows, the land was everywhere harmless, able to support neither serpents nor dormice, while the broad farms of the peasants, the splendid buildings, the immunities which had been granted, the serene sky and general harmony had all increased my love of the estate."[66]

Here and there on the farm, he set up small inscriptions expressing his enthusiasm for the place. None of the physical inscriptions survives, but the texts are found in Bartholin's collection of poems from 1668. One of them reads:

Figure 163. The Dutch rescue fleet under the command of Admiral Jacob van Wassenaar Obdam (1610–1665) passes the Sound off Kronborg Castle on 29 October 1658 with the Swedish warrior king Carl X Gustav (1622–1660) as a reluctant spectator (the Swedish flag). After a minor, victorious naval battle, the Dutch managed to slip past the Swedish fleet and come to the rescue of besieged Copenhagen. The Dutch war aid and supplies were crucial to the liberation of Copenhagen and the survival of the Kingdom of Denmark-Norway. Painting by F. C. Lund (1826–1901) 1872. Oil on canvas. 100 × 142 cm. National Gallery of Denmark.

"All the gifts of Nature and culture you find together in this place. And it does not matter that the farm lacks an alchemist's laboratory, for the barns store a golden harvest, and the stables contain a better goldmine than that of the alchemist."[67]

Hagestedgaard was also the ideal place for Bartholin's children to grow up, in safe and healthy rural surroundings and with teaching by the same private house teacher from the Latin school in Roskilde for 11 years from 1665 to 1676.[68] At the time of the purchase the farm, Bartholin and his wife Else had six children, three of each sex: Caspar (b. 1655), Christoffer (b. 1657), Thomas (b. 1659), Margrethe (b. 1652), Anna Margrethe (b. 1660) and Else Cathrine (b. 1662). In 1665 another son, Hans, was born. However, regular visits to Copenhagen were necessary to take care of the continuing commitments at the university and attending special events, such as when Niels Stensen returned home for a short stay in 1664.

Niels Stensen in Copenhagen 1664

Stensen's return to Copenhagen was not only welcomed by his family, but also very much by the university. After his appointment as honorary professor in 1661, first professor of medicine and Dean of the Faculty of Medicine Thomas Bartholin had largely withdrawn from practical teaching and devoted himself to his writings and administrative obligations. Certainly, he continued his medical research unabated and in that connection, and in the educational context, still undertook private dissections, but since in 1656 he had ceased to perform public dissections in the Anatomical Theatre in the Anatomy House, this theatre had essentially been abandoned and was, in Bartholin's own words in the letter to Ole Borch on 30 November 1661, "still covered in dust." And, as he added: "the art of medicine is on vacation at the university."

When reading the letter, you can almost hear his sighs in the background. His skilled prosector from the golden years of the discovery of the lymph vessels, Henrik à Møinichen, had returned home the year before with high qualifications and a doctorate in his luggage after six years of medical travel in Germany, Italy and France. He had been offered the professorship of anatomy, which had been vacant since Bartholin left it in 1654, but had declined and accepted a more convenient and probably better paid position as court physician to Frederik III.

Instead, the university had hired Stensen's childhood friend, Simon Paulli's son Jacob Henrik Paulli in the position in 1662, although so far it was only a designation with permission for some years of further medical studies abroad.[69] Paulli had carried these out in 1658–1660 and was now travelling to continue them, but came home after only a year to defend his earlier mentioned doctoral thesis, *Anatome Anatomiæ Bilsianæ*[70] – his only anatomical work – and take the professor's oath (Figure 164).[71]

He never started teaching in anatomy, and, like his father, he never lost his soul to this subject. He was more interested in poetry, history and law, and in 1663 took the philosophical master's degree in Copenhagen. In 1664 he was appointed vicarious professor of history for Rasmus Vinding (1615–1684), but at the same time retained his designated professorship in anatomy. From both professorships he received the same year leave for three years of study abroad and never again returned to anatomy.[72] In 1668 he became a royal historiographer and later a diplomat.[73] Second professor of medicine Christian Ostenfeld was more concerned with legal, political and administrative chores, while third professor of medicine Rasmus Bartholin's main interests were mathematics and physics.

It is therefore not surprising that Stensen aroused enthusiasm, admiration and hope for a new heyday for anatomy at the University of Copenhagen when he, returned as a new anatomical star, gave samples of his skills at private dissections in his home in the early summer of 1664 for a small circle of invited men, including his former teachers and colleagues of his own age. Thomas Bartholin, who was among those invited, came in from his manor to participate and

Figure 164. Portrait engraving of Jacob Henrik Paulli (1637–1704) by an unknown artist 1663. 12.5 × 10 cm.

The Latin text in the frame, "IACOBUS HENRICUS PAULLI. SIM. F. ANATOM. IN ACADEM. HAFNIENSI PROFESS. REGIUS. ÆTAT. XXVI. AN. MDCLXIII," reads: "Jacob Henrik Paulli. Son of Simon. Royal Professor of Anatomy at Copenhagen University. Age 26. Year 1663."

The text below the portrait reads: "Famed by virtue of his forefathers and father: No one was more worthy to take Bartholin's place." Signed I.W.

Jacob Henrik Paulli published his only independent medical work, *Anatome Anatomiæ Bilsianæ*, in Copenhagen the same year, i.e. 1663. It was dedicated to Frederik III. It came in a second edition in Strasbourg in 1665 which included a letter on anatomical doubts from the Swiss physician, pathologist and pharmacologist Johann Jakob Wepfer (1620–1695), who must be the man behind the initials I.W. above. Wepfer's statement was undoubtedly not shared by Thomas Bartholin, who was certainly not impressed by the young Paulli, but fought in vain to get his beloved, admired and ingenious student Niels Stensen to replace him in the anatomy professorship at the University of Copenhagen. Due to his father's close relationship with the king and without delivering so much as a dissection or a lecture, Paulli blocked Niels Stensen's access to the professorship for five crucial years from 1662–1667. When Paulli finally resigned in the autumn of 1667 to become a historian, Stensen had shortly before converted to the Catholic faith and had thus himself forever blocked his possibilities of an appointment as a professor in Copenhagen.
The Royal Library in Copenhagen.

IACOBVS HENRICUS PAULLI, SIM. F. ANATOM. IN ACADEM. HAFNIENSI PROFESS. REGIUS. ÆTAT. XXVI. AN. MDCLXIII
Clarus Avis, et Patre suo: non dignior alter
Qui BARTHOLINO sufficeretur erat.
I.W.

describes the event in two of his later publications of 1666, *De Hepatis Exautorati Desperata Causa* and *De Medicina Danorum Domestica* (On Danish domestic medicine). In the former work, Bartholin writes that the sources of saliva for centuries remained hidden in the mouth, and that no one had thought of them before salivary ducts, the descendants of the lymph vessels, were revealed by Wharton. With him, Bartholin says, our Stensen deserves to share the credit. Stensen, who according to Bartholin is born for anatomical studies, gains added credit by describing with diligence the duct of the parotid gland, with which Blasius has also dealt. Bartholin did not want to interfere in the strife between his friends, but as can be seen, he no longer hides to whom he assigns the priority to the discovery. He briefly describes the dispute, but then adds:

"Concerning the teachings of our friend Blasius, which is witnessed by so many published writings, we do not doubt, but of what his learned hand manages during an anatomical examination, I must necessarily be ignorant, as I have never seen the famous man, to whom I owe esteem, dissect. Our Stensen's practised hand and luck in making extraordinarily many discoveries are not only attested by his published observations, but I have seen him with my own eyes. For when the famous Stensen in 1664 due to his mother's death stayed in the king's city of Copenhagen,[74] in his home he undertook various dissections, with which he entertained his expert friends. Since I happened to have returned from my estate by then, he also invited me to be present along with the famous doctors Dr. Simon Paulli, Dr. Willum Worm and Dr. Henrik à Møinichen to see the new salivary ducts, which he had prepared and found. I received the invitation with friendly courtesy and looked forward to seeing new anatomical advances with a student to whom I had already long ago handed this torch. I would like to recall in this appropriate place what I saw then. On a calf's head he showed us the external salivary duct, that extends from the parotid gland to the lower corner of the upper jaw and from there up to the cheeks. Next, the small salivary ducts of the cheeks were presented in the small glands that fill the entire cheek region. These small salivary ducts give access to a pig's hair, and thanks to that you can watch them. Otherwise, they are hidden, unless you squeeze their liquid out with your fingers. On the surface of the palate, he also showed us countless small orifices, which receive salivary ducts from rounded glands. Moreover, he revealed to us, without cutting, the very small salivary ducts under the tongue. You can see them on the sides of the tongue within the tooth row. On the head of an African rooster, which he held in his hand, he showed, that the whole cheek, between the root of the tongue and the opening of the larynx, as well as on the sides of the tongue and under the tongue, hid glands, and that it was pierced by the openings of the salivary ducts. He demonstrated another duct, which is located at the front within the teeth and from the nose goes down through the palate (Figure 165), as well as small ducts piercing the cartilage of the larynx. The new duct found because of his dexterity, and which above all aroused applause, directs the fluids of the eyelids away from the tear points and into both nostril cavities. The other similar presentations, which Stensen carried out publicly with skill and great probity, provoked great enthusiasm among the audience. In order not to pass over our lymph vessels in silence, he made much of showing their real mouth into the veins. [...] We must not forget the chicken that Stensen brought with him. On it he found a special excretory duct from the yolk membrane to the intestines, and through it the yolk is pushed into the intestines to nourish the chicken.[75] This puts an end to many disputes in anatomy. Towards the end, with his investigation of the heart's new function and structure, he got our admiration to rise further. Thrilled that so many observations were being

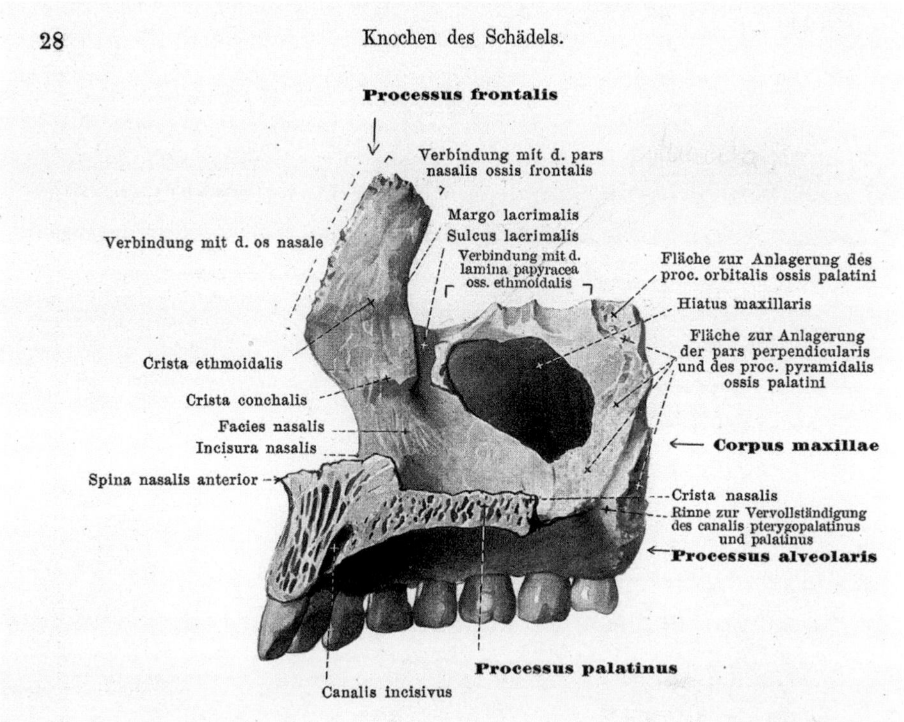

Figure 165. *Canalis insicivus*. The small canal in the upper jawbone above the front teeth connects to the nasal cavity. It was first described by Niels Stensen. From Werner Spalteholz: *Handatlas der Anatomie des Menschen*, Vol. 1, Leipzig 1910, p. 28.
Private ownership.

made public by another mouth, we left congratulating ourselves with such a significant pupil, who already surpassed his own teachers, and we congratulated not least the fatherland on such an outstanding prosector. Before the others, I thought of placing this great ornament of anatomy among the stars and in the council of Nature, because he has brought out things that were unknown to the ancients, and he has taken it upon himself to continue working with our lymph vessels, and it is not in vain, but of great benefit to men, who thereby become both more learned and more skilled at healing patients."[76]

In the other work mentioned, Bartholin mentions Stensen as "my former pupil, now a precious friend, an unusually diligent and lucky man, from whom one can expect the most complete."[77] After a list of all Stensen's discoveries so far Bartholin gives a slightly more detailed description of Stensen's demonstration of the heart for the small circle of invited doctors in Stensen's home:

"In the circle of anatomists, our Stensen has also presented and described many excellent things, first of all the external salivary ducts, the cheeks' salivary ducts, the small salivary ducts under the tongue, the salivary ducts of the palate, the anterior canal from the nose to the palate, the ducts of the larynx and nose, the connections from the eyelids to the nose, the tear ducts of the eyelids, the insertions and mouths of the lymph vessels, the yolk entry into the intestines of chickens, and daily his tireless hand and inquisitive nature make new discoveries. We have seen him demonstrate each finding himself, and we have congratulated the fatherland with the happy progress of anatomy. In addition, we have seen with him the perfectly new and elegant construction of the muscles and not without admiration his dissection of a heart, which he opened so skilfully that we could observe that the whole collection of fibres encircled each other like a snail to rise upwards again from the tip of the heart. With Hippocrates, therefore, we did not hesitate to call the heart a real muscle."[78]

Stensen's work on muscles, heart and glands and the lost professorship of anatomy

During the summer of 1664, Stensen in Copenhagen and Amsterdam published the acclaimed work on muscles and glands, *De Musculis et Glandulis Obervationum Specimen. Cum Epistolis duabus Anatomicis* (Test of observations of muscles and glands. With two anatomical letters), which summarizes his anatomical observations from the years in Holland. The book is dedicated to Frederik III, "Most serene and mighty principal and lord,My most gracious lord, to whom

Figure 166. The frontispiece of the Copenhagen edition of Niels Stensen's *De Musculis et Glandulis Observationum Specimen. Cum Epistolis duabus Anatomicis*, Copenhagen 1664.

Stensen has here abbreviated the title of the work to *Nicolai Stenoni's Observationum Anatomicarum Specimen*. The engraving illustrates, scattered between the flowers, several of Stensen's research areas: muscles, tongue, heart, lymph vessels and eggs (reproduction). The Royal Library in Copenhagen.

I hereby humbly offer this small summary, which touches on what I have observed so far in anatomy," signed "Most devoted subject Niels Stensen" (Figure 166). In the treatise Stensen says the famous words:

> "The heart has been made the home for innate heat, the throne of the soul, yes by some even the soul itself. The heart has been greeted as the actual sun, even the king, but if you examine it properly, you will find that it is nothing but a muscle."[79]

In the treatise, by comparing skeletal muscles with the heart, Stensen provides the final evidence that the heart is a muscle, and he explains the contraction mechanism of muscle fibres based on a preliminary geometric model. In addition, based on the course of the heart muscle fibres, he accounts for the movements of the heart. On these questions he disagrees with both Harvey's and Descartes' explanations, and his famous words quoted above are directly addressed to Harvey, who in his *De Motu Cordis et Sanguinis* had attributed to the heart the mentioned qualities.[80] Stensen's dedication of the book to Frederik III was certainly not without ulterior motives. There are many indications that Stensen was ready to take up a professorship in anatomy in Copenhagen at this early stage of his career,[81] and this had hardly prevented him from further education abroad, as leave for studies abroad – even for several years, as exactly shown in the case of Jacob Henrik Paulli – was quite common at the University of Copenhagen in Stensen's time.

As mentioned, the problem was that the professorship in anatomy was already occupied by his childhood friend Jacob Henrik Paulli, who had occupied it since 1662, and who, in spite of his lack of practical function in the professorship and his fundamental lack of interest in anatomy, had his leave extended for a further three years in 1664, that is, until 1667. In other words, the young Paulli blocked the far better qualified Stensen's access to the professorship for five years from 1662 to 1667. How this came about is revealed in a letter from Thomas Bartholin to Ole Borch on 31 May 1662, that is, at the very time when Bartholin and Borch had chosen Stensen as the next man in the professorship, and when Bartholin had tried to prepare Stensen's way to it by presenting the king and chancellor with Stensen's extraordinary abilities as a researcher.

Bartholin writes soberly but undoubtedly wistfully to Borch: "Our dear friend Simon Paulli has obtained the Anatomical Theatre for his son with a salary of 400 rix-dollars."[82] Neither Bartholin's wording for Borch nor his lack of praise of Jacob Henrik Paulli suggests that he himself, despite his status as first professor of medicine and Dean of the medical faculty, had had any material influence on the decision, or that his influence at the highest level could trump that of Simon Paulli. When barely a year later, on 7 April 1663, Bartholin again writes to Borch in Leiden, it does not appear that his enthusiasm for the young Paulli in the post as professor of anatomy and head of the Anatomy House has grown: "Your anatomical observations last winter were very joyful. Our own theatre is still quiet and will continue to be so until it is brought to life by Jacob Henrik Paulli. He is currently busy with his inaugural lecture."[83]

And Simon Paulli was not just anyone. As mentioned, Christian IV had summoned him to Denmark from Rostock in 1639 to an extraordinary professorship in anatomy, surgery and botany and supported his creation and leadership of the Anatomy House. Later they had collaborated on Paulli's *Flora Danica*. During the reign of Frederik III, Paulli took care to maintain his close relationship with the king and the royal family. Besides being Bartholin's colleague, Paulli was first court physician and later the king's personal physician. In this capacity he was close to the king, and undoubtedly also closer than Bartholin. And Paulli worked hard during these years to make his relationship to Frederik III even closer (Figure 167). Under these circumstances Stensen had to give up his wishes for a professorship of anatomy in Copenhagen for the time being. In the autumn of 1664, therefore, he travelled to Paris.

Figure 167. Simon Paulli kneels before Frederik III and presents him with the work *Quadripartitum Botanicum* (Botany in four parts), Copenhagen and Strasbourg 1667. Paulli dedicated the work to God and the king (DEO. REGI), as can be seen from the posted page. Engraving by Albert Haelwegh between the title page and the dedication. 18.9 × 14.7 cm.

The Latin text on the canopy above the king can be translated as follows: "Remember your mission, Frederik, is to govern and command the peoples (that is your art) and impose the way of peace; be sparing with your subjects and batter down the proud" (Adapted from Virgil's *Aeneid*, VI, pp. 851–53). The text at the bottom left: "I myself, decked with a chaplet of clipped olive leaves, shall bear gifts" (Virgil: *Georgics*, III, pp. 21–22). The text at the bottom right: "Seated on the threshold of the shining temple of Phoebus, he inspects the nation's gifts" (Virgil: *Aeneid* VIII, pp. 720–21). Source of the translations: Bruun (ed.) 2015, p. 141. The other people on the engraving are the king's counsellors, from left Christoffer Gabel (1617–1673), Erik Krag (1620–1672), Hans Schack (1609–1676) and Peder Reedtz (1614–1674). As can be seen, the king's personal physician Simon Paulli was skilled at positioning himself to win the king's favour. He utilized this to get his son employed in his own former anatomy professorship. It was fateful for Niels Stensen. The Royal Library in Copenhagen.

Work on comets

In late 1664 and early 1665, a large comet (C/1664 W1), which could be seen with the naked eye, appeared in the sky over most of the globe. In Copenhagen, it appeared in the sky in December 1664. Like many other scholars of the time, when the legacy of Tycho Brahe was still very much alive, both Thomas Bartholin and his brother Rasmus were interested in astronomy and took part in the interest in the comet. Rasmus Bartholin made thorough observations of it,[84] which he sent to the exiled Polish scholar Stanislav Lubieniecki (1623–1675),[85] who in these years lived in Copenhagen.

As always, when Thomas Bartholin dealt with areas of natural science other than medicine, it was the tangential medical aspects of these areas that occupied him. He thus published in 1665 at Peder Haubold's Publishing House a 154-page book *De Cometa, Consilium Medicum, Cum Monstrorum Nuper in Dania natorum Historia* (Medical consultation on the comet and an account about monsters recently born in Denmark), which is dedicated to Royal Professor of Medicine and Mathematics Rasmus Bartholin, and which ends with a letter from Thomas Bartholin to Stanislav Lubieniecki entitled *De Abcessu Coeli seu Cometa* (On the abscess in the sky or the comet). Both in the general population and among many scholars, comets were associated with various kinds of superstition. Many perceived comets as an omen sent from God of an impending punishment in the form of sickness, war, or disaster. In other words, they believed that man's deeds on Earth could affect the celestial bodies.

Bartholin was no exception. He writes in the book that comets never appear innocently in the sky. He thinks they portend evil and refers to the fact that a severe plague epidemic and the protracted German war (The Thirty Years' War) occurred after a comet had appeared in the sky in 1618. After the appearance of another comet in 1652, infectious diseases and wars in Poland and Denmark followed.[86]

Bartholin perceived comets as a kind of boil in the sky. He believed that they consisted of a substance that could harm the bodies of the Earth's inhabitants and cause disease, and that they could carry toxins from the sky to the Earth. In one section of the book, he describes his thoughts on two comets that followed consecutively in different places in the arc of the sky:

"I think of the comet almost as a boil in the sky, which has its increase and decrease; but whence the substance comes that collects in the tumour, I am not in a position to decide. Whether it is effusions that rise from our stinking Earth, or it is discharge from the celestial bodies, which are in a morbid state, is unknown, although the last assumption is most easily applauded by astronomers, who admit, that in the sky both destruction and new creation take place, but who will not so gladly admit that effluents from the Earth can reach up to the distant sky. But no matter how we assume that comets arise, they undoubtedly consist of a substance that can dissolve and return to us, whether hot or dry, or whatever other property it has, whereby our bodies and other things on Earth are either ignited by the heat or dried by the dryness. For I dare not predict anything auspicious from this heavenly boil, which either has burst or is about to burst. But after a tough winter it has brought us febrile illnesses and deadly pain in the side and an unusual mortality among the cattle. Therefore, it suggests a morbid condition in the universe, especially because a boil has failed to drain the disease substance, since another comet followed in Pegasus. The size and duration of both of them announce the same thing; for they indicate an excess of malignant matter, which, as it cannot be destroyed or distributed, returns to us and carries disease material to us. But which accidents should be feared is not so clear, because we know nothing of the nature of the distant boil: whether it threatens the earthlings with doom through dryness or heat or both. Particularly, I think, it applies to those who are born under Corvus, Hydra, Argo Navis, Canis Major, Lepus, Eridanus, Cetus, which celestial signs the first comet has passed through, or Pegasus, where the second comet has been seen beaming; but above all, those that lie under Ari have cause for anxiety, for here the two comets are almost joined. However, I would not emphasize these omens too much if the traces did not frighten me. Besides, other bad signs are connected with the time of both comets; for fire has fallen down from the sky, the snow is dyed red by bloody drops, ghosts have been observed, monsters have been born, numerous miscarriages have been seen, and a thousand other signs that indicate that Nature has been damaged are mentioned. I would judge more leniently about all this, if the hideous vices of men did not exceed the numbers of the warnings. Therefore, I must give up the hope of a happy omen and predict bad things for both the good and the evil ones. God grant that we may become better, so that the warnings may not harm."[87]

As a means of avoiding the harmful effects of comets, Bartholin recommended to eat as little meat as possible: "The eating of meat causes a lot of decay and many illnesses."[88]

The comet of 1664/1665 was, according to Bartholin, the explanation for an unusually high number of cases of bovine contagious abortion and abortions in general in 1665, and he therefore thanked God that his son Hans was healthy and well, even though he was born "under these ominous signs in heaven and earth."[89] Bartholin was also the first to describe meteorite falls, "rocks falling down from the sky," in Denmark.[90] He did so in 1657. However, he did not think that the meteorites originated from the sky, but that they were pieces of mountain tops carried by the wind.

Work on Danish medicine and Danish medicines

In 1666, Bartholin published at Peder Haubold's Publishing House the work *De Medicina Danorum Domestica*, which is of great cultural and medical historical value. In 10 chapters and an appendix, a total of 527 pages, Bartholin describes the history of medicine and medicine's conditions in Denmark, Danish food and drinks, the Danish way of life, Danish customs and the benefit and usefulness of Danish medicines. He dedicated the book to Crown Prince Christian. The topics of the 10 chapters are respectively the native Danish medical art, Danish doctors' discoveries, the treatment of Danish diseases, home remedies against Danish folk diseases, the Danish pharmacopoeia, the domestic Danish way of life (especially regarding diet), the Danes' drinking habits, the dignity of Danish doctors, the Danes' erroneous opinions regarding medicines and diet as well as the Danes' erroneous opinions regarding doctors.

Bartholin wanted to correct inappropriate lifestyles and guide the Danes in leading a healthy and appropriate life, among other things by avoiding immoderate consumption of food and drink. The book thus picks up the thread from the translated book by Luigi Cornaro. Bartholin criticizes the composition of the Danish diet. On his long trip in Europe, he had been accustomed to a completely different diet and way of life, and he describes how, after returning home, the sweat broke out on his forehead, when he saw or smelled the Danes' excessively salty food preparations. Except for cabbage, vegetables were rarely or never used.[91] On the other hand, Bartholin was careful to prescribe special diets for his patients. He noticed that when he and his wife wanted to serve delicious dishes to sick peasants in the area around Hagestedgaard, they did not want to eat them. They preferred porridge and clipfish in the belief that it was healthier for them. This made Bartholin conclude that both the healthy and the sick ought to stick to the diet they were used to, and doctors should not be so eager to prescribe diets: "I have more than once received thanks from convalescents for allowing them to eat the kind of food they were used to and wanted, but which other doctors forbade them."[92]

In the same way, Bartholin found that the peasants preferred to try to treat their illnesses themselves, rather than calling in the doctor: "Our peasants are empiricists. They do not use a pharmacy but collect medicines in their gardens and on their fields."[93] One of the plants the peasants frequently used as medicine was elder, but Bartholin wondered why the only thing from this plant that was not used were the berries, which he himself praised and called "wild grapes": "They are left to the birds, which seem more worthy of this divine food. I wonder why in this country they are not put on the table as dessert instead of foreign grapes, since they are both healthy for the stomach because of their acidity and suit our temperament."[94] In another place he writes: "If the East does not want to send us its balsamic drugs, the elder bush has everything, we need."[95] Bartholin recommends elder in several of his other writings as a medicinal plant, among other places in the *Dispensatorium Hafniense*, where he mentions elderberry jam, elderflower sugar and elderberry spirits. Given the knowledge of the present time, one has to say that he had a point. Elderberries, with their high vitamin C content, were able to both prevent and treat one of the serious diseases of the 17th century, scurvy, which was caused by a lack of vitamin C due to an unbalanced diet without fruit and vegetables.

Bartholin recommends fermented Danish beer rather than imported wine. It is nutritious, strengthens the stomach, keeps the joints free of gout, expels kidney stones and prevents constipation. Danish honey should be used as a sweetener instead of imported sugar, which is consumed in far too large quantities. Bread baked from Danish rye and other Danish cereals is healthier than bread baked from imported cereals. Porridge made of barley, wheat, oats or buckwheat is healthy and nutritious, and instead of making fun of the natives of Funen for their buckwheat porridge, you should honour them for it. Cabbage is praised by Bartholin. Homemade spices made from the herbs of the garden should be preferred to the new bad habit of importing eastern spices, from which only the merchants and the inhabitants of India benefit, and which have arousing effects and damage the health: "They are brought home from the East and from the West Indies at a great loss to our purse, but to even greater harm to our health.[96] [...] Foreign luxuries, foreign dishes, foreign spices, which our ancestors in their happy time did not know, light a fire in our body and change our nature. Because we indulge in foreign lifestyle and nutrition, we also need foreign medicines."[97]

Bartholin finds springs that rise from Denmark's soil just as healing as highly praised foreign springs, and meat from Danish vipers must be just as suitable as an ingredient in the compound drug theriac as the usually required Italian viper meat. Overall, Bartholin's dietary philosophy and pharmaceutical philosophy have the same starting point, namely that the Danish climate, flora and fauna offer the necessary basis for putting together a healthy diet and producing good medicines, so that you do not need to import any of them. He refers to Saxo's Nordic legendary hero, the bard and warrior Stærkodder: "He followed the customs of his fathers, subsisted poorly on the food of the fatherland and lived three times as long as other people."[98]

It is remarkable that the Nordic diet to which Bartholin pays tribute in this writing is based on the same principles, using available local ingredients, as the modern Nordic dogma diet, which is now – 350 years later – observed by Denmark's leading gourmet restaurants and wins so many international awards and honours. However, it is difficult to imagine that Bartholin himself, in the circles in which he moved, and with the experience he must have had with good French and Italian food from his study trip, should have preferred Stærkodder's Nordic dogma diet. Likewise, many of the medicaments he had recommended in his pharmacopoeia in 1658 were extracted from medicinal plants originating from exotic countries. His recommendations about Danish diet and Danish medicines should rather be seen as an expression of patriotism at a time when the country was economically and militarily repressed, and as an attempt to strengthen the physicians' position vis-à-vis that of the apothecaries.

The peasants around Hagestedgaard also frequently handled surgical problems themselves. Bartholin describes, among other things, a peasant who had accidentally chopped off a finger almost completely, so that only the stub hung in the skin. The peasant refused to summon a barber surgeon but healed the wound himself with an ointment consisting of wax, butter and turpentine.[99] Bartholin also deals with inappropriate behaviour associated with increased risk of infection, for example when the cup went around the table from guest to guest due to lack of glasses.[100]

Finally, mention can be made of Bartholin's reflections on the Danes' sexual behaviour, reproduction and birth assistance. As far as sexual behaviour is concerned, he believes that "there are fewer delusions here in our fatherland than elsewhere,"[101] and he explains this by the character of the people and the way of living: "Public brothels are not allowed, because people live more honourably by our sea."[102] On the other hand, the state suffers from the fact that the age difference between the sexes is often too large in married couples. When old men marry young women, it causes, according to Bartholin, fewer births and weaker children, and when both sexes or the woman are too old at the time of the marriage, there might come no children at all. Bartholin has, for example, seen a young man marry an old toothless woman of 90, who had to be led into the church with the support of others, and a very beautiful young girl marry a man of 70. To prevent such marriages, the state should, according to Bartholin, favour families with many children and marriages between young people.

Bartholin also sees a need for better education of the population about birth and maternity, and he expresses concern about midwives' lack of qualifications. Too often people seek help from a quack instead of a physician. It was a problem that male physicians could not appear at births for reasons of modesty. Bartholin himself had seen many births with an unhappy outcome, because they were assisted by incapable midwives without training. Even at Hagestedgaard he had witnessed how a midwife during a difficult birth had scratched the child most horribly with her nails, "and then these ignorant people claim afterwards that it is a miscarriage that has been born."[103] To this is added the previously mentioned tragedy, in which Bartholin lost his sister-in-law and her newborn child, according to Bartholin, due to an inexperienced midwife. Informed by such experiences, Bartholin later took the initiative of introducing an exam for midwives as part of the medicinal ordinance of 4 December 1672, *Forordning om Medicis oc Apotecker etc.*, which will be explained later.

Bartholin himself did not have much experience as an obstetrician. He puts it in these words: "This art of birth assistance I have neither practised nor wished to practise."[104] He mentions births throughout his writings, but apart from the previously mentioned book on midwifery in ancient times his only book particularly about birth was the earlier-mentioned book dealing with births outside the uterus, *De Insolitis Partus Humani Viis Dissertatio Nova*, published in Copenhagen in 1664.

Bartholin publishes his speeches and poems

In 1668, Bartholin followed up his large publication of letters with a selection of his speeches, *Orationes varii argumenti* (Speeches with various contents), which was published at Daniel Paulli's Publishing House in Copenhagen. The book has 384 pages in addition to the dedication to Frederik III's son-in-law, Duke Christian Albrecht of Schleswig-Holstein-Gottorp (1641–1695), preface and table of contents. The book contains 53 of Bartholin's speeches in chronological order over a 20-year period from 1645–1665, starting with his promotional speech in Basel about monsters in Nature. Most of the speeches are Bartholin's professorial lectures on various medical subjects from the teaching of the medical students in the university's large auditorium and the Anatomical Theatre, but there are also long memorial speeches for Ole Worm and Henrik Fuiren and speeches given in ecclesiastical and official contexts in Bartholin's capacity as Rector and Dean.

Throughout this book, examples have been given of several of the speeches. Here only the last one, number 53, *De Nihilo Medico* (On Medicine's Nothing), must be mentioned. It is an undated academic speech that Bartholin gave to the medical students at the University of Copenhagen. It is a philosophical speech that plays with the word "nothing" with focus on the field of medicine, where Bartholin elegantly balances satire and seriousness. He distances himself from Aristotle's statement that nothing can be created from nothing and begins the speech with the reverse statement: "Nothing in the world is more important or older than the nothing from which everything that has been created is created."[105] From this starting point, he continues his speech with diligent references to the philosophers and physicians of antiquity. For the physician the following applies:

"Among the most important duties of a physician is to do nothing but nothing, not to act, but to wait. Nothing is truer than this. [...] The physician often defeats the disease by doing nothing. As for the prognosis, he deserves our trust by making no prognosis, since there exist absolutely no certain prognoses for acute illnesses. Concerning the treatment, he refrains from acting during the crisis, when Nature is in full swing, so that it is not hindered in its work by the physician's inappropriate interference. [...] In that situation everything should be left to Nature, while the doctor watches and observes, whether the fever goes down, and other symptoms decrease, if the patient gets a fresher colour, if the pulse becomes stronger, and the patient stands up with recovered strength to finish the rest himself. There are febrile diseases, especially mild ones, which follow their natural course, and which have good qualities, against which no treatment is needed. In these cases, I have often acted as an observer, not a physician, while our colleagues have been puzzled that the disease is missed or neglected. A physician can do a lot, if he, like the empiricists, wants to drive the patient to the hearse, if he finds pleasure in apothecaries who have become rich at the expense of the sick, if he wants to act so as to feel that he has acted. [...] Through this 'nothing' the doctor can retain his good reputation and the patient his money."[106]

The following year, 1669, Bartholin published his poems, *Carmina varii argumenti* (Poems with various content) (Figure 168), again with Daniel Paulli. This work spans 239 pages, in addition to the dedication to Frederik III, the preface and an index of names. Before the title page there is an impressive frontispiece, in its symbolism anything but modest, with an etching made by Albert Haelwegh. It shows in the foreground a large eagle flying over the sea off Copenhagen, whose towers, churches and castles can be seen in the background under a sky with clouds, stars and a crescent moon with a face. The eagle holds in its claws a laurel wreath surrounding Thomas Bartholin's name.

The work consists of 8 small books (Liber I–VIII), which contain more than 30 years of poems from the period 1637–1668 about, to, and to a lesser extent

Figure 168. Albert Haelwegh's title etching for Thomas Bartholin's publication of his poems on various themes, *Carmina varii argumenti*, Copenhagen 1669. 12.8 × 8 cm.

In the foreground is a large eagle flying over the sea off Copenhagen, whose towers, churches and castles are seen in the background under a sky with clouds, stars and a crescent moon with a face. In its talons, the eagle holds a laurel wreath, which encircles Thomas Bartholin's name. The Royal Library in Copenhagen.

from many the persons with whom Bartholin has occupied himself – beginning with God and Jesus – and about the subjects that have occupied him. There are poems for kings, queens, other royals, nobles and high officials in Denmark, dukes and important persons abroad, family members, theologians as well as physicians and other scholars at home and abroad. Even female artists like the author and translator Birgitte Thott (1610–1662) and the Dutch painter, engraver, writer and scholar Anna Maria van Schurman (1607–1678) are addressed with Bartholinian poems (Figure 169).

Many of the poems are for special occasions, for example births, weddings and deaths (epitaphs and memorial poems), but there are also many poems for Bartholin's own books and for books written by Danish and foreign colleagues. The subject poems include, in the most literal sense, everything between heaven and earth, from celestial bodies and snow over India to stimulants, unicorn horns and other medical topics such as drugs and their composition, diseases, rarities, skeletons in the Anatomical Theatre, lymph and dissections.

Liber I–IV each contain around 50 of Thomas Bartholin's own poems, while Liber V contains 67 poems by others for Thomas Bartholin, with Vitus Bering as the major contributor. Liber VI contains 11 poems by Thomas Bartholin in Danish. Five of these have been written both with runes and Gothic letters and include, among other things, a memorial poem for Ole Worm and a couple of poems for Peder Hansen Resen's (1625–1688) edition of Snorri Sturluson's (1179–1241) *Edda* from 1665, one of which is reproduced below. The rest of Liber VI contains

an appendix with 32 poems by Thomas Bartholin and others, while Liber VII and VIII each contain 100 short poems by Thomas Bartholin with different contents.

"Edda rested under Bencke
Hidden in darkness/
No one could think of
Changing the wheel of happiness.
Sæmund wrote and Snorri more/
Resen kindled her light.
This brings Runa Honour/
Resen is given double price."[107]

The alchemist Borri

In the autumn of 1667, a long-awaited guest arrived at Copenhagen Castle: a tall, beautiful and well-dressed man in his prime with long black curly hair (Figure 170). He had hardly arrived before Frederik III provided him with a residence at the castle, horses, a carriage, four servants and a nobleman to wait on him.[108] Thomas Bartholin received him with a tribute poem in Latin.[109]

As discussed in the chapter on Bartholin's father, alchemy or goldsmithing, that is to say chemical production of gold from other, less valuable metals, was widespread and recognized in the Renaissance and the 17th century. Christian IV was, as mentioned, an alchemist himself, and Caspar Bartholin the Elder believed

Figure 169. Anna Maria van Schurman (1607–1678) painted by Jan Lievens (1607–1674) in 1649. Oil on canvas. 87 × 68.6 cm.

That Thomas Bartholin admired and felt attracted to women other than his two wives is evidenced by a few poems to foreign women in his poetry collection, *Carmina varii argumenti*, Copenhagen 1669. One of them is the Dutch Anna Maria van Schurman, who was famous in her time for both her high intellect and her wide-ranging artistic abilities. She was a painter, engraver, sculptor and, as shown in this portrait, an author. She reportedly mastered 14 languages, was the first woman in Europe to study at a university (Utrecht) and was known for her vast knowledge in theology, philosophy and medicine. She was also a feminist, who fought for women's rights to study and research. She corresponded with many learned men and women of the time, including René Descartes, Queen Christina of Sweden and the Danish author Birgitte Thott (1610–1662). Thomas Bartholin's epigram to Schurman (*Carmina*, Lib. IV, no. XXII, p. 98) is written for a portrait of her. Bartholin states that she surpasses her own sex in beauty and the opposite sex in spirit. Since the epigram is dated Utrecht 1646, it must be assumed that Bartholin met the nine-years-older Schurman there during his brief stay in the Netherlands before returning to Denmark.
National Gallery in London. Photo: Wikimedia Commons.

in alchemy, wrote about it and advised his students to read literature about it. Thomas Bartholin's peer, the nobleman Valdemar Daa (1616–1691), whose story Hans Christian Andersen (1805–1875) has described in the fairy tale *Vinden fortæller om Valdemar Daae og hans Døttre* (The wind tells about Valdemar Daa and his daughters),[110] squandered his entire fortune on alchemical experiments and reportedly took his chemicals, which could not tolerate frost, to bed in the winter because of lack of fuel. Many royal and princely courts employed alchemists. Thomas Bartholin and Frederik III also believed in alchemy.

Quackery was widespread in Bartholin's time, but as a university-educated physician, he was of course a strong opponent of the phenomenon and usually also quick to reveal it. But if you were a quack with a charming exterior, possessed special abilities to flatter with influential scholars, kings and princes and were skilled at persuading others about your abilities and excellence, you could still get far in Bartholin's time. One such man was the newly arrived guest, Giuseppe Francesco Borri (1627–1695), an Italian charlatan. He was the son of a physician of prominent Milanese nobility. As a young man, he had been employed at the Roman court and received instruction in alchemy, but due to attempts to create his own church and state, he was summoned in 1659 by the Inquisition in Rome and had to flee his homeland. In 1661 he was excommunicated, condemned as a heretic and burned in effigy. After a stay in Dresden with the Elector of Saxony, Johan Georg II (1613–1680), he travelled on in 1660 to Amsterdam, where he quickly gained a reputation as a miracle maker. Among those who were impressed by him was Ole Borch, who even before meeting him had written home to Bartholin about him in laudatory terms on 31 March 1661;[111] this was elaborated in further letters to Bartholin, after Borch and Niels Stensen had made Borri's acquaintance and had close contact with him.[112] At first Bartholin was sceptical, but after Borri, through Borch, had flattered him by mentioning his great genius, this scepticism evaporated.

One of Borri's experiments in particular, an eye operation, impressed Bartholin. Borch related that Borri cut through the cornea on a goose and forced the eye's fluids as well as the lens out through the wound, whereby the goose lost the sight of the eye. He then dripped Chelidonium extract in the eye and bandaged it. When, after some time, he removed the bandage, the wound was healed, and the goose could see with its eye again. Borch himself had attended the experiment twice. Bartholin saw completely new perspectives for medical science in this experiment, which he called the eighth wonder of the world, and Borch suggested inviting Borri to Copenhagen.[113] This idea seems to have come closer to being realized after Borri had made the acquaintance of Frederik III's son Ulrik Frederik Gyldenløve (1638-1704) and national treasurer Hannibal Sehested

(1609–1666) in Amsterdam, and after Borri himself told Borch in March 1663 that he would send a gift to the Danish king. Borri convinced Borch about his abilities as an alchemist by presenting gold ducats that he claimed to have manufactured. On 14 August, the Amsterdam physician Paul Barbette wrote to Thomas Bartholin that Borri, as a charlatan and deceiver, was preparing to leave Amsterdam.[114]

However, this did not happen before the end of 1666, when Borri fled Holland with ill-gotten money and jewels, and creditors and letters chasing him. After a short stay with Prince Rudolf August von Braunschweig-Wolfenbüttel (1627–1704) he came to Hamburg, where he became acquainted with Queen Christina of Sweden. In this city he was also received with hospitality by the English king's envoy. In return, he conveyed to the latter a heavy box with valuables for the king. It turned out, however, that it was filled with stones, but when this was revealed, Borri had already moved on to Copenhagen.

Here he came into Frederik III's service as an alchemist. To begin with, he performed his alchemical experiments together with Ole Borch in the old laboratory in Rosenborg Garden but quickly got permission to build a furnace and a laboratory behind the stock exchange, Børsen. At the same time, he acted as physician to the capital's notabilities and although most of his cures were ineffective, he was lucky enough to save the life of Ulrik Frederik Gyldenløve during a severe illness in January 1668, where other doctors had given up. This caused his reputation with Frederik III to rise to unimaginable heights. The poet and journalist Anders Bording (1619–1677) wrote in his newspaper *Den Danske Mercurius*,[115] Denmark's first newspaper – which was in verse – a tribute to Borri for this healing:

> "One has recently seen a glorious test of healing power,
> as Borri did on the noble Gyldenløve,
> where he was sick and helpless until death,
> yes, with one foot firmly stood in Charon's boat.
> His noble art, next to heaven, must be attributed,
> that the most noble blood escaped and remained alive.
> No: such a valiant hero and precious man of war
> should not go to the dark land of death so soon.
> What will Hippocrates, what will Galen now say?
> Here Borri with his Art much higher knows to scream [...]."[116]

In 1669, Borri's furnace and laboratory building were with great difficulty moved to what is now Rigensgade, near Nyboder. Borri was to live here and manufacture gold from silver from Kongsberg Silver Mill in Norway. Borri himself gave the

Figure 170. The Milanese charlatan, alchemist and miracle monger Giuseppe Francesco Borri (1627–1695) who, with his charm and trickery, completely fooled Thomas Bartholin, Ole Borch and several other Danish and Dutch physicians as well as Frederik III and other royals and courtiers. Engraving from 1675 by Pieter van Schuppen (1627–1702) from a painting by the Danish pupil of Rembrandt, Jürgen Ovens (1623–1678). 35.2 × 26.8 cm. The common theme of the engraving's text and illustrations is the wonders of Nature. Bottom left: Nature's omens. Top left: Luck in play. Top right: The art of miracles. Bottom right: Courage. Below Borri is the family coat of arms with the text Burrus, which in Latin means red, and an illustration of a bull, which could therefore be a red bull. Archive.org.

building the name Guldhuset (The Gold House), which has continued in the names of later buildings on the site. This removal was also set in verse by Bording in his newspaper:

"The House that Borri had set up behind Børsen,
to the use of his arts in work and operation,
He moved by ship and by land, just as it stood,
towards Østerport unbroken [...]."[117]

In Copenhagen, Borri sometimes performed his eye experiment on geese under observation by, among others, Thomas Bartholin and the king's surgeon Henrik Skriver. Skriver himself repeated the experiment without the use of Borri's Chelidonium extract and showed that the extract made no difference. With today's knowledge, this is partly due to the fact that the lens is not necessary for vision, and partly to the fact that the sclera of birds' eyes contains a bone shell, which makes it impossible to squeeze out more than a part of the eye's fluids. Thomas Bartholin later wrote about Borri's eye experiment in his journal *Acta Medica & Philosphica Hafniensia*,[118] where he expresses disappointment that Borri did not want to perform the experiment on humans. For example, he had hesitated about this when a royal page had had one of his eyes destroyed in an accident. When Borri finally tried to treat the man with his extract, it was too late. In another case, a peasant from Hagestedgaard had had his eye kicked to pieces by a horse, but Borri could not save the eye, as the optic nerve was destroyed.

Ole Borch, Henrik Møinichen and Henrik Skriver had in a couple of cases tried to treat injured eyes with Borri's extract without success. Thomas Bartholin had in vain tried, with drops of the extract, to restore the sight of the university's bursar Hans Mule (1605–1669), who had been blinded by a stroke that he suffered on his way from his own to Bartholin's official residence.

Borri's time in Guldhuset was short. As in Holland, he had, because of his favour with the king, his extravagant lifestyle, his deceitful behaviour and his heavy consumption of the kingdom's finances to no good use, become the source of indignation and envy in wide circles and made himself many enemies, including reportedly the future king, Crown Prince Christian. Therefore, when Frederik III died on 9 February 1670, Borri disappeared out of the country as quickly as he had come, followed to the borders of Copenhagen by an excited mob. Via Sweden he travelled to Germany. When he reached Moravia, he was recognized and arrested and sent to Vienna, and thence to Rome, where, after abjuring his old heresy, he was sentenced to life imprisonment and ended his days in the prison fortress Castel Sant'Angelo in 1691.

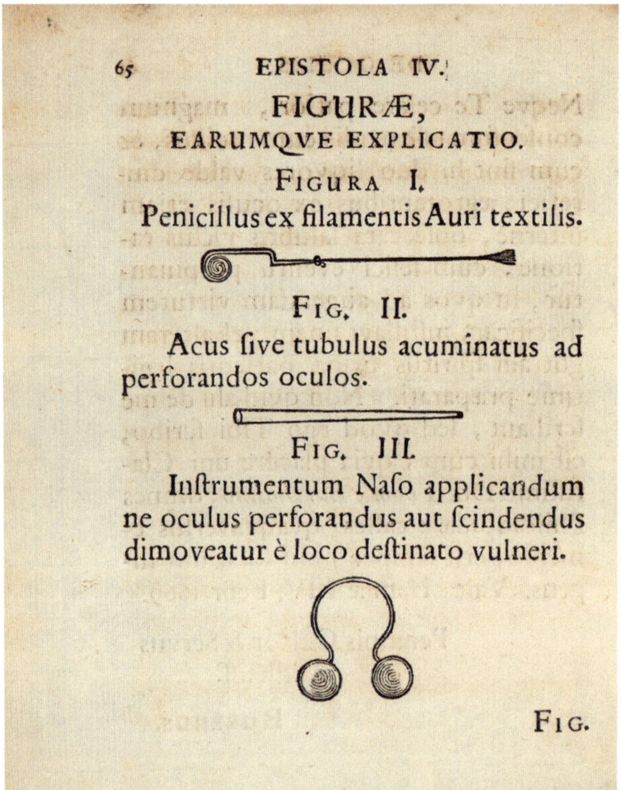

65 EPISTOLA IV.

FIGURÆ,

EARUMQVE EXPLICATIO.

FIGURA I.

Penicillus ex filamentis Auri textilis.

FIG. II.

Acus ſive tubulus acuminatus ad perforandos oculos.

FIG. III.

Inſtrumentum Naſo applicandum ne oculus perforandus aut ſcindendus dimoveatur è loco deſtinato vulneri.

FIG.

Figure 171. Instruments that Borri used for his questionable eye surgeries. From Thomas Bartholin's publication of two letters from Borri, *Francisci Josephi Burrhi Epistolae Duæ*, Copenhagen 1669. Figure legend:
Fig. I: Brush woven from ear hair. Fig. II: Needle or small pointed tube to perforate the eye.
Fig. III: An instrument to be placed on the nose to prevent the eye to be pierced or cut from moving from the intended site of the wound.
The Royal Library in Copenhagen.

Thomas Bartholin, who, as it appears, ended up sceptical towards Borri's eye experiment, retained, like Borch, veneration for him and published at Daniel Paulli's Publishing House in Copenhagen in 1669 a 68-page document with his letter correspondence with Borri.[119] From Hagestedgaard, Bartholin had on 10 March 1668 and 30 January 1669 written to Borri about the eye experiment and other issues, which Borri discusses in his respective responses on 24 January and 14 February 1669. The book is dedicated to Frederik III and ends with two poems, one by Ole Borch about the eye experiment and one by Borri to Frederik III. Borri's last letter included illustrations of his instruments (Figure 171). The subject's attention in Europe is illustrated by the fact that the publication was reviewed in both of Europe's scientific journals, *Journal des Sçavans* in Paris,[120] and the Royal Society's *Philosophical Transactions* in London.[121]

With the death of Frederik III, Bartholin had lost his great patron.[122] More accidents awaited him in the time to come.

Hagestedgaard burns to the ground

On 6 May 1670, Thomas Bartholin's teacher from boyhood, Royal Physician Poul Moth, died in Copenhagen. Bartholin had been involved in the treatment of the sick, old man and was present at his deathbed.[123] The funeral was to take place in Sankt Petri Church in Copenhagen on 18 May. Bartholin wanted to honour Moth and left for Copenhagen to speak for him on the day of the funeral and to set up a memorial tablet with a Latin inscription that read, among other things: "[...] in the service of the well-being of others, broken down by sickness and old age, he waits for eternity."[124]

While this was going on, a violent fire broke out on 17 May at Hagestedgaard. The main building with Bartholin's library burned to the ground, leaving only the cellars from Chancellor Johan Friis' (1494–1570) time. Moreover, the nearby parsonage and eight farms burned down, while Hagested Church escaped the flames.[125] The biggest loss was the library. Not the hundreds of printed books that Bartholin had collected or published throughout his life, which could be obtained in other places, but a long series of unprinted works that he had been working on for years, which were now lost forever.

On 1 July of the same year, he wrote to his sons Caspar, Christopher and Thomas a small 114-page book, *De Bibliothecæ Incendio Dissertatio ad Filios* (Dissertation for my sons on the burning of the library), which was published at Peder Haubold's Publishing House.[126] In the introduction he writes:

> "It was disastrous to be so respectful. For it was while I was paying full respect to my deceased teacher, the revered Moth, that I was interrupted by the dire message that my dwelling at Hagestedgaard had been destroyed by fire, and together with the rest of the household goods the library, a collection of the most esteemed physicians and philologians, who had lived in repose with me and daily consumed my leisure. Deprived of both preceptor and books, I sought to allay these misfortunes by the constancy of an unbroken spirit, and while such unforeseen disasters tried my spirit, they did not prevail."[127]

Despite the fact that the memorial in the Church of Our Lady had recently been prepared for Bartholin's own death, he did not want to follow the bad example of Titus Labienus (c. 100–45 BC), who no longer wanted to live after his writings had been burned.[128] Bartholin tries to console himself with the thought that the fire was God's will. God has not wished to punish him for his library contained no books or scripts that might hurt God, offend the king, or neglect to instruct and entertain the world. On the contrary, God wanted to spare him from the large workload that he would have had to complete in his old age, i.e. the many unfinished manuscripts that went up in smoke. God has shown him that books, which he thought immortal, are also mortal and perishable, and God has spared him, his wife and his children: "Let only the books vanish, as long as the children are safe."[129]

Bartholin looks forward to Hagestedgaard being rebuilt even more beautiful and better than it was, and he is delighted that the barn, the land and the flowering forests, meadows and fields have been spared; only eight of the estate's approximately 50 farms are burned, and the king, the newly acceded Christian V (1646–1699), has with indulgence granted the estate tax exemption.[130] In addition, Bartholin was supplied with building materials for the rebuilding of the farm, and this work progressed so rapidly that, in the preface to the fourth revised edition of the anatomy book on 21 December 1671, he wrote, "from my recently rebuilt farm Hagested". The only part remaining today from this building with Bartholin as the builder is the middle part of the west-facing tower on the garden side of the main building with the vaulted ceiling tower room (Figure 172)[131]. Above the window of this room is a small stone with the Latin inscription on the outer wall: VETERIS VESTIGIA TROIÆ TH. BARTHOLIN REPARAVIT 1672 (In 1672 Thomas Bartholin rebuilt the ancient Troy from its ruins) (Figure 173).

Bartholin continues in the writing about the fire:

"God ruined my purse, so that he might render it more ample. A very indulgent king favours me with his kindness. Friends at home and elsewhere condole and watch for opportunities to compete in services, so that they prove themselves to be true friends."[132]

Shortly before the Frederik III's death, he had on 31 January 1670 appointed Thomas Bartholin to "archiater honorarius," i.e. honorary court physician, based on the great diligence and industriousness he had displayed through all his publications, which had made him famous. According to the king, the appointment was to help Bartholin to be able to publish the other works he had planned, better and unhindered. On 8 September 1670, Christian V enriched Bartholin's post with an annual salary of 600 rix-dollars, and it was even arranged that Bartholin only had to serve, when the king summoned him (Figure 174).[133]

As further consolation, Bartholin reviews numerous examples of authors in antiquity whose works have been consumed by flames, and he mentions his grandfather Thomas Fincke's similar loss of famous medical and mathematical books to the fire god Vulcan – a loss that could always bring forth tears in the otherwise not very emotional man. He even goes so far as to mention the joy, benefit and symbolic value that fire has brought with it in different cultures and religions, including Jews, Romans, Persians and Egyptians. Bartholin believes that, with fire, God signals a desire that in the future he should spend more time on spiritual pursuits

Figure 172. The tower room on the ground floor of Hagestedgaard is the only preserved part of the building from Thomas Bartholin's rebuilding of the manor after the fire in 1670. Photo: Jesper Brandt Andersen.

Figure 173. On the memorial plaque on the wall above the tower room of Hagestedgaard, Thomas Bartholin wrote the following inscription: "VETERIS VESTIGIA TROIÆ TH. BARTHOLIN REPARAVIT 1672," which reads: "In 1672 Thomas Bartholin rebuilt ancient Troy from its ruins." Photo: Jesper Brandt Andersen.

CHRISTIANUS V. DEI GRATIA REX DANIÆ ET NORUE &c.
PIETATE ET IUSTITIA
H. Schaten. Sculp.

rather than trivial literary studies. If this really was God's plan with Bartholin, one must soberly state that the plan failed, for there are no signs that Bartholin during the last 10 years of his life devoted himself more to spiritual pursuits than before, and he by no means laid down his industrious pen before he took his last breath.

Bartholin ends the book's long introduction with a poem for his lost books, which will live on in his memory:

> "You spectator, who pass by,
> see and mourn for the erudite embers.
> These learned ashes mixed with profane ashes do not cease to speak,
> although they lack both letters and tongues.
> They of whom posterity speaks,
> struggling even with their own ashes,
> of which the flame was able to consume the content,
> never in the world, never out of memory.
> Go swiftly lest the dust be moved and vainly offend my eyes,
> which occupied themselves so pleasantly with the dust of these books,
> before they were ashes."[134]

This is followed by Bartholin's review of 27 works, that were planned for publication, but where the manuscripts were lost in the fire, or as he formulates it: "Here the reader will see what he may desire in the future but will never read."[135] It concerns both works written by Bartholin himself, and works written by others, which he planned to publish with or without his own additions. However, two of the mentioned works were in fact published. One, the revised fourth edition of the anatomy book, had been sent to Holland before the fire and was thereby saved. One must wonder why Bartholin includes it in this list. The second, the second edition of the work on the unicorn, was, as mentioned, published in 1678 through the intervention of his son Caspar.

Figure 174. Christian V (1646–1699). Engraving by Hubert Schaten (active 1671–1697). 23.7 × 17 cm.

The model for the engraving appears to be Abraham Wuchter's (1610–1682) oil portrait of Christian V in the Museum of National History at Frederiksborg Castle, which is dated to around 1675. The absolute king is depicted in armour with an ermine robe and a wide ribbon around him. Around the neck he wears a shirt frill. The Latin text on the ribbon, "CHRISTIANUS V. DEI GRATIA REX DANIÆ ET NORUE ETC.," reads: "Christian V, by the grace of God King of Denmark and Norway etc." The Latin text on the plate below the portrait, "PIETATE ET JUSTITIA," reads: "Piety and justice," the king's motto.

Due to the unexpected death of Frederik III, his son Christian V was relatively unprepared when he became king at the age of 23 in February 1670. The moderately gifted, but open and realistic Christian, therefore in his early years largely left the kingdom's important decisions to his advisors, which, until 1676, essentially meant Peder Schumacher Griffenfeld.
National Gallery of Denmark.

Among the most important lost works are the mentioned remaining three centuries of Bartholin's letters to and from the scholars of Europe. Two other significant works that were lost in the fire, and of which only one is on the list, are particularly interesting. The first is a large, planned work on pathological anatomy, which many had dealt with, but none had previously published a work about. Bartholin writes that he had spent many years on this useful work, where he did not follow the usual procedure describing the anatomy from dissections of corpses of healthy people. Instead, he described it from private and public dissections of people who died of disease. This included patients in hospitals that he had visited on his long journey abroad. During the dissections he had described in detail the various organs' appearance in numerous diseases, and he had combined these observations with reading both ancient and contemporary physicians' descriptions of dissections of corpses from people who died of disease.

Bartholin's plan had been that the work, which was to be entitled *Anatome Practica, Ex Morbo Defunctis, Pro Sede Affecta Inquirenda* (Practical Anatomy based on dissections of deceased patients to examine the diseased place), was to be included as part of his fourth revised edition of his father's textbook of anatomy, and he was sure that the work would have had great significance for the treatment of diseases, if he had been given the opportunity to complete it.

As a plaster on the wound, he published in 1674 at Peder Haubold's Publishing House the earlier-mentioned small, 48-page treatise *De Anatome Practica, Ex cadaveribus morbosis adornanda, Consilium, Cum Operum Autoris hactenus editorum Catalogo* (Consultation on the practical anatomy based on dissections of the corpses of sick people, with a catalogue of the author's published works to date). The thesis itself extends over the first 42 pages, while the last six pages are a chronological list of Bartholin's own 75 works published so far (see appendix 2) as well as a chronological list of works or letters written by others but published by Thomas Bartholin with his foreword.

It appears from the introduction that the reason for Bartholin's publication of this small work is that, the day before, he received information about the Swiss physician Théophile Bonet's (1620–1689) work on pathological anatomy, *Sepulchretum sive Anatomia Practica Ex Cadaveribus Morbo denatis* (The burial or the practical anatomy from corpses of sick people told in detail), which was published in two volumes in Geneva in 1679. With its review of approximately 3,000 dissections, it is considered the first complete work on pathological anatomy.[136] Bartholin is delighted that Bonet has turned his practised hand to this very subject, but at the same time he does not hide that part of the purpose with his own small printing is to make the whole world aware that he himself had been working for 20 years on such a work, which was almost complete, but now lost forever in the fire of his farm. The other part of the purpose is to act as an adviser to Bonet and others by explaining the sources of aid that he used for his own lost work.

The second significant work lost in the fire is curiously not included in Bartholin's list of lost works, but only listed as number 83 on the subsequent list of 129 "works hitherto published in my name or dedicated to me, which Vulcan consumed with the rest, but with less harm to me, since they are available elsewhere."[137] It is "copper engravings for the unfinished work, *Anatome Augusta*" (The Exalted Anatomy),[138] which Bartholin mentions in *Domus Anatomica Hafniensis* in the list of objects in his museum:

> "Pictures of most parts of the human body, internal and external, of natural size and shape, based on dissections conducted by Thomas Bartholin. They were first executed in natural colours by Karel van Mander, the royal court painter, and then engraved by the royal copperplate engraver, Albert Haelwegh; the illustrations are intended for the *Anatome Augusta* but have not yet reached their final completion."[139]

In other words, it was a large anatomical atlas, which would fulfil Bartholin's old plan from his letter to Ole Worm on 8 December 1642 regarding the second revised edition of the Bartholinian anatomy book: "I will omit the old illustrations and have new ones painted on the spot, while the dissections take place, and then engraved." Only after his return to Denmark did Bartholin, as professor of anatomy and head of the Anatomy House in Copenhagen, get the opportunity to complete the project with the king's support.

In an open royal letter of 9 March 1653, it appears that "Albrett Halweg" was to be granted an extra 150 rix-dollars annually for two years, where he had to engrave and complete "some anatomical figures painted by our dear Carl van Mander" at the "request" of "our dear, honourable and highly learned D: Thomas Bartholino medicinæ doctore and anathomices professore publico here at the University of Copenhagen."[140]

At the same time, the treasurers were ordered to provide Haelwegh with the necessary copper plates, and 100 rix-dollars were set aside for this purchase. Although Haelwegh, in return for the pay rise, was required to show diligence and accomplish as much as possible of the work within those two years, the project, as it appears, became delayed for many years, which proved fatal. The fact that, according to Bartholin, the burnt engravings for the work were available elsewhere suggests that a significant number had already been printed in multiple copies. However, such engravings have been neither mentioned nor found since, which suggests that they perished during one of the fires of

Copenhagen. Van Mander left behind at his death in 1670, i.e. the same year as the fire of Hagestedgaard, nine anatomical paintings with stretchers as well as the beginnings of another anatomical painting. These were later stored at Copenhagen Castle and then incorporated into the university's Natural History Museum on the first floor in the Anatomy House, where they were used in the teaching of the students.

A later inventory of the museum's losses in the Copenhagen fire in 1728 mentions "ten anatomical tableaux, which showed all the human organs and muscles and met the students' requirements to a high degree. These were superbly depicted on the spot by that once highly famed Royal Court Painter Karel van Mander, and the representation was so realistic that no human industry could have expressed it better."[141]

Den Kongelige Kobberstiksamling (The Royal Engraving Collection) at Statens Museum for Kunst (The National Gallery of Denmark) in Copenhagen owns eight black chalk drawings of various parts of the human skeleton, which, according to the museum's registry, were made by Karel van Mander in the period 1660–1663 for "an apparently unpublished work by Thomas Bartholin on the human skeleton" (Figure 175). Accordingly, the drawings in the bottom right corner are signed C. v. Mander. Since Thomas Bartholin does not mention such a work anywhere, the question is whether these drawings were in fact intended for *Anatome Augusta* and thus are the only surviving illustrations for this work.

Among the lost works, one finds further – as a follow-up to Bartholin's discovery of the lymph vessels – a work, *Liber de Morbis Lymphaticis* (Book on Diseases in the lymphatic system), referring to "all diseases that are caused by disturbances in the lymphatic system," including their medical treatment. Furthermore, there is in the list a work on medical practice using Danish medicines, *Medicina Danica Inquilinis Remedis Proprios Morbos Sanans*, of which Bartholin writes the following to illustrate his work as a doctor for the general population in the countryside:

"It is a pleasure to concern myself with our common people rather than the learned, although I may seem to be entering into the bad graces of the pharmacists. Furthermore, I have believed that I am more obligated to the public good of the nation than to the financial advantage of merchants growing rich from foreign commodities sought from afar. God granted domestic remedies for the treatment of domestic disorders, which anyone far removed from the shops of the pharmacists may employ, as well as he who does not have physicians he may consult the pouch of the pilgrim's remedies. An association arose on my farm for writing and testing, so that I might assist neighbouring friends, the destitute or domestics seeking aid from me. They do not always have the opportunity to go calmly to the city and to the shops, and I have considered it beyond the dignity of physicians and the honour of domestic nature to seek in distant parts and at great expense for remedies that are to be found everywhere at our feet at home and without cost. So recently, while I was dwelling on my farm, two girls were brought to me defiled by malignant little scores on the jaws, accompanied by creeping gangrene and threatening unsightliness to the whole face. Immediately I stayed the progress of the malignancy solely with sea water, in which absinthe[142] had been boiled, and delivered them from deformity. By means of this same easily acquired remedy I restrained a threatening gangrene in the foot of a man. When another displayed to me a paralysis of the left side, I ordered that aquavit, which our people employ in place of brandy, with extract of the herb of paralysis abounding in our woods, be applied to his spine. I cured the commonly occurring pleurisies solely with a decoction of barley, when the patients were averse to venesection. I happily restored not a few bed-ridden scorbutics by means of our aquatic trefoil decocted in that old beer that we preserve in our cellars for its singular fragrance. I purged dropsies with the more tender leaves of the elder tree in place of the leaves of senna, and everywhere I substituted the domestic for the foreign remedy. I omit innumerable other treatments."[143]

The last of the lost works, which must be mentioned here as a small curiosity, is Simon Paulli's *Observationes Crocodili* (Observations of crocodiles), a small, planned work to describe the "dexterous and praiseworthy" Simon Paulli's dissection of two crocodiles, which, according to Bartholin, not long ago were sent to Frederik III in Copenhagen. Distracted by other affairs, Paulli had entrusted Bartholin with his observations and illustrations from the dissections and asked him to publish an account of them augmented by his own additions. But Bartholin states that, although the crocodiles were previously accustomed to the waters of the Nile, they have now learned to tolerate fire.[144]

As the third accident in 1670, in the autumn Bartholin lost his brother Caspar (1618–1670), the lawyer, who owned the manor Kornerupgaard near Roskilde (Figure 176).

The medicinal ordinance of 4 December 1672

Bartholin's call to the king in the *Dispensatorium Hafniense* to help the doctors carry out reforms to ensure more orderly conditions in the Danish healthcare and medicinal system was, as might be expected, mothballed until the war against Sweden was brought to an end, and the new autocratic rule in Denmark

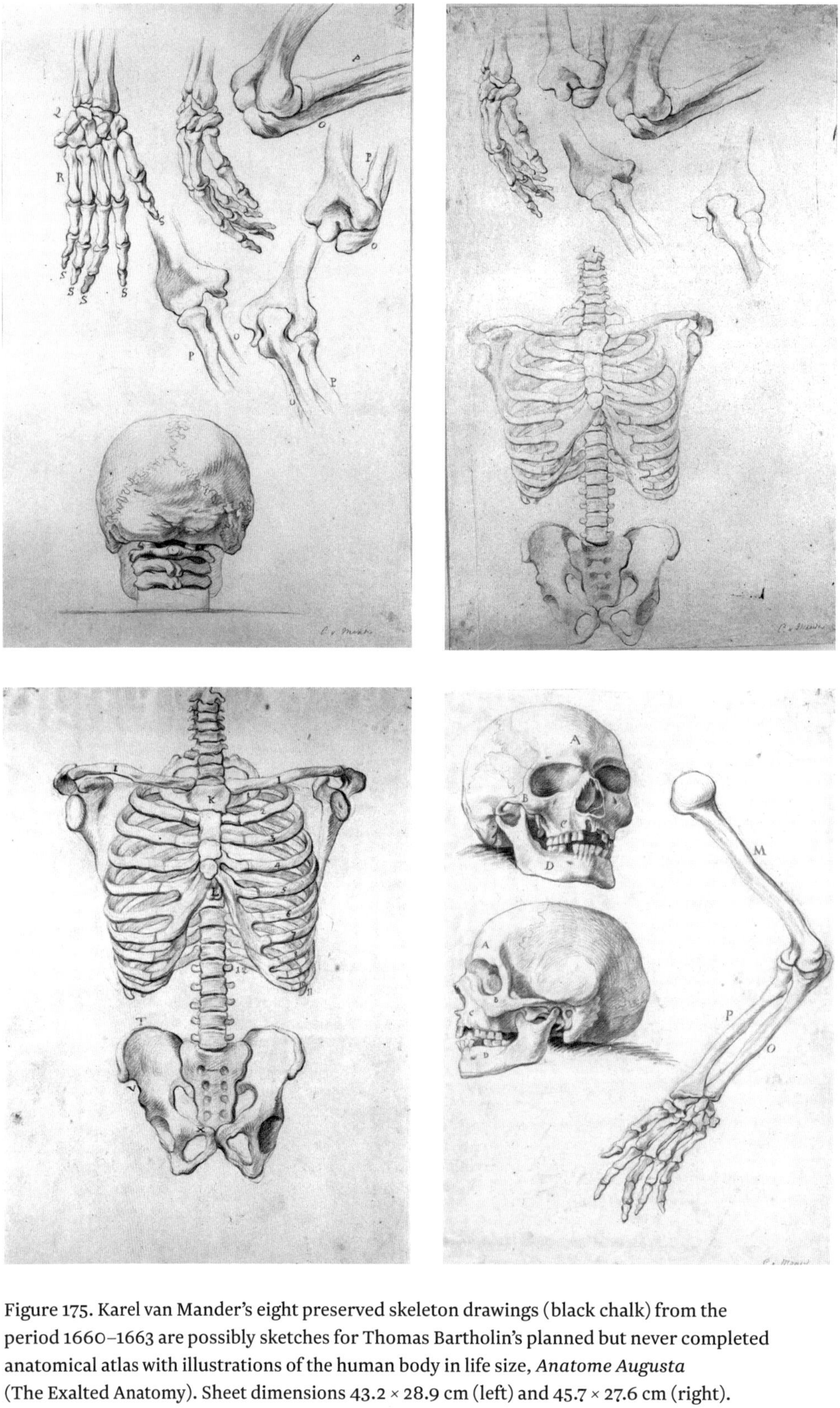

Figure 175. Karel van Mander's eight preserved skeleton drawings (black chalk) from the period 1660–1663 are possibly sketches for Thomas Bartholin's planned but never completed anatomical atlas with illustrations of the human body in life size, *Anatome Augusta* (The Exalted Anatomy). Sheet dimensions 43.2 × 28.9 cm (left) and 45.7 × 27.6 cm (right). The drawings are signed C. v. Mander in the lower right corner.

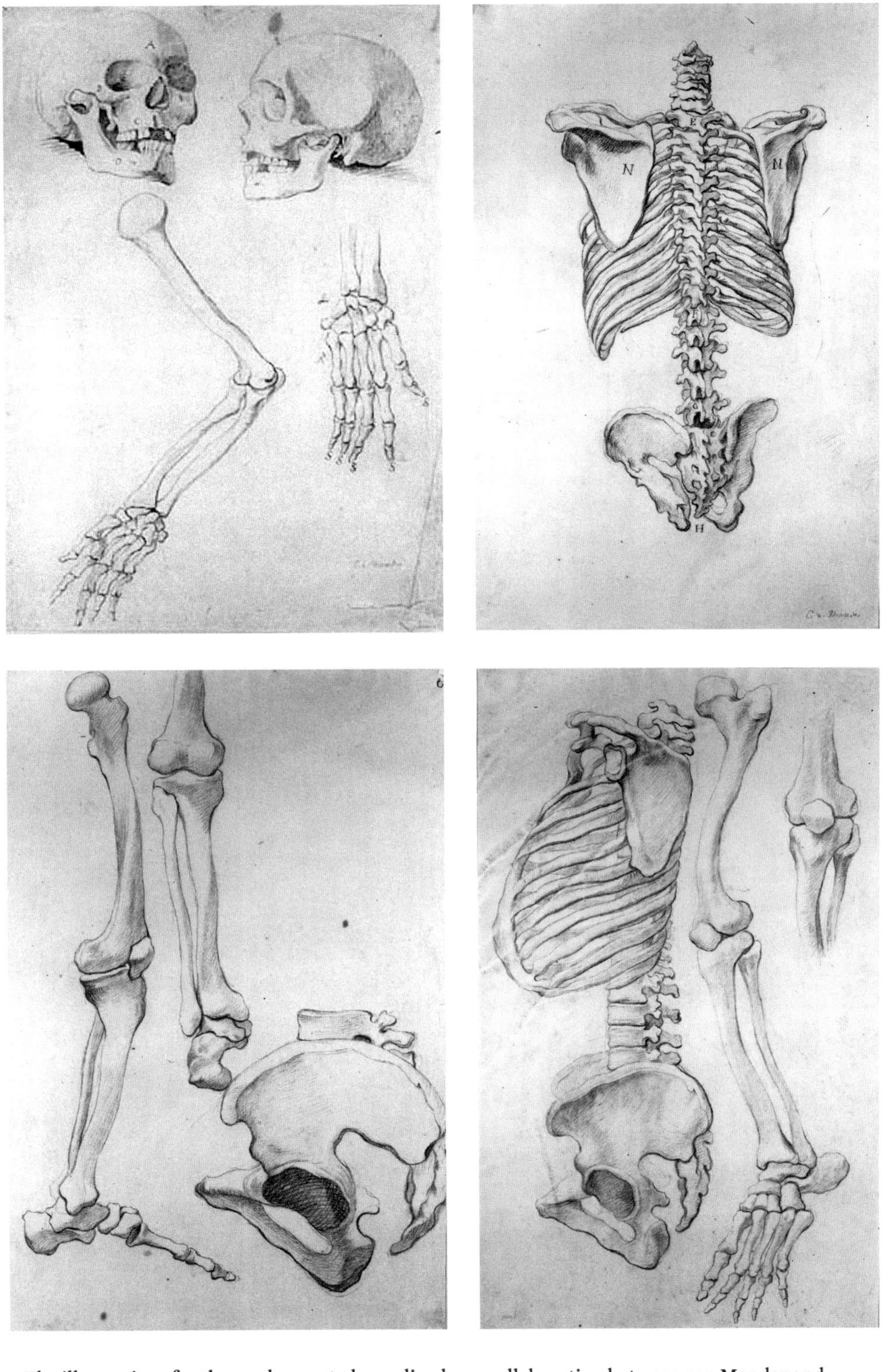

The illustrations for the work were to be realized as a collaboration between van Mander and the engraver Albert Haelwegh. Ten of van Mander's paintings for the work survived the fire of Hagestedgaard in 1670, but they later vanished in the flames of the Copenhagen fire of 1728. The destiny of Haelwegh's copper plates for the work is unknown.
The Collection of Engravings. National Gallery of Denmark. Photo: Jakob Skou-Hansen.

Figure 176. Doctor of Law and Cancelli Counsellor Caspar Bartholin (1618–1670). Portrait engraving by Albert Haelwegh after painting by Abraham Wuchters. 32.9 × 24.4 cm. Dated 1670–1673. With a Latin poem by Vitus Bering under the portrait.

Thomas Bartholin's lesser-known brother owned the smaller manor Kornerupgaard 30 kilometres from Hagestedgaard. As a young man, he travelled for ten years from 1639 to 1649 in the Netherlands, France, Italy, Switzerland and Germany. Some of the time he travelled with his brother Thomas, including partly on foot from Paris to Padua in the spring of 1641. His only literary production was a six-page Greenlandic-Latin dictionary in the first volume of *Acta Medica & Philosophica Hafniensia*. In 1674, 3½ years after his death, he and his children were ennobled with the name von Bartlin by Christian V for his efforts during the war against the Swedes 1658–1660. However, this first ennobled branch of the family died out in 1730 in the second generation on the male side.
National Gallery of Denmark.

introduced. But hereafter the process was sped up. In 1660, Rasmus Bartholin had started giving lectures in pharmacological topics such as *De proportionibus medicamentorum* (On the proportions of medicaments), and the following year Christian Ostenfeld gave lectures on *De remediis popularibus* (On the people's medicines) at the University.[145]

On 30 January 1661, a royal letter was finally issued to Dean Thomas Bartholin and the other professors and physicians associated with the Faculty of Medicine, inviting them to submit their proposals on how to put an end to the many complaints about illegal attempts by empiricists, pharmacists and others to carry out the work of physicians.[146] The result of this invitation was a proposal from the Faculty of Medicine for a medicinal ordinance formulated in 30 paragraphs, which was first published in Thomas Bartholin's *De Medicina Danorum Domestica* in 1666. In addition to Thomas Bartholin, the signatories to the proposal were the professors Christian Ostenfeld and Rasmus Bartholin, the royal physicians Simon Paulli, Poul Moth and Henrik à Møinichen as well as the general practitioners Christian Fabricius (1601–1666) and Ahasverus Payngk (1608–1667). In the book, Bartholin still expresses himself with great distance and almost contempt for apothecaries, whom he characterizes as follows:

> "A kind of people who often seem to be more of a nuisance than a benefit to the sick. For in the preparation of medicines they have only taste and stateliness in mind and empty their drawers to fill their purses with the sweat and money of the sick. There is good cause for complaint, since these pompous artists have scarcely set foot within the borders of our courts and cities before they must be kept in check by laws and forced to remain within the limits of their business. They are not to be trusted."

The main force behind the 30 paragraphs was undoubtedly Thomas Bartholin himself, which is confirmed by the fact that the content was virtually identical to the content of the 30 paragraphs of the final regulation of 1672, of which Bartholin was also the prime mover and mastermind.[147]

In the meantime, a minor power struggle took place between Thomas Bartholin and Simon Paulli, which shows that their relationship was not the warmest. Without the cause being known with certainty, in 1669 Paulli wrote his own comprehensive proposal for an ordinance, which he submitted to the Faculty of Medicine.[148] It filled 80 pages and was divided into seven chapters with a total of 95 paragraphs. In the preface Paulli complains that a former version of his was swept almost entirely off the table by the faculty (headed by Dean Thomas Bartholin), who barely allowed a single paragraph in the first two chapters to pass and rejected the other five chapters. Although the simple and justified explanation for the faculty's harsh judgement could be that Paulli had a few years earlier been a signatory to the faculty's proposal, which has essentially the same content as Paulli's first two chapters, and that the copious proposal was not practicable, Paulli was very upset. Again, he tried to exploit his close relationship with the king and his closest officials, Chancellor and Patron of the university Peder Reedtz (1614–1674) and State Governor Christoffer Gabel (1617–1673). In the preface, Paulli asked Reedtz and Gabel to appoint two censors to judge between him and the faculty, and he himself proposed two named members of the state college, whom he knew in advance would be favourably disposed towards his proposal. Frederik III granted Paulli's request, but unfortunately for Paulli, the king died soon after, and Reedtz and Gabel, under the new king, Christian V, quickly fell from the pinnacle of power in favour of Thomas Bartholin's faithful support and former pupil, the young up-and-coming statesman Peder Schumacher (later Griffenfeld).[149]

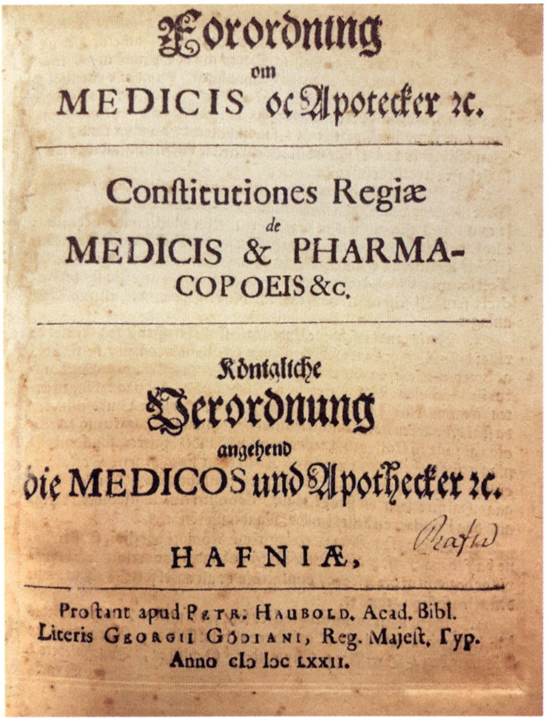

Forordning
om
MEDICIS oc Apotecker ꝛc.

Conſtitutiones Regiæ
de
MEDICIS & PHARMA-
COPOEIS &c.

Königliche
Verordnung
angehend
die MEDICOS und Apothecker ꝛc.

HAFNIÆ,

Proſtant apud PETR: HAUBOLD, Acad. Bibl.
Literis GEORGII GÖDIANI, Reg. Majeſt. Typ.
Anno cIↄ Iↄc LXXII.

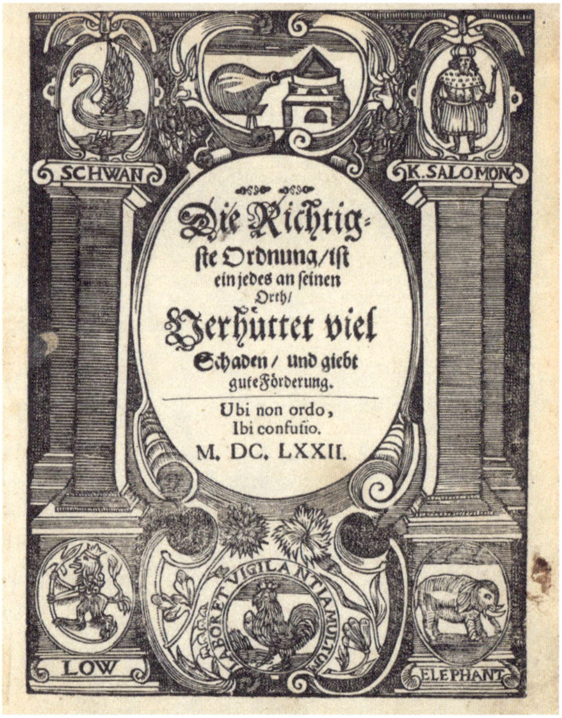

Figure 177. a. The title page of Denmark's first medicinal ordinance of 4 December 1672, *Forordning om Medicis oc Apotecker etc.*, behind which Thomas Bartholin was the driving force. The title is written in Danish, Latin and German from top to bottom.
Wikimedia Commons.

b. The frontispiece of the same. In the corners are the emblems of Copenhagen's four pharmacies, the Swan, Lion, Elephant and King Solomon pharmacies. The German text on the oval plate says: "The best arrangement is each one in his place. Prevents much damage and gives good support." The Latin text below reads: "Where there is no order, there is confusion." The Latin text around the cock in the bottom reads: "Work and great vigilance."
The Royal Library in Copenhagen.

Paulli's proposal was completely dismissed, and with the approval of first the State College and then Christian V, the faculty's proposal with the 30 paragraphs published by Thomas Bartholin in *De Medicina Danorum Domestica* in 1666 was established on 4 December 1672 as Denmark's first comprehensive medical ordinance covering the entire kingdom, *Forordning om Medicis oc Apotecker etc.* (Ordinance concerning physicians and apothecaries), which was accompanied by a new drug tariff (Figure 177). The ordinance formed the basis for the Danish healthcare system for approximately the next 250 years, until the next ordinance came in 1915.

In a modern context, the ordinance can be seen as the first significant attempt to undertake quality assurance of the entire Danish healthcare system, which at this time included 15 physicians and four pharmacies in Copenhagen, and in the whole country approximately 30 physicians and 21 pharmacies.[150] The overall

c. The corresponding tariff catalogue with its almost 3000 numbers was finished by 1671. Its frontispiece was made by the German engraver Cornelius Nicolaus Schurtz (1630–1690) in Nuremberg.

The catalogue contained significantly more numbers than previous tariffs, and there was a particular increase in the number of chemical medicaments, showing the growing importance of chemistry in medicine. The four corner illustrations show: upper left: the importance of the sun for the growth of the medicinal plants; upper right: the collection of the medicinal plants; bottom left and right: the preparation and the sale of the medicaments in the pharmacy, respectively. Copenhagen's four pharmacies are mentioned in the four bubbles, and the two men in the middle are probably the two fathers of medicine, Hippocrates and Galen. The Royal Library in Copenhagen.

objective of the ordinance was to ensure more qualified assistance for the population in case of illness and childbirth. Its 30 points deal with authorization of physicians, prescription rights of physicians, physicians' supervision of pharmacies, physicians' teaching, examination and supervision of midwives, physicians' treatment of plague and other infectious diseases, physicians' prescriptions, physicians' reporting of rare diseases to the medical faculty, physicians' payment, pharmacy license, apothecaries' prescription rights, apothecaries' authorization, pharmacies' medicaments, pharmacies' opening hours, pharmacies' storage of various liquids and substances, pharmacies' changes to physicians' prescriptions, pharmacies' professional secrecy, the number of pharmacies, physicians' and surgeons' manufacture of medicaments, unauthorized manufacture and sale of medicaments (pharmacies' monopoly) and unauthorized therapists. The tariff has almost 3000 entries, of which approximately 800 are from plants, 150 from animals and 1000 from minerals. There are a little more than 200 tinctures, and a relative increase in chemical preparations compared to earlier tariffs.

Niels Stensen returns to Copenhagen in 1672

During his ten-month stay in Paris from November 1664 to September 1665, Niels Stensen achieved real fame as an anatomist and almost rose to stardom. In Paris scientific development was no longer – as in Bartholin's time – driven by the university, but by the learned environment of the private scientific academies. With his sublime dissection technique and analytical ability, Stensen took this world by storm. This applies not least to his famous lecture on the anatomy of the brain, which he gave in the spring of 1665 to a crowd of learned men in the academy of his host, the orientalist and patron of science Melchisédech Thévenot (1620–1692). It was published by Thévenot in French in 1669 with the title *Discours de Monsieur Stenon sur l'Anatomie du Cerveau à Messieurs de l'Assemblée, qui se fait chez Monsieur Thévenot* (Mr. Stensen's discourse on the anatomy of the brain to the gentlemen of the assembly, which took place at Mr. Thévenot's house). The small book of 60 text pages and eight pages with illustrations of the brain in five different cuts (Figure 178) is a rhetorical and natural philosophical masterpiece, in which the 27-year-old Stensen enters the international stage as a scientific authority. In front of Descartes' own countrymen, Stensen presents in a friendly but firm manner the great philosopher's many mistakes regarding the anatomy and functions of the brain and the failure of his own method of exploring nature. He points out, among other things, that the very limited knowledge about the brain is due to the fact that anatomists throughout history have

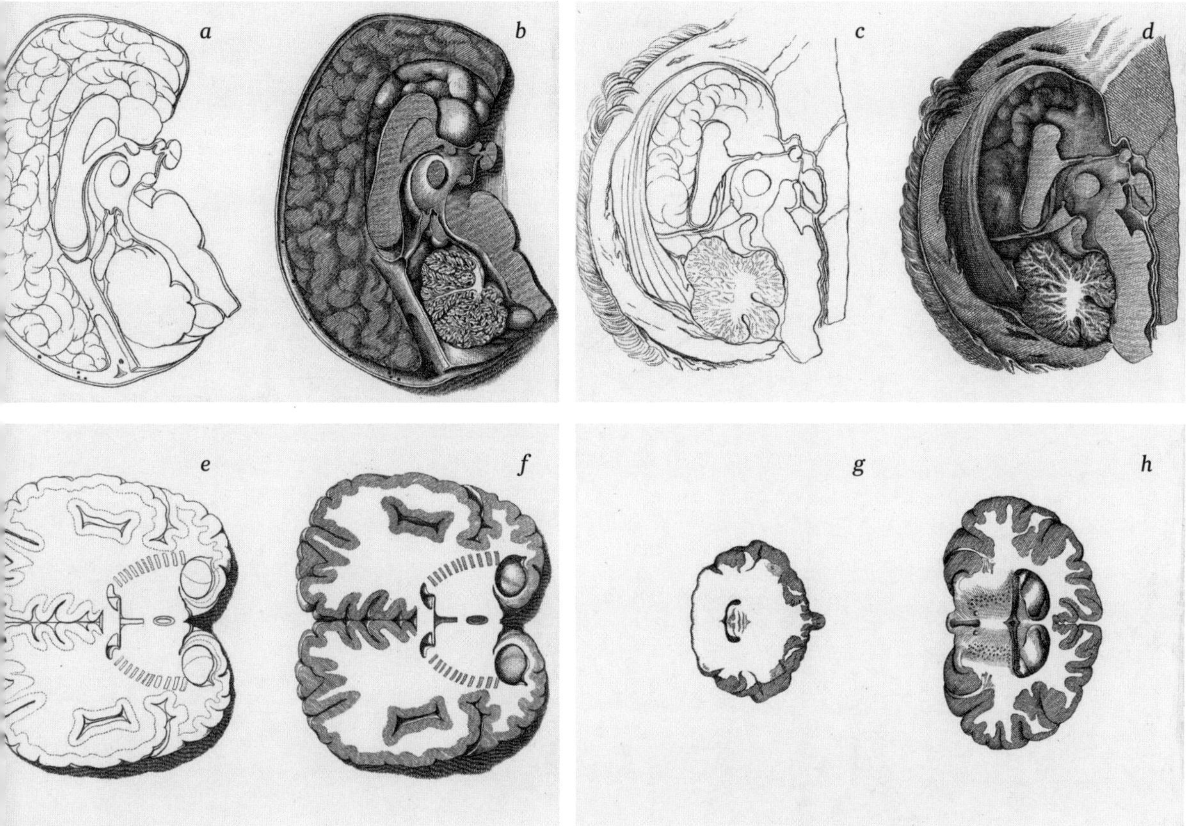

Figure 178. Niels Stensen's illustrations of the brain in *Discours de Monsieur Stenon sur l'anatomie du cerveau à messieurs de l'assemblée, qui se fait chez Monsieur Thévenot*, Paris 1669.

The lecture was held in the spring of 1665. The figures show: a and b: Longitudinal section of the brain in the median plane, i.e. the same plane as René Descartes' illustration in Figure 36. In Stensen's illustration, the structure of the brain (cerebrum, cerebellum, cerebral hemisphere, third and fourth ventricles, brain stem, hypothalamus and pituitary gland) are easily recognizable and correctly depicted. This also applies to the pineal gland which is seen in a horizontal position on the anterior part of the upper surface of the brainstem just in front of the cerebellum. c and d: Longitudinal section of the brain drawn by Stensen after Descartes' description of the brain. e and f: Cross-section through the centre of the brain. g and h: Two cross-sections, one at the back of the cerebrum (h) and one in the brain stem (g).
The Royal Library in Copenhagen.

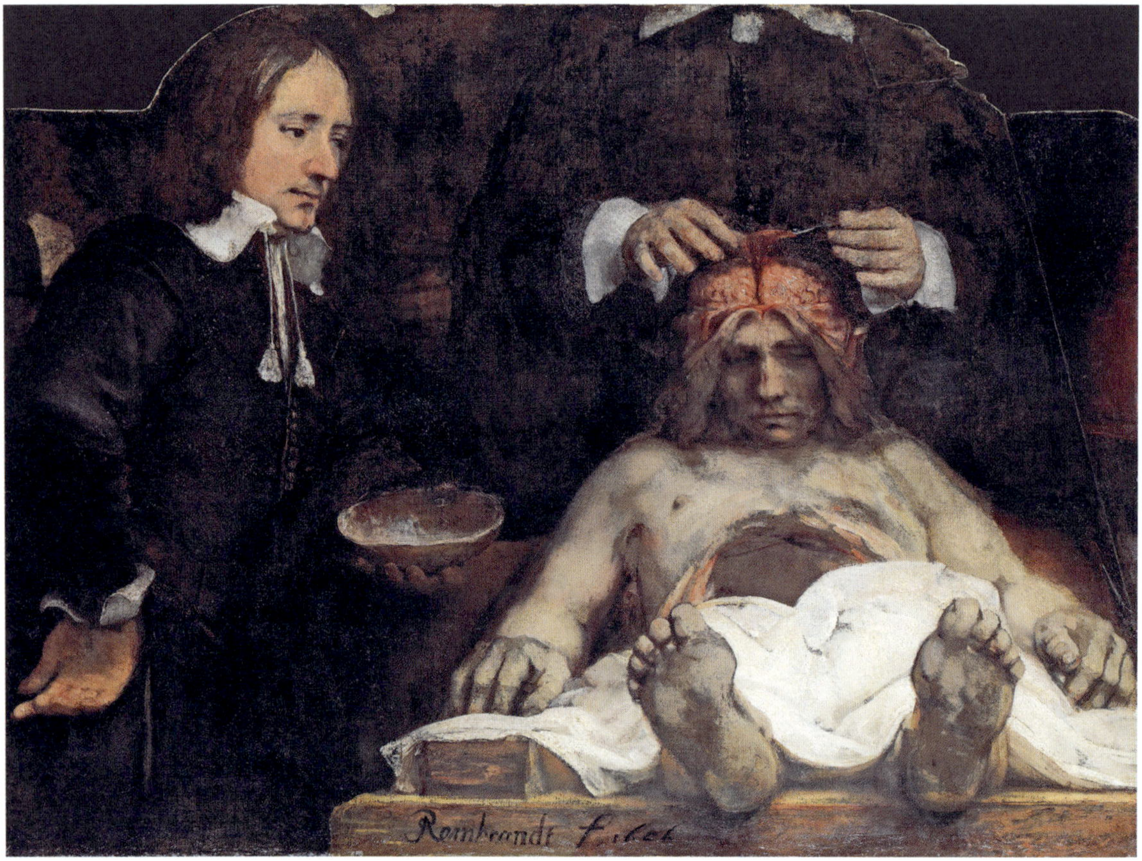

Figure 179. Rembrandt Harmenszoon van Rijn: Dr. Jan Deijmans anatomy lecture. Signed on the edge of the dissection table "Rembrandt f. 1656." Oil on canvas. 134 × 100 cm.

Jan Deijman (1619–1666) was Nicolaes Tulp's successor as an anatomy lecturer of the surgeons' guild in Amsterdam. On 28 January 1656, he performed his first public dissection on the executed 22-year-old Flemish tailor Joris Fonteijn. Deijman cuts his way through the two outermost meninges, the dura mater and arachnoid, while his assistant, surgeon Gijsbert Calkoen (1621–1664), stands with the top of the skull in his left hand. Rembrandt painted the dead man in an abbreviated version, so that it seems as if the feet and dissection table continue out of the image towards the viewer. Curious citizens could gain access to witness the dissection for four cents, and as the total revenue was 187 guilders, it is known that hundreds took advantage of the opportunity every day of the dissection. The painting was commissioned by the surgeons' guild and burned in its care in 1723, which is why only this fragment is preserved. In Rembrandt's surviving sketch for the painting, another seven doctors participate in the dissection.
Amsterdam Museum.

used incorrect methods when dissecting it, and he indicates the right method (Figure 179).

From Paris, Stensen travelled via Montpellier to Italy. With the recommendations of Thomas Bartholin,[151] he settled in Florence as a researcher under the Grand Duke of Tuscany, Ferdinand II de' Medici, and his experiment-based Academy of Sciences, Accademia del Cimento. Later, he served under Ferdinand's son Cosimo III de' Medici (1642–1723). In Florence he further developed his geometric muscular theory, in which, by presenting the muscle fibre as a parallelepiped (Figure 180), he explains why the volume of muscles does not increase, when they contract, presented in the work *Nicolai Stenonis Elementorum Myologiæ Specimen seu Musculi Descriptio Geometrica* (Specimen of Niels Stensen's muscle doctrine, or geometric description of the muscles), Florence 1667.

As mentioned earlier, Stensen also threw himself into geological research with groundbreaking results. The background to this was his dissection of the head a great white shark, a *Canis Carcharias*, in Florence in October 1666. Stensen was particularly interested in the shark's teeth, which he found to bear a striking resemblance to the Maltese tongue stones that his teacher Thomas Bartholin had dealt with so extensively and recently written about in the last volume of *Historiarum Anatomicarum Rariorum*.[152] In his dissertation *Canis Carchariæ Dissectum Caput* (The dissection of the head of a great white shark),[153] which was published together with the mentioned muscle dissertation in Florence in 1667, Stensen thus became the first to prove that tongue stones really are fossilized teeth from sharks from the past, and that they have been washed up on land by floods during the history of the Earth. As a result of periodical changes in the surface of the Earth, the fossils had, according to Stensen, become embedded in layers of earth and rock. As mentioned, the dissection of the shark's head led Stensen to further paleontological and stratigraphical studies that ended with his publication of the small geological dissertation *De Solido Intra Solidum Naturaliter Contento Dissertationis Prodromus* in Florence 1669,[154] and made him the founder of modern geology and crystallography.

During his stay in Italy, Stensen was captivated by the Catholic faith. On 4 November 1667, he took the Catholic Creed, and on 8 December of the same year he renounced the Protestant faith, thus converting to Catholicism.[155] On the same day he received – no doubt at Bartholin's instigation – a royal letter of recall from Copenhagen dated 19 October 1667.[156] Jacob Henrik Paulli's three-year leave from the professorship of anatomy was about to expire,[157] and the university now wanted Stensen home to take over. Unfortunately, it was just too late, since professors at the University of Copenhagen were obliged to submit to

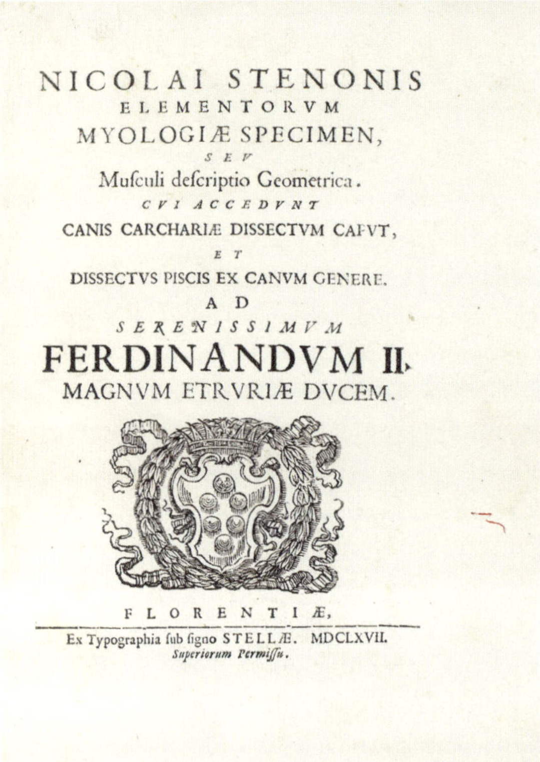

NICOLAI STENONIS
ELEMENTORVM
MYOLOGIÆ SPECIMEN,
SEV
Muſculi deſcriptio Geometrica.
CVI ACCEDVNT
CANIS CARCHARIÆ DISSECTVM CAPVT,
ET
DISSECTVS PISCIS EX CANVM GENERE.
AD
SERENISSIMVM
FERDINANDVM II.
MAGNVM ETRVRIÆ DVCEM.

FLORENTIÆ,
Ex Typographia ſub ſigno STELLÆ. MDCLXVII.
Superiorum Permiſſu.

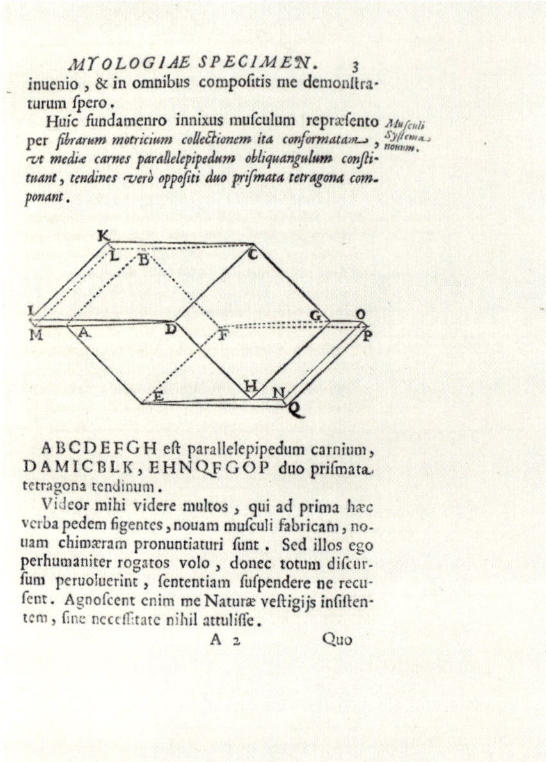

MYOLOGIAE SPECIMEN. 3

inuenio, & in omnibus compoſitis me demonſtraturum ſpero.

Huic fundamenro innixus muſculum repræſento per *fibrarum motricium collectionem ita conformatam, vt mediæ carnes parallelepipedum obliquangulum conſtituant, tendines verò oppoſiti duo priſmata tetragona componant.* *Muſculi Syſtema nouum.*

ABCDEFGH eſt parallelepipedum carnium, DAMICBLK, EHNQFGOP duo priſmata tetragona tendinum.

Videor mihi videre multos, qui ad prima hæc verba pedem figentes, nouam muſculi fabricam, nouam chimæram pronuntiaturi ſunt. Sed illos ego perhumaniter rogatos volo, donec totum diſcurſum peruoluerint, ſententiam ſuſpendere ne recuſent. Agnoſcent enim me Naturæ veſtigijs inſiſtentem, ſine neceſſitate nihil attuliſſe.

A 2 Quo

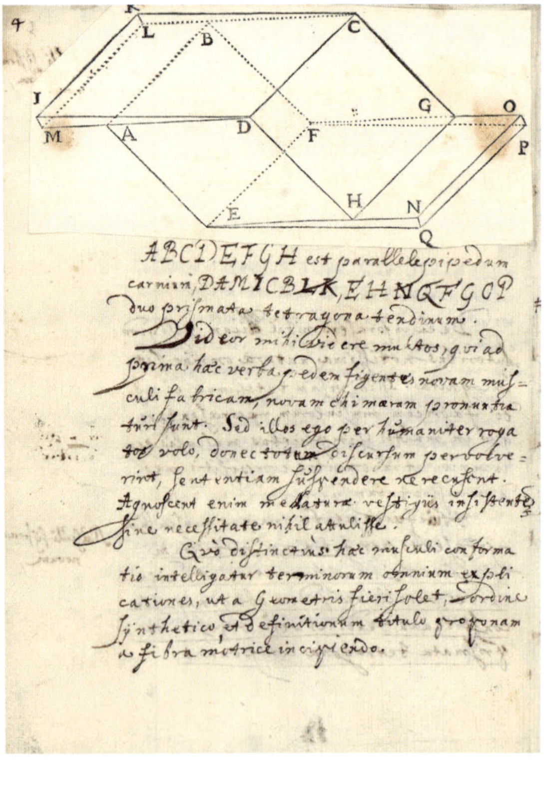

4

ABCDEFGH est parallelepipedum carnium, DAMICBLK, EHNQFGOP duo prismata tetragona tendinum. Videor mihi videre multos, qui ad prima hæc verba pedem figentes novam musculi fabricam, novam chimæram pronunciaturi sunt. Sed illos ego perhumaniter rogatos volo, donec totum discursum pervolverint, sententiam suspendere ne recusent. Agnoscent enim me Naturæ vestigijs insistentem sine necessitate nihil attulisse.

Quò distinctius hæc musculi conformatio intelligatur terminorum omnium explicationes, ut a Geometris fieri solet, ordine synthetico et definitionum titulo proponam a fibra motrice incipiendo.

Figure 180. a. The title page of *Nicolai Stenonis Elementorum Myologiæ Specimen seu Musculi Descriptio Geometrica*, Florence 1667, which also includes the treatises *Canis Carchariæ Dissectum Caput* and *Dissectus Piscis ex Canum Genere*. The Royal Library in Copenhagen.

b. Stensen´s *illustration* of the parallelepiped with which he explains the contraction of the muscle fibre. Ibid., p. 3. The Royal Library in Copenhagen.

c. Stensen's 206-page (103 leaves) handwritten manuscript for the book is filled with corrections and additions between the lines and in the margins, and the muscle treatise contains interspersed geometric figures like on this page, which shows the parallelepiped mentioned above.
The Royal Library in Copenhagen, Vesterlandske håndskrifter, NKS 4019 kvart.

the Augsburg Confession.[158] Therefore, it was not possible to employ a Catholic as a professor at the university. In other words, Stensen, with his conversion, had permanently cut himself off from the possibility of a professorship in Copenhagen, and he himself was fully aware of the problem. Hence, he responded to the recall by announcing his conversion and asking whether, on that basis, his return was still desirable.[159] More than four years passed before Stensen received an answer to that question in the form of a new royal recall letter dated 13 February 1672 from the new king, Christian V.[160] To circumvent the university's rules, a position had been created for Stensen as a royal anatomist (*Anatomicus Regius*) independent of the university and with the salary paid directly from the king's coffer. The men behind this arrangement were undoubtedly Thomas Bartholin and his former pupil,[161] now the king's closest adviser, the scientifically interested Peder Schumacher, who was soon to be chancellor of the realm and Denmark's most powerful man.[162]

Here, towards the end of Bartholin's third term as the university's rector in 1671/1672, the status of the professorships at the medical faculty was that he himself was first professor (*Medicus Primus*), Rasmus Bartholin was second professor (*Medicus Secundus*), and Villum Worm was third professor (*Medicus Tertius*), having acceded on 31 August 1671 after the death of Christian Ostenfeld.[163] The supernumerary professorship in anatomy, which had been occupied by Jacob Henrik Paulli, had not been filled, and this was also the case for the supernumerary professorship of medicine (fourth professor or *Medicus Quartus*), which Villum Worm had held from its creation in 1665 until 1671.[164] Bartholin had not given up hope of bringing Niels Stensen home to the University of Copenhagen, and now that his friend Peder Schumacher was high on the pinnacle of power, and when he himself was the rector of the university, it was a matter of striking while the iron was hot. With the increasing burden of age and kidney disease, he must have felt that it was the last call to secure the legacy of all that he had built up. With his unique qualifications, there could be no better heir than Stensen, and the only one of Bartholin's own sons who could stand as a candidate was Caspar, who was still only 16 years old.

Stensen asked his patron Cosimo III for permission to return home and serve his king and fatherland, and he got it. He arrived in Copenhagen on 3 July 1672 and settled down with his sister in his childhood home. From this date he was granted an annual salary of 400 rix-dollars, which was roughly equivalent to a professor's salary.[165] In addition, he was granted "free religious practice", or in other words, freedom to practise his own religion.[166] It was during the university's summer holidays and at a time of the year when it was too hot to perform dissections. The circumstances surrounding his new position and the duties that came with it, had not yet been clarified, and Bartholin was busy with many tasks at this time. Besides having to look after a professorship, a deanship and until 6 June 1672 a rectorship, Bartholin was, as mentioned, engaged in preparing Denmark's first medicinal regulation of 4 December 1672. In addition came his great work collecting, editing and writing articles himself for the first volume of *Acta Medica & Philosophia Hafniensia* and a great work preparing the publication of the fourth, greatly increased, richly illustrated and updated edition of the Bartholinian anatomy book, *Anatome Renovata*. To this must be added three more smaller works: a speech on the dignity due to rectors of the university, *De Rectoris Academici Dignate Oratio*, a preliminary treatise on the confection of the drug alkermes, *De Confectione Alkermes Dissertatio Preliminaries*, and finally a small book on diseases in the Bible, *De Morbis Biblicis Miscellanea Medica* (On the diseases in the Bible. A medical miscellany),[167] the first of its kind in the world.

Thus, it was no wonder that Stensen wrote the following about Bartholin in a letter to Cosimo III on 6 August 1672: "As far as my position is concerned, nothing has yet been decided, since the person who has taken it upon himself to find out what I should do, has a lot of work on his hands."[168] In contrast to Bartholin, Stensen in other words had no immediate duties to perform. Bartholin had high expectations for Stensen's return, and this was also the case for his brother Rasmus, who on 14 August 1672 wrote to the secretary of the Royal Society in London, Henry Oldenburg (1619–1677): "You should know that our Stensen has returned from Italy, and that his insight and his knowledge now enriches us for the benefit of the studies."[169] But the brothers Bartholin and Stensen's other old acquaintances were soon to learn that it was a different and changed Stensen who had returned to his homeland this summer. He was no longer the young ambitious and aspiring anatomist, who left the country in 1664, and who had kept religious issues at bay, because – in his own words – "the study of Nature and travel drew my interest in other directions" and "the study of the natural sciences, in which I utterly buried myself, did not allow me serious reflections on religion."[170] On the contrary, his letters this summer and autumn show that Stensen's sharp brain was now more concerned with religious issues than with natural sciences, and for the first few months after his arrival he spent most of his time familiarizing himself with the conditions of the Catholic congregation in Copenhagen. The recipients of the letters are exclusively Italian Catholics and clergymen.

Stensen's letters in particular to Cosimo III, Cardinal Leopoldo and the papal nuncio (messenger) of Florence, Opizio Pallavicini (1632–1700), show that he even assumed the role of envoy to the Vatican concerning the conditions for Catholics in Denmark. But Stensen did not restrict himself to the role of neutral announcer. He also became a missionary. He talks about them and us, ours and theirs. He calls his countrymen heretics and describes them as wretched, and he prays that the gracious and merciful God will come to their aid and drive them out of their darkness and awaken them from their deep and carefree sleep.[171]

The sources of Niels Stensen's dissections in Copenhagen

Niels Stensen's dissections during his stay in Copenhagen from 1672–1674 can be traced through mainly two sources. One is his pupil, Thomas Bartholin's nephew Holger Jacobsen's handwritten and, in large part, illustrated records in the Royal Library in Copenhagen with the title *Sectiones anatomicæ publicæ et privatæ prosectoris incomparabilis Nicolai Stenonis* (The incomparable prosector Niels Stensen's public and private anatomical dissections),[172] of which there are extracts in Vilhelms Maar's *Nicolai Stenonis Opera Philosophica.*[173] The second source is Thomas Bartholin's *Acta Medica & Philosophica Hafniensia.* Some of the dissections are also mentioned in Stensen's letters.[174] Danish summaries of a wide selection of the dissections with Holger Jacobsen's accompanying illustrations can be found in my 2021 biography of Niels Stensen, *Niels Stensen. Kongelig anatom og fyrstelig geolog* (Niels Stensen. Royal anatomist and princely geologist).[175] Most of the dissections are made on a wide variety of animals from Christian V's royal menagerie. There are only two dissections on humans, one on each gender, carried out in the Anatomy House.

The purpose of Stensen´s dissections in Copenhagen was both research and teaching. Some of the dissections were performed by Stensen himself, while at others he had the role of supervisor and teacher. Many of the dissections were attended by Stensen's two most important pupils, the young Caspar Bartholin, who turned 17 on 10 September 1672, and his five years older cousin, Holger Jacobsen. Thomas Bartholin attended many of Stensen's dissections and Ole Borch a few. In 1666 Jacobsen began his studies in theology at the University of Copenhagen with Professor Rasmus Vinding as private preceptor but switched in 1670 to medical studies under Thomas Bartholin's and Ole Borch's guidance. On 12 October 1671, he travelled to Holland with a view to further studies at the University of Leiden, where he received, among other things, teaching by Franciscus Sylvius. Already in August 1672 his studies were, however, interrupted by the French invasion of Holland, and he had to return to Copenhagen, where he thus arrived about the time, when Stensen began his dissections.

The royal anatomist begins dissecting

In September 1672, Stensen set about dissecting different mammals of different sizes. It mainly took place in private settings, among other places in his own home, but also in some cases in public auspices in the Anatomy House. The participants were typically medical students, colleagues and interested notables. Holger Jacobsen was present at many of the dissections and made drawings in his notebook, which Bartholin added to the articles about the respective dissections in *Acta Medica*.

On 3 September, Stensen and Borch dissected two hedgehogs, one of each gender, focusing on the genitals and the circularly arranged musculature in the skin around the spines and its function when, in defence, the hedgehog curls up into a ball of bristly spines. Borch subsequently gave a thorough description of the dissection in *Acta Medica* with the title *Anatome herinacei nostratis, seu echini terrestris* (Dissection of a Danish hedgehog, or a land hedgehog) (Figure 181).[176]

On 20 September, Stensen dissected a male bear donated by the French envoy Hugues de Terlon (1620–1690). Stensen focused on the layers of skin in the bear's front paws and demonstrated the inlets of the thoracic duct in the jugular veins.

On the 23–28 September, Stensen had his first public performance in the Anatomical Theatre. The task was to dissect the body of a soldier who had been hanged for theft. Stensen explained that the organs work like sieves. The kidneys secrete urine, the skin secretes sweat through its pores, the lungs emit breath, the liver bile. During the dissection of the brain, Stensen demonstrated, among other things, the convolutions and blood vessels on the surface and the nerves to the eye via the crossing point *chiasma opticum*. He demonstrated the structure of the skeletal muscles with a central fleshy part and two tendinous ends as well as the course of the muscle fibres of the heart. The following month Stensen dissected a fallow deer, a carp, a dormouse and a squirrel.[177]

It was a major event when Stensen, on 21 October, dissected a reindeer in the Anatomical Theatre. The animal was provided by Peder Schumacher Griffenfeld himself. Bartholin writes in *Acta Medica*[178] that Griffenfeld, out of love for the people, had made two young reindeer available for Stensen's public dissection in the Anatomical Theatre. It was, according to Bartholin, a unique opportunity, as reindeer are only rarely transported to Denmark from Norway, and when they are, it is rare that they survive the journey. But these two reindeer, a male and a female, were raised on Griffenfeld's farm, where they had become ill, as they could not stand the Danish air. Only the male animal was dissected this day, while the female animal was dissected on 26 November. Bartholin begins his report thus: "Among the animals that the very famous anatomist Niels Stensen, who

Figure 181. a. Holger Jacobsen's drawing of the genitals of a male hedgehog dissected by Ole Borch with Niels Stensen's participation on 3 September 1672 in Jacob Kitzerow's house in Studiestræde in Copenhagen. At the top are the two testicles, while the elongated organ at the bottom centre is the penis. (The two lower figures on either side of the penis do not relate to the dissection of the hedgehog. They are two water bladders found in the lungs of a monkey). From Ole Borch's dissection description in Thomas Bartholin's *Acta Medica & Philosophica Hafniensia Ann. 1671 & 1672*, Copenhagen 1673, Obs. 90, pp. 175–78. The illustration is found between p. 214 and p. 215. The Royal Library in Copenhagen.

b. Holger Jacobsen's coloured drawing of the same hedgehog in his notebook in the Royal Library in Copenhagen.

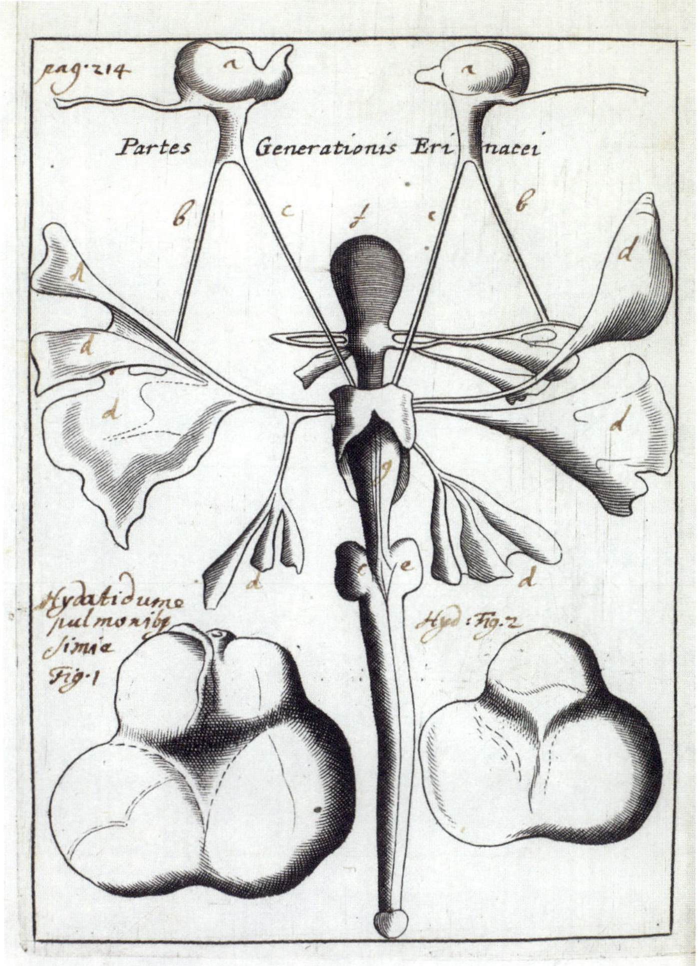

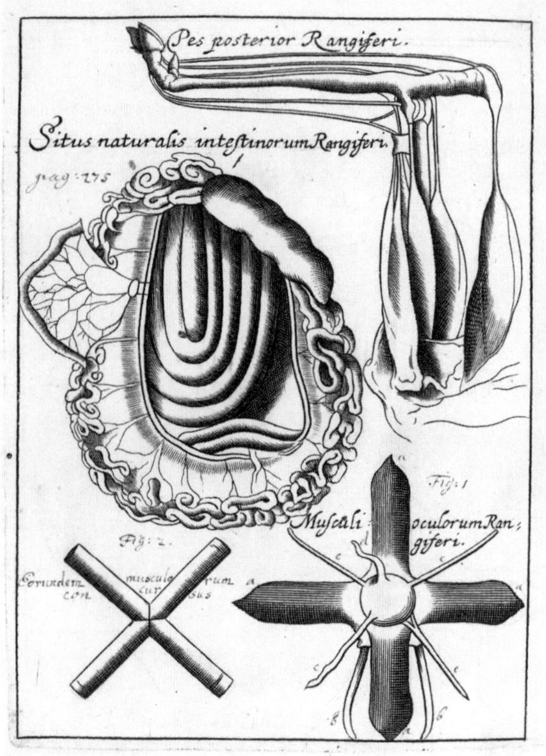

a

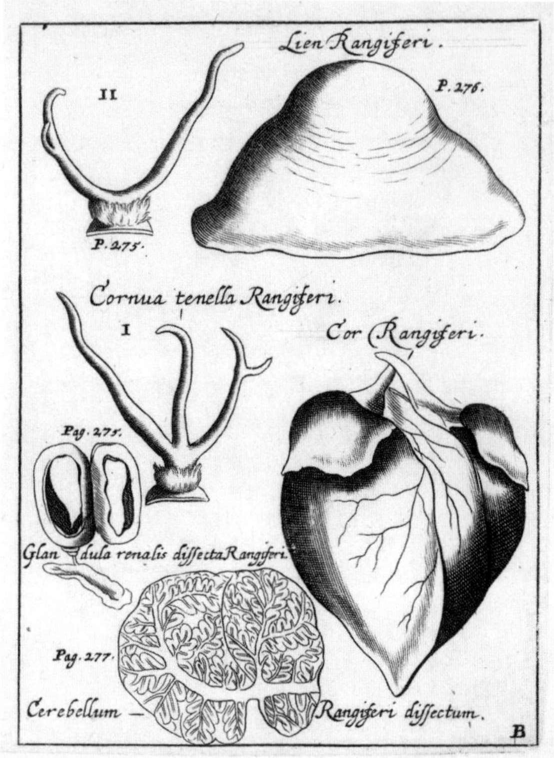

b

has come back to us, has dissected in public and in private with his tireless hand and displayed to spectators, the first place deserves to be awarded to the reindeer, who have never before been exposed to an anatomist's knife."

Then Bartholin describes in detail in 19 points Stensen's dissection of the male animal (Figure 182), and he concludes that it was affected by disease in the abdominal cavity, where watery blisters, swellings on the lymphatic vessels, abscesses and pus were found. On the same day, "the commendable Stensen" dissected a female bear (Figure 183), which according to Bartholin had been bestowed by the French envoy Terlon, in whose courtyard the dissection company had first seen the bear being strangled. Bartholin compares several of the dissection findings from the brains of the two animals. He concludes the report by stating that "our leader Stensen," when his time permits it, will convey other observations in the two reindeer, just as each day he passes similar observations on other animals to the interest of the students, and just as he has done it on a human being, whom he recently dissected in public. After this, Bartholin reproduces a poem on the dissection of the reindeer and the bear that the poet Georg Huber (d. 1686) wrote for *Den Danske Mercurius*.

c

Figure 182. Holger Jacobsen's drawings of the reindeer buck that Niels Stensen dissected in the anatomical theatre in the Anatomy House on 21 October 1672. From Thomas Bartholin's *Acta Medica & Philosophica Hafniensia Ann. 1671 & 1672*, Copenhagen 1673, between p. 274 and p. 275.
The Royal Library in Copenhagen.

a. The panel shows the reindeer's hind legs at the top, and below the intestinal system in its natural position. At the bottom are drawings of the eye muscles.

b. The panel shows at the top left two drawings of the reindeer's antlers and to the right the spleen and the heart. In the centre to the left is an adrenal gland, and at the bottom left is the cerebellum.

c. Holger Jacobsen's coloured drawing of the same reindeer buck. From Jacobsen's notebook in the Royal Library in Copenhagen.

Bartholin was at this time fully aware of Griffenfeld's increasing influence in Denmark and knew that it was important to maintain a good and close relationship with him to ensure the best opportunities for anatomy in Denmark in general and for Stensen in particular. This was not least important because, among influential Lutheran priests, there was opposition to Stensen's appointment as royal anatomist. The powerful head of the Danish Lutheran Church, Bishop of Zealand, professor of theology and Dean of the Faculty of Theology, Hans Wandall (1624–1675), was a decided Catholic hater, and his successor as bishop, Hans Bagger (1646–1693), called Stensen "a proselyte of Satan" and wrote about him, among other things:

"Deservedly we mourn the journeys of this man, in which he has been carried away to the ugly and shameful intercourse with the Babylonian whore."[179]

This is only one of many examples which, as expressed by Ian Herbert Porter, must have "confirmed his [Stensen's] impression that his religion was not welcome in official academic circles" in Copenhagen.[180] Therefore, Bartholin dedicated the second volume of *Acta Medica* to Griffenfeld, "the University of Copenhagen's

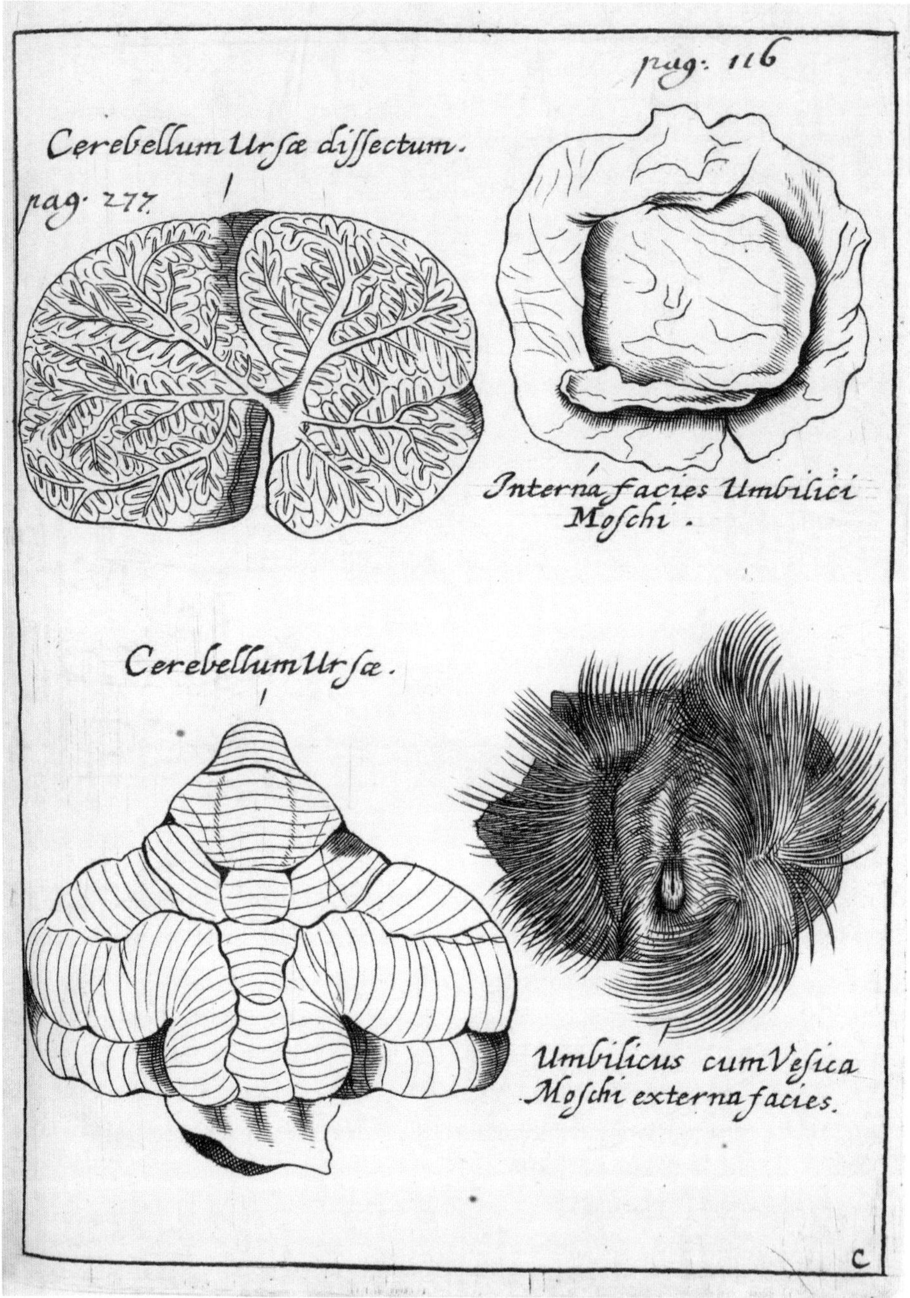

Figure 183. Holger Jacobsen's drawings (in the left side of the figure) of the cerebellum of the French envoy Terlon's strangled female bear, which Niels Stensen dissected in the anatomical theatre in the Anatomy House on 21 October 1672. From Thomas Bartholin's *Acta Medica & Philosophica Hafniensia Ann. 1671 & 1672*, Copenhagen 1673, between p. 214 and p. 215.
The Royal Library in Copenhagen.

preserver, maecenas and best patron." In the six-page dedication, which concludes with a poem, Bartholin emphasizes the importance that Griffenfeld has had for medical science in Denmark. Among other things, he writes the following words, which documents his collaboration with Griffenfeld to secure and support Stensen's employment as a royal anatomist in Copenhagen:

> "The Anatomical Theatre came to life again under your auspices, when Stensen, this great and famous prosector, was helped forward by your favour, in order that he, with his accurate hand and moderate speech, could thoroughly go through animal and human internal organs."

On 31 October 1672, Stensen dissected, together with Caspar Bartholin and other medical candidates, a hare, and Thomas Bartholin recorded the observations following the son's oral transmission, with diligent references to ancient authors (Figure 184).[181] The now 17-year-old Caspar Bartholin was well underway in the process of medical and anatomical studies. As a 12–13-year-old in 1668 he had already been introduced to the learned world in a new edition of his father's small treatise on the anatomy and song of the swan, *Dissertatio de Cygni Anatome, ejusq; Cantu*, dedicated to archbishop Hans Svane with astute allusions to his surname.[182]

According to Holger Jacobsen's notebook, "the excellent prosector Niels Stensen" dissected a dog and a calf in Ole Borch's house on 7 January 1673. Among other things he showed on both animals the joint confluence of the lymph vessels from the head and upper body with the thoracic duct at the left shoulder vein. Jacobsen explains in detail Stensen's method of dissection on the calf.[183]

Niels Stensen's public dissection of a female corpse in the Anatomical Theatre, January–February 1673

By the end of January 1673, at the peak of the dissection season in the middle of the winter, Bartholin finally had the necessary time to introduce Stensen to the learned society of Copenhagen. It happened in the form of a large-scale public dissection of an executed female in the Anatomical Theatre during the days from 29 January to 8 February – Stensen's inaugural lecture. The day before, on 28 January, the Dean of the Faculty of Medicine, royal life physician and honorary medical professor Thomas Bartholin, under the auspices of the Faculty of Medicine, issued a flashy invitation program (Figure 185), in which he picturesquely and poetically praised Stensen, the hope of the expectant students, to the skies.[184]

In the program, Bartholin rejoices that, by the grace of God and the king, the anatomy theatre has now begun to recover after the dust and silence that for

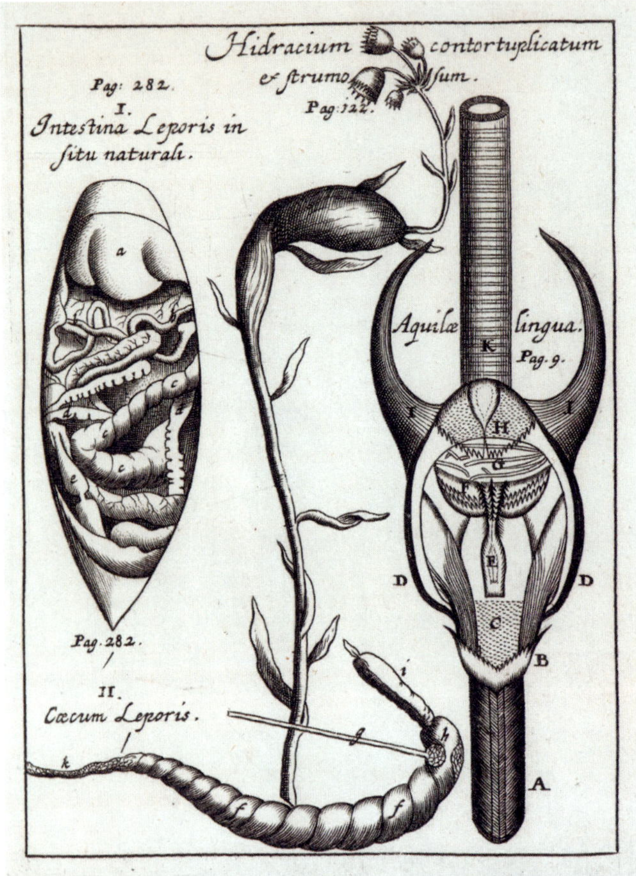

Figure 184. a. Holger Jacobsen's drawings from Niels Stensen's dissection of a hare in Copenhagen together with Caspar Bartholin and other medical graduates. According to Thomas Bartholin, the dissection took place on 31 October 1672, while according to Holger Jacobsen's notebook it was on 4 November 1672.

Top left (I) shows the intestinal system in its natural position. *a*. Liver. *b*. Small intestine. *c*. Appendix. *d*. Large intestine. *e*. Rectum. Bottom left (II) shows *f*. The removed appendix. *g*. Small intestine. *h*. Glandular tissue at the transition between the small intestine and the coecum. *i*. Large intestine. *k*. The glandular end of the appendix. From Thomas Bartholin's *Acta Medica & Philosophica Hafniensia Ann. 1671 & 1672*, Copenhagen 1673, between p. 282 and p. 283. The Royal Library in Copenhagen.

Anatome IX D. 4 Novemb.
141
Leporis

1. Pancreas non modo ab intestinis sub stomacho sinistrorsum ferebatur et lieni ultra mediam longitudinem ejus adhærebat, sed et inter intestina ipsa longo tractu descendebat supra ramum venæ mesentericæ, ubi ad extremitatem ejus a stomacho remotissimam egrediebatur ductus pancreaticus, ultram circiter ab insertione Ductus biliarii remotus. Ipsius pancreatis substantia non erat in unum corpus coacta sed per mesenterium circa venarum ramificationem sparsa.

Intestina in situ naturali. Intestinum Cæcum.

Cæcum ultra ulnam longum erat. Externus ultra duas tertias totius longitudinis spiralem lineam exhibens ut in cuniculis.

Tota portio intestini ab insertione Ductus biliarii usque ad insertionem ductus pancreatici habebat vasa sanguinea externus admodum turgentia sanguine ita quidem ut plurima ex capillaribus conspicerentur velut in inflammatione corneæ.

b. The same drawings, the dissection description and Holger Jacobsen's coloured drawing of the hare in his notebook in the Royal Library in Copenhagen.

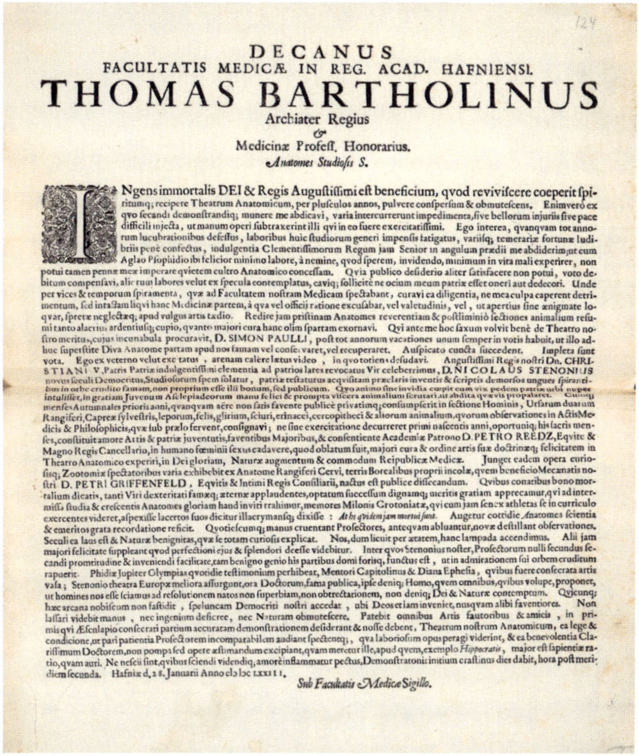

DECANUS
FACULTATIS MEDICÆ IN REG. ACAD. HAFNIENSI.
THOMAS BARTHOLINUS
Archiater Regius
&
Medicinæ Profess. Honorarius.
Anatomes Studiosis S.

INgens immortalis DEI & Regis Augustissimi est beneficium, qvod reviviscere coeperit spiritumq; recipere Theatrum Anatomicum, per plusculos annos, pulvere conspersum & obmutescens. Enimverò ex qvo secandi demonstrandiq; munere me abdicavi, varia intercurrerunt impedimenta, sive bellorum injuriis sive pace difficili injecta, ut manum operi subtraxerint illi qvi in eo fuere exercitatissimi. Ego interea, qvanqvam tot annorum lucubrationibus defessus, laboribus huic studiorum generi impensis fatigatus, variisq; temerariæ fortunæ ludibriis penè confectus, indulgentia Clementissimorum Regum jam Senior in angulum prædii me abdiderim, ut cum Aglao Psophidio ibi felicior minimo labore, à nemine, qvod sperem, invidendo, minimum in vita mali experirer, non potui tamen pennæ meæ imperare qvietem cultro Anatomico concessam. Qvia publico desiderio aliter satisfacere non potui, voto debitum compensavi, aliorum labores velut ex specula contemplatus, caviq; sollicitè ne ocium meum patriæ esset oneri aut dedecori. Unde per vices & temporum spiramenta, qvæ ad Facultatem nostram Medicam spectabant, curavi ea diligentia, ne mea culpa caperent detrimentum, sed intactam liqvi hanc Medicinæ partem, à qva vel officij ratione excusabar, vel valetudinis, vel, ut apertius sine ænigmate loqvar, spretæ neglectæq; apud vulgus artis tædio. Redire jam pristinam Anatomes reverentiam & postliminiò sectiones animalium resumi tanto alacrius ardentiusq; cupio, qvanto majori cura hanc olim spartam exornavi. Qvi ante me hoc saxum volvit benè de Theatro nostro meritus, cujus incunabula procuravit, D. SIMON PAULLI, post tot annorum vacationes unum semper in votis habuit, ut illo adhuc superstite Diva Anatome partam apud nos famam vel conservaret, vel recuperaret. Auspicato cuncta succedent. Impleta sunt vota. Ego ex veterno velut excitatus, arenam calere lætus video, in qvo totiens desudavi. Augustissimi Regis nostri Dn. CHRISTIANI V, Patris Patriæ indulgentissimi clementia ad patrios lares revocatus Vir celeberrimus, D. NICOLAUS STENONIUS novus seculi Democritus, Studiosorum spem solatur, patriæ testaturus acqvisitam præclaris inventis & scriptis demorsos ungues spirantibus in orbe erudito famam, non proprium esse illi bonum, sed publicum. Qvo animo sine invidia coepit cum vix pedem patriæ urbi nuper intulisset, in gratiam Juvenum Asclepiadeorum manu felici & prompta viscera animalium scrutari, ut abdita qvævis propalaret. Cumq; menses Autumnales prioris anni, qvanqvam aëre non satis favente publicè privatimq; consumpserit in sectione Hominis, Ursarum duarum Rangiferi, Capreæ sylvestris, leporum, felis, glirium, sciuri, erinacei, cercopitheci & aliorum animalium, qvorum observationes in Actis Medicis & Philosophicis, qvæ sub prælo fervent, consignavi; ne sine exercitatione decurreret primi nascentis anni, oportuniq; his sacris menses, constituit amore Artis & patriæ juventutis, faventibus Majoribus, & consentiente Academiæ Patrono D. PETRO REEDZ, Eqvite & Magno Regis Cancellario, in humano fœminii sexus cadavere, quod oblatum fuit, majori cura & ordine artis suæ doctrinæq; felicitatem in Theatro Anatomico experiri, in Dei gloriam, Naturæ augmentum & commodum Reipublicæ Medicæ. Junget eadem opera curiosisq; Zootomiæ spectatoribus varia exhibebit ex Anatome Rangiferi Cervi, terris Borealibus proprii incolæ, qvem beneficio Mecænatis nostri D. PETRI GRIFFENFELD, Eqvitis & Intimi Regis Consiliarii, nactus est publice dissecandum. Qvibus conatibus bono mortalium dicatis, tanti Viri dexteritati famæq; æternæ applaudentes, optatum successum dignamq; meritis gratiam apprecamur, qvi ad intermissa studia & crescentis Anatomes gloriam haud inviti trahimur, memores Milonis Crotoniatæ, qvi cum jam senex athletas se in curriculo exercentes videret, aspexisse lacertos suos dicitur illacrymansq; dixisse: *At hi qvidem jam mortui sunt.* Augetur cottidie Anatomes scientia & emeritos grata recordatione reficit. Qvotiescumq; manus cruentant Prosectores, anteqvam abluantur, novæ destillant observationes. Seculi ea laus est & Naturæ benignitas, qvæ se totam curiosis explicat. Nos, dum licuit per ætatem, hanc lampada accendimus. Alii jam majori felicitate suppleant qvod perfectioni ejus & splendori deesse videbitur. Inter qvos Stenonius noster, Prosectorum nulli secundus secandi promtitudine & inveniendi facilitate, tam benigno genio his partibus domi forisq; functus est, ut in admirationem sui orbem eruditum rapuerit. Phidiæ Jupiter Olympias qvotidie testimonium perhibeat, Mentori Capitolinus & Diana Ephesia, qvibus fuere consecrata artis vasa; Stenonio theatra Europæ meliora assurgunt, ora Doctorum, fama publica, ipse deniq; Homo, qvem omnibus, qvibus volupe, proponet, ut homines nos esse sciamus ad resolutionem natos non superbiam, non obtrectationem, non deniq; Dei & Naturæ contemptum. Qvicunq; hæc arcana nobiscum non fastidit, speluncam Democriti nostri accedat, ubi Deos etiam inveniet, nusqvam alibi faventiores. Non lassari videbit manus, nec ingenium deficere, nec Naturam obmutescere. Patebit omnibus Artis fautoribus & amicis, in primis qvi Æsculapio consecrati partium accuratam demonstrationem desiderant & nosse debent, Theatrum nostrum Anatomicum, ea lege & condicione, ut pari patientia Prosectorem incomparabilem audiant spectentq;, qva laboriosum opus peragi viderint, & ea benevolentia Clarissimum Doctorem, non pompa sed opere æstimandum excipiant, qvam meretur ille, apud qvem, exemplo *Hippocratis*, major est sapientiæ ratio, qvam auri. Ne nescii sint, qvibus sciendi videndiq; amorè inflammatur pectus, Demonstrationis initium crastinus dies dabit, hora post meridiem secunda. Hafniæ d. 28. Januarii Anno cIↄ Iↄc LXXIII.

Sub Facultatis Medicæ Sigillo.

Figure 185. Thomas Bartholin's invitation program from 28 January 1673 to Niels Stensen's public dissection of a female corpse in the anatomical theatre in the Anatomy House in Copenhagen January–February 1673. Stensen's dissection began with his introductory lecture on 29 January and ended on 8 February. The Royal Library in Copenhagen.

several years had descended upon it. With joy, he now sees life stirring again in the arena, where he himself has so often worked. He refers to Stensen as "the new Democritus of the century […] who is left behind no one in his dexterity of dissection and in his ease in making new discoveries, and who has captured the learned world's admiration. He is exalted by Europe's finest anatomical theatres and the mouths of scholars." Stensen now wanted to show the structure of the human body in such a way that the audience should understand that we humans are born to degrade, not to haughtiness or envy, least of all to contempt for God and Nature.

On the next day, before inserting his *culter anatomicus* into the body of the executed woman, Stensen gave his legendary philosophical introductory lecture, *Prooemium Demonstrationum Anatomicarum in Theatro Hafniensi Anni 1673* (Introductory lecture to the anatomical demonstrations in the Anatomical Theatre in Copenhagen in 1673) with the famous words: "Beautiful is what we see, more beautiful is what we know, by far the most beautiful is what we are ignorant of (Pulchra sunt, qvæ videntur, pulchiora, qvæ sciuntur, longe pulcherima, qvæ ignorantur)." Thomas Bartholin published the lecture as the last article in the second volume of *Acta Medica*.[185]

Considered in isolation, the famous words quoted appear wise, also from a scientific standpoint, but behind them are new thoughts with a direction away from science and into a more spiritual and religious world. The most remarkable thing about the lecture is not the famous words, but the radical transformation that had taken place with Stensen in just a few years, and which comes out clearly when comparing this lecture with the brain lecture in Paris in 1665. The brain lecture is largely cleansed of faith and religion, and in the 60-page lecture Stensen only mentions the Creator once, in passing. It is the natural scientist and the natural philosopher Niels Stensen who speaks and argues convincingly for the necessity of using the senses to make one's own observations and experiences and in order to discover Nature's hidden truths through certain evidence – just as Bartholin had taught him. It was this procedure that Stensen himself had followed in his research – perhaps more rigorously and uncompromisingly than any other natural scientist of the time – and with which he had achieved his epoch-making results. In the brain lecture, he criticizes and chastises anatomists who neglect these principles and instead try to guess and reason themselves to the scientific truths, and not least does this apply to Descartes, whom Stensen criticizes for failing his own method for exploring Nature. Like Descartes, Stensen compares the human body to a machine, and instead of God he speaks of "the master who built it." In the introductory lecture here eight years later, it is theologian-Stensen, who has taken over and turns the whole thing upside down. He now warns with examples about the danger of the deception of the senses, appearances and desires and advocates for the use of the intellect, the reason, and the divine soul that God has bestowed upon man:

"To judge of the perceptible we have our reason, thanks to which a reliable approach to the imperceptible is given us. It is far from us no longer to be men and put ourselves below the beasts. [...] So let us not dwell further on the senses, but with the eyes of the mind, through the eyes of the body, as through a window of a very artistic palace, let us look out over the most delightful meadow, in which there are as many species as there are flowers, as many particles as there are miracles. [...] Yes, who would not, when all the errors of the senses had been refuted, repeat: beautiful is what is accessible to the senses without dissection, more beautiful is what the dissection draws forth from the hidden interior, but by far the most beautiful is what escapes the senses and yet with the help of the sensuous is recognized through reason."

In the lecture, which is only seven pages long, the Creator is mentioned again and again, and only with his help can the anatomist discover the truths of Nature. However, there are still remnants left of the natural scientist, when Stensen emphasizes the necessity of examining the anatomy of plants, animals and humans right down to the smallest detail, even further down than the senses allow, which must be understood as an invitation to use technical aids like the microscope.[186] These considerations are entirely in line with Thomas Bartholin. Anatomists must go into the cave of Democritus, that is, right down to the smallest parts, to find the truth. But Stensen goes further. For him it would be best if you could look right into the soul, for it is the soul he speaks of, when he says that by far the most beautiful thing is, what we are ignorant of:

"If a very small part of the human surface is so beautiful and so highly affects the viewer, what beauties would we not get to see, what kind of pleasure would we not feel, if we could contemplate the art of the whole body's arrangement and the soul, that has so many and such elaborate tools available?"

For Stensen, the very purpose of anatomical research was no longer that, which Thomas Bartholin explained so well the following year in his little treatise on the practical anatomy: to get acquainted with the normal anatomy in order to understand the pathological anatomy and thereby acquire knowledge that can form the basis for prevention and treatment of diseases in humans. Or, as he had written to Stensen in the letter of 5 August 1662:

"Live well, my Stensen, and keep on pursuing the study of anatomy that you have embarked upon, and to which you seem to have been born, with customary zeal, for the benefit of sick human beings."

Remember that in December 1649, in the introduction to the first of his 18 disputations at the University, Bartholin had emphasized to his students the importance of both purposes for the study of anatomy: the understanding of human health and sickness, and God's creative power. For Bartholin, the most important purpose throughout his career was undoubtedly the former. For Stensen, on the contrary, the true goal of anatomical research had become to get closer to the Creator himself by acquiring knowledge and insight into his creations:

"And this is the true goal of anatomy: through the surprising artwork of the body to lift up the spectators to an understanding of the dignity of the soul, and consequently, through the wonders of the body and soul, to get to know them, and to love the Creator. The object of anatomy is the animal and especially the human body, insofar as they can be split into parts accessible to the senses. Such a great and obvious beauty cannot fail to evoke admiration and awaken the desire to gain knowledge of what escapes the senses. In this way, reason is lifted from the consideration of the individual parts and from the mutual comparison of the different parts to track down the Creator of such great miracles. [...] When organized for all the spectators, the true anatomy is the way in which God, by the hand of the anatomist, leads us to the knowledge of first the animal body, and then himself."

Stensen then even takes Thomas Bartholin and his more rational and modern views on anatomy and medicine to task:

"Thus, people who make anatomy a servant that simply has to prevent or cure diseases are mistaken and treat anatomy beneath its dignity. Admittedly, it is useful in these areas, but not as much as we think. Knowledge about the pathological state cannot surpass the knowledge about the healthy state, and since the knowledge about the healthy state is still very limited, the knowledge about the pathological state could not significantly expand its scope."

Time would soon prove Stensen fundamentally wrong on this point, since just six years later Théophile Bonet's mentioned fundamental work of pathological anatomy was published. Hereby the discipline on which fundament the understanding of modern medicine and treatment of diseases rests was founded. It was this radical transformation of Stensen that led his acquaintance and admirer as a scientist, G. W. Leibniz, to remember him with sadness in 1710:

"He was a great anatomist and extremely skilled in the study of Nature, but he regrettably abandoned the exploration of it, and from being a great physician, he became a mediocre theologian."[187]

Whether it was precisely the fact that on this day in January 1673 Stensen used his anatomical pulpit as a theological pulpit, which sealed his fate as royal anatomist in Copenhagen, is unknown, but it is most likely, since in the lecture he recited several messages that could easily be interpreted as Catholic[188] in front of an audience which included already sceptical and powerful Lutheran theologians. Anyhow, it is remarkable that Stensen's inaugural lecture also became his farewell lecture, for after the dissection of the female corpse was completed a few days later, Stensen never again appeared in public in the Anatomical Theatre.

Bartholin's grandiose and expectant words from the invitation program about a new heyday for anatomy in Copenhagen with the skilled Stensen at the helm were erased, almost before the ink had dried. But who erased them? Certainly not Bartholin, and most certainly not Stensen himself either. A letter found in 2018 from Stensen in Copenhagen to his friend in Florence, the Italian mathematician Vincenzo Viviani (1622–1703), dated 18 April 1673,[189] shows that Stensen's situation in Copenhagen at this time was unresolved. Stensen trusts that Bartholin, Griffenfeld and two other high-ranking officials can help him. He awaits a meeting with Griffenfeld, who has promised to see what can be done. Stensen hopes to be granted leave to return to Florence in the holy year of 1675 to act as tutor to Cosimo III's son.[190] In the letter, Stensen again forsakes the sensual and seeks comfort in the divine. He speaks of God's goodness and God as the originator of all joy and justice. How can we love his gifts without loving him? When man loves the sensual, he becomes a monster.

The result of the meeting with Griffenfeld was apparently that Stensen, during his continued stay in Copenhagen, was to act as a private tutor for Holger Jacobsen and Caspar Bartholin. It appears from Holger Jacobsen's notes and Thomas Bartholin's *Acta Medica* that, with few exceptions, it was Holger Jacobsen or Caspar Bartholin who held the knife in all the later dissections in Copenhagen where Stensen was present, and that the dissections were carried out on animals in a private setting.

The dissection of the female corpse brought no news, but Stensen made from the dissection a drawing of the circulation of the blood between the large organs in the abdominal cavity and chest, made clear by separating the chambers of the heart. Thomas Bartholin used the drawing in his fourth and last revised edition of the Bartholinian anatomy book (see later).

However, it must have made an indelible impression on the audience to watch the small and slender, almost fragile, but at the same time charismatic man with the gentle, calm and intelligent face, while with soul and commitment and with his beautiful and elegant language he explained the anatomical details and at the same time handled the instruments with impressive dexterity (Figure 186). A few letters from Stensen's stay in Paris 1664–1665 give an impression of this. When in the spring of 1665 he dissected in Thévenot's home in Paris, the physician André Graindorge (1616–1676) wrote to his friend the theologian Pierre-Daniel Huet (1630–1721) on 19 May 1665:

"This Mr. Stensen attracts attention. This evening, we saw the eye of a horse. To tell you the truth, next to him we are only apprentices. I asked him to show me the heart tomorrow morning, which he promised me with exceptional kindness. He is always in action. He has an unimaginable patience, and by practice he has acquired a skill beyond the ordinary. There is no such thing as a butterfly or a fly that avoids his zeal. He counts the legs of a flea, if it has legs."[191]

In a new letter to Huet on 30 May, Graindorge continues:

"When I called us apprentices in comparison with Mr. Stensen, I was right, for I have never seen dexterity like his; without using anything but the eye, scissors

Figure 186. Undated, but possibly original portrait of Niels Stensen in civic clothes with a thick book under his right arm. He is depicted with a pious and contemplative look and a sharp gaze towards the observer. The portrait may have been painted during his stay in Copenhagen 1672–1674.

Unknown artist. Oil on canvas. Size without frame: 68.9 × 55.8 cm, size with frame: 77.8 × 65 cm.
Medicinsk Museion in Copenhagen.
Photo: Thomas Howalt Andersen.

and another small instrument in his hand, which he always turns towards the audience, he lets us see everything that is worth noticing in the structure of the eye. [...] I have seen the course of the fibres of the heart, and the position that they retain in the musculature, which clearly shows the mechanics that Nature uses for muscle movement."[192]

The animal dissections in which Stensen participated during his stay in Copenhagen did not bring any significant news either. An exception, however, is his interesting dissection of all the muscles of an eagle on 4 April 1673. It was published in *Acta Medica* with the title *Historia Musculorum Aqvilæ D. Nicolai Stenonis* (Doctor Niels Stensen's account of the eagle's muscles).[193] and is the only dissection description that Stensen himself certainly provided during his employment as royal anatomist in Copenhagen. It systematically goes through all the muscles of the eagle from head to tail and feet, i.e. the origin, course and insertion (attachment) of the individual muscles. Stensen draws comparisons with the corresponding muscles in other animals and man, mentions the function of some of the muscles and bring out distinctive characteristics of the eagle and other birds. The treatise is thus a follow-up to his own invitation from the Florentine muscle treatise on a mapping of the body's various muscles and their detailed anatomy, combined with a thorough mechanical analysis of the importance of the individual muscles and muscle parts for the body's various movements. This invitation was later taken up and completed for the human being by his grandnephew Jacob Winsløw (1669–1760), another anatomist of the Bartholinian School. But also in this, Stensen's last small visionary anatomical treatise, which could give hope for his continued acting as an anatomist, the theologian shines clearly through. More than in any of his previous treatises, he highlights the ingenious devices of the Creator and emphasizes the importance of looking for God in the marvels of Nature. His expressed desire to complete his muscle doctrine of the various animals "in order to advance the understanding of the true causes of the differences in the external form of the different animals and to shed light on the mechanical capability" unfortunately never came to fruition.

Two more dissections performed by Stensen deserve mention because of their special character. The first is Stensen's study of a snake egg in July 1673. Thomas Bartholin reports about it in *Acta Medica*:

"In the month of July, I brought from my estate a pile of snake eggs, so that they could be examined by our Stensen. The snake eggs were found by our friend and relative Hans Haagen in a hollow tree root in the countryside. When little by little the eggs were opened, live snake cubs were found with a heart that clearly pulsed. Of the adherent yolk the placenta, the vessels of which had a firm consistency, was in the process of being formed, and via a string it ended in the navel, which is not far from the tail, as my beloved sister's son Holger Jacobæus has neatly outlined it in the illustration"[194] (Figure 187).

The second is Stensen's dissection of the scent glands at the tail of a fox. It took place on 15 February 1674 and was the last dissection in Copenhagen where Stensen himself guided the knife. The reason was that Christian V's head huntsman Vincents Hahn (1632–1680) had once told Thomas Bartholin about the peculiar violet scent from the fox's tail, which had given Bartholin the wish to investigate the phenomenon thoroughly, whenever he got the opportunity. He gave his son Caspar the task, and Caspar was given the opportunity on the said day to participate in the dissection of a fox in Henrik à Møinichen's house, executed by "the great anatomist Niels Stensen" with "his expert hand." The

Figure 187. Holger Jacobsen's coloured drawing of a baby snake that has just emerged out of one of the snake eggs that in June or July 1673 were found in a hollow tree root in the countryside near Thomas Bartholin's manor house Hagestedgaard and brought by him to Copenhagen for Niels Stensen's closer study. As described by Bartholin, Jacobsen's drawing shows the young snake's continued connection with the egg yolk and placenta via the umbilical cord. Thomas Bartholin writes that the event took place in July 1673, while Holger Jacobsen dates the examination to 21 June 1673. From Holger Jacobsen's notebook in the Royal Library in Copenhagen.

question of the scent glands was examined, and Caspar subsequently sent a description of the dissection from Copenhagen to his father at Hagestedgaard. This description was lost, however, and Thomas Bartholin instead included in *Acta Medica*[195] an extract from Caspar's more detailed description in his work *Exercises Miscellaneæ. Varii argumenti inprimis Anatomici* (Mixed exercises of various contents, but aimed at anatomy), published in Leiden 1675.[196]

Probably Stensen's last official appearance in the Anatomical Theatre took place on 30 September 1673. Together with Thomas Bartholin, Henrik à Møinichen, Ole Borch and city physicist Caspar Kölichen he participated as supervising physician at the first midwifery exam, which was a result of the mentioned new medicinal ordinance. 15 midwives were examined with the two older midwives Kristine Trømmer and Elisabeth Hansdatter as censors. The examined midwives' limited practical experience in the profession could to some extent be compensated by their personal experience with the births of their own children. 13 of them had each given birth to 12 children, one of them "only" to nine, and the last to 17. Their knowledge was generally assessed as being satisfactory, but as this did not apply to their anatomical knowledge, they were instructed to attend Niels Stensen's anatomical exercises the following winter. Likewise, they were encouraged to notify him, when they learned of deaths in the maternity bed among the poor population, so that he could without payment offer his teaching. Of course, Bartholin documented the historical event in *Acta Medica*.[197]

The rest of Stensen's stay in Copenhagen was, in addition to teaching Holger Jacobsen and Caspar Bartholin, marked by his continued work as an agent for the Papal States and a promotor of the Catholic mission in Denmark, his attempts to convert even his family members to the Catholic faith and his resulting conflicts with both the family and Lutheran theologians.[198] Under these circumstances, even his most faithful and influential supporters, including Thomas Bartholin and Peder Schumacher Griffenfeld, could not help him. It was probably primarily a combination of this domestic resistance and his longing to return to the more friendly co-religionists in Florence that, on 26 May 1674, made Stensen apply to Griffenfeld for royal permission to travel to Florence and act as tutor to Cosimo III's son, Grand Prince of Tuscany Ferdinando de' Medici (1663–1713).[199] On 4 July, he received permission from Christian V,[200] and shortly afterwards he left Denmark. In April 1675 he was ordained a Catholic priest in Florence. In 1677 he again worked for some months as Holger Jacobsen's and Caspar Bartholin's anatomy teacher during their stay in Italy, but thereafter he was no longer engaged in natural sciences. In the same year he was appointed titular bishop of Titiopolis and apostolic vicar of North Germany and the Nordic kingdoms. In this role he served from 1677–1680 in Hannover, 1680–1683 in Münster, 1683–

1685 in Hamburg, and after a ten-day wistful visit to Copenhagen in August 1685 he ended his days as Catholic bishop at the court in Schwerin, where he died 25 November 1686. Six months after Stensen's departure from Denmark, Thomas Bartholin wrote on 15 January 1675 to his friend and colleague Johann Georg Volkhamer (1616–1693) in Nürnberg:

"What we now dare expect from our Stensen, who has already returned to Florence, remains to be seen."[201]

It is hard to believe that Axel Garboe (1886–1970) did not hit the nail on the head, when he wrote about Thomas Bartholin's thoughts on the matter:

"It is as if he is casting a long, questioning and sad look after this eminent scientist, whose human and Christian development made him more and more alien to his countrymen and to the science of anatomy whose continuation at the University of Copenhagen was so close to Thomas Bartholin's heart."[202]

Niels Stensen was forever lost to anatomy in Copenhagen. Bartholin now had to rely on his 18-year-old son Caspar, who on 30 May 1674,[203] just four days after Stensen's letter to Griffenfeld, along with his 16-year-old brother Christopher and their cousin Holger Jacobsen, had begun their study trip to the Netherlands, France and Italy. In his letter to Volkhamer, Thomas Bartholin writes the following about Caspar:

"From my son Caspar, who has been entrusted with the position of anatomist at our university,[204] the medical world can also expect something good when he returns to us from his study trip."

Fourth revised edition of the anatomy book

Thomas Bartholin worked for many years on a fourth and last revised edition of the Bartholinian anatomy book, *Anatome Renovata* (The Renewed Anatomy Book). In the preface Bartholin states that, thanks to divine providence, the book was spared from "the greedy flames" of the fire at Hagestedgaard. Before the horrific fire, Bartholin had sent the manuscript to Gerard Blasius in Amsterdam with a view to the latter's further negotiations with the printer Hacke in Leiden about the organization and equipment of the book. According to the plan, the book was to be published in Leiden in 1673. Bartholin writes in the preface that 20 years have passed since the improved anatomy book, *Anatomia Reformata*, with his account of Harvey's doctrine of the circulation of the blood, came into the hands of the scholars. Since then, the book has been published in reprints in various places in numerous languages, and these reprints have become more and more full of errors, omissions and added words, so that Bartholin can almost no longer recognize the original.

At the same time, both he and the book had become so much older that it was time for the book to be "reborn as a new fruit". As examples of new topics in the book, Bartholin mentions Pecquet's discovery of the thoracic duct in animals and his own discovery of it in man, as well as his discovery of the lymph vessels. He describes in the book all the new things he has found himself, and as a builder erecting a new building from broken-down walls, he edits the old text from his father's anatomy book with his own and others' annotations in accordance with Harvey's doctrine of the circulation of the blood. All the new age's discoveries and experiences up to the date of publication are added and collated. The book is supplied with newly produced copper engravings, which no one has seen before, and even more from "various writings of excellent scholars." Bartholin regrets the loss to the flames of the previously mentioned, almost completed work on practical (pathological) anatomy, which he calls *Anatome Utilis seu Practica* (The Useful or Practical Anatomy). He had planned for this useful work, which described all the parts of the human body, and all the places where diseases strike, and whose purpose it was to describe both the sick and the healthy body, to be included in the fourth edition of the anatomy book.

The structure of the book, the number of chapters and their subjects are still unchanged, but it is a considerably enlarged and revised version, in terms of both text and illustrations. The book, which is dedicated to Christian V, has 807 pages and 116 anatomical plates, of which 12 are on folded leaves. Thomas Bartholin's text is, unlike the earlier editions, integrated into his father's text and not just added to this with brackets. The book is, to an even higher degree than the 1651 edition, Thomas Bartholin's own work, where the new text and the new illustrations are essentially derived from his own and contemporary anatomists' new discoveries and works after 1650.

This concerns, among other things Thomas Bartholin's, Niels Stensen's, Jean Pecquet's, Olof Rudbeck's, Johannes van Horne's and Frederik Ruysch's (1638–1731) research and discoveries regarding the lymphatic system, Marcello Malpighi's regarding the pulmonary capillaries (Figure 149), Reinier de Graaf's (1641–1673) and Jan Swammerdam's regarding the reproductive organs (Figure 188), de Graaf's regarding the pancreas (Figure 189), Niels Stensen's, Gerard Blasius', Thomas Wharton's and Francis Glisson's regarding the salivary glands (Figure 190), the latter's regarding the liver, Niels Stensen's regarding the lacrimal gland (Figure 191), muscles (Figure 192), heart (Figure 155) and blood circulation, Heinrich Meibom's (1638–1700) and Konrad Viktor Schneider's (1614–

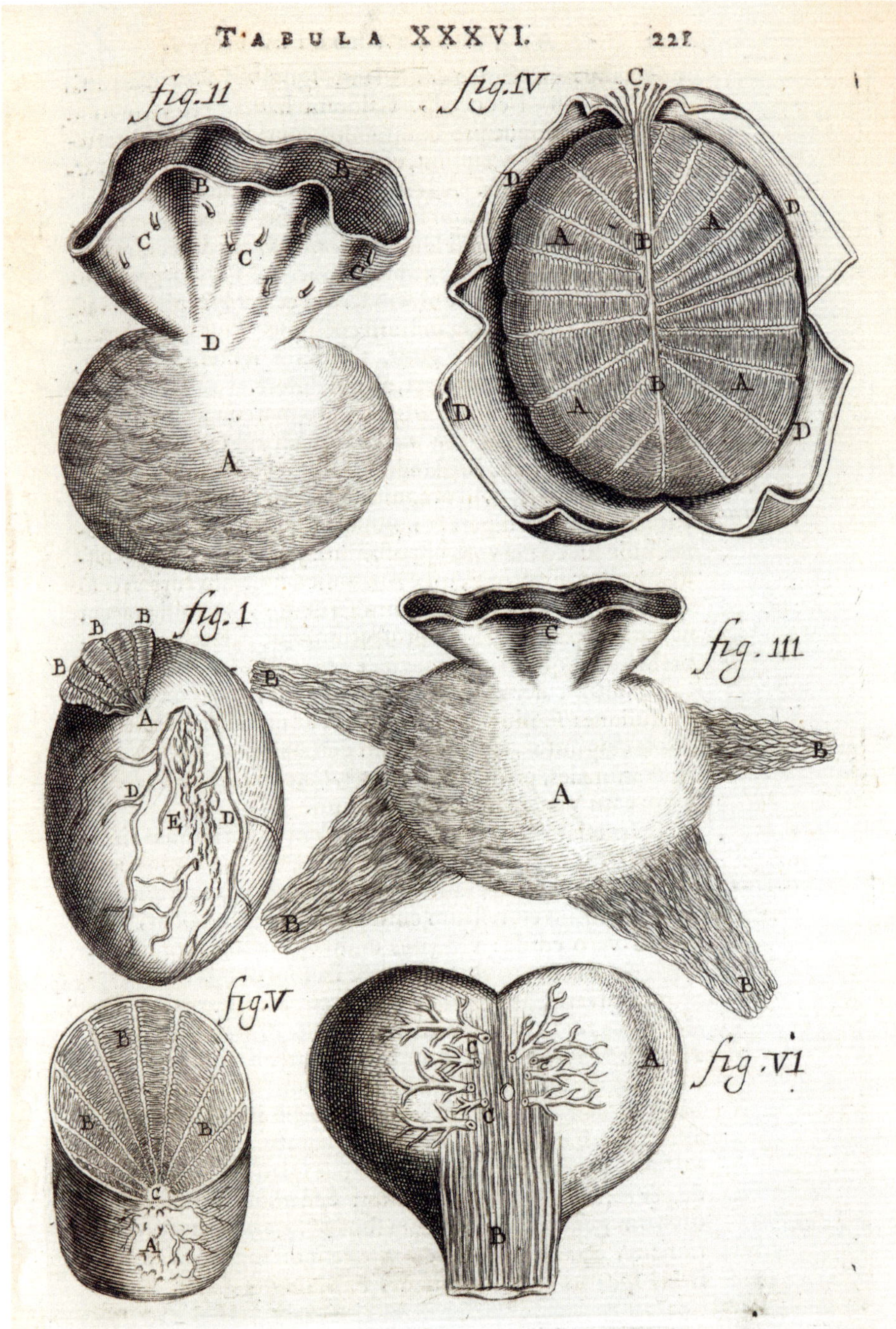

Figure 188. The Dutchman Reinier de Graaf's (1641–1673) illustrations of the male testicles dissected in different ways. From Thomas Bartholin's *Anatome Quartum Renovata*, Leiden 1673, p. 221.
The Royal Library in Copenhagen.

Figure 189. a. Reinier de Graaf was the first to examine the pancreatic juice of a living dog by inserting a permanent fistula into the pancreatic duct. He found that it was acidic and established that it was important for the digestive process by separating the different nutrients in the food.

Fig. 1 shows the fistula operation, and Fig. II shows the dog with a container mounted under the abdomen to collect the pancreatic juice. From Thomas Bartholin's *Anatome Quartum Renovata*, Leiden 1673, p. 122.

b. Reinier de Graaf's illustration of the pancreas with an accurate depiction of its excretory duct. From Thomas Bartholin's *Anatome Quartum Renovata*, Leiden 1673, p. 120.
The letters mark (C is missing on the figure):
A. Pancreatic duct (*ductus pancreaticus* or *ductus Wirsungianus*). B. Branches of the same duct.
D. Common biliary duct. E. Exit of both ducts.
F. Inner surface of pancreas. G. The same, separated from the biliary duct. H. Stomach pulled upwards.
I. Pylorus. K. Duodenum. L. and M. Small intestines.
N. Mesentery liberated from the small intestine.
O. Mesenteric veins. P. Mesenteric artery.
Q. Bile duct. R. Hepatic duct. S. Opened duodenum.
The Royal Library in Copenhagen.

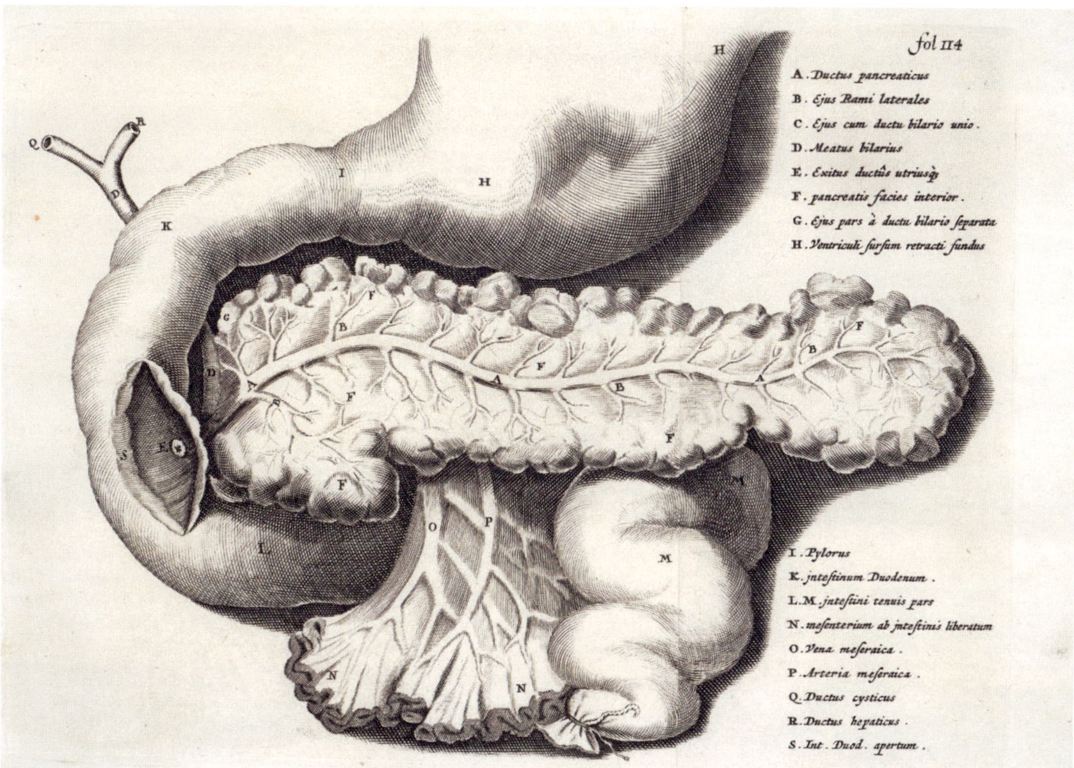

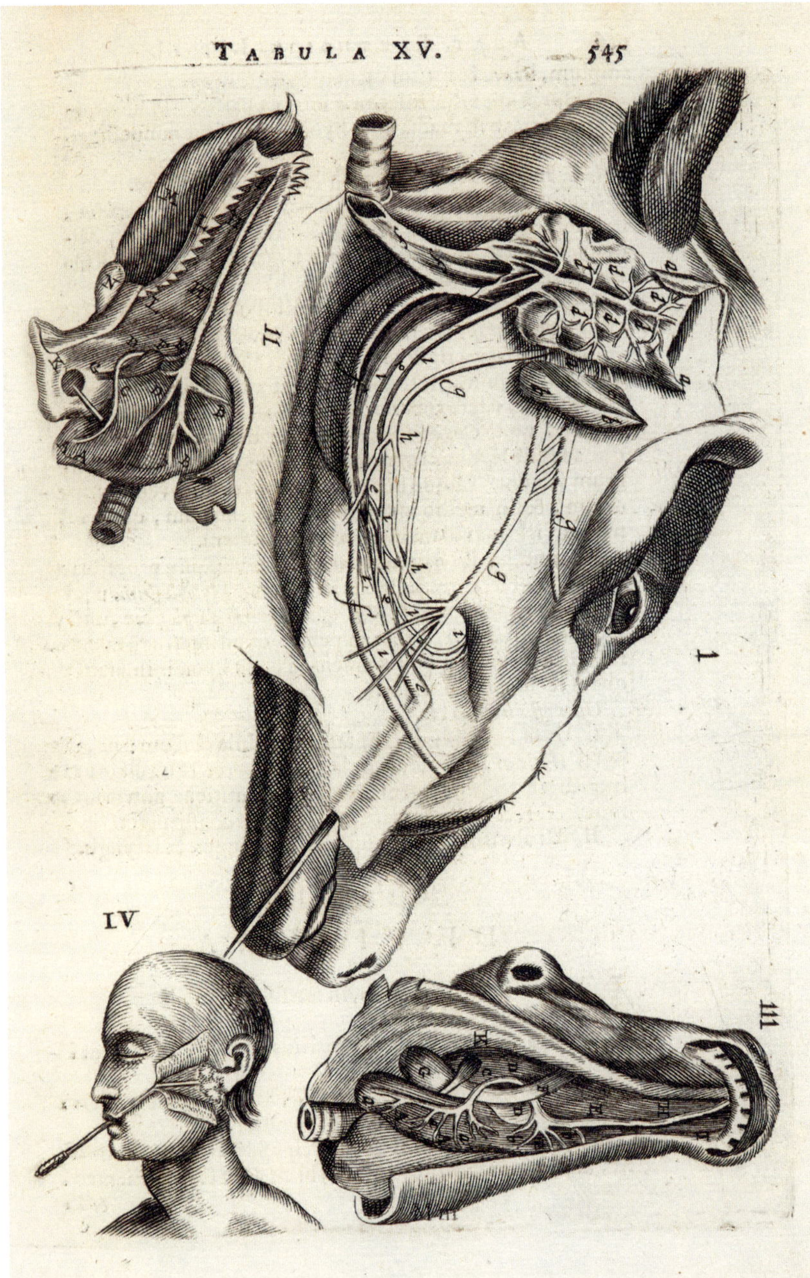

Figure 190. Salivary glands in Thomas Bartholin's *Anatome Quartum Renovata*, Leiden 1673, p. 545.

Bartholin has collected illustrations of salivary glands in calves and a human from the works of three other authors. Figure I shows Niels Stensen's illustration of the parotid gland of a calf (Figure 146 in this book). Figures II and III show Thomas Wharton's illustrations of the maxillary salivary gland of a calf (Figure 141 in this book). Figure IV shows Gerard Blasius' illustration of the parotid gland of a human with a probe through the main salivary duct (Stensen's duct). The Royal Library in Copenhagen.

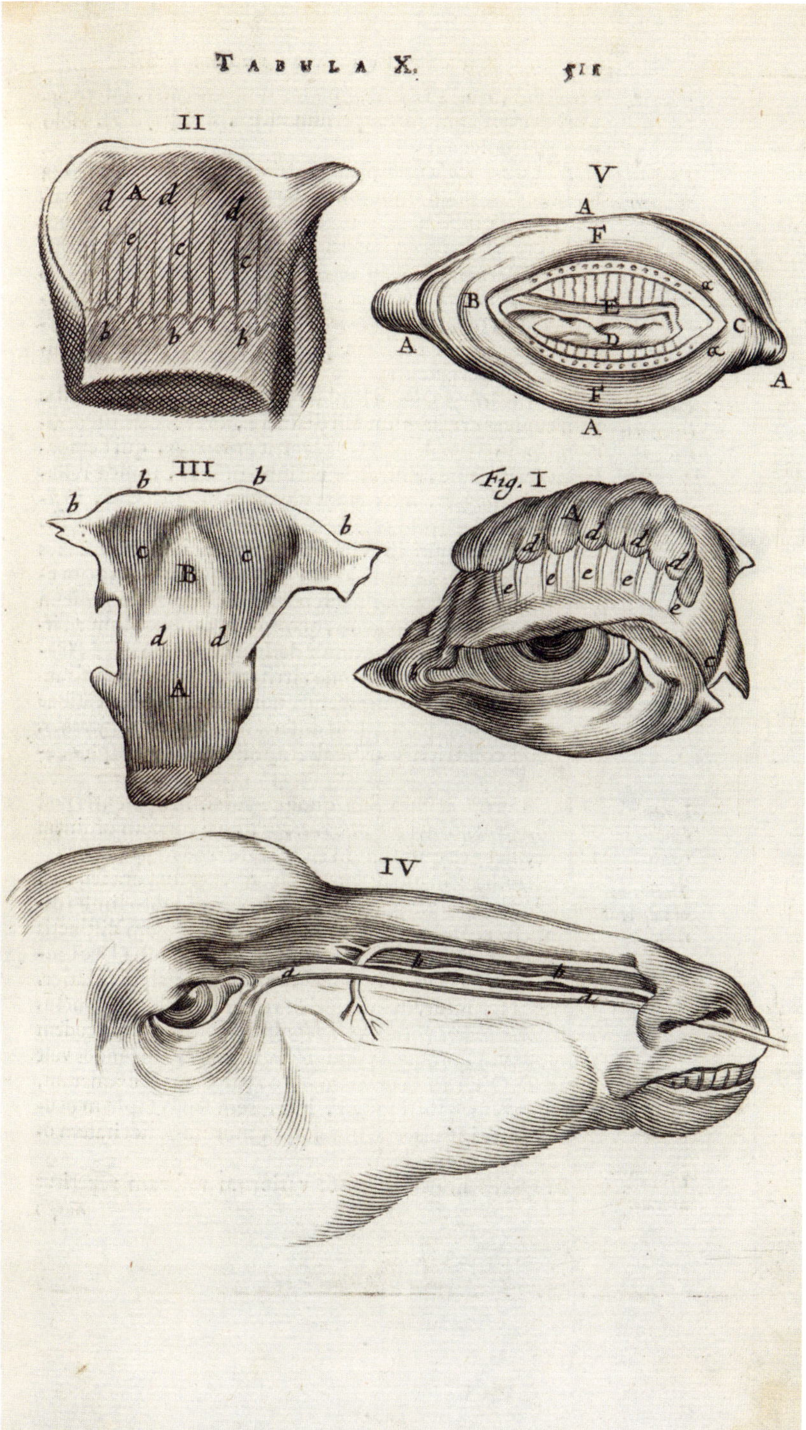

Figure 191. The lacrimal glands and the glands of the eyelids of the calf in Thomas Bartholin's *Anatome Quartum Renovata,* Leiden 1673, p. 511.

Again, Bartholin has collected illustrations from other authors' works. Figures I–IV show Niels Stensen's illustrations of the lacrimal gland and lacrimal ducts of a calf (Figure 147 in this book). Figure V shows Heinrich Meibom's (1638–1700) illustration of the sebaceous glands of the eyelids (*glandulae tarsales* or the *Meibomian glands*) from his work on their discovery, *De Vasis Palpebrarum Novis Epistola* (Letter on the new eyelid vessels), Helmstedt 1666.

In Figure V the letters mark: A. The skin of the eyelids cut off the head. B. Small eye corner. C. Large eye corner. D. Lower eyelid turned over, so that the lines of the ducts and the points of the ducts' mouths become clear on the edge. E. Upper eyelid, in which similar lines of the ducts and points of their openings are clear. F. Eyelid hairs. α. Tear points already known. The Royal Library in Copenhagen.

1680) regarding the glands of the eyelid (Figure 191) and the nasal mucosa respectively, Nathaniel Highmore's concerning the sinuses, Gerard Blasius' concerning the spinal cord and Thomas Willis' (1621–1675) and Richard Lower's (1631–1691) concerning the brain and the heart respectively. Among other anatomists whose illustrations Bartholin uses in the book can be mentioned Vesalius, Casserius, Aselli, Harvey, Vesling, Walæus, Tulp, Theodor Kerckring (1638–1693), Walter Needham (1631–1691) and Jacob Henrik Paulli.

The new frontispiece bears the year 1674 (Figure 193), while the title page bears the date 1673 (Figure 194), the book's official publication year. It was not an uncommon publisher's practice to advance the date of a publication, but the page following the title page testifies that the book can hardly have been on the street before the end of the year 1674. On this page there are two declarations of

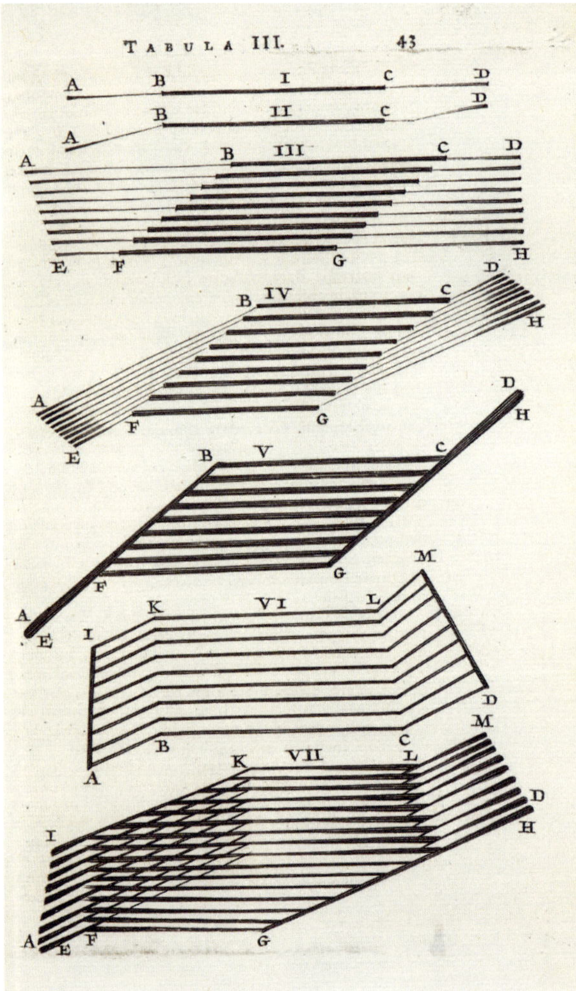

Figure 192. Geometric muscle models from Niels Stensen's *Elementorum Myologiæ Specimen seu Musculi Descriptio Geometrica*, Florence 1667, reproduced in Thomas Bartholin's *Anatome Quartum Renovata*, Leiden 1673, p. 43. In this fourth and last revised edition of the Bartholinian anatomy book, Thomas Bartholin unreservedly endorses Niels Stensen's new geometrical muscle theory and explains its elements in a pedagogical way. The Royal Library in Copenhagen.

Figure 193. The frontispiece of Thomas Bartholin's *Anatome Quartum Renovata*, which was officially published in Leiden in 1673, but in reality, as stated here, not until 1674. 15.5 × 9 cm. The copper engraving is made by the Flemish engraver Gonzales Appelmans (active 1666–1689).

The scene on the frontispiece is set in an anatomy hall with ancient physicians, where the teacher, who appears to be Hippocrates, the father of medicine, lectures over a male corpse with an opened abdomen, while a skeleton and the head of Thomas Bartholin, surrounded by two palm branches, watch in the background. On the wall above the teacher's head are dissection instruments. A seated man in the foreground records the teacher's words of wisdom. The Royal Library in Copenhagen.

exclusive right (summa privilegii) to printing of *Anatomiam Bartholinianam* in Germany/Austria/The Duchy Burgundy for six years and in Holland/Westfriesland for 15 years, respectively, signed by Emperor Leopold I (1640–1705) in Vienna 1671 and Johan de Witt (1625–1672) in The Hague 1668, respectively. The declaration from Vienna is countersigned by a certain Christoph Beüer on 1 August 1674, and so the book must have hit the streets after this date.[205] After that comes Bartholin's dedication to Christian V, the preface, a list of authors, the table of contents, poems to Thomas Bartholin from Vitus Bering, Joan Leonardszoon Blasius, Joh. Hopnerus and the sons Caspar, Christopher and Thomas. Then follows a new portrait engraving of Thomas Bartholin made by the Flemish copper engraver Gonzales Appelmans (Figure 195), after the German-born Danish court painter and engraver Heinrich Dittmers' (1625–1677) oil portrait (Figure 196), and after that begins the book itself (and the pagination) with an introduction.

What is not clear from the table of contents is that the book has three pages (805–807) before the closing index with a note from Thomas Bartholin to Gerard

THOMÆ BARTHOLINI
ANATOME
Ex omnium Veterum Recentiorumque
Obſervationibus
Inprimis Inſtitutionibus b.m. Parentis
CASPARI BARTHOLINI,
AD
CIRCULATIONEM HARVEJANAM,
ET
VASA LYMPHATICA
QUARTUM
RENOVATA.
Cum Iconibus novis, & Indicibus.

LUGDUNI BATAVORUM,
Ex Officinâ HACKIANA,
CIↃ IↃ C LXXIII.

Figure 194. Title page of Thomas Bartholin's *Anatome Quartum Renovata*, Leiden 1673.

Thomas Bartholin's fourth and final revised, updated and richly illustrated edition of his father Caspar Bartholin's anatomy book is a compendium and a treasure trove of 17th-century medical research and development. The illustration of the eagle taking off over a hilly landscape with towns, the sea and a ship sailing away, as well as the text MOVENDO (which is to be moved), seems to be an appropriate symbolism for the book.
The Royal Library in Copenhagen.

Blasius, and an associate drawing representing the circulation of the blood between the major organs in the abdominal cavity and thorax, made clear by separation of the heart chambers (Figure 197). The drawing was made by Niels Stensen based on his mentioned public dissection of the female corpse in the anatomy theatre in Copenhagen at the end of January and the beginning of February 1673. In the note to Blasius, Bartholin writes:

"Since the publication of my new anatomy has been brought to conclusion, thanks to your efforts, I have, to avoid a blank page, decided to add an illustration showing the blood circulation, on which my whole new anatomy is based. It is my good friend, Royal Anatomist Niels Stensen, who, as a courtesy, has provided the illustration. Please include it in the book and receive my best thanks for it, not to mention everything else, that brings me in debt to you."

Thomas Bartholin had with this final edition of the Bartholinian anatomy book, 62 years after his father's first edition, not just created a substantively impressive and visually beautiful anatomy book, but also a compendium of the most important anatomical discoveries of the 17th century. The book is thus a diamond in the history of medicine and deserves to be published in English. Until that happens, we have to make do with the Latin original or the German translation in Gothic script, which Bartholin had arranged by the Austrian physician, Doctor of Medicine and philosophy in Vienna, Elias Wallner.[206] It was published in Nuremberg in 1677 and was dedicated by Bartholin to Christian V's German-born queen, Charlotte Amalie (1650–1714). In the dedication, Bartholin mentions all the languages – French, Italian, English, Dutch and German – into which the different versions of the Bartholin anatomy book have been translated. A great mogul, driven by a special desire to learn about the human body, has even arranged for a foreign doctor to translate the book into an Indian language.[207]

But that was not the end. This most complete version of the Bartholinian anatomy book reached distribution as far away as China, where it was not a great mogul, but an emperor who wanted to enrich his knowledge with western medicine. Together with text from the French surgeon and anatomist Pierre Dionis' (1643–1718) anatomy book *L'Anatomie de l'Homme suivant la Circulation du Sang, & les dernières Découvertes, demontrée au Jardin Royal* (Human anatomy following the circulation of the blood, and the latest discoveries demonstrated in The Royal Garden), Paris 1690, a number of the illustrations from *Anatome Renovata* were used in a 90-page handwritten Manchurian anatomy book entitled *A Complete account of the parts of the body, produced by imperial order*, which it took the French Jesuit priest, missionary and geographer Dominique Parrenin (1665–1741) five years to produce for the Chinese Kangxi Emperor (1654–1722), and of which three copies were completed in 1700 (Figure 198).[208] Whether the copy owned by the Royal Library in Copenhagen is one of these three, a fourth, which was sent to France, or an atlas with explanatory text from the work on the book, is unknown.[209]

Bartholin establishes the world's first journal of medical science

After licking his wounds following the fire of Hagestedgaard, Bartholin found new courage and new energy and started another major literary project. He established the world's first medical journal *Acta Medica & Philosophica Hafniensia* (Medical and natural philosophical reports from Copenhagen). It was published in five volumes at Peder Haubold's Publishing House in Copenhagen from 1673 until Bartholin's death in 1680. Bartholin himself was both its publisher and its editor. The first volume (316 pages) containing the years 1671 and 1672 was published in 1673 (Figure 199), the second volume (376 pages) containing the year 1673 was published in 1675, the third (174 pages) and fourth (216

Figure 195. Portrait engraving of Thomas Bartholin from 1671 in his 56th year, made by Gonzales Appelmans from an oil portrait by Heinrich Dittmers (Figure 196). 15 × 10 cm. From Thomas Bartholin's *Anatome Quartum Renovata*, Leiden 1673, opposite page 1. The Royal Library in Copenhagen.

Figure 196. Thomas Bartholin portrayed by Henrich Dittmers (1625–1677). Oil on canvas. 100 × 70 cm.

The painting probably dates from 1671 or shortly before, since it served as a model for Gonzales Appelmans' portrait engraving from that year in *Anatome Quartum Renovata* (Figure 195). The painting shows the learned, experienced, authoritative and serious professor wearing a short-sleeved dark blue velour jacket with gilded buttons over a white shirt, and shirt frill on the outside. The long black hair is his own, as he was opposed to wigs. He is flipping through a thick book with the right hand, while you can imagine him holding aside with his left hand the blue curtain, which otherwise covers the thick parchment-bound books on the shelf behind. In this way, the diligent author invites the viewer into the cave of his literary works.
The Museum of National History at Frederiksborg Castle. Photo: Lennart Larsen.

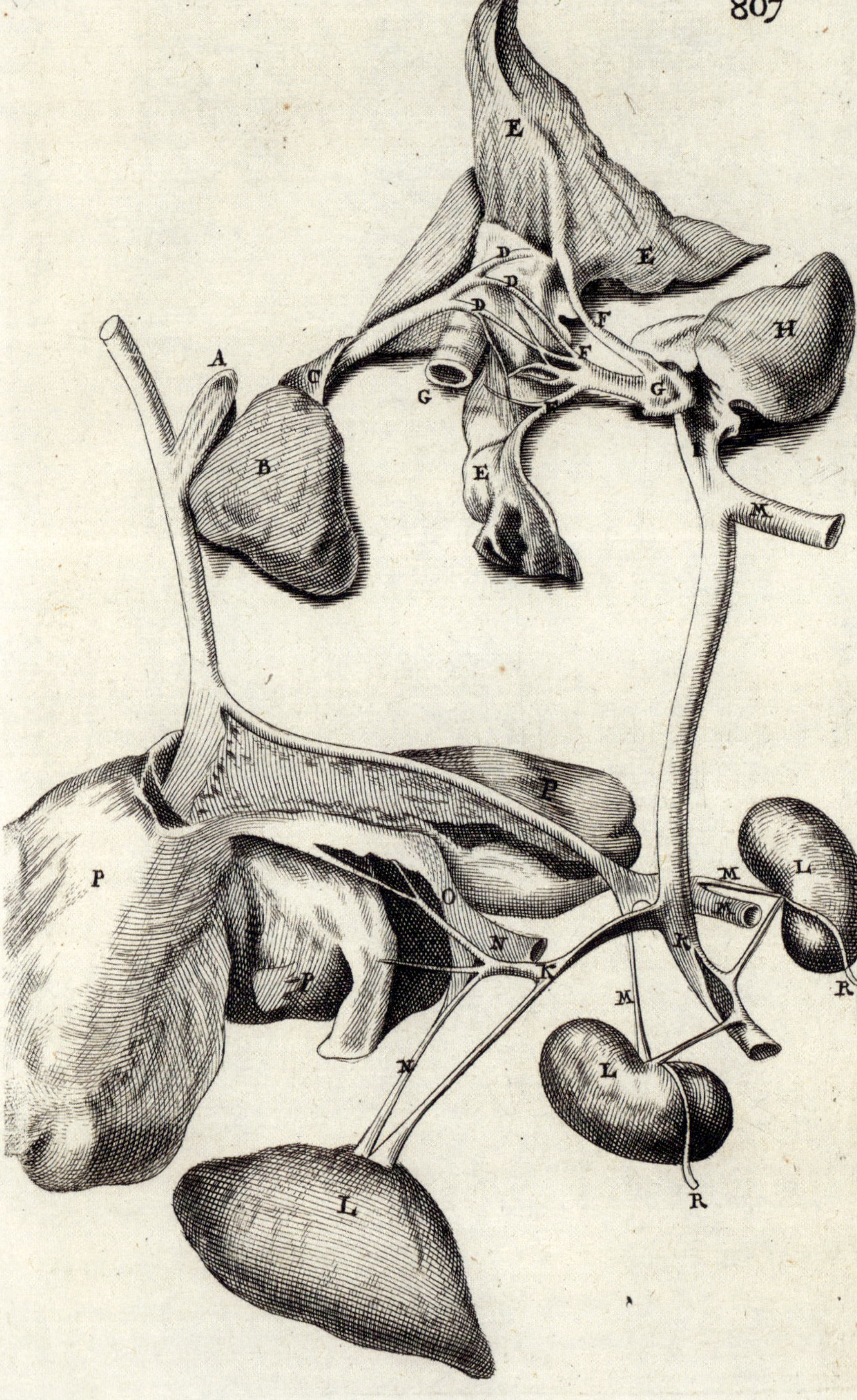
E
E
D
D
D
F
F
A
C
G
G
H
B
E
I
M
P
P
L
M
M
O
N
K
K
P
M
N
L
R
L
R

Figure 197. Niels Stensen's drawing of the blood circulation between the major organs in the abdominal cavity and the thorax clarified by the separation of the heart chambers. The drawing was made during Stensen's public dissection of a female in the anatomical theatre in the Anatomy House in Copenhagen on 7 February 1673.
From Thomas Bartholin's *Anatome Quartum Renovata*, Leiden 1673, p. 807.
The Royal Library in Copenhagen.

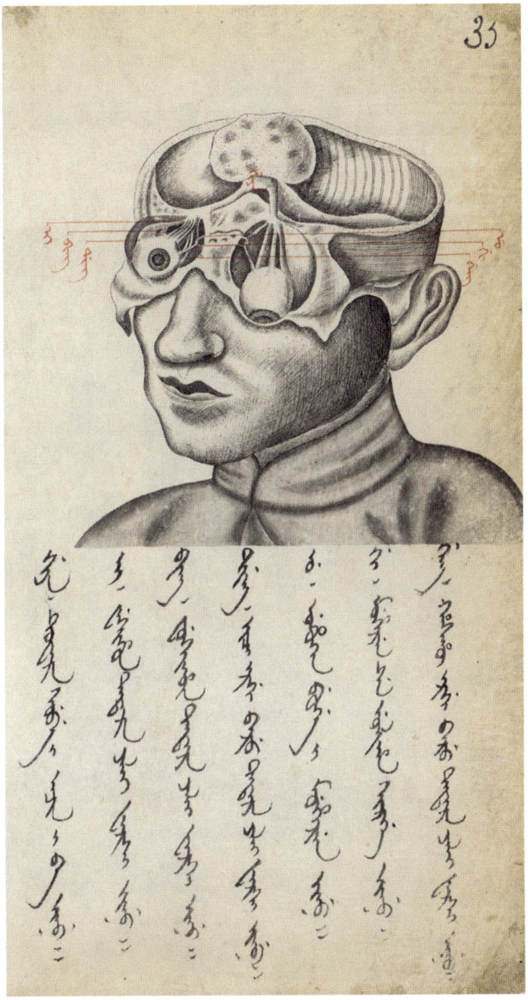

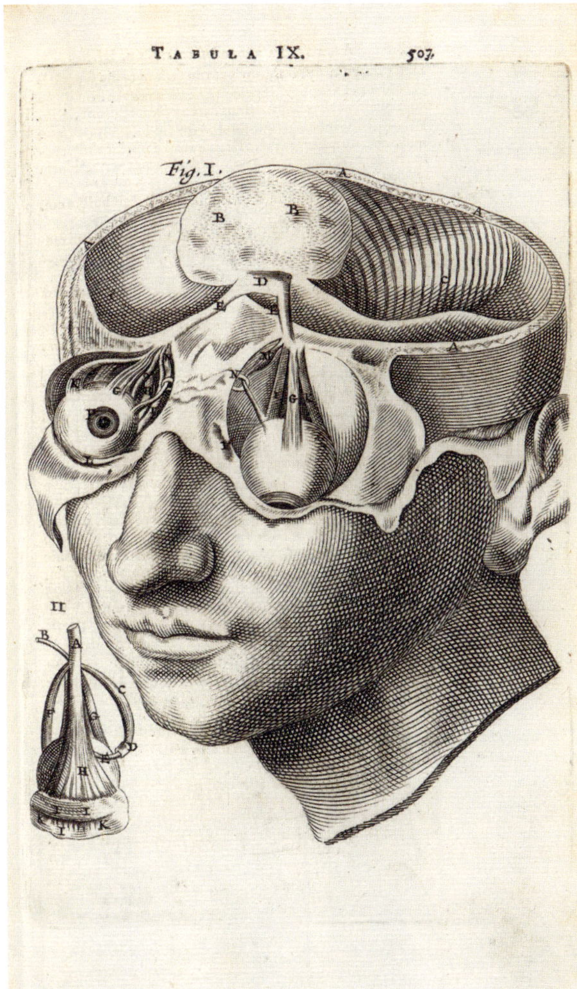

Figure 198. Left: Illustration of the eye muscles in a handwritten Manchurian anatomy book from 1700.
The Royal Library in Copenhagen.
Right: The model in Thomas Bartholin's *Anatome Quartum Renovata*, Leiden 1673, p. 507.
The Royal Library in Copenhagen.

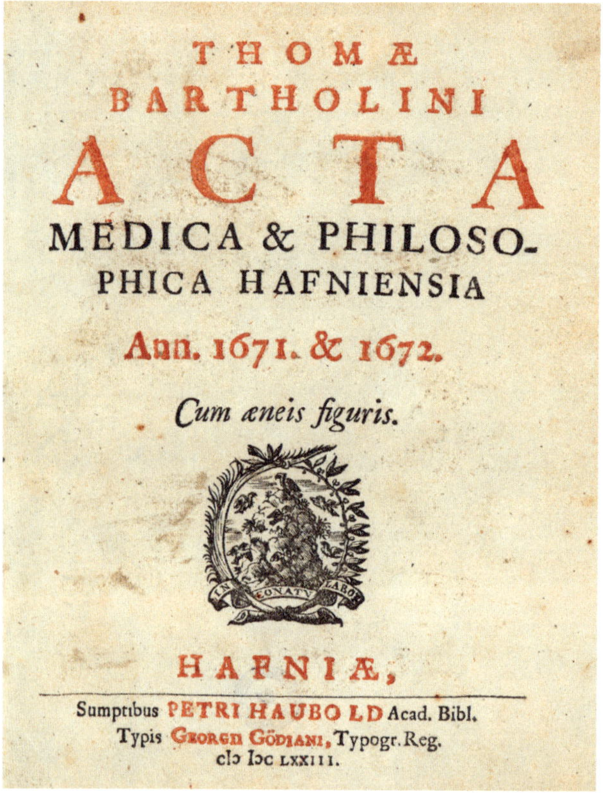

THOMÆ
BARTHOLINI
ACTA
MEDICA & PHILOSO-
PHICA HAFNIENSIA
Ann. 1671. & 1672.
Cum æneis figuris.

HAFNIÆ,
Sumptibus PETRI HAUBOLD Acad. Bibl.
Typis GEORGII GÖDIANI, Typogr. Reg.
ciↄ Iↄc LXXIII.

Figure 199. Title page of the first volume of Thomas Bartholin's medical and natural philosophical journal, *Acta Medica & Philosophica Hafniensia*, Copenhagen 1673. The journal was the first of its kind in the world and was published in five volumes comprising a total of nine years (1671–1679) at Peder Haubold's publishing house in Copenhagen from 1673 to Bartholin's death in 1680. Ten of Niels Stensen's medical writings were published in this journal. The Royal Library in Copenhagen.

pages) volumes containing the years 1674 and 1675 (third volume) and the year 1676 (fourth volume) were published together in 1677, and the fifth volume (342 pages) with the years 1677, 1678 and 1679 was published in 1680. All volumes are illustrated.[210]

The first volume was dedicated to the Triune God, "my creator, my saviour, my sanctifier," while the second volume was dedicated to Peder Griffenfeld (Figure 200), who in 1674 had become the university's patron. As a former medical student and as such Thomas Bartholin's pupil, he continued to have an affection for this science. For *Acta Medica* he had a particular importance by urging physicians and others, who experienced something remarkable in the medical and natural sciences, to report it to the Dean of the Medical Faculty, Thomas Bartholin.[211] The third volume was dedicated to Grand Chancellor Frederik Ahlefeldt (1623–1686) of Langeland.

Bartholin had undoubtedly found the inspiration for the journal in France and England. In Paris, the increasing need for communication, information and exchange of knowledge between learned people had, since around 1620, already led to the establishment of private scientific academies, a kind of club where scholars and scientists could meet at regular intervals and exchange knowledge and opinions and in certain cases perform experiments for each other.[212] This culture became widespread in Paris around the middle of the century.

One of the most important of the academies, Académie de Montmor, was established in 1657 by the Cartesian and man of letters Henry Louis Habert de Montmor (1600–1679) and was based in his house, Hôtel Montmor, in Rue du Temple (then rue Sainte-Avoye).[213] Thomas Bartholin's good friend Guy Patin was a member of this academy. Another of Bartholin's acquaintances, Pierre Bourdelot, had his own private academy in Rue de Tournon in the Latin Quarter just south of the Seine. Montmor's academy was dissolved in 1664 due to disagreements. Some of the members were supporters of Pierre Gassendi and thus wanted more emphasis on experimental science in the form of practical experiments and observations, while others, who were followers of René Descartes, wanted the emphasis placed on discussions and metaphysical theories.[214] The former group established a new academy, which had its seat in the previously mentioned scientific patron Melchisédech Thévenot's house in Rue de la Tannerie in the Marais-quarter just north of the Seine. One of the most valued members of this academy was Niels Stensen, who, on Bartholin's recommendation, lived with Thévenot during his 10-month stay in Paris in 1664–1665, and it was in this house for this academy of learned men that the 27-year-old Stensen held his famous lecture on the anatomy of the brain, where among other things he pointed out Descartes' many anatomical mistakes and confronted his pineal gland theory.

The experiment-oriented private academies' increasing needs for royal funds and capital for suitable premises and laboratory equipment led in 1666 to Louis XIV's (1638–1715) foundation of the Royal Academy of Natural Sciences, Académie des Sciences, which still exists today, on the recommendation of the statesman Jean-Baptiste Colbert (1619–1683). Initially it had its seat at the Louvre. Before this happened, the need for communication and exchange of knowledge and scientific results between scholars in the rest of France and the rest of Europe led to the establishment of the world's first academic journal, *Journal des Sçavans* (Journal of Scholars), which was first published in Paris on 5 January 1665. Niels Stensen is believed to have contributed to the birth of this journal, and in no. 12 on 23 March 1665 one could read a three-page review of his thesis on muscles and glands, *De Musculis & Glandulis Observationum Specimen.*

It was in the same way that in London Royal Society of London for Improving Natural Knowledge had been founded in 1660, and five years later the society gave birth to the world's first peer-reviewed scientific journal, *Philosophical Transactions.*

Figure 200. a. Peder Schumacher Griffenfeld (1635–1699) painted by Abraham Wuchters c. 1673. Oil on copper.

The talented Schumacher utilized his great talent and close relationship with the inexperienced and moderately gifted Christian V to become the country's true ruler until his downfall in 1676. His reform-mindedness, not least in religious matters, paved the way in 1672 for the convert Niels Stensen's recall to Copenhagen to a position as royal anatomist.
The Museum of National History at Frederiksborg Castle. Photo: Ole Haupt.

b. Portrait engraving from 1675 of Peder Schumacher Griffenfeld executed by Daniel Paulli (1640–1684) from a model by Conrad Woumans (1619–1681). 26.5 × 18 cm. From Thomas Bartholin's *De Eqvestris Ordinis Danebrogici ab augustissimo Rege Dn. Christiano V. Dan. Norweg. &c. Monarcha nuper instaurati Origine ad illustrissimum et excellentissimum Heroem Dn. Petrum Comitem de Griffenfeld, Tonsberg &c. Magnum Regni Daniæ Cancellarium utrisque Ordinis Eqvitem auratum &c. Dissertatio Historica* (Historical treatise on the origin of the Order of the Dannebrog, which was recently restored by King Christian V, who reigns supreme over Denmark, Norway, etc. The treatise is dedicated to the famous, excellent and superb Mr. Peder, Count of Griffenfeld, Tonsberg etc., Grand Chancellor of the Kingdom of Denmark, Knight of both orders), Copenhagen 1676. The engraving shows Chancellor of the Realm Griffenfeld (his age is stated on the engraving to be 38 years, but it must have been 39 or 40) at the pinnacle of power, the year before his fall. He is wearing an ermine robe and shirt frill. On his chest are the Order of the Elephant and the Order of the Dannebrog in separate bows.

The Latin text in the frame can be translated as follows: "Mr. Peder, Count of Griffenfeld and Tonsberg, Grand Chancellor of the Kingdom of Denmark, Knight of the Order of the Elephant and the Order of Dannebrog etc."

Underneath the portrait, Thomas Bartholin has written: "Depicted here is only his face. But his marvellous spiritual abilities, which have struck the world with admiration, no one can depict." (The epigram is translated by senior researcher Niels W. Bruun). In his younger years, Griffenfeld studied medicine and anatomy with Thomas Bartholin, and the two maintained a warm relationship throughout their lives. Bartholin took advantage of this when, in 1672, he got Griffenfeld to recall Niels Stensen from Italy to the position of royal anatomist in Copenhagen. Bartholin nowhere comments on Griffenfeld's dramatic fall in 1676 with a death sentence for corruption, abuse of office and treason, which ended in a last-minute pardon and life imprisonment.

Niels Stensen, on the other hand, describes it as follows in an undated letter to Cosimo III in the same year: "From my homeland, Father Sterck writes to me that on the same day that the sentence against Mr. Schumacher was to be carried out, and he was already prepared to receive the fateful blow, it happened that the king sent word at seven o'clock in the morning that he had granted him his life and changed the death penalty to life imprisonment. His coat of arms was removed by the executioner. However, the brother remains in the nobility."

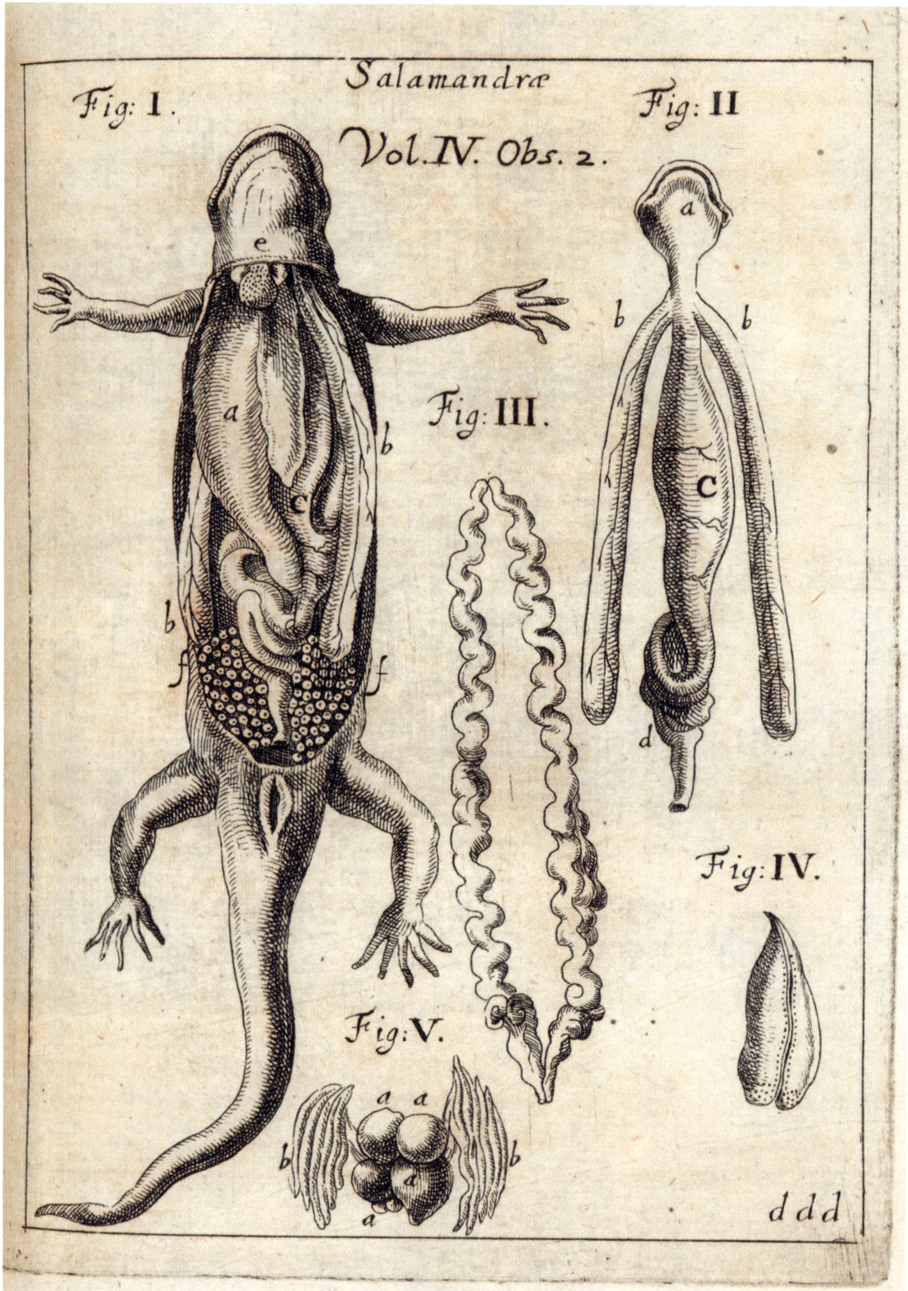

Figure 201. Illustration from Holger Jacobæus' dissection of a newt in Paris in February 1676. *Acta Medica & Philosophica Hafniensia Ann. 1674, 1675, 1676*, Copenhagen 1677, Obs. 2, pp. 5–9.
The Royal Library in Copenhagen.

Copenhagen had no scientific academy or society – it first came with the foundation of Det Kongelige Danske Videnskabernes Selskab or Academia Scientiarum Danica Regia (The Royal Danish Academy of Sciences) in 1742. Instead, Copenhagen had Thomas Bartholin, and it is in the light of the developments in Paris and London that one must see his initiative for the establishment of *Acta Medica*. Hitherto, the Danish physicians' theses, notices and research results had been published in a very scattered way. With *Acta Medica* Bartholin created a unified place of publication.

Like *Historiarum Anatomicarum Rariorum, Acta Medica & Philosophica Hafniensia* has a very broad and mixed content with a focus primarily on medical and secondarily natural history subjects. There are articles with a serious scientific content in the form of anatomical descriptions, descriptions of diseases and their treatment, case reports, descriptions from dissections of plants, insects, birds, fish, small and large mammals as well as people, botanical and zoological descriptions, descriptions of chemical and pharmaceutical processes etc. (Figure 201).

You will also find Bartholin's description of various Danish antiquities, his account of the rare nature of the Faroe Islands based on the priest and cartographer Lucas Jacobsøn Debes' (1623–1675) description from 1673,[215] his reflections on a remedy for scurvy among the Greenlanders, mention of the Icelandic ice, rocks and fossils as well as a Greenlandic-Latin dictionary made by Caspar Bartholin the Younger. There are reports from dissections of various animals from Frederiks III's royal menagerie, including a horse, an Indian baboon and an African lion. The dissection of the latter even took place in the Anatomical Theatre. Deformities, malformations and diseases in animals and humans are recurring topics (Figure 202). There are numerous articles on plants, including domestic, exotic, marine, medicinal and luminescent plants, many of which include associated illustrations (Figure 203).

The articles are mainly written by members of Thomas Bartholin's family and by the circle of scientists around him. Besides himself it is especially his son Caspar, his brother Rasmus, his two nephews Holger and Matthias Jacobsen, Niels Stensen, Ole Borch, Simon Paulli, Henrik Møinichen and Caspar Kölichen. But recognized foreign scientists, for example professor of medicine at the University of Kiel, Johann Ludwig Hannemann (1640–1724), professor of medicine at the university in Jena, Georg Wolfgang Wedel (1645–1721), the previously mentioned Johann Daniel Horst and the English physician Edward Tyson (1651–1708) also contributed, and one finds correspondences, reviews, mention of significant new releases of scientific publications, university programs and the first book indexes in Denmark in the form of annual lists of new Danish books.[216]

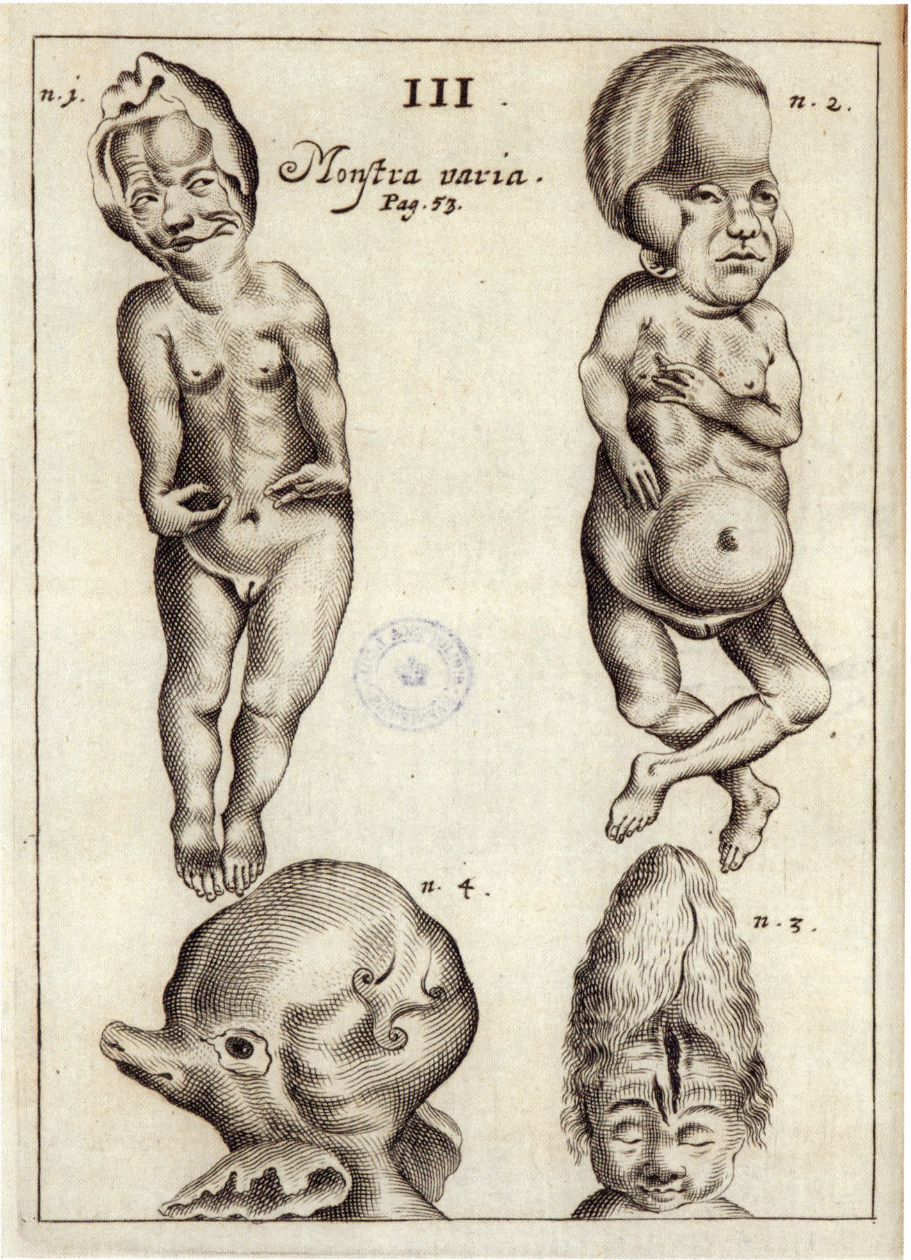

Figure 202. Illustrations of various deformities – a topic that was the subject of much attention in Thomas Bartholin's time. *Acta Medica & Philosophica Hafniensia Ann. 1671 & 1672*, Copenhagen 1673, Obs. 24, pp. 53–54.
The Royal Library in Copenhagen.

Figure 203. Illustration of an Asian tea plant sent from Batavia (Jakarta) to Simon Paulli with a letter from the physician André Clyer on 20. February 1674. *Acta Medica & Philosophica Hafniensia Ann. 1674, 1675, 1676*, Copenhagen 1677, Obs. 1, pp. 1–5.
The Royal Library in Copenhagen.

The journal's international importance and attention is emphasized by the fact that part of it was translated into French in *Collections Académiques* in the 1750s and 1760s,[217] and that its contents were continuously discussed in *Philosophical Transactions*. In remarkable contrast to the serious contents, you find in the usual Bartholinian manner imaginative and colourful anecdotes about for example a Norwegian monster with a grotesque skull-shape (Figure 204),[218] a little girl throwing up frogs and other animals, a cow pregnant with 30 puppies, and an excited male Indian baboon (Figure 205).

Among the many serious medical articles can be mentioned Niels Stensen's elegant, detailed and impressive description of the dissection of a malformed fetus in Paris in 1665, where he found the heart defect, tetralogy of Steno-Fallot, which was partially named after him. However, this eponym was not introduced until 1942, when the Danish cardiologist Erik Warburg (1892–1969) became aware of Stensen's description in *Acta Medica*.[219]

Mention must also be made of Caspar Bartholin the Younger's illustrated account of the olfactory organ,[220] and his uncle Rasmus Bartholin's account of how he, as the first in Denmark, cured himself of the form of malaria known as quartan fever with powder of China-bark.[221] China-bark contains quinine, which kills the malaria parasites. Quinine is still used as one of the most effective medicines against malaria all over the world. China-bark does not, as the name suggests, originate from China, but is an Inca Indian drug, that was introduced to Europe from Peru with the Spanish conquistadors in the 1630s and 1640s, and it is therefore also called Peruvian bark. It is the bark of the *Cinchona* tree (Figure 127). It is considered the first effective drug against a specific disease.[222] As mentioned earlier, in 1661 Thomas Bartholin had given the first account of the treatment of malaria with China-bark in Denmark, but he did not have enough of the costly powder to cure the patient.

Thomas Bartholin's own entries in *Acta Medica* decrease with the years, apparently as a result of his increasing sickness and fatigue, and the journal was obviously so much his personal work, that it could not be continued by others after his death.

Bartholin's interest in geological topics

Even though *Acta Medica* was primarily a medical journal, it also includes several geological articles by Thomas Bartholin, which illustrate the mutual influence between him and his pupil Niels Stensen concerning this science. Bartholin's interest in tongue stones (*glossopetrae*) and other kinds of fossils during his travels to South Italy and Malta has already been mentioned. In articles in *Acta Medica*, he describes his research in amber and his ideas about Icelandic fossils.

After the Carl Gustav wars with Sweden in 1657–1660, the construction of a fortification with ramparts and moats had taken place. The extensive work was carried out by soldiers over a period of many years. Huge amounts of earth were excavated and carried away, and temporary mills were established to pump away the intruding water. During the work, layers of sand, amber, wood etc. were penetrated.[223] During the 1660s, Bartholin eagerly studied the amber findings from the excavations of the moats.[224] Some of the finest findings were acquired by the mayor of Christianshavn,[225] Iver Casparsen Schøller, whom Bartholin visited to study his collection. The very finest pieces ended up in the King's Collection (Kongens Kunstkammer), which Bartholin also visited. According to his usual practice Bartholin was interested in the possible use of amber in medicine, especially its possible use as an ingredient in the mixed medicament theriac.[226] He was sceptical about the prevailing idea of the time, that amber was made in the earth from the oil substance bitumen. He was convinced that it was a plant substance, which he tried to show in a series of experiments published in the first volume of *Acta Medica*.[227] He had noticed that at the excavations of the moats, amber was often found together with bark and pieces of wood, and that the amber was often attached to the bark. However, he found it difficult to decide whether the amber was made in trees at the finding place, "or if it has been led hereto by streams in the Baltic Sea, which had then submerged a great part of Copenhagen, and at which coasts either a kind of oak trees or other trees with resin has secreted these tears."[228] He refers to the fact that great pieces of amber have been found during the ploughing of Reichs-Chancellor Peder Reedtz's grounds in the hilly landscape of central Zealand. Since he believes that seawater is needed for the formation of amber, he concludes that in ancient times there must have been sea in these areas.

As a conclusion of his investigations, he writes: "I have no doubt that it has been a resin or a tear which has floated out from a tree, and which by the influence of salt or just by the influence of time has hardened into amber at the shores of the sea."[229] Later he adds: "It is likely that amber is a plant substance. Everything points in the direction, that it is resin from a coniferous tree, which grows in a forest by the sea not far away from our costs, and that it has hardened under the influence of the salt of the sea."[230] The latter assumption Bartholin based on the facts that, during a visit at Iver Schøller, he had seen a spruce cone embedded in a piece of amber, and that in areas with many pines and other resin-producing trees you often find pieces of resin that look like amber. That the

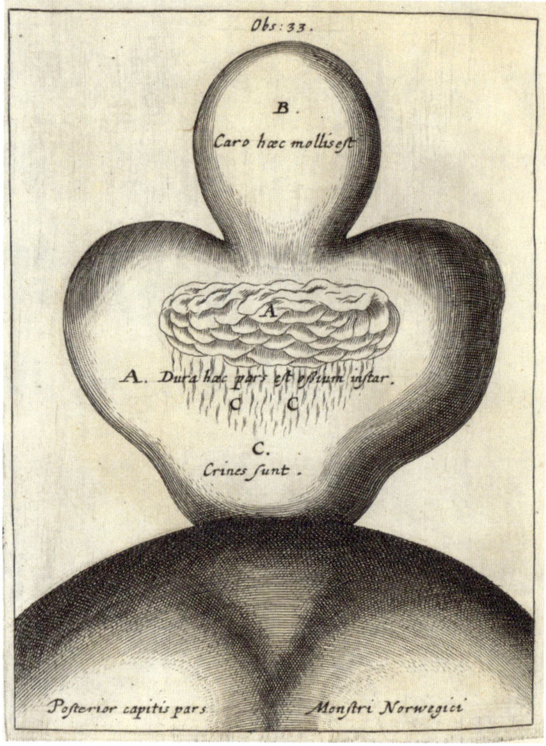

a

b

Figure 204. a, b and c. In the second volume of *Acta Medica & Philosophica Hafniensia* from 1675 concerning the year 1673, Obs. 33, pp. 80–81, Holger Jacobsen gives an account of "a shapeless female monster with a terrible appearance," who was born to Frederikke Hall in Norway on 30 May 1673. The poor mother arranged for the girl to be baptized with the name Catharina by emergency baptism at home. Holger Jacobsen writes about the child's appearance and fate:

"The fingers were very broad and flattened, the nails split down the middle. The skull consisted of a shapeless, soft, dark brown mass, shiny and transparent like glass. The back of the head, labelled with the letter A, had a peculiar contracted and wrinkled skin, beneath which the small bones, which also had a peculiar appearance, were covered with disfiguring scattered hairs. The eyes were completely protruding outside the eye sockets, creating a horrible appearance, as when white colour is mixed with bloody colour, and crystalline or white liquid is covered by dark, dull light. The nose and its bone avoided Nature's mistakes. The deformed and fleshy eyeballs were flaccid and spongy around their entire circumference. On the feet, the two front toes [the big toe and the neighbouring toe] were not separate, but joined and, on closer inspection, fused together. The breastbone was short and raised. But this monster did not lament her birth, her own misfortune or accident; she never uttered a sigh or a moan. She survived three days on nothing but milk, all of which she threw up through her nostrils before she died, undoubtedly exhausted from hunger."

The case report is accompanied by three illustrations:

a. Back of the head.
A. This part of the important entrance is hard.
B. This flesh is soft.
C. This is hair.
The Royal Library in Copenhagen.

b. Face.
The Royal Library in Copenhagen.

c. Hand and foot.

The Royal Library in Copenhagen.

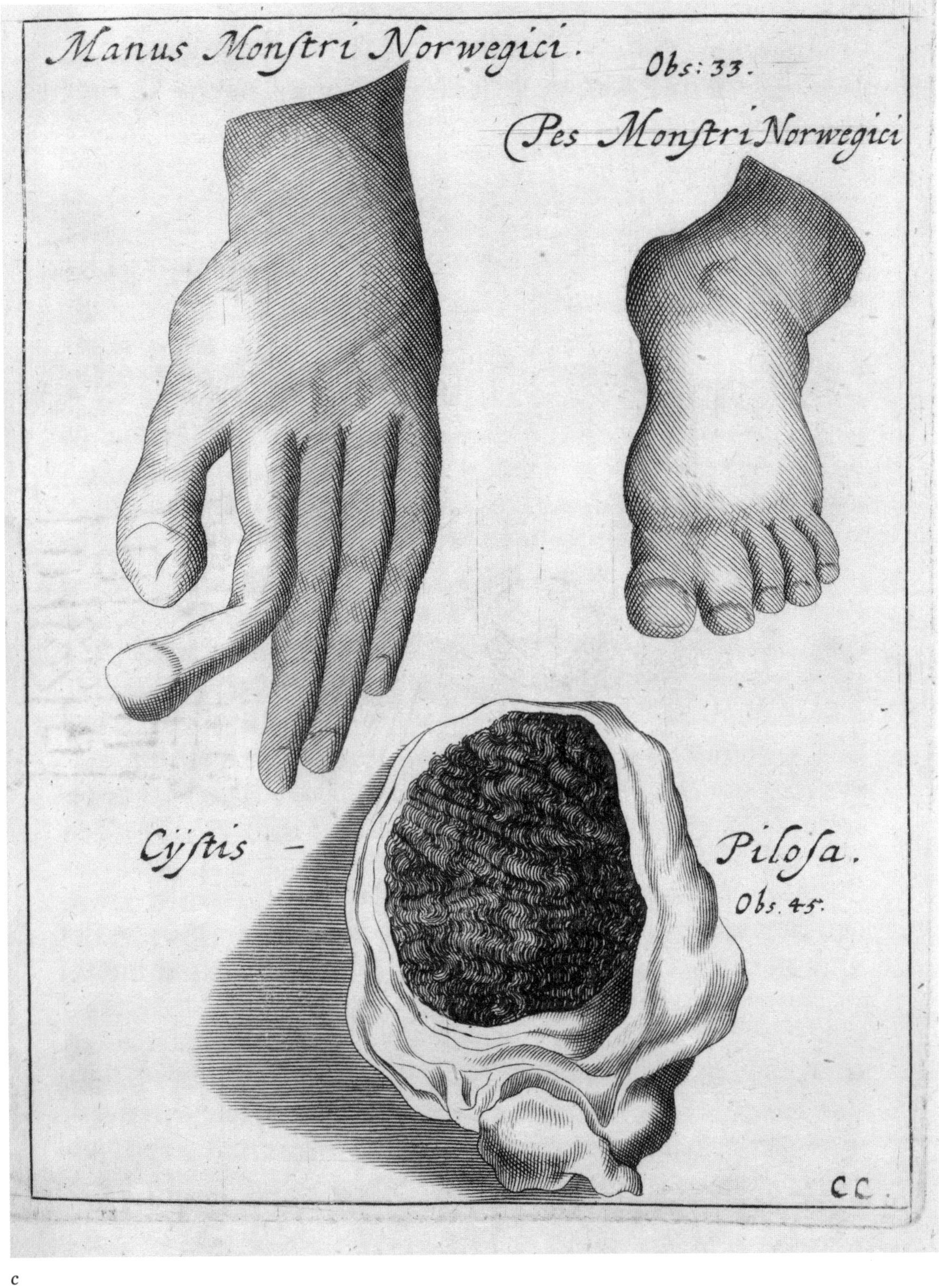
Manus Monstri Norwegici.
Obs: 33.
Pes Monstri Norwegici
Cystis - Pilosa.
Obs. 45.
CC.

c

Figure 205. Excited baboon from Frederik III's royal menagerie. When the baboon died, royal surgeon Henrik Jacob Scriver dissected it in the presence of Henrik Møinichen, Ole Borch, Matthias Moth and Caspar and Thomas Bartholin. Thomas Bartholin wrote a detailed dissection report of the findings. Subsequently, the baboon's skeleton was incorporated into his natural history collection. *Acta Medica & Philosophica Hafniensia Ann. 1671 & 1672*, Copenhagen 1673, Obs. 36, pp. 67–68. The Royal Library in Copenhagen.

amber had once been liquid he concluded partly from amber pieces' contents of insects, small animals and the like, and partly from his observations of another piece in Schøller's collection, which contained a drop that changed its shape between round, oblong and flat when you moved the piece. To confirm his assertion that amber is solidified resin, Bartholin made the court apothecary Johann Gottfried Becker (Figure 206) carry out a series of physical and chemical examinations.[231]

Almost two centuries had to pass before Thomas Bartholin's amber research and correct opinion about the formation of amber was recognized, and another half a century before this recognition was published.[232]

Bartholin must have told Niels Stensen about the amber findings in connection with the excavations in Copenhagen and shown him pieces of them, for just a month after his arrival in Copenhagen in 1672, Stensen writes about the subject in a letter to Cosimo III dated 6 August 1672. It is yet another beautiful example of how Stensen draws inspiration from Bartholin – once again in a geological subject – and then refines and further develops Bartholin's thoughts on the subject. Stensen writes to Cosimo:

> "When establishing moats around the city, two layers have been found below ground level, between which there was a clean layer of sand. Both layers are filled with pieces of wood and coal, among which there is a fairly large amount of amber or ambergris in different colours. I have seen pieces with flies and other insects and, among other things, a large piece in which you could see a drop of moving air surrounded by a liquid, in the same way as you see it in the crystal. From this, one can draw many conclusions, including the one that the whole island on which Copenhagen is built is formed by sea deposits in a place where in former times there was no sea, but forest, and that amber is a substance that has occasionally arisen in the forests, be it as plant sap, as turpentine, be it as another substance, which has changed under the ground as a result of a fluid contained in the ground, so that the amber found under Pomerania has not formed there, but has been uncovered as the soil, when washed away by the water, brings for the day what it has kept hidden before."[233]

In another article, Bartholin presents his speculations concerning a couple of Icelandic fossils.[234] An elephant tusk had been brought to Denmark from Iceland by the Danish lawyer and historian Peder Hansen Resen, who sent it to Bartholin and asked him to exhibit it in the university museum. Bartholin speculates about the nature and the origin of the tusk, which has the length of two forearm bones and the width of four fingers. Either, he thinks, it must be a real tusk from an elephant, which has been transformed into stone or flint after staying for a long time in the ground, or it must have been made from a substance in the ground, which in a natural way has hardened with time and by coincidence taken the shape of an elephant tusk.

In relation to the latter idea, Bartholin mentions that, in his own collection of rare natural objects, he also has different stones resembling animal parts. Playful Nature simply produces stones with shapes and contents resembling animals. On the other hand, he also owns stones in his museum that contain animal parts exactly ("exacte") resembling real animals. As an example, he mentions a stone he received a long time ago from the coast of Jutland. It is round like a ball and has the exterior shape of a loaf of bread, and Frederik III displays it as such, but inside it, in the middle of it, a snake can be seen in the whole circumference of the stone. Another softer stone, which Bartholin was recently given for his museum by Henrik à Møinichen, displays in its surface a small human being, but in the middle of its hollowness, it reproduces exactly a fish skeleton. Unlike Niels Stensen, Bartholin is doubtful about these things and asks:

D.O.M. S.
et memoriae
JOANNIS GOTFRIEDI BECKER
A. MDCLXI
FRIDERICO III.
SOPHIAM
GOTFRIEDUM
BECKER
A MDCLXXXIX
HELENAM MARGARETAM
MUNCK
JOANNES
GOTFRIEDUS
Tandem
A aetatis LXXI

Figure 206. a. Court and Travelling Apothecary Johann Gottfried Becker's (1639–1711) impressive baroque epitaph in St. Peter's Church (Sankt Petri Kirke) in Copenhagen, reflecting his status as a wealthy and important man. It was made in 1715 by the German sculptor Andreas Gercken the Elder (d. 1717). Photo: Jesper Brandt Andersen.

b. Becker, who had made study trips to Germany, Italy and Vienna, was appointed Court and Travelling Apothecary by Frederik III in 1668 on the recommendation of Giuseppe Francesco Borri. The following year, he established Copenhagen's first pharmacy, the Elephant Pharmacy, in Store Købmagergade. This contemporary painting by an unknown artist shows its interior around 1670, looking into the pharmacy from the street. On the end wall facing the courtyard on the opposite side, you can see Frederik III's and Christian V's nameplates, marking that the pharmacy was founded under the former in 1669 and inaugurated under the latter in 1670. On the table in the centre are weights and in the foreground two mortars.

In his younger days, Thomas Bartholin was sceptical of the apothecary profession. He perceived apothecaries as poorly educated, incompetent and greedy, with a tendency to encroach on doctors' expertise. In his older days, he became more co-operative and contributed significantly, not least through the Medicinal Ordinance of 1672, to improving the education of apothecaries, the quality and assortment of their goods and the consistency of their prices. Becker, who was a skilled chemist, was highly respected by Bartholin. In Becker's laboratory, they carried out together several chemical and pharmaceutical experiments, which Bartholin subsequently published in *Acta Medica & Philosophica Hafniensia*, and they collaborated on the drafting of the Medicinal Ordinance of 1672. Medicinsk Museion in Copenhagen. Photo: Ruben Blædel.

"But is it credible, when we see fish and snakes enclosed in hard stones, or previous soft material enclosed in stone, like in amber?"

The other finding brought to Bartholin from Iceland was a walrus tusk of the same kind as the elephant tusk transformed to the hardest flint. Bartholin believes that it must have been transformed into stone in the Icelandic Sea, in which it has been submerged. Whether it is related to the elephant tusk he dares not confirm nor deny. He is in doubt as to whether sulphur and pitch have been included in the formation of these fossils, since the elephant tusk smells of skinned meat exposed to fire. But how has the elephant tusk come to Iceland, he asks, if it has not been abducted there with the wave from the universal flood spoken of by different merchant seamen?

Furthermore, Bartholin mentions a very special stone with the shape of a small house put together by artificial water drops, which he has been given by the Faroese priest, topographer and State Governor Lucas Jacobsøn Debes, whose topographical description of the Faroe Islands from 1673 Bartholin includes in excerpts in *Acta Medica*.[235] The stone's interior and exterior parts are seen to consist of compact drops. In the interior hollowness, seven crusty columns are seen. They look like Turkish wheat put together by as many water-drops. The floor consists of thin layers. Round columns penetrate the floor and take root in the bottom. In the sides of the hollow is an opening resembling a door. The whole mass, which has the size of a fist, looks like a beautiful small house.

It should be mentioned that, many years earlier, and long before Niels Stensen began to occupy himself with fossils, Bartholin had also examined petrified wood from Iceland and had the same thoughts about it as about the elephant tusk and the walrus tusk. In the fifth and penultimate of the centuries of *Historiarum Anatomicarum Rariorum* from 1661 he reports about petrified wood found in Umbria, which exactly looks like Moorish cedar wood.[236] In that connection, he refers to a Latin version from 1641 of the Italian polymath Francesco Stelluti's (1577–1652) illustrated work about petrified wood, *Trattato Del Legno Fossile Minerale Nuovamente Scoperto*, Rome 1637. Bartholin speaks about the formation and nature of petrified wood and reports about petrified wood with a black and burnt appearance excavated in Iceland. He speculates whether the fossils are made of wood, which was first melted or made soft by the subterranean fire and then parched by subterranean heat. He refers to further reading about Icelandic fossils in Ole Worm's *Museum Wormianum*.[237]

Bartholin was also interested in petrification phenomena in human beings, which besides kidney stones, gallstones, salivary stones, tooth stones and the like also include the mentioned stone children (*lithopaedion*), i.e. unborn fetuses

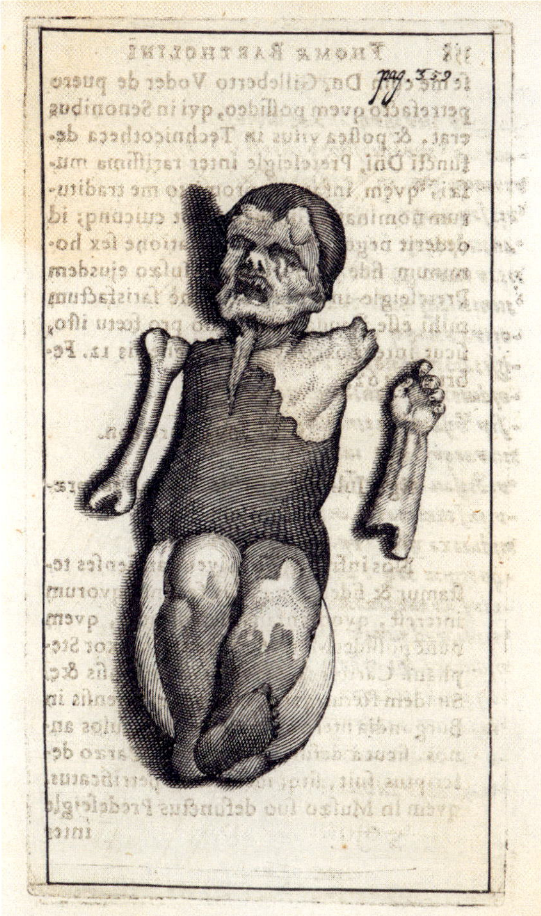

Figure 207. The stone child from Sens, as it looked when Thomas Bartholin examined it in Copenhagen in 1653. Engraving by Albert Haelwegh in the first volume of Thomas Bartholin's *Historiarum Anatomicarum Rariorum*, Copenhagen 1654, Cent. 2, Hist. 100, p. 359. The Royal Library in Copenhagen.

which, after many years' stay in the uterus, undergo petrification. A particularly famous one from the 1600s was the stone child from the French town Sens which, through Thomas Bartholin's intervention, was bought by Frederik III and became included in the King's Collection. Bartholin includes an illustrated report about the child in *Historiarum Anatomicarum Rariorum* (Figure 207).[238]

Study guides in medical travels and the art of reading books

Shortly before the three young men Caspar and Christopher Bartholin and Holger Jacobsen left on their study trip, Thomas Bartholin had, inspired by his father's *De Studio Medico*, published a small 73-page guide to medical travels, *De Peregrinatione Medica* (On medical travel), dedicated to the three travellers. The book, which is based on Bartholin's own experiences, is a tribute to medical travels and an underlining of their necessity. Bartholin advocates for the importance of making good professional contacts abroad and to learn broadly about the nature, culture, customs, diseases and treatments of foreign countries:

"In our age such great usefulness redounds to the physician from his travels that no one puts much faith in the authority of a physician who has not set foot outside his native land, and although each may have at home in abundance those things that are necessary for medical instruction, nevertheless they ought to be strengthened or increased by a comparison with things abroad. There is a vast delight and pleasure in gazing upon foreign lands and fields, mountains and rivers, observing the benignity of Nature's variety everywhere, the different conditions of the sick in homes and in hospitals with their great number of beds, which can readily be seen here and there, examining the methods for treating the patients, enjoying the conversation of the learned men and calling forth their experiences, and visiting the laboratories, the furnaces of the chemists, the pharmacies and unguent shops.[239] [...] We must gather those things that serve to enrich the medical store house from all sides, so that like Argonauts we may bring back the golden fleece from our journey. Everything which occurs attracts the eye of the physician: the air of different regions, the nature of the lands, the salubrity of the waters, the influence of the sky, the temper of the inhabitants, the arrangement of diet, the diseases, the domestic and foreign remedies, and the rest of Nature's gifts such as the plants, animals, medicinal waters, juices, medicinal earths, metals and other things of particular regions. From the physicians we learn the method of treatment peculiar to each place, the types of diseases and epidemics, the singular experiments and rare observations; from the surgeons, venesection, cupping, leeching and cautery; from the pharmacists, their medicinal compounds or those of the physicians; from the chemists, the method of fashioning mercury, potable gold and philosopher's stone; from the common people, common things, from kitchen-gardens, useful things; from all, something, since rarely do we enter hospitals without learning the use of a domestic remedy or the account of a rare disease. [...] Nothing ought to be neglected. [...] Although matters which deserve attention are to be found occurring in two ways, yet various advances of the art are to be observed in certain places, just as in the stars in their orbits; here the fundamentals may very readily be imbibed, there the practice of medicine flourishes, elsewhere there is a very full development of materia medica, the culture of Flora flourishes, works of pharmacy are exited, the fumes of chemicals are aroused, natural experiments are carried out, anatomical demonstrations are held, surgical operations are numerous; elsewhere the conclusions of the ancient physicians are discussed either by the followers of Hippocrates, the supporters of Galen or the disciples of the Arabs. The travelling physician must apply himself to these things individually, seeking out the merchandise in its own market, so that he emerges perfected by the erudition and experience of many."[240]

In a summary of his journey on the southern Italian mainland and especially Sicily, Bartholin includes numerous examples of all this, including descriptions of various geological natural phenomena, plants, fish, production of salt, cane sugar, ash tree manna, wine, honey, cheese, other foods and medicine, and he mentions healing spring waters as well as thermal steam and sulphur baths. He instructs his sons and nephew always to travel with an open and welcoming attitude, focus on the good and essential and try to make friends and avoid making enemies. He urges them to take care of their health by drinking clean water, eating a healthy diet and dressing properly. He gives concrete advice concerning the individual countries' and cities' best offers:

"To speak more clearly and without enigma, the principles of the art are taught both at home in our native land and in the neighbouring Low Countries; medical training and anatomical dissection flourish in Paris and Padua, and botany is exalted at Montpellier and in Montauban; Germany teaches pharmacy and chemistry, experiments are made in London, Florence, etc. Paris and Italy stand forth in dexterity of the surgical art. The Parisians and Romans search the oracles of Hippocrates, the Paduans follow in the footsteps of Galen, and those of Montpellier boast themselves as the heirs of Arab doctrine carried to them by the disciples of Avicenna[241] and Averroes,[242] who flourished for many centuries in the Academy of Cordoba through the diligence of the Moors. The Germans, followers of all sects, even join the medicine of Paracelsus to the rest in a friendly union. However, the British, fortunate in the fruitfulness of talented men and the essential supply of subsidies, exert themselves strenuously in new things wholly founded on experiments."[243]

Bartholin also makes his recommendations regarding the professional authorities of individual countries and cities, the most important of which have already been mentioned.[244] He narrows down his own choices to the following:

"I chose Walæus in Leiden, Patin in Paris, Sala in Padua, Severino in Naples and Castelli in Messina, whom I followed as guides in the treatment of the sick."[245]

Bartholin concludes about study tours:

"We give our sons, and we receive theirs from abroad. We teach some and we learn from others. We demonstrate to the world by example and the world is an example for us. Why, then, do we lament the loss of our sons? Why when they are about to see and to learn noteworthy things, should we draw them back by our contrary desires? They depart that they may return, and they journey abroad that they may dwell with us. They were not born for us, but for their native land and for the world, for, said Democritus, the whole world is the native land of a powerful mind."[246]

In continuation of the guidance in study travels, Bartholin published as one of his last books a guide for the students in reading books. The guide *De Libris Legendis Dissertationes VII* (7 dissertations on reading books) was published at Daniel Paulli's Publishing House in 1676. The book has 306 pages, including an appendix of poems to Thomas Bartholin from various persons, including some on his wife's death and a single one from himself to Christian V. The book is dedicated to Peder Griffenfeld. Bartholin took the opportunity to introduce the scholarly world to the middle of his five sons, the later historian Thomas (1659–1690), who also serves as the book's publisher and author of the dedication to Griffenfeld. The book consists of seven lectures, which Thomas Bartholin gave over seven Wednesdays in the spring of 1672, after he had been elected as the university's librarian on 13 September 1671.[247] The listeners were young students at the University of Copenhagen.

The book contains at once literature and scholarly history, book history, handbook knowledge and advice on writing and study techniques, and it discusses important examples of treasures among the university library's 35,000 books and manuscripts, with an emphasis on ancient, Renaissance and contemporary European literature in a wide range of sciences and subjects, with Latin as the primary language.[248]

Bartholin highlights the revolutionary importance of the art of printing for literary studies and once again emphasizes the scientific necessity that literary studies and personal experience through practical experiments should go hand in hand. He argues that all the world's peoples have the same natural ability to study, and as part of the gender discussions of the time, he is a strong advocate for women's access to study. In the position of university librarian, Bartholin also introduced several reforms within this area.[249]

It is of course no coincidence that Bartholin, towards the end of his life, when he really felt the weight of weariness, illness and old age, chose to prioritize his energy into works aimed at educating future generations. Teaching, training and passing on his own and others' knowledge permeates his entire life's work and is reflected in the unique scientific environment and the growth that he was able to create within his field at the University of Copenhagen, and which has rightly been named The Bartholinian School. The science historian C. C. A. Gosch characterizes it in this way:

"There was in Bartholin's school a mutual spirit of cooperation, and at the same time the teacher's respect for the pupils' independent ambitions and the disciples' esteem for the teacher's experience and scholarship. Furthermore, there was a spirit of freedom, which testified that it was a school animated by a true spirit of science."[250]

Caspar Bartholin the Younger's study trip. New times are heralded

The Bartholin brothers and their cousin largely followed Thomas Bartholin's travel guide and footsteps. Their first destination was the University of Leiden. Here they were matriculated on 23 June 1674. As a harbinger of new times, Caspar Bartholin here made acquaintance with Anton van Leeuwenhoek. In August 1674, Caspar wrote home to his father that this gifted but unlearned craftsman had shown him the red blood cells in one of his microscopes.[251] While in Leiden, the 19-year-old Caspar Bartholin was appointed professor of anatomy in Copenhagen on 27 October 1674.[252]

After a year in Leiden, the journey continued to Paris, where Caspar studied with the anatomist Joseph-Guichard Duverney (1648–1730), a later pioneer within the anatomy, physiology and diseases of the ear. At the entrance to the vagina on a cow he showed the young Bartholin a pair of small glands, one on each side. Caspar Bartholin, who like his teacher Niels Stensen was very interested in glands and muscles, hereafter described these newly discovered glands in a treatise on the diaphragm, *Diaphragmatis Structura Nova* (The new structure of the diaphragm) (Figure 208), published in Paris in 1676 with a portrait engraving (Figure 209) and a strategic dedication to his next host, Stensen's patron in Florence, Grand Duke Cosimo III of Tuscany.

The three students arrived at this destination in the spring of 1676, and Caspar Bartholin now studied further with Stensen for a year.[253] During a short stay in Rome, he published on 12 January 1677 at the request of the Roman anatomist and surgeon Giovanni Guglielmo Riva (1621–1677) a small dissertation on the female ovaries, *De Ovariis Mulierum*, designed as a letter to Riva. Here he once again describes his finding of the said glands at the entrance of the vagina on the lower third of the labia majora, but this time on a human female body, which he had dissected at a hospital in Florence. These ½–1-centimetre red-

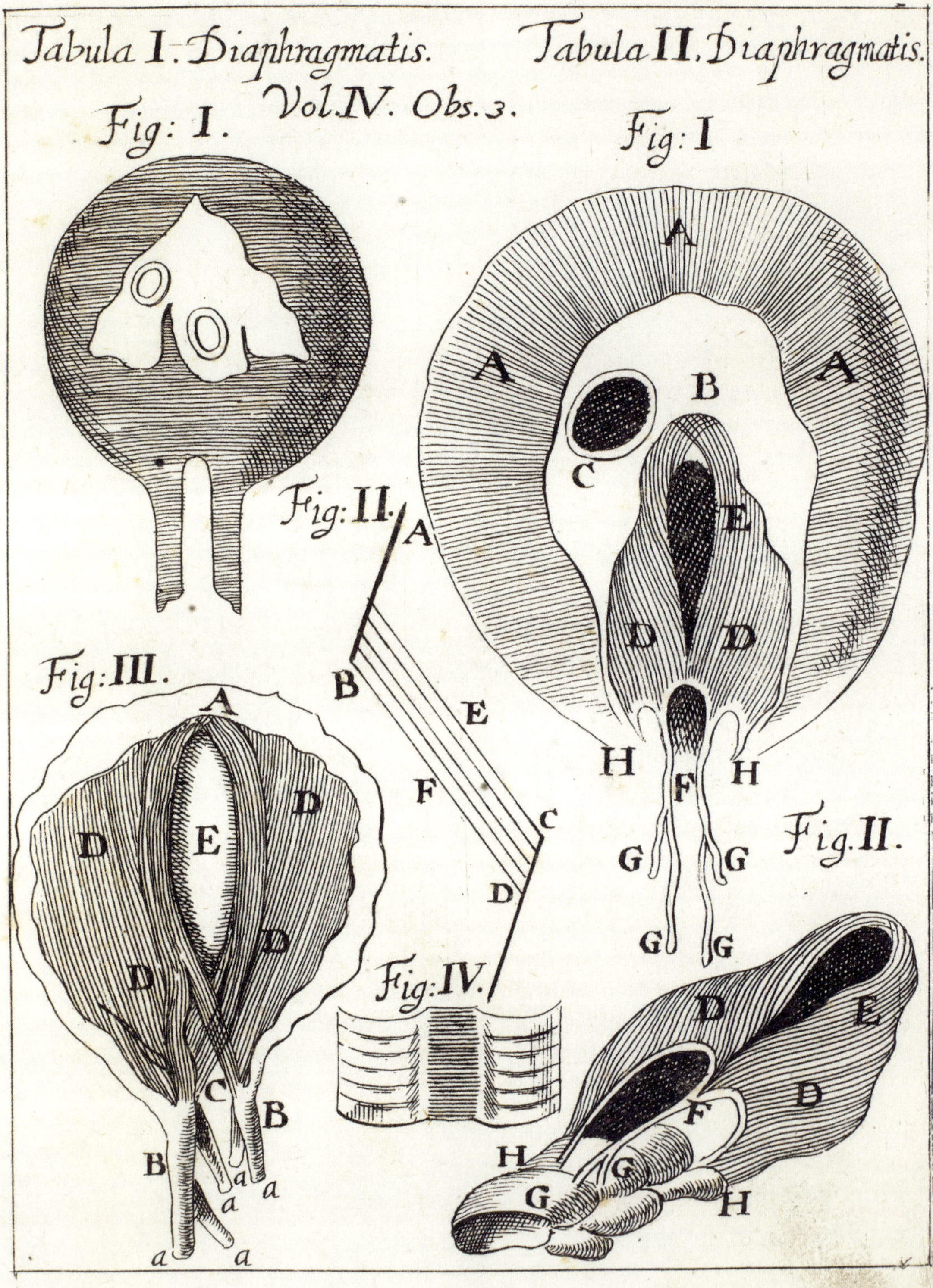
Tabula I. Diaphragmatis.
Tabula II. Diaphragmatis.
Vol. IV. Obs. 3.
Fig: I.
Fig: I
Fig: II.
Fig: III.
Fig: IV.
Fig. II.

dish-yellow glands, *glandulae vestibulares majores* (the large glands of the vaginal entrance), the 2-centimetre-long excretory ducts of which open at the back of the vaginal entrance, and whose purpose it is to release a lubricating secretion when the woman is sexually stimulated, were henceforth named the Bartholinian glands, *glandulae Bartholinianae*, after Caspar Bartholin, and not, as is often stated, after Thomas Bartholin. Inflammation of the Bartholinian glands occurs relatively frequently and is termed bartholinitis.

During his stay in Italy, Caspar Bartholin also had the opportunity to carry out studies of plant anatomy with Marcello Malpighi in Bologna, and he thereby acquired further knowledge about the microscope's potential as a technical aid in natural and medical science research. Undoubtedly inspired by Malpighi, he became one of the warmest advocates of his time of the use of the microscope in anatomical research. In the work *Administrationum Anatomicarum Specimen* (Sample on the practice of the anatomical disciplines), Frankfurt 1679, which he dedicated to Malpighi and Duverney, and in which he diligently mentions Stensen's and Malpighi's research, he states:

"Though others too rashly condemn the use of microscopes in anatomy and show their failing judgement, by means of microscopes there are found various things about which we previously spoke like the blind man speaks of the colours, so I, on the contrary, argue that without the proper use of the microscope the exercise of the anatomical disciplines is exceedingly limited."[254]

Figure 208. The illustrations in Caspar Bartholin's *Diaphragmatis Structura Nova* collected on a single page in Thomas Bartholin's *Acta Medica & Philosophica Hafniensia Ann. 1674, 1675, 1676*, Copenhagen 1677, Obs. 3, pp. 14–17.

At Niels Stensen's request, his student Caspar Bartholin has described and drawn the detailed structure of the diaphragm after his own dissections on a dog and a human. The drawings show both the upper and lower layers of the muscle and both the fleshy and tendinous parts of the muscle. Also as suggested by Niels Stensen, Caspar Bartholin has shown the course of the fleshy fibres around the hole for the passage of the oesophagus through the muscle. Figure 1 shows the diaphragm from above (from the thoracic cavity) of a dog. A is the fleshy fibres, B is the tendinous part, and C is the hole for the vena cava. D is the lower muscle layer with a passage E upwards for the oesophagus and a passage F downwards for the aorta. G is the tendons of the muscle that attach to the vertebrae's transverse processes H, as shown in figure II in the lower right corner. Figure III shows the course of the fibres in the diaphragm seen from below (from the abdominal cavity) of a human. A is the tendinous part called aponeurosis. B is the lower tendons which attach to the vertebrae with several branches *a*. E is the gap between the tendons, where the aorta passes through the diaphragm. D is the fleshy fibres. C is the opening for the passage of the oesophagus, where the fleshy fibres of the diaphragm close so much downwards that the stomach cannot slide up into the thorax. The Royal Library in Copenhagen.

CASPARVS BARTHOLINVS
THOM. FIL. CASP. NEP.
Anatomes Professor.
A. C. MDCLXXVI. Ætat. XXII.

Presumably out of fear that Stensen's attempt to convert them to Catholicism would succeed,[255] the three young men suddenly broke up, and by August 1677 they were back in Denmark.

During his three-year stay abroad, Caspar Bartholin had also published his father's works on bangles and birth assistance in ancient times, as well as an independent work on antique earrings,[256] to which was added a chapter on nose rings from Thomas Bartholin's book on diseases in the Bible. The book on antique earrings was dedicated to Thomas Bartholin with thanks that he had placed him (Caspar) in Minerva's bosom.[257] At home, Thomas Bartholin had followed his son's development through letters and was delighted about it. On New Year's Day 1675 he wrote to him:

"It fills me with exceedingly great joy, my son, to see you step into my footsteps. Nothing could be dearer to me than to have a son who is full of the same spirit as myself, and who is occupied with studies that go in the same direction as those I undertook in my youth, and which are now part of the joy of my old age. I am particularly pleased that you include history and language research in a sincere and happy union with medical science."[258]

Figure 209. Portrait engraving of the young anatomy professor Caspar Bartholin the Younger (1655–1738), Thomas' son, Caspar's grandson, in his 22nd year (21 years old), executed by the Flemish engraver Gérard Edelinck (1640–1707) in Paris 1676. 14.5 × 9 cm. From Caspar Bartholin's *Diaphragmatis Structura Nova*, Paris 1676.

Except for the number of stars, the coat of arms shown is identical to the one that Caspar and his only surviving brother, Professor of Theology Hans Bartholin (1665–1738), adopted as their coat of arms when they were ennobled by Christian VI on 17 April 1731. Its shield is parted horizontally in two equal-sized fields. The upper field is white with a black eagle in silver with a red beak and red claws. The lower field is sky-blue with a reclining yellow crescent moon encircling twelve yellow six-pointed stars arranged in three columns. On the helmet is a crown made of two vessel horns, one of which is black and silver, and the other blue and golden. Between the horns are once again the crescent moon and stars as described above.

The reason for the ennoblement was that his uncle Caspar's ennobled branch of the family had died out in 1730. Therefore, Christian VI found reason "by royal grace for the Bartholin family, which for almost two centuries has been known and respected in this country, and especially for those of the family, who are descended from the learned and widely famous Doctor Thomas Bartholin, to elevate his sons and their legitimate children and their legitimate offspring of both sexes to the nobility."
The Royal Library in Copenhagen.

The last years

While Bartholin's sons were abroad, their mother died of tuberculosis on 17 March 1675, and of his eight children Bartholin was now living at Hagestedgaard with the five youngest, aged 6 to 14 years. The youngest son Albert was born in 1668. Bartholin was struck by great grief and expressed it in places where one would not expect it. As Dean of the Faculty of Medicine, it was his duty to conduct regular supervisions of the four privileged pharmacies in Copenhagen. Before he had to perform this work in October 1675, he published a program in which he wrote:

"Actually, I didn't want to visit the pharmacies anymore, since in the year we are now wearing out, such a great sorrow has come to me that all the contents of the ointment jars fail to heal that wound. [...] From my own experience I have learned how badly we are doing with our pharmacies, which are certainly provided with all kinds of medicines, but could not save my wife's life. In the middle of her prime, she was taken away. Not without tears do I manage to speak of her, the mother of our lovely children, the joy of her husband, the ornament of her family, missed by all, half the life of my soul, loved while she lived, deeply missed after her death. So deep is the sadness that it is almost impossible to survive. [...] Her words relieved worries, she was good to get advice from, her cheerfulness dispersed sorrows, the mere sight of her made brought joy."[259]

In the same year, Bartholin entered into a strange dispute with a number of Copenhagen's theologians, not least the earlier mentioned bishop of Zealand Hans Wandal.[260] The reason was that, based on discussions with Professor Rasmus Vinding, Bartholin had published a small treatise on the wrong in consuming food that contains blood.[261] According to the Old Testament (Genesis and Leviticus) it was forbidden for Jews to eat meat with blood in it, and Bartholin was now of the opinion that this prohibition should also apply to Christians: "God, who is the keeper of our health, wants it to be so."[262] In a dissertation planned to be presented at a big promotion party for 11 new theological doctors at the University of Copenhagen in the spring of 1675, Wandal attacked Bartholin's statement, which he elsewhere had characterized as a deplorable delusion in a man "whom we all love."[263]

However, Wandal suddenly died on 1 May 1675, and the promotion party was postponed until June. Instead of Wandal, his colleague, professor of theology Jørgen Witzleben (1616–1676), spoke aggressively against Bartholin's proposed ban on consuming blood. Bartholin was not present, but stuck to his position, and in the continued controversy on the subject during the following years he included medical questions. Among other things, he distanced himself from the claim that epilepsy should be treatable by consuming other people's blood, and from the derived behaviour where epileptics showed up at places of execution to drink the blood of beheaded criminals. "We reject cannibalism. Why, then, do we not loathe the people who pollute their throats with human blood?"[264] wrote Bartholin. He also addressed the controversial issue of blood transfusions, which he considered an overly risky procedure.[265]

Another subject that occupied Bartholin in these years was the possibility of transmission of diseases from human beings to other living creatures or inanimate things; that is, the question of whether people could get rid of certain diseases by transmitting them to, for example, a specific animal. Bartholin discussed this topic with his colleague Hermann Grube (1637–1698),[266] who was, for good reason, sceptical.[267] Thus, an attempt to transmit diabetes from Frederik III's physician Peter Bülche to a dog by giving the dog salt made from distillation of Bülche's urine did not succeed. The dog died, and Bülche did not get better.[268]

In 1675, Bartholin was appointed supreme court assessor, which came to occupy part of his time in the last years of his life, where the main occupation otherwise seems to have been his work on the five volumes of *Acta Medica*. That same year he was shown the great honour of being offered a professorship of anatomy at the University of Padua but had to refuse because of his domestic obligations and his age.[269] He had passed the anatomical torch on to his son Caspar, who was on his way to Italy.

The autumn of 1678 offered a few happy events for Bartholin. On 3 October, Caspar Bartholin defended his doctoral dissertation on the structure and function of the heart,[270] which built on Niels Stensen's research, and which Thomas Bartholin, as Dean, had signed with a statement that it could be read with benefit by intelligent readers. In this disputation, Caspar Bartholin continued his father's Harvey-reception in Denmark by being the first in this country to advocate that all the blood from the right ventricle passes through the lungs, whereas, as previously discussed, his father continued to believe that the finer part of the blood passes around the lungs from the right to the left ventricle through invisible pores in the septum.

On his 62nd birthday on 20 October 1678, Bartholin married the 56-year-old Magdalene Rhode (1622–1708), who was a childless widow of the reputable and wealthy Copenhagen merchant Cort Heinrich Mercker (1607–1671). With this marriage, Bartholin became owner of Fjellebro Manor on Funen, which Magdalene Rhode had inherited from Mercker. On 12 December of the same year, Christian V made Bartholin the patron of the neighbouring Vanting parish.[271] It is doubtful, however, whether Bartholin ever visited his new possessions on Funen.[272]

During the marriage, Bartholin's health went steadily downhill because of the old kidney stone disease, which became increasingly troublesome. Nevertheless, he continued his work tirelessly. When in May 1680 a new rector had to be elected, it ended up being Bartholin for the fourth time, although he tried to avoid it because of his old age, weakness and several other obstacles and assignments.[273] As soon as July, however, he had to relinquish his supervision of the university's forests due to ill health. This included pointing out trees for felling, supervision of the delivery of pigs and the like.[274]

Bartholin continued, however, looking after his other work as rector right up until a week before his death. In September he sold Hagestedgaard to Laurits Jacobsen, assessor in the Admiralty College, who took it over on Bartholin's 64th birthday on 20 October 1680. Bartholin therefore ended his days in his stately professor's residence in the churchyard of the Church of Our Lady in Copenhagen with its large rooms with expensive French windows. The persistent efforts of seven colleagues with poultices, ointments, diuretics, etc. could not prevent the progress of the disease.[275] On 30 November, urination ceased, and on 4 December Bartholin took his last breath.

On the evening of 23 December, he was buried next to his first wife in the family burial place, which he himself had arranged in the chapel of the saint of the plague, Saint Rochus, in Our Lady's Church, next to the sepulchral monument of the family of his good friend, the law professor Rasmus Vinding, who died four years later. Many years earlier they had planned for a walled arch to connect the two graves, and on the arch, there was to be a verse saying, that "when they who love one another here on Earth connected in a sincere covenant of friendship are parted by death, their joined graves will cover them."[276] Monuments were erected with portraits and epitaphs, some of which had been authored almost 20 years before Bartholin's death.[277] How many of the epitaphs for Thomas Bartholin were actually included on his grave monument is unknown, as the chapel and church burned down in the Copenhagen fire in 1728. However, a good guess can be made from the texts that Willum Worm included in the publication of his memorial speech (see below).[278] In his travel book, written in the period 1671–1692, Holger Jacobsen depicts the joined grave monument of the families Bartholin and Vinding. He describes it as a majestic and superb artwork, which attracts the eyes of people passing by (Figure 210).[279] The most famous Danish priest of the Baroque, Thomas Kingo (1634–1703), who had been a pupil of Thomas Bartholin's brother Albert at Frederiksborg School,[280] wrote a rhymed memorial poem in Danish for the two professors, concerning their virtue, wisdom, modesty and close relation.[281]

Figure 210. Holger Jacobsen's drawing of the no longer extant united epitaphs of Thomas Bartholin and his friend and professor colleague Rasmus Vinding (1615–1684) in the St. Rochus Chapel in the Church of Our Lady in Copenhagen. On Bartholin's epitaph (right), he is depicted with two women behind him, one of whom must be his first wife Else Christophersdatter, née Hansen (1634–1675), whose name the epitaph also bears. From Holger Jacobsen's handwritten travel book in the Royal Library in Copenhagen.

On 4 January, the university held a memorial service for Thomas Bartholin in Auditorium Majus, where Willum Worm (Figure 211), Holger Jacobsen and Bartholin's successor as the university's rector, bishop of Zealand Hans Bagger, gave memorial speeches of panegyric character. Worm's and Bagger's speeches were published by Worm in *Oratio In Excessum Viri Incomparabilis Thomæ Bartholini. In Suprema Justitiæ Curia Assessoris, Archiatri Regii, Professoris Honorarii, & Facultatis Medicæ Decani. Publicé In Academia Hafniensi. IV.*

Januar. dicta. (Memorial speech for the incomparable man Thomas Bartholin, Assessor of the Supreme Court, Royal Physician, Honorary Professor and Dean of the Medical Faculty. Held in public at the University of Copenhagen on 4. January), Copenhagen 1681. This publication also includes memorial words by Vitus Bering, a memorial poem by Ole Borch and inscriptions from Bartholin's sepulchral monument. The latter two are included here for the first time in English translation by Peter Fisher:

On the tomb
OF THOMAS BARTHOLIN,
Which
Within the Church of the Blessed Virgin in Copenhagen
Is in the Chapel of Saint Roch
The following inscriptions survive.
THOMAS BARTHOLIN
Honorary Professor of Medicine
In the hope of future rest, which he had in mind,
While alive and desired for his body,
So that it should remain for his posterity,
Surviving for himself and his kindred;
He erected this memorial in 1663.

Figure 211. President of the Supreme Court Willum Worm (1633–1704) portrayed by an unknown artist (possibly Heinrich Dittmers) c. 1670. Oil on canvas.

Ole Worm's highly gifted son, from his second marriage, first attended Vor Frue Skole (the School of Our Lady) and then from 1650–1653 the University of Copenhagen with Thomas Bartholin as preceptor during his busy years exploring the lymphatic system. During Worm's subsequent 10-year study travel to the Netherlands, France, Italy and Germany, his course was set for a career in medicine. He became Doctor of Medicine in Padua in 1657. On his return, he was first a professor of physics, but from 1665 he also became a professor in a newly created professorship in medicine (Medicus Quartus). On Christian Ostenfeld's death in 1671, Worm was promoted to Medicus Tertius, on Thomas Bartholin's death in 1680 to Medicus Secundus and on Rasmus Bartholin's death in 1698 to Medicus Primus, a position he retained until his death. For two periods he was the university's rector. At the same time, he occupied several other high positions, including librarian of the Royal Library from 1671, chief inspector of the King's Art Chamber from 1673, assessor of the Supreme Court in 1679, cancelli counsellor from 1684 and president of the Supreme Court from 1690. As Ole Worm's son and Thomas Bartholin's pupil and later medical professor colleague for 15 years, he had a close and confidential relationship with Thomas Bartholin. However, he did not contribute anything significant to medical science.
The Museum of National History at Frederiksborg Castle.
Photo: Iben Kaufmann.

Higher up next to each effigy:
Although their forms are depicted in sacred marble,
And golden veils signify new titles,
Yet all would be condemned to be transitory, except for Christ's grace,
By which our names should be inscribed in the abodes of Heaven.
T.B.D.E.C.F.

On the tombstone:
Under this stone
The remains of
THOMAS BARTHOLIN
AND
ELISA CHRISTOPHERSDATTER
A married couple
With five sons and four daughters
Await eternity
With renewed life after their deaths.

On the Pillar:
In This Urn
Are Buried the Ashes
of The Most Prosperous Man
THOMAS BARTHOLIN,
Who after long travels through the world,
Famous among the greatest names in his Fatherland,
Happy with both his wives and desired offspring,
He lived on earth with God.
He deserved well of the world through loyalty his King,
Friendship with his neighbours,
And from his kindness.
Only this you would not wish: he was able to die
And be buried in this location.
But he ended his mortality, not his life,
Which he exchanged for eternal life in Heaven
By laying down here his mortal remains.
He left behind the longing of his kin, an example
For everyone and Eternity for himself.
Born in Copenhagen on 20 October 1616.
Died in Copenhagen on 4 December 1680.

To the Blessed Shade
OF DOCTOR THOMAS BARTHOLIN,
Now passed beyond mortal renown.
So, do you flee and reject our northern climes,
Realms long undeserving of your merits.
Bartholin, glory of physicians?
Did your wishes and your public lectures not lead to your success?
Did it not please you to have been alive, when errors were overcome?
There is now no Lernaean[282] marsh to survive you.
Nowhere does Cacus[283] live, nowhere the Stymphalian birds[284] of prey,
You have purged the world; deep peace reigns, and you now still heresy.
Ignorance itself is prostrate, yielding to your weapons,
It lies mute and numb, fetid in an everlasting prison.
We have lived unknown to ourselves, nor after viewing so many entrails,
Have the soothsayers revealed to the Greeks sufficient for themselves.
Let confidence, safe for no one, satisfy no one, there is a labyrinth everywhere.
Everywhere belief existed before discovery.
The republic of the body alone is uncertain,
Gloomy among so many Acquapendentes![285]
The water of the lymph vessels has been waiting,
Until you revealed the divine motions of Roscius,[286]
And renewed the journey with everlasting praise.
Why should I speak of the liver that has been drawn from its realm,
And that has sprung over the exercise ground to the proper home of its blood?
Nor has it given you the chance,
To make an attempt with your talent, sword and hand,
Unless your eagerness should immediately proceed against many:
For a host of lights has been lit from the light of BARTHOLIN,
For many that was a sufficient oracle to bring renown.
I call to witness the Bogdans and the famous Lyser,
And you, elegant glory of Torun, the knife of Seger,
These there is a hindrance in referring to; starting out from the anvil,
Above the clouds and the shining stars came Stensen.
In another part you are a sufficient remedy, Pæon,[287] for sick veins,
And you delight limbs floating next to a small boat:
O! How often have you overcome the thread of Lachesis[288] with your
Medicinal nectar and the headlong journey to death?

Dropsy has withdrawn, so has tuberous sclerosis and the belly's bloody
Torments and feverish threats have succumbed.
Nor have any fewer prizes been any less transferred from other contests,
And nothing has remained to be kept back by Phoebus[289] to keep back.
Whether there were any royal sceptres for Sophoclean[290] tragedy,
And deeds to be made known through the brush of Apelles,[291]
Or odes destined for the parental Academy,
Obligations due to distinguished men for their services,
This pen is equal to the task, bowed down in no other activity,
Nor seeming worn by the fault of long existence:
No, it has rather glowed with more excellence as its years have grown older,
And has earned distinction with a more florid crown.
Add encampments on Parnassus[292] of new female deities,
A delight to youth and a delight to old age.
So, you may go beyond this stage, a holy faultless dwelling,
Holy conduct and integrity call upon you:
That name of friendship, cherished by the people,
That openness without pretence, and that trust,
Once a rare bird in the East, but even rarer in the South,
And now hardly well known in northern climes.
They have chosen no place more certain in a heart,
Here they have found virtues, here a home for themselves.
Rage lies far from this temple, hidden dissension far from here.
No place is given to hate, or to quarrels or envy.
Sometimes, I admit, it has produced hostile giants,
That band has openly destroyed Antaeuses.[293]
But this is Jove's[294] doing, so that Olympus[295] may be ruled peaceably,
Refusing a place for Gyas,[296] Tityus[297] or Enceladus[298]
He has produced the men like Epictetus,[299] Antoninus[300] and Numa;[301]
Burdensome to none with their voice and pen.
A type of Church shining in an apostolic age,
When it has banished all evil far into exile,
Such was that home, fashioned by your brain,
Such the eminent progeny following your example.
Everything disposed and responding to your wishes,
Penelope[302] the last one, and Penelope the previous.

Machaon's[303] sceptres have tended the royal city,
The noble order of the Areopagus[304] has seen an equal.
Even like the most extreme marvels of Garama[305] and India,
Your writings, immortal name and honour flourish.
Only one thing was left: heaven, BARTHOLIN.
You came from the fatherhood of heaven.
And to heaven you must be restored.

OLE BORCH[306]

Holger Jacobsen published his speech the same year: *Oratio In Obitum Summi Daniæ Æsculapii Thomæ Bartholini. Accessit Operum Th. Bartholini Editorum Catalogus* (Memorial speech for the leading Danish physician Thomas Bartholin. Added a catalogue of Thomas Bartholin's published works), Copenhagen 1681. On 30 December 1680, Bartholin's former student, physician and professor of ethics and eloquence at Odense Gymnasium, Jørgen Hahn (1647–1699), had given a speech in his memory at the Gymnasium.[307]

Perhaps Bartholin's fear of comets had not been without reason. Scarcely three weeks before his death, the German astronomer Gottfried Kirch (1639–1710) discovered for the first time in history a large comet through a telescope, and in the following weeks and months this "frightening and unusually large" comet, Kirch's comet (Great Comet of 1680 or C/1680 V1) (Figure 212), could be seen in the sky over northern Europe with the naked eye, for example over Odense on 18 December[308]. At this time the Nordic countries and Latvia were attacked by cold fevers (malaria), and in Halmstad in Sweden typhoid fever killed a large part of the town's residents. The plague raged in Germany and threatened to invade Denmark.[309]

Apart from all his books, not many physical relics of Thomas Bartholin remain from his own time. Mentioned and depicted is the commemorative plaque from the university's first natural history museum, the tower room and stone plate at Hagestedgaard, Heinrich Dittmers' oil portrait at the Museum of National History at Frederiksborg Castle and Frederik III's ivory skeleton at the National Museum. Karel van Mander's oil portrait of Thomas Bartholin from 1651, the cover illustration of this book, was recently rediscovered in the Amsterdam Museum after being thought lost for 373 years. Otherwise, one must say that Bartholin's motto "semper ardentius" (always more ardently) in more than one point and in the most literal sense fits him to an astonishing degree, just as was said at the commemoration on the 300th anniversary of his birth:

"The professor's residence, where Thomas Bartholin was born, is burnt down. The residence in which he lived when he was a professor himself is burnt down. Burnt down is Domus Anatomica, where he made his discoveries. Burnt down is his manor. Burnt down is his library. Burnt down is the church where his grave was. Burnt are all the paintings of him with the exception of one [...] and that one has twice been on the verge of burning.[310] It is as if everything earthly that could remind us about Thomas Bartholin had to be wiped out."[311]

Posterity has, to a small extent, made up for all these losses to Vulcan. In the memorial grove at Jægerspris Castle, there is a memorial pillar (erected 1780) for Thomas Bartholin and his son, the historian and antiquarian of the same name (Figure 213); at Sorø Art Museum is a profile medallion in plaster of Thomas Bartholin, made by the artist Johannes Wiedewelt (1731–1802) (Figure

Figure 212. The Great Comet (Kirch's Comet, C/1680 V1) was one of the brightest comets of the 17th century and the first to be discovered through a telescope. The discoverer was the German astronomer Gottfried Kirch (1639–1710), who saw the comet in his telescope on 14 November 1680, perhaps as an omen of Thomas Bartholin's imminent death. This painting by the Dutch painter Lieve Verschuier (1627–1686) shows the comet over Rotterdam.
Museum Rotterdam.

Figure 213. Left: Johannes Wiedewelt's (1731–1802) memorial to Thomas Bartholin and his son, the historian Thomas Bartholin the Younger (1659–1690), in the forest at Jægerspris Castle. Etching by J.F. Clemens (1748–1831) c. 1780. 17.8 × 10.2 cm. National Gallery of Denmark.

Right: The memorial (no. 53) in the forest at Jægerspris Castle. Height 344 cm. The memorial is made of marble from Gjellebæk and was erected in 1780. In 2011, it was completely renewed in Turkish marble from Marmare. As a symbol of Thomas Bartholin's profession, a rod of Asclepius is seen on the front of the memorial, and at its base a book and a scroll of paper are seen, symbolizing his son's antiquarian studies. Photo: Jesper Brandt Andersen.

214); and in the great hall at Christiansborg Palace, the artist Bjørn Nørgaard (born 1947) has honoured Thomas Bartholin with a portrait in the upper right corner of Queen Margrethe's tapestry "The Reign of the Nobility" (Figure 215).[312]

At Aarhus University in Jutland is Bartholinbygningen (The Bartholin Building). It was erected between 1971–1974 and is part of the Institute of Biomedicine. Besides the building is Bartholins Allé (Bartholin's Avenue). At Rigshospitalet in Copenhagen is Bartholininstituttet (The Bartholin Institute), which was established in 1989 and carries out research in the areas of cancer, diabetes and diseases in the immune system. Close to Rigshospitalet, at the

Figure 214. Portrait medallion in plaster of Thomas Bartholin, made by Johannes Wiedewelt in 1779. 55.5 cm. Sorø Art Museum. Photo: Anders Sune Berg.

southeast corner of Sortedams Sø, is Bartholinsgade (Bartholin's Street), and in Gislinge close to Hagestedgaard is a road named Thomas Bartholins Have (Thomas Bartholin's Garden). In Aabenraa in southern Jutland is Thomas Bartholins Vej (Thomas Bartholin's Road).

In 1813, the Scottish botanist Robert Brown (1773–1858) named a genus of orchid after Thomas Bartholin. The genus, which is also called spider orchid, consists of the species *Bartholina pectinata* (or *Bartholina burmanniana*) and *Bartholina etheliae* (Figure 216). They grow in South Africa and Namibia. Brown named the orchid genus after Thomas Bartholin "in memory of the great Danish anatomist and physician Thomas Bartholin, whose different descriptions of plants in the old Copenhagen journal [*Acta Medica*] entitles us to embellish the history of science with his truly brilliant name."[313]

On the 400th anniversary of Thomas Bartholin's birth, from 3 October to 3 November 2016 there was an exhibition at the Royal Library in Copenhagen about his life and work,[314] and from 16–17 December of the same year, a workshop about him was held at the University of Padua,[315] which, as mentioned and depicted, also has a memorial plaque in stone and a painted portrait of Thomas Bartholin.

Figure 215. Bjørn Nørgaard's (b. 1947) tapestry portrait of Thomas Bartholin, which can be seen in the top-right corner to the left of the shield with the three lions. The tapestry "Adelsvældet" (Nobility), c. 1990.
Queen Margrethe's Tapestries, The Great Hall at Christiansborg Castle.
Photo: Thomas Brinkløv Grafisk.

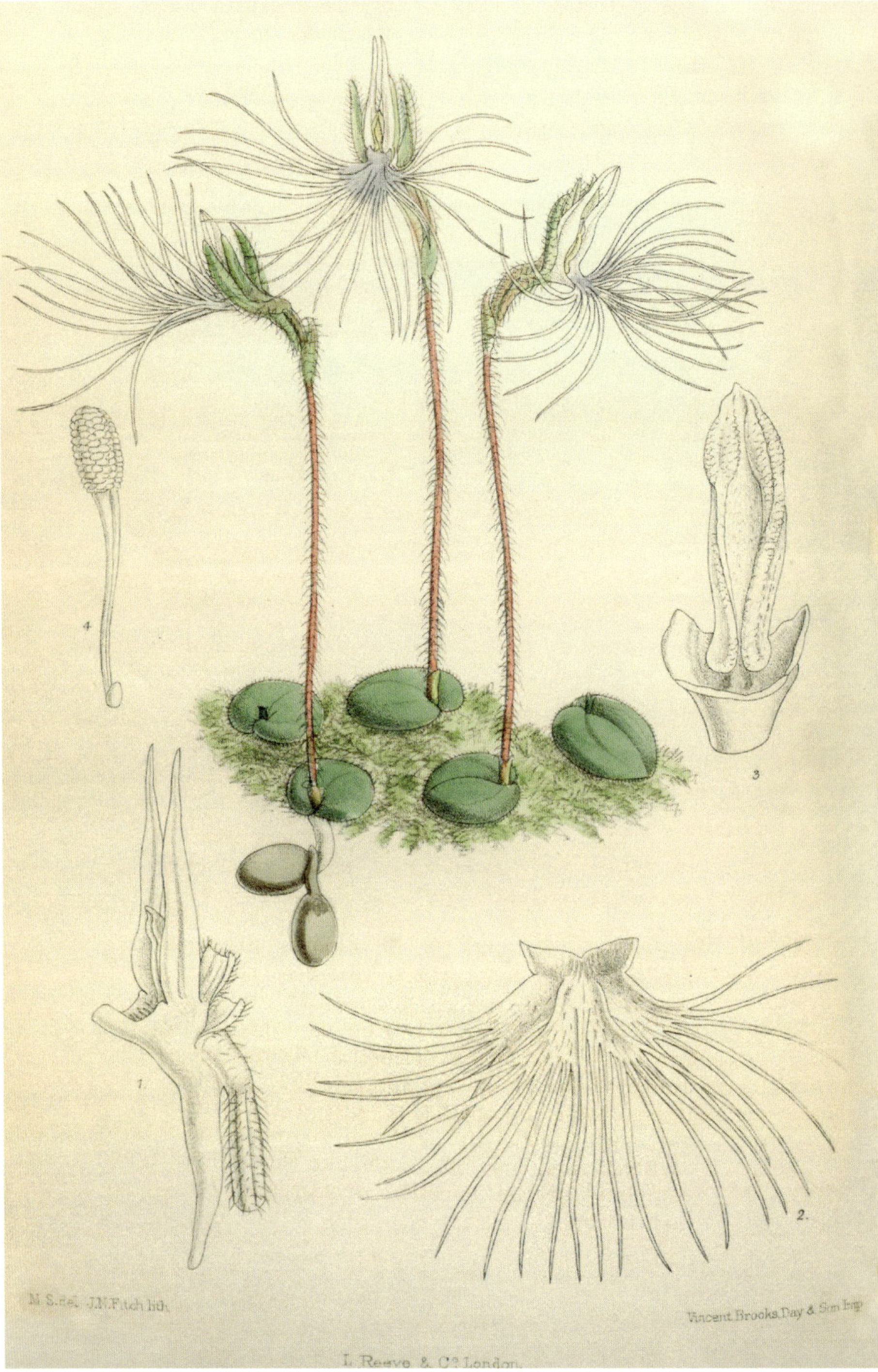

Figure 216. The spider orchid *Bartholina pectinata* or *Bartholina burmannia* which Scottish botanist Robert Brown (1773–1858) named after Thomas Bartholin in 1813. From *Curtis's Botanical Magazine*, Vol. 121 (Ser. 3, no. 51), London 1895, plate 7450. Wikimedia Commons.

Thomas Bartholin's legacy and significance for his own time and posterity

8

"Human beings waste their labour in seeking to understand Nature through the chatter of the sophists and the word-battle of the disputations. One must learn to know Nature by personal experience, not by disputing about it."[1] Thomas Bartholin, 1659

Based on national archivist A.D. Jørgensen's biography of Niels Stensen from 1884, an image was created that Thomas Bartholin's despicable and boundless nepotism, "which must considerably weaken the esteem for him as a human being," was responsible for Niels Stensen's genius being lost for Denmark. With his description of Bartholin as a self-satisfied braggart, a man who did not want to share power and honour with others, a poor and naive researcher, who with a hypocritical excuse about ailing health withdrew from his research in the lymphatic system at the most important moment, the medical historian Julius Petersen continued painting this ungraceful picture with a broad brush.[2]

Even at the memorial speech at the University of Copenhagen on the celebration of the 300th anniversary of Thomas Bartholin's birth, the same picture was once again drawn by another Stensen researcher, the physician and medical historian Vilhelm Maar, who further attempted to belittle Bartholin's discoveries in the field of the lymphatic system.[3] Right up to theologian Axel Garboe's biography of Thomas Bartholin in 1949–1950 this tendentious, distorted and unjustified description of Thomas Bartholin continued.[4] One must almost regret on behalf of the authors concerned, that they did not gain something better from their time spent with Thomas Bartholin. They might have asked themselves: Who created the medical environment at the University of Copenhagen from which Niels Stensen grew? Whom did Niels Stensen himself choose as his teacher? Who inspired him? Who secured connections for him to Europe's leading scientists? Who followed and supported him eagerly? Who discussed important questions with him and gave him new input and new knowledge through letters in his first years abroad? Who quickly became aware of his unique talent and recognized his research with enthusiasm when others doubted? Who praised him time and time again to the skies? Who recommended him to the chancellor and the king, describing him as a student who surpassed his teachers? Who tried to get Stensen home to Copenhagen in 1667 and presumably also took the initiative for his recall in 1672? Who provided for the publication of a significant part of his research, and so on? These questions reflect precisely the importance of Thomas Bartholin, not only for Niels Stensen, but also for his other pupils and for his entire epoch.

Bartholin's greatest achievement in the field of research, his discovery and naming of the lymph vessels, a completely new vascular system, is counted among the absolute greatest medical discoveries in the 1600s. As mentioned, Bartholin's thoughts about the functions and role of the lymph vessels in diseases also represented significant new knowledge. Such a significant discovery cannot but inspire others, and it certainly did. The next generation of anatomists, with Niels Stensen and his Dutch friends Jan Swammerdam and Frederik Ruysch (Figure 217) at the head, continued Bartholin's research in the lymph vessels and their valves, made new discoveries and provided new knowledge (Figure 218).[5]

But although the discovery of the lymph vessels was of great importance, and although this is, above all, what has made Thomas Bartholin famous, others made the same discovery at about the same time. His greatest significance for his own time and posterity is to be found in a completely different place. Bartholin understood – perhaps more than anyone else during his time – the importance of the discipline of disseminating knowledge. Moreover, he mastered and practised this discipline himself, perhaps better and more enthusiastically than anyone else. His enormous literary production of more than a hundred works over a span of 40 years (see Appendix 2), which was hardly surpassed by others in his lifetime, his extensive correspondence with Europe's leading scholars, his swift publication of own and others' research, his updated editions of Europe's most widely used textbook of anatomy and, as the climax of it all, his medical journal – the first of its kind in the world – all had the common purpose of communicating the his knowledge and that of others to the world. Time and time again, he says himself: What is knowledge and science worth if it is not communicated to others? Even at a very young age he had this understanding, and it was no coincidence that at the age of 24 he became the publisher of the first textbook of anatomy, which, through the addition of Johannes Walæus' letters, supported Harvey's controversial theory of the blood circulation. As soon as Bartholin himself had become convinced of the truth of this theory, he recognized the importance of communicating it, as well as his and others' support for it, as quickly and widely as possible. He became one of the first to dismiss publicly the ancient supposition that the liver was the blood-forming organ.

Figure 217. Jan van Neck (1634–1714): Dr. Frederik Ruysch's (1638–1731) anatomy lesson 1683. Oil on canvas. 141 × 203 cm.

Ruysch belonged to the generation of younger anatomists who, after Thomas Bartholin, continued research into the lymph vessels and their valves. Here, Ruysch has opened the abdominal cavity of a newborn child in front of his pupils and perhaps his colleagues. On the table in front of the child is the placenta. The child holds the umbilical cord with the right hand, while Ruysch himself pulls it up with the index finger and thumb of his left hand to demonstrate it while looking at the painter or viewer of the painting. On a cloth behind the child is a pair of surgical scissors, and on the table in front of the child is another surgical instrument. A very young assistant shows a child's skeleton as a model for the teaching. Remarkably, only two of the seven people in the painting are looking at the dissected child, while the others are either looking at each other or at the painter/viewer of the painting.
Amsterdam Museum.

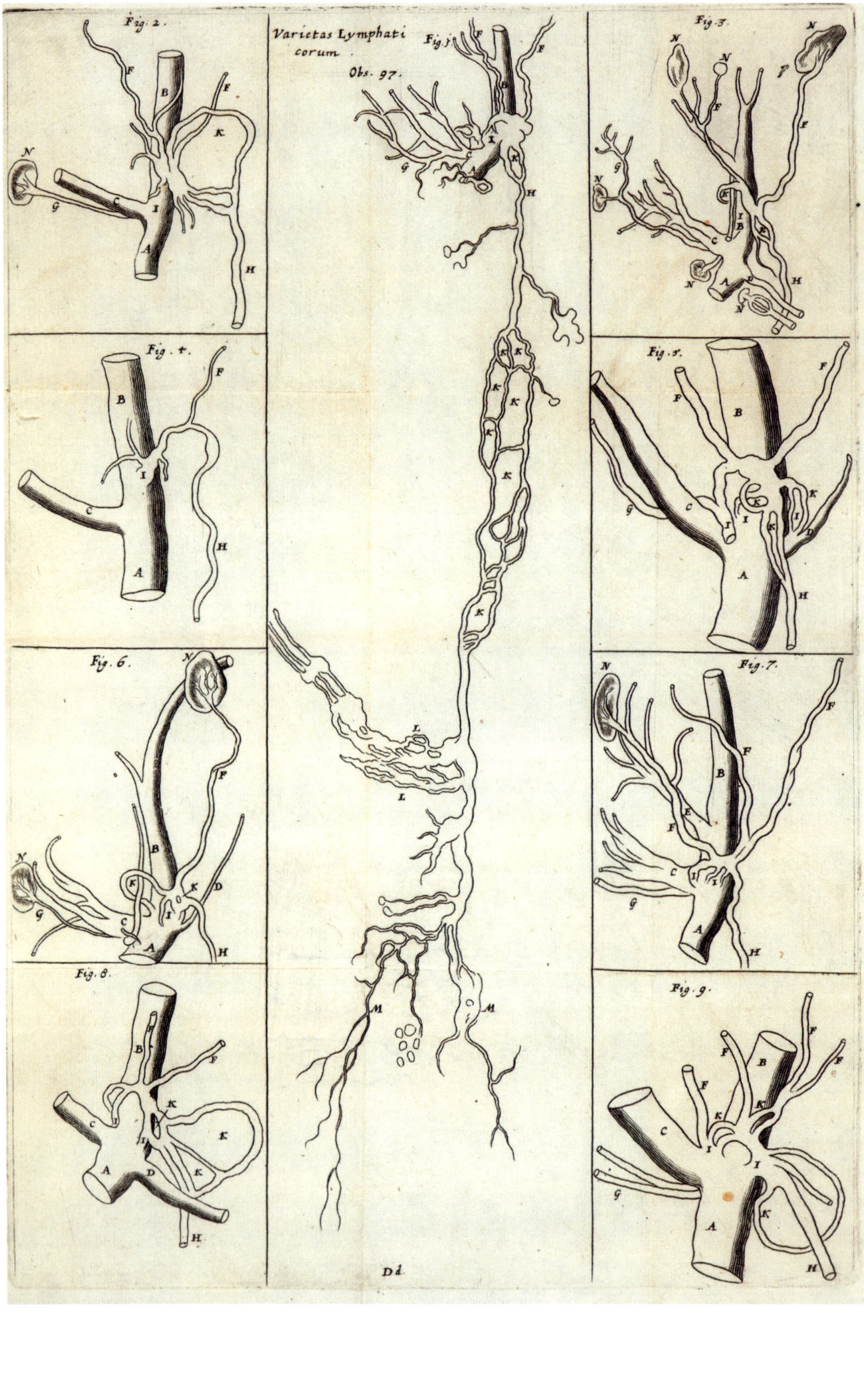
Varietas Lymphaticorum.
Obs. 97.
Fig. 1.
Fig. 2.
Fig. 3.
Fig. 4.
Fig. 5.
Fig. 6.
Fig. 7.
Fig. 8.
Fig. 9.

In the same way he became the first to publish a comprehensive work on luminescent phenomena in living organisms, the first to publish a scientific treatise on aneurysms, the first (together with Ole Worm) to convey the message that the suspected unicorn horns were in fact tusks from a Greenland whale, the narwhal, the first Dane to write a scientific treatise on a paediatric subject (a diphtheria epidemic among children in southern Italy), the first to describe the Bartholin-Patau malformation syndrome (trisomy 13), the first to provide a detailed description of the pancreas and its recently discovered excretory duct, the first after Malpighi himself to publish his findings of the pulmonary capillaries, the first to describe the importance of the discipline of pathological anatomy, the first to describe refrigeration anaesthesia – a modern method of anaesthesia, which has found application in the medical profession for the first time in recent decades – and so on. In addition to writing, Bartholin must have spent much of his time reading other people's works and publications. His own works testify to an incredibly well-read man with a huge overview of both ancient and contemporary literature within virtually all areas of the sciences. Again and again, he refers to the works, statements, theories and achievements of others.

But his books are not just enlightening. The language is beautiful, easy to understand and read, and Bartholin knows how to entertain his reader. But, as mentioned, this is precisely where you encounter his greatest weakness, and

Figure 218. Niels Stensen's illustrations of the varying inlets of the thoracic duct and its branches into the neck and shoulder veins of dogs. From the small treatise *Lymphaticorum Varietas* (Diversity of lymphatic ducts) in Thomas Bartholin's *Acta Medica & Philosophica Hafniensia Ann. 1673*, Copenhagen 1675, Obs. 97, pp. 240–41.

As an explanation for the illustration, Stensen states: "The diversity observed in dogs near the union of the lymphatic ducts with the vena cava on the left side of the neck is sometimes found in the branches of the insertions, I, which are sometimes many, sometimes few, sometimes in the small rings, K, which are either completely absent or present to some extent in more, fewer in others, sometimes narrow, sometimes very wide, as it appears in the figures. Fig. 1 is due to the diligence of my best friend Mr. Swammerdam, Fig. 2 & 3, which I have prepared, are a little different, as far as the insertion is concerned. I have concluded that it was superfluous to give specific explanations, since one common explanation for everything is sufficient.

Of the vena cava: A. Trunk above the heart. B. Branch of the jugular vein. C. Branch of the axillary vein. D. Branch extending towards the back of the neck. E. Branch extending towards the interior of the neck. F. Lymphatic ducts of the neck. G. Lymphatic ducts of the left front paw. H. Common trunk of the lymphatic ducts below the neck, called *Thoracicus*. I. Branches by which the lymphatic ducts continue into the vena cava. K. Rings of lymphatic ducts over which sometimes a lot of vessels sometimes a lot of nerves pass. L. Lymphatic ducts from the intestines and the liver. M. Lymphatic vessels from the loins and the hind paws. N. Conglobate glands."

The Royal Library in Copenhagen.

where people like Jean Riolan have rightly criticized him. In his eagerness to entertain, Bartholin time and time again puts his credibility to the test with stories in which it is difficult to imagine that he himself believed, and where you as his reader sit back and wish for better judgement. In some of these cases, however, one must of course be careful not to judge him on present critical terms, but remember, just as the title of this book suggests, that he belonged to an era when science moved from faith and superstition towards tangible, experimental observations. If you also include Bartholin's great interest in antiquity and his wide range of interests, you could say that he stood with his heels in the Renaissance, his toes in the Enlightenment and the rest of his feet solidly placed in the Baroque period. His great understanding of and ability to disseminate knowledge can be seen as a foretaste of the Enlightenment, and it was precisely in the motherland of the Enlightenment, France, that he was credited during this epoch, more specifically in 1766, for this effort. That year in *Collections Académiques*, the following was written about him:

> "Moreover, this physician has enjoyed great fame, not only in Denmark, but even throughout learned Europe, and this fame is justified by his skill in anatomy, his extensive knowledge in all fields of his science, his eagerness to collect interesting discoveries and useful observations, and his attention to their dissemination to the public."[6]

Bartholin himself was fully aware of his skills as a written communicator. For example, in his poetry collection *Carmina varii argumenti*, he begins a poem about a huge stone on the grounds of Hagestedgaard by saying that the stone was brought to this place across the sea by the power of a giant. This is followed by:

> "So that the stone will not always lie motionless on the ground,
> it will learn to fly across the sky, because of my writing.
> For my swift pen can do more with its words,
> than a mighty and giant strong hand."[7]

There is no doubt that Bartholin was also an excellent teacher. His scholarly, articulate and often humorous speeches to his students testify to this, and so does the University of Copenhagen's choice of him as a speaker at numerous official events. In his memorial speech for Bartholin, Willum Worm declares that no one surpassed him in eloquence.[8] The combination of the above-mentioned qualities and an almost unimaginable diligence and working ability must be the reason why Bartholin – of course also helped by his family, relations and the benevolence of successive kings and chancellors – was able to personify the revival of the medical faculty at the University of Copenhagen in the 1600s and create international awareness of the university and himself.

Finally, Thomas Bartholin's administrative importance for his own epoch and posterity in the fields of training and quality assurance should be mentioned. As the author of Denmark's first pharmacopoeia and the main force behind Denmark's first pharmaceutical regulations he made significant contributions to the pharmaceutical field in the form of rules for the harmonization of medicinal products and their prices, and the training and examination of apothecaries. Moreover, he established training and examination of midwives.

At Bartholin's death there was no one to carry on his legacy at the same level. As far as his most talented pupil Niels Stensen is concerned, his path to the professorship in anatomy at the University of Copenhagen was blocked for the most crucial five years (1662–1667) by the Paulli family, and when the chair finally became vacant, Stensen had shortly before converted to the Catholic faith and thus prevented his appointment as professor in Copenhagen for good. Thomas Bartholin could do nothing about this, and when Stensen returned to Copenhagen in 1672–1674, it was officially as a royal anatomist, but in reality as a missionary agent of the Catholic Church.

Although the four eldest of Bartholin's five sons became professors of various subjects,[9] none of them had a calibre approaching that of their father. In the long run, this unfortunately also applied to the otherwise promising young professor of anatomy and Stensen's student, Caspar Bartholin the Younger, who over the years became more interested and engaged in law and administration than in medicine (Figure 219). Thomas Bartholin's gifted youngest brother Rasmus, who became famous for the discovery of the double refraction of light in the mineral Icelandic calcite,[10] and who survived him by 18 years, also failed to continue the legacy. The heyday of the University of Copenhagen in the 1600s, and particularly of its medical faculty, began and ended with Thomas Bartholin.

Figure 219. Miniature of Caspar Bartholin the Younger made by Nicolai Christian Moritz (1684–1732). Gouache. 13 × 10 cm.

Thomas Bartholin's eldest son gave his name to the Bartholinian glands. As a student of Niels Stensen, in his younger years he gave rise to expectations of a glorious career in anatomy, but in the long run he was unable to fulfil his father's legacy as anatomy professor at the University of Copenhagen. Thomas Bartholin personified the beginning and the end of the university's heyday and international top position in anatomy and medicine in the 17th century. The latter part of Caspar's life was mainly devoted to administrative and legal tasks. In addition to 34 years as Dean of the medical faculty and several periods as Rector of the university, he was appointed to the Supreme Court in 1691, counsellor of justice in 1694, counsellor of state in 1708, and in 1719 he was appointed procurator general by the king (Frederik IV). In 1722 he became a conference counsellor, and in 1731 he and his family were ennobled. During the period 1712–1733 he was ephor of Borchs Kollegium.
National Gallery of Denmark. Photo: Jacob Skou-Hansen.

APPENDIX 1

Timeline of important events in Thomas Bartholin's life

1616	He is born in Copenhagen, Denmark on 20 October.
1629	His father Caspar Bartholin the Elder dies. His uncle Ole Worm takes over the role of guardian.
1631–1634	He is placed in the Fuiren family home and receives private lessons in medicine and anatomy from the physician Poul Moth.
1634	He matriculates at the University of Copenhagen.
1637	He begins a 9-year study trip in Europe. He matriculates at the University of Leiden on 11 July.
1638	He decides to study medicine.
1639	He becomes acquainted with the lacteal veins through dissections of dogs in Leiden.
1640	He leaves Leiden in June and arrives in Paris in July/August.
1641	He arrives in Padua after a short stay in Montpellier. His first revised edition of his father's textbook of anatomy is published in Leiden with the addition of two letters written by Johannes Walæus in support of William Harvey's theory on the circulation of the blood.
1642	In Padua, he participates in the dissection where the pancreatic duct in humans is discovered.
1643	Short stays in Naples and Rome.
1644	Stays in Campania, Sicily and Malta. He is offered a professorship in medicine at the University of Messina but declines.
1645	On 26 September in Basel, he becomes Doctor of Medical Science with a thesis on pleurisy.
1646	He arrives in Denmark after a return journey via Paris and Leiden.
1647	He is appointed professor in the Faculty of Philosophy at the University of Copenhagen.
1648	He is appointed professor of anatomy at the University of Copenhagen and takes over the responsibility for the Anatomy House in Copenhagen and the teaching here.
1649	He is appointed third professor of medicine (*Medicus Tertius*) at the University of Copenhagen. He marries Else Christophersdatter Hansen, daughter of the mayor of Copenhagen, Christopher Hansen.
1650	He begins his exploration of the lymphatic system and examines Aselli's lacteal veins in the human mesentery. He becomes a father for the first time.

1651	He detects the thoracic duct and the chyle bladder (*receptaculum chyli*) in a dog according to Jean Pecquet's instruction. On 15 December in Copenhagen, he demonstrates lymph vessels in a dog for the Frenchman Pierre Bourdelot and others.
1652	On 19 February in Copenhagen, he and his assistant Michael Lyser are the first to detect the thoracic duct in a human being. On 5 May in Copenhagen, he publishes a dissertation on this discovery. On 28 February, together with Lyser, he discovers new aqueous vessels in the abdominal cavity and forelegs of a dog.
1653	On 1 May in Copenhagen, he is the first to publish a paper on the new water-bearingvessels that were found on the dog on 28 February the previous year, as well as on several other dogs. He names them lymph vessels (*vasa lymphatica*).
1654	Together with his assistant Henrik Møinichen, he is the first to demonstrate the lymph vessels in a human being and publishes a treatise on this discovery in Copenhagen on 10 May. He becomes rector of the University of Copenhagen for the first time and second medical professor (*Medicus Secundus*) after the death of Ole Worm.
1656	He becomes first medical professor (*Medicus Primus*) of the University of Copenhagen after the death of his grandfather Thomas Fincke and ceases performing public dissections and demonstrations in the Anatomy House. He becomes Niels Stensen's preceptor, teacher and mentor.
1657	He establishes Denmark's first museum of natural science in Copenhagen.
1658	He publishes Denmark's first pharmacopeia.
1660	He publishes the first known book on cryoanalgesia.
1661	He becomes honorary professor at the University of Copenhagen.
1662	He buys the manor Hagestedgaard in North Zealand.
1670	He is appointed by Frederik III as the king's honorary physician. Hagestedgaard with his library burns down.
1671	He becomes the university's librarian.
1672	He rebuilds Hagestedgaard.
1673	He begins publishing the world's first medical journal *Acta Medica & Philosophica Hafniensia*.
1674	His oldest son, Caspar, is appointed professor of anatomy at the University of Copenhagen.
1675	His wife dies. He is appointed assessor of the Supreme Court.
1678	He marries Magdalene Rhode and becomes the owner of Fjellebro Manor on Funen.
1680	He sells Hagestedgaard. He dies in Copenhagen on 4 December.

APPENDIX 2

Holger Jacobsen's catalogue of Thomas Bartholin's published, adopted and unpublished works[1]

Catalogue of Thomas Bartholin's published works[2]

1. Anatomia Parentis C. Bartholini Novis Observationibus & Figuris locupletata. Lugd. Batavor. 1641, in 8vo. (The father Caspar Bartholin's anatomy book enriched with new observations and illustrations. Leiden 1641).
2. Anatomia Secundum Locupletata. Lugd. Bat. 1645, in 8. Gallicam versionem divulgavit. Abr. Prateus. Parisiis 1646, in 4. Italicis versibus reddidit Hostil. Contalgenus. Florentiæ 1651, in 12. (Second enriched edition of the anatomy book. Leiden 1645. French version published by Abraham du Prat. Paris 1646. Italian version in verse form by Hostilius Contalgenus. Florens 1651).[3]
3. Anatomia Tertium ad Sangvinis Circulationem Reformata, cum novis Iconibus. Lugd. Batav. 1651 & 1669, in 8. Hagæ Comit. 1655, 1660, 1666, in 8. In Belgicam Lingvam transtulit Th. Staffart. Lugd. Batav. 1653, 1668, in 8. Hagæ Comit. 1658, in 8. Anglicè edita, Londini, in 8. (Third edition of the anatomy book reformed according to the blood circulation, with new illustrations. Leiden 1651 & 1669. The Hague 1655, 1660, 1666. Translated into Dutch by Thomas Staffard. Leiden 1653, 1668. The Hague 1658. Published in English, London).[4]
4. Anatomia Quartum ad Sanguinis Circulationem & Vasa Lymphatica Renovata cum Figuris Novis & Idicibus. Lugd. Batav. 1673, in 8. Eadem edita, Lugduni in Gallis 1675, in 8. Germanicè edita ab Elia Wallnern. Noriberg 1677, in 4. (Fourth edition of the anatomy book renewed according to the blood circulation and the lymph vessels, with new figures and indexes. Leiden 1673. The same published in Paris in French 1675. Published in German by Elias Wallner. Nürnberg 1677).
5. Anatomica Aneurysmatis dissecti Historia. Panormi Siculorum 1644, in 8. (Anatomical account of the dissection of an aneurysm. Palermo, Sicily 1644).
6. De Unicornu Observationes Novæ. Patavii 1645, in 8. Amstelodami 1678, in 12. (New observations regarding the unicorn. Padua 1645. Amsterdam 1678).[5]
7. De Pleuritide Paradoxa. Basil 1645, in 4. (On paradoxes concerning pleurisy. Basel 1645).
8. De Monstris in Natura & Medicina. Basil 1645, in 4. (On Monsters in Nature and Medicine. Basel 1645).[6]
9. De Angina Puerorum Campaniæ Siciliæque Epidemica Exercitationes. Parisiis 1646, in 8. Neapol. 1663, in 8. (Investigations into a diphtheria epidemic among children in Campania and Sicily. Paris 1646. Naples 1663).
10. De Latere Christi aperto Dissertatio. Lugd. Batav. 1646, in 8. (A dissertation on the pierced side of Christ. Leiden 1646).
11. Antiquitatum Veteris Puerperii Synopsis. Hafniæ 1646, in 8. Edita cum Commentario Filii Caspari Bartholini. Amstelod. 1676, in 12. (A survey of midwifery in ancient times. Copenhagen 1646. Published with a commentary by his son Caspar Bartholin in Amsterdam 1676).

12. De Luce Animalium Libri III, admirandis Historiis, rationibusque novis referti. Lugd. Bat. 1647, in 8. Secunda Edito Auctior Hafniæ 1669, in 8. (Three books filled with wonderful accounts and new explanations of light emitted by living creatures. Leiden 1647. Second enlarged edition. Copenhagen 1669).
13. De Armillis Veterum præsertim Danorum. Hafniæ 1648, in 8. Edito Secunda Figuris auctior, per Casp. Bartholinum. Amstelodami 1676, in 12. (On bangles in ancient times, especially in Denmark. Copenhagen 1648.[7] Second edition enriched with figures, by Caspar Bartholin. Amsterdam 1676).
14. Anatomicæ Vindiciæ, Casparo Hofmanno aliisque oppositæ. Hafniæ 1648, in 4. (Anatomical counterclaims directed against Caspar Hofman and others. Copenhagen 1648).
15. Animadversiones in Anatomica Hofmanni. Ibid. (Comments on Hofman's anatomical remarks. The same place).
16. De Variis Reipublicæ Christianæ morbis, & placidis illorum Remediis Dissertatio Oratoria. Hafniæ 1649, in 4. (Oratorical dissertation on different diseases in the Christian society and the gentle remedies against them. Copenhagen 1649).[8]
17. Collegium Anatomicum Disputationibus XVII adornatum. Hafniæ 1651, in 4. (Anatomy lecture adorned with 17 disputations. Copenhagen 1651).[9]
18. De Cruce Christi Hypomnemata Quatuor: 1. De sedili medio. 2. De Vino Myrratho. 3. De Corona Spinea. De sudore sanguineo. Hafniæ 1651, in 8. (Four records on the cross of Christ. 1. On the seat in the middle. 2. On wine flavoured with myrrh. 3. On the of thorns. 4. On the bloody sweat. Copenhagen 1651).
19. De Lacteis Thoracicis in homine brutisque Observatis Historia Anatomica. Genuæ 1654, in 8. Lugd. Batav. 1654, in 12. Ultraject 1654, in 12. Heidelbergæ 1659, in 8. (An anatomical account of the thoracic ducts observed in man and animals. Genoa 1654. Leiden 1654. Utrecht 1654. Heidelberg 1659).[10]
20. De Lacteis Thoracicis Dubia Anatomica, & An Hepatis funus immutet medendi methodum. Hafniæ 1653, in 4. Paris. Leid. Heidelb. & c. (Anatomical doubts concerning the thoracic ducts, and the question of whether the abandonment of the liver changes the method of treatment. Copenhagen 1653. Paris. Leiden. Heidelberg etc.).[11]
21. Vasa Lymphatica in Animalibus inventa, & Hepatis Exseqviæ. Hafniæ 1653, in 4. Paris. Genuæ. Leid. Heidelberg & c. (The lymph vessels discovered in animals and the abandonment of the liver. Copenhagen 1653. Paris. Leiden. Heidelberg etc.).[12]
22. Vasa Lymphatica in Homine inventa. Hafniæ 1654, in 4. (The lymph vessels discovered in man. Copenhagen 1654).[13]
23. Defensio Vasorum Lacteorum & Lymphaticorum adversus Jo. Riolanum. Hafniæ 1655, in 4. (Defence of the lacteal and lymph vessels against Jean Riolan. Copenhagen 1655).[14]
24. Examen Lacteorum contra Harvejum & Riolanum. Hafniæ 1655, in 4. (Examination of the lacteal vessels against Harvey and Riolan. Copenhagen 1655).[15]
25. Spicilegium Primum ex Vasis Lymphaticis. Hafniæ 1657, in 4. (First aftermath of the lymph vessels. Copenhagen 1657).[16]
26. Spicilegium Secundum ex Vasis Lymphaticis. Hafniæ 1660, in 4. Bina hæc Spicilegia uno volumine recusa sunt Amstelodami 1661, in 12, editore D. Gerhardo Blasio (Second aftermath of the lymph vessels. Copenhagen 1660. Both these aftermaths were reissued in one volume in Amsterdam 1660[17] by Dr. Gerard Blasius).[18]
27. Paralytici Novi Testamenti Medico & Philologico Commentario illustrati. Hafniæ 1653, in 4. Basileæ 1662, in 8, procurante D. Georgio Segero. (Informative medical and philological commentaries on the paralytics in the New Testament. Copenhagen 1653. Basel 1662, provided by Dr. Georg Seger).

28. Oratio in Obitum D. Olai Wormii. Hafniæ 1655, in 4. (Memorial speech for Dr. Ole Worm. Copenhagen 1655).[19]
29. Oratio in Obitum D. Henrici Fuiren. Hafniæ 1659, in 4. (Memorial speech for Dr. Henrik Fuiren. Copenhagen 1659).[20]
30. Dispensatorium Hafniense. Hafniæ 1658, in 4. (The Copenhagen pharmacopoeia. Copenhagen 1658).
31. Historiarum Anatomicarum Rariorum Centuria Prima & Secunda. Hafniæ 1654, in 8. Hagæ Comitis, Roterodami, Amstelodami, 1654, in 8. Germanicè edita à D. Georgio Segero. Francofurt, 1657, in 8. Belgicè, Amstelodami 1657, in 8. (Rare medical reports,[21] first and second century. Copenhagen 1654. Den Haag, Rotterdam, Amsterdam, 1654. Published in German by Dr. Georg Seger. Frankfurt 1657. In Dutch, Amsterdam 1657).
32. Historiarum Anatomicarum Rariorum Centuria Tertia & Quarta. Hafniæ 1657, in 8. (Rare medical reports, third and fourth century. Copenhagen 1657).
33. Historiarum Anatomicarum Rariorum Centuria Quinta & Sexta. Hafniæ 1660,[22] in 8. (Rare medical reports, fifth and sixth century. Copenhagen 1660).
34. Panegyricus Regi Friderico III. Primo Regnorum Hæredi, publicè Academiæ nomine dictus. Hafniæ 1660, in fol. (Eulogy for Frederik III, first heir of the kingdoms, given publicly on behalf of the University. Copenhagen 1660).[23]
35. De Nivis Usu Medico Observationes variæ. Hafniæ 1660,[24] in 8. (Various observations regarding the medical use of snow. Copenhagen 1660).[25]
36. Disputationes variæ Medicæ Hafnienses. Hafniæ,[26] in 4. (Various medical disputations given in Copenhagen. Copenhagen).
37. Cista Medica Hafniensis, variis Consultationibus, Casibus, vitisque Medicorum Hafniensium repleta. Hafniæ 1661,[27] in 8. (The archive chest of the Medical Faculty in Copenhagen, full of various consultations, cases and biographies of different physicians of Copenhagen. Copenhagen 1661).[28]
38. Domus Anatomicæ Hafniensis Descriptio. Hafniæ 1662, in 8. (Description of the Anatomy House in Copenhagen. Copenhagen 1662).[29]
39. Responsio ad Novas Observationes Bilsianas contra Zassium. Hafniæ 1661, in 8. Amstelodam. 1661, in 12. Belgicè edidit Jo. Blasius Amstelodami, in 12. (Reply to Zas regarding Bils' new observations. Copenhagen 1661. Amsterdam 1661. Published in Dutch by Johannes Blasius in Amsterdam).[30]
40. Castigatio Epistolæ maledicæ Bilsii, ubi Bilsianæ artes deteguntur & Professoria dignitas vindicatur. Hafniæ 1661, in 8. Amstelod. 1661, in 12. (Correction of Bils' malicious letter, where the frauds of Bils are revealed, and the professorial dignity is vindicated. Copenhagen 1661. Amsterdam 1661).[31]
41. De Hepate defuncto Dissertatio, novis Bilsianorum experimentis opposita. Hafniæ 1661, in 8. (Dissertation on the abandoning of the liver against Bils' new experiments. Copenhagen 1661).[32]
42. De Pulmonum Substantia & Motu Diatribe. Hafniæ 1663, in 8. Lugd. Batav. 1672, in 12. (Treatise on the substance and movement of the lungs. Copenhagen 1663. Leiden 1672).
43. De Insolitis Partus Humani Viis Dissertatio Nova. Hafniæ 1664, in 8. (New dissertation about unusual birth paths in humans. Copenhagen 1664).
44. De Cometa Consilium Medicum, cum Monstrorum nuper in Dania natorum Historia. Hafniæ 1665, in 8. (Medical consultation on the comet and an account about monsters recently born in Denmark. Copenhagen 1665).

45. De Medicina Danorum Domestica Dissertationes X, cum Vindiciis & Additamentis. Hafniæ 1666, in 8. (Ten dissertations on Danish domestic medicine, with vindications and appendices. Copenhagen 1666).
46. De Hepatis Exautorati Desperata Causa. Hafniæ 1666, in 8. (On the abandoned liver's hopeless cause. Copenhagen 1666).[33]
47. Epistolarum Medicinalium Centuria I & II. Hafniæ 1663, in 8. (Medical letters Century I & II. Copenhagen 1663).
48. Epistolarum Medicinalium Centuria III. Historiis Medicis aliisque ad rem Medicam spectantibus plena. Hafniæ 1667, in 8. (Medical letters Century III full of medical accounts and other things concerning medicine. Copenhagen 1667).
49. Epistolarum Medicinalium Centuria IV. variis observationibus Curiosis & utilibus referta. Hafniæ 1667, in 8. (Medical letters Century IV filled with various careful and useful observations. Copenhagen 1667).
50. Orationum Varii argumenti Volumen. Hafniæ 1668, in 8. Orationum Volumen Secundum prælum expectat, Opus Posthumum. (A volume of speeches with various contents. Copenhagen 1668. A second volume with speeches awaits the press, a posthumous work).
51. De Flammulâ Cordis. Hafniæ 1667, in 8. (On the small flame of the heart. Copenhagen 1667).
52. De Cygni Anatome ejusque Cantu, auctior edit. à Casparo Bartholino filio. Hafniæ 1668, in 8. (On the anatomy and song of the swan, enlarged edition by my son Caspar Bartholin. Copenhagen 1668).
53. Carminum Libri VIII. Hafniæ 1669, in 8. (Eight books with poems. Copenhagen 1669).
54. De Medicis Poëtis Dissertatio. Hafniæ 1669, in 8. (Dissertation on medical poets. Copenhagen 1669).
55. Opuscula Lactea & Lymphatica, uno Volumine auctiora, & emendatoria edita. Francofurt 1670, in 8. (Writings on the milk and lymph vessels, an enlarged and improved edition in one volume. Frankfurt 1670).[34]
56. De Bibliothecæ Incendio Dissertatio ad Filios. Hafniæ 1670, in 8. (Dissertation for my sons on the burning of my library. Copenhagen 1670).[35]
57. De Medico Perfecto Oratio. Hafniæ 1670, in 4. (Speech on the perfect physician. Copenhagen 1670).
58. Oratio in Obitum Jo. Mulenii Georgii fil. Hafniæ 1670, in 4. (Memorial speech for Hans Jørgensen Mule. Copenhagen 1670).
59. De Flagrorum usu in Re Medica. Hafniæ 1670, in 8. (On the use of whips in medicine. Copenhagen 1670).
60. Quæstiones Nuptiales. Hafniæ 1670, in 4. (Wedding questions. Copenhagen 1670).
61. De Morbis Biblicis Miscellanea Medica. Hafniæ 1671, in 8. (On the diseases in the Bible. A medical Miscellany. Copenhagen 1671).[36]
62. De Theriaca Dissertationes Duæ. Hafniæ 1671, in 4. (Two dissertations on theriac. Copenhagen 1671).
63. De Rectoris Academici Dignitate Oratio. Hafniæ 1672, in 4. (Speech about the dignity of the rector of the University. Copenhagen 1672).
64. De Confectione Alkermes Dissertatio Præliminaris. Hafniæ 1672, in 4. (Preliminary dissertation on Alkermes confectionery. Copenhagen 1672).
65. De Visitatione Pharmacopæorum Programmata Septem. Hafniæ 1672. 1673. 1674. 1675. 1676. 1678 & 1679, in 4. (Seven programs for the supervision of apothecaries. Copenhagen 1672–1679).

66. De Transplantatione Morborum. Hafniæ 1673, in 8. (On the transfer of diseases. Copenhagen 1673).[37]
67. De Cerebri substantia & oculorum suffusione Epistolæ Duæ. Hafniæ 1669, in 4. (Two letters on the substance of the brain and the lubrication of the eyes. Copenhagen 1669).
68. De Sanguine Vetito Disquisitio Medica. Francof. 1673, in 8. De Sanguine Vetito Disquisitio Uberior. Francof. 1676, in 8. (Medical examination of a ban on blood. Frankfurt 1673. Full examination of a ban on blood. Frankfurt 1676).
69. Acta Medica & Philosophica Hafniensia, Anni 1671 & 1672. Hafniæ 1673, in 4. (Medical and natural philosophical reports from Copenhagen, years 1671 and 1672. Copenhagen 1673).[38]
70. Acta Medica & Philosophica Hafniensia, Anni 1673. Hafniæ 1675, in 4. (Medical and natural philosophical reports from Copenhagen, year 1673. Copenhagen 1675).
71. Acta Medica & Philosophica Hafniensia, Anni 1674, 1675, 1676. Hafniæ 1677, in 4. (Medical and natural philosophical reports from Copenhagen, years 1674–1676. Copenhagen 1677).
72. Acta Medica & Philosophica Hafniensia, Anni 1677, 1678, 1679. Hafniæ 1680, in 4. (Medical and natural philosophical reports from Copenhagen, year 1677–1679. Copenhagen 1680).
73. Epistola Consolatoria ad Fratriam super obitum D. Casp. Bartholini ICti. Hafniæ 1673, in 8. Danicè (Letter of consolation to my sister-in-law on the death of Doctor of Laws Caspar Bartholin. Copenhagen 1673. In Danish).[39]
74. De Simplicium Medicamentorum Inquilinorum Facultatibus cognoscendis. Francof. 1671, in 8. (How to know the properties of uncompounded and naturalized medicines. Frankfurt 1671).
75. De Anatome Practica Consilium. Hafniæ 1674, in 4. (Plan for the practical anatomy. Copenhagen 1674).[40]
76. De Legendis Libris Dissertationes VII. Hafniæ 1674, in 8. (Seven lectures on reading books. Copenhagen 1674).
77. De Peregrinatione Medica Dissertatio ad Filios. Hafniæ, in 4. (Dissertation on medical travel for my sons. Copenhagen 1674).[41]
78. De Peregrinatione ad Theodorum Fuiren Propempticon. Hafniæ 1674, in 4. (Valedictory for Theodor Fuiren on his peregrination. Copenhagen 1674).
79. Gratulatio Dn. Petro Com. in Griffenfeld publicè Academiæ nomine facta. Hafniæ 1674, in fol. (Congratulatory speech for Master Peter, Count of Griffenfeld, given publicly on behalf of Copenhagen University. Copenhagen 1674).
80. De Puerperio Veteri ad Filium Casparum Bartholinum Epistola. Hafniæ 1675, in 4. Amstelod. 1676, in 12. Romæ 1677, in 8. (Letter on ancient childbirth for my son Caspar Bartholin. Copenhagen 1675. Amsterdam 1676. Rome 1677).
81. De Equestris Ordinis Danebrogici Origine, Dissertatio Historica. Hafniæ 1676, in fol. (Historical dissertation on the origin of the order "Knight of Dannebrog." Copenhagen 1676).[42]
82. Panegyricus Regi Christiano Quinto post Victoriam Wandalicam & expugnationem Wismariæ Triumphatori, publicè Academiæ nomine dictus. Hafniæ 1676, in fol. (Eulogy for King Christian VI, who after the victory in North Germany and the conquest of Wismar emerged as a triumphator, given publicly on behalf of Copenhagen University. Copenhagen 1676).

The following works are not authored by Thomas Bartholin but published by him with his preface

83. Elenchus Operum M. Aurelii Severini. Neapoli 1643, in 4. (List of Marco Aurelio Severino's works. Naples 1643).
84. Jo. Walæi Epistolæ duæ de motu chyli & sanguinis ad Th. Barthol. Lugd. Bat. 1641 & c. (Johannes Walæus' two letters to Thomas Bartholin on the motion of the chyle and blood. Leiden 1641 etc.).
85. De Aureo Cornu Olai Wormii Eruditorum Iudicia. Patav. 1645, in 8. (Judgements of scholars on the gold horn of Ole Worm. Padua 1645).
86. De Hyssopo in Cruce Christi Cl. Salmasii Epistolæ 3 ad Th. Barthol. Lugd. Bat. 1646, in 8. (Three letters from Claude Saumaise to Thomas Bartholin regarding hyssop on the cross of Christ. Leiden 1646).
87. Observationes Anatomicæ P. Pawii selectæ posthumæ. Hafniæ 1656, in 8. (Posthumously selected anatomical observations of Pieter Pauw. Copenhagen 1656).
88. Joh. Rhodii Mantissa Anatomica ad Th. Barthol. Hafniæ 1660, in 8. (Johan Rhode's additional anatomical remarks for Thomas Bartholin. Copenhagen 1660).
89. Cornarii vita sobria ad usum vulgarem accomodata. Hafniæ 1657, in 12. Danicè. (Cornaro's sober life suited for common use. Copenhagen. In Danish.).[43]
90. Joh. Mulenii Numismata Danica. Hafniæ 1670, in 4. (Hans Mule's Danish coins), Copenhagen 1670).
91. Observationes Jo. Veslingii & Epistolæ Anatom. & Medicæ posthumæ. Hafniæ 1664, in 8. (Johannes Vesling's observations and posthumous letters on anatomy and medicine. Copenhagen 1664).
92. Meditationes Bonæ, linguâ vernaculâ versæ ab Elis. Tottia. Hafniæ 1664, in 12. Danicè (Useful meditations translated into Danish by Elisabeth Thott. Copenhagen 1664. In Danish).
93. Michaelis Lyseri Culter Anatomicus, cum duplici præfatione. Hafniæ 1665, in 8. (Michael Lyser's The anatomical knife, with two prefaces. Copenhagen 1665).
94. Observationes Medicæ Variorum, Lyseri, Møinichenii, Bogdani, Seidelii. Hafniæ 1665, in 8. (Various medical observations by Lyser, Møinichen, Bogdan, Seidel. Copenhagen 1665).
95. Alberti Bartholini de scriptis Danorum liber posthumus. Hafniæ 1666, in 8 (Albert Bartholin On Danish writings, a posthumous book. Copenhagen 1666).
96. Conradus Gesnerus De lunariis. Hafniæ 1669, in 8. (Conrad Gesner: On moon plants. Copenhagen 1669).
97. Lisseti Benanci de fraudibus pharmacopæorum ex Gallico declaratio. Francof. 1669 & 1671, in 8. (Lisset Benancio, Declaration of the fraud of the chemists, translated from French, Frankfurt 1669 & 1671).
98. Lodettus de iisdem ex Italico versus. ib. (Giovanni Antonio Lodetto on the same subject, translated from Italian. Ibid.).
99. Jo. Rhodii de Aciâ Dissertatio, secundâ editione auctior. Hafn 1672, in 4. (Johan Rhode's dissertation on the thread, more enlarged than the second edition. Copenhagen 1672).
100. J. Rhodii posthumum opusculum de ponderibus & mensuris Medicorum Vet. & Iudicia Eruditorum de Acia Rhodianâ. Hafniæ 1672, in 4. (Johan Rhode's small posthumous work on weight and measuring of veterinary medicine and scholars' judgements of the Rhodian thread. Copenhagen 1672).

101. Catalogus & Taxa medicamentor. officinalium. Hafniæ 1672, in 4. (Catalogue and tariff for pharmacies' medicaments. Copenhagen 1672).
102. Hugo Grotius de veritate religionis Christianæ, Danicè, operâ Th. Bartholini. Hafniæ 1678, in 12. (Hugo Grotius on the truth of the Christian religion, in Danish, translated into Danish by Thomas Bartholin. Copenhagen 1678).[44]

Titles of further works by Thomas Bartholin, which were ready to go press, but perished in the fire of Hagestedgaard in 1670[45]

1. Celsus de Medicina, Notis Variis Medicis, Philologicis & Antiquariis illustratus, cum Variis Lectionibus & Lexico Cl. V. Joh. Rhodii (Celsus on medicine, illustrated by various medical, philological and antiquarian notes, with various readings and the lexicon of the famous man Johan Rhode).
2. Anatome Practica, ex morbo defunctis pro sede affecta Inquirenda (Practical anatomy based on dissections of people who have died of disease, in order to examine the diseased site).
3. Antiquitates Puerperii Variarum gentium, in primis Romanorum (Various ancient people's birth care, primarily that of the Romans).
4. Anatome Pagana (Pagan anatomy).
5. Epistolarum Medicinalium Centuriæ Tres (Three centuries of medical letters).
6. Liber de Morbis Lymphaticis (Book on diseases in the lymphatic system).
7. Medicina Danica inquilinis remediis proprios morbos sanans (Danish medicine cures the Danes' own illnesses).
8. Bibliotheca Anatomica (Anatomical library).
9. Observationes de Unicornu Auctæ (Enlarged observations regarding the unicorn).
10. De Pygmæis Analecta (Additional remarks on pygmies).
11. Apologia pro Pharmacopæa Danorum (Defence of the Danish Pharmacopoeia).
12. De Glossopetris Melitensibus Dissertatio (Dissertation on Maltese tongue stones).
13. De Morbis Fungosis Opusculum (A small work on fungous diseases).
14. Antiquitates homini Medico necessariæ (Human antiquities necessary for the physician).
15. Instrumenta Medica (Medical instruments).
16. Miscellanea Medica (Mixed medical issues).
17. Arcanorum Medicorum Volumen (A volume on secret medicines).
18. De Animalibus in Norwegia (On the animals in Norway).
19. Notulæ Hippocraticæ in Aphorismos Hippocratis (Small Hippocratic comments on Hippocrates' aphorisms).
20. Notæ & Emendationes in Cælium Aurelianum (Notes and corrections to Caelius Aurelianus).
21. Strabus Gallus Notulis & Lectionibus Illustratus (Walafrid Strabo illustrated by means of small notes and readings).

APPENDIX 3

Bartholin versus Rudbeck

UBVA has been requested by paediatrician Jesper Brandt Andersen to make a statement concerning the historic contest between the Danish anatomist Thomas Bartholin (1616–1680) and his Swedish colleague Olof Rudbeck (1630–1702). UBVA has been requested to evaluate which of the two, according to modern estimation, must properly be credited with having discovered the lymph vessels. In the request it is mentioned that it is a contest for priority, which is classic in the history of Danish science and never properly resolved. It is requested that UBVA's statement be based on current standards, rather than those of the time.

It is first necessary to make the proviso that an actual decision about such a question demands historical studies, the undertaking of which does not lie in UBVA's sphere of work. An assessment of the matter must therefore be carried out based on the available material, which in this case means a chapter of a book manuscript, written by Jesper Brandt Andersen, who sent this to UBVA in an email of 22 April 2017. Jesper Brandt Andersen points out in the communication that in reading through the accessible literature on the subject, he has noted that a thorough presentation and analysis of the documents and known data of the subject is lacking. He has thus procured them for the chapter in hand. In this connection it should be observed that the UBVA has no way of either verifying or completing this presentation, and the UBVA's judgement must therefore stand or fall according to whether the information contained there is correct and thorough. In the following, the relevant data are obtained from the presentation, and for further details you are referred to this. On this basis it is possible to draw up the following chronology.

1650	Rudbeck observes, according to his own information, clear liquid and vessels in connection with the slaughter of a calf; they run from the aorta to the receptacle just under the diaphragm.
1651	Bartholin mentions the lacteal veins (chyle vessels) in the third edition of his Anatomy Book.
27 January 1651	Rudbeck observes, according to his own information, serum-filled vessels behind the large intestine and the rectum, which empty their contents into the chyle receptacle of a dog.
8 February 1651	Rudbeck points out the thoracic duct in a calf and a sheep, according to his own information.
Later in 1651	During the dissection of a cat and systematic ligation, Rudbeck finds out, according to his own claims, that the chyle in the intestine of the cat also empties into a receptacle lying upwards and backwards in the abdominal cavity, and that the tube emptied into the bloodstream at the shoulder vein's entry into the neck vein. This discovery is neither made public verbally nor published in written form at this point.
December 1651	Bartholin undertakes the vivisection of a dog according to Jean Pecquet's instruction. He points out lymph vessels in a dog for Pierre Bourdelot, Ole Worm and Henrik Fuiren in the Anatomy House in Copenhagen.

28 February 1652	Bartholin and Michael Lyser conclusively demonstrate the lymph vessels in a dog.
6 March 1652	Rudbeck finds, according to his own claims, serum-filled vessels in the thoracic cavity of a sheep.
16 March 1652	Bourdelot writes to Bartholin, saying that he has informed the Swedish queen Christina about Bartholin's work and experiments.
30 April 1652	Bartholin mentions the lymph vessels in a letter to Frederik Arnisæus.
April or May 1652	Rudbeck, at a dissection observed by the Swedish queen and several others, demonstrates the chyle receptacle. According to Rudbeck Professor Olaus Stenius gives an account of the matter at a lecture on this occasion. According to his own claims, Rudbeck also demonstrates the water-clear excretory ducts of the liver. Rudbeck finds out that Pecquet has published a book on the thoracic duct and the chyle receptacle, to which Rudbeck gains access in June. A short while afterwards he gains access to Johannes van Horne's treatise, published in May, on the thoracic duct.
5 May 1652	Bartholin publishes *De Lacteis Thoracicis* which describes the thoracic ducts in human beings.
18 May 1652	Bourdelot writes to Bartholin that a professor in Uppsala (actually the medical student Rudbeck) has made the same observations of the chyle vessels in the chest as Bartholin, without knowing Pecquet's name. The queen of Sweden has told the physicians in Uppsala about Bartholin and invited them to undertake vivisections.
September 1652	According to his own statement, Rudbeck gains access to Bartholin's treatise on the discovery of the thoracic duct in human beings.
19 October 1652	According to his own claims, Rudbeck finds serum-filled vessels in the abdominal cavity of a cat.
30 October 1652	According to his own claims, Rudbeck finds in the dissection of a dog tube connections between two glands in the mediastinum.
27 April 1653	According to his own claims, Rudbeck finds small glands behind the breastbone with serum-filled small branches to the thoracic duct.
1 May 1653	Bartholin publicises his treatise, *Vasa Lymphatica* etc., on the lymph vessels in a dog.
Summer 1653	In Vesterås, Rudbeck publishes a 24-page treatise on the discovery of the lymph vessels in animals with his own illustrations of the thoracic duct and the chyle receptacle, together with a description of two types of water-clear vessels. He cites Bartholin's discoveries of the lacteal veins from 1651.
May 1654	Bartholin publishes a treatise on the lymph vessels in human beings.

It is a well-known phenomenon that both inventions and discoveries are often made more or less in parallel or at the same time by several persons, but quite independently of one another. It is a natural consequence of the development of collected knowledge. When common understanding has reached a given stage, many researchers will work on that foundation and attempt to extend human understanding further. It is therefore inevitable that a good many researchers will work in parallel with similar problems. When essential progress is made in this way, doubt can arise as to which of several researchers should be credited with having made a particular discovery. This question is most often more of a scientific and ethical than a legal nature, since making a discovery itself seldom, if ever, has legal effect in current law. Discoveries themselves cannot be patented, for it is required for this purpose that there should be a reproducible and efficacious technical discovery that can be used industrially. They cannot enjoy copyright protection either, since for this purpose it is demanded that there should be a literary or artistic work, and the discovery itself will not constitute this. It will typically be described in a work which enjoys protection, but this protection will not prevent others writing in their own words a new and independent description of the same discovery. However, the question can have great practical significance, not least for the researcher's continuing career, scholarly reputation and allocation of honourable prizes.

It follows from this that the decision about who has made a discovery must be based on considerations of scientific theory rather than legal ones, but nevertheless such a judgement includes elements that are regularly used in the legal world: first and foremost, the evaluation of documentation and proofs. However, regarding scientific discoveries there exists a source of law in an international scheme, namely The Geneva Treaty on the International Recording of Scientific Discoveries (in the following referred to as the WIPO-treaty), which under the auspices of the World Intellectual Property Organisation was agreed on 3 March 1978.

It is not uninteresting in this connection that the treaty contains several definitions, of which it is reasonable to mention the following in particular:

"For the purposes of this Treaty:
(i) 'scientific discovery' means the recognition of phenomena, properties or laws of the material universe not hitherto recognized and capable of verification;
(ii) 'discoverer' means the natural person, who has made a scientific discovery himself, through observation, study, experimentation or reasoning, and in a manner decisive for arriving at the recognition thereof; where, in the making of a scientific discovery, several natural persons have jointly fulfilled the said requirements, any reference to a discoverer shall be construed as a reference to all of them [...]
(vi) 'Date of discovery' means the date on which the scientific discovery was, for the first time, published or communicated to the public [...]."

Although the treatise in question has not taken effect, it demands attention, first and foremost because here there is mention of an international forum of experts, which has agreed to definitions for use in a treatise. This normally happens with considerable preparations and afterthoughts, and definitions of this kind are sought as far as possible to be suitable to commonly used concepts and language in the relevant circles. It is also of special interest that the definition of the date of the discovery is not set at the point in time when the researcher in question obtains personal understanding of the relevant subject matter, but at the time of the first publication or announcement to the public.

The inclusion of the release or publication of discoveries to the establishment makes good sense. These explicit actions serve to ensure both a demonstrable and definitive dating. Many discoveries appear gradually, as the researcher makes observations, puts them together and recognizes their coherence or significance. The subjective "discovery" can therefore take place over a period of time, and as such one cannot meaningfully fix it to some particular date. Furthermore, in a scientific context, publication, reproducibility, verification and falsification are completely central, core concepts and conditions that underlie real research. It therefore goes together logically with the scientific theoretical understanding of when something can be said to be scientifically proved (discovered) that one dates a discovery to the moment it is published for testing and criticism, and not to the moment when it is subjectively understood by the discovering researcher. Furthermore, the issuing or publication of a research result is an event that is better suited to being dated objectively than by subjective recognition, which typically appears privately or within a closed circle, and where the subsequent dating can be influenced by many factors, including defective memories or wishes to influence one's own or others' profile.

The concepts mentioned, published and communicated to the public – or a publication – are well known within the copyright world, and since the treatise in question is agreed to within the framework of WIPO, which is also UN's "specialized agency" for copyright, it is natural to seek them illuminated by means of the understanding of the concept of copyright. The Bern Convention for the protection of literary and artistic works (most recently revised in Paris in 1971) contains in Article 3, Section 3, the following definition:

"The expression "published works" means works published with the consent of their authors, whatever may be the means of manufacture of the copies, provided that the availability of such copies has been such as to satisfy the reasonable requirements of the public, having regard to the nature of the work. The performance of a dramatic, dramatic-musical, cinematographic or musical work, the public recitation of a literary work, the communication by wire or the broadcasting of literary or artistic works, the exhibition of a work of art and the construction of a work of architecture shall not constitute publication."

The latter sentence in this definition is perhaps less relevant here, in that the relevant date for the discovery in the WIPO-treaty includes the day of the first announcement to the public as an alternative, but the sentence contributes to the understanding of what constitutes "statement to the public," namely any form of statement, talk, radio or television broadcast, exhibition or other function, by which a broader sphere than the researcher himself or his or her closest circle of family, friends or colleagues gains access to the relevant information.

When these rules are to be applied to the case in question, it is important to realize that the answer always depends on the question posed – in this case the discovery of the lymph system, in other words the system of vessels conveying a clear fluid around in the body. Part of what is today included in this system, the chyle system or the "lacteal veins," was undisputedly first discovered by the Italian Gaspare Aselli, while the thoracic duct and the chyle container or receptacle in the abdominal cavity was undisputedly fist discovered by the Frenchman Jean Pecquet.

It must further be noted that a printed publication of research results without further dissemination must be regarded as sufficient for claiming priority. For as far as concerns any other means of publicity, it must be claimed that this is testified by sources that are independent of the discoverer in question. If one only has the word of the latter that the discovery has been publicly presented, there will not exist any greater degree of proof than occurs when researchers maintain independently that they have observed something or other on a given date.

The first date of possible relevance for the publication of the discovery of the lymph system occurs on 15 December 1651 where Bartholin, during the vivisection of a dog, according to Pecquet's instructions, points out lymph vessels in a dog in the presence of Pierre Bourdelot, Ole Worm and Henrik Fuiren. However, it is unclear to what extent this event has been publicly accessible and witnessed by the people concerned. Even if reliable, confirmed witness statements from the named participants could be found, it must be pointed out with a reasonable degree of likelihood that it was a public demonstration. The available information gives no clear basis for assuming that this was the case, and therefore this date cannot with any certainty to be assumed to be evidence for publication of the discovery.

The next possible date is in April or May 1652, when Rudbeck presents the chyle container in a public dissection. The Queen and others are present, and Professor Stenius gives a lecture on the subject. From this information it must be assumed that it was a public presentation. Less clear, however, is what exactly was being presented. It seems undisputed, however, that Rudbeck displayed the chyle container on this occasion, but we apparently have only his own word, and that much later, that he displayed the lymph vessels too around the liver of a live dog. With the lack of trustworthy, confirmatory testimony from impartial eyewitnesses it cannot be assumed from the available sources that on this occasion the discovery of the lymph system was made public.

Bartholin's publication of 5 May 1652 describes the thoracic duct in the human being, which is a chyle duct and thus not a part of the system that conveys aqueous fluid forward in the body. Thus, this publication does not give Bartholin the status of having discovered the lymph system.

The conclusive date after this seems to be 1 May 1653, when Bartholin publishes his treatise on the lymph vessels in a dog. With this publication, Bartholin precedes Rudbeck, who first publishes his illustrated treatise of the lymph vessels in animals later that summer.

Finally, in May 1654, Bartholin publishes his treatise on the lymph vessels in human beings. There does not seem to be any debate about whether Bartholin gains the honour of that discovery.

Consequently, it must be concluded that, according to modern criteria whereby a discovery is dated by its publication or proclamation, not by its subjective perception, Thomas Bartholin must be regarded as the discoverer of the lymph system. However, it must be added that this conclusion is based on the available information, and that new source studies, including not least documentation from persons who were present at the public dissections etc. when the discovery was allegedly presented, will possibly be able to disrupt the foundation for it.

It must also be added that the principal criterion, the publication/proclamation, is very formal, and as such serves a purpose, if a decision is to be made. However, it seems clear that Bartholin and Rudbeck each gave a considerable and honourable contribution to the field in question, and the act of formally pointing out one of them as the official discoverer must not take away any part of the other's well-earned honour and reputation.

On behalf of UBVA
Jørgen Blomqvist
Associate professor, lic. jur.

Notes

Introduction

pp. 11–26

1. Thomas Bartholin in Copenhagen in a letter to Ole Borch (1626–1690) in Leiden on 19 April 1661. Thomas Bartholin: *Epistolarum Medicinalium* (Medical letters), Copenhagen 1667, Centuria 3, Epistola 90, pp. 388–90.
2. The work is translated into English by Robert Willis with the title An Anatomical Disquisition on the Motion of the Heart and Blood in Animals in *The Works of William Harvey*, London 1847, pp. 1–86. Danish translation by Valdemar Meisen: *William Harvey's Bog om Opdagelsen af Blodets Kredsløb*, Copenhagen 1929.
3. The full title is *Gaspare Aselli, De Lactibus sive lacteis venis quarto vasorum mesaraicorum genere, novo invento Gasparis Asellii Cremonensis anatomici Ticinensis, dissertatio, qua sententiæ anatomicæ multæ vel perperam receptæ convelluntur, vel parum perceptæ illustrantur* (A dissertation on the lacteal vessels or lacteal veins, a fourth type of mesenteric vessels, which are owed to a discovery by Gaspare Aselli of Cremona and anatomist in Tessino. In this treatise many anatomical opinions are analysed, some of them mistaken, while those which are not completely understood, are made clear), Milan 1627.
4. A supernova is a powerful explosion from a dying star. Tycho Brahe's supernova today carries the name SN 1572. It exploded 7500 years before Brahe saw it, since it is located 7500 light years from the Earth.
5. Tycho Brahe: *Tychonis Brahe Dani de nova et nullius ævi memoria prius visa Stella, iam pridem Anno à nato CHRISTO 1572, mense Novembrj primùm Conspecta, contemplatio mathematica* (The Dane Tycho Brahe's mathematical consideration over a new star never seen at any time before, but now for the first time observed in November 1572 AD), Copenhagen 1573.
6. A parallax is a change in an object's apparent position, when it is observed from two different places, i.e. from two different angles. The further away the object is located, the smaller the angle and thus the parallax between the object and the two observation points. That a star has no parallax means that it is located very far away.
7. Fixed star is an older designation for a star located so far away from the Earth that its position in the sky seems unchangeable, in contrast to planets and comets.
8. Wootton 2015, pp. 1, 12–14, 192–99, 456.
9. Brahe, Tycho: *Astronomiæ Instauratæ Mechanica* (The new mechanical astronomy), Wandesburg 1598.
10. Wootton 2015, p. 13.
11. Tycho Brahe: *Tychonis Brahe Dani Epistolarum astronomicarum libri* (Books with the Dane Tycho Brahe's astronomical letters), Uranienborg 1596.
12. H. D. Schepelern (ed.): *Breve fra og til Ole Worm I–III* (Letters from and to Ole Worm 1–3), Copenhagen 1965–1968, 3, Epist. 890, no. 1545, pp. 284–85.
13. Theriac is one of the most famous and esteemed medicaments throughout history. It originally contained 40–70 substances, later reduced to 5–6. It was used as a universal drug against snake bites, poisoning, plague and other grave diseases. Its composition changed from time to time. Typical ingredients were viper meat, castoreum, rhubarb, aloe, myrrh, saffron etc. mixed with honey and wine. Opium was also a usual ingredient. Theriac was initially mentioned in the first century BC and was still available in Danish pharmacies in the beginning of the 20th century.
14. Bartholin 1658, p. 254. Jesper Brandt Andersen: *Niels Stensen. Kongelig anatom og fyrstelig geolog* (Niels Stensen. Royal anatomist and princely geologist), Copenhagen 2021, pp. 18–19.
15. Bartholin 1669, p. 208 and Brandt Andersen 2021, p. 22. When in 1597 Tycho Brahe left Hven and Denmark after controversies with Christian IV and his civil servants, primarily Reichshof-Master Christoffer Valkendorf (1525–1601), he took all of his belongings, including his astronomical instruments, with him. His celestial globe ended up in the Jesuit college Neisse in Silesia (in nowadays Poland), where the son of Christian IV, Ulrik of Denmark (1611–1633), found it in 1632 as colonel in Saxon military service, took it as spoils of war and had it sent home to Denmark. He gave it to the university, and in this way it initially came to Rosenborg Castle, then to the university's upper auditorium (Auditorium superius) in Collegium novum, and finally to the observatory in the Round Tower, where it burned during the great fire of Copenhagen in 1728. Ole Worm mentions Ulrik's finding and conquest in a letter to bishop Thomas Wegner (1588–1654) in Stavanger on 23 April 1633: "I have no news to tell that is worthy of you, except that our famous Prince Ulrik, after the conquest of Neisse in Silesia, among the best spoils he won from the Jesuits reacquired Tycho Brahe's celestial globe of copper and of amazing size. He has sent it here and given it to the university. It is a gift that cannot be valued in money." Schepelern (ed.) 1965–1968, 1, Epist. 382 (no. 467), pp.279–80.
16. Bartholin refers to the German monk and alchemist Berthold Schwarz, who lived in the 14th century.
17. Bartholin 1662, p. 39. Niels W. Bruun (ed.): *The Anatomy House in Copenhagen*, Copenhagen 2015, pp. 94–95.

18. As his assistant in this work, Rasmus Bartholin profited from his later son-in-law, the astronomer Ole Rømer (1644–1710).
19. Thomas Bartholin's Parisian friend, the French physician Guy Patin (1601–1672) mentions the biography in a letter to Thomas Bartholin on 12 September 1655, shortly before Gassendi's death: "Our own Gassendi, who has described the life of that outstanding hero, your compatriot Tycho Brahe, is in a very poor condition here, suffering from pneumonia; he remains in greater danger because of the fact that he is averse to bloodletting, which is necessary in the treatment of such a serious disease." Thomas Bartholin: *Epistolarum Medicinalium* (Medical letters, hereafter abbreviated *Epist. Med.*), Centuria 2, Copenhagen 1663, Epistola 67, pp. 618–20. Loic Capron (ed.): *Correspondance complète et autres écrits de Guy Patin*, https://www.biusante.parisdescartes.fr/patin/.
20. The full title is *Discours de la Méthode pour bien conduire sa raison, et chercher la vérité dans les sciences* (Discourse on the Method of Rightly Conducting One's Reason and of Seeking Truth in the Sciences).
21. Pierre Borel: *Vitæ Renati Cartesii* (The life of René Descartes), Paris 1656, p. 4.
22. August Ziggelaar (ed.): Chaos. Niels Stensen's Chaos-manuscript, Copenhagen 1659. Complete edition with introduction, notes and commentary by August Ziggelaar. *Acta Historica Scientiarum Naturalium et Medicinalium* Vol. 44, Copenhagen 1997, p. 86.
23. Ibid., p. 474.
24. Letter from Niels Stensen to Lorenzo Magalotti, Florence, 18 May 1668. In: Harriet M. Hansen (ed.): *Niels Stensen's korrespondance i dansk oversættelse I–II* (Niels Stensen's correspondence in Danish translation I–II), Copenhagen 1987, 1, E 35c, pp. 123–24.
25. Brandt Andersen 2021, p. 564.
26. Thomas Bartholin: *De Lacteis Thoracicis in homine brutisq; nuperrimé observatis, Historia Anatomica* (An anatomical account of the thoracic ducts recently observed in man and animals), Copenhagen 1652. English translation: *The Anatomical History of Thomas Bartholinus, Doctor and Kings Professor: Concerning the Lacteal Veins of the Thorax, Observ'd by him lately in Man, and Beast. Publickly proposed by Michael Lyserus answering*, London 1653.

CHAPTER 1
Caspar Bartholin the Elder
pp. 17–48

1. Thomas Bartholin in 1652 in *De Lacteis Thoracicis*. Gabriel Tryde (ed.): Thomas Bartholin. Skrifter om Opdagelsen af Lymfekarsystemet i Udvalg. Paa Dansk ved G. Tryde (Thomas Bartholin. Selection of writings about the discovery of the lymphatic system. Danish translation by G. Tryde). *Klassisk dansk Medicin. Vol. III*, Copenhagen 1940, p. 15.
2. C. C. A. Gosch: *Udsigt over Danmarks zoologiske Literatur* (A review of Denmark's zoological literature), Copenhagen 1873, 2, 1, p. 12.
3. The full title of the book is *Anatomicæ Institutiones Corporis Humani, utriusque sexus, Historiam & Declarationem, exhibentes cum plurimis novis Observationibus & Opinionibus, nec non illustriorum quæ in Anthropologia occurrunt controversiarum decisionibus, cum Indice Capitum et rerum locupletissimo* (The anatomical arrangement of the human body, of both sexes, a description and exposition presented with a great many new observations and conjectures, and furthermore the setting out of the more famous disputes in anthropology. Furthermore, a list of chapters and a copious index of subjects), Wittenberg 1611.
4. Older name for Padova. The city is named after the River Po (Padana).
5. Where no other source is mentioned, biographical details of Caspar Bartholin are derived from the following sources: V. Ingerslev: *Danmarks Læger og Lægevæsen fra de ældste Tider indtil Aar 1800* (Denmark's physicians and medical authorities from the earliest times up to the year 1800), Copenhagen 1873, 1, pp. 270–74, Gosch 1873, pp. 10–23, and Valdemar Meisen and Julius Petersen: Caspar Bartholin. In: *Danish Biographical Dictionary*, Copenhagen 1933, 2, pp. 193–98.
6. In connection with the establishment of gymnasia (Latin schools), in 1619 Bartholin was asked to write new textbooks on the aforesaid subjects, and these books were by royal decree made obligatory in schools and gymnasia. See Morten Fink Jensen: *Fornuften under troens lydighed. Naturfilosofi, medicin og teologi i Danmark 1536–1636* (Reason under the obedience of belief. Natural philosophy, medicine and theology in Denmark 1536–1636), Copenhagen 2004, pp. 231–32.
7. The quotation comes from Vesalius' *De Humani corporis fabrica* but is here translated from Harald Moe: *Den anatomiske billedkunst i Renæssancen og Barokken* (The art of anatomical illustration in the Renaissance and the Baroque), Copenhagen 1993, pp. 15–16.
8. Translated from Moe 1993, p. 33.
9. The Latin text on the title page reads: *Pentæstheseion, Hoc est, De Quinque Sensibus Liber, Organorum Fabricam variis iconibus fideliter & ad viuum ære incisis illustratam, Nec non actionem & usum discursu Anatomico & Philosophico accuratè explicata continens.* (Pentæstheseion, this is a book about the structure of the five senses and is illustrated with various detailed copper engravings according to the organs in their natural state. It also describes the function and use of the organs, which are carefully made clear in an anatomical and physiological explanation).
10. Thomas Bartholin: *Anatomia ex Caspari Bartholini parentis institutionibus, omniumque recentiorum et propriis observationibus tertium ad sanguinis circulationem reformata, cum iconibus novis accuratissimis* (Third edition

of the *Anatomy* based on his father Caspar Bartholin's *Anatomy* and all more recent ones together with his own observations and adjusted according to the circulation of the blood. With new and more accurate illustrations), Leiden 1651, the preface.

11. Moe 1993, p. 64. The Latin text reads: *Qui non indicij acriore sensu, Casseri hæc tibi sensit exiisse, Cassus indicio ipsus atque sensu. Qui non artificem sagace dextram, cum mente hic penè sentit, atque palpat, Gustat, olfacit, audit, intuetur; Tactus sit stupidus, sit aure surdus, Nasi ab officio sit orbus impos, Linguæ, à lumine cæculusque utroque.*

12. Ejvind Slottved: *Lærestole og lærere ved Københavns Universitet 1537–1977* (Chairs and teachers at the University of Copenhagen 1537–1977), Copenhagen 1978, p. 150.

13. Ibid., p. 136.

14. Ibid., p. 107.

15. According to Fink-Jensen 2004, p. 243, Bartholin's appointment as professor of medicine took place with support from the theologian Hans Poulsen Resen (1561–1638), Jesper Brochmand (1585–1652) and Cort Aslaksen (1564–1624), and a letter from the physician Christen Lauritsen Bording (1578–1640) to Ole Worm in London on 2 February 1613 indicates that, in a wider circle, people considered Bartholin as an obvious choice: "His [Gellius Sascerides'] successor is still not named, but according to everyone's opinion and conjecture it will be Bartholin, nor can anyone better qualified be found." Schepelern (ed.) 1965, 1, Epist. 2 (no. 13), p. 6.

16. Slottved 1978, p. 222.

17. Ibid., pp. 73–74.

18. Ibid., pp. 76, 107. Fink-Jensen 2004, pp. 241–47, looks sceptically on the evidence from the earlier sources that Bartholin's shift to become a theologian was due to his vow to God, while he was suffering a crisis in his health, that he would dedicate all his studies and works to theology if he survived the illness. Fink-Jensen puts forward the idea that the change in career may have been Bartholin's own wish, since in 1619 he had discussed a possible theological career with Holger Rosenkrantz (1574–1642), who had great influence with the chancellor Christian Friis (1581–1639) in university matters. Fink-Jensen also airs the possibility that the appointment could have rested upon a wish from the university to occupy the chair with a natural philosopher to meet the challenges with which science confronted the church. This possibility is supported by a letter from Ole Worm to the physician Hans Andersen Skovgaard (1600–1627) in Wittenberg on 20 May 1624: "For since Dr. Cort Aslaksen, who was sick from a malicious and incurable gangrene in his right leg, has passed away, my brother-in-law Bartholin has taken over his chair, so that he now has to occupy himself with theology and consider becoming a doctor at the same faculty. The chair in medicine has been given to me." Schepelern (ed.) 1965, 1, Epist. 144 (no. 151), p. 86.

19. Slottved 1978, p. 222. For his whole life Caspar Bartholin had struggled with failing health (rheumatism in his joints, kidney pains and pains caused by kidney stones are mentioned by Ole Worm) and he had several stays at a health resort (1622 and 1624) in Karlsbad in Bohemia to restore himself. Ingerslev, 1873, 1, p. 273, establishes that Bartholin journeyed to Sorø in July 1629 to visit his children, who were placed there in Dr. Joachim Burser's (1583–1639) house to protect them from the plague, which was ravaging Copenhagen. According to Ingerslev, Bartholin died in Burser's house of his "old illness", while F. W. Mansa: *Folkesygdommenes og Sundhedspleiens Historie i Danmark fra de ældste Tider til Begyndelsen af det attende Aarhundrede* (The history of people's illness and hygiene in Denmark from the earliest times to the beginning of the eighteenth century), Copenhagen, 1873, p. 299, states that the cause of death was the plague. A letter from the well-informed Ole Worm in Copenhagen to Søren Olufsen (1582–1636) in Ørsted on 13 July 1629 (the date of Caspar Bartholin's death) indicates that Mansa was right: "The plague rages here, and within fourteen days I have lost my father, my sister and my brother-in-law." Schepelern (ed.), 1965, 1 (no. 280), pp. 173–174. It i s true that Thomas Bartholin also states in *Cista Medica Hafniensis* (The archive chest of the medical faculty in Copenhagen), Copenhagen 1662, p. 301, that his father died "tormented by colic, arthritis and a kidney stone," but one must take into account that it would undoubtedly have been considered a loss of prestige or perhaps even a loss of honour for a physician to have been contaminated with the plague, which could have been avoided with appropriate preventive behaviour. In the same way several sources prove that Ole Worm died of a bladder disease, even if he died on 31 August 1654 while the plague was ravaging Copenhagen. As late as in his last letter a couple of weeks before his death, he describes how people around him are dying like flies of the plague, and that he and his family are in the greatest danger. Furthermore, it should be considered that both in the Middle Ages and the Renaissance a large proportion of the population regarded the plague as God's punishment on mankind, which obviously implied that those who were overtaken by the sickness had made themselves well deserving of this retribution.

20. Fink-Jensen, 2004, p. 236.

21. For example, *Problematum philosophicorum et medicorum nobiliorum et rariorum miscell. exercitationes ad disputandum in illustri Albiaca propositæ* (Various disputation exercises in philosophical problems and more famous and rarer problems in medicine displayed in the famous Albiaca), Wittenberg, 1611; *Astrologia seu De Stellarum Naturâ, affectionibus, & effectionibus* (Astrology or about the stars' nature, impacts and their exercise hereof), Wittenberg 1612; and the section *Pharmacopœa Spagyrica seu Pyrotechnia* (Alchemical or pyrotechnic pharmacopœa)

in *De Studio Medico* (On the Study of Medicine) (see below), Copenhagen 1628. A thesis on unicorns *De unicornu eiusque affinibus et succedaneis* (About the unicorn and its relatives and successors) is included as the first thesis in *Opuscula quatuor singularia* (Four works in one volume), Copenhagen 1628.

22. For example, Ole Worm writes on 23 June 1617 to Jesper Brochmand: "When the said rector [Cort Aslaksen] attempted to impose on us, as legally agreed, the rule that some among the philosophers have adopted, namely that the students are not free to choose their own tutor, Dr. Bartholin declared that it was particularly decided for the purpose of being used against you and him. [...] When at the last exam there were some students who had made clear that they would not accept any other tutor than Dr. Bartholin, the rector had actually on the grounds of his official authority forbidden them this with threats and scolding; they complained about this and much more to Dr. Bartholin, and in the academic council the rector himself was compelled to admit the truth of their grievances. However, he maintained the opinion that he had acted properly within his authority in this matter." Scheperlen, 1965, 1, Epist. 34 (no. 29), pp. 21–22.

23. For a detailed explanation of this, see Fink-Jensen 2004, pp. 231–368.

24. Ibid. p. 356.

25. Morten Fink-Jensen (ed.): *The Foundation and Regulations of the University of Copenhagen 1539*, Copenhagen 2020.

26. Ibid. pp. 327–28, which refers to Caspar Bartholin's book about the nephrite stone: *De lapide nephritico opusculum. Cum præcipuis ad plerosque morbos amuletis* (A book about the nephrite stone. With excellent amulets against several diseases), Copenhagen, 1628. The nephrite stone was a precious stone, which people thought could cure a stone in the kidneys when worn on the loin.

27. Ibid. p. 292 and pp. 353–54.

28. Axel Garboe: Thomas Bartholin. Et Bidrag til dansk Natur- og Lægevidenskabs Historie i det 17. Aarhundrede (A contribution to the history of Danish natural and medical science in the 17th Century). *Acta Historica Scientiarium Naturalium et Medicinalium*, Vol. V, Copenhagen 1949, 1, p. 16.

29. Translated from Sten Ebbesen and Carl Henrik Koch: Dansk filosofi i renæssancen (Danish philosophy in the Renaissance). In: Sten Ebbesen & Carl Henrik Koch (eds.): *Den danske filosofis historie* (The history of Danish philosophy), 1, Copenhagen 2003, p. 257. For Peder Sørensen and *Idea Medicinæ Philosophiæ* see further Hans Skov & E. Bastholm: Petrus Severinus og hans Idea medicinæ philosophicæ. En dansk paracelsist (Petrus Severinus and his Ideal of Philosophical Medicine. A Danish Paracelsian). *Acta Historica Scientiarum Naturalium et Medicinalium*, Vol. 32, Copenhagen/Odense 1979.

30. Caspar Bartholin: *De Studio Medico inchoando, continuando & absolvendo. Pro accurato & supra vulgus future Medico, Consilium breve atque extemporaneum. Ad optime ingenuatum & paternarum virtutum* æmulum Juvenem Petrum Charisium Jonæ filium, Med. *& Philos. Stud. Amicum & affinem suum: Nec non ad filiolos suos Bartolum, Thomam, Casparum, Albertum, Jacobum & Erasmum* (On starting, continuing and completing the study of medicine. On behalf of an exact and future physician beyond the ordinary, brief and extemporaneous advice. To the excellent young Peder Charisius, follower of the noble and ancestral virtues, son of Jonas, student of medicine and philosophy, and his friend and relative. Furthermore, to his dear sons Bertel, Thomas, Caspar, Albert, Jacob and Rasmus), Copenhagen 1628. This study guide was also included as the fourth of the dissertations in Bartholin's *Opuscula quatuor singularia*, in the same place and year. (The first, second and third dissertations in this work deal respectively with the unicorn, the nephrite stone/amulets and pygmies.) The enduring nature of this study guide is underlined by the fact that Hermann Conring (see later) included it in his *Introductio in artem medicam* (Introduction to the art of medicine), Helmstedt 1687, which was reprinted in Halle/Leipzig in 1726. Bartholin also published in 1628 in Copenhagen a similar study guide, *De Studio Theologico* (On the study of theology) for theological students. Both study guides are treated in detail in Fink-Jensen 2004, pp. 231–359.

31. A detailed survey and analysis of these may be found in Fink-Jensen 2004, pp. 285–359.

32. Disputation (Latin, disputare – to debate): A formal academic debate at a university in which one or more claims (theses) are presented and defended by one party (the presenter) and attacked by the other (the opponent).

33. He gives examples of which subject areas of medicine one may profitably study at the different universities in European cities. Spagyric alchemy can be studied in Rostock, and botany in Giessen, Padua, Pisa, Bologna and Montpellier, or with Dr. Joachim Burser at the new academy in Sorø. Admission to public dissections of human beings (which at this time were not performed in Copenhagen) can be obtained in Padua, Bologna, Rome, Montpellier, Paris and Leiden, and practical training in medicine and surgery for medical students is held more often in Padua and Florence than in any other part of Europe.

CHAPTER 2
Study Travel in Europe
pp. 49–186

1. Thomas Bartholin in his cousin Thomas Fuiren's guest book in Leiden in 1637. Thott'ske Samling 561 8°, The Royal Library in Copenhagen.

2. In a letter from Leiden to Ole Worm in Copenhagen on 6 March 1638, Thomas Bartholin expresses huge gratitude for all the support Ole Worm has given to him since his childhood. For example, he writes: "But what shall I say about your services towards me, highly regarded Worm? Where shall I begin, and where shall I end? Even if I was given a hundred tongues in a hundred mouths, I would not be able to describe all the charities by which you have bound me to you in high esteem since my childhood. When my blessed father, through the envy of death, had ceased to exist, I got you – and that I reckon as the greatest luck – in my father's place as my guardian; and you have fulfilled all of a father's and a guardian's duties to such an extent, that I am in doubt as to whether I have been dearest to my natural father or the object of the greatest care from you. To such an extent have I never felt any limits to your favourable feelings towards me, that you (next to God) have been the source of all of my joys." Schepelern (ed.) 1967, 2, Epist. 644 (no. 702), p. 58.

3. Garboe 1949, 1, p. 20.

4. After graduation, a written testimonial from the university about a student's skill, qualifications and behavior.

5. Ejnar Hovesen: *Lægen Ole Worm* (The physician Ole Worm), Aarhus 1987, pp. 203–4. Ole Peter Grell: *The World of Worm: Physician, Professor, Antiquarian, and Collector, 1588–1654*, Abingdon and New York 2022, p. 101.

6. Probably the printer Jacob Marcus (1605–1654) in Leiden.

7. Marcus Tullius (106–43 BC), Roman statesman, lawyer, scholar, philosopher and author.

8. Letter from Thomas Bartholin in Leiden to Ole Worm in Copenhagen on 10 April 1640. Schepelern (ed.) 1967, 2, Epist. 659 (no. 837), pp. 176–77.

9. In the above-mentioned letter of 6 March 1638, Thomas Bartholin writes to Ole Worm: "That your agency has come about, that I have gained a share of my grandfather's generosity, I neither will nor should conceal," and in a later letter of 23 August 1638 to Worm: "Through your mediation and authority you will contrive that I am supported in the expense of the Fincke scholarship, until I can have another assigned to me." Ibid., Epist. 646 (no. 739), p. 94. To the latter, Worm answers on 9 October 1638: "With your grandfather I will speak more seriously about an extension of your scholarship, until it is possible for you to receive the royal scholarship." Ibid., Epist. 647 (no. 748), p. 103. In addition, Worm writes on 7 January 1639: "Your grandfather is so much more keenly determined that you receive an extension of his scholarship, as he has noted with satisfaction, that you show such great diligence and progress in your medical studies. The royal medical scholarship has been available for four years here. I shall try to provide you with the funds that have been made available, and I have the best hope that the outcome will correspond to my wishes. Thus, you will be able to enjoy 200 rix-dollars per year and maybe with supplements." Ibid., Epist. 649 (no. 762), p. 115.

10. Hans-Otto Loldrup: Anatomiske teatre i Europa (Anatomical theatres in Europe). In: Niels W. Bruun & Hans-Otto Loldrup (eds.): Thomas Bartholin. *Anatomihuset i København* (The Anatomy House in Copenhagen), edited and translated with commentary by Niels W. Bruun, Copenhagen 2007, p. 226.

11. Ibid., pp. 226–27.

12. Schepelern (ed.) 1967, 2, Epist. 644 (no. 702), p. 59. Since all the letters to and from Ole Worm referred to in this book are to be found in chronological order, arranged according to date and year in Schepelern's three-volume work, only the date and year of each single letter will from now on be given in the body text. In Schepelern's work, the first volume from 1965 covers the years 1607–1636, the second volume from 1967 the years 1637–1643, and the third volume from 1968 the years 1644–1654. For each single letter, Schepelern refers to the corresponding letter in Thomas Bartholin's medical letters, *Epistolarum Medicinalium*.

13. In a letter of 19 June 1638, Ole Worm gives Bartholin advice as to how to solve his health problems: "You are right to make careful considerations concerning your health, which is already fragile. I would not want you to exhaust yourself with too many drugs, which break down the innate heat and weaken your strength. You will help your stool problems if, on a fasting heart in the early morning hours, you eat a handful of black beans together with a drink of soup, or, if this does not help, then provide yourself with some pills that you know, and which work by supporting Nature and generating an opening, for example pills of Ammonium Quercetan, of which you, if necessary, should take three or four of a drachmas weight twice a week before noon. You will easily see, what will come out useful. But be aware not to inhibit Nature´s own urge to defecation. [...] It would be beneficial if, every morning, before you sit down to do your studies, as well as in the evening, when you are going to bed, you try to provoke defecation. Even if you do not succeed the first or second time, Nature will, when it is frequently encouraged, finally do its duty. Be carefully aware of its movements and do not cease to be ready, when the right moment comes."

14. Otto van Heurne (1577–1652), Dutch physician, professor of medicine in Leiden.

15. Adrian van Valckenburg (1581–1650), Dutch physician, professor of medicine in Leiden.

16. On 23 August 1638, Bartholin writes from Leiden to Worm: "My beloved mother has told me in a letter that my intention concerning medical studies sounds strange to her ears, and she is eager to make me change my mind by presenting different reasons."

17. On 31 March 1640, Bartholin writes to Worm: "The famous Golius sends his greetings. He guides us faithfully to the sources of the Arabs."

18. Letter from Thomas Bartholin to Ole Worm on 31 August 1639. Late in his life, Bartholin writes about this striving to obtain a broad education, and how he found time for it:

"Avid to learn, I travelled for ten years enthusiastically undertaking not only very long trips to farthest Melita—not Thule —but also because of my wide interest I spent many hours designed for necessary studies to the Hebrew, Aramaic and Coptic languages, to lectures on law, to the delights of the poets, to studies of the critics, to Platonic subtleties and to the curiosities of antiquity. I have forgotten a large part of all those things, but I do not regret that I undertook such a burden of no lucrative value. Whatever it may be, in addition to medical science, in which we seek pleasure in order to live happily, let us pursue it with diligence and care; and since a pure physician is commonly called a pure ass, one can devote those occasional hours, which others spend in games, unseemly love-making, drinking parties, gambling and adornment, to the pleasant studies of philology and antiquity, by which the physician can render himself agreeable and cultivated." Thomas Bartholin: *De Peregrinatione Medica* (On Medical Travel), Copenhagen 1674. Cited from the English translation by Charles D. O'Malley (ed.): *Thomas Bartholin. On the Burning of his Library and On Medical Travel*, Kansas 1961, pp. 56–57.

19. Adrianus Spigelius or Adriaan van den Spiegel (1578–1625), Dutch anatomist, botanist and professor of surgery and anatomy in Padua.

20. Jean Fernel or Joannis Fernelius (1497–1558), French court physician. The father of physiology.

21. Daniel Sennert or Daniele Sennerto (1572–1637), German physician, alchemist, chemist and professor of medicine in Wittenberg.

22. Cæsar Cremonius or Cecare Cremonini (1550–1631), Italian physician, philosopher and professor in Padua.

23. Johann Sperling (1603–1658), German physician, zoologist and professor of medicine in Wittenberg.

24. Antonio Ponce de Santa Cruz (1561–1632), Spanish court physician and professor at the University of Valladolid.

25. Schepelern (ed.) 1967, 2, Epist. 646 (no. 739), p. 94.

26. Adolphus Vorstius or Adolf van Voorst (1597–1663), Dutch physician and botanist.

27. Turtles.

28. See further H. D. Schepelern: *Museum Wormiamum: dets Forudsætninger og Tilblivelse* (*Museum Wormianum: its prerequisites and genesis*), Aarhus 1971. The full title of Worm's catalogue is *Museum Wormianum seu Historia Rerum Rariorum, Tam Naturalium, quam Artificialium, tam Domesticarum, quam Exoticarum, quæ Hafniæ Danorum in ædibus Autor servantur. Adornata ab Olau Worm, Med. Doct., & in Regia Hafniensi Academia, olim Professore publico. Variis et accuratis Iconibus illustrata* (Worm's Museum or the report about rare things, natural as well as artificial, domestic as well as exotic and Danish from Copenhagen in this author's custody. Adorned by Ole Worm, physician and former public professor at Copenhagen University. Illustrated with various accurate images).

29. Garboe 1949, 1, p.27.

30. Willem Piso (1611–1678), Dutch physician and scientist. In 1637–1644 he took part in a Dutch expedition to Brazil as the expedition's physician. In 1648 he published in Leiden and Amsterdam the work *Historia Naturalis Brasiliae* (An account of Brazil's natural history) on Brazil's fauna and flora. Apart from that he was one of the founding physicians in the subject of tropical medicine. Maurice of Orange (1567–1625) was chief magistrate of the provinces of the Dutch Republic.

31. Thomas Bartholin: *De Anatome Practica ex cadaveribus morbosis adornanda, consilium, cum operum autoris hactenus editorum catalogo* (On practical anatomy, advice and counsel, to be provided from diseased corpses, with a catalogue of the author's works published so far), Copenhagen, 1674. Danish translation by Hartvig Frisch: Den praktiske anatomi: Et råd og en vejledning. *Bibliotek for Læger* 1918, p. 95.

32. For a short summary of Galen's teachings, see Roy Porter: *The Greatest Benefit to Mankind – A Medical History of Humanity from Antiquity to the Present*, London 1997. Danish translation by Bent Bjerre: *Ve og Vel. Medicinens Historie fra Oldtid til Nutid*, Copenhagen 2000.

33. Malpighi, Marcello: *De Pulmonibus Observationes Anatomicæ* (Anatomical observations of the lungs), Bologna 1661.

34. Garboe 1949, p. 20. In 1670 Thomas Bartholin wrote a biography of Poul Moth with the title *Medicus perfectus ex vita beati senis D. D. Pauli Moth archiatri regii informatus* (The perfect physician illustrated from the life of the blessed old man Poul Moth, chief royal consultant), where Bartholin states that it was Moth who first gave him the incentive to study anatomy.

35. Vopiscus Fortunatus Plempius or Foppe Plemp (1601–1671), Dutch pharmacist and professor of medicine at the University of Leuven.

36. Vopiscus Fortunatus Plempius: *Fundamenta seu Institutiones Medicinæ* (Basic principles or instructions in medicine), Leuven 1638.

37. Twelve years later in his third revised edition of 1651 of his father's anatomy book (see later), Bartholin explains the phenomenon by stating that the liver draws extra chyle to itself from the intestines through the lacteal veins when an animal or a human experiences fear and pain. But when the strangling caused by hanging puts an end to the fear and pain, the liver suddenly stops drawing chyle to itself, with the result that chyle accumulates in the lacteal veins: "And this can thus be the reason why the lacteal veins of a man who was hanged in Amsterdam and dissected by Dr. Tulp remained visible for many days afterwards." Thomas Bartholin: *Anatomia tertium ad Sangvinis Circulationem Reformata, cum novis iconibus* (Third edition of the textbook of anatomy, revised according to the circulation of the blood, with new illustrations), Leiden 1651, p. 425.

38. Johann Friedrich Meckel: Über die Devertikel am Darmkanal. *Archiv für die Physiologie*, Halle 1809, 9, pp. 421–53.

39. J. Sagar, V. Kumar, D. K. Shah: Meckel's diverticulum: a systematic review. *Journal of the Royal Society of Medicine* 2006, 99, pp. 501–5.
40. www.whonamedit.com. Meckel's diverticulum.
41. Bartholin's letter to Ole Worm of 27 March 1640 was not included in Bartholin's medical letters, *Epistolarum Medicinalium*, published in Copenhagen 1663-1667, but was first published with Ole Worm's letters, *Olai Wormii et ad eum Doctorum Virorum Epistolæ* (Letters of Ole Worm and to him from learned men), Copenhagen 1751.
42. A book printer in Leiden.
43. Epistola Johannis Walæi De Motu Sanguinis: Ad Thomam Bartholinum, Casp. Filium (Letter on the motion of the blood from Johannes Walæus to Thomas Bartholin, son of Caspar). Dated Leiden, 10 October 1640. Bartholin 1641, pp. 385–408. Altera Epistola Johannis Walæi, De Motu Sanguinis, Ad Eundem (Second letter on the motion of the blood from Johannes Walæus to the same). Dated Leiden, 1 December 1640. Ibid., inserted between pp. 408 and 409. Walæus' two letters are included in all four of Thomas Bartholin's revised editions of his father's textbook of anatomy, all the way to the last edition *Anatome Renovata* from 1673, though in revised versions with the necessary updates in accordance with new knowledge, especially concerning the circulation of blood and chyle. In *Anatome Renovata* the title of the first letter has been changed to Epistola Prima De Motu Chyli Et Sanguinis: Ad Thomam Bartholinum, Casp. Filium (First letter on the motion of the chyle and blood to Thomas Bartholin, son of Caspar), and three illustrations have been added. Bartholin 1673, pp. 761–93. The second letter has its title unchanged and is found on pp. 794–804.
44. Ludvig Choulant: *History and bibliography of anatomic illustration in its relation to anatomic science and the graphic arts.* Translated and edited with notes and a biography by Mortimer Frank, Chicago 1920, p. 246.
45. Massimo Collice & Roasa Collice, Alessandro Riva: Who discovered the sylvian fissure. *Neurosurgery*, 63, 4, 2008, pp. 623–28.
46. Bartholin 1641, p. 624.
47. Bartholin 1651, p. 319.
48. Bartholin 1641, pp. 205–6.
49. James Primrose (1600–1659), English physician. Persistent defender of the ancient Greek physicians' teaching and a profiled opponent of Harvey. In 1630 he published a refutation of Harvey's teaching about the circulation of the blood, which is what Ole Worm is alluding to. James Primrose: *Exercitationes et Animadversiones in Librum de Motu Cordis et Circulatione Sanguinis adversus Gulielmum Harveum* (Exercises in and commentaries on the book about the motion of the heart and the circulation of the blood directed against William Harvey, London 1630. Primrose called Harvey "a producer of trivialities," and his discovery "ridiculous" and "without the least benefit to medicine." See George Frederic Still: *The History of Paediatrics. The Progress of the Study of Diseases of Children up to the End of the XVIIIth Century*, London 1931, p. 250.
50. Annie Bithol-Hespériès: Cartesian Philosophy. In: Stephen Gaukroger, John Shuster and John Sutton (eds.): *Descartes' natural Philosophy*, London and New York 2000, p. 355.
51. Ibid., p. 362.
52. Ibid., p. 353.
53. Ibid.
54. Descartes had to postpone and adapt his publication of *Le Monde* in view of the Inquisition's proceedings against Galileo Galilei after his publication of the work *Dialogue Concerning the Two World Systems* in 1632, which therefore resulted in *Le Monde* being published for the first time fourteen years after Descartes' death. Thomas Bartholin's brother, Rasmus Bartholin, who personally knew Descartes, mentions the work to Ole Worm on 2 April 1650, scarcely two months after Descartes' death: "I certainly know that in his mind he had prepared a dissertation on the human being, for he has declared by word of mouth that he had an extremely clear and distinct perception of the whole nature and structure of a human being and could present it just as clearly as Euclid's proofs, but I fear that it has gone to the grave with him, since he has scarcely left anything tangible in written presentation. He was not accustomed to commit anything to paper before he had conceived the whole thing at once in his mind and thought it through thoroughly." In 1647, Descartes published another work about the human body, *La description du corps humain, et de toutes ses fonctions* (Description of the human body and all its functions).
55. Annie Bithol-Hespériès 2000, pp. 357–58.
56. René Descartes: *Om Metoden. Discours de la Méthode.* Danish translation by Viggo Brøndal and Valdemar Hansen, Copenhagen 1937, p. 58.
57. Meisen 1929, p. 5.
58. Ibid., p. 93.
59. Ibid., p. 95.
60. Ibid., p. 37.
61. Ibid., p. 78.
62. Brøndal and Hansen 1937, p. 50.
63. Ibid, p. 53.
64. Ibid., p. 58.
65. *Epist. Med.* 1663, 1, Epist. 46: De editione Anatomia, ductus Pancreaticus etc. (On the edition of the anatomy book, the pancreatic duct etc.), pp. 195–98.
66. Bartholin 1651, pp. 253–54.
67. Ibid., pp. 232–73.
68. Harriet M. Hansen (ed.) 1987, 1, E 13, pp. 67–71. *Epist. Med.* 1667, Cent. 4, Epist. 70: *Nova Musculorum & Cordis fabrica* (A new structure of the muscles and the heart), pp. 414–21. Vilhelm Maar (ed.): *Nicolai Stenonis Opera Philosophica, Vol. I-II*, Copenhagen 1910, 1, OPH No. 14, pp. 155–60. Gustav Scherz (ed.): *Nicolai Stenonis Epistolae et Epistolae ad eum datae* (Letters to and from Niels Stensen), Copenhagen and Freiburg 1952, *Tomus I–II*, E 13, pp. 175–79. English translation in Troels Kardel & Paul

Maquet (eds.): *Nicolaus Steno. Biography and Original Papers of a 17th Century Scientist*. Second Edition, Berlin 2018, pp. 543–47.
69. Niels Stensen: *De Musculis & Glandulis Observationum Specimen. Cum Epistolis duabus Anatomicis* (A specimen of observations of muscles and glands. With two anatomical letters), Copenhagen 1664. Maar (ed.) 1910, 1, OPH No. 15-17, pp. 161–218. English translation in Kardel & Macquet (eds.) 2018, pp. 549–77. The two letters to Willem Piso and Paul Barbette concerning the dissection of a ray and the passage of the yolk into the intestines of the chick, respectively, can further be found in Danish translation in Harriet M. Hansen (ed.) 1987, 1, E 15, pp. 73–86, and E 16, pp. 87–95.
70. Sten Ebbesen & Carl Henrik Koch: Dansk filosofi i renæssancen 1537–1700 (Danish philosophy in the Renaissance 1537–1700). In: Sten Ebbesen & Carl Henrik Koch (eds.): *Den danske filosofis historie* (The history of Danish philosophy), 2, Copenhagen 2003, p. 317.
71. Niels Stensen: *Discours de Monsieur Stenon sur l'Anatomie du Cerveau à Messieurs de l'Assemblée, qui se fait chez Monieur Thévenot* (Mr. Stensen's discourse on the anatomy of the brain held for an assembly of scientifically interested men at Mr. Thévenot's place), Paris 1669. Maar (ed.) 1910, 2, OPH No. 18, pp. 1–35. English translation in Kardel & Maquet (eds.) 2018, pp. 603–29.
72. The pineal gland or epiphysis is a small cone-shaped endocrine gland weighing approximately 0.15 grams, which is located on the anterior part of the upper side of the brainstem. It produces, among other things, the sleep hormone melatonin, which is released into the blood and has an impact on the circadian rhythm.
73. Ever since the ancient physicians there had been a general opinion that in living creatures there existed a life spirit, *spiritus animalis*, which was a kind of volatile substance uniting the soul and the body.
74. Bartholin 1651, p. 336.
75. *Epist. Med.* 1667, Cent. 3, Epist. 85, pp. 360–69. Brandt Andersen 2021, p. 192.
76. *Epist. Med.* 1667, Cent. 3, Epist. 86, pp. 369–74. Brandt Andersen 2021, p. 193. Similarly, Bartholin comments many years later in his *De Libris Legendis* from 1676 (see later) on Archbishop Anselm of Canterbury (1033–1109), who 76 years old on his deathbed reportedly regretted that he could not live a little longer so that he could resolve the difficult question of the origin of the soul. Bartholin states that he strongly doubts, that the archbishop would have succeeded to solve the question even if he had been allowed to live for another 76 years (Bartholin 1676, pp. 163–64).
77. Florentius Schuyl (1619–1669), Dutch physician and botanist.
78. *Epist. Med.* 1667, Cent. 4, Epist. 26: Observationes Anatomicæ in avibus et cuniculus (Anatomical observations in birds and rabbits), pp. 103–13. Maar (ed.) 1910, 1, OPH No. 9, pp. 113–20. Scherz (ed.) 1952, 1, E 9, pp. 158–64. Harriet M. Hansen (ed.) 1987, 1, E 9, pp. 51–56. Kardel & Maquet (eds.) 2018, pp. 511–16.
79. Niels Stensen *Discours de Monsieur Stenon sur l'Anatomie du Cerveau a Messieurs de l'Assemblée, qui se fait chez Monsieur Thévenot* 1669, p. 2.
80. Bartholin 1651, p. 453.
81. Thomas Bartholin: *Oratio in obitum clarissimi viri D. Henrici Fuiren medicinæ apud Hafniensis doctoris* (Memorial speech for that very famous man, Henrik Fuiren, Doctor of Medicine in Copenhagen), Copenhagen 1659, p. 12. Cited from Garboe 1949, 1, p. 203.
82. Thomas Bartholin in *De Lacteis Thoracicis*, Copenhagen 1652, p. 10.
83. Stensen 1669, pp. 48–49.
84. Vilhelm Maar: *Lidt om Descartes og Danmark* (A little about Descartes and Denmark), Copenhagen 1931, pp. 12–17. In a letter from Rasmus Bartholin in Leiden to Ole Worm in Copenhagen on 10 April 1648, he mentions Descartes many times and among other things writes: "René Descartes is thinking of travelling back to his homeland now in the coming month of May in order finally to comply with his king's requests. He has great schemes concerning the creation of mankind, but since there is still a question where he cannot find a satisfactory standpoint, he doesn't want this work to see the light of the day, something he has told me himself in person." Schepelern (ed.) 1968, 3, Epist. 916 (no. 1572), pp. 307–9.
85. Letter from Rasmus Bartholin in Leiden to Ole Worm in Copenhagen of 2 November 1649. Schepelern (ed.) 1968, 3, Epist. 920 (no. 1663), pp. 406–8.
86. Descartes 1649, p. 318. Brandt Andersen 2021, p. 125.
87. Maar refers to Charles Adam & Paul Tannery: *Oevres de Descartes*, Paris 1897–1910, 3, p. 254, note a.
88. Maar 1931, p. 10. Thomas Bartholin and René Descartes both stayed in Leiden from April to June 1640. During the remaining part of Bartholin's first stay in Holland in 1637–1640, Descartes lived at the coast outside Harlem, west of Amsterdam, and during Bartholin's second stay in Holland in 1646, Descartes lived in the area around Egmond, north-west of Amsterdam. See Chronological Table in Stephen Gaukroger: *Descartes. An Intellectual Biography*, New York 1995.
89. Athanasius Kircher: *Prodromus Coptus sive Ægyptiacus* (Coptic or Egyptian preface), Rome 1634.
90. Georgius Agricola: *De animantibus subterraneis* (On subterranean creatures), Bale 1561.
91. The reissue was apparently not definitively abandoned before the fire of Bartholin's estate Hagestedgaard and its library in 1670. In his book about the fire, he mentions among the lost works the planned book about pygmies, *De Pygmæis Analecta* (Analects about pygmies) – his father's book enlarged with his own little bits of read and observed material.

92. J. Lévy-Valenci: *La médicine et les médecins francais au XVIIe siècle*, Paris 1933, p. 220.
93. Ibid., pp. 488–92.
94. Moreover, Guy de la Brosse published in 1640 second edition of his botanical work *De la Nature, vertu, et utilité des Plantes. Divisé en cinq Livres*. Bartholin awaits this publication in his letter to Ole Worm of 8 September 1640.
95. Christen Thomesen Sehested (1590–1657), nobleman and, from 1640, the king's and the university's chancellor.
96. The book has three letters from Saumaise to Bartholin and two from Bartholin to Saumaise. Moreover, it has two letters to and from Adolphus Vorstius, respectively, and two letters to and from the Dutch physician and writer Johan van Beverwijck (1594–1647), respectively. These letters are all from 1646. Finally, there are two letters from 1645 to and from the Italian theologian Hieronymus Bardi, respectively.
97. Johann Schioldann-Nielsen & Kurt Sørensen (eds.): Thomas Bartholin: On diseases in the Bible. A Medical Miscellany 1672. *Acta Historica Scientiarum Naturalium et Medicinalium*, Vol. 41, Copenhagen 1994, Introduction, pp. 9–18. Translated from the Latin edition *De Morbis Biblicis Micellanea Medica*, Copenhagen 1671, second corrected edition, Frankfurt 1672.
98. The chosen prince, Christian of Denmark (1603-1647), son of Christian IV and Anna Cathrine of Brandenburg. He was chosen to succeed his father as King of Denmark, but died the year before his father and therefore never became king.
99. Henny Glarbo: Danske og norske studerende i Padua (Danish and Norwegian students in Padua). *Personalhistorisk Tidsskrift*, 1928, Vol. 49, series 9, 1, p. 111.
100. O'Malley (ed.) 1961, p. 92.
101. Helge Kragh: Da fosfor kom til Danmark (When phosphorus came to Denmark), *Dansk kemi*, 2004, Vol. 84, 3, pp. 34–35 and same author: Phosphors and phosphorus in Early Danish Natural Philosophy, *Historisk-filosofiske Meddelelser*, 2003, Vol. 88.
102. E. Newton Harvey: *A History of Luminescence. From the Earliest Times Until 1900*, Philadelphia 1957, p. 307. The full title of la Gala's work is *De Phænomenis in Orbe Lunæ novi Telescopii usu a Galileo Galilei nunc iterum suscitatis Physica disputatio* (Physical treatise on phenomena within the moon's circumference now again brought about using Galileo Galilei's new telescope).
103. E. Newton Harvey 1957, p. 311.
104. Garboe 1949, 1, pp. 41–42.
105. E. Newton Harvey 1957, p. 115.
106. Ibid., pp. 107–15 goes through the work, which is also discussed in Gosch 1873, pp. 54–55 and in Helge Kragh (ed.): Fra Middelalderlædom til Den Nye Videnskab (From the learning of the Middle Ages to the new Science), 1000–1730. *Dansk Naturvidenskabs historie* (The History of Danish Natural Science), 1, Aarhus 2005, pp. 340–41.
107. E. Newton Harvey 1957, p. 116.
108. Ibid., p. 18. Bartholin writes: "The light that it has absorbed from the lit-up surrounding air it soon emits in a beautiful way in the dark, glittering as coal set on fire, and it preserves this ability to create light for many years. With a new calcification it is proofed again."
109. Ibid., p. 15. In his much later work *De Morbis Biblicis Miscellanea Medica* (On the diseases in the Bible. A medical miscellany), Copenhagen 1671, second edition, Frankfurt 1672, Bartholin ascribes the light on Moses' face as having merely symbolic meaning. See Schioldann-Nielsen and Sørensen (eds.) 1993, pp. 42–48.
110. E. Newton Harvey 1957, pp. 110–11 and Bartholin 1647, p. 57.
111. E. Newton Harvey 1957, pp. 278–79.
112. Ibid., p. 540.
113. Ibid.
114. Ibid., p. 541.
115. Ibid., p. 108.
116. In addition, see Thomas Bartholin: *Historiarum Anatomicarum Rariorum* (Rare medical reports, hereafter abbreviated *Hist. Anat.*), *Centuria I & II*, Copenhagen 1654, 2, Hist. XII. Carnes in macello lucentes (Self-illuminating meat in the meat market), pp. 191–93 and Thomas Bartholin: *De flammulâ cordis epistola, cum Jacobi Holsti eiusdem argumenti dissertatione accessit de carnibus lucentibus Danielis Puerarii responsio* (Letter about the flame of the heart, together with Jacob Holst's treatise about the same topic. Moreover, Daniel Puerarius' answer concerning the self-illuminating meat), Copenhagen 1667 (for a report of this work, see E. Newton Harvey 1957, pp. 465–68) and Garboe 1949, 1, p. 43.
117. Kragh 2004, p. 35.
118. Egill Snorrason: *Dansker en Johan Rhode i det 17' århundredes Padua* (The Dane Johan Rhode in 17th century Padua), Copenhagen 1965, p. 22. In 1630 there were more than 1000 foreign students in Padua.
119. According to Ingerslev 1873, 1, p. 483 Ostenfeld stayed in Padua from 1651–1655 and became Doctor of Medicine there on 19 May 1655, while Glarbo 1928, p. 106 writes that Ostenfeld was matriculated at the University of Padua on 9 August 1653.
120. The goddess of justice.
121. Snorrason 1965, pp. 23–24.
122. Università degli Studi di Padova: *A visit to Palazzo Bo*, Padua 2019.
123. Garboe 1949, 1, pp. 50–51.
124. Frisch (ed.) 1918, p. 94.
125. Jerry Stannard: Prospero Alpini. In: Charles Coulston Gillispie (ed.): *Dictionary of Scientific Biography*, New York 1981, 1, pp. 124–25. The only survivor from the Danish expedition was the German mathematician and cartographer Carsten Niebuhr (1733–1815) who gathered Forsskål's botanical notes and published them in Copenhagen 1775 in the work *Flora Ægyptiaco-Arabica, sive descriptiones*

plantarum, quas per Ægyptum inferiorem et Arabiam felicem detexit (Flora of Egypt and Arabia, or descriptions of plants discovered in the lower Egypt and the happy Arabia).

126. Thomas Bartholin: *De Cometa* (On the comet), Copenhagen 1665, p. 21.

127. Snorrason 1965, p. 26.

128. Personal information from the guide in Palazzo Bo in Padua, July 2019.

129. Università degli Studi di Padova: *A visit to Palazzo Bo*, Padua 2019. For the history of the University of Padua, see also Piero del Negro (ed.): *The University of Padova. Eight Centuries of History*, Padua 2003. Padua was placed under the administration of the Venetian Republic until the fall of the latter in 1797.

130. Snorrason 1965, p. 14.

131. Christian Bruun: Johan Rhode: Født i Kjøbenhavn i 1587. Død i Padua i 1659. In: *Paa Hundrede-aarsdagen efter at Det store Kongelige Bibliotek blev erklæret for at være et offentligt Bibliotek ved Kongelig Reskript af 15. November 1793* (Johan Rhode. Born in Copenhagen 1587. Dead in Padua in 1659. In: On the hundred-year anniversary after the great Royal Library was declared to be a public library by royal rescript of 15 November 1793), Copenhagen 1893, p. 86.

132. Rhode had in 1651 in Padua published a new edition of Johannes Vesling's *Syntagma anatomicum* (Anatomical treatise), Padua 1647, and among Rhodes' legacies was a small work by Vesling, which Bartholin published as an appendix to his book *De Insolitis Partus Humani Viis Dissertatio Nova, accedunt Johannis Veslingii de pullitie Ægyptiorum et aliæ ejusdem Observationes anatomicæ & Epistolæ medicæ posthumæ* (New treatise about unusual birth paths in human beings. Moreover, Johannes Vesling: About hatching of eggs by the Egyptians. By the same: Anatomical observations and posthumous letters), Copenhagen 1664. See Bruun 1893, p. 44. Rhode published in 1639 in Padua a book *De Acia dissertatio ad Cornelii Celsi mentem* (Treatise about the thread according to Cornelius Celsus) about the meaning of the word "Acia" in the Roman encyclopaedist Aulus Cornelius Celsus' (c. 25 BC–c. 50 AD) work *De Medicina* from c. 26 AD. On his death, Rhode left behind a copy of the book with attached notes with the purpose of publishing a new edition. This was fulfilled by Bartholin in 1672 with the book *De acia dissertatio ad Cornelii Celsi mentem, accredit De ponderibus et mensuris ejusdem autoris dissertatio, et vita Celci, qua simul universa fibulæ ratio explicatur, secundis curis ex autographo autoris auctior et emendatior, cum judiciis doctorum èdita a Th. Bartholino* (Treatise about the thread according to Cornelis Celsus, furthermore the same author's treatise about measure and weight and a biography of Celsus, where at the same time the whole of the fibulas' idea is investigated. Increased and improved second edition building upon the author's manuscript together with the opinions of the learned world, edited by Thomas Bartholin). See Bruun 1893, pp. 54–55. In 1657, Rhode published a book about different diseases based on concrete case reports: Johannes Rhodii: *Observationum Medicinalium Centuria Tres* (Medical observations in three centuries), Padua 1657. Several of the case reports from this book Bartholin later republished in his *Hist. Anat.* and in his journal of medicine and natural history, *Acta Medica & Philosophica Hafniensia* (hereafter abbreviated *Acta Medica*). See Bruun 1893, pp. 62–63. To the third and last volume of *Historiarum Anatomicarum Rariorum* from 1661 Bartholin added Rhode's *Mantissa Anatomica* (Anatomical addition). Finally, it should be mentioned that Rhode also left behind a small study guide for medical students, which Bartholin published in Copenhagen in 1662 as part of his *Cista Medica Hafniensis* and with the title *Johannis Rhodii de artis Medicæ exercitatione, Consilia tria, Amicis intimæ tantum admissionis hactenus communicate* (Johan Rhode's three directions to education in medicine, so far only reported to his close friends). Danish translation in Niels W. Bruun & Hans Otto Loldrup (eds.): *Thomas Bartholin. Cista Medica Hafniensis*, Copenhagen 1982, pp. 86–94. See also Bruun 1893, p. 63.

133. According to Saxo Grammaticus' *Gesta Danorum*, Denmark was during the time of Christ ruled by King Frode Fredegod (Frode III). After the numerous war victories of this king and during his subsequent reign, there was peace and order in Denmark for 30 years. The king showed harshness towards robbery and theft, and to challenge the honesty of "the people in Jutland, which so to say was the head of his realm, he suspended a very heavy bangle on a public road, to test the righteousness he had demanded, by putting on display such a valuable prey. Even though it tempted the dishonest souls and tormented the evil ones, the fear of the inevitable danger prevailed. So great was Frode's royal authority, that even gold, displayed in such a way that everyone could take it, was as sure as if it had been kept behind the strongest locks. This unusual idea granted its inventor great fame." See Saxo Grammaticus: *Danmarks Krønike*, translated by Frederik Winkel Horn, Copenhagen 1898, 1, p. 201.

134. The *vena cava* is the largest vein of the human body. It has an upper part (*vena cava superior*) in the upper chest, draining blood from the head, neck, arms and upper chest, and a lower part (*vena cava inferior*) in the abdomen and the lower chest, draining blood from the legs, the abdomen and the lower chest. They join in the right atrium of the heart.

135. *Arteria venosa* = the lung veins, i.e. the two large veins leading oxygenated blood from the lungs to the left atrium of the heart.

136. *Vena arteriosa* (*truncus pulmonalis*) or the pulmonary artery is the large artery leading deoxygenated blood from the right ventricle of the heart into the lungs.

137. The Venetian anatomist Cecilio Folli (1615–1660), in the work *Sanguinis a dextro in sinistrum cordis ventriculum defluentis facilis reperta via* (On the discovered channel for the blood flowing from the right to the left heart chamber), Venice 1639, had erroneously claimed that the oval hole (*foramen ovale*), which is present between the right and left

atrium of the heart during fetal life, is also responsible for the transportation of the blood from the right to the left half of the heart after birth, which would mean that the blood would bypass the lungs. Under normal conditions, the *foramen ovale* closes at birth, and only in certain congenital heart malformations and pathological conditions with changed pressure in the heart and lungs at birth, which must have been the case in the patient mentioned by Bartholin, does the *foramen ovale* remain open after the birth. Henrik Fuiren mentions Follis' work and theory in a letter to Ole Worm from Padua on 1 December 1639. Bartholin demonstrates the *foramen ovale* in a couple of illustrations (p. 194) in his second revised edition from 1645 of his father's textbook of anatomy and mentions Follis' mistaken theory in the caption.

138. Roman feast for Saturn, which is celebrated during the days around 17 February.

139. O'Malley (ed.) 1961, pp. 83–84.

140. His grandfather Thomas Fincke's scholarship and the royal medical scholarship.

141. Axel Garboe: Enhjørningen (The unicorn). *Medicinsk-Historiske Smaaskrifter ved Vilhelm Maar*, 12, Copenhagen 1915, p. 19.

142. Ibid., p. 20.

143. Bruun (ed.) 2015, pp. 100–1 and pp. 118–19.

144. Helmer Kofod: Narhvalen og enhjørningen (The narwhal and the unicorn). *Tidsskriftet Grønland* 1985, pp. 101–16. Note 41, p. 116 refers to *Forth Book of Moses* 23, 22 and 24, 8; *Fifth Book of Moses* 33, 17; *Book of Job* 39, 12–15 and *Book of Psalms* 22, 22; 29, 6 and 92, 10.

145. Garboe 1915, pp. 14–15.

146. Ibid, p. 12.

147. Ibid., p. 33.

148. Ibid., p. 32.

149. Ibid.

150. Kaj Birket-Smith: *Jens Munk's rejse og andre danske ishavsfarter under Christian IV* (Jens Munk's travel and other Danish Arctic travels during the reign of Christian IV), Copenhagen 1929, p. 75.

151. Garboe 1915, p. 36 and Gosch 1873, p. 27.

152. Worm's corollarium is found on pp. 98–102, and the drawing of the narwhal skull and tusk on p. 103.

153. Bartholin 1662, Chapter 37, pp. 394–98.

154. For a detailed examination of this topic, see Schepelern 1971.

155. Bartholin 1662, Chapter 37: Examination of Greenlandic unicorn tusk, pp. 393–94, Garboe 1915, p. 59, Ingerslev 1873, pp. 426–27.

156. Garboe 1915, pp. 62–63 gives several examples and prices.

157. De Thou 1622, book 123, p. 1518.

158. Jesper Brandt Andersen: Sælsomme beretninger i lægen og anatomen Thomas Bartholins værker (Curious reports in the physician and anatomist Thomas Bartholin's works). *Fund og Forskning i Det Kongelige Biblioteks Samlinger* 2018, 57, pp. 85–91. Jan Bondeson: *The Two-headed Boy and Other Medical Marvels*, New York 2004, pp. 51–63. Bartholin mentions the story about Franciscus Troviluvius in both the first and second edition of *De Unicornu Observationes Novæ*, p. 18 and p. 2, respectively. The phenomenon of horned people truly exists. It is caused by tumours in the skin, cutaneous horns (*cornu cutaneum*), which grow out from the keratin layer of the skin.

159. The learned authors with whom Bartholin and Worm correspond in the appendix about the golden horn are Fortunio Liceti, Henrik Fuiren, Johan Rhode, the Neapolitan physician and surgeon Marco Aurelio Severino (1580–1656) and the German Jesuit and polymath Athanasius Kircher.

160. Bartholin 1645, Chapter 15: De Unicornu Gronlandico (About the Greenlandic unicorn), pp. 94–106.

161. Bruun 1893, pp. 46–47.

162. Gosch 1873, p. 23.

163. In 1657, Bartholin writes about the Greenlandic unicorn: "The unicorn has been very famous among the ancient writers. It was believed to be a four-footed terrestrial animal with one single horn in the forehead. They praised the horn for its power to neutralize poison, for which reason it had such a value among the magnates of the past that in Rome it was sometimes bought for 90,000 rix-dollars and was considered a treasure. But in our days the price of the unicorn horn has been much reduced, after the supply has grown because of the merchants' growing commerce and the discovery of the animal. For our merchants who frequently sail around Iceland and Greenland have found out that the unicorn, which the ancient trumpeted so much in India and the East, is a sea animal of the whale species. By the Icelanders it is called the narwhal. In the upper jaw it has one single long and thick tusk, which is twisted in circles and has stripes reaching from the upper root to the lower tip. This is the tooth that everybody has previously sold as real unicorn horn in the belief that it had fallen off a terrestrial animal. Thus, I can say for sure that all unicorn horns that have previously had this name originate from the narwhal's tusk. I have at various places seen pieces, and I have seen whole tusks in treasure chambers, in Utrecht in St. Dionysii Church and elsewhere. [...] I will not add more about this horn, since I have examined this tooth with the necessary diligence elsewhere, in my book *De Unicornu Observationes Novæ*. I will briefly touch upon the powers that I have found in this Greenlandic narwhal's tusk in experiments. When it was delivered to us in sufficient amounts from northern Greenland by the Davis Strait with Henrik Müller's ships in 1652 and 1653, our capital's physicians have luckily observed that it possessed much power to drive out sweat in the then present malignant fever, so that the fever ceased with a great amount of sweat. Taken as a powder from a scruple to half a drachm in *Cnicus Benedictus*-water or elder-water, or in beer, it pressed out a great amount of sweat. The excellent physicians Dr. Worm and Dr. Moth have experienced the same power in their sick and sincerely assured me about it. And not

only did it help taken as powder but hung around the neck in small pieces it served as an amulet. By request I sent a piece of it to the famous physician Albert Kyper, professor in Leiden, my friend. He used it last year with great luck in powder, given with an adequate amount of water to those who were hit by the plague that ravaged all Holland, so that either they were quickly cured, or they more easily survived the disease, or certain death was postponed for some days. In earlier times, the bishops in Denmark had their crosiers made from this Greenlandic unicorn's horn. In the chapel of the Cathedral of Roskilde I have seen such a stick of suitable size made of unicorn horn. This treasure has belonged to bishop Lucius and at that time cost a great sum of money. Now it has been given to our king's museum." *Hist. Anat.* 1657, 4, Hist. 4: Unicornu Grönlandicum (The Greenlandic unicorn), pp. 220. Odin Wolff: Synderlige anatomiske og medicinske Tildragelser (Special anatomical and medical incidents). *Journal for Politik, Natur- og Menneskekund-skab*, 1828, 1, pp. 99–101. Minister for finance (contemporary Danish: Rentemester) Henrik Müller (1609–1692) was the father of Thomas Bartholin's brother Caspar's wife Anna Cathrine. Müller was a very rich merchant, who built up his fortune on the basis of monopoly on trade in Iceland and Greenland. He gave his manor Kornerupgaard at Roskilde to his son-in-law Caspar Bartholin (See Garboe 1950, 2, pp. 117–18). Albert Kyper (1614–1655) was a German physi-cian born in Königsberg. From 1650 he was professor of medicine at the University of Leiden. Thomas Bartholin had a letter correspondence with Kyper in 1654–1655: *Epist. Med.* 1663, 2, Epist. 52: De Unicornu viribus (About the power of the unicorn's horn) from Kyper in Leiden to Bartholin in Copenhagen on 23 November 1654, pp. 577–78, and Epist. 53: De Unicornu usu & utendi modo (About the use of unicorn horn and the way to use it) from Bartholin in Copenhagen to Kyper in Leiden on 14 June 1655, pp. 578–80.

164. Garboe 1949, 1, p. 48.

165. John M. Howard, Walter Hess, William Traverso: Johan Georg Wirsung (1589–1643) and the Pancreatic Duct: The Prosector of Padua, Italy. *Journal of the American College of Surgeons*, 1998, 187 (2), p. 202.

166. Alfred Max Schirmer: *Beitrag zur Geschichte und Anatomie des Pankreas*, Basel 1893, p. 20.

167. Bartholin 1645, p. 78.

168. Howard et al. 1998, p. 203.

169. The other receivers were Jean Riolan in Paris, Caspar Hofman in Altdorf, Marco Aurelio Severino in Napoli, Paul Marquart Schlegel (16051653) in Hamburg, Werner Rolfink (1599–1673) in Jena and Johann Georg Volkhamer (1616–1693) in Nürnberg. Ibid., p. 206.

170. Irvin M. Modlin et al.: The history of the pancreas. In: Hans Berger et al. (eds.): *The Pancreas. An Integrated Textbook of Basic Science, Medicine, and Surgery*, Massachusetts, Oxford, Victoria, second ed. 2008, pp. 10–11.

171. Ibid., p. 10.

172. The Dutch anatomist and physiologist Reinier de Graaf (1641–1673), who was a pupil of Franciscus Sylvius in Leiden, was the first to examine the pancreatic juice by creating a permanent fistula in the pancreas' excretory duct on a live dog. He found that it was sour and concluded that it was of importance for the digestive process by separating the different nutrients in the food (see Evan R. Ragland: Experimenting with Chymical Bodies: Reinier de Graaf's Investigations of the Pancreas. *Early Science and Medicine*, 2008, 13 (6), pp. 615–664). De Graaf published his results in *De succi Pancreatici Natura et Usu Exercitatio Anatomica Medica* (Anatomical and medical investigation of the nature and function of the pancreatic juice), Leiden 1664. In the fourth and last edition of his textbook of anatomy, *Anatome Quartum Renovata*, Leiden 1673, Thomas Bartholin goes through de Graaf's exploration of the pancreas in text and illustrations (pp. 113–23). Two hundred years later, Graaf's results were confirmed by the French physiologist Claude Bernard (1813–1878), who proved the importance of the pancreatic juice in the digestion and the uptake of fat and sugar (Claude Bernard: *Mémoires sur le pancréas et sur le role du sac pancréatique dans les phénomènes digestifs, particulièrement dans la digestion des matières grasses neutres*, Paris 1856. The hormone-producing cell islands, the islets of Langerhans, were discovered by the German physician and anatomist Paul Langerhans (1847–1888) in 1869, and in 1889 the Lithuanian physician Oskar Minkowsky (1858–1931) and the German physician Joseph von Mehring (1849–1908) discovered that dogs developed diabetes when their pancreas was removed. They concluded that the pan-creas produces one or more substances that regulate the blood sugar. The English physiologist Edward Albert Sharpey-Schafer (1850–1935) was of the opinion that this substance was produced in the islets of Langerhans, and he gave the substance the name insulin (lat., *insula* = island). It was not until 1921 that the Canadian physician Frederick Banting (1891–1941) and his assistant Charles Best (1899–1978) succeeded in treating a diabetic dog with injections of insulin isolated from the islets of Langerhans in another dog. The following year they treated in the same way the first human being, a 14-year-old boy, for diabetes. See Porter 1997. Niels Stensen's research in glands during his stay in Holland in 1660–1664 was also inspired by Wirsung's discovery, t hrough the influence of Thomas Bartholin.

173. Howard et al. 1998, p. 205–6.

174. Claudio Bassi, Giuseppe Malleo: The unsolved mystery of Johann Georg Wirsung and of (his?) pancreatic duct. *Surgery* 2011, 149, pp. 153–55. Bartholin has a somewhat different version of Wirsung's death in *De Peregrinatione Medica* (1674): "Wirsung was shot down on the doorstep of his own house in Padua by a stray cannonball." Where the cannonball and the story came from, Bartholin does not tell. See O'Malley (ed.) 1961, p. 53. In *Anatomia Reformata*, p. 78, Bartholin writes, that Wirsung was "killed by a cruel destiny."

175. Bartholin 1673, p. 113.
176. A vein on the surface of the heart, which collects the venous blood from the heart muscle and leads it to the right atrium of the heart.
177. Francesco I d`Este (1610–1658).
178. Lennart Torstenson (1603–1651), Swedish count, councillor of the realm and field-marshal. He invaded Jutland in December 1643 and commanded the Swedish army in the Danish-Swedish war 1643–1645 (The Torstenson War), which ended with The Piece in Brömsebro in August 1645. Denmark had to give up Halland for 30 years, the islands Gotland and Øsel in the Baltic Sea, as well as Jæmtland and Hejredalen in Norway. Moreover, the Swedes gained exemption from duty in Øresund (the Sound).
179. Besides the attack on Jutland under Torstenson's command, the Swedes attacked Scania from the north with an army under the command of field marshal Gustaf Horn (1592–1657). Thus, the Swedish tactic was a pincer manoeuvre. The last element in the Swedish plan, to put in land troops on Zealand, failed because of the superiority of the Danish fleet.
180. Garboe 1949, 1, p. 45.
181. Alessandro Bichi (1596–1657).
182. Anne of Austria (1601–1666). As the widow of Louis XIII she ruled France from 1643–1651 on behalf of her son Louis XIV (1638–1715).
183. Cassiano dal Pozzo (1588–1657), Italian scholar, art patron and secretary of cardinal Francesco Barberini (1597–1679).
184. Ole Worm: *Danicorum Monumentorum Libri Sex* (Danish memorials in six books), Copenhagen 1643. This work mentions, on page 438, Thomas Bartholin's two letters to Ole Worm concerning the earlier-mentioned Florentine horns.
185. Plinius the Elder (23–79 AD), Roman historian, natural philosopher and military leader, especially known for the encyclopaedic work *Naturalis Historia*. As the leader of the fleet, he became a witness to Vesuvius' all-destroying eruption in 79 AD, which buried the ancient cities of Pompeii and Herculaneum. He died himself from inhaling the poisonous gases from the eruption.
186. O'Malley (ed.) 1961, p. 69.
187. Marcus Aurelius Severinus: *De recondita Abscessuum Natura Libri VIII* (8 books about the hidden nature of boils), second edition, Frankfurt 1643 and *Vipera Pythia. Id est, De Viperæ Natura Venero, Medicina, Demonstrationes, et Experimenta nova* (The Pythic viper, i.e. demonstrations and new experiments with the viper's nature, poison and medicine), Padua 1643.
188. O'Malley (ed.) 1961, p. 63.
189. Garboe 1949, 1, p. 62.
190. Ibid.
191. Ibid, pp. 62–63.
192. Thomas Bartholin: *Anatomica Aneurysmatis dissecti Historia. Accedit Johannis van Horn eiusdem argumenti Epistola* (Anatomical account of the dissection of an aneurism. With the addition of a letter of argumentation about the same topic by Johannes van Horne), Palermo 1644.
193. Ian H. Porter: The Nineteenth-Century Physician and Cardiologist Thomas Bevill Peacock (1812–88). *Medical History* 1962, 6 (3), p. 244.
194. Henrik Fuiren writes from Padua to Ole Worm on 19 May 1644: "After Thomas Bartholin left Naples, where he during some time lived together with my brother in Severino's house, he sailed to Sicily, where he had shortly before been called by envoys to a well-paid professorship in medicine – without accepting it, however." Schepelern (ed.) 1968, 3, Epist. 564 (no. 1201), pp. 18–19.
195. Scylla and Charybdis were two sea monsters who, according to Greek mythology (Homer's *Odyssey*), lived in the Strait of Messina between the Italian mainland and Sicily, posing great danger to passing seamen. Bartholin writes in his book from 1674 on medical travel that: "Scylla and Charybdis of the ancients lie in the narrows of the sea between Calabria and Sicily. [...] Scylla is a rock, and Charybdis the sea, each a menace to those ships driven towards them. However, the rock did not appear above the sea, and Charybdis clashing its waves together against us in a playful mood, carried us rapidly to Messina in the bay of Pelori." O'Malley (ed.) 1961, p. 73.
196. Pietro Castelli: *Hortus Messanensis* (The Garden of Messina), Messina 1640.
197. Bartholin gives the recipe for this viscous laxative in 1674 in his book on medical travel: "Among the common diseases afflicting the inhabitants was flatus of the hypochondria, which Castelli happily dissipated with his electuary diatartar, of which he employed one ounce an hour before meals with a resultant large expurgation, and he prescribed it on alternate days for those suffering from lues venerea. Since this very useful electuary of Castelli has reached the notice of the poor and is suitable even for our country, I shall not hesitate to repeat its description, published in Messina, 1644, by his nephew Giovanni Pietro Corvino. Take half an ounce each of very finely ground and sifted tartar of white wine and choice senna, one ounce each of choice manna and brown sugar, one drachm each of ginger, anise, cinnamon and galanga minor, five ounces of syrup of roses; mix, and the electuary is made. Four ounces of meat juice ought to be drunk afterward so that it will work more readily. All the shops of Messina sold this electuary at a reasonable price, and in the whole of Sicily and nearby Calabria the diatartar of Castelli was so highly praised that it was preferred to all other cathartics." O'Malley (ed.) 1961, p. 75.
198. The word angina was in Bartholin's era used as a general term for inflammation of the throat as well as more specifically for diphtheria. Here he uses it for diphtheria.
199. One wonders why Bartholin believes that swimmers' feet are cut off by the shark's tail and not by its teeth.
200. O'Malley (ed.) 1961, p. 64.

201. Thomas Bartholin: *De Angina Puerorum Campaniæ Siciliæque Epidemica Exercitationes. Accedit de Laryngotomia Cl. V. Renati Moreau Paris. Med. & Profess. Regij, Epitola* (Investigations into a diphtheria epidemic among children in Campania and Sicily, accompanied by a letter by the famous Parisian physician and Royal Professor René Moreau about laryngotomy), Paris 1646.
202. Henrick Smid's (1495–1563) *Fierde Urtegaard flittelig tilsammen lest aff lerde Lægers Bøger faar Quindfolckernis och unge Spædbørns Siugdommer*, Copenhagen 1557, included in *Henrick Smids Lægebog*, Copenhagen 1577, and Niels Mikkelsen Aalborg's (1562–1645) Om Spæde Børns Siuger, which is the fifth book in the same author's *Medicin eller Læge-Boog*, Copenhagen 1635, are both translations of German works, and P.C. Abildgaard's (1740–1801) *Underretning om Børne-Sygdomme og deres Helbredelse til Nytte for Landfolket*, Copenhagen 1769, is a translation with remarks and explanations of the Swedish physician Nils Rosén von Rosensteins (1706–1773) *Underrättelser om Barn-Sjukdomar och deras Bote-Medel*, Stockholm 1764. In Danish literature, the obstetrician Carl Edvard Marius Levy's (1808–1865) posthumous published lectures *De almindeligste Sygdomme hos spæde børn*, Copenhagen 1868, have so far been considered the first independent work on paediatrics. See Erik Thamdrup: *Pædiatri i Danmark. Træk af den historiske udvikling*, Copenhagen 1994, pp. 20–21.
203. Paul of Aegina (625–690), Greek physician, best known for writing the medical encyclopaedia *Medical Compendium in Seven Books*.
204. Lilith was, according to a Jewish tradition from the Middle Ages, Adam's first wife. Since she refused to submit herself to Adam, discord arose between them. After she had said the holy name of God, she escaped from Eden and started to commit adultery with demons. Adam complained to God over her escape, and God sent the three angels Senoi, Sanesenoi and Samangelof to bring Lilith back. Because she refused, God let her demon-children kill. However, she was given the power to kill boy children until they were eight days old (the number of days that have to pass before Jewish boys should be circumcised), and girl children until they were 20 days old. But Lilith swore to the angels that if she saw their names written, she would not have power to kill the infants, and therefore some Jewish societies have had the tradition to write the names of the angels over the doors to infants' homes. Lilith, which comes from a Semitic root meaning night, came to symbolize the horrors of the night and infant death.
205. Alyssum is a plant genus in the Mediterranean area with approximately 50 species.
206. Chiron was, according to the Greek mythology, a wise and medically knowledgeable centaur, who raised Achilles and Asklepios on the mountain Pelion in Thessaly.
207. *Epist. Med.* 1663, 1, Epist. 80: De Laryngotomia (On laryngotomy), pp. 331–36.
208. The Belgian-American historian of science George Sarton (1884–1956) calls *De medicamentis* "an unusual mixture of traditional knowledge, popular medicine and superstition." See George Sarton: *Introduction to the History of Science*, Baltomore 1927, 1, p. 391. *De medicamentis* was republished by Maximillian Niedermann under the title "Marcelli de medicamentis liber" in *Copus Medicorum latinorum*, 5, Leipzig and Berlin 1916.
209. The letter is included in *Epist. Med.* 1663, 1, Epist. 81: De eadem Laryngotomia Responso (Answer concerning the mentioned laryngotomy), pp. 336–49.
210. One of the writings to which Moreau refers is Severino's *Pædanchone Loimodes, seu de pestilente ac præfocante Pueros Abcessu* (Contagious whooping cough or about a contagious disease choking children), which Severino published together with the second edition of his book about boils, *De recondita Abcessuum Natura* (On the secret behind the nature of boils), Frankfurt 1643, and to which Bartholin refers diligently in *De Angina Puerorum*.
211. George Frederic Still: *The History of Paediatrics. The Progress of the Study of Diseases of Children up to the End of the XVIIIth Century*, London 1931. Thomas Bartholin's *De Angina Puerorum* is described in Chapter 23, pp. 230–34.
212. It is surprising that Still translates with the word "virus." Viruses were not known until 1892, and diphtheria is caused by a bacterium. Bartholin uses the words "aërem inquinatum" (polluted air). Bartholin 1646, p. 8.
213. The cribriform plate or *lamina cribrosa* is a thin bony plate that constitutes the ceiling of the nasal cavity and separates the nasal cavity from the brain. It contains small holes, which are penetrated by the olfactory nerve.
214. Marco Aurelio Severino: *Therapeuta Neapolitanus Seu Veni Mecum Consultor Curandarum febrium & internorum omnium morborum. Inclusa Pædanchone affectu pestilente, ac pueros præfocante. Cum Commentario Cl. Thomæ Bartholini Regis Danorum Archiatri* (The Neapolitan treater or handbook for the treatment of fever diseases and all internal diseases. Included is the child suffocator, a lethal disease that suffocates children. With a commentary by Thomas Bartholin, physician to the king of Denmark), Naples 1653.
215. Bartholin uses the Latin word *centuria* for a hundred, in this case a hundred letters.
216. The five-page letter *De Plantis circa Æthnam nascentibus* (On the plants growing around Etna) is addressed to Carolo Avantio Rhodigino in Padua and is sent from Siracusa on 25 April 1644.
217. Tansy.
218. Melita is used as the name of both the island (country) Malta and an ancient town on the island. Bartholin probably means the island.
219. O'Malley (ed.) 1961, p. 79.
220. Bartholin 1644, p. 6, Garboe 1949,1, p. 64.
221. Bartholin 1646, the preface, Garboe 1949, 1, p. 65.
222. Bartholin 1647, p. 10, Garboe 1949, 1, p. 65.

223. Bartholin 1661, p. 4. In a letter a couple of years later Bartholin repeats this explanation of Iceland's volcanos and hot springs: "By the external coldness in the Nordic regions the entrails of the Earth are heated up." Garboe 1949, 1, p. 66.
224. Bartholin 1661, p. 6. Garboe 1949, 1, pp. 65–66.
225. Brandt Andersen 2021, pp. 588–94. Letter from Niels Stensen in Milano to Cosimo III in Florence, summer 1671. In: Angelo Fabroni: *Lettere inedite di uomini illustri* (Unpublished letters from famous people), Vol. 2, Florence 1775, letter no. 141, pp. 318–21. Letter from Niels Stensen i n Milano to Cosimo III in Florence of 19 August 1671. Ibid., letter no. 142, pp. 321–27, (The original letter in Italian language and associated drawings can be found in Bibliotheca Nationale in Florence, Gal. 287, 58r–61r. Maar (ed.) 1910, 2, OPH No. 29: On the grotto above Gresta, pp. 239–42 and OPH No. 30: On the grotto of Moncodine, pp. 243–48. Scherz (ed.) 1952, 1, E 62, pp. 238–40, and E 63, pp. 241–46. Kardel & Maquet (eds.) 2018, pp. 839–48.
Harriet M. Hansen (ed.) 1987, 1, E 62, pp. 164–66, and letter E 63, pp. 166–70.
226. Garboe 1949, 1, p. 66.
227. Latinization of Guillaume Rondelet (1507–1566), French physician and professor of medicine at the university in Montpellier.
228. *Hist. Anat.* 1661, 6. Hist. 1. Glossopetrarum Melitensium Usus Medicus (On the medical use of Maltese tongue stones), pp. 193–201. Wolff 1928, 4, pp. 1–4.
229. Garboe 1949, 1, pp. 71–72.
230. Bartholin 1670, pp. 73–74. O'Malley (ed.) 1961, p. 28.
231. Niels Stensen: *Canis Carchariæ Dissectum Caput* (Dissection of a white shark's head), Florence 1667. Maar (ed.) 1910, 2, OPH No. 23, pp. 113–45. Kardel & Maquet (eds.) 2018, pp. 699–731. Brandt Andersen 2021, pp. 435–64.
232. Stensen's foundational geological work, *De Solido Intra Solidum Naturaliter Contento Dissertationis Prodromus* (Forerunner to a dissertation on solids naturally embedded in other solids), was published in Florence in 1669. A planned, more comprehensive geological work was never published, but *De Solido Intra Solidum* was, despite its brevity, so convincing that it is generally considered the founding work of geology. It deals with stratigraphy, palaeontology and crystallography. Among other things, Stensen accounts for the principle of superposition, i.e. that the age of the Earth can be determined based on its layers, where the most superficial layers are the youngest, and the most profound the oldest. Furthermore, Stensen accounts for the formation of fossils, the growth of crystals and for the basic crystallographic law that the opposite faces are always parallel in all kinds of crystals.
233. Garboe 1949, 1, p. 66.
234. Ibid., p. 73 and *Epist. Med.* 1663, 1, Epist. 54, pp. 227–36.
235. Ibid., pp. 72–73 and *Epist. Med.* 1663, 1, Epist. 55: De vomitu in calculo (On vomiting in kidney stone disease), pp. 236–39.
236. The earlier-mentioned Felix Plater.
237. Realdo Colombo (1515–1559), professor in anatomy in Padua.
238. *Hist. Anat.* 1654, 2, Hist. 45. Os in Corde Urbani VIII (A bone in Urban VIII's heart), pp. 243–44. Bartholin also mentions the finding in *Anatome Quartum Renovata*, Leiden 1673, p. 394.
239. Garboe 1949, 1, pp. 77–78. On 24 June, Worm writes to the physician and botanist Johan Caspar Bauhin (1605–1685) in Basel: "Two of our excellent young compatriots, who are skillful in the art of medicine, namely my relatives Henrik Fuiren and Thomas Bartholin (the latter known through his writings) now come to you from Italy, or maybe they have already arrived. From you they intend to bring home trophies for the efforts they have gone through. If you in the same way as your father can support them with favour and advice, I will consider it done for me." Rhode writes on 16 November 1645 from Padua to Worm, with the expectation that Bartholin is already back in Copenhagen: "I congratulate you that the Bartholins have been given back to you. I have recommended Thomas and Henrik Fuiren to Stupanus [Emmanuel Stupanus (1587–1664), Swiss physician and professor of medicine in Basel] and Bauhin."
240. Endoxa (plural of the Greek endoxon) are, according to Aristotle, opinions that one can agree with after careful examination of arguments both for and against it, with the former emerging stronger, i.e. a kind of consensus. The full title of the disputation is *De Pleuritide paradoxa endoxa. Superiorum permissu. Pro Summa Med. Laurea. In illustri Rauracorum Lycæo, horis consuetis, Publicè ventilanda proponet* (Thomas Bartholin will with the permission of the authorities and to achieve the highest medical distinction present *Acknowledged paradoxes concerning pleurisy* publicly at the usual time at the famous University of Basel).
241. Thomas Bartholin: *De Monstris in Natura & Medicina Oratio, in Brabeuterio Academico Augustæ Rauracorum publicè dicta* (Speech about deformed creatures in Nature and in medicine held for an academic audience in Brabeuterium), Basel 1645. The speech is reprinted in Thomas Bartholin: *Orationes varii argumenti* (Speeches with various contents), Copenhagen 1668 (hereafter abbreviated *Orationes*), pp. 1–11. Augusta Raurica, which was in Bartholin's era used synonymously with Augst and possibly also with Basel, which Bartholin's title indicates, refers to the ancient Roman city with the same name in Augst approximately 20 km east of Basel. This ruined city with its amphitheatre was considered Basel's mother city and was, from its establishment by the University of Basel in 1460, an object of great interest from the university. It was a favourite excursion destination for both archaeological and botanical studies. The latter was practised by Caspar Bartholin the Elder's teacher, the physician and botanist Caspar Bauhin the Elder (1560–1624), and his students. See Edgar Bonjour: *Die Universität Basel von den Anfängen bis zur Gegenwart 1460–1960*, Basel 1960, pp. 154 and 181.

242. Albrecht Burckhardt: *Geschichte der Medizinischen Fakultät zu Basel 1460–1900*, Basel 1917, p. 419. This book also mentions Bartholin's disputation in 1645 and includes him among "famous men who in these years picked up the Baselian doctor's hat," p. 181. Martin Möhle: Alte Universität Basel. *Schweizerische Kunstführer GSK*. Serie 80, No. 795, Bern 2006.
243. Henrik Fuiren became Doctor of Medical Science at the University of Basel on the same day as Thomas Bartholin. Garboe 1949, 1, p. 78.
244. Bartholin 1645, p. 1.
245. *Institutions Anatomiques de Gasp. Bartholin, Docteur et Professeur du Roy de Dannemarck, Augmentées & enrichies pour la seconde fois, tant des Opinions & Observations novelles des Modernes, dont la plus grande partie n'a j amais esté mise en lumiere, que de plusieurs Figures en talille-douce, par Thomas Bartholin Docteur en Medicine, Fils de l'Auteur; et Traduictes en François par Abr. Du Prat, Docteur en Medecine*, Paris 1647. On the initiative of Thomas Bartholin's predecessor as professor in anatomy in Copenhagen, the German Simon Paulli (1603–1680), the second edition was also translated into German by Paulli's pupil Michael Kirstein (1620–1678) and published in Copenhagen in 1648 with the title *Institutiones Anatomicæ, Das ist künstliche Zerlegung Menschlichen Leibes, Des Hochgelahrten und Weitberühmten Herrn D. Caspari Bartholini, Itzo durch Anordnung D. Simonis Paulli, Med. Anat. ac Botan. P.P. In der Königlichen Academien Kopenhagen, Mit fleiss Allen Wundartzen zu Nutz ins Deutsche übersetzet. Und Mit dreyen unterschiedlichen weitläufftigen Registern, zu dero sonderbaren Nutzen versehen; Davon an seinen Orte der günstige Leser absonderlich berichtet wird. Gedruckt zu Kopenhagen, Im Jahr 1648*. About this translation, Worm writes on 24 January 1646 to Thomas Bartholin in Paris: "Our Simon Paulli plans a German translation for the sake of surgery, but rewritten in a shorter version, as far as I understand."
246. However, there could be another contributing reason for this. A letter from Johannes Walæus in Leiden on 11 February 1644 indicates that Bartholin did not receive the promised help from Johannes Vesling, who wished to use the illustrations from his dissections for a new edition of his own *Syntagma Anatomicum* (Anatomic Survey), which had been published in Frankfurt in 1641. The second edition with new illustrations was published in Leiden in 1647. Walæus writes in the letter to Thomas Bartholin: "Your father's and your anatomy book has now been sold out and brought to honour. The printer has fewer than ten copies left. [...] He therefore prepares a new edition, which, if we should believe him, shall consist of better paper, typeface and illustrations. But I warned him that you had promised the publication of a new edition in Padua, provided with new illustrations drawn from Vesling's anatomical dissections. Nevertheless, he answered that he could produce finer prints than the Italians, for he knew that they were uncertain about living up to their guarantees. Indeed, I was recently stricken with fear, when I read a letter from Vesling written to your acquaintance Vorstius, where he declared that he was entirely committed to enriching and increasing his *Syntagma Anatomicum* and wanted to be sure that there were new drawings from the dissections, with which he wanted to beautify it; but I understand that he has been stopped from carrying out these dissections." *Epist. Med.* 1663, 1, Epist. 46: De editione Anatomia, ductus Pancreaticus etc. (On the edition of the anatomy book, the pancreatic duct etc.), pp. 195–98.
247. The new illustrations are the opened coecum with the point of entry of the small intestine into the large intestine (*ostium ileocoecale*) and the valve (*valva ileocoecale*) that confines this opening (p. 62); the pancreas with the newly found secretory duct (*ductus pancreaticus*) (p. 70); the heart seen from right and left side, respectively, and with a probe through *foramen ovale* (the open oval connection between the atria of the heart during fetal life) (p. 194); the aorta's branches to the head and neck (p. 206); the vein valves of the forearm and the lower leg (p. 310); the spinal cord (*medulla spinalis*) (p. 376); the bones of the cranial base seen from inside (p. 396); the bones of the top of the cranium seen from inside (p. 398); the cranium seen from behind (p. 402); the eardrum, the small bones of the middle ear, the bones of the inner ear (*cochlea*) and the surrounding bones of the cranium (p. 408); two illustrations from Walæus' dissections on dogs concerning the blood circulation (pp. 452 and 458); and the mesentery with Aselli's lacteal veins (p. 474). Three pages with gynaecological illustrations from the first edition, among which two illustrations are redrawn, are now confined to a single page (p. 144).
248. However, in a letter to Bartholin in Leiden on 18 April 1640, Worm had already mentioned the six cranial bones: "When I publicly gave a lecture here on your father's dissertation with the help of our skeleton, I found in our skeleton's cranium in sutura lambdoidea six small bones, three at its right edge, and three at its left, visible in both shells of the cranium, both the outer and the inner, but both larger and more visible in the inner than in the outer plate. I don't remember that they have been seen or described by others. On the vaulted surface they look smaller and are interwoven in the suture's larger extended teeth, so that they are only seen when you look thoroughly for them. The lower of them is in this cranium placed at the mamillar protrusion, the middle barely half a finger's distance above, the third is somewhat further away from the second. Their forms are different, triangular, oval, indefinite. They are larger on the left seam of the lambdoidea than on the right. The largest does not exceed a thumbnail in size. I found them for the first time in 1628, when your father was still alive, and I showed them to him. He wondered why they had not been noticed by him or others, and since we were in doubt

as to whether this was something special for this particular cranium, we examined others and found them, but different in size, number and form."
249. *Epist. Med.* 1663, 1, Epist. 63: Auris interna delineatio (Drawing of the inner ear), pp. 255–63.
250. The portrait engraving is also mentioned and shown in Otto Amdrup: Om Portræter af Thomas Bartholin (On portraits of Thomas Bartholin) in Vilhelm Maar (ed.): *Thomas Bartholin 1616, 20. Oktober, 1916. Mindeskrift paa 300-Aarsdagen for hans Fødsel* (Thomas Bartholin 1616, 20 October 1916. Commemorative words on the 300th anniversary of his birth), Copenhagen 1916, pp. 55–57, 62 and 64. Garboe 1949, 1, pp. 54–56. Bruun & Loldrup (eds.) 2007, pp. 246–47 and 272. Bruun (ed.) 2015, pp. 10 and 213. P.B.C. Westergaard: *Danske Portræter i Kobberstik, Litografi og Træsnit 1–2* (Danish portraits as engravings, lithographs and woodcuts), Copenhagen 1930–1933, 1, no. 628, p. 90.
251. Bartholin includes in his mentioned book about the fire in his library from 1670 this translation as no. 5 on his list of lost published works in the fire: "The same anatomy in Italian verses, translated by Hostilius Contalgenus, Florence 1651." Bartholin 1670, p. 86. O'Malley (ed.) 1961, p. 33.
252. Caspar Hofman: *Institutionum medicarum libri sex* (Medical doctrines in six books), Lyon 1645.
253. Gosch 1873, p. 61–66 goes through several of the points of criticism and Bartholin's answers to them.
254. pp. 210–12 and p. 191 in the mentioned editions, respectively.
255. pp. 265–67 and p. 399–402 in the mentioned editions, respectively.
256. Bartholin 1645, p. 201.
257. Gunde Rosenkrantz (1604–1674), councillor of the realm, son of Holger Rosenkrantz the Learned.
258. *Epist. Med.* 1663, 1, Epist. 74: De Dubis in Anatome nostra (On questions of doubt in our anatomy book), pp. 308–10.
259. Ibid., Epist. 76: De Respiratione Cartesiana (On the breathing mechanism according to Descartes), pp. 315–17.
260. Ibid., Epist. 77, p. 317–22.
261. In a letter to Bartholin in Basel on 28 October 1645, Worm writes: "There are two professorships vacant with us, one in rhetoric, the other in philosophy. The former is sought by your brother Bertel, who, I hope, will get it. The latter hasn't a single suitor, but I hope politely, that it will be given to you as your bride. If you agree with me on this, I advise you to come home as soon as possible. I have negotiated about this case with our high Lord Chancellor. He is favourably disposed towards your endeavours and misses nothing else than your presence, so that the university will not be harmed by waiting too long. I have given my word that you will be with us before next spring. Make sure you do not leave me in the lurch, since it is highly in your own interest."
262. *Epist. Med.* 1663, 1, Epist. 94, pp. 385–97.
263. According to Ingerslev 1873, 1, p. 477, Bartholin travelled home through Germany and arrived to Copenhagen on 14 October 1646, but according to a letter from the French physician and philosopher Samuel de Sorbière (1615–1670) in Paris to Thomas Bartholin in Copenhagen on 1 February 1647, the journey home took place by ship. *Epist. Med.* 1663, 1, Epist. 97: De Cartesio judicium, et alia (A judgement of Descartes and other things), p. 408.

CHAPTER 3
Professor of anatomy in Copenhagen
pp. 187–220

1. Thomas Bartholin in *Vasa Lymphatica, Nuper Hafniæ in Animantibus inventa, Et Hepatis exseqviæ* (The lymph vessels, which were recently discovered in animals in Copenhagen, and the abandonment of the liver), Copenhagen 1653, p. 10. G. Tryde (ed.) 1940, p. 95.
2. Jakob Matthiesen was married to Thomas Bartholin's sister Anna. The couple is depicted on an epitaph in Aarhus Domkirke (The cathedral of Aarhus).
3. *Epist. Med.* 1663, 1, Epist. 98, pp. 410–11.
4. Professor of anatomy Simon Paulli (see later).
5. Christian Fabricius (1601–1666), physician, the son of Royal Physician Jacob Fabricius.
6. In normal medical terminology, the word fontanelle refers to the soft spot in the top of the cranium of a baby. The meaning of the word here is unclear.
7. E. Ingerslev: *Fragmenter af Fødselshjælpens Historie* (Fragments of the history of obstetrics), Copenhagen 1906–1907, 1, pp. 46–47.
8. O'Malley (ed.) 1961, pp. 23–24.
9. Ingerslev 1906, 1, p. 48.
10. Thomas Bartholin: *De armillis veterum, presertim Danorum, Schedion, accessit Olai Wormii De aureo cornu ad F. Licetum responsio* (On bangles in ancient times, especially in Denmark, followed by Ole Worm's reply to Fortunio Liceti concerning the Golden Horn), Copenhagen 1647. It also came in a 114-page second edition in Amsterdam in 1676, dedicated to Ole Worm's son, librarian at the Royal Library and professor of medicine Willum Worm (1633–1704) with the addition of new illustrations. The title is added: *Editio novissima, figuris æneis illustrata* (Latest edition illustrated with copper engravings).
11. Carl Neergaard: Thomas Bartholin and ancient research in Denmark in the 17th century. In: Maar (ed.) 1916, p. 43.
12. Ibid., p. 45.
13. Bartholin gives this planned book the title *Epistolarum & Epigrammatum variis in locis editorum Volumen* (A volume of letters and epigrams published in various places). The book was never published. Instead, Bartholin published his entire scientific correspondence in four centuries (four volumes of each 100 letters) entitled *Epistolarum Medicinalium*

(Medical letters), of which the first two centuries appeared in Copenhagen in 1663 and the last two in 1667. Furthermore, in 1669 in Copenhagen he published *Carmina varii argumenti* (Poems with various contents, hereafter abbreviated as *Carmina*). (For both works, see later).
14. Garboe 1949, 1, pp. 89–90 and letter from Ole Worm in Nykøbing Falster to Thomas Bartholin in Copenhagen, 20 February 1647, and vice versa on the same day. The planned apology by Bartholin titled *De Cruce Apologia ad Nihusium* (Apology against Niehus concerning the cross), Leiden 1647, has been lost, but according to Garboe it is possibly identical to *Hypomnema I* (first record), in Thomas Bartholin: *De Cruce Christi hypomnemata IV* (Four records of the cross of Christ), Copenhagen 1651.
15. *Epist. Med.* 1663, 1, Epist. 100, pp. 414–16. Loic Capron (ed.): *Correspondance complète et autres écrits de Guy Patin*, https://www.biusante.parisdescartes.fr/patin/.
16. *Acta Consistorii*, 31 July 1647, Rigsarkivet (Danish National Archives). Henning Matzen: *Kjøbenhavns Universitets Retshistorie 1479–1879* (Legal History of the University of Copenhagen 1479–1879), Copenhagen 1879, 2, p. 47. Slottved 1978, p. 105.
17. *Acta Consistorii*, 16 October 1647. Matzen 1879, 2, p. 371. Garboe 1949, 1, p. 93. Slottved 1978, s. 152.
18. Thomas Bartholin: *Historia paralytici primi ex qvinto Joannis evangelio* (The Story of the first paralytic man from John chapter 5), Copenhagen 1647. Published in Thomas Bartholin: *Paralytici Novi Testamenti Medico & Philologico Commentario illustrati* (Informative medical and philological commentary on the paralytics in the New Testament), Copenhagen 1653, pp. 1–13. Respondent to the disputation was the theologian, later professor at the University of Copenhagen 1652-1668 and bishop of Zealand 1668–1675, Hans Wandal (1624–1675).
19. *Acta Consistorii*, 15 December 1647. Slottved 1978, p. 152.
20. Garboe 1949, 1, p. 93. The speech is published in *Orationes*, pp. 12–18.
21. Gosch 1873, p. 66.
22. *Orationes*, pp. 18–39. The speeches (and the disputations) were held on 22 March, 15 April, 27 May, 28 June and 22 July 1648, respectively.
23. Ibid. 39–43.
24. Bruun & Loldrup (eds.) 2007, pp. 253–54.
25. Ibid., p. 27.
26. Ingerslev 1873, 1, pp. 146–47.
27. Ibid, p. 147.
28. Bruun & Loldrup (eds.) 1982, p. 71.
29. Frisch (ed.) 1918, p. 95.
30. Ibid, p. 98.
31. Bruun (ed.) 2015, p. 77.
32. Ibid.
33. Ibid., pp. 75–81.
34. Ibid., p. 151.
35. Ibid., p. 89.
36. Ibid.
37. Mansa 1873, p. 331.
38. V.A. Secher: Københavns Rets- og Kulturhistorie i Kristian IV.s og Frederik III.s Tid 1624-1663. VII. Livsstraffene og deres Fuldbyrdelse (Copenhagen's legal and cultural history in the time of Christian IV and Frederik III 1624–1663. The life sentences and their execution). *Historiske Meddelelser om København 1907–08*, pp. 574–78.
39. Ingerslev 1873, 1, pp. 226–27 and 283. Matzen 1879, 2, p. 94. Garboe 1949, 1, pp. 93–94. Slottved 1978, pp. 108–9.
40. Ingerslev 1873, 1, p. 227. Matzen 1879, 2, p. 94.
41. Ingerslev 1873, 1, p. 283.
42. Ibid., p. 227. Garboe 1949, 1, p. 96. Bruun (ed.) 2015, p. 89. Slottved 1978, p. 109.
43. Slottved 1978, p. 109. Garboe 1949, 1, p. 96, Matzen 1879, 2, p. 20 and pp. 94–95.
44. Matzen 1879, p. 20 and pp. 94–95. Garboe 1949, 1, p. 96.
45. Garboe 1949, 1, p. 97.
46. E. Snorrason: Illustrationerne i anatomen Thomas Bartholins publikationer 1641–1661 (The illustrations in the anatomist Thomas Bartholin's publications 1641-1661). *Fund og Forskning i Det Kongelige Biblioteks Samlinger* 1973, 20, note 44, p. 105. Snorreson informs us that the precise location of the professor's residence was Klædebo Kvarter no. 239.
47. Bruun (ed.) 2015, p. 89.
48. https://natmus.dk/historisk-viden/temaer/nationalmuseets-historie/kunstkammeret/
49. Bruun (ed.) 2015, p. 91.
50. Ibid., p. 137.
51. *Hist. Anat.* 1654, 1, Hist. 5: Anatome solennis 1649, pp. 15–16.
52. Jens Lauridsen Wolf: *Encomion Regni Daniæ*, Copenhagen 1654, p. 357.
53. Bruun (ed.) 2015, pp. 55–57.
54. Niels W. Bruun: Kongens stol i Domus Anatomica Hafniensis (The king's chair in Domus Anatomica Hafniensis). *Fund og Forskning i Det Kongelige Biblioteks Samlinger* 2012, 51, pp. 131–47.
55. Bruun (ed.) 2015, p. 139. The method is described by Simon Paulli in *Acta Medica Ann. 1673*, Copenhagen 1675, chapter 18. Modus dealbandi ossa pro Sceletopæia (Method of making bones for skeletons), pp. 42–46 and chapter 113. Observationes in coctura ossium, presertim sterni (Observations on boiling bones, in particular sternum), pp. 279–88.
56. *Hist. Anat.* 1657, 3, Hist. 87: Corij humani usus & preparationio (Use and preparation of human skin), pp. 177–79.
57. Bruun (ed.) 2015, p. 91.
58. Wolf 1654, p. 357.
59. Bruun (ed.) 2015, pp. 91–93.
60. *Hist. Anat.* 1654, 1, Hist. 14: Anatome puella nuper nate Anni 1650 (Dissection of a newborn girl in 1650), pp. 27–30.

61. Bartholin refers to the oval hole of fetal life, *foramen ovale*, located between the right and left atrium, which together with the small *ductus arteriosus* between the pulmonary artery, *truncus pulmonalis*, and the body artery, *aorta*, which Bartholin mentions in the next sentence, ensure that the blood passes through the heart outside the lungs during fetal life, when the lungs are not in use. Under normal circumstances, both connections close at birth; when the baby screams, the lungs unfold, and the pressure, flow and oxygenation conditions in the heart change.
62. *Orationes*, Oratio 28: Anatome fæminini cadaveris (Dissection of a female corpse), pp. 149–52. See also Bruun (ed.) 2015, p. 159, who refers to Bartholin's second edition, Copenhagen 1665, of Michael Lyser's *Culter Anatomicus* (The anatomical knife), Copenhagen 1653, p. 3: "However, it is also necessary to cut open an infant, so that you can recognize its difference from an adult. At times even embryos that have been removed by a miscarriage are exposed to the sharp edge of an anatomical scalpel."
63. Garboe 1949, 1, pp. 104–7.
64. *Hist. Anat.* 1654, 2, Hist. 67: In Porcis Observationes raræ (Rare observations in pigs), pp. 292–94 and the same place, Hist. 32: In Boum Anatome observationes (Observations on dissection of cows), pp. 222–25.
65. Ibid., 1, Hist. 99: Galli ovipari Anatome (Dissection of an egg-laying cock), pp. 162–65. Garboe 1949, 1, pp. 110–12. Brandt Andersen 2019, pp. 72–74.
66. *Epist. Med.* 1667, 3, Epist. 52: De Scriptis Schoockianis. De gallo oviparo (On the writings of Martin Schoock. On the egg-laying cock), pp. 212–15.
67. Axel Garboe: Om fabeldyret basilisken (On the fabulous creature, the basilisk). In: *Teologisk Tidsskrift for den danske Folkekirke* 1917, 3, 8, pp. 17–30.
68. As another example, Bartholin mentions in his paper about unusual birth paths in man, that Jean Riolan in his *Anthropographia*, Paris 1618, so as not to dismiss the court gossip, mentions an ectopic pregnancy of the French queen's washer maid, although, according to his colleague Guy Patin (1601–1672), Riolan himself did not believe in the story. See Ingerslev 1907, pp. 330–31.
69. *Epist. Med.* 1663, 2, Epist. 92: De Lue Regiomontana & Basilico (About the Epidemic in Königsberg and on the basilisk), pp. 704–7. Letter from Ludwig Kepler in Lübeck to Thomas Bartholin in Copenhagen on 2 October 1657. Garboe 1949, 1, p. 112.
70. *Hist. Anat.* 1661, 6. Hist. 87: Vermes in ovo (Worms in an egg), pp. 348–52. Wolff 1928, 4, pp. 12–15.
71. Garboe 1949, 1, p. 111. The experiment is described in *Acta Medica Ann.* 1673, Copenhagen 1675, no. 10: Ova Galli & Serpentum (Cock eggs and snake eggs), pp. 14–16.
72. H.C. Bering Liisberg: *Kunstkammeret. Dets Stiftelse og ældste Historie* (The King's Art Chamber. Its foundation and oldest history), Copenhagen 1897, p. 161.
73. Ibid., p. 158.
74. *Orationes*, pp. 67–68. Garboe 1949, 1, p. 107.
75. Thomas Bartholin: *De Variis Reipublicæ Christianæ Morbis et placidis eorum remidiis, Dissertatio Oratoria* (On the Christian community's illnesses and about the lenient measures that are found to counter them), Copenhagen 1649. Bartholin later published the speech in *Orationes*, pp. 42–73.
76. Thomas Bartholin: *Collegium Anatomicum Disputationibus XVIII adornatum* (Anatomical lecture decorated with 18 disputations), Copenhagen 1649–1651. 10 of the opening speeches of the disputations are also found in *Orationes*, pp. 73–102.
77. *Orationes*, pp. 73–77: De Morte Veslingi & Walæi (On the death of Vesling and Walæus).
78. Ibid. De ureteribus (On the ureters), pp. 94–96.
79. *Thomæ Bartholini Casp. F. Anatomia ex Caspari Bartholini parentis Institutionibus, omniumque recentiorum et propriis observationibus tertium ad sanguinis circulationem reformata cum iconibus novis accuratissimis* (Third edition of the anatomy book based on the anatomy book of his father Caspar Bartholin and all recent and own observations and adjusted according to the blood circulation. With new and very accurate illustrations), Leiden 1651.
80. Gosch 1873, p. 77.
81. Bartholin 1651, p. 395.
82. Ibid., p. 419. Gosch 1873, p. 77, also highlights the drawings of the adrenal glands in man and certain animals (ox, porpoise (whale) and lamb) on page 123 of the work as originals.
83. Gosch 1873, p. 77. Frank (ed.) 1920, p. 246.
84. The portrait engraving and copies thereof are mentioned and depicted in Andrup 1916, pp. 57–61, Garboe 1949, 1, p. 91, Bruun (ed.) 2007, pp. 261 and 272 and Bruun (ed.) 2015, pp. 27 and 213. The engraving is mentioned in Westergaard 1930–33, 1, no. 630, p. 90.
85. Thomas Bartholin: *Anatomia, often, Ontledinge des Menschelijcken Lichaems, soon, onlanghs uyt sijn Vaders, als oock vele nieuwe Schrijvers, en eyghene ondervindingke, in't latijn t'samen ghestelt, hier zijn by gevoegt twee Brieven van Joh. Walaeus, raeckende de bewegende des Gijls ende Bloets, als mede, een map verhael van een nieuw ghevonden wegh des Gijls naer het Herte*, Leiden 1653.
86. *Epist. Med.* 1663, 2, Epist. 6: Gratulatio Nuptialis (Wedding Congratulations), pp. 428–30.
87. Garboe 1949, 1, p. 115. Some of the poems, by Vitus Bering among others, are found in Thomas Bartholin's *Carmina*, pp. 137–42.
88. Garboe 1949, 1, p. 116. *Hist. Anat.* 1654, 2, Hist. 65: Diarrhoea, & Dysenteria Epidemica (Diarrhoea, and dysentery epidemic), pp. 287–90 and same work 1657, 4, Hist. 31: Prognosticum lethale (A harbinger of death), pp. 294–95.

CHAPTER 4
The discovery of the thoracic duct and the lymph vessels
pp. 221–306

1. Thomas Bartholin: *Spicilegium Secundum Ex Vasis Lymphaticis* (Second Aftermath of the Lymph Vessels), Copenhagen 1660, p. 1.
2. G. Tryde (ed.) 1940, p. 20.
3. Ibid. Later in the treatise (p. 64), Bartholin writes. "The publicly dissected child killer, who, about five hours before he was hanged, had been filled up with a heavy meal and with wine, still had some food and thick and thin chyle left in the stomach."
4. Secher 1907–08, p. 569.
5. Frisch (ed.) 1918, p. 94.
6. Secher 1907–08, pp. 335–36.
7. Ibid., p. 337.
8. *Hist. Anat.* 1654, 1, Hist. 53: Lacteæ thoracicæ, in Anatome virili 1652, primum quæsita & inventæ (The chest ducts were first searched and found at the dissection of a male person in 1652), pp. 79–84. The text up to the quoted section can be translated as follows: "In our anatomical theatre on February 19, the body of a child killer was brought in. He had actually been sentenced to the breaking wheel, but by the consent of our High Royal Majesty, Frederik the Third, our most gracious Lord, he was left to the gallows, so that we could more easily observe the new milk ducts. To this end, five hours before the hanging he had been given food and drink."
9. Scurvy is a disease of the connective tissue caused by a lack of vitamin C in the diet. Symptoms are bleeding of the skin, the subcutaneous tissue, muscles, joints and gums, loose and decayed teeth, swollen joints, fatigue, decreased appetite, weight loss and, in the worst case, death. The disease was most prevalent in cold areas and among sailors, where vitamin C-containing fruits and vegetables were not part of the daily diet. In 1747 the Scottish naval physician James Lind (1716–1794) realized that scurvy could be prevented and cured by the consumption of citrus fruits (*A treatise of the scurvy*, Edinburgh 1753).
10. Consumption is an older term for serious illness that causes weight loss and general impairment. The term was often used for tuberculosis.
11. Buboes are swollen lymph glands caused by infection.
12. Serum is blood without blood cells.
13. The omentum is a vessel containing fold of connective tissue in the abdominal cavity.
14. Pus is yellow inflammatory fluid.
15. In Bartholin's time, the terms lacteal veins or chyle vessels were used for the lymph vessels of the intestine, since the liquid chyle (Greek: chylos = juice) that they contain has a milky appearance due to its high fat content. For the same reason, the lymph glands in the intestine were called chyle glands.
16. For further information about Karel van Mander, see Povl Eller: *Kongelige portrætmalere i Danmark 1630–82* (Royal portrait painters in Denmark 1630–82), Copenhagen 1971, pp. 219–65.
17. Thomas Bartholin: *De Lacteis Thoracicis in homine brutisq; nuperrimé observatis, Historia Anatomica* (An anatomical account of the thoracic ducts recently observed in man and animals), Copenhagen 1652. English translation in: *The Anatomical History of Thomas Bartholinus, Doctor and Kings Professor: Concerning the Lacteal Veins of the Thorax, Observ'd by him lately in Man, and Beast. Publickly proposed by Michael Lyserus answering*, London 1653. Danish translation by G. Tryde (ed.) 1940, pp. 1–76.
18. Tryde (ed.) 1940, p. 76.
19. Ibid., p. 7.
20. Eviscerated here refers to a dog whose entrails have been removed.
21. Ibid, p. 27.
22. Tryde (ed.) 1940, p. 21.
23. Ibid.
24. Ibid., p. 75.
25. Democritus (c. 460–400 BC), Greek philosopher and the founder of atomism. Natural sciences in the 1600s were significantly influenced by ancient atomism. Bartholin must surely here mean that the dissector has to dissect himself down to the smallest structures to find the truth.
26. Ibid., p. 76.
27. Michael Lyser: *Culter Anatomicus. Hoc est Methodus brevis facilis ac perspicua artificiose & compendiose humana incidendi cadavera. Cum nonnullorum instrumentorum Iconibus* (The anatomical knife, that is, a concise, simple, and obvious method of artful and clear dissection of human corpses. With illustrations of some instruments), Copenhagen 1653. About Michael Lyser, see also Axel Garboe: Kongelig Majestæts Læge for Lolland-Falster Michael Lyser (Royal Majesty's physician for Lolland-Falster Michael Lyser). In: *Lolland-Falsters Historiske Samfunds Årbog* 1952, 3, 1, pp. 83–93. Bartholin cites among the works lost in the fire of Hagestedgaard a work of his own, *Instrumenta Medica*, in which he had described all medical instruments invented by ancient and modern doctors. See O'Malley (ed.) 1961, p. 29.
28. Bruun (ed.) 2015, p. 89. The book's official year of publication, which appears on the title page, was 1653.
29. *Epist. Med.* 1663, Cent. 2, Epist. 24: De Infantis ex chylo nutritione (On the fetus' nutrition from the chyle), pp. 478–81.
30. Nathaniel Highmore: *Corporis Humani Disquisitio Anatomica, in qua Sanguinis Circulationem in quavis Corporis particula plurimis typis novis, ac Ænygmatum Medicorum succincta dilucidatione ornatam prosequutus est* (An anatomical study of the human body, in which the author has given a representation, set with many new types, of the circulatory system in any part of the body, as well as an illustration of the medical riddles), The Hague 1651.

31. Willum Worm joined the two noblemen and statesmen Erik Rosenkrantz (1612–1681) and Peder Reedtz (1614–1674) on a journey to England in 1652.
32. English translation by Howard B. Adelmann in: *The Embryological Treatises of Hieronymus Fabricius of Aquapendente*, Ithaca, New York 1942, 1, pp. 235–334.
33. English translation in Willis (ed.) 1847, pp. 143–586.
34. At this early stage, before the microscope gained ground in medical research, one had no knowledge about the microcirculation of the placenta, and the exchange of nutrients and gases between the capillaries of the mother and fetus there. Such knowledge first began to form in the Enlightenment some 100 years later with the Scottish anatomist William Hunter's (1718–1783) beautifully illustrated work on the pregnant uterus, *Anatomia uteri humani gravidi tabulis illustrata* (Anatomy of the pregnant uterus illustrated with tablets), Birmingham 1774. For further reading on this topic, see Joseph Needham: *A history of embryology*, New York 1959 and Charles D. O'Malley and J.B. de C.M. Saunders: *Leonardo da Vinci on The Human Body*, New York 1982, pp. 208–16.
35. William Harvey: *Exercitatio Anatomica de Circulatione Sanguinis: ad Joannem Riolanum filium Parisiensem, medicum peritissimum anatomicoru coryphæum, in Academia Parisiensi Anatomes & Herbariæ Professorem Regium & egregium atque decanum, Reginæ matris Lodovici XIII medicum primarium* (Reply to Jean Riolan concerning an anatomical exercise of the blood circulation), Cambridge 1649. English translation in Willis (ed.) 1847, pp. 87–141.
36. Robert Tigerstedt (ed.): *Dokument ur Medicinens Historia under sextonde och sjuttonde Århundradet* (Documents from the history of medicine during the sixteenth and seventeenth century), Stockholm 1921, p. 362.
37. Willis (ed.) 1847, pp. 604–10. It has not been possible to identify a Parisian doctor by the name R. Morison or to establish whether by "the Parisian physician R. Morison" Harvey possibly means the Scottish botanist Robert Morison (1620–1683), who at this time stayed in Paris.
38. Ibid., 612–15.
39. Ibid., p. 612.
40. Ibid., p. 614.
41. *Epist. Med.* 1663, Cent. 2, Epist. 57: Dubia quædam de Lacteis thoracicis (Some doubts concerning the thoracic ducts), pp. 592–95.
42. Ibid., Epist. 63: Pro Venis Lacteis ad Harveum responsio (A response to Harvey about the lacteal veins), pp. 606–8.
43. Ibid., Epist. 62: Harvey dubia de Thoracicis (Harvey's doubts about the thoracic ducts), pp. 603–5.
44. Garboe 1949, 1, p. 128. Thomas Bartholin: *Collegii Anatomici Disputatio Tertia. De Mesenterio et Pancreate* (On the mesentery and pancreas), Copenhagen 1649 and *Collegii Anatomici Disputatio Quarta: De Duobus reliquis Mesenterii Vasis, id est, De Novis Asellii et Notis Veterum Venis* (On the other two vessels of the mesentery; that is, about Aselli's new vessels and about the old ones), Copenhagen 1650.
45. Thomas Bartholin: *Cista Medica*. Apertio cadaveris lingva bubula extincti (Autopsy of the corpse of a person who was choked by an ox tongue), pp. 524–25. Bruun and Loldrup 1982, p. 169.
46. The statement seems a bit strange. We have just been told that the good Sven Olsen on that very day had stuffed himself with food and drink, so apparently Bartholin practises cynical humour and means by "frugal dinner" the beef tongue that never reached the stomach. From the abundant food intake earlier in the day, such a long time had apparently passed that many of the lacteal veins were no longer visible.
47. *Hist. Anat.* 1654, 1, Hist. 11. Anatome Olai à lingva bubula suffocati (Dissection of Olsen, who was strangled by an ox tongue), pp. 23–25.
48. Bartholin 1651, p. 425.
49. Ibid., pp. 87–94.
50. Ibid., p. 90.
51. Ibid., p. 91.
52. Otto Sperling (1602–1681), physician and botanist, born in Hamburg. His plant lists from a botanical survey tour in Scania, Halland, Blekinge and Gotland in 1623 were published by Thomas Bartholin in *Cista Medica*. Sperling's autobiography in German has been translated into Danish and published in excerpts by S. Birket-Smith, Copenhagen 1885, entitled: *Dr. med. Otto Sperlings Selvbiografi 1602-1673, oversat fra tysk i Uddrag efter Originalhaandskriftet med særligt Hensyn til Forfatterens Ophold i Danmark og Norge samt Fangenskab i Blaataarn.*
53. Ludvig Pouch (1620–1681), provincial physician in Ribe.
54. Tryde (ed.) 1940, pp. 55–56.
55. Bartholin 1651, pp. 420–21.
56. Ibid., p. 421.
57. Ibid, p. 425.
58. Ibid., pp. 418–20.
59. Jean Pecquet: *Experimenta Nova Anatomica, quibus incognitum hactenus Chyli Receptaculum, et ab eo per Thoracem in ramos usque Subclavios Vasa Lactea deteguntur, eiusdem Dissertatio Anatomica de Circulatione Sanguinis et Chyli Motu, Accedunt Clarissimorum Virorum perelegantes ad Authorem Epistolæ* (New Anatomical Experiments exposing the hitherto unknown vessel for chyle and lacteal vessels, extending from it to branches under the collarbone, to which are added the very elegant letters of great celebrities to the author), Paris 1651. English translation including Thomas Bartholin's *De Lacteis Thoracicis: New Anatomical Experiments of John Pecquet of Deip. By which the hitherto unknown Receptacle of the Chyle, and the Transmission from thence to the Subclavial Veins by the now discovered Lacteal Chanels of the Thorax, is plainly made appear in Brutes. As also an Anatomical Dissertation of the Motion of Blood and Chyle. Together with the further Description of the same Lacteal Chanels newly discovered in the Body of Man as well as Brutes. Being an Anatomical*

Historie, publickly propos'd by Thomas Bartoline, Dr. and Reg. Professor both in Physick and Anatomy, to Michael Lysere, Answering, London 1653.
60. Ibid., p. 3.
61. Ibid.
62. Ibid, pp. 2–3.
63. Tryde (ed.) 1940, pp. 18–19.
64. Ibid., p. 19–20.
65. Ibid., s. 56.
66. Ibid.
67. Peter Englund: *Silvermasken – en kort biografi över drottning Kristina* (The silver mask. A short biography of Queen Christina), Stockholm 2006, p. 32.
68. Cornelis van Hoghlande (1590–1651), Dutch physician and philosopher.
69. Garboe 1949, 1, pp. 118–19.
70. *Epist. Med.* 1663, 2, Epist. 26: De valetudine sua & itinere (On his health and his journey), p. 485–87.
71. J. Lévy-Valensi 1933, p. 586.
72. Tryde (ed.) 1940, p. 20.
73. Hermann Conring: *De Sanguis Generatione et motu naturalii Opus novum* (A new work on the generation and natural movement of the blood), Helmstedt 1643. The book was a support for William Harvey's circulatory theory and contributed significantly to its spread and recognition in Germany. In the book, Conring theorizes that thin chyle, which is formed by the drink, flows through some hitherto unknown vessels to the spleen – a theory that Bartholin refutes in both *Anatomia Reformata* and in *De Lacteis Thoracicis*. See Tryde (ed.) 1940, pp. 12–13.
74. *Epist. Med.* 1663, 2, Epist. 13: De novo chyli Receptaculo et Hepatis usu (On the new chylus container and the function of the liver), pp. 444–47.
75. *Epist. Med.* 1663, 2, Epist. 14: De Lacteis Pequetianis responsio et dubia (An answer and doubt concerning Pequet's lacteal veins), pp. 447–52.
76. Bartholin agrees with this finding in *De Lacteis Thoracicis* and adds that the thoracic duct also on its way up through the ribcage sends small branches to the oesophagus. See Tryde (ed.) 1940, p. 43.
77. *Epist. Med.* 1663, 2, Epist. 15: De Generatione Harvei (About Harvey's book on procreation), pp. 452–55.
78. See Tryde (ed.) 1940, pp. 61–62.
79. *Epist. Med.* 1663, 2, Epist. 20: De Regina Christina (About Queen Christina), pp. 466–67.
80. Ibid., Epist. 21: De Lacteis, Aquæ potu, & semine muliebri (About lacteal veins, drinking of water and the female sperm), pp. 468–70.
81. Ibid., Epist. 22, De Aquæ potu post medicamentum & de Harvei opinione (On drinking of water after medication intake and on Harvey's view), pp. 470–72.
82. Ibid., Epist. 18: De Pequeti inventi Utilitatibus (On the usefulness of Pequet's discovery), pp. 461–64.
83. Juno = Roman goddess, wife of Jupiter, after whom the month of June is believed to be named.
84. By this stream, Bartholin must mean receptaculum chyli.
85. Ibid., Epist. 19: Cur lacteas in hepate Pequetis non annotarite? (Why has Pequet not listed lacteal veins in the liver?), pp. 464–66. The same view of the circulation of the chyle and division of the blood formation between the liver and the heart, which Bartholin puts forward in this letter, he repeats in *De Lacteis Thoracicis*. See Tryde (ed.) 1940, pp. 53–62.
86. Tryde (ed.) 1940, pp. 26–27.
87. Ibid., p. 45.
88. Ibid., pp. 56–57. In his first treatise on the lymph vessels *Vasa Lymphatica, Nuper Hafniæ* etc. mentioned below, Bartholin explains further: "Admittedly, we have often at the dissection of dogs seen some water-carrying vessels coming from the liver, but since we considered them to be chyle vessels, we paid no further attention to them and did not take into account anything extraneous to us." See Tryde (ed.) 1940, pp. 93–94.
89. Hydatid = water bladder.
90. Tryde (ed.) 1940, pp. 94–95.
91. *Ramus axillaris* = branch of the axillary artery.
92. Tryde (ed.) 1940, p. 95.
93. Ibid., p. 100.
94. Ibid., p. 101.
95. Ibid., pp. 101–2.
96. Ibid., pp. 103–4.
97. Ibid., p. 104.
98. *Epist. Med.* 1663, 2, Epist. 25: De Anatome Helmstadiana (On the Anatomy of Helmstedt), pp. 481–85. This letter is dated Wittenberg, 15 June 1652, and this date has hitherto been taken as evidence for the assumption that Lyser left Denmark as early as the first half of 1652 (see for example Ingerslev 1873, 1, p. 529, and Axel Garboe: Michael Lyser, a 17th-century anatomist. *Acta Medica Scandinavica* 1952, suppl., p. 66), but in connection with the preparation of this book, three things have drawn my attention to the fact that the letter must be misdated, presumably by one year, so the correct dating should be 15 June 1653. The three circumstances which prove the misdating are: 1) that Lyser mentions in the letter the death of Schlegel, which took place on 20 February 1653; 2) that it appears from the letter that Lyser on his journey is carrying specimens of *Vasa Lymphatica, Nuper Hafniæ in Animantibus inventa, Et Hepatis exseqviæ* (see below), which did not appear until 1 May 1653; and 3) that Thomas Bartholin on 4 April 1653 addresses a letter to Lyser in Copenhagen (*Epist. Med.* 1663, 2, Epist. 29: De Sectionibus Anatomicis (On anatomical dissections), pp. 491–96). Michael Lyser published on 3 September 1653, at the University of Leipzig, a short 16-page dissertation on the ear (*Dissertationem physicam de auditu* (Physical treatise on hearing). The letter from Wittenberg must therefore be from the period between 1 May and 3 September 1653, which fits with 15 June 15, 1653. This makes it likely that Lyser was Bartholin's assistant (prosector) for one more year than previously thought.

99. Ibid., Epist. 50: De Lymphaticis, morte Slegelij ect. (About the lymphatic vessels, Schlegel's death etc.), 14. September 1654, pp. 567–71.
100. Thomas Bartholin: *Opuscula Nova Anatomica De Lacteis Thoracicis Et Lymphaticis Vasis, Uno Volumine Comprehensa. Ab Auctore aucta & recognita* (New anatomical writings on the thoracic ducts and lymphatic vessels, published in one volume, augmented and reviewed by the author), published at Daniel Paulli's Publishing House in Copenhagen and Frankfurt 1670 and sold in Amsterdam at Johannes Bleau as well.
101. Bartholin 1670, p. 94.
102. It has not been possible to verify the reference in question, but the story of the earwig is reproduced uncritically by many later sources, for example Christian Wilhelm Kestner: Medicinisches Gelehrten-Lexicon, Jena 1740, p. 792: "Slegelius starb zu Hamburg 1653 in 48. Jahre seines Alters, nachdem er kurz zuvor bey einer öffentlichen Anatomie von dem zurück gebogenen und unversehens wieder vorgefallenen Arme des Cadaveris eine Ohrfeige bekommen und hierdurch in nicht geringes Schrecken gesetzet worden."
103. O'Malley (ed.) 1961, p. 53. For the story about Schlegel see also Brandt Andersen 2018, pp. 69–72.
104. Tryde (ed.), p. 104.
105. Thomas Bartholin: *Vasa lymphatica, Nuper Hafniæ in Animantibus inventa, Et Hepatis exseqviæ*, Copenhagen 1653. English translation in: *The Anatomical History of Thomas Bartholinus, Doctor and Kings Professor: Concerning the Lacteal Veins of the Thorax, Obser'd by him lately in Man, and Beast. Publickly proposed by Michael Lyserus answering*, London 1653. Danish translation in Tryde (ed.) 1940, pp. 77–119. Republished by Vilhelm Maar in Copenhagen in 1916 "On the tercentenary of the birth of Thomas Bartholinus" with a biographical introduction and notes in English. Bartholin's dedication to Riolan reads: "Riolan, the world's and the city of Paris' greatest anatomist, dedicated by Thomas Bartholin."
106. Bartholin 1653, the last page.
107. Bartholin 1651, p. 213.
108. Tryde (ed.) 1940, p. 70.
109. Ibid., p. 71.
110. Ibid, p. 129.
111. Ravn was bailiff in Copenhagen from 1638 to 1660. See O. Nielsen: *Kjøbenhavns Historie og Beskrivelse* (History and Description of Copenhagen), Copenhagen 1881, 3, pp. 251–56. Thanks to Bjørn Westerbeek Dahl for help with the identification of Ravn.
112. Byfogedregnskaber Sjælland og Møn København 1648–1653, Rigsarkivet, Copenhagen. See also Secher 1907–08, p. 590.
113. *Hist. Anat.* 1654, 1, Hist. 92: Anatome puerperæ (Dissection of a woman who had given birth to a child), pp. 147–52.
114. Thomas Bartholin: *Dubia Anatomica de Lacteis Thoracicis & An Hepatis funus immutet Medendi Methodum* (Anatomical doubts concerning the thoracic ducts, and the question of whether the abandonment of the liver changes our method of treatment), Copenhagen 1653. Danish translation in G. Tryde (red.) 1940, pp. 121–60.
115. Tryde (ed.) 1940, p. 131.
116. Ibid., p. 170.
117. *Hist. Anat.* 1657, 3, Hist. 2: Phthisici Anatome in eoque Vasa Lymphatica detecta (Dissection of a tuberculous male person, in whom the lymphatic vessels were found), pp. 5–10.
118. Ole Worm writes on 29 April 1654 to his son Willum in Leiden: "Our Bartholin has – while I in my illness encouraged him to do so—discovered the lymphatic vessels of one who had died here of shrinkage and has given an account of this matter to the printing house, perhaps he will now send it to you too."
119. Tryde (ed.) 1940, pp. 174–75.
120. Ibid, p. 173.
121. Bartholin's father-in-law, the prosperous merchant and mayor Christopher Hansen (1598–1679), owned several properties in Copenhagen, so it cannot be said with certainty which one of these it is about. According to Erik Pontoppidan: *Den Kongelige Residens-Stad Kiøbenhavn*, Copenhagen 1760, p. 37, Christopher Hansen lived on the site where St. Clement's Church (demolished in 1530) had been located, i.e. on the present Strøget at Rådhuspladsen, more specifically Frederiksberggade 38, Copenhagen K. According to O. Nielsen: *Kjøbenhavns Historie og Beskrivelse*, Copenhagen 1881, 3, 1, p. 218, Christopher Hansen owned and inhabited no. 44 on Gammel Strand, and he also owned Vimmelskaftet (in the 1600s Skoboderne) no. 36, eight apartments in Mikkel Bryggers Gade, a garden with eight apartments at Vestervold between Lille S. Klemensstræde and Vestergade, seven apartments in Didrik Bartskærs Gang and a fishpond at Store Ravnsborg.
122. Tryde (ed.) 1940, p. 175.
123. *Hist. Anat.* 1657, 3, Hist. 60: Pestilentia magna 1654 (The Great Plague 1654), pp. 118–20. Plague, which is usually divided into bubonic plague and pneumonic plague, is caused by the bacterium *Yersinia pestis* and can today be cured with a simple cure of penicillin. On the plague in Copenhagen 1654–1655 and the precautions for combating it, see also Carl Bruun: *Kjøbenhavn. En illustreret Skildring af dets Historie, Mindesmærker og Institutioner* (Copenhagen. An Illustrated Depiction of Its History, Memorials and Institutions), Copenhagen 1890, Part 2, pp. 49–50 and Mansa 1873, pp. 391–403.
124. Garboe 1949, 1, p. 177.
125. Tryde (ed.) 1940, p. 177.
126. Ibid., p. 178.
127. Danish translation in Tryde (ed.) 1940, pp. 161–98.
128. Bruun (ed.) 2015, p. 89. Møinichen was on a six-year study trip in Europe from 1654 to 1660. From the spring of 1655 to the spring of 1660, he studied in Padua. On

28 April 1660, on his way back to Denmark through France and Holland, he obtained his medical doctorate in Angers, France. When he returned to Denmark in the autumn of 1660, he was offered the vacant professorship of anatomy after Bartholin, but declined and instead became court physician to Frederik III and later Christian V. See Ingerslev 1873, 1, pp. 508–9 and H. Ehrencron-Müller 1929, 6, pp. 33–34. He published, among other things, *Observationes medico-chirurgicæ missæ ad Th. Bartholinum* (Medical-surgical observations sent for Thomas Bartholin) in Copenhagen 1665 and some contributions to Thomas Bartholin's *Acta Medica* and *Epist. Med.*

129. Frisch (ed.) 1918, p. 98.
130. Ibid., p. 96. Alexipharmaca = antidotes.
131. Oleum rutaceum = a kind of citrus oil. Rutaceae are dicotyledonous tropical plants of the Rude family.
132. Acetum calendulæ = calendula vinegar.
133. Frisch (ed.) 1918, p. 96.
134. Ibid., p. 97.
135. Ibid., p. 98.
136. Ibid., p. 97.
137. Tryde (ed.) 1940, p. 30.
138. The parenchyma is the functional tissue in an organ, in this case the liver cells.
139. Tryde (ed.) 1940, p. 51.
140. Ibid.
141. Ibid., pp. 52–53.
142. Ibid., p. 111.
143. Ibid., p. 113.
144. Anasarca and oedema are fluid accumulation in the tissue outside the bloodstream.
145. Ascites is abnormal accumulation of fluid in the abdominal cavity, which can be seen in a variety of diseases, for example cancers, cirrhosis of the liver and inflammation.
146. Tryde (ed.) 1940, p. 114.
147. Hydrops is abnormal water retention in the body.
148. Tryde (ed.) 1940, pp. 183–84.
149. Ibid., pp. 190–91.
150. Ibid., p. 194.
151. Porter 2000, p. 10 and pp. 219–20.
152. Ibid., p. 341.
153. Ibid., p. 446.
154. Barry Cooper: The origins of bone marrow as the seedbed of our blood: from antiquity to the time of Osler. *Baylor University Medical Center Proceedings* 2011, 24, 2, pp. 115–18.
155. Niels Stensen: *De musculis et glandulis observationum specimen*, Amsterdam 1664, pp. 42. See also, Maar (ed.) 1910, 1, OPH No. 15, pp. 163–92 and English translation (Specimen of Observation on Muscles and Glands) in Kardel & Maquet (eds.) 2018, pp. 549–77. For a review of the history of sweat, see E.T. Renbourn: The history of sweat and the sweat rash from the earliest times to the end of the 18th century. *Journal of the History of Medicine and Allied Sciences*, April 1959, 14, 2, p. 217.
156. *Epist. Med.* 1663: Epist. 54: De semita ad mammas (On the path to the breasts), pp. 581–82.
157. Ibid., Epist 75: Pro lymphatico spicilegio gratiarum actio (Thanks for *Aftermath of the Lymph Vessels*), pp. 642-44.
158. Jean Riolan: *Opuscula Anatomica Nova. Quæ nunc primùm in lucem prodeunt. Instauratio magna Physicæ et Medicinæ, per novam Doctrinam de Motu Circulatorio Sangvinis in Corde. Access Notæ in Johannis Wallei duas Epistolas de Circulatione Sanguinis* (New anatomical opuscules, which are now for the first time published. A great renewal of physiology and medicine through the new teachings about the blood circulation in the heart. In addition, some comments on Johannes Walæus' two letters on the blood circulation), London 1649.
159. Gosch 1873, p. 67. For a summary of Riolan's attack on Bartholin, see ibid., pp. 66–77.
160. The criticisms are formulated in the section Animadversiones in Caspari Bartholini Doctoris Medici, Theologi, ac Professoris Regii Institutiones anatomicas (Remarks on Doctor of Medicine and Theology and Royal Professor Caspars Bartholin's Book of Anatomy), pp. 379–428.
161. Gosch 1873, pp. 73–75.
162. *Anatomia Reformata* 1651, p. 59. See also Gosch 1873, p. 75, and E. Ingerslev 1907, 2, p. 330.
163. Porter 2000, p. 564.
164. Bartholin 1651, p. 125.
165. Gosch 1873, p. 74. The French physician Charles Édouard Brown-Séquard (1817–1894) became in 1856 aware of the vital function of the adrenal glands, as his removal of the adrenal glands in laboratory animals brought about their deaths. He concluded that the adrenal glands must supply the blood stream with one or more vital substances (hormones). See Porter 2000, p. 563.
166. Garboe 1949, 1, p. 126.
167. *Epist. Med.* 1663, 2, Epist. 8, 3 September1650: De Hofmanno & Riolano (On Hofman and Riolan) pp. 431–33.
168. Tryde (ed.) 1940, p. 145.
169. Ibid., pp. 135–36.
170. Ibid., p. 136.
171. Ibid.
172. Ibid., p. 139.
173. Ibid., pp. 145–52.
174. Aretaeus of Cappadocia (second century), ancient Greek physician.
175. Ibid., p. 154. Stygian refers to the river Styx, which in Greek mythology formed the boundary to Hades.
176. *Epist. Med.* 1663, 2, Epist. 31: De Riolani Opusculus (On the small writings of Riolan), pp. 498–500. French translation in *Correspondance complete de Guy Patin et autres écrits, édités par Loïc Capron*: www.biusante.parisdescartes.fr/patin.
177. Ibid., Epist. 32: De Ejusdem Riolani dissidiis (On discrepancies with the said Riolan), pp. 500–2. French translation in *Correspondance complete de Guy Patin*

et autres écrits, édités par Loïc Capron: www.biusante.parisdescartes.fr/patin.
178. Jean Riolan: *Anthropographia et Osteologia, Judicium novum de venis lacteis tam mesentericis quam thoracicis adversus Th. Bartholinum. Lymphatica vasa Bartholini refutata. Animadversiones Secundæ ad Anatomiam Reformatam Bartholini. Ejusdem Dubia Anatomica De Lacteis Thoracicis resoluta. Hepatis funerati et ressuscitati vindiciæ* (Anthropography and osteology. A reassessment – directed against Thomas Bartholin – of the lacteal veins in both the mesentery and the thoracic duct. Rejection of Bartholin's lymphatic vessels. For the second time critical remarks on Bartholin's revised anatomy. Resolution of Bartholin's doubts regarding the thoracic duct. Apology of the buried and resurrected liver), Paris 1653.
179. Thomas Bartholin: *Defensio Vasorum Lacteorum et Lymphaticorum adversus Joannem Riolanum Celeberrimum Lutetiæ Anatomicum. Accedit Cl. V. Guilielmi Harvey de venis Lacteis sententia expensa ab oedem Th. Bartholino* (Defence of the lacteal and lymph vessels against Jean Riolan, the famous anatomist in Paris. In addition, William Harvey's opinion on the lacteal veins assessed by Thomas Bartholin), Copenhagen 1655.
180. Gosch 1873, pp. 98–99.
181. Ibid., p. 101.
182. Thomas Bartholin: *Spicilegium Ex Vasis Lymphaticis: ubi cl. v. Glissonii & Pecqueti sententiæ expenduntur: publicè in Reg. Acad. Haffn. propositum; Repondente [sic] M. Johanne Christiano Stolbergk, 28 Nov. 1655* (Aftermath of the Lymph Vessels, in which the views of the famous men Glisson and Pequet are assessed, publicly presented at the University of Copenhagen with Hans Christian Stolbergk as respondent), Copenhagen 1655.
183. Gosch 1873, p. 102.
184. *Epist. Med.* 1663, 2, Epist. 75: Pro Lymphatico spicilegio gratiarum actio (Thanks for *Aftermath of the Lymph Vessels*), pp. 642–44.
185. Swedish translation in Tigerstedt (ed.) 1921, pp. 363–403, Danish translation in Tryde (ed.) 1940, pp. 199–229, and English translation by Aage Nielsen: A Translation of Olof Rudbeck's Nova Excercitatio Anatomica Announcing the Discovery of The Lymphatics. *Bulletin of the History of Medicine* 1942, 11, 1, pp. 30–39.
186. Garboe 1949, 1, p. 156 and Gunnar Eriksson: *Rudbeck 1630–1702. Liv, lärdom, dröm i barockens Sverige* (Rudbeck 1630––1702. Life, learning, dream in Sweden during the baroque), Stockholm 2002, p. 60.
187. Olaus Martini Stenius (1597–1660), professor of mathematics, astronomy and medicine and rector of Uppsala University.
188. Johannes van Horne: *Novus Ductus Chyliferus. Nunc primum Delineatus, Descriptus & Eruditorum Examini expositus* (The new chyle duct, now for the first time illustrated, described and presented to the study of the experts' survey), Leiden 1652.
189. Tryde (ed.) 1940, p. 211.
190. *Epist. Med.* 1663, 2, Epist. 20: De Regina Christina (About Queen Christina), pp. 466–67.
191. Ibid., Epist. 21: De Lacteis, Aquæ Potu, & semine muliebri (About the lacteal veins, drinking water and the female sperm), pp. 468–70.
192. Tryde (ed.) 1940, pp. 218–19.
193. Ibid., p. 215.
194. Ibid., p. 226.
195. *Epist. Med.* 1663, 2, Epist. 34: De ductibus serosis (On the serous ducts), pp. 505–11.
196. Ibid., Epist. 47: Pro Lymphaticorum prima inventione (In defence of the first discovery of the lymph vessels), p. 559: "Pondering over the question of the author I could not figure out anything before around Christmas time, when the bookseller's messenger brought the printed work to our house, whereby I got to know both the author, whom I had never heard or read about before, and his project." See longer quote from the letter later.
197. Sibold Hemsterhuis: *Messis aurea triennalis, exhibens anatomica novissima et utilissima experimenta. Ex. editione Siboldi Hemsterhuis* (A 3-year golden harvest of new and useful anatomical experiments. Published by Sibold Hemsterhuis), Leiden 1654. A copy of the second edition from 1659 is owned by The Royal Library in Copenhagen.
198. Gosch 1873 writes on page 105, that "Bartholin had sent copies of his treatises on the lacteal and the lymph vessels to Hemsterhuis in Leyden to be published by him", which could give the impression, that Bartholin was behind the release. However, it does not seem likely, that Bartholin should have been behind the publication of a collection, which also contained the treatises of Pecquet and Rudbeck, and no other sources support this interpretation. It is likely, however, that Bartholin sent the said copies of his treatises at the request of Hemsterhuis, in accordance with Hemsterhuis' words in the preface to the book, that if any readers may have theses, that relate to the topic of the book, they are welcome to send them to him.
199. *Epist. Med.* 1663, 2, Epist. 36, 2. February 1654: De Receptaculo Lacteo in Homine (On the chyle container in man), pp. 514-516: "Your own works (as well as those of Pecquet and the Swede) are now finished with the printer, and your name has been added to them. I sent them yesterday to Your Excellency with the son of the Ottonian consul Johannes Christian Kóllebrun."
200. Bartholin writes in a letter to Willum Worm in Leiden on 10. March 1654, that he sends him the first two centuries of the work, which have just been published. See *Epist. Med.* 1663, 2, Epist. 37: De Receptaculo Lacteo Responsio (Answer concerning the chyle container), pp. 516-17.
201. Bartholin repeats these dates again on page 621 of the fourth revised edition of his anatomy book, *Anatome Quartum Renovata* from 1673.

202. *Hist. Anat.* 1654, 2. Lymphaticorum Vasorum Inventio prima (The first discovery of the lymph vessels), pp. 248-58.
203. *Epist. Med.* 1663, 2, Epist. 47: Pro lymphaticorum prima inventione (In defence of the first discovery of the lymph vessels), pp. 547-62.
204. Gosch 1873, p. 107.
205. *Epist. Med.* 1663, 2, Epist. 37: De Receptaculo Lacteo Responsio (Answer concerning the chyle container), pp. 516-17.
206. Gosch 1873, pp. 107-8.
207. Ibid., p. 108.
208. *Epist. Med.* 1663, 2, Epist. 47: Pro Lymphaticorum prima inventione (In defence of the first discovery of the lymph vessels), pp. 547-62.
209. *Olai Rudbeckii Sveci ad Thomam Bartholinum Danum Epistola, qua sibi primam inventionem Vasorum Serosum Hepatis, invictis rationibus, contra Bogdani cuiusdam impudentissimas Calumnias, sub Apologiæ Nomine pro Bartholini editas, asserit*, etc. (A letter from the Swede Olof Rudbeck to the Dane Thomas Bartholin, in which, for irrefutable reasons directed against a certain Bogdan's utterly shameless accusations (allegedly published in defence of Bartholin), he claims to be the first to discover the liver's serous vessels), Uppsala 1657.
210. Eriksson 2002, p. 58.
211. Gosch 1873, p. 108. Eriksson 2002, p. 68.
212. *Epist. Med.* 1667, 4, Epist. 37, 20 October 1662, pp. 213–17.
213. Ibid., Epist. 42, 25 November 1662, pp. 240–49.
214. "I have recently read Rudbeck's *Atlantica* and marvelled at the same man's capacity to invent and present hidden and previously unknown things, by which he has acquired great esteem to the Nordic countries and granted them with the honour that Greek and Roman writers either by malice or ignorance have deprived them. He has with irrefutable arguments proven that many things that ancient writers hold to have happened in the Orient, Asia and Greece, really happened in the Nordic countries." *Ludvig Holberg's Væker*, 11, Epistola CXCIII, pp. 167–72. Arkiv for Dansk Litteratur (www.adl.dk).
215. Rudbeck 1689, 2, pp. 38–39.
216. Olof Rudbeck: *Hortus Upsaliensis Academiæ* (The Garden at the University of Uppsala), Uppsala 1666, which was a reprint of his *Catalogus plantarum* (Plant catalogue), Uppsala 1658. In 1664 came *Horticultura nova* Upsaliensis (Uppsala's new horticulture).
217. In the section on his pupils and assistants in *Domus Anatomica Hafniensis*, Bartholin writes in 1662 about Bogdan: "Martin Bogdan, who at Copenhagen in 1653 wrote two defences [the publication year for both was 1654], in which he reclaimed and reappropriated the lymphatic vessels for their discoverer." Bruun (ed.) 2015, p. 89.
218. George Joyliffe (1621–1658). English physician who in June 1652 described to his teacher Francis Glisson that, by dissections of animals, he had observed in several places on their bodies a kind of aqueous vessels that were not veins, arteries or nerves. He himself never published his observations, but Glisson mentioned them in his *Anatomia Hepatis*, London 1654.
219. Bartholin 1673, p. 622.
220. Eriksson 2002, p. 61.
221. Tryde (ed.) 1940, p. 215.
222. See for example Johannes G. Rudbeck: Biografiske anteckningar om Olof Rudbeck den äldres anatomiske skrifter och striden med Thomas Bartholin (Biographical notes on Olof Rudbeck the Elder's anatomical writings and the dispute with Thomas Bartholin), *Samleren* 1904, 25, pp. 37–57, Tigerstedt (ed.) 1921, p. 359 and Sten Lindroth. *Svensk lärdomshistoria*, 2, Stormaktstiden, Stockholm 1975, pp. 414 and 416–17.
223. Eriksson 2002, p. 68.
224. Ibid., p. 58.
225. C. D. O'Malley: Thomas Bartholin. In: Gillespie (ed.) 1981, 1, pp. 482–83. O'Malley writes: "Although there was extended controversy, there is now little doubt of Bartholin's priority." See also Roy Porter 2000, pp. 219–20, who writes: "Aselli put the [chyle] vessels in connection with the lymph nodes in the groin, and Jean Pecquet followed them on to the ductus thoracicus (a chyle-carrying duct, that leads the chyle into the bloodstream) and thus contributed to the rejection of Galen's liver-centred physiology. At about the same time, the Dane Thomas Bartholin demonstrated the presence of corresponding chyle-carrying vessels in man and further explained the lymphatic system in the *Vasa Lymphatica Nuper Hafniæ in Animantibus inventa, Et Hepatis exseqviæ* (The lymph vessels recently discovered in Copenhagen in animals and the abandonment of the liver) from 1653."
226. Jan Bondeson: *The Two-Headed Boy and Other Medical Marvels*, New York 2004, p. 59.
227. *Epist. Med.* 1663, 2, Epist. 48: De Vasis Rudbeckii (On Rudbeck's vessels), pp. 562–63.
228. Thomas Bartholin: *Spicilegium Secundum Ex Vasis Lymphaticis* (Second Aftermath of the Lymph Vessels), Copenhagen 1660, p. 1.

CHAPTER 5
Administrator and author
pp. 307–340

1. *Orationes*, Copenhagen 1668.
2. Slottved 1978, p. 107.
3. Mansa 1873, p. 401.
4. Ibid and Garboe 1949, 1, p. 183.
5. *Orationes*, pp. 162–211. The preface to the speech is found at the same place, pp. 159–61.
6. Garboe 1949, 1, p. 185.
7. www.natmus.dk. See also Schepelern 1971 and H.C. Bering Liisberg 1897.
8. Schepelern 1971, pp. 226–27.

9. Ibid, p. 226.
10. Charles Lyell: *Principles of Geology*, London 1830, p. 27.
11. C.E. Werlauff & L. Engelstoft: *Udsigt over Københavns Universitets-Bygnings Historie, fra Universitetets Stiftelse indtil 1836* (Review of the history of the university buildings in Copenhagen, from the establishment of the university until 1836), Copenhagen 1836, p. 21.
12. Svend Dahl: *Universitetsbibliotekets Bygninger gennem Tiderne. En historisk Fremstilling* (The buildings of the university library through the ages. A historical presentation), Copenhagen 1932, p. 21.
13. Werlauff & Engelstoft 1836, p. 21.
14. Garboe 1949, 1, p. 205.
15. Werlauff & Engelstoft 1836, p. 22.
16. Ibid.
17. Vilhelm Maar: En Mindestens Historie (The history of a memorial tablet). In: Vilhelm Maar (ed.) 1916, pp. 69-73 and Bruun & Loldrup (eds.) 2007, pp. 148–50.
18. *Orationes*, pp. 250–77. Garboe 1949, 1, p. 203.
19. Slottved 1978, p. 105.
20. Ibid., p. 107.
21. Garboe 1949, 1, p. 209.
22. Slottved 1978, p. 109.
23. *Hist. Anat.* 1657, 4, Hist. 6: Anatome Hominis sani (Dissection of a healthy human), pp. 221–24.
24. *Orationes*, no. 36: Anatome ultima publica (The last public dissection), pp. 228–33. Garboe 1949, 1, p. 210.
25. Ibid.
26. "Rare anatomical cases" would be a more direct translation, but the translation "Rare medical cases" better reflects the content of the work.
27. Klaus Patau, David W. Smith, Eeva Therman, Stanley L. Inhorn and Hans P. Wagner: Multiple congenital anomaly caused by an extra autosome. *Lancet* 1960, 7128, pp. 790–93. See also Jesper Brandt Andersen & Niels W. Bruun: Steno-Fallots tetralogi og Bartholin-Pataus syndrom. En hjertemisdannelse og et misdannelsessyndrom første gang beskrevet af danske anatomer i 1600-tallet (Steno-Fallot's tetralogy and Bartholin-Patau syndrome. A heart malformation and a malformation syndrome first described by Danish anatomists in the 17th century). *Fund og Forskning i Det Kongelige Biblioteks Samlinger* 2015, 54, pp. 197–233.
28. Ibid. and www.whonamedit.com.
29. *Hist. Anat.* 1657, 3, Hist. 47: Monstrum sine oculis, p. 95.
30. Contemporary illustrations, however, including the one presented by Bartholin himself, seem to show that the fusion between the two brothers was rather localized to the upper part of the abdomen, as also described in the newborn brothers by the Italian physician Augustino Pinceto (see later).
31. *Hist. Anat.* 1654, 1, Hist. 66: Frater pectori fratis connatus (A brother who was born on his brother's breast), pp. 116–17. Wolff 1827, 3, pp. 254–55.
32. Bondeson 2004, p. vii.
33. Ibid., pp. vii–xix.
34. Liceti 1665, pp. 273–92. The report about the two conjoined brothers is found on pp. 273–75 with Bartholin's illustration on p. 274.
35. Ibid, p. 114–15.
36. Ibid., p. 117.
37. Zacchia 1688, question 9, p. 601: Novæ Monstrosorum aliquæ historiæ, & alia circa hanc materiam scitu digna (Reports about some new monsters and other similar topics, which are worth knowing about).
38. Henri Sauval: *Historie et Recherches des Antiquités de la Ville de Paris*, Paris 1724, pp. 564–65.
39. John Spalding: *The History of the Troubles and Memorable Transactions in Scotland from the Year 1624 to 1645*, 2, Aberdeen 1792, p. 4.
40. Mansa 1873, pp. 384–87 and *Hist. Anat.* 1654, 2, Hist. 65: Diarrhoea & Dysenteria Epidemica (Epidemic diarrhoea and dysentery), pp. 287–90.
41. *Hist. Anat.* 1654, 2, Hist. 56. Febris Tertiana Epidemica (An epidemic of third day fever), pp. 272–73. Wolff 1828, 4, pp. 217–18. Worm furthermore describes the third day fever, by which he himself is afflicted, in a letter to Pierre Bourdelot in Stockholm on 29 May 1652. See Schepelern (ed.) 1968, 3, pp. 477–78.
42. Girolamo Bardi (1600–1667), Italian physician, theologian and philosopher.
43. Opus cit. 1661, 5, Hist. 50: Pulvis Febrifugus Peruvianus (Peruvian antipyretic powder), pp. 107–16. See also Vilhelm Maar: *Den første Anvendelse af Kinabark i Danmark* (The first use of China bark in Denmark), Leiden 1925, pp. 14–17 and 27–28.
44. Bartholin refers to the wars against Sweden, the so-called Carl Gustav wars 1657-1660 (see later).
45. Philip Hacquart (1616–1698), French surgeon, who came to Denmark from Lothringen/Lorraine during the reign of Christian IV. In 1651 he was granted support for a study trip in Italy and other countries. From 1 October 1655 he was a surgeon at Holmen (several small islands in Copenhagen previously serving as the headquarters of the Danish fleet), Tugthuset (a prison) and Provianthuset (the House of Provisions). From 15 January 1658 he was personal surgeon for Frederik III and from 13 February 1671 for Christian V. See K. Carøe: Livkirurg Philip Hacquart og hans Sønner (Royal Surgeon Philip Hacquart and his sons). *Personalhistorisk Tidsskrift* 1920, rk. 7, bd. 5, 150–67.
46. *Hist. Anat.* 1661, 5, Hist. 100: Cancri in mammis exulcerati excisio (Removal of an ulcerated breast cancer), pp. 185–92. Danish translation in Niels Nørskov-Lauritsen & Niels W. Bruun: En celeber kræftoperation. Philip Hacquarts fjernelse af et mammakarcinom fra Lisbeth Lunge, København den 16. juni 1658 (A fashionable cancer operation. Philip Hacquart's removal of a mammary carcinoma from Lisbeth Lunge, Copenhagen, 16 June 1658). *Bibliotek for Læger* 1995, 187, 2, pp. 103––22.
47. Carøe 1920, p. 150.

48. *Hist. Anat.* 1657, 2, Cent. 3, Hist. 15: Motus sine corde (Movement without the heart), pp. 36–37.
49. Bacon 1623, pp. 390–91. Quoted after the English edition with the title *The Historie of Life and Death*, London 1638, pp. 270–71.
50. Schepelern (ed.) 1968, 3, Epist. 790 (no. 1430), pp. 194–95.
51. *Hist. Anat.* 1654, Cent. 2, Hist. 11: Sirenis seu Marini Hominis Anatome (The dissection of a mermaid or a sea human), pp. 186–91.
52. Ibid., p. 188.
53. Charles Darwin: *On the Origin of Species by Means of Natural Selection or the Preservation of favoured Races in the Struggle for Life*, London 1859.
54. Ibid., p. 191, and Bruun & Loldrup (eds.) 2007, p. 164. For the whole section about the dissection of the mermaid, see Brandt Andersen 2018, pp. 98–101.
55. Thomas Bartholin: *Domus Anatomica Hafniensis brevissime descripta* (The Anatomy House in Copenhagen briefly described), Copenhagen 1662, p. 52. Bruun (ed.) 2015, pp. 110–11 and 186-87.
56. Bartholin 1651, p. 59.
57. Gosch 1873, p. 75.
58. Salmuth 1648, Cent. 3, no. 94, pp. 156–57. See also E. Ingerslev: *Fragmenter af Fødselshjælpens Historie* (Fragments of the history of birth care), Copenhagen 1907, 2, p. 330.
59. Riolan 1653, p. 15, and Gosch 1873, p. 75.
60. *Hist. Anat.*, Cent. 5, Hist. 98: Felis per os pariens (Cat gives birth through the mouth), pp. 180–83. Thal Jantzen: Af Thomas Bartholins "anatomiske Historier." Special edition of *Bibliotek for Læger*, 8, 5, 1904, p. 5.
61. Ibid.
62. Bartholin 1664, Chapter IX: Ex ore Nativitas (Birth through the mouth), pp. 55-64.
63. Bartholin 1673, p. 80. The section about birth through the mouth is translated from Brandt Andersen 2018, pp. 67–69.
64. *Hist. Anat.*, 1654, 2, Hist. 38: Caseus, Placenta, globulus per vomitum rejecta (Cheese, placenta, bullet excreted through vomiting), pp. 230–32. Wolff 1828, pp. 165–66.
65. Letter from Thomas Bartholin in Copenhagen to the city medical officer in Christiania (now Oslo) Ambrosius Rhode (1605–1696) on 25 August 1655. *Epist. Med.* 1663, 2, Epist. 66: An Diabolus in substantia malus (Is the devil evil by nature?), pp. 613–17. For the subject, see also Garboe 1949, 1, p. 188.
66. *Acta Medica Ann. 1671 & 1672*, Copenhagen 1673, pp. 102–3. Garboe 1949, 1, p. 188.
67. German translation: *Dess Höckstberühmten Anatomici, ungewohnlicher Anatomischer Geschichten Erstes und Zweytes Hundert, Auff Vieler Begehren auss dem Lateinischen ins Teutsche versetzt, und mit darzu dienlichen Registern-verfertiget, von Georg Seger*, Frankfurt 1657. Georg Seger was born in Torun (then Prussia, now Poland). He studied for some years during the 1650s in Copenhagen, where his publication list extends from 1652 to 1658 (see H. Ehrencron-Müller 1929, 7, pp. 325–25).

According to *Kjøbenhavns Universitets Matrikel* (The register of Copenhagen University, see S. Birket-Smith 1890, 1: 1611–1667, p. 238) he was matriculated on 7 August 1652 with Thomas Bartholin as his preceptor after previous studies in Wittenberg, Leipzig and Königsberg. He was a pupil of both Ole Worm and Thomas Bartholin and a friend of Niels Stensen, with whom he studied in Copenhagen from 1656 to 1658. In 1653 he published a synopsis of Worm's collection of rarities and during the mentioned years a few anatomical writings, mainly about the lymphatic system. After the stay in Copenhagen, he worked some years in Basel, where he obtained his medical doctorate, and then some years as a physician in Torun, before in 1675 he became professor of physics and medicine in Gdansk, where he died in 1678. Seger is among other things known for the first description of the disease cystic fibrosis in a three-year-old Polish girl in 1673. In his book about the Anatomy House in Copenhagen from 1662, Thomas Bartholin writes the following about Seger in the section about his pupils: "Georg Seger, who, after the detection of the lymph vessels in Copenhagen in 1653, brought out *A Triumph for the Heart*, which was later reprinted in Basel, and undertook some clever dissertations on Bartholin's lymphatic system in Copenhagen in 1656." See Bruun (red.) 2015, p. 89.
68. *Twee hondert Getal van Thomas Bartholyns seltsame ondervindingen ofte Geschiedenissen, omtrent de Anatomia ofte Ontledinge*, Dordrecht 1657.
69. Concerning the latter, see for example Gosch 1873, pp. 65–66 and pp. 133–36, P.L. Panum: Bidrag til Kundskab om vort medicinske fakultets Historie med hensyn til dets Betydning for Naturvidenskabernes og Lægevidenskabens udvikling i Danmark fra Frederik den 3dies Tronbestigelse indtil Frederik den 5tes Død (1648–1766) – *Indbydelsesskrift til Kjøbenhavns Universitets Aarsfest til Erindring om Kirkens Reformation* (Contribution to the knowledge about the history of our medical faculty concerning its significance for the development of the natural sciences and the medical science in Denmark from the time of the coronation of Frederik III to the death of Frederik V (1648–1766) – Invitation program to the university's annual celebration of the reformation of the church), Copenhagen 1880, pp. 28–29, H.C. Bering Liisberg 1897, pp. 30–32, Julius Petersen: *Bartholinerne og Kredsen om dem* (The Bartholins and the circle around them), Copenhagen 1898, p. 57, Thal Jantzen 1904, p. 2, E. Ingerslev 1907, p. 329 and 335, and Maar (ed.) 1916, pp. 7–8.
70. Brandt Andersen 2018, p. 114 and English summary p. 115.
71. *Epist. Med.* 1667, 2, Cent. 3, Epist. 83: De narrationibus Fabulosis (On fabulous narrations), pp. 354–57. For the section concerning Patin's criticism, see Brandt Andersen 2018, pp. 112–13.

72. *Hist. Anat.* 1661, 3, Cent. 6, Hist 61: Ex imaginatione natus (Born by imagination), pp. 296–304.
73. Ibid., Hist. 92: Foetus extra uterum in abdomine inventus (Fetus found in the abdominal cavity outside the uterus), pp. 359–64.
74. Ibid., pp. 362–63.
75. *Epist. Med.* 1667, 2, Cent. 3, Epist. 84: De mirabilibus in partubus humanis (On wonders in human births), pp. 357–59.
76. Ibid.
77. Jean Berryat et al. (eds.): *Collection Académique. Tome Septiéme de la Partie Etrangere, Et le Premier de la Médecine séparée,* Dijon and Paris 1766, the preface.
78. E.g. Maar (ed.) 1916, p. 7.
79. Kristian Carøe: *Medicinalordningens Historie indtil Sundhedskollegiets Oprettelse 1803*, Copenhagen 1917, pp. 7–8.
80. Ibid., p. 9.
81. Ibid., pp. 10–12.
82. A facsimile edition with an added Danish and English text by Kurt Bærentsen about the work in the light of history was published in Copenhagen in 1966 by Dansk Farmacihistorisk Selskab.
83. Bartholin 1658, p. 254.
84. Ibid., pp. 128–34.
85. *Epist. Med.* 1663, 2, Epist. 1: De pharmacopeis (On apothecaries), pp. 417–20. Loïc Capron (ed.): *Correspondance complète de Guy Patin et autres écrits.* https://www.biusante.parisdescartes.fr/patin/.
86. Bartholin 1661, pp. 83–84.
87. Garboe 1950, 2, p. 25.
88. *Orationes*, Copenhagen 1668, Oratio 45: Panegyricus Augustissimo Regi Friderico III (Eulogy to the highest king, Frederik III), p. 285–307.

CHAPTER 6
Niels Stensen – the ingenious pupil
pp. 341–416

1. Said by Niels Stensen during a demonstration of the anatomy of the brain in Paris in 1665. Published in the same place in 1669 with the title *Discours de Monsieur Stenon sur l'Anatomie du Cerveau à Messieurs de l'Assemblée, qui se fait chez Monsieur Mr. Thévenot.*
2. According to the university's register, Stensen was matriculated on 27 November 1656 under his Latin name Nicolaus Stenonis. See Birket-Smith 1890, pp. 264–65.
3. Eilersen and Borch were professors at the Faculty of Philosophy from 1654 and 1660, respectively. See Slottved 1978, pp. 210 and 157.
4. S. Birket-Smith (ed.): *Etatsraad og Landkomissær Johan Monrad's Selvbiografi (1638–1692). Udgivet efter Original-manuskriptet af S. Birket Smith* (Councillor of state and country commissioner Johan Monrad's autobiography (1638–1692). Published from the original manuscript by S. Birket Smith), Copenhagen 1888, p. 42.
5. Holger Rørdam: *De danske og norske Studenters Deltagelse i Kjøbenhavns Forsvar mod Karl Gustav. Et Bidrag til Kjøbenhavns Universitets Historie 1658–60* (The participation of the Danish and Norwegian students in the defence of Copenhagen against Carl Gustav. A contribution to the history of the University of Copenhagen), Copenhagen 1855, p. 159.
6. Garboe 1950, 2, pp. 22–23 documents, that Thomas Bartholin and his family left Copenhagen and sought safety in Roskilde in August or the first days of September 1658, and on p. 24 documents, that Bartholin was at least back in Copenhagen on 26 May 1659, where he gave a eulogy at the university for his cousin Henrik Fuiren, who had died on 8 January of the same year.
7. *Hist. Anat.* 1657, Cent. 4, Hist. 100: Leonis Anatome (Dissection of a lion), pp. 424–30.
8. *Hist. Anat.* 1661, Cent. 5, Hist. 1: Undecim costæ in thorace (Eleven ribs in thorax), pp. 1–3.
9. Ibid., Hist. 49: Civettæ feminæ Anatome (Dissection of a female owl), pp. 105–7.
10. Slottved 1978, pp. 109 and 152.
11. Brandt Andersen 2021, pp. 122–28.
12. Rasmus Bartholin: *De Studio Lingvæ Danicæ* (On the study of the Danish language), Copenhagen 1657. In: Rasmus Bartholin: *De Naturæ Mirabilibus Questiones Academicæ* (Academic questions on the marvels of Nature), Copenhagen 1674, pp. 146–70. No. 11 of the book's 13 questions. Danish translation by Carl Berend with the title "Rasmus Bartholins Tale om det danske Sprog." In: *Studier fra Sprog og Oldtidsforskning*, no. 97, 1914. Reprinted in Henrik Bertelsen (ed.): *Danske Grammatikere fra Midten af det syttende til Midten af det attende Aarhundrede*, reprinted in Copenhagen 1979 (original edition 1915–1929), 6, pp. 35–64. The quoted piece is on p. 48. The Latin text is in both editions in Vol. 1, pp. 3–33.
13. Brandt Andersen 2021, pp. 560–63.
14. Ziggelaar (ed.) 1997, col. 27, pp. 88–90 and 455.
15. *Philosophical Transactions*, London 1670, no. 67, pp. 2039–48.
16. A correspondence between Rasmus Bartholin and Christiaan Huygens can be found in *Oeuvres complètes de Christiaan Huygens*, The Hague 1888, 1, No. 352, pp. 515–16, No. 361, pp. 528–29 and 1889, 2, No. 445, pp. 113–14.
17. André Authier: *Early Days of X-ray Crystallography*, Oxford 2013, p. 298.
18. For further reading on this subject, see Axel Garboe: Nicolaus Steno (Niels Stensen) and Erasmus Bartholin. Two 17th Century Danish Scientists and the Foundation of exact Geology and Crystallography, Special edition of *Danmarks Geologiske Undersøgelse* 1954, IV. Række, Vol. 3, No. 9, pp. 1–48.
19. August Ziggelaar (ed.): Chaos. Niels Stensen's Chaos-manuscript, Copenhagen 1659. Complete edition with Introduction, Notes and Commentary by August Ziggelaar, *Acta Historica Scientiarum Naturalium et Medicinalium* Vol. 44, Copenhagen 1997. For a synopsis and analysis, see Brandt Andersen 2021, pp. 134–50.

20. Ziggelaar (ed.) 1997, pp. 461–69. Brandt Andersen 2021, p. 138.
21. Surprisingly, the letter does not appear in *Epist. Med.*, but in the previously mentioned letter from Paris on 17 April 1656, which is found in *Epist. Med.*, Pecquet justifies his dedication to Bartholin: "When I dedicated the latest fruits of my labours to you, I did not imitate those who, with their wretched little gifts, try to obtain the generosity of influential people. On the other hand, I hoped you would taste them with your open-minded palate and evaluate them in the same way of thinking that you use as you work to find the undivided truth." It appears from the letter that Pecquet became happy with Bartholin's support for the work, and that he perceives Bartholin as an important ally in the dispute with the perfidious Riolan.
22. On 31 July 1654, Thomas Bartholin's faithful friend in Paris, Guy Patin, had notified him about Pecquet's upcoming book: "Pecquet lives here and is preparing for the release of another expanded edition of his work with several new exercises to explain and confirm the content" (*Epist. Med.* 1663, Cent. 2, Epist. 46: De Riolano & Pecqueto (On Riolan and Pecquet), pp. 545–47. Loic Capron (ed.): *Correspondance complète et autres écrits de Guy Patin*, https://www.biusante.parisdescartes.fr/patin/. On 13 December of the same year, Pecquet sent a copy from Paris of the book to Thomas Bartholin: "Excellent sir, my little book has long been prepared for its journey and awaited favourable winds to reach you, to whom it is dedicated. [...] Therefore, I thought it was best not to delay it anymore, but to assure you of my devotion as soon as possible." (*Epist. Med.* 1663, Cent. 2, Epist. 54: De semita ad mammas (On the path to breasts), pp. 581–82).
23. According to the handwritten catalogue in the Royal Library's archive E2 Joachim Gersdorff's library contained books by, among others, Tycho Brahe, Pierre Borel, René Descartes, Galileo Galilei, Pierre Gassendi, Athanasius Kircher, Simon Paulli, Matthias Untzer and Francis Bacon. For a more detailed list, see H.D. Schepelern (ed.): Niels Stensen. En dansk student i 1659 og noterne i hans Chaos-manuskript (A Danish student in 1659 and the notes in his Chaos manuscript). *Skrifter udgivet af Universitetsbiblioteket 2. Afdeling, Danmarks natur- og lægevidenskabelige Bibliotek*, Copenhagen 1986, p. 71, which source also mentions (p. XIV) that the library contained approximately 7000 volumes concerning all kinds of subjects and sciences. Like other students, Stensen presumedly also had access to other of the city's private libraries (for a mention of these, see ibid., pp. XIII–XIV) and from 7 July 1657 to the newly inaugurated university library on the first floor of the recently built Trinity Church, a couple of hundred metres from his home.
24. Dirk van Miert: *Humanism in an age of Science. The Amsterdam Athenaeum in the Golden Age 1632–1704*, Leiden 2009, p. 92.
25. Brandt Andersen 2021, p. 155.
26. In a letter to Thomas Bartholin on 25 March 1661, Gerard Blasius' brother, Joan Blasius, writes: "But because I have seen that my only brother has made contact with you through letters of introduction [...]." *Epist. Med.* 1667, Cent. 3, Epist. 19, pp. 73–76.
27. In a letter from Gerard Blasius to Thomas Bartholin dated 16 July 1661 (see later), Blasius writes: "It is well over a year since I extended hospitality to Mr. Niels Stensen, who came furnished with your letters of recommendation."
28. *Epist. Med.* 1667, Cent. 3, Epist. 24: De prima ductus salivalis exterioris inventione, & Bilsianis experimentis (On the first Discovery of an External Salivary Duct and the Experiments of Bils), pp. 86-95. Maar (ed.) 1910, 1, pp. 3–7. Scherz (ed.) 1952, 1, E1, pp. 133–37. English translation in Kardel & Maquet (eds.) 2018, pp. 419–23. Danish translation in Harriet M. Hansen (ed.) 1987, 1, E1, pp. 27–32 and Christensen, R.E., Hansen, Axel and Larsen, Knud (eds.): Niels Stensen's (Stenonis) Værker i Oversættelse (works in translation). In: *Klassisk dansk Medicin*, Vol. II, Copenhagen 1939, pp. 3–10.
29. See below.
30. For the most recent letter registrant, see Helge Clausen: Niels Stensens brevregistrant. Nicolaus Steno. Index of letters. *Ælnoths skriftserie Nr. 18*, 2019.
31. Borch's very detailed and informative travel diary from the first almost five years of his travel (the diary from the last year's travel in Italy has not been preserved) was published by H.D. Schepelern in four volumes in London and Copenhagen in 1983 under the title *Olai Borrichii Itinerarium 1660–1665. The Journal of the Danish Polyhistor Ole Borch*, I–IV. Volume I contains Borch's short autobiography in English translation.
32. *Epist. Med.* 1667, Cent. 3., Epist. 85, pp. 360–69.
33. Wharton 1656, Chapter 21: De glandulis maxillaribus (On the salivary glands of the upper jaw), pp. 12–37. Illustration on p. 132.
34. Ibid., Chapter 20: De parotidibus (On the parotid glands), pp. 124–27.
35. Ibid., p. 127.
36. Ibid, Chapter 18: De glandulis thyroidæis (On the thyroid glands), p. 122.
37. Further, Wharton refers to Bartholin's section about the glands of the intestine and peritoneum, the adrenal glands, the thymus, the glands of the oesophagus, the pituitary gland, the placenta and the uterus.
38. Bartholin 1660, the introduction to chapter 5 on Wharton's new discoveries and opinions on the lymph vessels.
39. William LeFanu: Thomas Wharton. In: Gillispie (ed.) 1981, 14, pp. 286–87.
40. Henricus Eyssonius (1620–1690), Dutch physician and anatomist. From 1660 professor of medicine at the University of Groningen. Joan Leonardszoon Blasius (1639–1672), Dutch poet, playwright, theatre director, translater and lawyer. The brother of Gerard Blasius.

41. Stensen's seven letters to Bartholin during the period from April 1661 to July 1663 were, as mentioned, published in the third and fourth century of Bartholin's *Epistolarum Medicinalium* in Copenhagen 1667, eight of Stensen's smaller treatises were published in vols. I and II of Bartholin's *Acta Medica et Philosophica Hafniensia* in Copenhagen 1673 and 1675, respectively, and Stensen's drawing of the blood circulation made clearer by separation of the ventricles was published in Bartholin's *Anatome Quartum Renovata* in Leiden in 1673. For a complete list of Stensen's 34 publications of natural science (anatomy and geology), see Brandt Andersen 2021, pp. 773–74.
42. The name of the University of Amsterdam at that time.
43. The disputation was, after about 300 years in oblivion, rediscovered in 1960 by Gustav Scherz in the Library Company of Philadelphia, Pennsylvania, USA, which owns the only known copy of it. See Gustav Scherz: Niels Stensen's First Dissertation. *Journal of the History of Medicine and Allied Sciences* 1960, 15, 3, pp. 247–64.
44. Borch left Denmark on 11 November 1660 and arrived in Amsterdam on 3 December of the same year. On 4 January 1661, after a month's stay with Blasius in Amsterdam, he travelled on via Haarlem to Leiden, where he joined Stensen.
45. *Epist. Med.* 1667, Cent. 3., Epist. 85, pp. 360–69.
46. G.A. Lindeboom: Johannes van Horne. In: Gillispie (ed.) 1981, 6, pp. 508–9.
47. The dedication is dated "Pridie Non. Maij," meaning the day before Non. Maij. Non. is an abbreviation of Nonas, and in the Latin dating rules "Nonas Maij" means the seventh of May. The date is therefore the day before the seventh of May, i.e. the sixth of May.
48. *Epist. Med.* 1667, Cent. 3, Epist. 86, pp. 369–74.
49. Ibid., Epist. 10, pp. 48–52.
50. Ibid., Epist. 87, pp. 374–77.
51. Blasius 1661, p. 24.
52. Ziggelaar (ed.) 1997, col. 12, 10. March 1659, p. 50.
53. The publication is dated Amsterdam 1661 and the parts separately Amsterdam 1660, although the first part was published in Copenhagen 1655 and the second part in Copenhagen 1660.
54. *Epist. Med.* 1667, Cent. 3, Epist. 20, pp. 77–79.
55. For Stensen's research on the lymph vessels during his stay in Leiden, see later.
56. *Epist. Med.* 1667, Cent. 3, Epist. 25, pp. 95–98. Scherz (ed.) 1952, 1, E2, pp. 138–40. Harriet M. Hansen (ed.) 1987, 1, E2, pp. 32–34.
57. *Epist. Med.* 1667, Cent. 3, Epist. 91, pp. 390–95.
58. Ibid., Epist. 94, pp. 405–9.
59. Seger refers to Olof Rudbeck's claim to the discovery of the lymph vessels.
60. Ibid., Epist. 35, pp. 125–30.
61. The Latin text is reproduced in Maar (ed.) 1910, 1, pp. 9–51. English translation in Kardel & Maquet (eds.) 2018, pp. 425–63. Danish translation in Christensen, Hansen and Larsen (eds.) 1939, pp. 11–77. All these sources reproduce the text of the second edition, which begins the collection *Nicolai Stenoni's Observationes Anatomicæ* (Niels Stensen's anatomical observations), Leiden 1662.
62. Kardel & Maquet (eds.) 2018, p. 426. D.D.D. is an abbreviation of dat, dicat, dedicat (gives, devotes, dedicates).
63. Ibid., pp. 427–28.
64. Ibid., p. 428.
65. Ibid., p. 436.
66. Ibid., p. 437.
67. Ibid., p. 438.
68. Ibid., p. 449.
69. *Epist. Med.* 1667, Cent. 3, Epist. 43: Examen Disputationis Stenonianæ de inventione Ductus salivalis exterioris (Study of Niels Stensen's treatise on the discovery of the external salivary duct), pp. 158–84.
70. *Epist. Med.* 1667, Cent. 3, Epist. 44: Reconsiliatio inter dissidentes amicos tentatur (Attempt at mediation between disputing friends), pp. 184–88.
71. In Greek mythology, Sinon is the name of the Greek warrior who persuaded the Trojans to draw the Trojan horse within the walls, so that the Greek soldiers hidden in it could open the city gates under cover of night.
72. Maar (ed.) 1910, 1, pp. 143–54. Kardel & Maquet (eds.) 2018, 533–42.
73. Letter from Thomas Bartholin in Copenhagen to Nicolas Hoboken in Utrecht on 4 April 1663 (see Gosch 1873, p. 160, who corrects to this date from Bartholin's given date of 4 June 1662 with reference to *Spicilegium Epistolarum* (Aftermath of the Letters) in Hoboken's *Anatomia Secundinæ Humanæ* (Second edition of the human anatomy), Utrecht 1669, Epist. 3, pp. 72–76), *Epist. Med.* 1667, Cent. 4, Epist. 4, pp. 15–19. Furthermore, Gosch 1873, p. 161, who refers to Hoboken 1669, Epist. 4, pp. 76–98.
74. Stensen 1664, p. 30. According to Epicurus, the universe is infinite and contains an infinite number of worlds. Rooms between the worlds of Epicurus are empty. Epicurus also believed that all knowledge is based on objective sensations, and that errors only occur when adding subjective opinions. Thus, his philosophy was fundamentally consistent with the experiment-based empirical research method practised by Stensen.
75. Greek, *hugros* = wet, humid; *ophthalmos* = eye.
76. *Epist. Med.* 1667, Cent. 3, Epist. 93, pp. 399–404.
77. Ibid., Epist. 94, pp. 405–9.
78. Niels Stensen: *De Glandulis Oculorum, Novisque earundem vasis Observationes Anatomicæ, Quibus Veri lacrymarum fontes deteguntur* (Niels Stensen's anatomical observations on the glands of the eyes and their new vessels, by which the true sources of the tears are revealed), Leiden 1662. Maar (ed.) 1910, 1, pp. 75–90. Kardel & Maquet (eds.) 2018, pp. 483–94.

79. Ibid., Epist. 57: Variæ in Oculis & Naso observationes novæ etc. (Various new observations in the eyes and nose etc.), pp. 224–30. Maar (ed.) 1910, 1, OPH No. 3, pp. 53–58. Scherz (ed.) 1952, 1, E 3, pp. 140–44. English translation in Kardel & Maquet (eds.) 2018, pp. 465–68. Danish translation in Harriet M. Hansen (ed.) 1987, 1, E 3, pp. 34–37.
80. For this experiment, see later.
81. *Epist. Med.* 1667, Cent. 3, Epist. 28: Usus Vasis novi in Naso (The use of the new vessel in the nose), pp. 231–34. Harriet M. Hansen (ed.) 1987, 1, E 4, pp. 38–39. Scherz (ed.) 1952, 1, E4, pp. 144–46.
82. Titus Maccius Plautus (254–184 BC): *Bacchides*.
83. On 18 November 1661, eight days after Bartholin's letter to Stensen, Stensen had written the following to the professor of medicine at University of Groningen, Anton Deusing (1612–1666): "At all places indeed, which the praised famous Gentleman [Konrad Viktor Schneider] holds as sources of the catarrhs, on page 554, I have observed and shown to friends fleshy glands even before I saw this third tome [...] and, in the mucous membrane of the nose, at the side of this [nasolacrimal] duct, which runs from the lacrimal points to the anterior aspects of the nose, I have recently observed another excretory vessel originating from these small glands, so that I concluded that the conglomerate glands provide us with the true springs of the catarrhs." Kardel & Maquet (eds.) 2018, pp. 477–78.
84. Maar (ed.) 1910, pp. 91–97. Kardel & Maquet (eds.) 2018, pp. 495–99.
85. Franciscus Sylvius used the word conglomerate for exocrine glands, like the salivary glands, and the word conglobate for lymph glands.
86. Kardel & Maquet (eds.) 2018: Origin of Sweat from Glands. On the Insertion and Valvula of the Thoracic Duct and of the Lymphatics, pp. 501–3. *Epist. Med.* 1667, Epist. 65: De insertione & valvula lactei thoracici & lymphaticorum, pp. 262–66. Maar (ed.) 1910, 1, OPH No. 7, pp. 101–3. Scherz (ed.) 1952, 1, E 5, pp. 146–49. Danish translation in Harrie M. Hansen (ed.) 1987, 1, E 5, pp. 40–42.
87. Peder Reedtz had become chancellor in 1661.
88. *Epist. Med.* 1667, Cent. 3, Epist. 95, pp. 409–13. For the defence against Deusing, see later.
89. Bartholin mentions his king-granted pension in a letter to Ole Borch on 29 March 1661: "My own mental faculties have grown dull and weary through these constant labours of mine, not to mention the weakness of my body. Therefore, to avoid becoming involved in new work, I have resolved to lay aside the weariness of my former life for the retirement granted me by his most serene Royal Majesty, and I shall follow the example of great men by granting myself rest from my labours." *Epist. Med.* 1667, Cent. 3, Epist. 88, pp. 378–81.
90. *Epist. Med.* 1667, Cent. 3, Epist. 96, pp. 413–16.
91. Ibid., Epist. 97, pp. 416–25.
92. Ibid., Epist. 66, pp. 266–69. Scherz (ed.) 1952, 1, E6, pp. 149-151. Harriet M. Hansen (ed.) 1987, 1, E 6, pp. 42–43.
93. Ibid. Epist. 98, pp. 425–29.
94. Stensen refers to his discovery of the excretory duct of the parotid gland, *ductus parotideus or ductus Stenonianus*.
95. Bartholin had, on the last page of his 84-page letter to Anton Deusing, *Dissertatio Anatomica De Hepate Defuncto* etc., Copenhagen 1661 (see later), mentioned Stensen. It was Bartholin's first, but certainly not last, mention of Stensen in a published work.
96. *Epist. Med.* 1667, Cent. 4, Epist. 1: Cur Nicotianæ pulvis oculos clariores reddat. De Lactea gelatina observatio (Why tobacco powder makes the eyes clearer. Observation on a milky gelatine), pp. 1–10. Maar (ed.) 1910, 1, OPH 8, pp. 105–11. Scherz (ed.) 1952, 1, E 7, pp. 151–56. English translation in Kardel & Maquet (eds.) 2018, pp. 505–9. Danish translation in Harriet M. Hansen (ed.) 1987, 1, E 7, pp. 44–48.
97. *Epist. Med.* 1667, Cent. 4, Epist. 2: Quare lumen intensum oculis officit? etc. (Why does bright light harm the eyes?), pp. 10–14. Harriet M. Hansen (ed.) 1987, 1, E 8, pp. 48–51. Scherz (ed.) 1952, 1, E 8, pp. 156–58.
98. *Epist. Med.* 1667, Cent. 4, Epist. 26: Observationes Anatomicæ in avibus et cuniculus (Anatomical observations in birds and rabbits) pp. 103–13. Maar (ed.) 1910, OPH No. 9, pp. 113–20. Scherz (ed.) 1952, 1, E 9, pp. 158–64. English translation in Kardel & Maquet (eds.) 2018, pp. 511–16. Danish translation in Harriet M. Hansen (ed.) 1987, 1, E 9, pp. 51–56.
99. *Epist. Med.* 1667, Cent. 4, Epist. 27, pp. 113–17. Scherz (ed.) 1952, 1, E10, pp. 164–66. Harriet M. Hansen (ed.) 1987, 1, E10, pp. 56–58.
100. Epistola Johannis Walæi De Motu Sanguinis: Ad Thomam Bartholinum, Casp. Filium (Letter on the motion of the blood from Johannes Walæus to Thomas Bartholin, son of Caspar). Dated Leiden, 10 October 1640. Bartholin 1641, pp. 385–408.
101. Thus, for example, the German physician and botanist Johann Theodor Schenck (1619–1671) writes on 6 May 1663 from Jena to Thomas Bartholin in Copenhagen after receiving Bartholin's book about the lungs with Malpighi's two letters: "Malpighi owes his fame to your expertise: without it, his own two letters would have been nothing but shadows, for they burst forth in a superficial sweat." *Epist. Med.* 1663, Cent. 4, Epist. 72: De Pulmonibus dubia (Doubts about the lungs), pp. 427–37.
102. Tigerstedt (ed.) 1921, pp. 343–45.
103. In antiquity, it was widely believed that the substance of lung tissue was fleshy, but Bartholin emphasizes here that Hippocrates did not share this view.
104. Giovanni Alfonso Borelli.
105. Lat.: *ad extispicinam*. The reference is to the practice in ancient Roman religion of examining the entrails of animals for prophetic signs.
106. William Harvey: *Exercitationes de Generatione Animalium, quibus açç. quædam de Partu: de Membranis ac Humoribus Uteri et de Conceptione* (Exercises on the generation of animals, to which are added some on parturition, on the membranes and fluids of the uterus, and on conception), London 1651.

107. Pierre Gassendi: *Viri Ilustris Nicolai Claudii Fabricii de Peiresc, Senatoris Aquisextiensis, Vita*, (The life of the illustrious Nicolas-Claude Fabri de Peiresc, Senator of the Parliament at Aix), Paris 1641.
108. Cornelis van Hogelande (1590–1662), Dutch physician educated and living in Leiden. A friend of Descartes.
109. *Epist. Med.* 1667, Cent. 3, Epist. 82: De Pulmonis Anatome nova (On the new anatomy of the lungs), pp. 350–54.
110. Luigi Belloni: Marcello Malpighi. In: Gillispie (ed.) 1981, 9, pp. 62–66.
111. Tigerstedt (ed.) 1921, p. 347.
112. The most powerful of van Leeuwenhoek's microscopes, now in the University Museum in Utrecht, has a linear magnification force of 266 and a resolution strength of 1.34 micrometres, which means that two objects with a distance of 0.00135 millimetres can be distinguished as two separate objects by the observer. This was first surpassed in the 19th century. See Porter 2000, pp. 224–25 and Johannes Heniger: Antoni van Leeuwenhoek. In Gillispie (ed.) 1981, 8, pp. 126–30. Caspar Bartholin the Younger visited van Leeuwenhoek in 1674 and had a demonstration of red blood cells in one of his microscopes. Caspar mentioned this in a letter to his father in August of that year, and Thomas Bartholin published the letter in *Acta Medica Ann. 1674, 1675 & 1676*, Copenhagen 1677, Obs. 3: Microscopiis detecti globuli in sangvine (Blood cells made visible with a microscope), pp. 7–8. Jan Swammerdam already saw red blood cells in one of his microscopes in 1666, and Marcello Malpighi described them in *De Polypo Cordis Dissertatio* (Treatise on cardiac polyps), which was published as an appendix to the work *De Viscerum Structura Exercitatio Anatomica* (Anatomical exercise in the structure of the viscera), Bologna 1666.
113. Carl Henrik Koch: *Spinoza. Kætter og filosof* (Spinoza. Heretic and philosopher), Copenhagen 2015, pp. 96–98.
114. Both the Hebrew Baruch and the Latin Benedict mean "the blessed one."
115. Schepelern (ed.) 1983, 1, p. 214. The quality of Spinoza's microscopes is testified to by a statement from the Dutch physician and anatomist Theodor Kerckring (1638–1693), who in his *Spicilegium Anatomicum* (Anatomical aftermath), Amsterdam 1670, writes: "My excellent microscope, made by the noble philosopher and mathematician Benedict Spinoza, and with which one can see the lymph vessels spreading in different strands, before they enter their conglobate glands and continue, until they depart again and gather in a single vessel." Kerckring then mentions further anatomical observations, that he has made with his "admirable instrument" (p. 178).
116. *Epist. Med.* 1667, Cent. 4, Epist. 51: De Malpighii s cripto, observationes variæ Anatomicæ (About Malpighi's book on various anatomical observations), pp. 328–36.
117. Ibid., Epist. 55: De Vesiculis in pulmone. Anatome cuniculi precursor. In pulmonibus experimenta. De lacteis mammarum. In cygno observationes, etc. (Vesicles in the lung. Dissection of a pregnant rabbit. Experiments in lungs. On the milk ducts of the breasts. Observations in a Swan, etc.), pp. 348–59. Maar (ed.) 1910, 1, pp. 129–36. Scherz (ed.) 1952, 1, E11, pp. 167–73. English translation in Kardel & Maquet (eds.) 2018, pp. 523–28. Danish translation in Harriet M. Hansen (ed.) 1987, 1, E 11, pp. 59–64.
118. Tigerstedt (ed.) 1921, p. 347.
119. *Epist. Med.* 1667, Cent. 4, Epist. 70: Nova Musculorum & Cordis fabrica (The new structure of the muscles and the heart), pp. 414–21. Maar (ed.) 1910, 1, pp. 155–60. Scherz (ed.) 1952, 1, E 13, pp. 175–79. English translation in Kardel & Maquet 2018, pp. 543–47. Danish translation in Harriet M. Hansen (ed.) 1987, 1, E 13, pp. 67–71.
120. Ibid., Epist. 76: De vesiculis pulmonum, ranarum Anatome, aëre in cor penetrante, vesicæ tunica, mode solvendi, calculos per aquam calcis vivæ (On the vesicles of the lungs, the dissection of frogs, air entering the heart, the tunic of the bladder, a method of dissolving stones with water of quicklime), pp. 446–53.
121. Bartholin 1663, p. 29.
122. Johannes Schreyer: *Erörterung und Erläuterung der Frage / Ob es ein gewiß Zeichen / wenn eines todten Kindes Lunge im Wasser untersincket / daß ein solches in Mutter-Leibe gestorben sey* (Discussion and explanation of the question / Whether it is a certain sign / When a dead child's lungs sink into the water / That such a child has died in its mother's womb), Zeitz 1690, pp. 5 and 15.
123. Stensen refers to his earlier mentioned investigations of the heart of the raven and the rabbits.
124. Finn Geneser: *Histologi*, Copenhagen 1981, pp. 246 and 258.
125. Letter from Niels Stensen in Hanover to G. W. Leibniz in the same place, November 1677. Scherz (ed.) 1952, E 143, pp. 366–69. Harriet M. Hansen (ed.) 1987, 1, E 143, pp. 283–85.
126. Bartholin 1651, p. 239 and Caspar Bartholin's *Editio Nova correcta & locupletata* (The new corrected and enriched edition), Goslar, p. 186.
127. Ibid.
128. Porter 2000, pp. 73–83.
129. Johannes Antonides van der Linden (1609–1664), Dutch physician, botanist, author and librarian. Professor of practical medicine at the University of Leiden from 1651 and head of the Leiden University Hospital. His *Medicina Physiologica* (Physiological Medicine) was published in 1653 in Amsterdam.
130. *Epist. Med.* 1667, Cent. 4, Epist. 71: De musculoso corde (On the muscular heart), pp. 421–27 (There are errors in the pagination; pp. 423–24 are missing). Scherz (ed.) 1952, 1, E 14, pp. 179–82. Harriet M. Hansen (ed.) 1987, 1, E 14, pp. 71–73.

131. Bartholin 1673, p. 361. Gustav Scherz is unjust to Thomas Bartholin when he writes that 17 centuries, including Stensen's teacher Thomas Bartholin, had joined Galen's rejection that the heart is a muscle (Scherz 1987, 1, p. 105). This was probably the case for Caspar Bartholin the Elder, but not for Thomas Bartholin. Scherz thus blames Thomas Bartholin for Caspar Bartholin's text in the Bartholinian anatomy book, but one should be aware that Thomas Bartholin, in his revised editions of his father's anatomy book, has only slightly altered his father's text. This also applies where the text was in dispute with Thomas Bartholin's own updated perceptions. The issue at hand is just one of many examples on this. Thomas Bartholin's renewal of the work consists primarily of additions of illustrations and new, updated text. On the advice of Ole Worm, he introduced his new text in square brackets in the editions of 1641, 1645 and 1651, although not entirely consistently. In the 1673 edition, the square brackets have been removed, and the text written together with Caspar Bartholin's text. The only sure way to distinguish Caspar Bartholin's original text from that of his son's later updated additions, therefore, is by compiling Caspar's last edition, *Editio Nova correcta & locupletata* (New corrected and enriched edition), Goslar 1632, with Thomas' four revised editions.

132. Linden, Johannes Antonides van der: *Medicina Physiologica* (Physiological Medicine), Amsterdam 1653, pp. 166–69.

133. The diploma of the University of Leiden is dated 4 December 1664. Scherz (ed.) 1952, 1, E 17, p. 183. Harriet M. Hansen (ed.) 1987, 1, E 17, pp. 95-96.

CHAPTER 7
Honorary professor and landowner
pp. 417–570

1. *Epist. Med.* 1667, Cent. 3, Epist 86: De sede anima (On the residence of the soul), etc., pp. 369–74, Brandt Andersen 2021, p. 193.

2. Ibid., Cent. 4, Epist. 21, 25 September 1662, pp. 91–96. Loic Capron (ed.): *Correspondance complète et autres écrits de Guy Patin*, https://www.biusante.parisdescartes.fr/patin/.

3. Garboe 1950, 2, pp. 36–37, Slottved 1978, p. 105.

4. Garboe 1950, 2, pp. 38–39.

5. Slottved 1978, p. 105.

6. Garboe 1950, 2, p. 37, who refers to documents in Sjællandske Registre 1661.

7. Slottved 1978, p. 224.

8. Garboe 1950, 2, p. 67, S. Birket-Smith: *Kjøbenhavns Universitets Matrikel* (The register of Copenhagen University), Copenhagen 1890, 1: 1611–1667 and 2: 1667–1740.

9. The whole controversy is reviewed in detail in Gosch 1873, pp. 117–30 and Brandt Andersen 2021, pp. 193–221.

10. For further information about Bils, see Luuc Koojimans: *Death Defied. The Anatomy Lessons of Frederik Ruysch*, Leiden 2010, pp. 18–32.

11. Gosch 1873, p. 117 and letter from Ole Borch in Leiden to Thomas Bartholin in Copenhagen on 3 March 1661. *Epist. Med.* 1667, Cent. 3, Epist. 85, pp. 360–69.

12. Ludovici de Bils: *Epistolica Dissertatio: Qua verus Hepatis circa Chylum, & pariter ductus Chiliferi hactenus dicti usus, docetur*, Rotterdam 1659.

13. *Epist. Med.* 1667, Cent. 3, Epist. 86, pp. 369–74.

14. Ibid., Epist. 87: De Bilsianis promissis, motu humorum in lymphaticis, materia liqvoris lymphaticis (On Bils' promises, the movement of the liquids in the lymph vessels, the origin of the liquid in the lymph vessels), pp. 374–77.

15. Thomas Rasmussen Walgensten (1627–1681), later physicist, city builder in Copenhagen and high court judge in Gotland, where he was born. He is, among other things, known for his manufacturing of one of the first usable slide cameras, a *laterna magica*.

16. P.A. Quernes, Norwegian student.

17. *Epist. Med.* 1667, Cent. 3, Epist. 88: Bilsius impugnatur. Lympha an ex Animalibus Spiritibus oriatur (Attack on Bils. Does the lymph originate from spiritus animalis?), pp. 378–81.

18. Ibid., Epist. 89: De motu humoris lymphatici, modo sine sangvine secandi, mumiis, sale mirabilia Glauberi, Burrhi solertia (On the movement of the lymph, a dissection method without bleeding, the miracle salt of Glauber, Borri's ingenuity), pp. 382–87.

19. According to several sources, the unhealthy indoor climate in Bils' home, induced by the stench of the embalming fluids and the decay of the dead bodies, contributed to his early death.

20. The Dutch translation was provided by Joan Blasius, according to a letter from him to Thomas Bartholin of 25 March 1661. *Epist. Med.* 1667, Cent. 3, Epist 19: De Bilsianis (About Bils), pp. 73–76.

21. Ibid.: "My brother Gerard Blasius sent you a short letter written by someone named Zas."

22. In connection with the war against Sweden, Krag was on a diplomatic journey with Godske Buchwald (1624–1700) from November 1659 to August 1660 to the Dukes of Braunsweig-Lüneburg, the general states in Haag and the new English king Charles II to obtain support for Denmark.

23. The tragedy *Hercules furens* (The Mad Hercules) by the Roman writer Lucius Annaeus Seneca (d. 65 AD).

24. *Epist. Med.* 1667, Cent. 3, Epist. 90, pp. 388–90.

25. Louis de Bils: *Brief van Louijs de Bils Heere von Koppensdamme etc. aan Th. Bartholijn Professor tot Koppenhaven*, dated 10 April 1661. According to Gosch 1873, p. 122, the letter was also published in Latin translation. Bartholin received this letter from Joan Blasius, who also offered to translate Bartholin's answer into Dutch. *Epist. Med.* 1667, Cent. 3, Epist. 19, pp. 7376.

26. Nicolai Stephani (Thomas Bartholin): *Castigatio Epistolæ maledicæ quam mense Aprili Nobilissimus Calumniator Ludovicus De Bils Scripsit ad Thomam Bartholinum* (Correction of the malicious letter from April from the famous bully Louis de Bils to Thomas Bartholin), Copenhagen and

Amsterdam 1661. The 39-page pamphlet is dated 7 May 1661 and is accompanied by a 6-page letter from Bartholin to Stephanus in Gdansk (that is, really from Bartholin to himself) dated 13 May 1661. This letter is reprinted in *Epist. Med.* 1667, Cent. 3, Epist. 32, pp. 112–18, while the pamphlet is reprinted in *Orationes*, Oratio 47: In Bilsium invectiva, de ejusdem artibus & Professoria dignitate (Introduction to Bils, to his art as well as to the esteemed professor), 7 May 1661, pp. 311–51. In the letter to Stephanus (himself), Bartholin keeps an ironic tone, where he thanks Stephanus for having fought for his cause, declares his own innocence, and regrets Bils' sensitivity for well-intended criticism. Referring to Cicero, Bartholin should not have uttered those words that Bils and his camp deserved to hear, but those which he could have spoken with honour: "I touched Bils' imperfections with the lightest hand. If he, as it turns out, cannot bear it, I will withdraw it with regret." With reference to both ancient and contemporary authors, and with address to Bils, Bartholin then draws a picture of what can be perceived as good manners, honour and dignity for doctors and other scholars. To want money for one's knowledge makes one a slave, is a worse disease than insanity, and should be cured by all doctors by a united effort. In ancient Rome, the emperor did not want art and science to be kept secret based on an envious attitude but to be spread to the general population. To ensure this, good salaries were introduced for the doctors, who were all considered as equal. In a letter to Thomas Bartholin on 23 July 1661 (*Epist. Med.* 1667, Cent. 3, Epist. 93, pp. 399–404) Ole Borch mentions that Bils is more upset than ever after receiving Stephanus' pamphlet. Borch is by many suspected to be Stephanus, and therefore he has received threats from Bils' friend Jordaan. In his answer of 19 August of the same year, Bartholin regrets this, but he does not reveal that he himself is Stephanus (*Epist. Med.* 1667, Cent. 3, Epist. 94, pp. 405–9).

27. In the mentioned letter to Stephanus, Bartholin also writes: "I will spare his [Bils'] work in the future, and because I have decided to refrain from any unfriendly dispute, so that I can live the rest of my life with a gentle mind, I will send Bils the old Roman divorce form: RES TUAS TIBI HABETO [Keep what belongs to you]."

28. For example, Borch writes in the letter to Bartholin on 20 March 1661: "When I was recently sitting next to the splendid van Horne in the so-called medical fair, he told me that he made great progress in his private work of copying Bils' mummies, and that he was very close to the success he was striving for." Borch continues in the letter to Bartholin on 21 April of the same year: "The excellent van Horne continues to work on his embalmed body. When I was with him recently, I saw a human arm that had been treated in such a manner without the use of conditioner that all the vessels were very visible, just as they would be in a fresh body. The natural colours of the body parts were preserved – a bit too pale, but he says he will be able to fix it soon – as well as the mobility of the joints. If he succeeds in finishing the rest of the body in the same way, Bils can just eat onion; he will lose 50,000 golden florins (the total sum he spends on his embalmed bodies) and will then perhaps be more careful in praising his own work, which seems to be beneath the skills of van Horne."

29. *Epist. Med.* 1667, Cent. 3, Epist. 24: De prima Ductus salivalis exterioris inventione, & Bilsianis experimentis, pp. 86–95. Maar (ed.) 1910, 1, OPH No. 1, pp. 1–7. Scherz (ed.) 1952, E 1, pp. 133–38. English translation (On the First Discovery of an External Salivary Duct and on the Experiments of Bils) in Kardel & Maquet (eds.) 2018, pp. 419–23. Danish translation in Harriet M. Hansen (ed.) 1987, 1, E 1, pp. 27–32.

30. In the 17th century, the body parts were divided into spermatic, sanguine and mixed parts. The spermatic parts, i.e. the parts believed to originate from the sperm, included bones, cartilage, ligaments, membranes, tendons, nerves, arteries and veins. See Thomas Bartholin: *Anatome Quartum Renovata*, Leiden 1673, p. 7.

31. The full title is *Exercitationes Physico-Anatomicæ De Nutrimenti in Corpore Elaboratione. Ubi De Chylificatione, & Chyli Motu; Sanguificatione; Depuratione Alimenti; Itemque Spiritibus. Quibus adjecta Appendix, In qua Examini ac Iudico aliorum Subjiciuntur varia De Chyli motu, & Nutrimenti in Corpore Elaboratione, Nec non De admiranda Anatome Nobiliss. Viri D. Ludovici Bils* (Physical-anatomical exercises concerning the processing of food in the body, including the formation and the movement of the chyle, the blood formation and the purification of the food as well as the spirits. Added an appendix, in which the movement of the chyle and the processing of food in the body are subjected to different other examinations and evaluations, but not by the admirable anatomist, the famous Doctor Louis de Bils), Groningen 1660.

32. Stensen does not exclude the latter possibility, but he points out that neither Bartholin, himself nor anyone else that he knows of has been able to confirm it in dissections.

33. At the end of the said treatise, Bartholin had put forward this invitation to Bils in his characteristic ironic tone: "But he [Bils] promises that he will one day demonstrate that the whole body by this (namely ros) [ros was Bils' word for chyle] is made suitable for nutrition by irrigation and fermentation. Although I have not been against ros in particular, if it can be distinguished from chyle, we will, since he has a different view on the matter, with great longing wait for the demonstrations, so that we, after the cover has been removed, under more favourable conditions can see what Nature has revealed to the famous man, and we ask as politely as possible, that he fulfil his word soon. We wish that luck may favour him in the task he has undertaken." Christensen, Hansen and Larsen (eds.) 1939, p. 107.

34. Everaerts 1661, pp. 17–18.

35. *Epist. Med.* 1667, Cent. 3, Epist. 93, pp. 399–404.

36. Ibid., Epist. 47: De Bilsianus Secretis judicum (Verdict on Bils' works), pp. 194–97.

37. Ibid., Epist. 57: Variæ in Oculis & Naso observationes novæ etc. (Various new observations in the eyes and nose etc.), pp. 224–30. Maar (ed.) 1910, 1, OPH No. 3, pp. 53–58.

Scherz (ed.) 1952, 1, E 3, pp. 140–44. English translation in Kardel & Maquet (eds.) 2018, pp. 465–68. Danish translation in Harriet M. Hansen (ed.) 1987, 1, E 3, pp. 34–37.
38. Ibid., Epist. 93, pp. 399–404.
39. It appears from a later letter of thanks from Deusing to Bartholin, dated 1 June 1662, that Bartholin sent the dissertation to Deusing. *Epist. Med.* 1667, Cent. 4, Epist. 9: Pro Hepate Defuncto gratiarum actio (Thanksgiving for *De Hepate Defuncto*), pp. 30–33.
40. Descartes 1644, pars secunda, chapter 33, pp. 50–52.
41. *Epist. Med.* 1667, Cent. 4, Epist. 17, 25. June 1662, pp. 70–78, and ibid., Epist. 51, 1. March 1663, pp. 328–36.
42. Ibid., Epist. 32, pp. 200–2.
43. Stensen refers to Bartholin's *Vasa Lymphatica, Nuper Hagniæ in Animantibus inventa, Et Hepatis exseqviæ* with its epitaph for the liver.
44. Stensen 1662, pp. 55–78. Maar (ed.) 1910, 1, OPH No. 4, pp. 59–73. English translation in Kardel & Maquet (eds.) 2018, pp. 469–81. Danish translation in Christensen, Hansen and Larsen (eds.) 1939, pp. 87–111.
45. For a detailed, annotated review, see Brandt Andersen 2021, pp. 213–17.
46. *Epist. Med.* 1667, Cent. 3, Epist. 28, pp. 231–34. Scherz (ed.) 1952, 1, E4, pp. 144–46. Danish translation in Harriet M. Hansen (ed.) 1987, 1, E4, pp. 38–39.
47. *Epist. Med.* 1667, Cent. 4, Epist. 1: Cur Nicotianæ pulvis oculos clariores reddat. De Lactea gelatina observatio (Why tobacco powder makes the eyes clearer. Observation on a milky gelatine), pp. 1–10. Maar (ed.) 1910, 1, OPH 8, pp. 105-111. Scherz (ed.) 1952, 1, E 7, pp. 151–56. English translation in Kardel & Maquet (eds.) 2018, pp. 505-9. Danish translation in Harriet M. Hansen (ed.) 1987, 1, E 7, pp. 44–48.
48. *Epist. Med.* 1667, Cent. 4, Epist. 26: Observationes Anatomicæ in avibus et cuniculus (Anatomical observations in birds and rabbits) pp. 103–13. Maar (ed.) 1910, OPH No. 9, pp. 113–20. Scherz (ed.) 1952, E 9, pp. 158–64. English translation in Kardel & Maquet (eds.) 2018, pp. 511–16. Danish translation in Harriet M. Hansen (ed.) 1987, 1, E 9, pp. 51–56.
49. They can all be found in *Epist. Med.* 1667, Cent. 3 and 4.
50. *Epist. Med.* 1667, Cent. 4, Epist. 27, pp. 113–17. Scherz (ed.) 1952, 1, E10, pp. 164–66. Harriet M. Hansen (ed.) 1987, 1, E10, pp. 56–58.
51. The full title is *Cista Medica Hafniensis, Variis Consiliis, Curationibus, casibus rarioribus, Vitis Medicorum Hafniensium, aliisq; ad rem Medicam, Anatomicam, Botanicam & Cymicam spectantibus referta. Accedit ejusdem Domus Anatomica brevissime descripta* (The archive chest of the Medical Faculty in Copenhagen, full of various pieces of advice, cures, rare cases, the lives of different Copenhagen physicians and others pertaining to the medical, anatomical, botanical and chemical professions. Added is The Anatomy House in Copenhagen briefly described). A Danish translation of parts of *Cista Medica Hafniensis* is found in Niels W. Bruun & Hans-Otto Loldrup (eds.): *Thomas Bartholin. Cista Medica Hafniensis*, Copenhagen 1982. A Danish translation of *Domus Anatomica brevissime descripta* is found in Niels W. Bruun & Hans-Otto Loldrup (eds.): *Anatomihuset i København*, Copenhagen 2007, and an English translation is found in Niels W. Bruun (ed.): *The Anatomy House in Copenhagen*, Copenhagen 2015.
52. Ozæna = fetid nasal catarrh.
53. Bruun & Loldrup (eds.) 1982, p. 170.
54. Bruun and Loldrup (eds.) 1982, pp. 151–56.
55. For other forensic reports of the period, see G. Tryde: *Retsmedicinske Erklæringer afgivet af Københavns Universitets Medicinske Fakultet i Aarene 1630–1662*, Copenhagen 1939.
56. Bartholin 1662, pp. 49–50, Bruun (ed.) 2015, pp. 107–9.
57. Especially Petersen 1898, pp. 46–73 and Maar (ed.) 1916.
58. Bartholin 1662, pp. 37–40. Quoted after Bruun (ed.) 2015, pp. 93–97.
59. *De Tranquillitate Animi* (On the tranquillity of the mind) is a Latin work by the Stoic philosopher Lucius Annaeus Seneca the Younger (4 BC–65 AD). The dialogue concerns the state of mind of Seneca's friend Annaeus Serenus, and how to cure Serenus of anxiety, worry and disgust with life.
60. Thomas Bartholin: *De Bibliothecæ Incendio Dissertatio Ad Filios* (Dissertation for my sons on the burning of my library), Copenhagen 1670, pp. 65–66. O'Malley (ed.) 1961, pp. 24–25.
61. Rigsarkivet (The National Archive in Copenhagen), Krigskollegiet, 1664. 5. Indkomne breve fra flere end 2. The letter is partially reproduced and transcribed in Garboe 1950, 2, pp. 84–85.
62. Svend Ellehøj: Christian 4.s Tidsalder 1596–1660 (The era of Christian IV 1596–1660), in: John Danstrup and Hal Koch (eds.): *Danmarks Historie* (The history of Denmark), Copenhagen 1964, 7, pp. 494–95 and Gunnar Olsen: Den unge Enevælde 1660–1721 (The young absolute monarchy 1660–1721), ibid., 8, p. 46.
63. The deed is dated 2 November 1663. Garboe 1950, 2, p. 43.
64. Gunnar Olsen: Hagestedgaard. In: Arthur G. Hassø (ed.): *Danske Slotte og Herregaarde. Ny Samling* (Danish castles and manors. New Collection), Copenhagen 1944, 1, pp. 330–35.
65. For granted properties, tithes and rights in the area around Hagestedgård Manor on Zealand and Fjellebro Manor on Funen, see F.J. West: *Kronens Skøder paa afhændet og erhvervet Jordegods i Danmark, fra Reformationen til Nutiden.* Andet Bind. 1648–1688, Copenhagen 1908, pp. 261–62, 276, 286, 426, 463–64. Thanks to Svend Erik Jensen for this reference.
66. Bartholin 1670, pp. 8-9, O'Malley (ed.) 1961, p. 4.
67. *Carmina*, p. 208. Garboe 1950, 2, pp. 48–49.
68. Garboe 1950, 2, 73–75.
69. Ibid., pp. 53–54.

70. The dissertation was dedicated to Frederik III and appeared in a second edition in Strasbourg in 1665 with the title *Anatomiæ Bilsianæ Anatome*, with two added letters, one from (dated 20 June 1664 in Schaffhausen) and one to (dated 11 July 1664 in Strasbourg) the Swiss physician, pathologist and pharmacologist Johann Jakob Wepfer (1620–1695) concerning anatomical doubts.
71. Garboe 1950, 2, p. 54.
72. Slottved 1978, pp. 109 and 155.
73. C.O. Bøggild Andersen: Jacob Henrik Paulli v. Rosenchild. In: Engelstoft, Povl & Dahl, Svend (eds.): *Dansk Biografisk* Leksikon, 2nd edition, Copenhagen 1933–1944, 20, p. 136.
74. Stensen's mother died in the summer of 1664, which prolonged Stensen's stay in Copenhagen.
75. *Ductus vitellinus* or vitelline duct.
76. Thomas Bartholin: *De Hepatis Exautorati Desperata Causa*, Copenhagen 1666, pp. 81–84.
77. Thomas Bartholin: *De Medicina Danorum Domestica*, Copenhagen 1666, p. 51.
78. Ibid., pp. 51–52.
79. Stensen 1664, p. 4.
80. Brandt Andersen 2021, pp. 293–310.
81. Ibid., pp. 333–39.
82. *Epist. Med.* 1667, Cent. 3, Epist. 100, p. 439.
83. Ibid., Cent. 4, Epist. 52, pp. 336–37.
84. Rasmus Bartholin: *De Cometis Anni MDCLXIV & MDCLXV Opusculum. Ex observationibus Hauniæ habitis adornatum* (Treatise on the comets in the years 1664 and 1665. Based on observations made in Copenhagen), Copenhagen 1665.
85. Lubiniecki was a Socinian and thus denier of the divinity of Christ.
86. Bartholin 1665, pp. 64–66.
87. Frank Hansen and Mogens Nørgaard Olesen: *Halleys Komet og vort verdensbillede* (Halley's Comet and our view of the world), Copenhagen 1985, pp. 110–13.
88. Garboe 1950, 2, p. 90.
89. Bartholin 1665, p. 125, Garboe 1950, 2, p. 91.
90. Axel Garboe: Meteoritfaldet på Fyen i Aaret 1654 (The meteorite fall in Fyen in the year 1654), *Meddelelser fra Dansk Geologisk Forening* 1959, 12, pp. 284–97.
91. Bartholin: *De Medicina Danorum Domestica* 1665, p. 217 and 391, Mansa 1873, p. 326.
92. Ibid., pp. 251–52, Garboe 1950, 2, p. 78.
93. Ibid., p. 31, Garboe 1950, 2, p. 79.
94. Ibid., pp. 309–10, Garboe 1950, 2, p. 79.
95. *Carmina*, p. 111, Garboe 1950, 2, p. 79.
96. Ibid., p. 302, Garboe 1950, 2, p. 95.
97. Ibid., p. 75, Garboe 1950, 2, p. 95.
98. Ibid., the preface, Garboe 1950, 2, p. 92.
99. Ibid., p. 31, Garboe 1950, 2, p. 80.
100. Ibid., p. 386, Mansa 1873, p. 327.
101. Ibid., p. 400–1, Garboe 1950, 2, p. 97.
102. Ibid.
103. Ibid., p. 405, Garboe 1950, 2, p. 71.
104. E. Ingerslev 1907, 2, p. 326.
105. *Orationes*, p. 376. Swedish translation with comments by Krister Östlund: De nihilo medico. Ett tal av Thomas Bartholin. *Bibliotheca NEOLATINA Upsalensis IV*, Upsala 2016, p. 11.
106. Ibid., pp. 379–81, Östlund (ed.) 2016, pp. 17–21.
107. Ibid., p. 176.
108. E.C. Werlauff: Efterretninger om italieneren J. F. Borro's Ophold ved det danske Hof i Aarene 1667–1670 (Information about the Italian J.F. Borro's stay at the Royal Court in Denmark during the years 1667–1670). In: J.D. Herholdt and F.V. Mansa (eds.): *Samlinger til den danske Medicinal-Historie*, Copenhagen 1835, 1, pp. 152–90. Where nothing else is stated, this source is the background for the information about Borri.
109. *Carmina*, p. 91.
110. H.C. Andersen: *Nye Eventyr og Historier*, 3, Copenhagen 1859.
111. *Epist. Med.* 1667, 3, Epist. 89, pp. 382–87.
112. Ibid., Epist. 93, pp. 399–404.
113. Ibid. 1667, 4, Epist. 17, 25. June 1662, pp. 70–78.
114. Ibid., Epist. 88, pp. 507–8.
115. Published for the first time on 1 August 1666 and then once a month until 1691.
116. Werlauff 1835, p. 177.
117. Ibid., p. 172.
118. *Acta Medica Ann. 1671 & 1672*, Copenhagen 1673, Obs. 132: De Oculi humoribus restituendis (On restoration of the eye's fluids), pp. 262-264, and Obs. 133: De sponte renatis in Ansere Oculorum humoribus observatio (On observation of spontaneous restoration of the fluids in goose eyes), pp. 264–67.
119. Thomas Bartholin: *Francisci Josephi Burrhi Epistolæ Duæ* (Two letters from Giuseppe Francesco Borri to Thomas Bartholin). *I. De cerebri ortu et Usu Medico* (On the origin of the brain and its use in medicine). *II. De Artificio oculorum humores restituendi* (On the art of restoring the eye's fluids). *Ad Th. Bartholinum*, Copenhagen 1669.
120. *Journal des Sçavans* 1669, p. 540.
121. *Philosophical Transactions* 1670, p. 2081.
122. On the occasion of Bartholin's 50th birthday in October 1666, Frederik III had granted him a gift of 4000 rix-dollars.
123. Garboe 1950, 2, p. 108.
124. Ibid., p. 109.
125. Ibid., p. 110.
126. English translation by Charles C.D. O'Malley: *Thomas Bartholin. On The Burning of his Library and On Medical Travel*, Kansas 1961.
127. Ibid., p. 4.
128. Ibid., pp. 11–12. Matzen 1879, 1, p. 234 refers to *Acta Consistorii* 28 January 1659: "Grave site for Th. Bartholin in the northern chapel at the choir west of Dr. Gellii grave, on which must be given him a deed, when he gives satisfaction therefore."
129. Garboe 1950, 2, p. 112.

130. Initially, the estate and its eight farms were given three years of tax exemption, but this was later extended. See Garboe 1950, 2, p. 115.
131. Personal information from landlord Anders Hvass.
132. O'Malley (ed.) 1961, p. 18.
133. Garboe 1950, 2. p. 117.
134. O'Malley (ed.) 1961, p. 21.
135. Ibid.
136. The Italian anatomist Giovanni Battista Morgagni (1682–1771) is generally considered the founder of modern pathological anatomy. He published in 1761 in Venice the five-volume work *De Sedibus, et Causis Morborum per Anatomen Indagatis Libri quinque* (Five books on the seats and causes of diseases sought out through dissections).
137. O'Malley (ed.) 1961, p. 32.
138. Bartholin 1670, p. 101.
139. Bruun (ed.) 2015, p. 123.
140. Sthyr 1965, p. 15. On 11 February 1654, Thomas Bartholin applied for release from his duties as Consistory Notary to work on *Anatome Augusta*, among other things. See Garboe 1649, 1, note 42a, p. 183, which refers to *Acta Consitorii*, same date, The Danish National Archives (Rigsarkivet), fol. 78.
141. Bruun (ed.) 2015, p. 200, Werlauff & Engelstoft 1836, p. 46. For Karel van Mander's interest in anatomy, see Povl Eller: *Kongelige portrætmalere i Danmark 1630–82. En undersøgelse af kilderne til Karel van Manders og Abraham Wuchters' virksomhed* (Royal portrait painters in Denmark 1630–82. A study of the sources of Karel van Mander's and Abraham Wuchters' activities), Copenhagen 1971, pp. 261–65.
142. Alcoholic drink with garden wormwood, fennel and aniseed.
143. O'Malley (ed.) 1961, pp. 26–27.
144. O'Malley (ed.) 1961, pp. 30–31.
145. Ingerslev 1873, 1, pp. 457–58.
146. Kristian Carøe: *Medicinalordningens Historie i ndtil Sundhedskollegiets Oprettelse 1803* (The History of The Medicinal Ordinance until the Establishment of the Health College in 1803), Copenhagen 1917, p. 13.
147. Ibid., p. 36.
148. Ibid, pp. 18–27.
149. In July 1671, Peder Schumacher was ennobled by Christian V with the name Griffenfeld.
150. Carøe 1917, p. 36.
151. During his own stay in Italy in the early 1640s, Bartholin had made acquaintance with Ferdinand II, and as mentioned elsewhere in this book, Bartholin dedicated *Cista Medica Hafniensis* to the Grand Duke in 1662.
152. *Hist. Anat.* 1661, 6. Hist. 1. Glossopetrarum Melitensium Usus Medicus (On the medical use of Maltese tongue stones), pp. 193–201.
153. Maar (ed.) 1910, 2, pp. 113–45. English translation in Gustav Scherz (ed): Steno. Geological Papers. *Acta Historica Scientarium Naturalium et Medicinalium*, Vol. 20, Odense 1969, pp. 65–131 and Kardel & Maquet (eds.) 2018, pp. 699–731. Danish translation in Valdemar Meisen & Knud Larsen (eds.): *Stenoniana*, Copenhagen 1933, pp. 50–99.
154. Maar (ed.) 1910, 2, pp. 181–226. English translation in Scherz (ed.) 1969, pp. 133–234 and Kardel & Maquet (eds.) 2018, pp. 763–825. Danish translation in August Krogh & Vilhelm Maar (eds.): *Nicolaus Steno: Foreløbig Meddelelse til en Afhandling om faste Legemer, der findes naturlig indlejrede i andre faste Legemer*. With introduction and notes, Copenhagen 1902.
155. Brandt Andersen 2021, note 894, p. 742.
156. Garboe 1950, 2, pp. 64–66 and 127–28. Axel Garboe & Knud Larsen: Niels Stensen. In: Engelstoft, Povl & Dahl, Svend (eds.): *Dansk Biografisk Leksikon*, 2nd edition, Copenhagen 1933–1944, 22, pp. 562–75.
157. According to Slottved 1978, p. 109, it expired on 5 December 1667.
158. In Denmark, the Augsburg Confession was recognized with the introduction of the Reformation in 1536. Since the introduction of "Kongeloven" (The Royal Act) in 1665, the Augsburg Confession has been the official expression of the faith of the Church of Denmark.
159. Harriet M. Hansen (ed.) 1987, 1, p. 13.
160. Scherz (ed.) 1952, E74, p. 261. Harriet M. Hansen (ed.), E74, pp. 188–89.
161. When Bartholin as newly joined professor of anatomy and medicine at the University of Copenhagen held his first lecture series of a total of 18 disputations on various anatomical topics with various respondents in the period from December 1649 to July 1651, the then only 14-year-old Schumacher was respondent to number eight, on 20 February 1650, on the ureters and bladder (Thomas Bartholin: *Collegii Anatomi Disputatio Octava. De Ureteribus et Vesica Urinaria*). Bartholin presented in his opening speech for the defence an apt prophecy, when he praised the talent and maturity of the young Schumacher, who reached far ahead of his age and promised the fatherland great things (*Orationes*, pp. 94–96). On 4 September of the same year, Schumacher disputed again, this time about the nerves (Thomas Bartholin: *Collegii Anatomici Disputatio Decima Sexta. De Nervis*). For the close relation between Thomas Bartholin and Peder Schumacher Griffenfeld, see Brandt Andersen 2021, pp. 596–600.
162. See letter from Stensen in Florence to Griffenfeld in Copenhagen on 26 April 1672. Scherz (ed.) 1952, 1, E75, pp. 261–62. Harriet M. Hansen (ed.) 1987, 1, E 75, p. 189.
163. Slottved 1978, p. 109.
164. Ibid.
165. Garboe 1950, 2, p. 128.
166. Garboe & Larsen 1942, p. 572.
167. English translation by James Willis: On Diseases in the Bible. A Medical Miscellany. Edited and with introduction by Johan Schioldann-Nielsen and Kurt Sørensen. *Acta Historica Scientarium Naturalium et Medicinalium*, Vol. 41, Copenhagen 1994.

168. Scherz (ed.) 1952, 1, E 82, pp. 268–70. Harriet M. Hansen (ed.) 1987, 1, E 82, pp. 196–97.
169. Scherz 1987, 1, p. 305.
170. Letter from Niels Stensen in Florence to Johannes Sylvius 12. January 1672. Scherz (ed.) 1652, 1, E 73, pp. 257–60. Danish translation in Harriet M. Hansen (ed.) 1987, 1, E 73, pp. 185–88.
171. Brandt Andersen 2021, pp. 613–14, gives examples of such letters.
172. The Royal Library, NKS 309aa 4°. The title of Jacobsen's collected lecture notes, in which the dissection descriptions are included, is *Exercitia Academica Hafniæ et Lugdun. Batavorum Olgeri Jacobæi* (Holger Jacobsens exercises at Copenhagen and Leiden University).
173. Maar (ed.) 1910, 2, pp. 297–310.
174. An overview can also be found in Scherz (ed.) 1952, 2, p. 926.
175. Brandt Andersen 2021, Chapter 9, pp. 611–77.
176. *Acta Medica Ann. 1671 & 1672*, Copenhagen 1673, Obs. 90, pp. 175–78.
177. Brandt Andersen 2021, pp. 626–29.
178. *Acta Medica Ann. 1671 & 1672*, Copenhagen 1973, Obs. 135: Anatome Rangiferi (Dissection of a reindeer), pp. 274–78. English translation by Paul Maquet in Troels Kardel & Paul Maquet: Anatomy of a reindeer dissected in Copenhagen in 1672 by Niels Stensen as reported by Thomas Bartholin. *Rangifer* 2012, 32, 1, pp. 49–55.
179. A.D. Jørgensen: *Niels Stensen. Et mindeskrift*, Copenhagen 1884, pp. 200-201. Brandt Andersen 2021, 671–72.
180. Ian Herbert Porter: Thomas Bartholin (1616–80) and Niels Stensen (1638–86), Master and Pupil. In: *Medical History* 1963, 7, 2, pp. 99–125. The quote is from p. 115.
181. *Acta Medica Ann. 1671 & 1672*, Copenhagen 1673, Obs. 136: Leporis Anatome (Dissection of a Hare), pp. 278–82. Thomas Bartholin indicates 31 October 1672 as the date of this dissection, while Holger Jacobsen and Gustav Scherz indicate 4 November 1672. Scherz (ed.) 1952, 2, p. 927. Wilhelm Maar (ed.) 1910, 2, p. 299.
182. The Danish word Svane means swan.
183. See Brandt Andersen 2021, pp. 637–38.
184. Det Kongelige Bibliotek (The Royal Library in Copenhagen), REX 34: 3, -79 2°. English translation by August Ziggelaar in Kardel & Maquet (eds.) 2018, pp. 849–52 and Danish translation by Niels W. Bruun in Brandt Andersen 2021, pp. 777–80.
185. *Acta Medica Ann. 1673*, Copenhagen 1675, Obs. 34 (misnumbered as Obs. 24), pp. 359–66. A simultaneous copy exists in Holger Jacobsen's mentioned records in the Royal Library, NKS 399aa 4°. Maar (ed.) 1910, 2, pp. 249–56. Kardel & Maquet (eds.) 2018, pp. 853–59. Danish translation by Knud Larsen in Valdemar Meisen & Knud Larsen (eds.): *Stenoniana*, Copenhagen 1933, pp. 100–11, by Aage Kragelund in *Den humanistiske renæssance og Antikken*, Copenhagen 1976, pp. 226–44, and by Jesper Brandt Andersen in *Niels Stensen. Kongelig anatom og fyrstelig geolog*, Copenhagen 2021, pp. 780–86. Annotated excerpts can be found in Jørgensen 1884, pp. 170–73.
186. Compare the sentence: "vel qui saltem microscopium cuti admovet" (or who simply puts the skin under a microscope), with which Stensen refers to Malpighi's microscopic studies of the skin in the work *De Externo Tactus Organo Anatomica Observatio* (Anatomical observation of the external sensory organ), Naples 1665. Knud Larsen and Aage Kragelund both erroneously translate "microscopium" as "magnifying glass" (respectively *Stenoniana*, p. 106, and Kragelund 1976, p. 233). When Stensen elsewhere in the speech touches on the microscopic aspects of the structure of plants: "But if he went even further and in a single plant examined the inner harmony between the small parts, which it consists of" (*Stenoniana*, p. 104), he also thinks of Malpighi's investigations with a microscope of the anatomy of plants in the work *Anatome Plantarum* (Anatomy of Plants), London 1675.
187. Ebbesen and Koch 2003, p. 317, referring to G. W. Leibniz: *Essais de Théodicée* I, §100, in: G. W. Leibniz: *Die Philosophischen Schriften* published by C.J. Gerhardt, Berlin 1885, 6, p. 158.
188. Brandt Andersen 2021, pp. 651–52, who among other things emphasizes that Stensen's speech draws threads to both the Italian Catholic philosopher Thomas Aquinas (1225–1274) and the Catholic Church Father Augustine of Hippo (354–430). Kragelund 1976, p. 236, notes, that Stensen draws on Augustine's *Confessiones*.
189. Det Kongelige Bibliotek, København, Håndskrift-samlingen (The Royal Library, Copenhagen, The Manuscript Collection), manuscript Acc. 2019/11.
190. Since 1300, the Catholic Church had celebrated every 25 years the papally proclaimed holy year (jubilee year), in which Catholic pilgrims make pilgrimages to Rome to seek absolution in the city's churches. The tradition is still maintained.
191. Det Kongelige Bibliotek (The Royal Library in Copenhagen), NKS 4660, 4°. Léon Tolmer: Une page d'histoire des sciences 1661–1669. Vingt-deux lettres inédites d'André de Graindorge à P.-D. Huet. *Mémoires de l'Académie Nationale des Sciences, Arts et Belles-Lettres de Caen*, Nouvelle Série, 1942, 10, pp. 245–337. Gustav Scherz: Da Stensen var i Paris (When Stensen was in Paris). *Fund og Forskning i Det Kongelige Biblioteks Samlinger* 1969, 16, pp. 43–52.
192. Ibid.
193. *Acta Medica Ann. 1673*, Copenhagen 1675, Obs. 127, pp. 320–45, Maar (ed.) 1910, 2, pp. 257–77. English translation in Kardel & Maquet (eds.) 2018, pp. 871–91. For an annotated review, see Brandt Andersen 2021, pp. 652–54.
194. *Acta Medica Ann. 1673*, Copenhagen 1675, Obs. 10: Ova Galli et Serpentum (Rooster and snake eggs), pp. 14–16. Brandt Andersen 2021, pp. 657–58.

195. *Acta Medica Ann. 1674, 1675, 1676*, Copenhagen 1677, Obs. 21: De caudæ Vulpinæ odore violaceo, & Ursi suctione (On the violet scent of the fox's tail and the bear's breath on his paw), pp. 32–35. See also mention in Gosch 1873, pp. 263–64 and Brandt Andersen 2021, pp. 663–67.
196. Caspar Bartholin 1675: Obs. 4: Unde odor violaceus in cauda Vulpina? (Where does the violet scent in the fox's tail come from?), pp. 63–79.
197. *Acta Medica Ann. 1673*, Copenhagen 1675, Obs. 22: Obstetricum Examen (Obstetric exam), pp. 53–55.
198. See for example Brandt Andersen 2021, pp. 660–62 for Stensen's dispute with the Lutheran Johan Brunsmand (1637–1707) and pp. 669–71 for his attempts to convert his family members and for Ole Borch's opposition to Stensen's conversion. Ole Borch was the son of a Lutheran priest, and, unlike Thomas Bartholin, he was not religiously tolerant. At the celebration of the 150th anniversary of the Reformation on 1 November 1667, Ole Borch gave a speech at Copenhagen University, *Olai Borrichii Oratio Jubilæa Evangelica* (Ole Borch's evangelical jubilee speech), which was strongly anti-Catholic, and in which he gives examples of Catholic follies, while on the other hand paying tribute to Luther and Lutheranism. See Jens Glebe-Møller: Lutheraneren Ole Borch. In: *Borchs Kollegium 1691–1991*, Copenhagen 1991, p. 40. When Stensen, during Borchs' rectorate 1675/1676, as a newly hatched Catholic priest touched on confessional issues in a letter, Borch urged him to return to his Fathers' Faith, which led Stensen to write a long emotional letter to Borch from Florence, in which he professes his love for his new true faith but acknowledges his failure of the fatherland. The formal form of the letter reveals the chilled relationship between the two friends who were previously so close. Det Kongelige Bibliotek (The Royal Library in Copenhagen), Bøllings Brevsamling, D 480 2°. Scherz (ed.) 1952, 1, E 108, pp. 308–10. Harriet M. Hansen (ed.) 1987, 1, E 108, pp. 233–35. Brandt Andersen 2021, pp. 669–70.
199. Scherz (ed.) 1952, 1, E 95, pp. 293–94. Harriet M. Hansen (ed.) 1987, 1, E 95, p. 219.
200. Scherz (ed.) 1952, 1, E 96, pp. 294–95. Harriet M. Hansen (ed.) 1987, 1, E 96, pp. 219–21.
201. *Acta Medica Ann. 1674, 1675, 1676*, Copenhagen 1677, Obs. 22, pp. 35–36.
202. Garboe 1950, 2, pp. 138–39.
203. *Holger Jacobæus' Rejsebog*, p. 75.
204. The 19-year-old Caspar Bartholin had been appointed professor of anatomy in Copenhagen in absentia on 17 October 1674.
205. In accordance with this, Bartholin writes in his mentioned small work on the practical anatomy from 1674: "*Anatome Nova*, which at the end of this year will be published in The Netherlands."
206. Thomæ Bartholini: *Neu-verbesserte künstliche Zerlegung dess Menschlichen Leibes/ In vier absonderliche Bücher eingetheilet. Darinnen Grund-richtig gehandlet wird/ I. Von dem Unter-Bauch/II. Von der mittlern Höle dess Leibes/ III. Von der obern Höle desselben/ IV. Von den äussersten Gliedmassen/Und zwar also und dergestalt/ dass jedes bemeldter Bücher wieder mit seinem absonderlichen Büchlein versehen ist/ Worinnen die Blut-Puls- und Spann-Adern/ samt den Gebeinen/ Krospeln und Sennen vorgestellet werden. Denen noch über das Johannis Walæi Zwey Send-Schreiben/ von der Bewegung dess Milch-Safftes und Geblüts, beygefüget sind [...] aus der Lateinischen Sprache in die Deutsche Sprache übersesset/ durch Eliam Wallnern/ Philosophiæ & Medicinæ Doctorem, einer hochloblichen Unter-Oesterreichischen Landschaft bestellten Physicum zu Baden*, Nuremberg 1677.
207. Caspar Bartholin the Younger writes in a letter from Paris on 2 August 1675: "Otherwise, nothing has happened that I can report, except that I recently spoke with a prominent French doctor named Bernier, who had returned from a trip in India where he had spent seven years at the Great Mogul's court. He said that he had made sure that my father's *Anatomy* was translated into an Indian language, and that the Great Mogul greatly appreciated it because it contained things they had not heard of, and which were completely new to them." See *Acta Medica Ann. 1674, 1675 & 1676*, Copenhagen 1677, Obs. 54: De Anatome in Indicam Lingvam versa (On the anatomy book in an Indian language), pp. 86–87.
208. Guan Xuelin: Vestlig medicinsk praksis ved Kangxi-kejserens hof med efterskrift af Helge Kragh (Western medical practice at the Khangxi-emperor's court with postscript by Helge Kragh). In: Ole Willumsen Krog (ed.): *Skatte fra kejserens Kina. Den Forbudte By og Det Danske Kongehus* (Treasures from the emperor's China. The Forbidden City and the Danish Royal House), Copenhagen 2006, pp. 126–31.
209. E. Ingerslev: Nogle Bemærkninger om Bartholins Anatomi i kinesisk Omklædning (Some remarks on Bartholin's Anatomy in Chinese clothing). Special edition of *Ugeskrift for Læger* no. 21, 1907.
210. The number of illustrated reports in the five volumes are 16, 18, 14, 6 and 14, respectively.
211. The preface of the second volume.
212. Harcourt Brown: *Scientific Organizations in Seventeenth Century France (1620–1680)*, New York 1934.
213. J. Schiller & J. Théodoridès: Stenon et les milieux scientifiques parisiens. In: G. Scherz (ed.): The historical aspects of brain research in the seventeenth century, proceedings of the international historical symposium on Nicolaus Steno and brain research in the seventeenth century held in Copenhagen 18.–20. August 1965. *Analecta Medica-Historica*, 3, Oxford 1968, p. 156.
214. Harcourt Brown 1934, pp. 135–42, J. Schiller & J. Théodoridès 1968, p. 157 and Nicolas Dew: *Orientalism in Louis XIV's France*, Oxford 2009, p. 91.
215. Lucas Jacobsøn Debes: *Færoæ & Færoa reserata: Det er Færøernis oc Færøeske Indbyggeris Beskrifvelse, udi hvilcken føris til Liuset adskillige Naturens Hemeligheder/ oc nogle Antiqviteter, som her til Dags udi Mørcket hafve været*

indelugt, oc nu her opladis / Alle curieuse til Velbehagelighed, sammenskrefven oc forklaret aff Lucas Jacobsøn Debes, Copenhagen 1673.

216. For a detailed description of these records, entitled *De Opusculis Medicis & Philosphicis editis Hafniæ vel ab Hafniensibus* (On medical and philosophical works published in Copenhagen or by Copenhageners), see Ingrid Ilsøe: Bøger og boghandlere under Christian V (Books and bookstores during the reign of Christian V). *Fund og Forskning i Det Kongelige Biblioteks Samlinger* 1981, 25, pp. 26–27.

217. Jean Berryat, et al. (eds.): *Collection Académique*, Partie Etrangere, Vol. 4, Dijon and Paris 1757, pp. 185–376, Vol. 6, Dijon and Paris 1761, pp. 362–440 and Vol. 7, Dijon and Paris 1766, pp. 140–386.

218. For Holger Jacobsen's report on the malformed Norwegian girl, who died three days old, see Brandt Andersen 2018, pp. 101–3.

219. Until then, the heart defect had been called tetralogy of Fallot after the French forensic pathologist Étienne-Louis Arthur Fallot (1850–1911), who described it in 1888. See Brandt Andersen and Bruun 2015, where Stensen's Latin text is translated into Danish with associated notes, analysis and perspectivation. The original description, Embryo monstro affinis Parisiis dissectus, is found in *Acta Medica* Ann. *1671 & 1672*, Copenhagen 1673, pp. 200–3. For Fallot's description, see Étienne-Louis Arthur Fallot: Contribution à l'Anatomie patologique de la Maladie bleue (Cyanose cardiaque). *Marseille Medical* 1888, 25, pp. 77–93, 138–58, 207–23, 341–54, 370–86, 403–20.

220. *Acta Medica Ann. 1677, 1678 & 1679*, Copenhagen 1680, Obs. 16: De Vero Olfactus Organo (The true Organ of smell), pp. 61–72.

221. *Acta Medica 1677, 1678 & 1679*, Copenhagen 1680, Obs. 112: Pulveris corticis Chinæ-Chinæ usus contra febrim qvartanam in Dania probatus (China-bark powder used against malaria in Denmark with favourable effect), pp. 287–90. Rasmus Bartholin's account is translated into Danish in Vilhelm Maar: The first use of China-bark in Denmark, Leiden 1925, pp. 29–32. This little writing also reviews the history of China-bark in general and in Denmark specifically.

222. Porter 2000, p. 230.

223. Georg F.L. Sarauw: Cromer-skovlaget i Frihavnen og trælevningerne i de ravførende sandlag ved København (The Cromer-woodlayer in Frihavnen and the wooden remains in the amber-containing sand layers at Copenhagen), *Meddelelser fra Dansk Geologisk Forening*, 1897, 1, 4, pp. 17–44.

224. Axel Garboe: Thomas Bartholin og Rav-studiet i Danmark. Et Blad af Mineralogisk Museums Historie (Thomas Bartholin and the amber study in Denmark. A page of the history of the Mineralogical Museum). Special edition of *Meddelelser fra Dansk Geologisk Forening*, 1946, 11, pp. 97–116.

225. Christianshavn is a part of Copenhagen.

226. Thomas Bartholin: *Th. Bartholini Decani, Regii Archiatri & Prof. Honorar. Dissertatio prima de Theriaca in Officina Christophori Heerford Sen. Pharmacop. Reg. 1. Febr. 1671 dispensata & Dissertatio secunda de Theriaca in Officina Joh. Gothofredi Becker Pharmacop. Reg. Aulic. 1. Febr. 1671 dispensata* (Thomas Bartholin, Dean, Royal Archiater & Honorary Professor. First dissertation on theriac dispensed in the pharmacy of Christopher Heerfordt the Elder, Royal Pharmacist, 1 February 1671 and Second dissertation on theriac dispensed in the pharmacy of Johann Gottfried Becker, Court Pharmacist, 1 February 1671), Copenhagen 1671. The passage about amber is in the second dissertation.

227. Thomas Bartholin: *Acta Medica Ann. 1671 & 1672*, Copenhagen 1673, Obs. 57: De succino experimenta (On experiments with amber), pp. 110–15.

228. Garboe 1946, p. 102.

229. Ibid., pp. 102–3.

230. Ibid., p. 103.

231. For further publications on amber written by Bartholin see *Acta Medica Ann. 1673*, Copenhagen 1675, Obs. 122: De Theriaca Danica & succini generatione (About Danish theriac and the formation of amber), pp. 306–11, and ibid., Obs. 123: De Succini resolutione, & viribus (About the resolution and strength of amber), pp. 312–14, and *Acta Medica Ann. 1674, 1675 & 1676*, Copenhagen 1677, 4 (1676), Obs. 26: De Succino Experimenta singularia (About an experiment with amber), pp. 73–74.

232. Christian Vaupell: En botanisk Undersøgelse af det fossile Træ, der findes i Leer- og Sandlagene i Danmark og af havet opkastes med Ravet på de danske Kyster (A botanical investigation of the fossilized wood, which is to be found in the clay and sand layers in Denmark, and which is thrown up on the Danish coasts together with the amber). Published by N. Hartz in *Meddelelser fra Dansk Geologisk Forening*, 1906, 2, 12, pp. 43–64. Vaupell's investigations date from the beginning of the 1850s.

233. Scherz (ed.) 1952, 1, E 82, pp. 268–70. Harriet M. Hansen (ed.) 1987, 1, E 82, pp. 196–97.

234. *Acta Medica Ann. 1671 & 1672*, Copenhagen 1673, Obs. 46: Petrificata animalia, eroumq; partes (petrified animals and parts hereof), pp. 83–84.

235. Lucas Jacobsøn Debes: *Færoe & Færoa reserata: Det er Færøernis oc Færøske Indbyggeris Beskrifvelse, udi hvilcken føris til Liuset adskillige Naturens Hemeligheder/ oc nogle Antiqviteter, som her til dags udi Mørcket hafve været indelugt, oc nu her opladis/ Alle curieuse til Velbehagelighed, sammenskrefven oc forklaret aff Lucas Jacobsøn Debes* (The Faroe Islands and the Faroese displayed: That is the description of The Faroe Islands and the Faroese inhabitants, in which several of nature's secrets and some antiquities, which hitherto have been hidden in the dark, are now revealed. All these curious things have for the pleasure been written down and explained by Lucas Jacobsøn Debes),

Copenhagen 1673. *Acta Medica Ann. 1671 & 1672*, Copenhagen 1673, Obs. 49: Rara Naturæ in Insulis Ferröensibus (The rarities of nature on the Faroe Islands), pp. 86–102, and mentioned in *Acta Medica Ann. 1673*, Copenhagen 1675, Obs. 19, pp. 369–72.

236. *Hist. Anat.* 1661, Cent. 5, Hist. 72: Lignum fossile (Petrified wood), pp. 145–47.

237. The first book in *Museum Wormianum* deals with fossils. It is divided into sections about soil types, stones and metals. For an overview, see Schepelern 1971, pp. 231–49, and English summary, p. 378. In his view on fossils, Worm was mainly inspired by the two mineralogists, the German Georgius Agricola (Georg Bauer) (1494–1555) and the Flemish Anselmus Boëtius de Boodt (1550–1632). Worm shared the general view of the time, that fossils were subterranean "growths", and that petrification was caused by "a petrifying juice" in the ground, in the same way as stones (kidney stones, gallstones etc.) were believed to be made in the human body. Like Thomas Bartholin, Worm was interested in the possible medical use of geology, and like Bartholin he was sceptical towards alchemy.

238. *Hist. Anat.* 1654, Cent. 2, Hist. 100: Homines Petrificati (Petrified human beings), pp. 353–60. For a full account of the stone child from Sens, see Brandt Andersen 2018, pp. 92–98.

239. O'Malley (ed.) 1961, p. 50–51.

240. Ibid., pp. 57–58.

241. Avicenna (980–1037), Persian physician, philosopher and politician. Follower of Aristotle.

242. Averroes (1126–1198), Moorish physician, philosopher and mystic. Follower of Aristotle.

243. O'Malley (ed.) 1961, p. 58.

244. Ibid., pp. 51–52.

245. Ibid., p. 83.

246. Ibid., p. 45.

247. Garboe 1950, 2, p. 125. In this position, Bartholin was paid 120 rix-dollars annually.

248. For a review of the book, see Karen Skovgaard-Petersen: Biblioteksfaglig vejledning I 1670'erne – om Thomas Bartholins forelæsninger som universitetsbibliotekar (Library guidance in the 1670s – about Thomas Bartholin's lectures as a university librarian). In: Lauridsen, John T. & Olsen, Olaf: *Umisteligt. Festskrift til Erland Kolding Nielsen*, Copenhagen 2007, pp. 137–52.

249. Ibid and Garboe 1950, 2, p. 125.

250. Gosch 1873, p. 190.

251. *Acta Medica Ann. 1674, 1675, 1676*, Copenhagen 1677, 3, Obs. 3: Microscopiis detecti globuli in sangvine (Red blood cells seen in a microscope), pp. 7–8.

252. Slottved 1978, p. 109.

253. For further information about this stay, see Brandt Andersen 2021, pp. 678–85.

254. Caspar Bartholin 1679, pp. 45–46.

255. Brandt Andersen 2021, p. 681.

256. Caspar Bartholin: De inauribus veterum syntagma. Accedit Mantissa ex Thomæ Bartholini miscellaneis medicis de annulis narium (Treatise on antique earrings. Added a chapter on noserings from Thomas Bartholin's book on diseases in the Bible), Leiden 1676.

257. Garboe 1950, 2, p. 152. Minerva is in Roman mythology the goddess of wisdom, knowledge, art and craftsmanship.

258. Ibid.

259. Thomas Bartholin: De visitatione officinarum pharmeceuticum programma IV (Program number 4 for visitation of the pharmacies) (27 October 1675), Copenhagen 1675. Garboe 1950, 2, pp. 144–45.

260. Garboe 1950, 2, pp. 157–60.

261. Thomas Bartholin: *De Sanguine Vetito Disquisito Medica, cum Cl. Salmasio Judicio* (Medical examination of a ban on blood, with Claude Saumaise as judge), Frankfurt 1673. The dissertation was dedicated to Rasmus Vinding.

262. Ibid, p. 61, and Garboe 1950, 2, p. 157.

263. Garboe 1950, 2, p. 157. Hans Wandal's mentioned dissertation from 1675 was published posthumously in Copenhagen in 1686 with the title *Dissertatio Theologica De Antistitum Ecclesiæ orthodoxæ legitima vocatione & ordinatione. Item De sangvinis & suffocatorum animalium esu, inter Christianos haud illicito.*

264. Ibid, p. 160.

265. Ibid.

266. Thomas Bartholin: *De transplantatione morborum dissertatio epistolica* (Letter concerning dissertation on the transmission of diseases), Copenhagen 1673. Danish translation by Sebastian Popp with the title *Afhandling om Sygdommes Overplantning eller om den Maade paa hvilken Sygdomme overføres fra det menneskelige Legeme til andre, saavel levende som livløse Ting. Forestillet i et Brev til den danske Læge Hermann Grube*, Copenhagen 1794.

267. Herman Grube: *De transplantatione morborum. Analysis nova* (On the transmission of diseases. A new analysis), Hamburg 1674.

268. Garboe 1950, 2, pp. 156–57.

269. Ibid., pp. 160–61.

270. Caspar Bartholin: *De Cordis Structura & Motu Disquisitio* (Investigation of the structure and movement of the heart), Copenhagen 1678. Thomas Bartholin's censorship is dated 22 September 1678.

271. F.J. West 1908, p. 463.

272. In the previously mentioned letter from Thomas Bartholin to Jørgen Hahn in Odense on 23 April 1679, Bartholin states that he is unable to visit Fjellebro due to court duties and must therefore decline Hahn's invitation. Instead, his wife will go with his son Caspar, who is looking forward to getting to know Hahn.

273. *Acta consistorii*, 29 May 1680, Garboe 1950, 2, p. 181.

274. *Acta consistorii*, 31 July 1680, Garboe 1950, 2, p. 181.

275. Garboe 1950, 2, p. 184.

276. Rasmus Vinding: *Regia Academia Hauniensis* (The Royal University of Copenhagen), Copenhagen 1665, p. 388. Peder Hansen Resen: *Inscriptiones Hafnienses Latinæ, Danicæ et Germanicæ* (Latin, Danish and German inscriptions in Copenhagen), Copenhagen 1668, pp. 94–95. Maar, Vilhelm (ed.): *Holger Jacobæus' Rejsebog* (Holger Jacobæus' Travel Book) 1671–1692, Copenhagen 1910, p. 41. The original handwritten travel book is in the Royal Library in Copenhagen: Holger Jacobsen: Forskellige skrifter (Various writings), NKS 309 aa kvart.
277. Vinding 1665, who also has a mini biography of Thomas Bartholin on pp. 378–81 and a list of his publications until 1665 on pp. 381–84, includes an epitaph for himself dated 1661 on p. 388 and one for Thomas Bartholin dated 1663 on p. 381. On pp. 379–80 he includes the earlier quoted panegyric tablet text about Bartholin by Christian Ostenfeld, which was published by Bartholin in *Domus Anatomica* in 1662.
278. Willum Worm 1681, pp. 117–20.
279. Maar (ed.) 1910, p. 40.
280. This brother's posthumous author's encyclopaedia *De Scriptis Danorum*, the first of its kind in Denmark, was published by Thomas Bartholin in Copenhagen in 1666 with additions from the book lists in the linguist Peder Syv's (1631–1702) work *Nogle Betenkninger om det Cimbriske Sprog* (Some Reports on the Cimbrian Language), Copenhagen 1663.
281. The poem is reproduced in Brandt Andersen 2017, p. 342 and in Thomas Kingo: *Samlede Skrifter* (Collected writings) 1–7. Published by Hans Brix, Paul Diderichsen and F.J. Billeskov Jansen, Copenhagen 1975, 1, p. 200. The poem is undated, and its aim has been discussed. See Carsten Bach-Nielsen: Tale over en tom grav. Thomas Kingos meditation over Thomas Bartholins og Rasmus Vindings fælles begravelse i Vor Frue Kirke (Speech over an empty grave. Thomas Kingo's meditation over Thomas Bartholin's and Rasmus Vinding's common burials in the Church of Our Lady). *Kirkehistoriske Samlinger* 2021, pp. 166–77.
282. In classical Greece, Lerna was a marshy region near the east cost of the Peloponnese, south of Argos.
283. Cacus is in Greek and Roman mythology a fire-breathing giant and the son of Vulcan. He was killed by Heracles/Hercules.
284. In Greek mythology, Stymphalian birds are man-eating birds with beaks of bronze, sharp metallic feathers, which they could launch at their victims, and poisonous dung.
285. The earlier-mentioned Italian anatomist Girolamo Fabrici d'Acquapendente (1533–1619).
286. Quintus Roscius (c. 126–62 BC) was a Roman actor.
287. Pæon or Pæan is in Greek mythology the physician of the Greek gods.
288. In Greek mythology, Lachesis is the middle of the Three Fates.
289. Phoebus Apollo is in Greek and Roman mythology the god of the sun, medicine, music, poetry, and the sciences.
290. Sophocles (c. 497/6–406/5 BC) was a Greek tragedian.
291. Apelles of Kos (4th century BC) was a renowned painter of ancient Greece.
292. Mount Parnassus is a mountain range in central Greece frequently mentioned in ancient Greek literature.
293. In Greek mythology, Antaeus is the half-giant son of Poseidon and Gaia. He was defeated by Heracles.
294. Jove or Jupiter is the god of the sky and thunder and king of the gods in ancient Roman religion and mythology.
295. Mount Olympus is the highest mountain in Greece and in Greek mythology the home of the Greek gods.
296. Gyas is in Greek mythology a giant with a hundred arms.
297. Tityos is a giant in Greek mythology, the son of Zeus and Elara.
298. Enceladus is a giant in Greek mythology, the offspring of Gaia and Uranus.
299. Epictetus (c. 50–135) was a Greek Stoic philosopher.
300. Antoninus Pius (86–161) was a Roman emperor.
301. Numa Pompilius (c. 753–672 BC) was the second king of Rome.
302. Penelope is in Homer's *Odyssey* the queen of Ithaca and the faithful wife of Odysseus.
303. Machaon is in Greek mythology a highly valued surgeon and physician.
304. Areophagus is a prominent rock located northwest of the Acropolis in Athens. In Greek mythology it is the location of the Athenian judicial council.
305. Germa, or in ancient times Garama, is an archaeological site in Libya. It was the capital of the Garamantian Kingdom.
306. Willum Worm 1681, pp. 113–20.
307. Jørgen Hahn: *ORATIO in Obitum VIRI Nobilissimi DN. D. THOMÆ BARTHOLINI, publicè habita, Ao MDCLXXX. Die 30. Decemb: in Regio Othiniensi Gymnasio, à GEORGIO HANNÆO, Ethices ac Eloqventiæ Professore*, Copenhagen 1680. In a letter from Thomas Bartholin to Hahn in Odense on 23 April 1679, Bartholin thanks Hahn for his invitation to visit him in Odense and regrets that he cannot come. He tells him that the solemn doctoral promotion is imminent, and that Hahn can let him know if he also wants to be promoted to doctor. However, Hahn did not accept the offer and did not receive his doctorate until 1684. The letter is an example of Bartholin's active work as a professor and his commitment to his students right up until his death. The letter is written in Latin and can be found in the National Archives for Funen (Landsarkivet for Fyn) in Odense, Hahn's archive, file No. 1, fol. 1rv. The letter was kindly made available by Niels W. Bruun.
308. Mansa 1873, p. 478, who refers to Jens Bircherod's (1658–1708) diaries, published by Christian Molbech in 1846.
309. Ibid., pp. 478–79.

310. Maar refers to the painted portrait by Heinrich Dittmers in The Museum of National History at Frederiksborg Castle (Figure 196 in this book). The painted portrait by Karel van Mander (this book's Figure 96a and cover illustration) was until recently considered lost but rediscovered in 2024. (see the figure text and Jesper Brandt Andersen: Forsvundet i 373 år. Om genopdagelsen af Karel van Manders malede portræt af Thomas Bartholin og samarbejdet imellem en berømt anatom og en feteret kunstner i 1600-tallets København (Lost for 373 years. The rediscovery of Karel van Mander's painted portrait of Thomas Bartholin, which initiated a fruitful collaboration between a famous anatomist and a celebrated artist in 17th-century Copenhagen). *Fund og Forskning i Det Kongelige Biblioteks Samlinger* 2024, 63, pp. 11-36).
311. Maar (ed.) 1916, p. 12.
312. Peter Michael Hornung (ed.): *Gobeliner til Danmarks Dronning. Bjørn Nørgaards Danmarkshistorie på Christiansborg, vævet i Manufactures nationales des Gobelins et de Beauvais* (Tapestries for the Queen of Denmark. Bjørn Nørgaard's History of Denmark at Christiansborg, woven in Manufactures nationales des Gobelins et de Beauvais). Guide, Copenhagen 2000, pp. 78–79, no. 12.
313. John Lindsay: *Edwards's Botanical Register: or, ornamental flower-garden and shrubbery: consisting of coloured figures of plants and shrubs, cultivated in British gardens; accompanied by their history, best method of treatment in cultivation, propagation, etc.*, London 1835, 7, plate 1653. See also J. Stewart and E.F. Hennessy: *Orchids of Africa. A select review*, London 1981, pp. 56–57.
314. https://pure.kb.dk/da/activities/udstill-ing-om-l%C3%A6gen-og-polyhistoren-thomas-bartholins-1616-1680-l and https://ugeskriftet.dk/nyhed/diamanten-fejrer-bartholins-400-ars-dag.
315. https://www.dissgea.unipd.it/sites/dissgea.unipd.it/files/Locandina%20convegno_V02.pdf.

CHAPTER 8
Thomas Bartholin's legacy and significance for his own time and posterity
pp. 571–577

1. Thomas Bartholin: *Oratio in obitum clarissimi viri D. Henrici Fuiren medicinæ apud Hafniensis doctoris* (Memorial speech for that very famous man, Henrik Fuiren, Doctor of Medicine in Copenhagen), Copenhagen 1659, p. 12.
2. Petersen 1898, pp. 46–74.
3. Vilhelm Maar: Mindetale holdt på Universitetet paa 300-Aarsdagen for Thomas Bartholins fødsel (Memorial speech held at the university at the 300th anniversary of Thomas Bartholin's birth. In: Maar (ed.) 1916, pp. 1–13.
4. Thus, for example, Gunnar Olsen 1944, pp. 331–32: "Foreign anatomists often attacked him hard and for good reason, since he was strongly addicted to boasting and self-aggrandizement, and he only reluctantly gave credit to others. At home in Denmark no one dared to attack him; he ruled almost autocratically at the university and used his power to occupy numerous teaching posts with his relatives," and Valdemar Meisen: Nicolaus Steno. In: Valdemar Meisen (ed.): *Prominent Danish Scientists through the Ages*, Copenhagen 1932, pp. 36–43: "In 1664 Steno returned to Copenhagen. Young, as he was, he was yet known as a man of science, and he was certainly entitled to a university position. But even though Thomas Bartholin had been much interested in the progress of Steno's studies abroad, he let family ties be decisive in this case, and the unusual gifts of Steno were thus lost to his native country."
5. *Epist. Med.* 1663, 3, Epist. 65: Sudorum origo ex glandulis. De insertione & valvula lactei thoracici & lymphaticorum (Sweat's origin from glands. On the inlet and valves of the thoracic duct and of the lymph vessels). Letter from Niels Stensen to Thomas Bartholin, Leiden 9. January 1662, pp. 262–66. Maar (ed.) 1910, 1, OPH No. 7, pp. 99–103. Scherz (ed.) 1952, E 5, pp. 146–49. English translation in Kardel & Maquet (eds.) 2018, pp. 501–3. Danish translation in Harriet M. Hansen (ed.) 1987, 1, E 5, pp. 40–42. Furthermore, *Acta Medica Ann. 1673*, Copenhagen 1675, Obs. 97: Lymphaticorum varietas (Diversity of lymphatic ducts), pp. 240–42. Maar (ed.) 1910, OPH No. 12, pp. 137–42. English translation in Kardel & Maquet (eds.) 2018, pp. 529–31. Frederik Ruysch: *Dilucidatio Valvularum in Vasis Lymphaticis et Lacteis. Cum Figuris æneis. Accesserunt Quædam Observationes Anatomicæ rariores* (Explanation of the valves in lymph vessels and lacteal veins. With copper engravings. Added some rare anatomical observations), The Hague 1665.
6. Jean Berryat et al. (eds.): *Collection Académique. Tome Septiéme de la Partie Etrangere, Et le Premier de la Médecine séparée*, Dijon and Paris 1766, the preface.
7. *Carmina*, p. 193.
8. Willum Worm 1681, p. 49.
9. Caspar (1655–1738) in medicine, anatomy and physics, Christopher (1657–1714) in mathematics, Thomas (1659–1690) in history and Hans (1665–1738) in theology and philology. Source: Slottved 1978. Only Albert Bartholin (1668–1703) did not become professor, but mayor of Copenhagen.
10. Rasmus Bartholin: *Experimenta crystalli Islandici* etc. (Studies of Icelandic crystal), Copenhagen 1669.

APPENDIX 2

Holger Jacobsen's catalogue of Thomas Bartholin's published, adopted and unpublished works

pp. 581–587

1. Holger Jacobsen: *Oratio In Obitum Summi Daniæ Æsculapii Thomæ Bartholini. Accessit Operum Th. Bartholini Editorum Catalogus* (Memorial speech for the leading Danish physician Thomas Bartholin. Added a catalogue of Thomas Bartholin's published works), Copenhagen 1681, pp. 28–36, which is an update of Bartholin's own catalogue in *Anatome Practica*, Copenhagen 1674, pp. 43–48. For a full overview of Thomas Bartholin's publications, see also H. Ehrencron-Müller: *Forfatterlexikon omfattende Danmark, Norge og Island indtil 1814*, Copenhagen 1924, 1, pp. 276–290. For Bartholin's contributions to the German journal *Miscellanea Curiosa Medico-Physica Academiæ Naturæ Curiosorum sive Ephemeridum Medico-Physicarum Germanicarum Curiosarum*, see ibid., p. 287.
2. English translation is added in parentheses after each work title. Jacobsen indicates the titles in Roman numerals. Note that Jacobsen has slightly altered some of Bartholin's original titles.
3. In addition, a German translation by Michael Kirstein was published in Copenhagen in 1648.
4. The English translation was done by Nicholas Culpeper (1616–1654) and published posthumously in London in 1668.
5. The second edition from Amsterdam 1678 was published by Caspar Bartholin the Younger.
6. Reprinted in *Orationes*, Copenhagen 1668, pp. 1–11.
7. The official year of publication was 1647.
8. Reprinted in *Orationes*, Copenhagen 1668, pp. 43–73.
9. The number XVII must be an error made by Jacobsen. The official number of disputations was XVIII = 18.
10. Very surprisingly and strangely, Jacobsen does not include the original edition from Copenhagen 1652 and the editions from London 1652 and Paris 1654. The book was reprinted in *Opuscula Nova Anatomica*, Copenhagen 1670, pp. 1–71. English translation, London 1653. Danish translation in G. Tryde (ed.), Copenhagen 1940, pp. 1–76.
11. Reprinted in *Opuscula Nova Anatomica*, Copenhagen 1670, pp. 113–47. Danish translation in G. Tryde (ed.), Copenhagen 1940, pp. 121–60.
12. Reprinted in ibid., pp. 73–112. Danish translation by G. Tryde (ed.), Copenhagen 1940, pp. 77–119.
13. Reprinted in ibid., pp. 149–84. Danish translation in G. Tryde (ed.), Copenhagen 1940, pp. 161–98.
14. Reprinted in ibid., pp. 185–382.
15. Reprinted in ibid., pp. 383–412.
16. Reprinted in ibid., pp. 413–62.
17. The official year of publication was 1661.
18. Reprinted in *Opuscula Nova Anatomica*, Copenhagen 1670, pp. 463–518.
19. Reprinted in *Orationes*, Copenhagen 1668, pp. 162–211.
20. Reprinted in ibid., pp. 250–77.
21. The literal translation would be "Rare anatomical stories", but the content of the work is better covered by the translation "Rare medical reports."
22. The official year of publication was 1661.
23. Reprinted in *Orationes*, Copenhagen 1668, pp. 285–308.
24. The official year of publication was 1661.
25. The work is added Rasmus Bartholin's illustrated dissertation on snow crystals and a catalogue over Thomas Bartholin's works hitherto.
26. The years of publication were 1656–1657.
27. The official year of publication was 1662.
28. Danish translation of parts of the work in Niels W. Bruun and Hans Otto Loldrup (eds.), Copenhagen 1982.
29. Danish translation by Niels W. Bruun in Niels W. Bruun (ed.) 2007. English translation by Peter Fisher in Niels W. Bruun (ed.) 2015.
30. Reprinted in *Opuscula Nova Anatomica*, Copenhagen 1670, pp. 519–48.
31. Reprinted in *Orationes*, Copenhagen 1668, pp. 311–51.
32. Reprinted in *Opuscula Nova Anatomica*, Copenhagen 1670, pp. 549–614.
33. Reprinted in ibid., pp. 615–726.
34. The work was also published in Copenhagen in the same year.
35. English translation in C. D. O'Malley (ed.) 1961, pp. 43–101.
36. Second corrected edition, Frankfurt 1672. English translation by James Willis in Johan Schioldann-Nielsen and Kurt Sørensen (eds.) 1994.
37. Danish translation, Copenhagen 1794.
38. For a full overview of Bartholin's own contributions to all volumes of this journal, see Ehrencron-Müller 1924, 1, pp. 287–89.
39. Danish title: *Trøste-Breff til sin Hierte kiere og høybedrøfvede Syster Anna Catharina Müller S. D. Caspari Bartholini Efterlefverske.*
40. Danish translation by Hartvig Frisch in *Bibliotek for Læger*, 1918, pp. 86–113.
41. English translation by Charles D. O'Malley (ed.) 1961, pp. 1–42.
42. Danish translation by G. L. Baden 1814.
43. Published under the pseudonym Tienst Berede with the title *Et Edrue Lefnets Gafn oc Nytte.*
44. Danish title: *Om den Christen Troes Sandhed: Alle sine Jefn-Christne/sær dennem som seile til Ost- oc Vest-Indien /til gaffn. Af det Latinske Ofversat oc mangestæds Forklaret Aff Th. Bartholin.*
45. Bartholin's own list in his script on the fire (no. 56) contains another five lost works, but they were authored by others, which may explain why Jacobsen has omitted them (Bartholin 1670, no. XXII–XXIV, pp. 80–83 and O'Malley (ed.) 1961, pp. 30–31).

Literature

Achen, Sven Tito: *Danske Adelsvåbener. En heraldisk nøgle*, Copenhagen 1973.

Adam, Charles & Tannery, Paul: *Oevres de Descartes*, Paris 1897–1910.

Agricola, Georgius: *De Ortu et Causis Subterraneorum* (On the origin and causes of underground phenomena), Basel 1546.

Agricola, Georgius: *De animantibus subterraneis* (About the subterranean creatures), Basel 1561.

Aiton, William Townsend: *Hortus Kewensis, or a catalogue of the plants cultivated in the Royal Botanic Garden at Kew*, London 1813.

Aldrovandi, Ulisse: *Monstrorum Historia* (History of monsters), Bologna 1642.

Alpini, Prospero: *De Plantis Ægypti* (On Egyptian plants), Venice 1592.

Andersen, Hans Christian: *Nye Eventyr og Historier*, 3, Copenhagen 1859.

Aselli, Gaspare: *De lactibus sive lacteis venis, quarto vasorum mesaraicorum genere* (About the lacteal vessels or lacteal veins, a fourth kind of mesenterial vessels), Milan 1627.

Authier, André: *Early Days of X-ray Crystallography*, Oxford 2013.

Bach-Nielsen, Carsten: Tale over en tom grav. Thomas Kingos meditation over Thomas Bartholins og Rasmus Vindings fælles begravelse i Vor Frue Kirke. *Kirkehistoriske Samlinger* 2021, pp. 166-177.

Bacon, Francis: *Historia Vitæ et Mortis* (The History of Life and Death), London 1623.

Barbette, Paul: *Aenmerckingen van Paulus Barbette, der Medicynen Doctor, Op d'Anatomische Schriften van Jonckheer Lodowyk de Bils*, Amsterdam 1659.

Bartholin, Albert: *De Scriptis Danorum, Liber Posthumus* (On the writings of the Danes, a posthumous book), Copenhagen 1666. Published by Thomas Bartholin.

Bartholin, Caspar the Elder: *Enchiridion logicum ex Aristotele* (Handbook in logic according to Aristotle), Strasbourg 1608.

Bartholin, Caspar the Elder: *Anatomicæ Institutiones Corporis Humani, utriusque sexus, Historiam & Declarationem, exhibentes cum plurimis novis Observationibus & Opinionibus, nec non illustriorum quæ in Anthropologia occurrunt controversiarum decisionibus, cum Indice Capitum et rerum locupletissimo* (The anatomical arrangement of the human body, of both sexes, a description and exposition presented with a great many new observations and conjectures, and furthermore the setting out of the more famous disputes in anthropology. Furthermore, a list of chapters and a copious index of subjects), Wittenberg 1611.

Bartholin, Caspar the Elder: *Problematum philosophicorum et medicorum nobiliorum et rariorum miscell. exercitationes ad disputandum in illustri Albiaca propositæ* (Various disputation exercises in philosophical and more famous and rarer problems in medicine displayed in the famous Albiaca), Wittenberg 1611.

Bartholin, Caspar the Elder: *Astrologia seu De Stellarum Natura affectionibus et effectionibus* (Astrology or concerning the nature of the stars, their actions and consequences), Wittenberg 1612.

Bartholin, Caspar the Elder: *Opuscula quatuor singularia* (Four unique little pieces), Copenhagen 1628.

Bartholin, Caspar the Elder: *De Studio Theologico* (On the study of theology), Copenhagen 1628.

Bartholin, Caspar the Elder: *De Studio Medico* (On the study of medicine), Copenhagen 1628.

Bartholin, Caspar the Elder: *De lapide nephritico opusculum. Cum præcipuis ad plerosque morbos amuletis* (A book about the nephrite stone. With excellent amulets against several diseases), Copenhagen, 1628.

Bartholin, Caspar the Younger: *Exercises Miscellaneæ. Varii argumenti inprimis Anatomici* (Mixed exercises of different content, but aimed at anatomy), Leiden 1675.

Bartholin, Caspar the Younger: De inauribus veterum syntagma. Accedit Mantissa ex Thomæ Bartholini miscellaneis medicis de annulis narium (Treatise on antique earrings. Added a chapter on noserings from Thomas Bartholin's book on diseases in the Bible), Leiden 1676.

Bartholin, Caspar the Younger: *Diaphragmatis Structura Nova* (The new structure of the diaphragm), Paris 1676.

Bartholin, Caspar the Younger: *De Ovariis Mulierum* (On the female ovaries), Rome 1677.

Bartholin, Caspar the Younger: *Expositio veteris in pueperio ritus* (Exposition of ancient customs and practices at births), Rome 1677.

Bartholin, Caspar the Younger: *De Cordis Structura & Motu Disquisitio* (Investigation of the structure and movement of the heart), Copenhagen 1678.

Bartholin, Caspar the Younger: *Administrationum Anatomicarum Specimen* (Sample on the practice of the anatomical disciplines), Frankfurt 1679.

Bartholin, Rasmus: *De Cometis Anni MDCLXIV & MDCLXV Opusculum. Ex observationibus Hauniæ habitis adornatum* (Treatise on the comets in the years 1664 and 1665. Based on observations made in Copenhagen), Copenhagen 1665.

Bartholin, Rasmus: *Specimen recognitionis nuper editarum Observationum Astronomicarum Nobilissimi Viri Tychonis Brahe* (Exploring sample of the nobleman Tycho Brahe's recently published astronomical observations), Copenhagen 1668.

Bartholin, Rasmus: *Experimenta Crystalli Islandici* (Experiments with Islandic crystal), Copenhagen 1669.

Bartholin, Rasmus: *De Studio Lingvæ* Danicæ (On the study of the Danish language), Copenhagen 1657. In: Rasmus Bartholin: *De Naturæ Mirabilibus Questiones Academicæ* (Academic questions on the marvels of Nature), Copenhagen 1674, pp. 146–70. No. 11 of the book's 13 questions. Danish translation by Carl Berend with the title "Rasmus Bartholins Tale om det danske Sprog." In: *Studier fra Sprog og Oldtidsforskning*, no. 97, 1914. Reprinted in Henrik Bertelsen (ed.): *Danske Grammatikere fra Midten af det syttende til Midten af det attende Aarhundrede*, reprinted in Copenhagen 1979 (original edition 1915–1929), 6, pp. 35-64. The Latin text is in both editions in Vol. 1, pp. 3–33.

Bassi, Claudio & Malleo, Guiseppe: The unsolved mystery of Johann Georg Wirsung and of (his?) pancreatic duct. *Surgery* 2011, 149, pp. 153–55.

Bauhin, Caspar: *Theatrum Anatomicum* (The anatomical theatre), Frankfurt 1605 and 1621.

Bering Liisberg, Henrik Carl: *Kunstkammeret. Dets Stiftelse og ældste Historie*, Copenhagen 1897.

Bernard, Claude: *Mémoires sur le pancréas et sur le role du sac pancréatique dans les phénomènes digestifs, particulièrement dans la digestion des matières grasses neutres*, Paris 1856.

Berryat, Jean et al. (eds.): *Collection Académique*, Partie Etrangere, Vol. 4, Dijon and Paris 1757, pp. 185–376, Vol. 6, Dijon and Paris 1761, pp. 362–440 and Vol. 7, Dijon and Paris 1766, preface and pp. 140–386.

Bils, Louis de: *Waarachtig Gebruik der tot noch toe gemeende Gijlbuis, beneffens de verrijzenis der Lever, voorheen zoo lichtvaardig in't graf* gestooten, Rotterdam 1658. Latin translation: *Epistolica Dissertatio: Qua verus Hepatis circa Chylum, & pariter ductus Chiliferi hactenus dicti usus, docetur*, Rotterdam 1659.

Bils, Louis de: *Aan alle ware Liefhebbers der Anatomie*, Rotterdam 1659.

Bils, Louis de: *Brief van Louijs de Bils Heere von Koppensdamme etc. aan Th. Bartholijn Professor tot Koppenhaven*, Rotterdam 1661.

Birket-Smith, Kaj: *Jens Munk's rejse og andre danske ishavsfarter under Christian IV*, Copenhagen 1929.

Birket-Smith, S: *Kjøbenhavns Universitets Matrikel*, 1: 1611–1667 and 2: 1667–1740, Copenhagen 1890.

Birket-Smith, S. (ed.): *Dr. med. Otto Sperlings Selvbiografi 1602-1673, oversat fra tysk i Uddrag efter Originalhaandskriftet med særligt Hensyn til Forfatterens Ophold i Danmark og Norge samt Fangenskab i Blaataarn, Copenhagen* 1885.

Birket-Smith, S. (ed.): *Etatsraad og Landkomissær Johan Monrad's Selvbiografi (1638-1692). Udgivet efter Originalmanuskriptet af S. Birket Smith*, Copenhagen 1888.

Bithol-Hespériès, Annie: Cartesian Philosophy. In: Stephen Gaukroger, John Shuster and John Sutton (eds.): *Descartes' natural Philosophy*, London and New York 2000.

Blaeu, Willem: *Stedeboeck der Nederlanden*, Amsterdam 1649.

Blasius, Gerard: *Medicina Generalis* (General Medicine), Amsterdam 1661.

Blasius, Gerard: *Anatome Medullæ Spinalis et Nervorum inde provenientium* (Anatomy of the spinal cord and the nerves emanating from it), Amsterdam 1666.

Bogdan, Martin: *Insidiæ Structæ Thomæ Bartholini Vasis Lymphaticis ab Olao Rudbeckio Sveco in suis Ductibus Hepaticis et Vasis Glandularum. Arosiæ editis* (The treacherous ambush on Thomas Bartholin's *Vasa Lymphatica* instigated by the Swede Olof Rudbeck, in his writing about the liver's excretory ducts and the glands' vessels, published in Västerås), Copenhagen 1654.

Bogdan, Martin: *Apologia pro Vasis Lymphaticis D. Thomæ Bartholini Medici Regij Professoris & Anatomici, conscripta a Martino Bogdano Drifna Marchico contra Insidias Secundo Structas ab Olao Rudbeck Sueco* (Apology in defence of Royal Professor of Medicine and Anatomy Dr. Thomas Bartholin's *Vasa Lymphatica*. The defence was composed by Martin Bogan from Driesen in Neumark and directed against the hidden ambush, which for the second time was instigated by the Swede Olof Rudbeck), Copenhagen 1654.

Bondeson, Jan: *The Two-headed Boy and Other Medical Marvels*, New York and London 2004.

Bonet, Théophile: *Sepulchretum sive Anatomia Practica Ex Cadaveribus Morbo denatis* (The burial or the practical anatomy from corpses of sick people told in detail), Geneva 1679.

Bonjour, Edgar: *Die Universität Basel von den Anfängen bis zur Gegenwart 1460–1960*, Basel 1960.

Borch, Ole (Benedictus Blottesandæus): *Deusingius Heautontimorumenos sive epistolæ selectæ eruditorum, quæ immaturis Antonii Deusingii medici Groningensis, scriptis larvam strictim sed, sincere detrahunt etc.* (The self-tormentor Deusing or letters by selected scholars, which degrade the immature physician in Groningen Anton Deusing's bewitched and superficial, but genuine writings etc.), Hamburg 1661.

Borch, Ole: *Olai Borrichii Oratio Jubilæa Evangelica* (Ole Borch's evangelical jubilee speech), Copenhagen 1667.

Bording, Anders: *Den Danske Mercurius*, Copenhagen 1668.

Borel, Pierre: *Vitæ Renati Cartesii* (The life of René Descartes), Paris 1656

Brahe, Tycho: *Tychonis Brahe Dani de nova et nullius ævi memoria prius visa Stella, iam pridem Anno à nato CHRISTO 1572, mense Novembrj primùm Conspecta, contemplatio mathematica* (The Dane Tycho Brahe's mathematical consideration over a new star never seen at any time before, but now for the first time observed in November 1572 A.C.), Copenhagen 1573.

Brahe, Tycho: *De mundi ætherei recentioribus phænomenis* (About the new phenomena in the heavenly world), Uranienborg 1588.

Brahe, Tycho: *Tychonis Brahe Dani Epistolarum astronomicarum libri* (Books with the Dane Tycho Brahe's astronomical letters), Uranienborg 1596.

Brahe, Tycho: *Astronomiæ Instauratæ Mechanica* (The new mechanical astronomy), Wandesburg 1598.

Brandt Andersen, Jesper & Bruun, Niels W.: Steno-Fallots tetralogi og Bartholin-Pataus syndrom. En hjertemisdannelse og et misdannelsessyndrom første gang beskrevet af danske anatomer i 1600-tallet. *Fund og Forskning i Det Kongelige Biblioteks Samlinger* 2015, 54, pp. 197–233.

Brandt Andersen, Jesper: *Thomas Bartholin. Lægen og anatomen. Fra enhjørninger til lymfekar*, Copenhagen 2017.

Brandt Andersen, Jesper: Sælsomme beretninger i lægen og anatomen Thomas Bartholins værker. *Fund og Forskning i Det Kongelige Biblioteks Samlinger* 2018, 57, pp. 61–115 (with English summary).

Brandt Andersen, Jesper: Da Niels Stensen tog Paris med storm. *Dansk Medicinhistorisk Årbog* 2018, pp. 27–71 (with English summary).

Brandt Andersen, Jesper & Bruun, Niels W.: Steno-Fallots tetralogi og Bartholin-Pataus syndrom. En hjertemisdannelse og et misdannelsessyndrom første gang beskrevet af danske anatomer i 1600-tallet. *Dansk Medicinhistorisk Årbog* 2018, pp. 73–105 (with English summary).

Brandt Andersen, Jesper: *Niels Stensen. Kongelig anatom og fyrstelig geolog*, Copenhagen 2021.

Brandt Andersen, Jesper: Forsvundet i 373 år. Om genopdagelsen af Karel van Manders malede portræt af Thomas Bartholin og samarbejdet imellem en berømt anatom og en feteret kunstner i 1600-tallets København. With English summary. *Fund og Forskning i Det Kongelige Biblioteks Samlinger* 2024, 63, pp. 11-36.

Brosse, Guy de la: *De la Nature, vertu, et utilité des Plantes. Divisé en cinq* Livres. 2nd edition, Paris 1640.

Brown, Hartcourt: *Scientific Organizations in Seventeenth Century France (1620–1680)*, New York 1934.

Bruun, Carl: *Kjøbenhavn. En illustreret Skildring af dets Historie, Mindesmærker og Institutioner*, Copenhagen 1890.

Bruun, Christian: Johan Rhode: Født i Kjøbenhavn i 1587. Død i Padua i 1659. In: *Paa Hundrede-aarsdagen efter at Det store Kongelige Bibliotek blev erklæret for at være et offentligt Bibliotek ved Kongelig Reskript af 15. November 1793*, Copenhagen 1893.

Bruun, Niels W. & Loldrup, Hans Otto (eds.): *Thomas Bartholin: Cista Medica Hafniensis*, Copenhagen 1982.

Bruun, Niels W. & Loldrup, Hans-Otto (eds.): *Thomas Bartholin: Anatomihuset i København* (The Anatomy House in Copenhagen), Copenhagen 2007.

Bruun, Niels W.: Kongens stol i Domus Anatomica Hafniensis. *Fund og Forskning i Det Kongelige Biblioteks Samlinger* 2012, 51, pp. 131–47 (with English summary).

Bruun, Niels W. (ed.): Thomas Bartholin*: The Anatomy House in Copenhagen*, Copenhagen 2015.

Brøndal, Viggo & Hansen Valdemar (eds.): René Descartes: *Om Metoden. Discours de la Méthode*. Danish translation, Copenhagen 1937.

Burckhardt, Albrecht: *Geschichte der Medizinischen Fakultät zu Basel 1460–1900*, Basel 1917.

Burman-Becker, Johan Gottfried: *Hof- og Reise-Apotheker Johan Gottfried Beckers Levnet*, Copenhagen 1864.

Carøe, Kristian: *Medicinalordningens Historie indtil Sundhedskollegiets Oprettelse 1803*, Copenhagen 1917.

Carøe, Kristian.: Livkirurg Philip Hacquart og hans Sønner. *Personalhistorisk Tidsskrift* 1920, 7, 5, 150–67.

Casserio, Giulio: *De Vocis Auditusque Organis Historia Anatomica* (An anatomical description of the vocal and auditory organs), Ferrara 1600.

Casserio, Giulio: *Pentæstheseion, Hoc est, De Quinque Sensibus Liber, Organorum Fabricam variis iconibus fideliter & ad viuum ære incisis illustratam, Nec non actionem & usum discursu Anatomico & Philosophico accuratè explicata continens.* (Pentæstheseion, this is a book about the structure of the five senses and is illustrated with various detailed copper engravings according to the organs in their natural state. It also describes the function and use of the organs, which are carefully made clear in an anatomical and physiological explanation), Venice 1609 and Frankfurt 1610.

Castelli, Pietro: *Hortus Messanensis* (The Garden of Messina), Messina 1640.

Cesalpino, Andrea: *De Metallicis Libri Tres* (Three books about metals), Rome 1596.

Choulant, Ludvig: *History and bibliography of anatomic illustration in its relation to anatomic science and the graphic arts.* Translated and edited with notes and a biography by Mortimer Frank, Chicago 1920.

Christensen, R. E., Hansen, Axel and Larsen, Knud (eds.): Niels Steensens (Stenonis) Værker i Oversættelse. In: *Klassisk dansk Medicin*, Vol. II, Copenhagen 1939.

Clausen, Helge: Niels Stensens brevregistrant. Nicolaus Steno. Index of letters. *Ælnoths skriftserie* No. 18, 2019.

Collice, Massimo, Collice & Roasa, Riva, Alessandro: Who discovered the sylvian fissure. *Neurosurgery*, 63, 4, 2008, pp. 623–28.

Colonna, Fabio: *De Glossopetris Dissertatio, qua ostenditur Melitenses Linguas Serpentinas, sive Glossopetras dictas, non esse lapideas, ut quidam asserunt, sed osseas & Carcariæ Lamiæ, sive Canicularum & similium dentes, maris æstu olim terra tenui & lutosa obrutos* (Dissertation about tongue stones. Here, it is shown, that the Maltese snake stones, or tongue stones, as they are called, are not made from stone, as some suggest, but from bone, and that they are teeth from great white sharks, smaller sharks or similar, which were once covered by a thin layer of earth and mud). In: *Lyncea Purpura*, Rome 1616, p. 31–39.

Conring, Hermann: *De Sanguis Generatione et motu naturalii Opus novum* (A new work on the generation and natural movement of the blood), Helmstedt 1643.

Conring, Hermann: *Introductio in artem medicam* (Introduction to the art of medicine), Helmstedt 1687.

Cooper, Barry. The origins of bone marrow as the seedbed of our blood: from antiquity to the time of Osler. *Baylor University Medical Center Proceedings* 2011, 24, 2, pp. 115–18.

Copernicus, Nicolaus: *De revolutionibus orbium coelestium* (On the Revolutions of the Heavenly Spheres), Nuremberg 1543.

Cornaro, Luigi: *Discorsi della vita sobria*, Padua 1558.

Culpeper, Nicholas: *The English Physitian OR An Astrologo-Physical Discourse of the Vulgar Herbs of this Nation. Being a Compleat Method of Physick, whereby a man may preserve his Body in Health; or cure himself, being sick, for three pence charge, with such things only as grow in England, they being most fit for English Bodies*, London 1652.

Dahl, Svend: *Universitetsbibliotekets Bygninger gennem Tiderne. En historisk Fremstilling*, Copenhagen 1932.

Dapper, Olfert: *Naukeurige Beschrijvinge der Afrikaensche Gewesten*, second edition, Amsterdam 1676.

Darwin, Charles: *On the Origin of Species by Means of Natural Selection or the Preservation of favoured Races in the Struggle for Life*, London 1859.

Debes, Lucas Jacobsøn: *Færoæ & Færoa reserata: Det er Færøernis oc Færøeske Indbyggeris Beskrifvelse, udi hvilcken føris til Liuset adskillige Naturens Hemeligheder/ oc nogle Antiqviteter, som her til Dags udi Mørcket hafve været indelugt, oc nu her opladis / Alle curieuse til Velbehagelighed, sammenskrefven oc forklaret aff Lucas Jacobsøn Debes*, Copenhagen 1673.

Descartes, René: *Discours de la Méthode*, Leiden 1637.

Descartes, René: *Principia Philosophiæ* (Principles of philosophy), Amsterdam 1644.

Descartes, René: *Geometria* (Geometry) Leiden 1649.

Descartes, René: *Les Passions de l'Ame*, Paris 1649.

Descartes, René: *Tractatus de Homine* (Treatise on man), Leiden 1662.

Descartes, René: *Le Monde*, Paris 1664.

Descartes, René: *L'Homme*, Paris 1664.

Deusing, Anton: *Exercitationes Physico-Anatomicæ De Nutrimenti in Corpore Elaboratione. Ubi De Chylificatione, & Chyli Motu; Sanguificatione; Depuratione Alimenti; Itemque Spiritibus. Quibus adjecta Appendix, In qua Examini ac Iudico aliorum Subjiciuntur varia De Chyli motu, & Nutrimenti in Corpore Elaboratione, Nec non De admiranda Anatome Nobiliss. Viri D. Ludovici Bils* (Physical-anatomical exercises concerning the processing of food in the body, including the formation and the movement of the chyle, the blood formation and the purification of the food as well as the spirits. Added an appendix, in which the movement of the chyle and the processing of food in the body are subjected to different other examinations and evaluations, but not by the admirable anatomist, the famous Doctor Louis de Bils), Groningen 1660.

Deusing, Anton: *Exercitationes Physico-Anatomicæ De Nutrimento Animalium Ultimo. Ubi De Sanguinis Usu, Ac Commento Nutritii Succi per Nervos influentis. Accessit Dissertatio Epistolica De Hepatis Officio, ad Celeberrimum Virum, D. Thomam Bartholinum, Medicum et Anatomicum Excellentissimum, Professorem Regium Hafniæ* (Physical-anatomical exercises concerning animals' last meals, including the function of the blood and considerations about the nutritional juice flowing from the nerves. Added a dissertation in the form of a letter about the function of the liver to the famous man, excellent physician and anatomist, professor at Copenhagen University, Doctor Thomas Bartholin), Groningen 1661.

Deusing, Anton: *Appendix Ad Dissertationem De Hepatis Officio Seu Vindiciæ Hepatis Redivivi, Leni correctione tangentes sequiorem Interpretationem Clarissimi Viri D. J. van Horne, Med. Doct. Anatomiæ & Chirurgiæ in Acad. Lugd. Bat. Professoris* (Appendix to the dissertation about the function of the liver or defence of the liver's resurrection, which considerately corrects the famous doctor and professor of anatomy and surgery at the University of Leiden Johannes van Horne's less good interpretation), Groningen 1661.

Deusing, Anton: *Disquisitio Ulterior De Chyli Motu Atque Officio Hepatis Ad Excellentiss. & Inclytum Virum D. Thomam Bartholinum, etc.* (Further investigation of the movement of the chyle and the function of the liver to the excellent and famous man Thomas Bartholin, etc.), Groningen 1662.

Deusing, Anton: *Resurrectio Hepatis Asserta; Contra Socium Larvatum Vincentium Slegium, sub Personati Blottesandæi cohorte furiosa Signiferum* (Defence for the resurrection of the liver against Vincent Slegel, the bewitched confederate of the hidden Blottesandæus, the leader of the mad mob), Groningen 1662.

Deusing, Anton: *Vindiciarum Hepatis Redivivi Supplementum* (Appendix to vindication for the resurrection of the liver). In: *Apologeticæ Defensionis Pro Oeconomia Corporis Animalis Prodromus, Quo Personata Cuidam Benedicto Blottesandeo Larva detrahitur* (Forerunner of a protective defence of a system of life expressions in the body of the living being, and where the personal mask is torn off a certain Benedictus Blottesandeus), Groningen 1662.

Dew, Nicolas: *Orientalism in Louis XIV's France*, Oxford 2009.

Digby, Kenelm: Two *Treatises in the one of which, The Nature of Bodies; in the other, The Nature of Mans Soule; is looked into: In Way of Discovery, of the Immortality of reasonable Soules*, Paris 1644.

Dionis, Pierre: *L'Anatomie de l'Homme suivant la Circulation du Sang, & les dernières Découvertes, demontrée au Jardin Royal*, Paris 1690.

Ebbesen, Sten & Koch, Carl Henrik: Dansk filosofi i renæssancen 1537–1700. In: Sten Ebbesen & Carl Henrik Koch (eds.): *Den danske filosofis historie*, 2, Copenhagen 2003.

Ehrencron-Müller, H.: *Forfatterlexikon omfattende Danmark, Norge og Island indtil 1814*, I–XII, Copenhagen 1924–1935.

Ellehøj, Svend: Christian 4.s Tidsalder 1596–1660. In: John Danstrup and Hal Koch (eds.): *Danmarks Historie*, Copenhagen 1964, 7.

Eller, Povl: *Kongelige portrætmalere i Danmark 1630–82. En undersøgelse af kilderne til Karel van Manders og Abraham Wuchters' virksomhed*, Copenhagen 1971.

Empiricus, Marcellus: *De Medicamentis* (On medicaments).

Engelstoft, Povl & Dahl, Svend (eds.): *Dansk Biografisk Leksikon*, 2nd edition, Copenhagen 1933–1944.

Englund, Peter: *Silvermasken – en kort biografi över drottning Kristina*, Stockholm 2006.

Eriksson, Gunnar: *Rudbeck 1630–1702. Liv, lärdom, dröm i barockens Sverige*, Stockholm 2002.

Everaerts, Antony: *Novus et Genuinus Hominis Brutique Animalis Exortus* (A new and genuine origin of humans and animals), Middelburg 1661.

Everaerts, Antony: *Lux e Tenebris Affulsa ex Viscerum Monstrosi Partus Enucleatione* (The light shines in the darkness, according to the explanation of the entrails of a deformed fetus), Middelburg 1662.

Fabrici d'Acquapendente, Girolamo: *De Visione, Voce et Auditu* (On the vision, voice and hearing), Venice 1600.

Fabrici d'Acquapendente, Girolamo: *De Formato Foetu* (On the formation of the foetus), Venice 1600 and 1620. English translation by Howard B. Adelmann in: *The Embryological Treatises of Hieronymus Fabricius of Aquapendente*, Ithaca, New York 1942, 1, pp. 235–334.

Fabrici d'Acquapendente, Girolamo: *De Formatione Ovi et Pulli* (On the formation of the egg and of the chick), Padua 1621. English translation by Howard B. Adelmann in: *The Embryological Treatises of Hieronymus Fabricius of Aquapendente*, Ithaca, New York 1942, 1, pp. 137–233.

Fabroni, Angelo: *Lettere inedite di uomini illustri*, Vol. 2, Florence 1775.

Fink-Jensen, Morten: *Fornuften under troens lydighed. Naturfilosofi, medicin og teologi i Danmark 1536–1636*, Copenhagen 2004, pp. 231–32.

Fink-Jensen, Morten (ed.): *The Foundation and Regulations of the University of Copenhagen 1539*, Copenhagen 2020.

Folli, Cecilio: *Sanguinis a dextro in sinistrum cordis ventriculum defluentis facilis reperta via* (About the found channel for the blood flowing from the right to the left heart chamber), Venice 1639.

Folli, Cecilio: *Nova auris internæ delineatio* (A new sketch of the inner ear), Venice 1640.

Fallot, Étienne-Louis Arthur: Contribution à l'Anatomie pathologique de la Maladie bleue (Cyanose cardiaque). *Marseille Medical* 1888, 25, pp. 77-93, 138-158, 207-223, 341-354, 370-386, 403-420.

Galilei, Galileo: *Siderius Nuntius* (Starry Messenger), Venice 1610.

Galla, Giulio Cesare la: *De Phænomenis in Orbe Lunæ novi Telescopii usu a Galileo Galilei nunc iterum suscitatis Physica disputatio* (Physical treatise on phenomena within the moon's circumference now again brought about using Galileo Galilei's new telescope), Venice 1612.

Garboe, Axel: Enhjørningen. *Medicinsk-Historiske Smaaskrifter ved Vilhelm Maar*, 12, Copenhagen 1915, p. 19.

Garboe, Axel: Om fabeldyret basilisken. In: *Teologisk Tidsskrift for den danske Folkekirke* 1917, 3, 8, pp. 17–30.

Garboe, Axel & Larsen, Knud: Niels Stensen. In: Povl Engelstoft & Svend Dahl (eds.): *Dansk Biografisk Leksikon*, 2nd edition, Copenhagen 1933–1944, 22, pp. 562–75.

Garboe, Axel: Thomas Bartholin og Rav-studiet i Danmark. Et Blad af Mineralogisk Museums Historie. Særudgave af *Meddelelser fra Dansk Geologisk Forening*, 1946, 11, pp. 97–116.

Garboe, Axel: *Thomas Bartholin*. Et Bidrag til dansk Natur- og Lægevidenskabs Historie i det 17. Aarhundrede, *Acta Historica Scientiarium Naturalium et Medicinalium*, Vol. V, 1–2, Copenhagen 1949–1950.

Garboe, Axel: Kongelig Majestæts Læge for Lolland-Falster Michael Lyser. In: *Lolland-Falsters Historiske Samfunds Årbog* 1952, 3, 1, pp. 83–93.

Garboe, Axel: Nicolaus Steno (Niels Stensen) and Erasmus Bartholin. Two 17th Century Danish Scientists and the Foundation of exact Geology and Crystallography. Special edition of *Danmarks Geologiske Undersøgelse* 1954, IV. Række, Vol. 3, No. 9, pp. 1–48.

Garboe, Axel: Meteoritfaldet på Fyen i Aaret 1654. *Meddelelser fra Dansk Geologisk Forening* 1959, 12, pp. 284–97.

Gassendi, Pierre: *Viri Ilustris Nicolai Claudii Fabricii de Peiresc, Senatoris Aquisextiensis, Vita* (The life of the illustrious Nicolas-Claude Fabri de Peiresc, Senator of the Parliament at Aix), Paris 1641.

Gassendi, Pierre: *Tychonis Brahei, Equitis Dani, Astronomorum Coryphæi Vita* (The life of Tycho Brahe, Danish knight, astronomical pioneer), Paris 1654.
Gaukroger, Stephen: *Descartes. An Intellectual Biography*, New York 1995.
Geneser, Finn: *Histologi*, Copenhagen 1981.
Gesner, Konrad: *De Lunariis Herbis, et Rebus Noctu Lucentibus* (On self-illuminating plants, and things which shine at night), Zurich 1555.
Gillispie, Charles Coulston (ed.): *Dictionary of Scientific Biography*, New York 1981.
Glarbo, Henny: Danske og norske studerende i Padua. *Personalhistorisk Tidsskrift*, 1928, Vol. 49, series 9, 1, pp. 32–44 and 103–15.
Glebe-Møller, Jens: Lutheraneren Ole Borch. In: *Borchs Kollegium 1691–1991*, Copenhagen 1991, pp. 33–44.
Glisson, Francis: *Anatomia Hepatis* (The Anatomy of the Liver), London 1654.
Gosch, Christian Carl August: *Udsigt over Danmarks zoologiske Literatur*, Copenhagen 1873.
Graaf, Reinier de: *De succi Pancreatici Natura et Usu Exercitatio Anatomica Medica* (Anatomical and medical investigation of the nature and function of the pancreatic juice), Leiden 1664.
Grell, Ole Peter: *The World of Worm: Physician, Professor, Antiquarian, and Collector, 1588–1654*, Abingdon and New York 2022.
Grube, Herman: *De transplantatione morborum. Analysis nova* (On the transmission of diseases. A new analysis), Hamburg 1674.
Gräf, Holger Thomas & Tacke, Andreas (eds.): *Von Augsburg nach Frankfurt. Der Kupferstecher Johann Philipp Thelott (1639–1671)*, Marburg 2022.
Guan Xuelin: Vestlig medicinsk praksis ved Kangxi-kejserens hof med efterskrift af Helge Kragh. In: Ole Willumsen Krog (ed.): *Skatte fra kejserens Kina. Den Forbudte By og Det Danske Kongehus*, Copenhagen 2006, pp. 126–31.
Hagedorn, Lea: Buch-Bildnisse: Thelotts Porträtarbeiten für Verlagshäuser. In: Holger Thomas Gräf & Andreas Tacke (eds.): *Von Augsburg nach Frankfurt. Der Kupferstecher Johann Philip Thelott (1639–1671)*, Marburg 2022, pp. 75–85.
Hahn, Jørgen: *ORATIO in Obitum VIRI Nobilissimi DN. D. THOMÆ BARTHOLINI, publicè habita, Ao MDCLXXX. Die 30. Decemb: in Regio Othiniensi Gymnasio, à GEORGIO HANNÆO, Ethices ac Eloqventiæ Professore* (Eulogy for the most noble doctor Thomas Bartholin delivered publicly on 30 December 1680 at the Royal Gymnasium in Odense by Jørgen Hahn, professor of ethics and rhetoric), Copenhagen 1680.
Hansen, Frank & Olesen, Mogens Nørgaard: *Halleys Komet og vort verdensbillede*, Copenhagen 1985.
Hansen, Harriet M. (ed.): *Niels Stensens korrespondance i dansk oversættelse I-II*, Copenhagen 1987.
Harvey, E. Newton: *A History of Luminescence. From the Earliest Times Until 1900*, Philadelphia 1957.
Harvey, William: *Exercitatio Anatomica de Motu Cordis et Sanguinis in Animalibus* (An anatomical exercise on the motion of the heart and blood in animals), Frankfurt am Main 1628. English translation in Robert Willis (ed): *The Works of William Harvey*, London 1847, pp. 1–86. Danish translation in Valdemar Meisen (ed.): *William Harvey's Bog om Opdagelsen af Blodets Kredsløb*, Copenhagen 1929.
Harvey, William: *Exercitatio Anatomica de Circulatione Sanguinis: ad Joannem Riolanum filium Parisiensem, medicum peritissimum anatomicoru coryphæum, in Academia Parisiensi Anatomes & Herbariæ Professorem Regium & egregium atque decanum, Reginæ matris Lodovici XIII medicum primarium* (Reply to Jean Riolan concerning an anatomical exercise of the blood circulation), Cambridge 1649. English translation in Willis (ed.) 1847, pp. 87–141.
Harvey, William: *Exercitationes de Generatione Animalium. Quibus accedunt quædam De Partu: de Membranis ac humoribus Uteri: & de Conceptione* (Exercises on the generation of animals, to which are added some on parturition, on the membranes and fluids of the uterus, and on conception), London and Amsterdam 1651. English translation in Robert Willis (ed): *The Works of William Harvey*, London 1847, pp. 143–586.
Hemsterhuis, Sibold: *Messis aurea triennalis, exhibens anatomica novissima et utilissima experimenta. Ex. editione Siboldi Hemsterhuis* (A 3-year golden harvest of new and useful anatomical experiments. Published by Sibold Hemsterhuis), Leiden 1654.
Highmore, Nathaniel: *Corporis Humani Disquisitio Anatomica, in qua Sanguinis Circulationem in quavis Corporis particula plurimis typis novis, ac Ænygmatum Medicorum succincta dilucidatione ornatam prosequutus est* (An anatomical study of the human body, in which the author has given a representation, set with many new types, of the circulatory system in any part of the body, as well as an illustration of the medical riddles), The Hague 1651.
Hildebrand, Reinhard: Anatomie und Medizin im 17. Jahrhundert. In: Holger Thomas Gräf & Andreas Tacke (eds): *Von Augsburg nach Frankfurt. Der Kupferstecher Johann Philip Thelott (1639–1671)*, Marburg 2022, pp. 139–57.
Hoboken, Nicolas: *Novus Ductus Salivalis Blasianus, in luceum protractus* (Blasius's new salivary duct brought to light), Utrecht 1662.
Hoboken, Nicolas: *Anatomia Secundinæ Humanæ* (Second edition of the human anatomy), Utrecht 1669.
Hofman, Caspar: *De thorace, eiusque partibus commentarius tripartitus* (Commentary in three parts on the thorax a nd its parts), Frankfurt 1627.
Hofman, Caspar: *Institutionum medicarum libri sex* (Medical doctrines in six books), Lyon 1645.
Holberg, Ludvig: *Ludvig Holbergs Værker*. Arkiv for Dansk Litteratur (www.adl.dk).

van Horne, Johannes: *Novus Ductus Chyliferus. Nunc primum Delineatus, Descriptus & Eruditorum Examini expositus* (The new chyle duct, now for the first time illustrated, described and presented to the examination of the learned), Leiden 1652.

van Horne, Johannes: *Disquisitio Anatomica de Ductibus Salivalibus* (Anatomical Examination of the salivary ducts), Leiden 1656.

van Horne, Johannes: *Waerschowinge aen alle Liefhebbers der Anatomie teegens de gepretendeerde Weetenschap derselver van Joncheer Louys de Bils*, Leiden 1660.

van Horne, Johannes: *Microcosm seu Brevis Manuductio ad Historiam Corporis Humani* (Microcosm or brief guide to the knowledge of the human body), Leiden 1662.

van Horne, Johannes: *Mikrotekne: seu methodica ad chirurgiam introduction* (Micro-techne or introduction to surgical method), 2nd edition, Leiden 1668.

Hornung, Peter Michael (ed.): *Gobeliner til Danmarks Dronning. Bjørn Nørgaards Danmarkshistorie på Christiansborg, vævet i Manufactures nationales des Gobelins et de Beauvais*. Guide, Copenhagen 2000.

Hovesen, Ejnar: *Lægen Ole Worm*, Aarhus 1987.

Howard, John M., Hess, Walter & Traverso, William: Johan Georg Wirsung (1589–1643) and the Pancreatic Duct: The Prosector of Padua, Italy. *Journal of the American College of Surgeons*, 1998, 187 (2), p. 202.

Huygens, Christiaan: *Traité de la Lumière*, Leiden 1690.

Huygens, Christiaan: *Oeuvres complètes de Christiaan Huygens*, The Hague 1888, 1, No. 352, pp. 515–16, No. 361, pp. 528–29 and 1889, 2, No. 445, pp. 113–14.

Hunter, William: *Anatomia uteri humani gravidi tabulis illustrata* (Anatomy of the pregnant uterus illustrated with tablets), Birmingham 1774.

Ilsøe, Ingrid: Bøger og boghandlere under Christian V. *Fund og Forskning i Det Kongelige Biblioteks Samlinger* 1981, 25, pp. 26–27.

Imperato, Ferrante: *Dell Historia Naturale Libri XXVIII*, Napoli 1599.

Ingerslev, V.: *Danmarks Læger og Lægevæsen fra de ældste Tider indtil Aar 1800*, Copenhagen 1873.

Ingerslev, E: *Fragmenter af Fødselshjælpens Historie*, Copenhagen 1907.

Ingerslev, E: Nogle Bemærkninger om Bartholins Anatomi i kinesisk Omklædning. Special edition of *Ugeskrift for Læger*, No. 21, 1907.

Jacobsen, Holger: *Oratio In Obitum Summi Daniæ Æsculapii Thomæ Bartholini. Accessit Operum Th. Bartholini Editorum Catalogus* (Memorial speach for the leading Danish physician Thomas Bartholin. Added a catalogue of Thomas Bartholin's published works), Copenhagen 1681.

Jantzen, Thal: Af Thomas Bartholins "anatomiske Historier." Special edition of *Bibliotek for Læger*, 8, 5, 1904, pp. 1–48.

Jørgensen, Adolf Ditlev: *Niels Stensen. Et mindeskrift*, Copenhagen 1884.

Kardel, Troels & Maquet, Paul: Anatomy of a reindeer dissected in Copenhagen in 1672 by Niels Stensen as reported by Thomas Bartholin. *Rangifer* 2012, 32, 1, pp. 49–55.

Kardel, Troels & Maquet, Paul (eds.): *Nicolaus Steno. Biography and Original Papers of a 17th Century Scientist*, 2nd edition, Berlin 2018.

Kerckring, Theodor: *Spicilegium Anatomicum* (Anatomical aftermath), Amsterdam 1670.

Kestner, Christian Wilhelm: *Medicinisches Gelehrten-Lexicon*, Jena 1740.

Kingo, Thomas: *Samlede Skrifter* 1–7. Published by Hans Brix, Paul Diderichsen & Fredrik Julius Billeskov Jansen, Copenhagen 1975, 1, p. 200.

Kircher, Athanasius: *Prodromus Coptus sive Ægyptiacus* (Coptic or Egyptian forerunner), Rome 1634.

Koch, Carl Henrik: *Spinoza. Kætter og filosof*, Copenhagen 2015.

Kofod, Helmer: Narhvalen og enhjørningen. *Tidsskriftet Grønland* 1985, pp. 101–16.

Koojimans, Luuc: *Death Defied. The Anatomy Lessons of Frederik Ruysch*, Leiden 2010.

Kragelund, Aage: *Den humanistiske renæssance og Antikken*, Copenhagen 1976.

Kragh, Helge: Da fosfor kom til Danmark. *Dansk kemi*, 2004, Vol. 84, 3, pp. 34–35.

Kragh, Helge: Phosphors and phosphorus in Early Danish Natural Philosophy. *Historisk-filosofiske Meddelelser*, 2003, Vol. 88.

Kragh, Helge (ed.): Fra Middelalderlædom til Den Nye Videnskab 1000–1730. *Dansk Naturvidenskabs historie*, 1, Aarhus 2005.

Krog, Ole Willumsen (ed.): *Skatte fra kejserens Kina. Den Forbudte By og Det Danske Kongehus*, Copenhagen 2006.

Krogh, August & Maar, Vilhelm (eds.): *Nicolaus Steno: Foreløbig Meddelelse til en Afhandling om faste Legemer, der findes naturlig indlejrede i andre faste Legemer*. With introduction and notes, Copenhagen 1902.

La Peyrère, Isaac: *Relation du Groenland*, Paris 1647.

Lauremberg, Peter: *Collegium anatomicum* (Anatomical lecture), Rostock 1636.

Lévy-Valensi, J.: *La médicine et les médecins francais au XVIIe siècle*, Paris 1933.

Liceti, Fortunio: *Litheosphorus sive de Lapide Bononiensi* (On the Luminous Stone or the Bologna Stone), Udine 1640.

Liceti, Fortunio: *De Anulis Antiquis Librum Singularem* (A unique book on antique armrings), Udine 1645.

Liceti, Fortunio and Blasius, Gerard: *De Monstris* (On monsters), Amsterdam 1665.

Lind, James: *A treatise of the scurvy*, Edinburgh 1753.

Linden, Johannes Antonides van der: *Medicina Physiologica* (Physiological Medicine), Amsterdam 1653.

Lindroth, Sten: *Svensk lärdomshistoria*, 2, Stormaktstiden, Stockholm 1975.

Lindsay, John: *Edwards's Botanical Register: or, ornamental flower-garden and shrubbery: consisting of coloured figures of plants and shrubs, cultivated in British gardens; accompanied by their history, best method of treatment in cultivation, propagation, etc.*, London 1835.

Loredan, Giovanni Francesco: Giovanni Francesco Loredan: *Le Glorie de gli Incogniti O vero gli Huomini Illustri dell' Accademia de' Signori Incogniti di Venetia*, Venice 1647.

Lusitano, Abraham Zacuto: *De praxi medica admiranda* (On the admirable practice of medicine), Amsterdam 1634.

Lyell, Charles: *Principles of Geology*, London 1830.

Lyser, Michael: *Culter Anatomicus. Hoc est Methodus brevis facilis ac perspicua artificiose & compendiose humana incidendi cadavera. Cum nonnullorum instrumentorum Iconibus* (The anatomical knife, that is, a concise, simple, and obvious method of artful and clear dissection of human corpses. With illustrations of some instruments), Copenhagen 1653. The second edition was published by Thomas Bartholin, Copenhagen 1665. English translation by G. Thomson: *The Art of Dissecting the Human Body, in a Plain, Easy, and Compendious Method. With the several Instruments necessary, curiously engraved; wherein all the different Parts of the Human Body are described, but particularly the Muscles, in a full and accurate Manner, and their Uses clearly explained*, London 1740.

Lyser, Michael: *Dissertationem physicam de auditu* (Physical treatise on hearing), Leipzig 1653.

Maar, Vilhelm (ed.): *Nicolai Stenonis Opera Philosophica* (Niels Stensen's natural philosophical works), Vols. I–II, Copenhagen 1910.

Maar, Vilhelm (ed.): *Holger Jacobæus' Rejsebog* 1671–1692, Copenhagen 1910.

Maar, Vilhelm (ed.): *Thomas Bartholin 1616, 20. oktober, 1916. Mindeskrift paa 300-Aarsdagen for hans Fødsel*, Copenhagen 1916.

Maar, Vilhelm: *Den første Anvendelse af Kinabark i Danmark*, Leiden 1925.

Maar, Vilhelm: *Lidt om Descartes og Danmark*, Copenhagen 1931.

Magnus, Olaus: *Historia de gentibus septentrionalibus* (The History of the Nordic People), Rome 1555.

Malpighi, Marcello: *De Pulmonibus Observationes Anatomicæ* (Anatomical observations of the lungs), Bologna 1661.

Malpighi, Marcello: *De Externo Tactus Organo Anatomica Observatio* (Anatomical observation of the external sensory organ), Naples 1665.

Malpighi, Marcello: *De Polypo Cordis Dissertatio* (Treatise on cardiac polyps), Bologna 1666.

Malpighi, Marcello: *De Viscerum Structura Exercitatio Anatomica* (Anatomical exercise in the structure of the viscera), Bologna 1666.

Malpighi, Marcello: *Anatome Plantarum* (Anatomy of Plants), London 1675.

Mansa, F.V: *Folkesygdommenes og Sundhedspleiens Historie i Danmark fra de ældste Tider til Begyndelsen af det attende Aarhundrede*, Copenhagen 1873.

Matzen, Henning: *Kjøbenhavns Universitets Retshistorie 1479–1879*, Copenhagen 1879.

Meckel, Johann Friedrich: Über die Devertikel am Darmkanal. *Archiv* für die *Physiologie*, Halle 1809, 9, pp. 421–53.

Meibom, Heinrich: *De Vasis Palpebrarum Novis Epistola* (Letter on the new eyelid vessels), Helmstedt 1666.

Meisen, Valdemar (ed.): *Prominent Danish Scientists through the Ages*, Copenhagen 1932.

Meisen, Valdemar & Larsen, Knud (eds.): *Stenoniana*, Copenhagen 1933.

Mercator, Gerardus: *Atlas Minor*, Amsterdam 1609.

van Meurs, Johannes: *Athenæ Batavæ. Sive, De Urbe Leidensi, & Academia, Virisque claris; qui utramque ingenio suo, atque scriptis, illustrarunt: Libri duo* (The Dutch Athens. Or, On the City and University of Leiden, and the famous men, who have distinguished it by their intellect and writings: In two books), Leiden 1625.

Miert, Dirk van: *Humanism in an age of Science. The Amsterdam Athenaeum in the Golden Age 1632–1704*, Leiden 2009.

Modlin, Irvin M. et al.: The history of the pancreas. In: Hans Berger et al. (eds.): *The Pancreas. An Integrated Textbook of Basic Science, Medicine, and Surgery*, 2nd edition, Massachusetts/Oxford/Victoria 2008.

Moe, Harald: *Den anatomiske billedkunst i Renæssancen og Barokken*, Copenhagen 1993.

Morgagni, Giovanni Battista: *De Sedibus, et Causis Morborum per Anatomen Indagatis Libri quinque* (Five books on the seats and causes of diseases sought out through dissections), Venice 1761.

Möhle, Martin: Alte Universität Basel. *Schweizerische Kunstführer GSK*. Serie 80, No. 795, Bern 2006.

Møinichen, Henrik: *Observationes medico-chirurgicæ missæ ad Th. Bartholinum* (Medical-surgical observations sent for Thomas Bartholin), Copenhagen 1665.

Needham, Joseph: *A history of embryology*, New York 1959.

Negro, Piero del (ed.): *The University of Padova. Eight Centuries of History*, Padova 2003

Niebuhr, Carsten & Forsskål, Pehr: *Flora Ægyptiaco-Arabica, sive descriptiones plantarum, quas per Ægyptum inferiorem et Arabiam felicem detexit* (Flora of Egypt and Arabia, or descriptions of plants discovered in lower Egypt and the happy Arabia), Copenhagen 1775.

Nielsen, O: *Kjøbenhavns Historie og Beskrivelse*, Copenhagen 1881.

Nørskov-Lauritsen, Bruun, Niels W.: En celeber kræftoperation. Philip Hacquarts fjernelse af et mammakarcinom fra Lisbeth Lunge, København den 16. juni 1658. *Bibliotek for Læger* 1995, 187, 2, pp. 103–22.

Olsen, Gunnar: Hagestedgaard. In: Arthur G. Hassø (ed.): *Danske Slotte og Herregaarde. Ny Samling*, Copenhagen 1944, 1, pp. 330–35.

Olsen, Gunnar: Den unge Enevælde 1660–1721. In: John Danstrup & Hal Koch (eds.): *Danmarks Historie*, Copenhagen 1964, 8, p. 46.

O'Malley, Charles D. (ed.): *Thomas Bartholin. On the Burning of his Library and On Medical Travel*, Kansas 1961.

O'Malley, Charles D.: Thomas Bartholin. In: Charles Coulston Gillispie (ed.): *Dictionary of Scientific Biography*, New York 1981, 1, pp. 482–83.

O'Malley, Charles. D. & Saunders, John Bertrand de Cusance Morant: *Leonardo da Vinci on The Human Body*, New York 1982.

Panum, Peter Ludvig: Bidrag til Kundskab om vort medicinske fakultets Historie med hensyn til dets Betydning for Naturvidenskabernes og Lægevidenskabens udvikling i Danmark fra Frederik den 3dies Tronbestigelse indtil Frederik den 5tes Død (1648–1766) – *Indbydelsesskrift til Kjøbenhavns Universitets Aarsfest til Erindring om Kirkens Reformation*, Copenhagen 1880.

Paré, Ambroise: *Des Monstres & Prodiges*, Paris 1573.

Patau, Klaus et. al.: Multiple congenital anomaly caused by an extra autosome. *Lancet* 1960, 7128, pp. 790–93.

Paulli, Jacob Henrik: *Anatome Anatomiæ, Bilsianæ* (An anatomical examination of Bils' anatomy), Copenhagen 1663.

Pecquet, Jean: *Experimenta Nova Anatomica, quibus incognitum hactenus Chyli Receptaculum, et ab eo per Thoracem in ramos usque Subclavios Vasa Lactea deteguntur, eiusdem Dissertatio Anatomica de Circulatione Sanguinis et Chyli Motu, Accedunt Clarissimorum Virorum perelegantes ad Authorem Epistolæ* (New anatomical experiments exposing the hitherto unknown vessel for chyle and lacteal vessels, extending from it to branches under the keybone, to which are added the very elegant letters of great celebrities to the author), Paris 1651. English translation including Thomas Bartholin's *De Lacteis Thoracicis: New Anatomical Experiments of John Pecquet of Deip. By which the hitherto unknown Receptacle of the Chyle, and the Transmission from thence to the Subclavial Veins by the now discovered Lacteal Chanels of the Thorax, is plainly made appear in Brutes. As also an Anatomical Dissertation of the Motion of Blood and Chyle. Together with the further Description of the same Lacteal Chanels newly discovered in the Body of Man as well as Brutes. Being an Anatomical Historie, publickly propos'd by Thomas Bartoline, Dr. and Reg. Professor both in Physick and Anatomy, to Michael Lysere, Answering*, London 1653.

Petersen, Julius: *Bartholinerne og Kredsen om dem*, Copenhagen 1898.

Piso, Willem: *Historia Naturalis Brasiliæ* (An account of Brazil's natural history), Leiden and Amsterdam 1648.

Plempius, Vopiscus Fortunatus: *Fundamenta seu Institutiones Medicinæ* (Basic principles or instructions in medicine), Leuven 1638.

Pontoppidan, Erik: *Den Kongelige Residens-Stad Kiøbenhavn*, Copenhagen 1760.

Porter, Ian Herbert: The Nineteenth-Century Physician and Cardiologist Thomas Bevill Peacock (1812–88). *Medical History* 1962, 6 (3), p. 244.

Porter, Ian Herbert: Thomas Bartholin (1616–80) and Niels Stensen (1638–86), Master and Pupil. *Medical History* 1963, 7, 2, pp. 99–125.

Porter, Roy: *The Greatest Benefit to Mankind – A Medical History of Humanity from Antiquity to the Present*, London 1997. Danish translation by Bent Bjerre: *Ve og Vel. Medicinens historie fra oldtid til nutid*, Copenhagen 2000.

Primrose, James: *Exercitationes et Animadversiones in Librum de Motu Cordis et Circulatione Sanguinis adversus Gulielmum Harveum* (Exercises in and commentaries on the book about the motion of the heart and the circulation of the blood directed against William Harvey), London 1630.

Ragland, Evan R.: Experimenting with Chymical Bodies: Reinier de Graaf's Investigations of the Pancreas. *Early Science and Medicine*, 2008, 13 (6), pp. 615–64.

Renbourn, Edward Tobias: The history of sweat and the sweat rash from the earliest times to the end of the 18th century. *Journal of the History of Medicine and Allied Sciences*, April 1959, 14, 2, p. 217.

Resen, Peder Hansen: *Inscriptiones Hafnienses Latinæ, Danicæ et Germanicæ* (Latin, Danish and German inscriptions in Copenhagen), Copenhagen 1668.

Rhode, Johan: *De Acia dissertatio ad Cornelii Celsi mentem* (Treatise about the thread according to Cornelius Celsus), Padua 1639.

Rhode, Johan: *De acia dissertatio ad Cornelii Celsi mentem, accredit De ponderibus et mensuris ejusdem autoris dissertatio, et vita Celci, qua simul universa fibulæ ratio explicatur, secunbdis curis ex autographo autoris auctior et emendatior, cum judiciis doctorum èdita a Th. Bartholino* (Treatise about the thread according to Cornelis Celsus, furthermore, the same authors' treatise about measure and weight and a Celsus biography, where at the same time the whole of the fibulas' idea is investigated. Enlarged and improved second edition building upon the authors' manuscript together with the opinions of the learned world, edited by Thomas Bartholin), Copenhagen 1672.

Rhode, Johan: *Observationum Medicinalium Centuria Tres* (Medical observations in three centuries), Padua 1657.

Riolan, Jean: *Anthropographia et Osteologia* (Anthropography and Osteology), Paris 1626.

Riolan, Jean: *Opuscula Anatomica Nova. Quæ nunc primùm in lucem prodeunt. Instauratio magna Physicæ et Medicinæ, per novam Doctrinam de Motu Circulatorio Sangvinis in Corde. Accessere Notæ in Johannis Wallei duas Epistolas de Circulatione Sanguinis* (New anatomical opuscules, which are now for the first time published. A great renewal of physiology and medicine through the new teachings about the blood circulation in the heart. In addition, some comments on Johannes Walæus' two letters on the blood circulation), London 1649.

Riolan, Jean: *Opuscula Nova Anatomica* (New anatomical opuscules), Paris 1653.

Riolan, Jean: *Anthropographia et Osteologia, Judicium novum de venis lacteis tam mesentericis quam thoracicis adversus Th. Bartholinum. Lymphatica vasa Bartholini refutata. Animadversiones Secundæ ad Anatomiam Reformatam Bartholini. Ejusdem Dubia Anatomica De Lacteis Thoracicis resoluta. Hepatis funerati et ressuscitati vindiciæ* (Anthropography and osteology. A reassessment – directed against Thomas Bartholin – of the lacteal veins in both the mesentery and the thoracic duct. Rejection of Bartholin's lymphatic vessels. For the second time critical remarks on Bartholin's revised anatomy. Resolution of Bartholin's doubts regarding the thoracic duct. Apology of the buried and resurrected liver), Paris 1653.

Riolan, Jean: *Responsiones Duæ: Prima, Ad Experimenta nova Joannis Pecqueti, Doctoris Medici Montpeliensis: Altera, Ad Pecquetianos duos Doctores Parisiensis, adversus sanguificationem in Corde*, etc. (Two answers. The first concerns the new experiments of Jean Pecquet, Doctor of Medicine in Montpellier. The second is against blood formation in the heart, addressed to two Parisian doctors and followers of Pecquet, etc.), Paris 1655.

Rudbeck, Johannes G.: Biografiske anteckningar om Olof Rudbeck den äldres anatomiske skrifter och striden med Thomas Bartholin. *Samleren* 1904, 25, pp. 37–57.

Rudbeck, Olof: *Nova Excercitatio Anatomica, Exhibens Ductus Hepaticos Aquosos, & Vasa Glandularum Serosa, nunc primum inventa*, æneisque *figuris delineata* (A new anatomical examination of the liver's limpid excretory ducts and the gland's serous vessels, now discovered for the first time and illustrated with sketches), Västerås 1653. Swedish translation in Robert Tigerstedt (ed.) 1921, pp. 363–403. Danish translation in Gabriel Tryde (ed.) 1940, pp. 199–229, and English translation by Aage Nielsen: A Translation of Olof Rudbeck's Nova Excercitatio Anatomica Announcing the Discovery of The Lymphatics. *Bulletin of the History of Medicine* 1942, 11, 1, pp. 304–39.

Rudbeck, Olof: *Insidiæ Structæ, Olai Rudbeckii Sueci Ductibus Hepaticis Aquosis, & Vasis Glandularum Serosis. Arosiæ editis, a Thoma Bartholino* Médico Regio *Professore in Academia Hafniensi* (The treacherous ambush on the Swede Olof Rudbeck's *The liver's crystal-clear excretory ducts and the glands' serous vessels*, published in Västerås, instigated by Thomas Bartholin, Royal Physician and Professor of Anatomy at Copenhagen University), Leiden 1654.

Rudbeck, Olof: *Olai Rudbeckii Sveci ad Thomam Bartholinum Danum Epistola, qua sibi primam inventionem Vasorum Serosum Hepatis, invictis rationibus, contra Bogdani cuiusdam impudentissimas Calumnias, sub Apologiæ Nomine pro Bartholini editas, asserit*, etc. (A letter from the Swede Olof Rudbeck to the Dane Thomas Bartholin, in which, for irrefutable reasons directed against a certain Bogdan's utterly shameless accusations (allegedly published in defence of Bartholin), he claims to be the first to discover the liver's serous vessels), Uppsala 1657.

Rudbeck, Olof: *De sero ejusque vasis* (On serum and its vessels), Uppsala 1661.

Rudbeck, Olof: *Hortus Upsaliensis Academiæ* (The Garden at the University of Uppsala), Uppsala 1666.

Rudbeck, Olof: *Atlantica*, Uppsala 1679–1702.

Ruysch, Frederik: *Dilucidatio Valvularum in Vasis Lymphaticis et Lacteis. Cum Figuris æneis. Accesserunt Quædam Observationes Anatomicæ rariores* (Explanation of the valves in lymph vessels and lacteal veins. With copper engravings. Added some rare anatomical observations), The Hague 1665.

Rørdam, Holger: *De danske og norske Studenters Deltagelse i Kjøbenhavns Forsvar mod Karl Gustav. Et Bidrag til Kjøbenhavns Universitets Historie 1658–60*, Copenhagen 1855.

Sagar, Jayesh et al.: Meckel's diverticulum: a systematic review. *Journal of the Royal Society of Medicine*, 2006, 99, pp. 501–5.

Salmuth, Philip: *Observationum Medicarum Centuriæ Tres* (Medical observations in three centuries), Braunschweig 1648.

Sarauw, Georg F.L.: Cromer-skovlaget i Frihavnen og trælevningerne i de ravførende sandlag ved København. *Meddelelser fra Dansk Geologisk Forening*, 1897, 1, 4, pp. 17–44.

Sarton, George: *Introduction to the History of Science*, Vol. 1, Baltimore 1927.

Sauval, Henri: *Historie et Recherches des Antiquités de la Ville de Paris*, Paris 1724.

Saxo Grammaticus: *Danmarks Krønike*, translated by Frederik Winkel Horn, Copenhagen 1898.

Schepelern, Henrik Ditlev (ed.): *Breve fra og til Ole Worm I–III*, Copenhagen 1965–1968.

Schepelern, Henrik Ditlev: *Museum Wormiamum: dets Forudsætninger og Tilblivelse*, Aarhus 1971.

Schepelern, Henrik Ditlev (ed.): *Olai Borrichii Itinerarium 1660–1665. The Journal of the Danish Polyhistor Ole Borch*, I–IV, Copenhagen and London 1983.

Schepelern, Henrik Ditlev (ed.): Niels Stensen. En dansk student i 1659 og noterne i hans Chaos-manuskript. *Skrifter udgivet af Universitetsbiblioteket 2. Afdeling, Danmarks natur- og lægevidenskabelige Bibliotek*, Copenhagen 1986.

Scherz, Gustav (ed.): *Nicolai Stenonis Epistolae et Epistolae ad eum datae* (Letters from and to Niels Stensen), Copenhagen and Freiburg 1952.

Scherz, Gustav: Niels Stensen's First Dissertation. *Journal of the History of Medicine and Allied Sciences* 1960, 15, 3, pp. 247–64.

Scherz, Gustav (ed): Steno. Geological Papers. *Acta Historica Scientarium Naturalium et Medicinalium*, Vol. 20, Odense 1969.

Scherz, Gustav: Da Stensen var i Paris. *Fund og Forskning i Det Kongelige Biblioteks Samlinger* 1969, 16, pp. 43–52.

Schiller, J. and Théodoridès, J: Stenon et les milieux scientifiques parisiens. In: Gustav Scherz (ed.): The historical aspects of brain research in the seventeenth century, proceedings of the international historical symposium on Nicolaus Steno and brain research in the seventeenth century held in Copenhagen 18.–20. August 1965. *Analecta Medica-Historica*, 3, Oxford 1968, p. 156.

Schioldann-Nielsen, Johann & Sørensen, Kurt (eds.): Thomas Bartholin: On diseases in the Bible. A Medical Miscellany 1672. *Acta Historica Scientiarum Naturalium et Medicinalium*, Vol. 41, Copenhagen 1994.

Schirmer, Alfred Max: *Beitrag zur Geschichte und Anatomie des Pankreas*, Basel 1893.

Schooten, Frans van: *Francisci à Schooten Principia Matheseos Universalis seu Introductio ad Geometriæ Methodum Renati des Cartes, edita ab Er. Bartholino, Casp. Fil.* (Frans van Schooten's general mathematical principles or introduction to René Descartes' geometric method, published by Rasmus Bartholin, son of Caspar), Leiden 1651.

Schreyer, Johannes: *Erörterung und Erläuterung der Frage / Ob es ein gewiß Zeichen / wenn eines todten Kindes Lunge im Wasser untersincket / daß ein solches in Mutter-Leibe gestorben sey*, Zeitz 1690.

Secher, Vilhelm Adolf: Københavns Rets- og Kulturhistorie i Kristian IV.s og Frederik III.s Tid 1624–1663. VII. Livsstraffene og deres Fuldbyrdelse. *Historiske Meddelelser om København 1907–08*, pp. 574–78.

Severino, Marco Aurelio: *De recondita Abscessuum Natura Libri VIII* (8 books about the hidden nature of boils), 2nd edition, Frankfurt 1643, which includes *Pædanchone Loimodes, seu de pestilente ac præfocante Pueros Abcessu* (Contagious whooping cough or about a contagious disease choking children).

Severino, Marco Aurelio: *Vipera Pythia. Id est, De Viperæ Natura Venero, Medicina, Demonstrationes, et Experimenta nova* (The Pythic viper, i.e. demonstrations and new experiments with the viper's nature, poison and medicine), Padua 1643.

Severino, Marco Aurelio: *Therapeuta Neapolitanus Seu Veni Mecum Consultor Curandarum febrium & internorum omnium morborum. Inclusa Pædanchone affectu pestilente, ac pueros præfocante. Cum Commentario Cl. Thomæ Bartholini Regis Danorum Archiatri* (The Neapolitan treater or handbook for the treatment of fever diseases and all internal diseases. Included is the child suffocator, a lethal disease that suffocates children. With a commentary by Thomas Bartholin, physician to the king of Denmark), Naples 1653.

Skov, Hans & Bastholm, Eyvind: Petrus Severinus og hans Idea medicinæ philosophicæ. En dansk paracelsist. *Acta Historica Scientiarum Naturalium et Medicinalium*, Vol. 32, Copenhagen/Odense 1979.

Skovgaard-Petersen, Karen: Biblioteksfaglig vejledning i 1670'erne – om Thomas Bartholins forelæsninger som universitetsbibliotekar. In: John T. Lauridsen & Olaf Olsen: *Umisteligt. Festskrift til Erland Kolding Nielsen*, Copenhagen 2007, pp. 137–52.

Slottved, Ejvind: *Lærestole og lærere ved Københavns Universitet 1537–1977*, Copenhagen 1978.

Snorrason, Egill: *Danskeren Johan Rhode i det 17' århundredes Padua*, Copenhagen 1965.

Snorrason, Egill: Illustrationerne i anatomen Thomas Bartholins publikationer 1641–1661. *Fund og Forskning i Det Kongelige Biblioteks Samlinger*, 20, 1973.

Spalding, John: *The History of the Troubles and Memorable Transactions in Scotland from the Year 1624 to 1645*, 2, Aberdeen 1792.

Spalteholz, Werner: *Handatlas der Anatomie des Menschen*, Leipzig 1909.

Spigelius, Adrianus: *De Humani Corporis Fabrica Libri Decem* (Ten books on the structure of the human body), Venice 1627.

Stelluti, Francesco: *Trattato Del Legno Fossile Minerale Nouvamente Scoperto*, Rome 1637.

Stensen, Niels: *Disputatio Physica de Thermis* (A physical disputation on hot springs), Amsterdam 1660.

Stensen, Niels: *Disputatio Anatomica de Glandulis Oris, & nuper observatis inde prodeuntibus Vasis* (Anatomical disputation on the glands of the mouth and their newly observed salivary ducts), Leiden 1661.

Stensen, Niels: *De Glandulis Oculorum, Novisque earundem vasis Observationes Anatomicæ, Quibus Veri lacrymarum fontes deteguntur* (Niels Stensen's anatomical observations on the glands of the eyes and their new vessels, by which the true sources of the tears are revealed), Leiden 1662.

Stensen, Niels: *Nicolai Stenonis Observationes Anatomicæ, Quibus Varia Oris, Oculorum, & Narium Vasa describuntur, novique salivæ, lacrymarum & muci fontes deteguntur, et Novum Nobilissimi Bilsii De lymphæ motu et usu commentum Examinatur & rejictur* (Niels Stensen's anatomical observations, where various new observations concerning the vessels of the mouth, the eyes and the nose are described, where the new origin of the saliva, the tears and the nasal mucus is exposed, and where the famous Bils' new conception of the movement and function of the lymph is examined and rejected), Leiden 1662.

Stensen, Niels: *Apologiæ Prodromus quo demonstratur, judicem Blasianum et rei anatomicæ imperitum esse, & affectuum suorum servum* (Introduction to a defence which shows that Blasius' judge is ignorant of anatomical matters and a slave to his passions), Leiden 1663.

Stensen, Niels: *De Musculis & Glandulis Observationum Specimen. Cum Epistolis duabus Anatomicis* (A specimen of observations of muscles and glands. With two anatomical letters), Copenhagen and Amsterdam 1664.

Stensen, Niels: *Canis Carchariæ Dissectum Caput* (Dissection of a white shark's head), Florence 1667.

Stensen, Niels: *Nicolai Stenonis Elementorum Myologiæ Specimen seu Musculi Descriptio Geometrica* (Specimen of Niels Stensen's muscle doctrine, or geometric description of the muscles), Florence 1667.

Stensen, Niels: *Discours de Monsieur Stenon sur l'anatomie du cerveau à messieurs de l'assemblée, qui se fait chez Monsieur Mr. Thévenot*, Paris 1669.

Stensen, Niels: *De Solido Intra Solidum Naturaliter Contento Dissertationis Prodromus* (Forerunner to a dissertation on solids naturally embedded in other solids), Florence 1669.

Stensen, Niels: *Prooemium Demonstrationum Anatomicarum in Theatro Hafniensi Anni 1673*, (Introductory lecture to the anatomical demonstrations in the Anatomical Theatre in Copenhagen in 1673). *Acta Medica Ann. 1673*, Copenhagen 1675, Obs. 34 (misnumbered as Obs. 24), pp. 359–66. A simultaneous copy exists in Holger Jacobsen's mentioned records in the Royal Library in Copenhagen, NKS 399aa 4°. Maar (ed.) 1910, 2, pp. 249–56. Kardel & Maquet (eds.) 2018, pp. 853–59. Danish translation by Knud Larsen in Valdemar Meisen & Knud Larsen (eds.): *Stenoniana*, Copenhagen 1933, pp. 100–11, by Aage Kragelund in *Den humanistiske renæssance og Antikken*, Copenhagen 1976, pp. 226–44, and by Jesper Brandt Andersen in *Niels Stensen. Kongelig anatom og fyrstelig geolog*, Copenhagen 2021, pp. 780–86. Annotated excerpts can be found in Jørgensen 1884, pp. 170–73.

Stensen, Niels: *Historia Musculorum Aqvilæ D. Nicolai Stenonis* (Doctor Niels Stensen's account of the eagle's muscles). *Acta Medica Ann.* 1673, Copenhagen 1675, Obs. 127, pp. 320–45, Maar (ed.) 1910, 2, pp. 257–77. English translation in Kardel & Maquet (eds.) 2018, pp. 871–91. For an annotated review, see Brandt Andersen 2021, pp. 652–54.

Stewart, Joyce & Hennessy, Esmé Frances: *Orchids of Africa. A select review*, London 1981, pp. 56–57.

Still, George Frederic: *The History of Paediatrics. The Progress of the Study of Diseases of Children up to the End of the XVIIIth Century*, London 1931.

Sylvius, Franciscus: *Disputationum Medicarum Decas* (Ten Medical Disputations), Amsterdam 1663.

Syv, Peder: *Nogle Betenkninger om det Cimbriske Sprog*, Copenhagen 1663.

Sørensen, Peder: *Idea Medicinæ Philosophicæ* (Ideal of Philosophical Medicine), Basel 1571.

Teisen, Jens Villiam & Bobé, Louis (eds.): Danmarks Adels Aarbog, Copenhagen 1932.

Thamdrup, Erik: *Pædiatri i Danmark. Træk af den historiske udvikling*, Copenhagen 1994, p. 20–21.

Thou, Jacques Auguste de: *Historichen Beschreibung deren Namhafftigsten, Geistlichen und Weltlichen Geschichten*, Frankfurt 1622.

Tigerstedt, Robert (ed.): *Dokument ur Medicinens Historia under sextonde och sjuttonde Århundradet*, Stockholm 1921.

Tolmer, Léon: Une page d'histoire des sciences 1661–1669. Vingt-deux lettres inédites d'André de Graindorge à P.-D. Huet. *Mémoires de l'Académie Nationale des Sciences, Arts et Belles-Lettres de Caen*, Nouvelle Série, 1942, 10, pp. 245–337.

Tomasini, Giacomo Filippo: *Gymnasium Patavinum* (The College in Padua), Utini 1654.

Tryde, Gabriel: *Retsmedicinske Erklæringer afgivet af Københavns Universitets Medicinske Fakultet i Aarene 1630–1662*, Copenhagen 1939.

Tryde, Gabriel (ed.): Thomas Bartholin. Skrifter om Opdagelsen af Lymfekarsystemet i Udvalg. Paa Dansk ved G. Tryde. *Klassisk dansk Medicin. Vol. III*, Copenhagen 1940.

Vaupell, Christian: En botanisk Undersøgelse af det fossile Træ, der findes i Leer- og Sandlagene i Danmark og af havet opkastes med Ravet på de danske Kyster. Published by N. Hartz in: *Meddelelser fra Dansk Geologisk Forening*, 1906, 2, 12, pp. 43–64.

Vesalius, Andreas: *De Humani corporis fabrica Libri septem* (On the structure of the human body in seven books), Basel 1543.

Vesling, Johannes: *Syntagma anatomicum* (Anatomical treatise), Padua 1641, and second edition 1647.

Vinding, Rasmus: *Regia Academia Hauniensis* (The Royal University of Copenhagen), Copenhagen 1665.

Wandal, Hans: *Dissertatio Theologica De Antistitum Ecclesiæ orthodoxæ legitima vocatione & ordinatione. Item De sangvinis & suffocatorum animalium esu, inter Christianos haud illicito*, Copenhagen 1686.

Werlauff, Erich Christian: Efterretninger om italieneren J. F. Borro's Ophold ved det danske Hof i Aarene 1667–1670. In: Johan Daniel Herholdt & Frederik Vilhelm Mansa (eds.): *Samlinger til den danske Medicinal-Historie*, Copenhagen 1835, 1, pp. 152–90.

Werlauff, Erich Christian & Engelstoft, Laurids: *Udsigt over Kiøbenhavns Universitets-Bygnings Historie, fra Universitetets Stiftelse indtil 1836*, Copenhagen 1836.

West, Frederik Jürgensen: *Kronens Skøder paa afhændet og erhvervet Jordegods i Danmark, fra Reformationen til Nutiden.* Andet Bind. 1648–1688, Copenhagen 1908.

Westergaard, Peder Basse Christian: *Danske Portræter i Kobberstik, Litografi og Træsnit I–II*, Copenhagen 1930–1933.

Wharton, Thomas: *Adenographia: sive glandularum totius corporis descriptio* (Adenography or description of the glands of the whole body), London 1656.

Willis, Robert (ed.): *The Works of William Harvey*, London 1847.

Wolf, Jens Lauridsen: *Encomion Regni Daniæ*, Copenhagen 1654.

Wolff, Odin: Synderlige anatomiske og medicinske Tildragelser. *Journal for Politik, Natur- og Menneskekundskab*, 1827–1829.
Wootton, David: *The Invention of Science: A New History of the Scientific Revolution*, London 2015.
Worm, Ole: *An os illud quod vulgo pro cornu Monocerotis venditatur verum sit Unicornu?* (Is that bone, which is commonly offered as the tusk of a unicorn, really the tusk of a unicorn?), Copenhagen 1638.
Worm, Ole: *Institutionum Medicarum Epitome* (Epitome of medical instructions), Copenhagen 1640.
Worm, Ole: *De Aureo Serenissimi Domini Christiani Quinti Daniæ, Norvegiæ, etc. electi Principis Cornu Dissertatio* (Treatise on the golden horn that belongs to His Royal Highness Christian V, hereditary Prince of Denmark, Norway, etc.), Copenhagen 1641.
Worm, Ole: *Danicorum Monumentorum Libri Sex* (Danish memorials in six books), Copenhagen 1643.
Worm, Ole: *Museum Wormianum seu Historia Rerum Rariorum, Tam Naturalium, quam Artificialium, tam Domesticarum, quam Exoticarum, quæ Hafniæ Danorum in ædibus Autor servantur. Adornata ab Olau Worm, Med. Doct., & in Regia Hafniensi Academia, olim Professore publico. Variis et accuratis Iconibus illustrata* (Worm's Museum or the report about rare things, natural as well as artificial, domestic as well as exotic and Danish from Copenhagen in this author's custody. Adorned by Ole Worm, physician and former public professor at Copenhagen University. Illustrated with various accurate images), Leiden and Amsterdam 1655.
Worm, Willum: *Oratio In Excessum Viri Incomparabilis Thomæ Bartholini. In Suprema Justitiæ Curia Assessoris, Archiatri Regii, Professoris Honorarii, & Facultatis Medicæ Decani. Publicé In Academia Hafniensi. IV. Januar. dicta.* (Memorial speech for the incomparable man Thomas Bartholin, Assesssor of the Supreme Court, Royal Physician, Honorary Professor and Dean of the Medical Faculty. Held in public at Copenhagen University on 4 January), Copenhagen 1681.
Zacchia, Paolo: *Quæstionum Medico-Legalium Tomi Tres* (Medico-legal questions in three volumes), Frankfurt 1688.
Zas, Nicolaas: *Epistola Apologetica Ad Magnum Th. Bartholinum Regium in Regia Hafniensi Professorem Anatomicorum Coryphæum De Calumniis Nobiliss. Ludovico Bilsio à Coppensdamme, Bonem etc. Arcanorum Naturalium Scrutatori Subtiliss. nec non vivorum ac mortuorum Anatomico Incomparabili, perperam impactis* (Apology letter to the great Thomas Bartholin, Royal Professor and leader of the anatomists at Copenhagen University, concerning the false accusations that are improperly directed against the noble Louis de Bils of Koppensdamme, Bonem etc., who is a very sharp investigator of Nature's secrets and an incomparable anatomist, who dissects living and dead), Rotterdam 1661.
Zeiler, Martin & Merian, Matthäus (eds.): *Topographia Germaniæ inferioris*, Frankfurt am Main 1659.
Ziggelaar, August (ed.): Chaos. Niels Stensen's Chaos-manuscript, Copenhagen 1659. Complete edition with introduction, notes and commentary by August Ziggelaar. *Acta Historica Scientiarum Naturalium et Medicinalium* Vol. 44, Copenhagen 1997.
Östlund, Krister (ed.): De nihilo medico. Ett tal av Thomas Bartholin. *Bibliotheca NEOLATINA Upsalensis IV*, Uppsala 2016.

Unpublished sources

Bartholin, Thomas: Letter to the mayor in Holbæk 25. April 1664, Rigsarkivet, Copenhagen, Krigskollegiet, 1664. 5. Indkomne breve fra flere end 2.
Bartholin, Thomas: Letter to Jørgen Hahn in Odense 23. April 1679. Landsarkivet for Fyn, Odense, Hahn's Archive, file no. 1, fol. 1rv.
Byfogedregnskaber Sjælland og Møn København 1648–1653, Rigsarkivet, Copenhagen.
Graindorge, André: Letters to Pierre Daniel Huet. Det Kongelige Bibliotek, Copenhagen, NKS 4660, 4º.
Jacobsen, Holger: *Sectiones anatomicæ publicæ et privatæ prosectoris incomparabilis Nicolai Stenonis* (The incomparable prosector Niels Stensen's public and private anatomical dissections). Det Kongelige Bibliotek, Copenhagen, NKS 309aa 4°. The title of Jacobsen's collected lecture notes, in which the dissection descriptions are included, is *Exercitia Academica Hafniæ et Lugdun. Batavorum Olgeri Jacobæi* (Holger Jacobsen's exercises at Copenhagen and Leiden University).
Stensen, Niels: Letter in Italian from Copenhagen 18 April 1673 to Vincenzo Viviani in Florence. Det Kongelige Bibliotek, Copenhagen, Håndskrift-samlingen, manuscript Acc. 2019/11.
Stensen, Niels: Letter from Florence to Ole Borch in Copenhagen, November 1675. Det Kongelige Bibliotek, Copenhagen, Bøllings Brevsamling, D 480 2°.

Index of Names

G

H

I

J

K

R

S